AF360712

Cyanotoxins in Bloom: Ever-Increasing Occurrence and Global Distribution of Freshwater Cyanotoxins from Planktic and Benthic Cyanobacteria

Cyanotoxins in Bloom: Ever-Increasing Occurrence and Global Distribution of Freshwater Cyanotoxins from Planktic and Benthic Cyanobacteria

Editors

Triantafyllos Kaloudis
Anastasia Hiskia
Theodoros Triantis

MDPI • Basel • Beijing • Wuhan • Barcelona • Belgrade • Manchester • Tokyo • Cluj • Tianjin

Editors

Triantafyllos Kaloudis
Laboratory of Organic
Micropollutants, Water Quality
Control Department
Athens Water Supply &
Sewerage Company, EYDAP SA
Athens
Greece

Anastasia Hiskia
Institute of Nanoscience and
Nanotechnology
NCSR Demokritos
Athens
Greece

Theodoros Triantis
Institute of Nanoscience and
Nanotechnology
NCSR Demokritos
Athens
Greece

Editorial Office
MDPI
St. Alban-Anlage 66
4052 Basel, Switzerland

This is a reprint of articles from the Special Issue published online in the open access journal *Toxins* (ISSN 2072-6651) (available at: www.mdpi.com/journal/toxins/special_issues/Cyanotoxins_Bloom).

For citation purposes, cite each article independently as indicated on the article page online and as indicated below:

LastName, A.A.; LastName, B.B.; LastName, C.C. Article Title. *Journal Name* **Year**, *Volume Number*, Page Range.

ISBN 978-3-0365-3922-5 (Hbk)
ISBN 978-3-0365-3921-8 (PDF)

Contents

Preface to "Cyanotoxins in Bloom: Ever-Increasing Occurrence and Global Distribution of Freshwater Cyanotoxins from Planktic and Benthic Cyanobacteria"

At present, cyanobacteria and their toxins (also known as cyanotoxins) constitute a major threat for freshwater resources worldwide. Cyanotoxin occurrence in water bodies around the globe is constantly increasing, whereas emerging, less studied or completely new variants and congeners of various chemical classes of cyanotoxins, as well as their degradation/transformation products are often detected. In addition to planctic cyanobacteria, benthic cyanobacteria, in many cases, appear to be important toxin producers, although far less studied and more difficult to manage and control. This Special Issue highlights novel research results on the structural diversity of cyanotoxins from planktic and benthic cyanobacteria, as well as on their expanding global geographical spread in freshwaters.

Triantafyllos Kaloudis, Anastasia Hiskia, and Theodoros Triantis

Editors

Editorial

Cyanotoxins in Bloom: Ever-Increasing Occurrence and Global Distribution of Freshwater Cyanotoxins from Planktic and Benthic Cyanobacteria

Triantafyllos Kaloudis [1,2,*], **Anastasia Hiskia** [2] **and Theodoros M. Triantis** [2]

1 Laboratory of Organic Micropollutants, Water Quality Control Department, Athens Water Supply & Sewerage Company, EYDAP SA, Flias 11, 13674 Menidi, Greece
2 Institute of Nanoscience and Nanotechnology, National Center for Scientific Research "DEMOKRITOS", Partiarchi Grigoriou E & Neapoleos 27 Str., Agia Paraskevi, 15341 Athens, Greece; a.hiskia@inn.demokritos.gr (A.H.); t.triantis@inn.demokritos.gr (T.M.T.)
* Correspondence: kaloudis@eydap.gr

Citation: Kaloudis, T.; Hiskia, A.; Triantis, T.M. Cyanotoxins in Bloom: Ever-Increasing Occurrence and Global Distribution of Freshwater Cyanotoxins from Planktic and Benthic Cyanobacteria. *Toxins* **2022**, *14*, 264. https://doi.org/10.3390/toxins14040264

Received: 26 March 2022
Accepted: 1 April 2022
Published: 8 April 2022

Publisher's Note: MDPI stays neutral with regard to jurisdictional claims in published maps and institutional affiliations.

Toxic cyanobacteria in freshwater bodies constitute a major threat to public health and aquatic ecosystems [1]. Cyanobacterial blooms are increasing in frequency, magnitude and duration globally, while eutrophication, rising CO_2 and climate change promote their global expansion [2,3]. Toxic cyanobacteria metabolites, known as cyanotoxins, comprise a wide range of compounds, including cyclic peptides (microcystins, nodularins) and alkaloids (cylindrospermopsins, anatoxins, saxitoxins) that can be hepatotoxic, cytotoxic, genotoxic or neurotoxic. In response to the risks associated to known cyanotoxins, the World Health Organization (WHO) has published guidelines for their monitoring and management, including provisional guideline values for exposure via drinking water and recreational activities [4]. Nevertheless, the high metabolic potential of cyanobacteria yields a plethora of secondary metabolites that are largely understudied. A recently developed database of cyano-metabolites reported in the literature (CyanoMetDB) contains more than 2000 molecules, including more than 300 microcystin congeners [5]. Still, research so far on the occurrence and impacts of cyano-metabolites has mostly focused on a small number of cyanotoxins, particularly on a few microcystin congeners.

This Special Issue aims to present novel research results on the presence and structural diversity of cyanotoxins and cyano-metabolites in freshwater bodies worldwide. We welcomed research and review papers that showcase the expanding global geographical spread of cyanotoxins, including reports from less-studied areas and on understudied cyanotoxins and cyano-metabolites. We particularly encouraged advances and novelties in the areas of cyanotoxin analysis and monitoring, structural elucidation of new cyano-metabolites, biotic and abiotic factors linked to cyanotoxin production and the role of benthic cyanobacteria as cyanotoxin producers.

A number of published papers reported the presence of toxic cyanobacteria and cyanotoxins using diverse monitoring techniques, in freshwater bodies encompassing Central and South European, Mediterranean, Southeast Asian and North American regions. Van Hassel et al. [6] reported results from monitoring of cyanobacterial blooms in lakes of Wallonia, Flanders and Brussels, Belgium, using LC-MS/MS, PCR and sequencing techniques, to assess the risks associated to recreational waters. More than 20% of samples exceeded the WHO guideline value for microcystins, while the *mcyE* gene was detected in 76% of samples. Fournier et al. [7] investigated the deep-water, red-pigmented biomass occurrences in Lake Constance, which is the third largest lake in Central-Western Europe that borders Germany, Austria and Switzerland. Using 16S rRNA gene-amplicon sequencing and LC-MS/MS they showed that these blooms were contributed by microcystin-producing *Planktothrix* spp. A one-year monitoring study of Slovenian waterbodies using qPCR (*mcyE, cyrJ, sxtA* genes) and LC-MS/MS (microcystins, cylindrospermopsin, saxitoxin) was conducted by Zupancic et al. [8]. Potentially toxic *Microcystis* and *Planktothrix* cells were detected by qPCR and microscopic analysis and a positive correlation between the numbers of *mcyE* gene copies and microcystin concentrations was observed. Furthermore, potential cylindrospermopsin and saxitoxin producers were detected by qPCR, showing the potential of molecular techniques to complement chemical and microscopic analysis in freshwater monitoring programs. Zervou et al. [9] reported the results of a 3-year monitoring of Lake Vegoritis in Northwestern Greece, which is used for irrigation, fishing and recreational activities. LC-MS/MS analysis showed the co-occurrence of cyanotoxins (seven microcystin congeners and cylindrspermopsin, at low levels, <1 µg/L) with other cyanobacterial peptides (anabaenopeptins, microginins). An investigation for the presence of cyanotoxins in Lake Karaoun, the largest artificial lake in Lebanon that serves multiple purposes, was conducted by Hammoud et al. [10], using complementary analytical techniques (LC-MS/MS, qPCR, ELISA and in vitro bioassays). A total of 11 microcystin congeners were detected in concentrations up to 211 and 199 µg/L for MC-LR and MC-YR, respectively. In addition, typical volatile and odorous cyanobacteria compounds were detected by GC-MS. Using a polyphasic approach, Ballot et al. [11] characterized cyanobacterial strains isolated from Meiktila Lake, a shallow reservoir close to Meiktila city in central Myanmar. The strains were classified morphologically and phylogenetically as *R. raciborskii*, and *Microcystis* spp. Cylindrospermopsins were detected by ELISA and LC–MS in 3 of the 5 *Raphidiopsis* strains, while *Microcystis* strains produced a wide range of microcystins, including 22 previously unreported congeners. Zastepa et al. [12] characterized nearshore deep chlorophyl layers from two embayments of Lake Huron, Canada. These layers were shown to be dominated by *Planktothrix* cf. *isothrix*. Microcystins, anabaenopeptins and cyanopeptolins were detected through the water column, along with the corresponding genes. The results also indicated that intersecting gradients of light and nutrient-enriched hypoxic hypolimnia are key factors in supporting deep chlorophyl layers in these embayments.

A serious incident involving dog deaths in Mandichosee, a mesotrophic reservoir of the River Lech, Germany, was investigated by Bauer et al. [13]. Anatoxin-a and dihydroanatoxin-a (dhATX) from benthic *Tychonema* sp. were detected by LC-MS/MS in the stomachs of two dogs in concentrations up to 1207 µg/L, while up to 68,000 µg/L anatoxins were present in lake samples containing large amounts of mat material. The findings of this study are extremely important as they underscore the role of less-studied benthic cyanobacteria in the production of potent toxins, such as the neurotoxic anatoxins.

Several studies reported results on new or less common cyanobacteria metabolites and cyanotoxin producers. Kust et al. [14] applied a molecular networking and dereplication approach in high-resolution mass spectrometry data using the open global natural product social networking (GNPS) web platform to putatively identify a wide range of cyanopeptides from eutrophic fishponds in the Czech Republic. Forty peptides belonging to the groups of anabaenopeptins, microcystins, cyanopeptolins, microginins, cyanobactins, radiosumins, planktocyclins and epidolastatins were identified. Zervou et al. [15] reported the occurrence and structural variety of anabaenopeptins in cyanobacterial blooms and cultured strains from Greek freshwaters using LC-MS/MS. Thirteen structures of anabaenopeptins were annotated based on interpretation of fragmentation spectra, including three structures not reported before. Cordeiro et al. [16] screened 157 strains from the Azorean Bank of Algae and Cyanobacteria (BACA) for cyanotoxin production (microcystins, saxitoxins and cylindrospermopsins) using qPCR, LC-MS/MS and 16S rRNA phylogenetic analysis. Cyanotoxin-producing genes were amplified in 13 strains, and 4 were confirmed as toxin producers by LC-MS/MS. Two nostocalean strains, possibly belonging to a new genus, were identified as new cylindrospermopsin producers, as they were positive for *cyrB* and *cyrC* genes and the presence of cylindrospermopsin was further confirmed by LC-MS/MS.

Two papers reported effects of nutrient and climate factors on the proliferation of cyanobacteria and the production of cyanotoxins. Barnard et al. [17] investigated the role of phosphorus and nitrogen limitation on microcystin and anatoxin production from *Microcystis* spp. and *Planktothrix* spp. in Western Lake Erie. The results showed the importance of reducing both nitrogen and phosphorus to limit cyanotoxin and cyanobacterial biomass production. Le Moal et al. [18] analyzed 13 years of eutrophication and climatic data of Lac au Duc, one of the largest shallow water bodies in Brittany, Western France, which is used as recreational and drinking water reservoir. Analysis showed interannual variability of cyanobacterial composition, with dominant species shifting from *Planktothrix agardhii* towards *Microcystis* sp. and then *Dolichospermum* sp. due to climatic pressures and nitrogen limitation.

Paleolimnological studies based on analysis of sediment cores for cyanobacteria and cyanotoxins, can contribute historic data on the prevalence of toxic cyanobacterial blooms. Weisbrod et al. [19] explored the spatial variability and historical cyanobacterial composition in sediment cores from Lake Rotorua in the South Island of New Zealand, focusing on the abundance of *Microcystis*, *mcyE* gene copy numbers and microcystins. The results showed that toxin producing *Microcystis* blooms are a relatively recent phenomenon in Lake Rotorua, initiated after the 1950s. In addition, results indicated that a single sediment core sampling used by most paleolimnological studies in small to medium-sized lakes can capture dominant microbial communities.

The Special Issue includes three review papers that present emerging areas of toxic cyanobacteria and cyanotoxins research. Metcalf and Codd [20] reviewed and discussed cases were cyanobacteria and cyanotoxins co-occurred with additional hazards such as algal toxins, microbial pathogens, metals, pesticides and microplastics. The authors discussed challenges in assessment of toxicity in such cases and identified further research needs in this field. Sundaravadivelu et al. [21] reviewed the current methodologies for the analysis of freshwater cyanotoxins and prymnesins with emphasis in samples other than water. The authors discussed their limitations, especially with respect to accurate quantitation and structural confirmation of various cyanotoxins, where mass spectrometric techniques are advantageous as they can potentially be applied for detection and unambiguous identification of multiple toxins. Lastly, Monteiro et al. [22] reviewed the existing knowledge on the less-studied, structurally diverse cyclic hexapeptides anabaenopeptins that are increasingly detected in freshwaters in elevated concentrations and possibly play important roles in aquatic ecosystems.

Funding: This research received no external funding.

Acknowledgments: The editors are grateful to all the authors who submitted and contributed their work to this Special Issue. Special thanks to the expert peer reviewers for the rigorous evaluations of all submitted manuscripts.

Conflicts of Interest: The authors declare no conflict of interest.

References

1. Brooks, B.W.; Lazorchak, J.M.; Howard, M.D.; Johnson, M.-V.V.; Morton, S.L.; Perkins, D.A.; Reavie, E.D.; Scott, G.I.; Smith, S.A.; Steevens, J.A. Are harmful algal blooms becoming the greatest inland water quality threat to public health and aquatic ecosystems? *Environ. Toxicol. Chem.* **2016**, *35*, 6–13. [CrossRef] [PubMed]
2. Huisman, J.; Codd, G.A.; Paerl, H.W.; Ibelings, B.W.; Verspagen, J.M.H.; Visser, P.M. Cyanobacterial blooms. *Nat. Rev. Microbiol.* **2018**, *16*, 471–483. [CrossRef] [PubMed]
3. Visser, P.M.; Verspagen, J.M.; Sandrini, G.; Stal, L.J.; Matthijs, H.C.; Davis, T.W.; Paerl, H.W.; Huisman, J. How rising CO_2 and global warming may stimulate harmful cyanobacterial blooms. *Harmful Algae* **2016**, *54*, 145–159. [CrossRef] [PubMed]
4. Chorus, I.; Welker, M. (Eds.) *Toxic Cyanobacteria in Water*, 2nd ed.; CRC Press: Boca Raton, FL, USA; On Behalf of the World Health Organization: Geneva, Switzerland, 2021.
5. Jones, M.R.; Pinto, E.; Torres, M.A.; Dörr, F.; Mazur-Marzec, H.; Szubert, K.; Tartaglione, L.; Dell'Aversano, C.; Miles, C.O.; Beach, D.G.; et al. CyanoMetDB, a comprehensive public database of secondary metabolites from cyanobacteria. *Water Res.* **2021**, *196*, 117017. [CrossRef] [PubMed]
6. Van Hassel, W.H.R.; Andjelkovic, M.; Durieu, B.; Marroquin, V.A.; Masquelier, J.; Huybrechts, B.; Wilmotte, A. A Summer of Cyanobacterial Blooms in Belgian Waterbodies: Microcystin Quantification and Molecular Characterizations. *Toxins* **2022**, *14*, 61. [CrossRef] [PubMed]
7. Fournier, C.; Riehle, E.; Dietrich, D.R.; Schleheck, D. Is Toxin-Producing *Planktothrix* sp. an Emerging Species in Lake Constance? *Toxins* **2021**, *13*, 666. [CrossRef]

8. Zupančič, M.; Kogovšek, P.; Šter, T.; Rekar, R.; Cerasino, L.; Baebler, Š.; Klemenčič, A.K.; Eleršek, T. Potentially Toxic Planktic and Benthic Cyanobacteria in Slovenian Freshwater Bodies: Detection by Quantitative PCR. *Toxins* **2021**, *13*, 133. [CrossRef]

9. Zervou, S.-K.; Moschandreou, K.; Paraskevopoulou, A.; Christophoridis, C.; Grigoriadou, E.; Kaloudis, T.; Triantis, T.; Tsiaoussi, V.; Hiskia, A. Cyanobacterial Toxins and Peptides in Lake Vegoritis, Greece. *Toxins* **2021**, *13*, 394. [CrossRef]

10. Hammoud, N.A.; Zervou, S.-K.; Kaloudis, T.; Christophoridis, C.; Paraskevopoulou, A.; Triantis, T.M.; Slim, K.; Szpunar, J.; Fadel, A.; Lobinski, R.; et al. Investigation of the Occurrence of Cyanotoxins in Lake Karaoun (Lebanon) by Mass Spectrometry, Bioassays and Molecular Methods. *Toxins* **2021**, *13*, 716. [CrossRef]

11. Ballot, A.; Swe, T.; Mjelde, M.; Cerasino, L.; Hostyeva, V.; Miles, C.O. Cylindrospermopsin- and Deoxycylindrospermopsin-Producing *Raphidiopsis raciborskii* and Microcystin-Producing *Microcystis* spp. in Meiktila Lake, Myanmar. *Toxins* **2020**, *12*, 232. [CrossRef]

12. Zastepa, A.; Miller, T.; Watson, L.; Kling, H.; Watson, S. Toxins and Other Bioactive Metabolites in Deep Chlorophyll Layers Containing the Cyanobacteria *Planktothrix* cf. *isothrix* in Two Georgian Bay Embayments, Lake Huron. *Toxins* **2021**, *13*, 445. [CrossRef] [PubMed]

13. Bauer, F.; Fastner, J.; Bartha-Dima, B.; Breuer, W.; Falkenau, A.; Mayer, C.; Raeder, U. Mass Occurrence of Anatoxin-a- and Dihydroanatoxin-a-Producing *Tychonema* sp. in Mesotrophic Reservoir Mandichosee (River Lech, Germany) as a Cause of Neurotoxicosis in Dogs. *Toxins* **2020**, *12*, 726. [CrossRef] [PubMed]

14. Kust, A.; Řeháková, K.; Vrba, J.; Maicher, V.; Mareš, J.; Hrouzek, P.; Chiriac, M.-C.; Benedová, Z.; Tesařová, B.; Saurav, K. Insight into Unprecedented Diversity of Cyanopeptides in Eutrophic Ponds Using an MS/MS Networking Approach. *Toxins* **2020**, *12*, 561. [CrossRef] [PubMed]

15. Zervou, S.-K.; Kaloudis, T.; Gkelis, S.; Hiskia, A.; Mazur-Marzec, H. Anabaenopeptins from Cyanobacteria in Freshwater Bodies of Greece. *Toxins* **2021**, *14*, 4. [CrossRef] [PubMed]

16. Cordeiro, R.; Azevedo, J.; Luz, R.; Vasconcelos, V.; Gonçalves, V.; Fonseca, A. Cyanotoxin Screening in BACA Culture Collection: Identification of New Cylindrospermopsin Producing Cyanobacteria. *Toxins* **2021**, *13*, 258. [CrossRef]

17. Barnard, M.A.; Chaffin, J.D.; Plaas, H.E.; Boyer, G.L.; Wei, B.; Wilhelm, S.W.; Rossignol, K.L.; Braddy, J.S.; Bullerjahn, G.S.; Bridgeman, T.B.; et al. Roles of Nutrient Limitation on Western Lake Erie CyanoHAB Toxin Production. *Toxins* **2021**, *13*, 47. [CrossRef]

18. Le Moal, M.; Pannard, A.; Brient, L.; Richard, B.; Chorin, M.; Mineaud, E.; Wiegand, C. Is the Cyanobacterial Bloom Composition Shifting Due to Climate Forcing or Nutrient Changes? Example of a Shallow Eutrophic Reservoir. *Toxins* **2021**, *13*, 351. [CrossRef]

19. Weisbrod, B.; Wood, S.A.; Steiner, K.; Whyte-Wilding, R.; Puddick, J.; Laroche, O.; Dietrich, D.R. Is a Central Sediment Sample Sufficient? Exploring Spatial and Temporal Microbial Diversity in a Small Lake. *Toxins* **2020**, *12*, 580. [CrossRef]

20. Metcalf, J.S.; Codd, G.A. Co-Occurrence of Cyanobacteria and Cyanotoxins with Other Environmental Health Hazards: Impacts and Implications. *Toxins* **2020**, *12*, 629. [CrossRef]

21. Sundaravadivelu, D.; Sanan, T.T.; Venkatapathy, R.; Mash, H.; Tettenhorst, D.; Danglada, L.; Frey, S.; Tatters, A.O.; Lazorchak, J. Determination of Cyanotoxins and Prymnesins in Water, Fish Tissue, and Other Matrices: A Review. *Toxins* **2022**, *14*, 213. [CrossRef]

22. Monteiro, P.; Amaral, S.D.; Siqueira, A.; Xavier, L.; Santos, A. Anabaenopeptins: What We Know So Far. *Toxins* **2021**, *13*, 522. [CrossRef] [PubMed]

Article

A Summer of Cyanobacterial Blooms in Belgian Waterbodies: Microcystin Quantification and Molecular Characterizations

Wannes Hugo R. Van Hassel [1,2,*], **Mirjana Andjelkovic** [3], **Benoit Durieu** [2], **Viviana Almanza Marroquin** [4], **Julien Masquelier** [1], **Bart Huybrechts** [1,†] and **Annick Wilmotte** [2]

[1] Toxins Unit, Organic Contaminants and Additives, Sciensano, Rue Juliette Wytsmanstraat 14, 1050 Brussels, Belgium; Julien.Masquelier@sciensano.be (J.M.); Bart.Huybrechts@sciensano.be (B.H.)

[2] InBios-Centre for Protein Engineering, Department of Life Sciences, University of Liège, Allée du Six Août 11, 4000 Liège, Belgium; benoit.durieu@uliege.be (B.D.); awilmotte@uliege.be (A.W.)

[3] Risk and-Health Impact Assessment, Sciensano, Rue Juliette Wytsmanstraat 14, 1050 Brussels, Belgium; Mirjana.Andjelkovic@sciensano.be

[4] Phytoplankton and Phytobenthos Laboratory, EULA Center, University of Concepcion, Barrio Universitario Box 160, Concepcion 3349001, Chile; valmanza@udec.cl

* Correspondence: Wannes.Vanhassel@sciensano.be

† The author has passed away.

Abstract: In the context of increasing occurrences of toxic cyanobacterial blooms worldwide, their monitoring in Belgium is currently performed by regional environmental agencies (in two of three regions) using different protocols and is restricted to some selected recreational ponds and lakes. Therefore, a global assessment based on the comparison of existing datasets is not possible. For this study, 79 water samples from a monitoring of five lakes in Wallonia and occasional blooms in Flanders and Brussels, including a canal, were analyzed. A Liquid Chromatography with tandem mass spectrometry (LC-MS/MS) method allowed to detect and quantify eight microcystin congeners. The *mcyE* gene was detected using PCR, while dominant cyanobacterial species were identified using 16S RNA amplification and direct sequencing. The cyanobacterial diversity for two water samples was characterized with amplicon sequencing. Microcystins were detected above limit of quantification (LOQ) in 68 water samples, and the World Health Organization (WHO) recommended guideline value for microcystins in recreational water (24 μg L^{-1}) was surpassed in 18 samples. The microcystin concentrations ranged from 0.11 μg L^{-1} to 2798.81 μg L^{-1} total microcystin. For 45 samples, the dominance of the genera *Microcystis* sp., *Dolichospermum* sp., *Aphanizomenon* sp., *Cyanobium/Synechococcus* sp., *Planktothrix* sp., *Romeria* sp., *Cyanodictyon* sp., and *Phormidium* sp. was shown. Moreover, the *mcyE* gene was detected in 75.71% of all the water samples.

Keywords: planktonic cyanobacteria; microcystin; blooms; monitoring; analysis; mass spectrometry; Liquid Chromatography with tandem mass spectrometry (LC-MS/MS)

Key Contribution: First assessment and comparison of microcystin content; the contribution of microcystin congeners; toxin-producing potential; and dominant genera with standardized methods in water samples from monitoring of five lakes and occasional sampling in the three Belgian regions.

Citation: Van Hassel, W.H.R.; Andjelkovic, M.; Durieu, B.; Marroquin, V.A.; Masquelier, J.; Huybrechts, B.; Wilmotte, A. A Summer of Cyanobacterial Blooms in Belgian Waterbodies: Microcystin Quantification and Molecular Characterizations. *Toxins* **2022**, *14*, 61. https://doi.org/10.3390/toxins 14010061

Received: 19 November 2021
Accepted: 12 January 2022
Published: 16 January 2022

Publisher's Note: MDPI stays neutral with regard to jurisdictional claims in published maps and institutional affiliations.

1. Introduction

A Belgian global picture of the diversity of cyanotoxins and the taxa producing them is currently lacking. This impairs a better understanding of their importance, their yearly variations and the need to prevent and mitigate blooms. Drinking water in Belgium is mostly provided by aquifers and barely depends on reservoirs [1,2]. Moreover, the water supply varies among the regions (Brussels regions, Flanders, and Wallonia). In Flanders, besides aquifers, the Meuse river and the Albert canal, only eight reservoirs supply water. If this is not sufficient, water is imported from Wallonia and neighboring countries [2]. The drinking water in Brussels originates mainly (>90%) from Wallonia [3]. Water exploitation in Wallonia is dependent on ground water for 80% and is supplemented with water exploitation from the Meuse, some old mining sites and six dams [1,4,5]. Further, recreational waters are a sensitive issue, as there is an increasing societal demand for such areas in summer. External sources of eutrophication, such as untreated sewage discharge or agriculture run off, in these waterbodies may promote cyanobacterial bloom formation [6,7]. However, little information about eutrophication sources is available for fresh waterbodies in Belgium.

Ingestion of cyanotoxin contaminated water has been shown to be detrimental to human and animal health [8–14]. Yet in Belgium, no causative link has so far been found between toxic blooms and associated symptoms in humans and animals, such as gastroenteritis, vomiting, liver damage or convulsions [15,16]. However, suspicious bird deaths have been reported, which coincided with a toxic *Microcystis* bloom [17].

Between 1994 and 2008, a few studies identified the morphological and toxin diversity in toxic cyanobacterial blooms in Belgian lakes and ponds [17–24]. The concentrations of total microcystin (MC) measured by Ultra High Performance Liquid Chromatography (UHPLC) reached 18 to 2651 μg g^{-1} dry weight (DW), and the bloom samples contained up to six variants [20,22]. *Microcystis* was found to be the most dominant genus, followed by *Planktothrix*. These results prompted a large scale study (BelSPO project B-BLOOMS2) from 2007 to 2010 [23,24]. During this study, 89% and 83% of the 162 samples tested showed the presence of *mcyE* and *mcyA* genes, respectively. *Microcystis*, *Anabaena* (now taxonomically identified as *Dolichospermum*), *Aphanizomenon*, *Planktothrix*, and *Woronichinia* were the primary bloom-forming cyanobacterial taxa. Furthermore, a quantitative toxin analysis of the samples showed that the total congeners concentration varied from 0.120 μg L^{-1} to 37500 μg L^{-1} total microcystin, analyzed with high performance liquid chromatography—photodiode array detection (HPLC-DAD) and ELISA, in parallel [23–25]. Based largely on the B-BLOOMS2 report, the public authorities started to take action by informing the citizens and including cyanobacterial blooms in monitoring studies. Presently, Flemish and Walloon environmental agencies perform limited monitoring of recreational ponds (where bathing or other activities involving water contact are allowed) but using different protocols. This approach precludes the possibility of obtaining a global overview of blooms in Belgium. Furthermore, the data is limited to a small number of waterbodies, as bathing in Belgian surface waters is very restricted. Our study will extend the data from the B-BLOOMS2 study by providing new toxin, molecular and cyanobacterial occurrence data for water samples after a 10-year hiatus, using a uniform protocol. Moreover, the MC quantification revealed the existence of new microcystin congeners.

MCs belong to the most common cyanotoxins group found worldwide and are produced by multiple taxa (e.g., *Microcystis aeruginosa*, *Planktothrix* sp., *Anabaena*/*Dolichospermum*, *Oscillatoria* and *Nostoc*) [26,27]. The MCs covalently bind the protein phosphatases 1 and 2A (PP1 and PP2A) in eukaryotes, inhibiting their functions and eventually causing cell death [28–30]. Upon ingestion by mammals, the congeners are primarily transported in the liver cells through specific organic anion transporting polypeptides (OATPS) [31–34], which results in a hepatotoxic effect, causing nausea, intestinal problems and liver damage [29,30,35–41]. These toxins can also effect other organs such as the lungs and kidneys [34,39,42]. Human exposure to the MCs through multiple routes have been described (e.g., drinking water, recreational exposure, cyanobacteria-based food supplements, contaminated crops, . . .) [14,16,39,41,43–49]. The most prevalent congener found in Europe is MC-LR, though it is rarely detected in isolation [21,27,50–55]. Therefore, MC-LR is also commonly used in toxicological assays [36,37,40,56,57]. However, other congeners have single or multiple modifications in their structure, such as different amino acids in positions two and/or four, or methylations at different positions [58–60]. The structurally different congeners interact differently with the OATPS, PP1 and PP2A, resulting in different toxicities [31,58,61–64]. The half maximal inhibitory concentration (IC$_{50}$) for PP2a and lethal dose for half of the test population (LD$_{50}$) in vivo for MC-RR are shown to be lower than MC-LR [61,62,64]. More efficient uptake for MC-LF and MC-WR than MC-LR is suggested to correspond to higher in vivo toxicity, while the PP-inhibiting capabilities are comparable [31,34,61,62]. Although differences in toxicity between the congeners are known, no uniform toxicity equivalency factors are available to adjust for the variation in their activity, as is the case for marine toxins [65,66]. An accurate risk assessment, when several congeners are present, is, therefore, difficult. Thus, the World Health Organization (WHO), Environmental Protection Agency (EPA) and other regulatory agencies use the sum of the concentrations for all MCs, described as MC-LR equivalent (MC-LR Equiv). The WHO published, in 1994, an initial tolerable daily intake (TDI) guideline of 0.04 μg kgbodyweight^{-1} day^{-1}, which translated to a concentration of 20 μg L^{-1} in surface waters used for recreational activities [67]. In 2020, the WHO updated its provisional guideline value to 24 μg L^{-1} [39]. The changed value results from a difference in calculations. The original value (20 μg L^{-1}) was based on the proposed TDI, the body weight of an adult and the involuntary ingestion of 100 mL of water during swimming activities [68]. The new guideline value is calculated from the no observed adverse effect level (NOAEL) (40 μg kg^{-1} bodyweight) [57], a ten times reduced uncertainty factor compared to the proposed TDI due to the short-term nature of recreational exposure, a volume of 250 mL of involuntarily ingested water, and taking into account the bodyweight of a child instead of an adult. The WHO guideline values are calculated to provide an adequate margin of safety [39]. The US E.P.A. also provided a new guideline for recreational waters in 2019. A reference dose (RfD) of 0.05 μg kg bodyweight^{-1} day^{-1}, the mean body weight of children between 6 and 10 years and an incidental ingestion factor were used to calculate 8 μg L^{-1} as the recommended value [69].

Since the B-BLOOMS2 study was finalized a decade ago, no studies with standardized protocols have been performed to monitor cyanobacterial diversity and toxins in water samples from all over Belgium. For the first time in Belgium, we utilized Ultra High Performance Liquid Chromatography with tandem mass spectrometry (UHPLC-MS/MS) to identify and quantify the most frequent microcystin congeners in the water samples. By also detecting the genetic potential for synthesizing MCs and the dominant species in the samples, we tried to obtain more insights concerning the bloom characteristics in Belgium. Our cooperation with the three regional environmental agencies in Belgium achieved a sampling on a wide spatial scale of 79 water samples, covering 23 aquatic ecosystems. The species and toxin profiling by a standard set of analyses reveal the importance of monitoring in space and time. Furthermore, by targeting various waterbodies, we aim to identify the locations where monitoring would be needed because there is a health risk. Additionally, the data could help to design more effective prevention and mitigation measures.

2. Results

2.1. Toxin Quantification

In 86.08% (68/79) of the water samples, at least one of the quantified MCs (MC-RR, MC-LA, MC-LF, MC-LR, MC-LY, MC-LW, MC-YR, MC-WR) could be found above the limit of quantification (LOQ = 12.5 μg kg^{-1} before correction for the total sample weight and volume). Moreover, 22.78% (18/79) of the samples contained toxin concentrations higher than 24 μg L^{-1} total microcystin [39]. A complete overview of the results can be found in Table S1 in the Supplementary Materials. The concentration range was between 0.11 μg L^{-1} and 2798.81 μg L^{-1} total microcystin.

Furthermore, 82.61% (19/23) of the waterbodies contained, at least once during the summer, a quantifiable concentration of toxin (>LOQ), whereas concentrations higher than 24 μg L^{-1} total microcystin were detected in 34.78% (8/23) of the waterbodies.

Interestingly, the canal samples (BV1, BV2, BV3) from Brussels contained congener concentrations higher than LOQ. One of the samples even reached 1831.32 μg L^{-1} total microcystin. These results are the first reports on blooms in Belgium waterways.

Validation results for the UHPLC-MS/MS method used for analysis of the water samples can be found in Tables S5 and S6 in the Supplementary Materials.

2.2. Toxin Congener Diversity

The detection frequency of MC-RR was the highest (84.81%), followed by the detection frequencies of MC-LR (81.01%) and MC-YR (50.63%).

MC-LR, MC-RR and MC-YR also contribute the most to the total MCs concentration in individual samples, compared to the other congeners. When comparing the proportions of the highest contributors in Belgium, the proportion of MC-RR is significantly higher than MC-LR, based on the Wilcoxon signed-ranked test ($\alpha < 0.05$). Proportionally, MC-LR is the second-highest contributor, followed by MC-YR. The Wilcoxon test shows a significant difference ($\alpha < 0.05$) between the proportional contributions of MC-YR compared to the other congeners (Figure 1a). Separate statistical analysis of samples containing a total MC concentration above or below the WHO guideline value for recreational water showed that the proportions of MC-YR are also significantly lower in relation to the proportions of MC-LR and MC-RR both above or below the 24 µg L^{-1} total microcystin (Figure 1b,c).

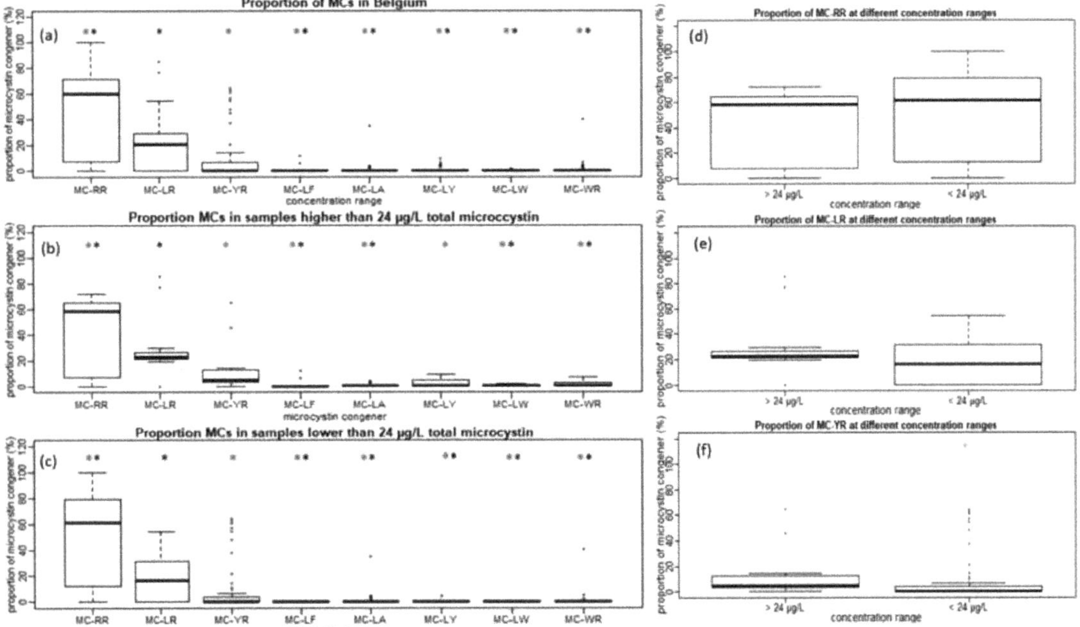

Figure 1. (**a**) The distribution of the proportion of microcystin congeners (MCs) calculated at an individual sample level for all Belgian samples. (**b**) Samples with concentrations higher than 24 µg L^{-1} total microcystin. (**c**) Samples with concentrations lower than 24 µg L^{-1} total microcystin. (**d**) Proportions of MC-RR are compared in samples below and above the World Health Organization (WHO) guideline value. (**e**) Proportions of MC-LR are compared in samples below and above the WHO guideline value. (**f**) Proportions of MC-YR are compared in samples below and above the WHO guideline value. * Proportion of MC is significantly different from MC-LR at $\alpha < 0.05$ using the Wilcoxon test. * Proportion of MC is significantly different from MC-YR at $\alpha < 0.05$ using the Wilcoxon test. ˙ Proportion of MC is significantly different from the proportion of MC at concentration range > 24 µg L^{-1} total microcystin with $\alpha < 0.05$ using the Wilcoxon test.

When comparing samples with a total MC concentration above or below the guideline value, several observations can be made. There was a significant difference in MC-YR (Figure 1f) but no significant difference in the proportions of MC-LR and MC-RR (Figure 1d,e). Comparing the two concentration ranges for the proportions of MC-LA,

MC-LY, MC-LF and MC-LW, a significant difference was shown using the Wilcoxon signed-rank test ($\alpha < 0.05$) (data not shown). A higher diversity of congeners contributed to the total MCs concentration when the concentration was above the WHO guideline value for recreational water (24 µg L^{-1} total microcystin) (Figure 1b,c).

Additionally, the water samples were screened for six other MCs (MC-HtyR, dm-MC-LR, D-asp- MC-LR, dm-MC-RR, D-asp-Dhb-MC-RR and MC-HilR), which are also commonly detected in other studies [20,70,71]. These toxins were not included in the initially designed validation process. However, due to their prevalence and possible toxicity, they were screened. The congeners are identified based on molecular mass, production ions and elution time with the UHPLC-MS/MS method. However, dm-MC-LR and D-asp-MC-LR as well as dm-MC-RR and D-asp-Dhb-MC-RR could not be separated based on these parameters and are reported together (Table 1). We further establish limits of detection for the congeners, shown in Table 1. Overall, dm-MC-RR/D-asp-Dhb-MC-RR were the most abundant congeners in the water samples, followed by dm-MC-LR/D-asp- MC-LR, MC-HilR and MC-HytR, sequentially. An overview of their detection frequency can also be seen in Table 1, as well as their frequency related to the total quantified microcystin concentration in the samples. A complete overview of the results per sample can be found in Table S7 in the Supplementary Materials.

Table 1. Overview of precursor ion, product ions and limit of detection for not validated microcystin congeners. Additionally, the table also includes the detection frequency of the congeners in the analyzed samples at different total microcystin concentrations.

Toxins	MC-HtyR	dm MC-LR/D-asp MC-LR	D-asp-Dhb MC-RR/dm MC-RR	MC-HilR
Precursor ion	1059.5	981.14	512.7	505.3
Product ions	106.9; 135.27	106.8; 135.07	103.2; 135.13	126.99; 134.92
Limit of Detection (µg L^{-1})	0.1	0.1	0.1	0.1
All samples	13.92%	53.16%	77.22%	34.18%
Samples < 1 µg L^{-1} total microcystin	0.00%	9.38%	50.00%	0.00%
Samples > 1 µg L^{-1} total microcystin	23.40%	82.98%	95.74%	57.45%
Samples < 24 µg L^{-1} total microcystin	25.00%	100.00%	100.00%	100.00%

2.3. Molecular Analysis of Water Samples

PCR amplification of the 16S rRNA fragment was attempted for 76 water samples. The fragment was successfully amplified for 45 samples. Some water samples were amplified twice (e.g., BL1.29, BV1.34, B04.29, I04.32), as can be seen in Table S3. Direct Sanger sequencing of the 16S rRNA fragment from the water samples resulted in 49 sequences of sufficient quality that could be analyzed with BLAST. They were of cyanobacterial origin, except for one plastid sequence in sample E04.32. The majority (42/49) of the analyzed 16S rRNA fragments showed 97% or higher similarities to sequences found in Genbank, as shown in Table S3. However, not all samples had a single dominant species, and 31 failed sequencings corresponded to mixtures of sequences that could not be analyzed. In four cases, the PCR with different primer pairs gave different dominant genera, and both are shown in Table S3. For all the samples analyzed with direct Sanger sequencing, *Microcystis* was the most dominant genus (12/76 or 15.79%), closely followed by *Dolichospermum* (11/76 or 14.47%). The third most abundant genus was *Aphanizomenon* (8/76 or 10.53%). Furthermore, the *Synechococcus* and *Planktothrix* genera were dominant in five and three samples, respectively. The *Cyanobium* genus was observed twice, while the *Cyanodictyon*, *Romeria* and, *Phormidium* genera were found once.

The amplicon sequencings by Illumina indicated a dominance of sequences from five OTUs belonging to the *Dolichospermum* genus for sample BL5.29 (71% of the reads), followed by *Microcystis* (20.5% of the reads) and a minor fraction of *Aphanizomenon* and *Cyanobium/Synechococcus*. This corresponds to the dominance of *Dolichospermum* inferred from the direct sanger sequencing. In contrast, sample VL1.36 was completely dominated by sequences of *Microcystis* (99.1% of the reads) followed by 0.6% of the reads affiliated to *Dolichospermum*. However, the direct Sanger sequencing did not give any readable sequence to compare (Table S4).

The dominance of *Microcystis* in VL1.36 coincides with a high diversity (seven microcystin congeners) and a high concentration of total microcystin (128.93 µg L^{-1}). In contrast, only two congeners and a lower total microcystin concentration (1.22 µg L^{-1}) were found in the *Dolichospermum* dominated bloom BL5.29 (Table S1).

The *mcyE* gene amplification was tested in 70 water samples. The *mcyE* gene was detected together with MCs in 71.43% of the samples. Moreover, 4.29% of the samples contained the *mcyE* gene though the presence of MCs could not be detected by UHPLC-MS/MS method. However, the presence of the *mcyE* gene only implies the potential to produce MCs. Microcystin will not be produced when *mcyE* or other genes of the *mcy* gene cluster are lacking, are silenced or contain mutations.

2.4. Cyanobacteria Dominance at Different MCs Concentrations

In the water samples, there was a dominance of the genera *Microcystis*, *Dolichospermum*, *Cyanobium/Synechococcus*, *Aphanizomenon* and *Planktothrix*. Sequences affiliated with *Cyanodictyon*, *Romeria*, and *Phormidium* were each observed in one of the samples. In samples with quantified MCs below the WHO guideline value, *Aphanizomenon*, *Dolichospermum*, *Microcystis*, *Synechococcus/Cyanobium*, *Planktothrix* and a plastid were identified as dominant (Figure 2a). For water samples that contained MCs concentrations above the WHO guideline value for recreational waters, primarily *Microcystis* or *Dolichospermum* could be identified as dominant species, based on the direct Sanger sequencing (Figure 2b). However, in both cases, it was not always possible to determine the dominant taxon, as the direct sequencing was not successful. When MCs were present in samples where *Dolichospermum* was dominant, concentrations ranged between 0.67 to 2420.91 µg L^{-1} total microcystin, while for samples with *Microcystis* as dominant species, concentrations ranged from 1.07 to 2798.81 µg L^{-1} total microcystin. Samples, dominated by *Aphanizomemon*, contained concentrations between 0.11 and 4.35 µg L^{-1} total microcystin. In two water samples from lake H02, where *Planktotrix* was dominant, total microcystin concentrations were 4.05 and 2.80 µg L^{-1}. In contrast, a concentration of 250.35 µg L^{-1} total microcystin was quantified in a water sample from lake I04 containing *Planktothrix*. However, the origin of the MCs can be debated as a week earlier, *Microcystis* was abundant in lake I04, and the bloom could have been in decline, as further shown in Figure 3a. Our results cannot support definitive conclusions that would link high toxin concentrations to a specific cyanobacterial taxon. However, higher total microcystin concentrations are observed when *Microcystis* or *Dolichospermum* are the dominant taxa.

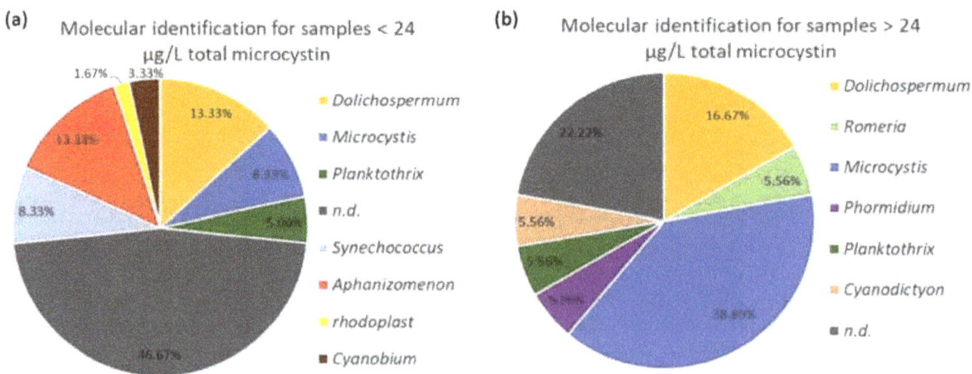

Figure 2. Identification of species using direct Sanger sequencing and BLAST analysis. Samples were

divided based on the WHO guideline value for recreational ponds (24 μg L^{-1} total microcystin (MC)). The "n.d." abbreviation refers to not exploitable 16S rRNA sequences. (**a**) Species distribution for samples containing total MCs concentration below the WHO guideline value. (**b**) Species distribution for samples containing total MCs concentration above the WHO guideline value.

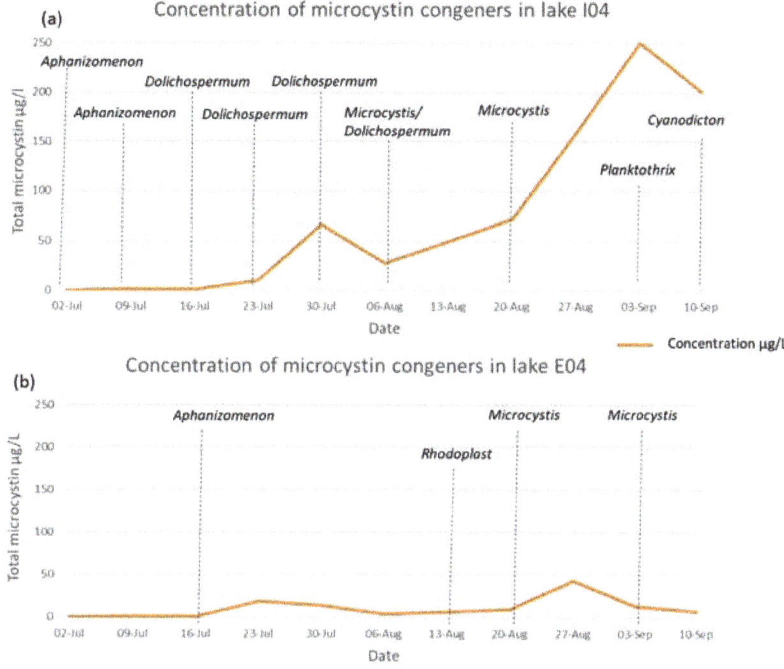

Figure 3. (**a**) Evolution of total microcystin concentrations in Lake I04 (lac de Bambois, Fosses-La-Ville) during the summer of 2019. Dominant genera detected in the samples are also indicated. (**b**) Evolution of total microcystin concentrations in Lake E04 (Grand large, Mons) during the summer of 2019. Dominant genera detected in the samples are also indicated.

2.5. Monitoring of Walloon Recreational Lakes

The weekly sampling of the Walloon lakes provided an opportunity to look at the evolution of the toxin concentration and dominant species with time. The samples from lake I01 (Falemprise) showed a total microcystin concentration higher than 1 μg L^{-1} (I01.31) only once, and when the direct Sanger sequencing was possible, the dominant cyanobacteria belonged to the unicellular *Synechococcus/Cyanobium* (Table S1). This lake was designated as a reference recreational lake during the B-BLOOMS2 study and has been regularly monitored since then. In lake B04 (Renipont plage), the concentration of MCs only rose slightly above 1 μg L^{-1} in three instances when the potentially toxic cyanobacteria genera, *Aphanizomenon* and *Planktothrix* were found (Table S1). MC concentrations in the samples from H02 (Saint Léger sport complex) never reached the WHO guideline for recreational use but were slightly increasing over the summer and peaked at 4.35 μg L^{-1} at the end of August, coinciding with the presence of the potentially toxic *Aphanizomemon* genus. However, the values decreased in the following weeks (Table S1). The two lakes where the WHO guideline value was exceeded were E04 (Grand Large, Mons) and I04 (Lac de Bambois). There was an increase in MCs over the summer, ending with a decrease in September in these samples. However, in the latter lake, the MC values were much higher, reaching 250.35 μg L^{-1}, and the decrease was more gradual. For both lakes, *Microcystis* sp. was prevalent in the samples just before the MC increases (Figure 3a,b). However, in lake

I04, a higher diversity of potentially toxic genera was detected by direct Sanger sequencing, *Aphanizomenon*, *Dolichospermum* and *Planktothrix* (Figure 3a).

3. Discussion

For the first time since the B-BLOOOMS2 study, microcystin congeners in Belgian surface waters have been reliably quantified by a standardized, state of the art analytical method in all three Belgian regions. Moreover, unmonitored waterbodies in the Brussels region were included in the study and provided additional proof of toxic blooms.

True monitoring data was achieved in Wallonia due to the weekly sampling of five recreational lakes during the summer of 2019. Three recreational waters (H02, I01, B04) showed MC values lower than the WHO guideline. Falemprise (I01) and Bambois (I04) had already served as a reference recreational lake with regular monitoring and a sporadically analyzed lake, respectively, during the B-BLOOMS2 study. Falemprise had shown quite a variable planktonic diversity, with a regular presence of potentially toxic taxa, the *mcy* genes and total microcystin concentrations ranging from 0.12 to 6.11 µg L^{-1}. In Bambois, *Anabaena* (now *Dolichospermum*) formed blooms with MCs concentrations lower than 2.6 µg L^{-1} [23]. In the present study, both lakes contained MCs, showing a persistent problem. Lake Bambois presented the highest values for the five Walloon lakes and exceeded the WHO guideline value for recreational use between the end of July to the middle of September (weeks 31 to 37). These results seem to indicate an increase in toxin concentration during the bloom events. *Aphanizomenon* and *Dolichospermum* were observed before the start of the MCs peaks.

Globally, 34.78% of the studied waterbodies contained total microcystin concentrations higher than the suggested guideline for recreational waters. The highest values were observed in waterbodies in Flanders. These samples also contained the highest fresh weights of filtered biomass (20.4 10^{-6} g L^{-1} average) compared to Brussels (4.6 10^{-6} g L^{-1} average) and Wallonia (2.3 10^{-6} g L^{-1} average). However, high concentrations of total microcystin could not always be linked to a high fresh weight of the biomass, as shown in Table S1. The total microcystin range (0.11 µg L^{-1} to 2798.81 µg L^{-1}) found in 68 of the 78 analyzed water samples is similar to the one found a decade ago during the B-BLOOMS2 study (0.120 µg L^{-1} to 37,500 µg L^{-1} total microcystin) by HPLC or ELISA [23]. The specific screening can explain the difference in certain MCs, resulting in underestimating the actual concentration, as the screening method only includes a selection of congeners and might miss certain ones. The ELISA assays, performed during the B-BLOOMS2 study, have the advantage of detecting all the possible hydrophobic adda groups specific for MCs (Figure S1) in the samples. In contrast, our triple quadrupole MS/MS method is very specific and will detect only the targeted toxins selected a priori in the detection method. This type of method is not well suited for a full scan analysis of samples. Worldwide, total microcystin concentrations can reach a maximum of 42.7 mg L^{-1}, although concentrations are usually lower [51–54,70,72–80]. For instance, during a sampling campaign in the United Kingdom, only 18% of the samples contained a concentration above 20 µg L^{-1} total microcystin [51].

In terms of congener diversity, MC-RR (84.81%), MC-LR (81.01%), and MC-YR (50.63%) were most prevalent in our study. Earlier, MC-LR was found to be the dominant congener (in 64% of analyzed water samples) in Wallonia and Luxemburg by Willame et al. [20]. MC-YR was the second most common congener, while MC-RR was not found [20]. Other reports confirm the presence of these most abundant congeners (MC-LR, MC-RR and MC-YR) in western Europe [26,51,53,54,70,71,81–83].

Moreover, five MCs (MC-LA, MC-LF, MC-WR, MC-LW and MC-LY) were also quantified in lower concentrations, while six other MCs (MC-HytR, dm-MC-LR, D-asp- MC-LR, dm-MC-RR, D-asp-Dhb-MC-RR and MC-HilR) were only screened and not quantified during our study. For the five quantified MCs, only a minor contribution to the total microcystins concentration at different concentration levels was shown. The MCs that were not quantified were primarily present in samples with total microcystin concentrations above the WHO guideline limit (24 µg L^{-1}). The lower contribution or presence of these eleven congeners at higher total microcystin concentrations are consistent in the samples from independent waterbodies and suggest that the biosynthesis dynamics of these toxins is different compared to the more abundant congeners (MC-LR, MC-RR, MC-YR). However, shifts in environmental factors, cyanobacterial strain diversity, growth phase and toxin production dynamics could influence the overall toxin diversity and quota in the blooms [84].

Overall, the minor contribution or presence of these eleven MCs (quantified or not) to the total MC concentration might be of minor importance for the degree of health risk compared to the more abundant MCs (MC-LR, MC-RR and MC-YR). In most cases, the sum of the MC-LR, MC-RR and MC-YR concentrations already exceeded the new WHO guidelines for microcystin, without the contributions of the other congeners (Table S1) [39]. However, different geographical origins, species and environmental factors might influence which congeners are more abundant. Analysis of the most common congeners during monitoring is therefore still advisable [20,53–55,71,76,77,85,86].

Incorporation of the toxic demethylated congeners in the quantification methodology would be similarly advisable in the future [16].

Besides toxins, *mcyE* was detected in almost 76% of the water samples. Comparing these results is difficult due to differences in the monitoring frequency and amplification method. For instance, the B-BLOOMS2 study found *mcyE* in 89% of their samples, but the number of samples was larger due to more regular monitoring in surface waters prone to bloom events and the addition of occasional bloom samples [23]. Similarly, Moreira et al. (2020) also found variations in *mcyE* presence in various lakes in Portugal. During their sampling from April to September, the *mcyE* detection rate varied from 33.3% to 83.3% for different lakes, but they used a different amplification method than our study [80]. A good correlation between toxin content and the *mcyE* compared to other genes has been shown in the multiple studies reviewed in Pacheco et al. [87]. The authors also pointed out that correlations between *mcyE* copy number and toxin concentration are still controversial. Nonetheless, this data supports the need for our approach of assessing the potential of toxin production by the detection of the *mcyE*.

With the direct Sanger sequencing approach, only the most abundant species would be detected if there was a strong dominance in the sample. Indeed, if the cyanobacteria populations were too heterogeneous, the direct sequencing would fail because the resulting chromatograms would not be interpretable. Therefore, only 45 out of 79 samples produced usable data due to species heterogeneity. Nevertheless, direct Sanger sequencing showed that *Microcystis, Dolichospermum/Anabaena, Synechococcus/Cyanobium, Planktothrix* and *Aphanizomenon* were generally observed in the samples, as earlier described in the B-BLOOMS2 study. However, *Woronichinia*, also present during B-BLOOMS2, was not detected. Our sampling season, which focused on the warmer months of the year, might have missed the occurrence of *Woronichinia*, which is more prevalent at temperatures below 15 °C [88]. Yet, they might also be present in the more heterogeneous samples but could not be detected with our current methods. Furthermore, the *mcyE* gene was found in samples containing *Cyanobium*, which raises the question of the possible toxicity of this genus, as discussed during B-BLOOMS [23]. Little information is available about the toxicity of picocyanobacteria in eutrophic waters. *Cyanobium* have previously been identified in freshwater ecosystems [89,90], although their toxicity has been put in question. Multiple studies showed low amounts of MCs being produced [15,90–92], while the sequenced genomes of *Cyanobium* strains seem to lack a complete NRPS/PKS gene clusters [93].

Amplicon sequencing by a high-throughput sequencing technique is an alternative to direct Sanger sequencing that allows the identification of species in complex populations, allowing the study of heterogeneous samples. However, amplicon sequencing protocol is more expensive and time-consuming. Thus, it was tested for in two water samples, of which one (BL5.29) was used for both sequencing types. This comparison shows that the dominant taxon was quasi-identical with both methods. Indeed, the Sanger sequence was 99.8% similar to the most similar hit for the dominant OTU72 (*A. ellipsoides* Ana HB).

The samples from waterbodies were either collected through one of the regional monitoring programs or following a sporadic bloom notice by the regional environmental agencies in an unmonitored waterbody. The samples from Wallonia and some of the samples from Flanders were obtained from already established monitoring programs for recreational waterbodies, whereas the samples from ponds in Brussels were obtained specifically for the study when a bloom notice was received. In the cases with monitoring programs, the situation is followed by the public authorities and the recreational waterbodies in both regions get regularly closed if (toxic) blooms are observed. In the not monitored ponds in the Brussels region, we found that 66.67% and 16.67% of the samples were above LOQ and 24 μg L^{-1} total microcystin, respectively [67,68]. By comparison, the total MC concentration rose above LOQ and 24 μg L^{-1} total microcystin in 86.08% and 22.78% of the Belgian waterbodies, respectively, at one point in time. Even though the numbers seem to be lower in Brussels compared to the rest of the country, this is artificial. In Brussels, only a small number of waterbodies was sampled. Furthermore, these results were obtained outside the standardized protocol and from only three or fewer time points per site. Therefore, we could have missed bloom peaks during the summer. As we reported, the canal in Brussels also contained a total of 1831.32 μg L^{-1} total microcystin at one sampling spot. Currently, there is a public dynamic in the region to make more waterbodies available for recreational use during the summer. Moreover, the unauthorized use of waterbodies as bathing areas was also common practice during hot summers. Without proper monitoring, this could create a public health risk for humans and domestic animals [9]. In addition, other usages of this water need to be considered. For instance, pond water could be used to irrigate urban vegetable gardens, and water from canals could be used in agriculture. Several research groups have already shown that plants could accumulate MCs when irrigated with contaminated water [45,47,94–96]. The connection between water usage and toxin accumulation in plants needs to be further investigated in the future.

Based on our results, it appears that monitoring of any potential bathing sites where unauthorized bathing occurs could be recommended. In the context of Belgium, this would be in the Brussels region. Monitoring could reveal links between public health issues and any potential hazard related to cyanobacterial blooms. The other two Belgian regions monitor their official bathing sites, using different sampling protocols and analysis techniques. Harmonizing the monitoring methods could provide insight into the species and toxin diversity in the blooms in Belgium. Moreover, it might help to uncover environmental drivers that promote the blooms. Techniques used during the monitoring could vary depending on the expertise and resources that are available. Cell counting and species identification are relatively low tech monitoring tools to determine the intensity of the blooms and quantify the potential toxin-producing species, but they are time-consuming and need taxonomic expertise. ELISA and *mcyE* amplification are fast, relatively cheap tests appropriate for screening MCs or MC-production potential, respectively, when a bloom is observed. However, these four techniques need to be supplemented with UHPLC-MS/MS to accurately determine MCs concentrations during and after the bloom to ensure public safety. When toxin equivalency factors become available for the different congeners, UHPLC-MS/MS approaches will also be crucial to accurately determine the risk.

More detailed information about the bloom incidences in a country can also benefit possible mitigation strategies in the future. Preventive strategies could be designed to reduce the influx of nutrients where this is feasible. This strategy requires information about the sources of eutrophication, which is still lacking for most fresh waterbodies in Belgium. During this study; the regional environmental agency listed agriculture, the discharge of purified sewage water, water influx by a canal and the feeding of fish during recreational fishing as potential sources of eutrophication in the Walloon lakes. For smaller ponds in Brussels, an overpopulation of waterfowl can cause nutrient loading due to an abundance of excrement. Another mitigation strategy is hydrogen peroxide treatments.

This treatment can eliminate the bloom and MCs but needs to be optimized based on bloom density and the quantity of toxins. For toxin quantification and assistance to prevention, analytical methods, such as the one presented in this paper, would be suitable [97–105]. Approaches such as flock lock or flock sink techniques could be a viable solution to prevent bloom incidence by capturing phosphorous on the bottom of larger waterbodies with a greater depth. However, the sediments of shallow recreational lakes might be too frequently disturbed for this approach to be effective [106–109]. External phosphorous loading in the waterbodies after treatment (e.g., sewage disposal, floods, …) will also undermine the effectiveness of these treatments. To ensure public safety, monitoring of waterbodies will always be necessary and toxin quantification should be included as a part of the monitoring techniques.

4. Conclusions

During this study, we validated and used a UHPLC-MS/MS method for the first time to analyze Belgian water samples from the three different regions with an identical method. The microcystin concentrations found clearly illustrate a persistent problem of toxic blooms throughout Belgium with a potential health impact. The three most abundant MCs (MC-RR, MC-LR, MC-YR) contributed the most to the total microcystin concentrations. Our fast and efficient method can be applied to monitoring programs in Belgium and other parts of the world. PCR amplification of the *mcyE* gene linked its occurrence to the toxin presence for 71.43% of the water samples. Moreover, the dominant blooming taxa were also determined in a number of samples. Interestingly, this study also characterized a cyanobacterial bloom in a Belgian canal for the first time. The abundance of water samples that contained MCs shows the need to enlarge the sampling of waterbodies where there could be a risk of human exposure and include them in existing or new monitoring programs.

5. Materials and Methods

UPLC/MS grade solvents (Biosolve B.V., Valkenswaard, The Netherlands) were used for extraction or basis for the mobile phase. The MCs standards were ordered as a solid powder from Enzo Life Sciences (Antwerp, Belgium)®, except for D-asp-Dhb-MC-RR from Cyano Biotech GmbH (Berlin, Germany) and Dm-MC-RR from Novakits (Nantes, France). They were initially dissolved in 100% methanol and used to prepare mixed stock solutions in 50% methanol with 1% acetic acid. The dissolved cyanotoxin standards were kept at −20 °C. Whatman GF/C grade filters were obtained from Sigma Aldrich (Overijse, Belgium).

5.1. Sampling

The sampling was performed from July until mid-September in 2019 at 23 different locations in the three Belgian regions: Wallonia (5 locations), Flanders (7 locations) and Brussels (11 locations). The sampling frequency was dependent on the region, type of waterbodies and access to the lakes (directly or via the environmental agency). Recreational waterbodies are defined as ponds and lakes where bathing is permitted. The water samples were either collected every week or only after a bloom was observed. Each sample was annotated by combining a three digit annotation of the sample site followed by a number giving the week of the year (e.g., XYZ.12). Names for the waterbodies can be found in Table S2 in Supplementary files.

In Wallonia, water samples were collected weekly in 5 recreational lakes (I01, I04, E04, B04 and H02), independently of the presence of a bloom following a standardized protocol. The environmental agency Institut Scientifique de Service Public (ISSEP) sampled the surface water with a 500 mL glass bottle at a fixed point. Samples were stored at 4 °C and later transported to Sciensano within three days of the collection for further processing.

In Flanders, water samples were taken from 3 recreational waterbodies (AN1, AN2 and AN3) by the environmental agency Vlaamse Milieu Maatschappij (VMM) only when a bloom was present. The samples were stored in plastic containers at 4 °C before further processing and analysis by Sciensano. Four ponds (GH1, VL1, VL2 and VL3) that were not used for recreation were sampled by Sciensano. GH1 is a sedimentation pond for wastewater, while VL1, VL2 and VL3 are shallow ponds in parks where fishing is allowed. The surface water was sampled with 500 mL sterilized glass bottles. They were processed the same day. In this case, public media had indicated the presence of the blooms. These waterbodies were only sampled a second time by Sciensano if a bloom was still present.

In Brussels, we performed samplings in ponds where bathing is not allowed and thus are not considered as recreational waterbodies. Each waterbody was initially sampled after a bloom notification by the regional environmental agency Environment.Brussels or Port.Brussels. The latter manages the port estate in the Brussels capital region. In total, 8 ponds (BL1-8) and 3 locations in the Brussels canal (BV1-3) were sampled. Each spot was sampled at least a second time independent of bloom presence, except for two spots in the canal. Samples were taken from the surface water with 500 mL sterilized glass bottles. They were processed the same day. An overview of all the sampling sites is shown in Figure 4. An overview of the sample sites with waterbody type can be found Supplementary Material (Table S2).

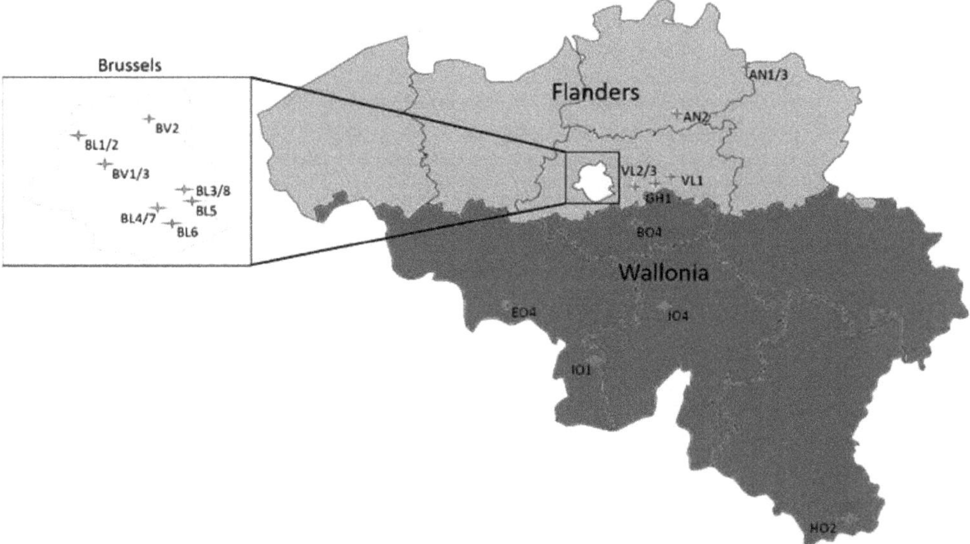

Figure 4. Map of Belgium showing the sample sites. The first three letters of the sample names are used as abbreviation. In Flanders, 7 sites were sampled (AN1-3, VL1-3 and GH1). In Wallonia, 5 recreational lakes were sampled (I01, I04, E04, B04 and H02). For clarity, the Brussels region is enlarged. Here 8 ponds were sampled (BL1-8), as well as the Brussels canal at 3 different sites (BV1-3). Place names for the waterbodies and their type can be found in Table S2 in Supplementary files.

In general, 150 mL of the sample was filtered on a GF/C Whatman® filter under vacuum to collect the biomass. Lower volumes were filtered due to clogging when dealing with high bloom density. The sample filters were weighed before and after filtration to determine the weight of the wet biomass. The sample filters were stored at $-20\,^{\circ}$C before analysis. Filtration was performed in duplicate. One filter was used for the quantification of the MCs, while the other was used for the molecular work. The filtrates were collected and stored at $-20\,^{\circ}$C to determine extracellular toxin concentration.

5.2. Quantitative Analysis of Microcystin Congeners

5.2.1. Intracellular and Extracellular Microcystin Extraction

Only the most common toxins in Europe (MCs) were selected for our quantification method. Earlier studies in Belgium suggested that these toxins are the most prevalent public health threat [20,23]. To properly validate the method, only commercially available MCs were selected.

The method used for analysis was validated in house. Results of the validation can be found in Tables S5 and S6 in the Supplementary Materials. The filters, containing biomass, used for toxin extraction underwent a freeze-thaw step and liquid extraction. When the filters were initially stored at $-20\,^{\circ}\mathrm{C}$, they only need to be defrosted. The filters were cut in half and weighted. For the liquid extraction, 4.5 mL 80% methanol was added together with the filter in 50 mL plastic tubes. Solvent and biomass contact was increased by regularly mixing during 1 h. The samples were centrifuged for 10 min at 3900 rpm.

The extract was filtered through a Phenomenex 0.2 μm RC syringe filter (Utrecht, The Netherlands) to remove debris. Samples were stored in a 15 mL plastic tube at $-20\,^{\circ}\mathrm{C}$. Samples with high concentrations of the MCs were diluted after the initial analysis to fit within the range of the calibration curve. The calibration curve was made in a blank matrix.

The sample filtrates (extracellular fraction) were also purified using a Phenomenex 0.2 μm RC syringe filter and analyzed separately through direct injection of 10 μL the Xevo TQ-S, similar to Turner et al., 2018 [51].

5.2.2. Detection and Quantification of Cyanotoxins

The detection and quantification parameters were identical for intra- and extracellular toxins analysis. A Waters Acquity UPLC H-class (Eten-Leur, The Netherlands) connected to a Waters XEVO TQ-S was used for the detection of the cyanotoxins. A 1.7 μm, 2.1 mm × 100 mm Waters Acquity BEH C18 column fitted with a Waters Acquity BEH C18 1.7 μM VANGUARD PRE-Col separated the toxins under the influence of a gradient elution program. The fraction of acetonitrile (B) in the eluent changed as followed: 0 min, 2% B; 1.00 min, 40% B; 7.00 min, 55% B; 7.20 min, 98% B; 8.00 min, 98% B; 9.00 min; 2% B; 12 min, 2% B. Both Organic and water phases were supplemented with 0.025% formic acid. The flow rate was 0.5 mL min^{-1}. The column was heated to $60\,^{\circ}\mathrm{C}$, and 10 μL of sample was injected.

Multiple reaction monitoring (MRM) was then used to detect the toxins by selectively quantifying compounds within complex samples. The triple quadrupole MS initially targeted the ions corresponding to the toxins of interest, referred to as the "precursor ion". Two product ions from the collision induced fragmentation were selected. One was used for quantification of the cyanotoxin, the other as a qualifier. The MS parameters were set according to the literature data and optimized to the instrument setting (Table 2).

Table 2. MS/MS parameters for eight microcystin congeners (MCs).

Toxins	Precursor Ion m/z	Quantifier Ion m/z	Collision Energy (eV)	Cone Voltage (V)	Qualifier Ion m/z	Collison Energy (eV)	Cone Voltage (V)
MC-LR	995.4	135.0	70	80	213.1	60	80
MC-RR	519.8	134.8	30	50	107.2	60	50
MC-YR	1045.5	135.3	80	60	212.9	60	60
MC-WR	1068.4	135.3	70	100	213.1	60	100
MC-LY	1002.4	135.4	60	50	213.0	50	50
MC-LA	910.3	135.1	60	50	107.1	80	50
MC-LF	986.3	135.0	60	70	213.1	60	70
MC-LW	1025.4	134.9	60	60	213.1	50	60

After quantification, the concentration for each cyanotoxin was recalculated to μg L^{-1}, corrected with the mass of the original mass of the filter and the filtered volume of the sample. The concentration of each MC in the filtrate (extracellular fraction) was added to the final concentration of the MC extracted from the biomass. Thereafter, the sum of all the congeners was calculated to provide a µg L^{-1} total microcystin value.

Congener proportions to the total MCs concentration were calculated in each sample. The differences in proportions for the separate congener were then compared for Belgium using the Wilcoxon test at $\alpha < 0.05$. Additionally, the same statistical test was used to compare the difference in congener proportions for samples containing MCs concentration higher and lower than 24 µg L^{-1} total microcystin, separately. The samples without MCs were excluded. Proportions of MC-LR, MC-RR and MC-YR were also compared between the two concentration ranges.

5.3. Molecular Analysis of the 16S rRNA and the mcyE Gene

5.3.1. DNA Extraction

First, 0.8 mL lysis buffer (40 mM EDTA 5, 50nM Tris-HCl, 0.75 M sucrose) was added to each sample filter (containing biomass), and a bead-beating step (at 30 m s^{-1} for 30 s) was performed. Then, a lysozyme (Sigma-Aldrich, St. Louis, MI, USA) (20 mg mL^{-1}) digestion for 30 min at 37 °C was followed by a treatment with 22.22 mg mL^{-1} proteinase K (Macherey-Nagel, Düren, Germany), supplemented with 80 µL SDS (100%), for 2 h at 55 °C. The lysate was transferred to a new Eppendorf tube. Subsequently, the filters were rinsed with 1 mL lysis buffer during a 10 min incubation at 55 °C. The second lysate was stored in another Eppendorf tube.

A Phenol-chloroform-isoamyl alcohol solution (25:24:1, pH 8) (VWR, Leuven, Belgium) was added in an equal volume to the extract volume (V:V) to both lysates. Next, the samples were centrifuged at 14,000× *g* for 15 min. The upper phase of each tube was transferred to a new Eppendorf tube, and chloroform-isoamyl alcohol (24:1, pH 8) was added V:V. The tubes were centrifuged again at 14,000× *g* for 15 min to collect the upper phase. For each sample, the two lysates were combined

Finally, the DNA was precipitated with 0.1 V:V of sodium acetate (3 M, pH 5.2) and 0.6 V:V of cold isopropanol. After centrifugation, the DNA was rinsed once with 300 µL ice-cold ethanol (Merck, Branchburg, NJ, USA) (100%) and once with 300 µL ice-cold ethanol (70%). The supernatant was removed, and the pellet was air dried. Finally, the DNA pellet was dissolved in 100 µL TE buffer (10mM Tris-HCl and 1mM EDTA, pH 8) and stored at −20 °C.

5.3.2. Gene Amplification of Partial rRNA and mcyE Gene Sequences

For the rRNA gene sequences, two protocols were tested. A long rRNA fragment was amplified with the cyano-specific primers 359F/23S30R [110] using the SuperTaq Plus$^{©}$ enzyme (HT Biotechnology, Cambridge, UK), buffer and dNTPs obtained from SpharoQ (NL). The amplification program was 95 °C—5 min, 35 times; 95 °C—30 s, 57 °C—45 s, 68 °C—1 min; followed by 69 °C—5 min, 16 °C—infinite. As a shorter PCR product could give a higher amplification efficiency, the primer pair 359F/781R [111] was tested later. However, the SuperTaq Plus$^{©}$ enzyme was no longer commercialized and was replaced by the Q5 High Fidelity polymerase (New England Biolabs, Ipswich, MA, USA) for the majority of the PCR reactions. The amplification program was: 98 °C—5 min, 35 times; 98 °C—30 s, 65 °C—45 s, 72 °C—1 min; 72 °C—5 min, 16 °C—infinite. The *mcyE* gene involved in the production of MCs was amplified with the primer pair mcyEF2/mcyER4 [112] using the SuperTaq Plus© enzyme. The amplification program was 94 °C—3 min, 30 times; 94 °C—30 s, 57 °C—45 s, 68 °C—1 min, followed by 68 °C—10 min and 16 °C infinite, as described in the final report of B-BLOOMS2 [23]. Amplifications were performed in a Thermal cycler T100 (Bio-Rad, Hercules, CA, USA). The presence of PCR products of the right size was visualized by electrophoresis on a 1.5% agarose gel during a 95 min run at 90 V.

5.3.3. Sanger Sequencing and Sequence Analysis

After PCR, the 16S rRNA amplicons were sent for Sanger sequencing with primers 359F, 781R or 23S30R at Giga Genomics (ULiege) [110,111]. Some sequences were of bad quality, which prohibited further analysis. These sequences probably resulted from a mixture of organisms without clear dominance by one taxon. The forward and reverse sequences were not obtained in all cases for each PCR product, and therefore, the individual sequences of a single strand were used for further analysis, admitting that some sequencing errors might be present but that the quality would be sufficient to determine the dominant genus. In three cases, the sequences obtained on different PCR products (short or longer ones) for the same sample were affiliated to different genera. The sequences used during the further analysis can be found in the supplementary files (Table S8).

The NCBI nucleotide BLAST (basic local alignment search tool) was used to identify the most closely related strain sequences for the 16S rRNA sequences, using individual sequences obtained by the different primers tested and the identification was based on this data, as shown in Table S3.

5.3.4. Amplicon Sequencing with the Illumina Technique

For samples BL5.29 and VL1.36, partial 16S rRNA gene sequences were obtained by PCR using the primer set CYA359F and CYA781Ra/CYA781Rb, which amplifies the V3-V4 region of the cyanobacterial 16S rRNA gene [111]. Primers were modified to include a 10-bp sample-specific barcode tag at the 5′ end to allow samples to be multiplexed for sequencing. PCR reactions were performed in triplicates in order to minimize the influence of amplification biases. These were pooled to equivalent concentrations and purified using the NucleoSpin® Gel and PCR Clean-up kit (Macherey-Nagel, Düren, Germany). Purified samples were sent to Genewiz (South Plainfield, NJ, USA), where sequencing adapters were ligated to the amplicons and sequencing was performed using Illumina MiSeq (Illumina, San Diego, CA, USA) using 2 × 300 bp paired-end libraries. The bioinformatic analysis is adapted from a validated method by Pessi et al. and consists of processing raw reads to remove chimeric sequences, followed by the clustering into an operational taxonomic unit (OTU) [113]. Briefly, paired-end reads were merged, filtered and only reads containing both barcodes in the 3′ and 5′ ends were kept. Two and zero mismatches were allowed to the primer and barcode sequences, respectively, and reads with a maximum expected error of more than 0,5 and a length of less than 370 bp were removed. Singletons were removed, and remaining quality-filtered sequences were denoised to remove chimaeras and sequencing errors using unoise3 [114]. The denoised operational taxonomic units (ZOTUs) obtained were then clustered into OTUs at a 99% similarity threshold [115]. The representative sequence of an OTU is the most abundant unique sequence of each OTU cluster. Taxonomic classification was performed by extracting from Genbank the most closely related sequences of each OTU using BLAST.

Supplementary Materials: The following are available online at https://www.mdpi.com/article/10.3390/toxins14010061/s1, Figure S1: Microcystin core structure with annotated adda group, Table S1: Overview of experimental data for the water samples, concentrations of MC congeners and total microcystin ($\mu g\ L^{-1}$), presence of genes coding for 16S rRNA and *mcyE*. molecular identification, coverage and identity % of the most similar hit by BLAST and primer used for Sanger sequencing. The first three characters of the samples annotation indicate the sample location, while the last two numbers indicate the week of the year when the sample was collected. Samples are grouped by origin, Flanders. Brussels and Wallonia. N.A. indicates samples for which no PCR was performed. "-" represent samples were no visible bands were obtained after gel electrophoresis of the PCR products or sequencing results were off too pore a quality to provide a reliable result. "?" represent inconclusive PCR results where the band was visible but was too faint or not at the correct height. The asterisk * indicates that the sequence was obtained on the basis of a PCR reaction produced with the primer pair 359F-781R, whereas the others sequences were obtained with 359F-32S30R. "ⁱ" denotes the samples that were analyzed using Illumina technology, Table S2: Overview of sampling sites in the different regions, water sample annotation, waterbody type and specific monitoring for cyanobacteria, Table S3: Detailed information for taxonomic identification based on BLAST analysis. * Samples for which PCR with different primers gave different dominant taxa, Table S4: Number of reads and BLAST analysis of the OTUs obtained with the Illumina amplicon sequencing, Table S5: Validation results for UHPLC-MS/MS method quantification method of 8 MCs and Nodularin in filtered cyanobacterial biomass. The different validation concentration levels Limit of detection (LOD) was set at lowest tested concertation where signal to noise ratio was higher than 3. Limit of Quantification (LOQ) was selected as lowest concentration for which the method was validated. Signal to noise ratio for LOQ should be above 20. Additionally, values for recovery, repeatability, reproducibility, measurement uncertainty (MU) and R^2 of the linear curve are presented, Table S6: Results of the ion ratio for the validation of the UHPLC-MS/MS method quantification method of 8 MCs and Nodularin in filtered cyanobacterial biomass. Acceptation criteria are based on European Decision 2002/EC/657, Table S7: Detection results for six additional microcystin congeners with a limit of detection (LOD) at 0.1 µg L^{-1}. Not detected is abbreviated by "n.d.". Detected MCs are annotated as > LOD, Table S8: Overview of the single sequences amplified by the Sanger method used for taxonomic identification based on BLAST analysis.

Author Contributions: Conceptualization, W.H.R.V.H., A.W., M.A. and B.H.; methodology, W.H.R.V.H., A.W. and B.H.; validation, W.H.R.V.H. and B.H.; formal analysis, W.H.R.V.H. and B.D.; investigation, W.H.R.V.H., A.W., B.D. and V.A.M.; resources, A.W., M.A., V.A.M. and J.M.; data curation, W.H.R.V.H.; writing—original draft preparation, W.H.R.V.H., A.W., B.D., B.H. and M.A.; writing—review and editing W.H.R.V.H., A.W., B.D., M.A., V.A.M. and J.M.; visualization, W.H.R.V.H.; supervision, A.W., M.A. and J.M.; project administration, W.H.R.V.H., A.W., M.A. and J.M.; funding acquisition, A.W. and M.A. All authors have read and agreed to the published version of the manuscript.

Funding: This research was funded by the Federal Agency for the Safety of the Food Chain (FAVV-AFSCA-FASFC) in the framework of the work of WVH towards his PhD. FAVV-AFSCA-FASFC cannot be held liable for the use of the data nor the conclusions that could be drawn from their treatment. Annick Wilmotte is Senior Research Associate of the Belgian Funds for Scientific Research (FRS-FNRS).

Institutional Review Board Statement: Not applicable.

Informed Consent Statement: Not applicable.

Conflicts of Interest: The authors declare no conflict of interest.

References

1. Direction des Eaux Souterraines; Direction de la Coordination des Données. Qualite Des Eaux Distribuees Par Le Reseau Public En Wallonie. Available online: http://environnement.wallonie.be/de/eso/eau_distribution/pdf/eau_distribution.pdf (accessed on 29 March 2021).
2. Vlaamse Milieu Maatschappij (VMM) Status Indicator Bevoorrading Leidingwater: Situatie op 24/09/2020. Available online: https://www.vmm.be/data/kraanwater-beschikbaarheid/indicator_bevoorrading_kraanwater (accessed on 30 March 2021).
3. Environnement Brussels Kwaliteit van Water Bestemd voor Menselijke Consumptie Periode 2014–2015–2016. Available online: https://leefmilieu.brussels/sites/default/files/user_files/rapport_drinkwater_2014-2015-2016.pdf (accessed on 25 March 2021).
4. Brahy, V.; Huart, M.; T'serstevens, J.-J. L'exploitation des Ressources en Eau de Surface. Available online: http://etat.environnement.wallonie.be/files/live/sites/eew/files/Publications/Rapport%20analytique%202006-2007/Chap04/3_ExploitationEauSurface/RES_EAU_02.pdf (accessed on 18 November 2021).
5. État de l'Environnement Wallon Production d'Eau de Distribution. Available online: http://etat.environnement.wallonie.be/contents/indicatorsheets/RESS%203.html (accessed on 18 November 2021).
6. Anderson, D.M.; Glibert, P.M.; Burkholder, J.M. Harmful Algal Blooms and Eutrophication: Nutrient Sources, Composition, and Consequences. *Estuaries* **2002**, *25*, 704–726. [CrossRef]
7. O'Neil, J.M.; Davis, T.W.; Burford, M.A.; Gobler, C.J. The Rise of Harmful Cyanobacteria Blooms: The Potential Roles of Eutrophication and Climate Change. *Harmful Algae* **2012**, *14*, 313–334. [CrossRef]
8. Edwards, C.; Beattie, K.A.; Scrimgeour, C.M.; Codd, G.A. Identification of Anatoxin-A in Benthic Cyanobacteria (Blue-Green Algae) and in Associated Dog Poisonings at Loch Insh, Scotland. *Toxicon Off. J. Int. Soc. Toxinol.* **1992**, *30*, 1165–1175. [CrossRef]
9. Foss, A.J.; Aubel, M.T.; Gallagher, B.; Mettee, N.; Miller, A.; Fogelson, S.B. Diagnosing Microcystin Intoxication of Canines: Clinicopathological Indications, Pathological Characteristics, and Analytical Detection in Postmortem and Antemortem Samples. *Toxins* **2019**, *11*, 456. [CrossRef]

10. Bauer, F.; Fastner, J.; Bartha-Dima, B.; Breuer, W.; Falkenau, A.; Mayer, C.; Raeder, U. Mass Occurrence of Anatoxin-a- and Dihydroanatoxin-a-Producing *Tychonema* Sp. in Mesotrophic Reservoir Mandichosee (River Lech, Germany) as a Cause of Neurotoxicosis in Dogs. *Toxins* **2020**, *12*, 726. [CrossRef]

11. Carmichael, W.W.; Azevedo, S.M.; An, J.S.; Molica, R.J.; Jochimsen, E.M.; Lau, S.; Rinehart, K.L.; Shaw, G.R.; Eaglesham, G.K. Human Fatalities from Cyanobacteria: Chemical and Biological Evidence for Cyanotoxins. *Environ. Health Perspect.* **2001**, *109*, 663–668. [CrossRef] [PubMed]

12. Zurawell, R.W.; Chen, H.; Burke, J.M.; Prepas, E.E. Hepatotoxic Cyanobacteria: A Review of the Biological Importance of Microcystins in Freshwater Environments. *J. Toxicol. Environ. Health B Crit. Rev.* **2005**, *8*, 1–37. [CrossRef] [PubMed]

13. Codd, G.A.; Morrison, L.F.; Metcalf, J.S. Cyanobacterial Toxins: Risk Management for Health Protection. *Toxicol. Appl. Pharmacol.* **2005**, *203*, 264–272. [CrossRef] [PubMed]

14. Codd, G.; Bell, S.; Kaya, K.; Ward, C.; Beattie, K.; Metcalf, J. Cyanobacterial Toxins, Exposure Routes and Human Health. *Eur. J. Phycol.* **1999**, *34*, 405–415. [CrossRef]

15. Bláha, L.; Babica, P.; Maršálek, B. Toxins Produced in Cyanobacterial Water Blooms—Toxicity and Risks. *Interdiscip. Toxicol.* **2009**, *2*, 36–41. [CrossRef]

16. Testai, E.; Buratti, F.M.; Funari, E.; Manganelli, M.; Vichi, S.; Arnich, N.; Fessard, V.; Sialehaamoa, A. Review and Analysis of Occurrence, Exposure and Toxicity of Cyanobacteria Toxins in Food. *EFSA Extern. Sci. Rep.* **2016**, *13*, 998E. [CrossRef]

17. Wirsing, B.; Hoffmann, L.; Heinze, R.; Klein, D.; Daloze, D.; Braekman, J.C.; Weckesser, J. First Report on the Identification of Microcystin in a Water Bloom Collected in Belgium. *Syst. Appl. Microbiol.* **1998**, *21*, 23–27. [CrossRef]

18. Van Hoof, F.; Castelain, P.; Kirsch-Volders, M.; Vankerkom, J. Toxicity Studies with Blue-Green Algae from Flemish Reservoirs. In Proceedings of the First International Symposium on Detection Methods for Cyanobacterial (Blue-Green Algal) Toxins, Claverton Down, UK, 27–29 September 1993; The Royal Society of Chemistry: Cambridge, UK, 1994; Volume 149, pp. 139–141.

19. Willame, R.; Hoffmann, L. Bloom-Forming Blue-Green Algae in Belgium and Luxembourg. *Algol. Stud. Hydrobiol.* **1999**, *94*, 365–376. [CrossRef]

20. Willame, R.; Jurczak, T.; Iffly, J.-F.; Kull, T.; Meriluoto, J.; Hoffmann, L. Distribution of Hepatotoxic Cyanobacterial Blooms in Belgium and Luxembourg. *Hydrobiologia* **2005**, *551*, 99–117. [CrossRef]

21. Willame, R.; Boutte, C.; Grubisic, S.; Wilmotte, A.; Komárek, J.; Hoffmann, L. Morphological and Molecular Characterization of Planktonic Cyanobacteria from Belgium and Luxembourg. *J. Phycol.* **2006**, *42*, 1312–1332. [CrossRef]

22. Wilmotte, A.; Descy, J.-P.; Vyverman, W. *Algal Blooms: Emerging Problem for Health and Sustainable Use of Surface Waters (B-BLOOMS)*; Belgian Science Policy: Brussels, Belgium, 2008; p. 85.

23. Descy, J.-P.; Pirlot, S.; Verniers, G.; Viroux, L.; Lara, Y.; Wilmotte, A.; Vyverman, W.; Vanormelingen, P.; Van Wichelen, J.; Van Gremberghe, I.; et al. *Final Report: Cyanobacterial Blooms: Toxicity, Diversity, Modelling and Management*; Research Programme Science for a Sustainable Development; Belgian Science Policy: Brussels, Belgium, 2011; p. 84.

24. Lara, Y.; De Wever, A.; Versniers, G.; Pirlot, S.; Viroux, L.; Leporcq, B.; Vanormelingen, P.; Van Wichelen, J.; Van der Gucht, K.; Peretyatko, A.; et al. Metatdata Compilation for the B-BLOOMS2 Dataset: Cyanobacterial Bloom Monitoring. *Freshw. Metadata J.* **2017**, *28*, 1–8. [CrossRef]

25. Descy, J.-P.; Leprieur, F.; Pirlot, S.; Leporcq, B.; Van Wichelen, J.; Peretyatko, A.; Teissier, S.; Codd, G.A.; Triest, L.; Vyverman, W.; et al. Identifying the Factors Determining Blooms of Cyanobacteria in a Set of Shallow Lakes. *Ecol. Inform.* **2016**, *34*, 129–138. [CrossRef]

26. Mantzouki, E.; Lürling, M.; Fastner, J.; de Senerpont Domis, L.; Wilk-Woźniak, E. Temperature Effects Explain Continental Scale Distribution of Cyanobacterial Toxins. *Toxins* **2018**, *10*, 156. [CrossRef]

27. Svirčev, Z.; Lalić, D.; Bojadzija Savic, G.; Tokodi, N.; Drobac Backovic, D.; Chen, L.; Meriluoto, J.; Codd, G. Global Geographical and Historical Overview of Cyanotoxin Distribution and Cyanobacterial Poisonings. *Arch. Toxicol.* **2019**, *93*, 2429–2481. [CrossRef] [PubMed]

28. Gehringer, M.M. Microcystin-LR and Okadaic Acid-Induced Cellular Effects: A Dualistic Response. *FEBS Lett.* **2004**, *557*, 1–8. [CrossRef]

29. MacKintosh, C.; Beattie, K.A.; Klumpp, S.; Cohen, P.; Codd, G.A. Cyanobacterial Microcystin-LR Is a Potent and Specific Inhibitor of Protein Phosphatases 1 and 2A from Both Mammals and Higher Plants. *FEBS Lett.* **1990**, *264*, 187–192. [CrossRef]

30. Runnegar, M.T.; Kong, S.; Berndt, N. Protein Phosphatase Inhibition and in Vivo Hepatotoxicity of Microcystins. *Am. J. Physiol. Gastrointest. Liver Physiol.* **1993**, *265*, G224–G230. [CrossRef]

31. Fischer, A.; Hoeger, S.J.; Stemmer, K.; Feurstein, D.J.; Knobeloch, D.; Nussler, A.; Dietrich, D.R. The Role of Organic Anion Transporting Polypeptides (OATPs/SLCOs) in the Toxicity of Different Microcystin Congeners in Vitro: A Comparison of Primary Human Hepatocytes and OATP-Transfected HEK293 Cells. *Toxicol. Appl. Pharmacol.* **2010**, *245*, 9–20. [CrossRef]

32. Steiner, K.; Zimmermann, L.; Hagenbuch, B.; Dietrich, D. Zebrafish Oatp-Mediated Transport of Microcystin Congeners. *Arch. Toxicol.* **2016**, *90*, 1129–1139. [CrossRef]

33. Altaner, S.; Jaeger, S.; Fotler, R.; Zemskov, I.; Wittmann, V.; Schreiber, F.; Dietrich, D.R. Machine Learning Prediction of Cyanobacterial Toxin (Microcystin) Toxicodynamics in Humans. *ALTEX* **2020**, *37*, 24–36. [CrossRef]

34. Feurstein, D.; Holst, K.; Fischer, A.; Dietrich, D.R. Oatp-Associated Uptake and Toxicity of Microcystins in Primary Murine Whole Brain Cells. *Toxicol. Appl. Pharmacol.* **2009**, *234*, 247–255. [CrossRef] [PubMed]

35. Funari, E.; Testai, E. Human Health Risk Assessment Related to Cyanotoxins Exposure. *Crit. Rev. Toxicol.* **2008**, *38*, 97–125. [CrossRef]
36. Heinze, R. Toxicity of the Cyanobacterial Toxin Microcystin-LR to Rats after 28 Days Intake with the Drinking Water. *Environ. Toxicol.* **1999**, *14*, 57–60. [CrossRef]
37. Humpage, A.R.; Falconer, I.R. Microcystin-LR and Liver Tumor Promotion: Effects on Cytokinesis, Ploidy, and Apoptosis in Cultured Hepatocytes. *Environ. Toxicol.* **1999**, *14*, 61–75. [CrossRef]
38. Yoshida, T.; Makita, Y.; Nagata, S.; Tsutsumi, T.; Yoshida, F.; Sekijima, M.; Tamura, S.I.; Ueno, Y. Acute Oral Toxicity of Microcystin-LR, a Cyanobacterial Hepatotoxin, in Mice. *Nat. Toxins* **1997**, *5*, 91–95. [CrossRef] [PubMed]
39. World Health Organization. *Cyanobacterial Toxins: Microcystins: Background Document for Development of WHO Guidelines for Drinking-Water Quality and Guidelines for Safe Recreational Water Environments*; WHO: Geneva, Switzerland, 2020.
40. Falconer, I.R.; Burch, M.D.; Steffensen, D.A.; Choice, M.; Coverdale, O.R. Toxicity of the Blue-green Alga (*Cyanobacterium*) *Microcystis aeruginosa* in Drinking Water to Growing Pigs, as an Animal Model for Human Injury and Risk Assessment. *Environ. Toxicol. Water Qual.* **1994**, *9*, 131–139. [CrossRef]
41. Falconer, I.R.; Humpage, A.R. Health Risk Assessment of Cyanobacterial (Blue-Green Algal) Toxins in Drinking Water. *Int. J. Environ. Res. Public. Health* **2005**, *2*, 43–50. [CrossRef] [PubMed]
42. Xu, S.; Yi, X.; Liu, W.; Zhang, C.; Massey, I.Y.; Yang, F.; Tian, L. A Review of Nephrotoxicity of Microcystins. *Toxins* **2020**, *12*, 693. [CrossRef] [PubMed]
43. Dietrich, D.; Hoeger, S. Guidance Values for Microcystins in Water and Cyanobacterial Supplement Products (Blue-Green Algal Supplements): A Reasonable or Misguided Approach? *Toxicol. Appl. Pharmacol.* **2005**, *203*, 273–289. [CrossRef] [PubMed]
44. Gilroy, D.J.; Kauffman, K.W.; Hall, R.A.; Huang, X.; Chu, F.S. Assessing Potential Health Risks from Microcystin Toxins in Blue-Green Algae Dietary Supplements. *Environ. Health Perspect.* **2000**, *108*, 435–439. [CrossRef]
45. Crush, J.R.; Briggs, L.R.; Sprosen, J.M.; Nichols, S.N. Effect of Irrigation with Lake Water Containing Microcystins on Microcystin Content and Growth of Ryegrass, Clover, Rape, and Lettuce. *Environ. Toxicol.* **2008**, *23*, 246–252. [CrossRef] [PubMed]
46. Gutiérrez-Praena, D.; Campos, A.; Azevedo, J.; Neves, J.; Freitas, M.; Guzmán-Guillén, R.; Cameán, A.M.; Renaut, J.; Vasconcelos, V. Exposure of *Lycopersicon esculentum* to Microcystin-LR: Effects in the Leaf Proteome and Toxin Translocation from Water to Leaves and Fruits. *Toxins* **2014**, *6*, 1837–1854. [CrossRef]
47. Llana-Ruiz-Cabello, M.; Jos, A.; Cameán, A.; Oliveira, F.; Barreiro, A.; Machado, J.; Azevedo, J.; Pinto, E.; Almeida, A.; Campos, A.; et al. Analysis of the Use of Cylindrospermopsin and/or Microcystin-Contaminated Water in the Growth, Mineral Content, and Contamination of *Spinacia oleracea* and *Lactuca sativa*. *Toxins* **2019**, *11*, 624. [CrossRef]
48. Chorus, I.; Welker, M. *Toxic Cyanobacteria in Water: A Guide to Their Public Health Consequences, Monitoring and Management*, 2nd ed.; Taylor&Francis: London, UK, 2021; ISBN 9781003081449.
49. Falconer, I.R.; Bartram, J.; Chorus, I.; Kuiper-Goodman, T.; Utkilen, H.; Burch, M.; Codd, G.A. Safe Levels and Safe Practices. *E FN Spon.* **1999**, 155–178.
50. Via-Ordorika, L.; Fastner, J.; Kurmayer, R.; Hisbergues, M.; Dittmann, E.; Komarek, J.; Erhard, M.; Chorus, I. Distribution of Microcystin-Producing and Non-Microcystin-Producing *Microcystis* sp. in European Freshwater Bodies: Detection of Microcystins and Microcystin Genes in Individual Colonies. *Syst. Appl. Microbiol.* **2004**, *27*, 592–602. [CrossRef]
51. Turner, A.D.; Dhanji-Rapkova, M.; O'Neill, A.; Coates, L.; Lewis, A.; Lewis, K. Analysis of Microcystins in Cyanobacterial Blooms from Freshwater Bodies in England. *Toxins* **2018**, *10*, 39. [CrossRef]
52. Roy-Lachapelle, A.; Duy, S.V.; Munoz, G.; Dinh, Q.T.; Bahl, E.; Simon, D.F.; Sauvé, S. Analysis of Multiclass Cyanotoxins (Microcystins, Anabaenopeptins, Cylindrospermopsin and Anatoxins) in Lake Waters Using on-Line SPE Liquid Chromatography High-Resolution Orbitrap Mass Spectrometry. *Anal. Methods* **2019**, *11*, 5289–5300. [CrossRef]
53. Pekar, H.; Westerberg, E.; Bruno, O.; Lääne, A.; Persson, K.M.; Sundström, L.F.; Thim, A.M. Fast, Rugged and Sensitive Ultra High Pressure Liquid Chromatography Tandem Mass Spectrometry Method for Analysis of Cyanotoxins in Raw Water and Drinking Water-First Findings of Anatoxins, Cylindrospermopsins and Microcystin Variants in Swedish Source Wa. *J. Chromatogr. A* **2016**, *1429*, 265–276. [CrossRef] [PubMed]
54. Christophoridis, C.; Zervou, S.-K.; Manolidi, K.; Katsiapi, M.; Moustaka-Gouni, M.; Kaloudis, T.; Triantis, T.M.; Hiskia, A. Occurrence and Diversity of Cyanotoxins in Greek Lakes. *Sci. Rep.* **2018**, *8*, 17877. [CrossRef] [PubMed]
55. Hummert, C.; Reichelt, M.; Weiß, J.; Liebert, H.P.; Luckas, B. Identification of Microcystins in Cyanobacteria from the Bleiloch Former Drinking-Water Reservoir (Thuringia, Germany). *Chemosphere* **2001**, *44*, 1581–1588. [CrossRef]
56. Dawson, R.M. The Toxicology of Microcystins. *Toxicon* **1998**, *36*, 953–962. [CrossRef]
57. Fawell, J.K.; Mitchell, R.E.; Everett, D.J.; Hill, R.E. The Toxicity of Cyanobacterial Toxins in the Mouse: I Microcystin-LR. *Hum. Exp. Toxicol.* **1999**, *18*, 162–167. [CrossRef]
58. Bouaïcha, N.; Miles, C.O.; Beach, D.G.; Labidi, Z.; Djabri, A.; Benayache, N.Y.; Nguyen-Quang, T. Structural Diversity, Characterization and Toxicology of Microcystins. *Toxins* **2019**, *11*, 714. [CrossRef]
59. Du, X.; Liu, H.; Yuan, L.; Wang, Y.; Ma, Y.; Wang, R.; Chen, X.; Losiewicz, M.D.; Guo, H.; Zhang, H. The Diversity of Cyanobacterial Toxins on Structural Characterization, Distribution and Identification: A Systematic Review. *Toxins* **2019**, *11*, 530. [CrossRef]
60. Welker, M.; Von Dohren, H. Cyanobacterial Peptides: Nature's Own Combinatorial Biosynthesis. *FEMS Microbiol. Rev.* **2006**, *30*, 530–563. [CrossRef]

61. Niedermeyer, T.H.J.; Daily, A.; Swiatecka-Hagenbruch, M.; Moscow, J.A. Selectivity and Potency of Microcystin Congeners against OATP1B1 and OATP1B3 Expressing Cancer Cells. *PLoS ONE* **2014**, *9*, e91476. [CrossRef] [PubMed]
62. Chen, Y.-M.; Lee, T.-H.; Lee, S.-J.; Huang, H.-B.; Huang, R.; Chou, H.-N. Comparison of Protein Phosphatase Inhibition Activities and Mouse Toxicities of Microcystins. *Toxicon* **2006**, *47*, 742–746. [CrossRef]
63. Monks, N.R.; Liu, S.; Xu, Y.; Yu, H.; Bendelow, A.S.; Moscow, J.A. Potent Cytotoxicity of the Phosphatase Inhibitor Microcystin LR and Microcystin Analogues in OATP1B1- and OATP1B3-Expressing HeLa Cells. *Mol. Cancer Ther.* **2007**, *6*, 587–598. [CrossRef] [PubMed]
64. Stoner, R.D.; Adams, W.H.; Slatkin, D.N.; Siegelman, H.W. The Effects of Single L-Amino Acid Substitutions on the Lethal Potencies of the Microcystins. *Toxicon* **1989**, *27*, 825–828. [CrossRef]
65. Food and Agriculture Organization; World Health Organization. *Technical Paper on Toxicity Equivalency Factors for Marine Biotoxins Associated with Bivalve Molluscs*; FAO: Rome, Italy, 2016; p. 180. ISBN 9789241511483.
66. European Food Safety Authority (EFSA). Scientific Opinion of the Panel on Contaminants in the Food Chain on a Request from the European Commission on Marine Biotoxins in Shellfish—Saxitoxin Group. *EFSA J.* **2009**, *7*, 1019. [CrossRef]
67. World Health Organization. *Cyanobacterial Toxins: Microcystin-LR in Drinking-Water*; WHO: Geneva, Switzerland, 1998; p. 2.
68. Chorus, I.; Bartam, J. *Toxic Cyanobacteria in Water: A Guide to Their Public Health Consequences, Monitoring and Managment*; WHO: New York, NY, USA; FN Spon: London, UK, 1999; ISBN 0419239308.
69. United States Environmental Protection Agency. *Recommended Human Health Recreational Ambient Water Quality Criteria or Swimming Advisories for Microcystins and Cylindrospermopsin*; Health and Ecological Criteria Division: Washington, DC, USA, 2019; p. 249.
70. Pitois, F.; Fastner, J.; Pagotto, C.; Dechesne, M. Multi-Toxin Occurrences in Ten French Water Resource Reservoirs. *Toxins* **2018**, *10*, 283. [CrossRef]
71. Zervou, S.-K.; Christophoridis, C.; Kaloudis, T.; Triantis, T.M.; Hiskia, A. New SPE-LC-MS/MS Method for Simultaneous Determination of Multi-Class Cyanobacterial and Algal Toxins. *J. Hazard. Mater.* **2017**, *323*, 56–66. [CrossRef] [PubMed]
72. Neumann, U.; Campos, V.; Cantarero, S.; Urrutia, H.; Heinze, R.; Weckesser, J.; Erhard, M. Co-Occurrence of Non-Toxic (Cyanopeptolin) and Toxic (Microcystin) Peptides in a Bloom of *Microcystis* sp. from a Chilean Lake. *Syst. Appl. Microbiol.* **2000**, *23*, 191–197. [CrossRef]
73. Ballot, A.; Krienitz, L.; Kotut, K.; Wiegand, C.; Pflugmacher, S. Cyanobacteria and Cyanobacterial Toxins in the Alkaline Crater Lakes Sonachi and Simbi, Kenya. *Harmful Algae* **2005**, *4*, 139–150. [CrossRef]
74. Tamele, I.J.; Vasconcelos, V. Microcystin Incidence in the Drinking Water of Mozambique: Challenges for Public Health Protection. *Toxins* **2020**, *12*, 368. [CrossRef]
75. Simiyu, B.M.; Oduor, S.O.; Rohrlack, T.; Sitoki, L.; Kurmayer, R. Microcystin Content in Phytoplankton and in Small Fish from Eutrophic Nyanza Gulf, Lake Victoria, Kenya. *Toxins* **2018**, *10*, 275. [CrossRef]
76. Mekebri, A.; Blondina, G.J.; Crane, D.B. Method Validation of Microcystins in Water and Tissue by Enhanced Liquid Chromatography Tandem Mass Spectrometry. *J. Chromatogr. A* **2009**, *1216*, 3147–3155. [CrossRef]
77. Dyble, J.; Fahnenstiel, G.L.; Litaker, R.W.; Millie, D.F.; Tester, P.A. Microcystin Concentrations and Genetic Diversity of *Microcystis* in the Lower Great Lakes. *Environ. Toxicol.* **2008**, *23*, 507–516. [CrossRef]
78. Xue, Q.; Steinman, A.D.; Su, X.; Zhao, Y.; Xie, L. Temporal Dynamics of Microcystins in *Limnodrilus hoffmeisteri*, a Dominant Oligochaete of Hypereutrophic Lake Taihu, China. *Environ. Pollut.* **2016**, *213*, 585–593. [CrossRef]
79. Trung, B.; Dao, T.-S.; Faassen, E.; Lürling, M. Cyanobacterial Blooms and Microcystins in Southern Vietnam. *Toxins* **2018**, *10*, 471. [CrossRef]
80. Moreira, C.; Gomes, C.; Vasconcelos, V.; Antunes, A. Cyanotoxins Occurrence in Portugal: A New Report on Their Recent Multiplication. *Toxins* **2020**, *12*, 154. [CrossRef]
81. Padisak, J. *Cylindrospermopsis raciborskii* (Woloszynska) Seenayya et Subba Raju, an Expanding, Highly Adaptive Cyanobacterium: Worldwide Distribution and Review of Its Ecology. *Arch. Hydrobiol.* **1997**, *107*, 563–593.
82. Kaloudis, T.; Zervou, S.-K.; Tsimeli, K.; Triantis, T.M.; Fotiou, T.; Hiskia, A. Determination of Microcystins and Nodularin (Cyanobacterial Toxins) in Water by LC-MS/MS. Monitoring of Lake Marathonas, a Water Reservoir of Athens, Greece. *J. Hazard. Mater.* **2013**, *263 Pt 1*, 105–115. [CrossRef] [PubMed]
83. Minasyan, A.; Christophoridis, C.; Wilson, A.E.; Zervou, S.-K.; Kaloudis, T.; Hiskia, A. Diversity of Cyanobacteria and the Presence of Cyanotoxins in the Epilimnion of Lake Yerevan (Armenia). *Toxicon Off. J. Int. Soc. Toxinol.* **2018**, *150*, 28–38. [CrossRef] [PubMed]
84. Orr, P.T.; Willis, A.; Burford, M.A. Application of First Order Rate Kinetics to Explain Changes in Bloom Toxicity—the Importance of Understanding Cell Toxin Quotas. *J. Oceanol. Limnol.* **2018**, *36*, 1063–1074. [CrossRef]
85. Zastepa, A.; Pick, F.R.; Blais, J.M. Fate and Persistence of Particulate and Dissolved Microcystin-LA from *Microcystis* Blooms. *Hum. Ecol. Risk Assess. Int. J.* **2014**, *20*, 1670–1686. [CrossRef]
86. Czyżewska, W.; Piontek, M.; Łuszczyńska, K. The Occurrence of Potential Harmful Cyanobacteria and Cyanotoxins in the Obrzyca River (Poland), a Source of Drinking Water. *Toxins* **2020**, *12*, 284. [CrossRef]
87. Pacheco, A.B.F.; Guedes, I.A.; Azevedo, S.M.F.O. Is QPCR a Reliable Indicator of Cyanotoxin Risk in Freshwater? *Toxins* **2016**, *8*, 172. [CrossRef]

88. Bucka, H.; Wilk-Woźniak, E. Ecological Aspects of Selected Principal Phytoplankton Taxa in Lake Piaseczno. *Oceanol. Hydrobiol. Stud.* **2005**, *34*, 79–94.

89. Callieri, C. Single Cells and Microcolonies of Freshwater Picocyanobacteria: A Common Ecology. *J. Limnol.* **2010**, *69*, 257–277. [CrossRef]

90. Śliwińska-Wilczewska, S.; Maculewicz, J.; Barreiro Felpeto, A.; Latała, A. Allelopathic and Bloom-Forming Picocyanobacteria in a Changing World. *Toxins* **2018**, *10*, 48. [CrossRef] [PubMed]

91. Domingos, P.; Rubim, T.K.; Molica, R.J.R.; Azevedo, S.M.F.O.; Carmichael, W.W. First Report of Microcystin Production by Picoplanktonic Cyanobacteria Isolated from a Northeast Brazilian Drinking Water Supply. *Environ. Toxicol.* **1999**, *14*, 31–35. [CrossRef]

92. Lincoln, E.P.; Carmichael, W.W. Preliminary Tests of Toxicity of Synechocystis Sp. Grown on Wastewater Medium. In *The Water Environment: Algal Toxins and Health*; Carmichael, W.W., Ed.; Springer: Boston, MA, USA, 1981; pp. 223–230. ISBN 9781461332671.

93. Shih, P.M.; Wu, D.; Latifi, A.; Axen, S.D.; Fewer, D.P.; Talla, E.; Calteau, A.; Cai, F.; de Marsac, N.T.; Rippka, R.; et al. Improving the Coverage of the Cyanobacterial Phylum Using Diversity-Driven Genome Sequencing. *Proc. Natl. Acad. Sci. USA* **2013**, *110*, 1053–1058. [CrossRef]

94. Li, Y.-W.; Zhan, X.-J.; Xiang, L.; Deng, Z.-S.; Huang, B.-H.; Wen, H.-F.; Sun, T.-F.; Cai, Q.-Y.; Li, H.; Mo, C.-H. Analysis of Trace Microcystins in Vegetables Using Solid-Phase Extraction Followed by High Performance Liquid Chromatography Triple-Quadrupole Mass Spectrometry. *J. Agric. Food Chem.* **2014**, *62*, 11831–11839. [CrossRef]

95. Corbel, S.; Bouaïcha, N.; Nélieu, S.; Mougin, C. Soil Irrigation with Water and Toxic Cyanobacterial Microcystins Accelerates Tomato Development. *Environ. Chem. Lett.* **2015**, *13*, 447–452. [CrossRef]

96. Díez-Quijada, L.; Guzmán-Guillén, R.; Prieto Ortega, A.I.; Llana-Ruíz-Cabello, M.; Campos, A.; Vasconcelos, V.; Jos, Á.; Cameán, A.M. New Method for Simultaneous Determination of Microcystins and Cylindrospermopsin in Vegetable Matrices by SPE-UPLC-MS/MS. *Toxins* **2018**, *10*, 406. [CrossRef]

97. Weenink, E.F.J.; Luimstra, V.M.; Schuurmans, J.M.; Van Herk, M.J.; Visser, P.M.; Matthijs, H.C.P. Combatting Cyanobacteria with Hydrogen Peroxide: A Laboratory Study on the Consequences for Phytoplankton Community and Diversity. *Front. Microbiol.* **2015**, *6*, 714. [CrossRef] [PubMed]

98. Häkkinen, P.J.; Anesio, A.M.; Granéli, W. Hydrogen Peroxide Distribution, Production, and Decay in Boreal Lakes. *Can. J. Fish. Aquat. Sci.* **2011**, *61*, 1520–1527. [CrossRef]

99. Barroin, G.; Feuillade, M. Hydrogen Peroxide as a Potential Algicide for *Oscillatoria rubescens* D.C. *Water Res.* **1986**, *20*, 619–623. [CrossRef]

100. Matthijs, H.C.P.; Visser, P.M.; Reeze, B.; Meeuse, J.; Slot, P.C.; Wijn, G.; Talens, R.; Huisman, J. Selective Suppression of Harmful Cyanobacteria in an Entire Lake with Hydrogen Peroxide. *Water Res.* **2012**, *46*, 1460–1472. [CrossRef] [PubMed]

101. Sorlini, S.; Collivignarelli, C.; Carnevale Miino, M.; Caccamo, F.M.; Collivignarelli, M.C. Kinetics of Microcystin-LR Removal in a Real Lake Water by UV/H2O2 Treatment and Analysis of Specific Energy Consumption. *Toxins* **2020**, *12*, 810. [CrossRef] [PubMed]

102. Spoof, L.; Jaakkola, S.; Važić, T.; Häggqvist, K.; Kirkkala, T.; Ventelä, A.-M.; Kirkkala, T.; Svirčev, Z.; Meriluoto, J. Elimination of Cyanobacteria and Microcystins in Irrigation Water—Effects of Hydrogen Peroxide Treatment. *Environ. Sci. Pollut. Res.* **2020**, *27*, 8638–8652. [CrossRef]

103. Piel, T.; Sandrini, G.; White, E.; Xu, T.; Schuurmans, J.M.; Huisman, J.; Visser, P.M. Suppressing Cyanobacteria with Hydrogen Peroxide Is More Effective at High Light Intensities. *Toxins* **2020**, *12*, 18. [CrossRef]

104. Cooper, W.J.; Lean, D.R.S. Hydrogen Peroxide Concentration in a Northern Lake: Photochemical Formation and Diel Variability. *Environ. Sci. Technol.* **1989**, *23*, 1425–1428. [CrossRef]

105. Drábková, M.; Matthijs, H.C.P.; Admiraal, W.; Maršálek, B. Selective Effects of H2O2 on Cyanobacterial Photosynthesis. *Photosynthetica* **2007**, *45*, 363–369. [CrossRef]

106. Van Oosterhout, F.; Waajen, G.; Yasseri, S.; Manzi Marinho, M.; Pessoa Noyma, N.; Mucci, M.; Douglas, G.; Lürling, M. Lanthanum in Water, Sediment, Macrophytes and Chironomid Larvae Following Application of Lanthanum Modified Bentonite to Lake Rauwbraken (The Netherlands). *Sci. Total Environ.* **2020**, *706*, 135188. [CrossRef]

107. Lürling, M.; van Oosterhout, F. Controlling Eutrophication by Combined Bloom Precipitation and Sediment Phosphorus Inactivation. *Water Res.* **2013**, *47*, 6527–6537. [CrossRef]

108. Waajen, G.; Van Oosterhout, F.; Douglas, G.; Lürling, M. Management of Eutrophication in Lake De Kuil (The Netherlands) Using Combined Flocculant—Lanthanum Modified Bentonite Treatment. *Water Res.* **2016**, *97*, 83–95. [CrossRef] [PubMed]

109. Noyma, N.P.; de Magalhães, L.; Miranda, M.; Mucci, M.; van Oosterhout, F.; Huszar, V.L.M.; Marinho, M.M.; Lima, E.R.A.; Lürling, M. Coagulant plus Ballast Technique Provides a Rapid Mitigation of Cyanobacterial Nuisance. *PLoS ONE* **2017**, *12*, e0178976. [CrossRef] [PubMed]

110. Taton, A.; Grubisic, S.; Brambilla, E.; De Wit, R.; Wilmotte, A. Cyanobacterial Diversity in Natural and Artificial Microbial Mats of Lake Fryxell (McMurdo Dry Valleys, Antarctica): A Morphological and Molecular Approach. *Appl. Environ. Microbiol.* **2003**, *69*, 5157–5169. [CrossRef]

111. Nübel, U.; Garcia-Pichel, F.; Muyzer, G. PCR Primers to Amplify 16S RRNA Genes from Cyanobacteria. *Appl. Environ. Microbiol.* **1997**, *63*, 3327–3332. [CrossRef] [PubMed]

112. Rantala, A.; Fewer, D.P.; Hisbergues, M.; Rouhiainen, L.; Vaitomaa, J.; Börner, T.; Sivonen, K. Phylogenetic Evidence for the Early Evolution of Microcystin Synthesis. *Proc. Natl. Acad. Sci. USA* **2004**, *101*, 568–573. [CrossRef] [PubMed]
113. Pessi, I.S.; Maalouf, P.D.C.; Laughinghouse, H.D.; Baurain, D.; Wilmotte, A. On the Use of High-Throughput Sequencing for the Study of Cyanobacterial Diversity in Antarctic Aquatic Mats. *J. Phycol.* **2016**, *52*, 356–368. [CrossRef]
114. Edgar, R.C. UNOISE2: Improved Error-Correction for Illumina 16S and ITS Amplicon Sequencing. *bioRxiv* **2016**, 081257. [CrossRef]
115. Edgar, R.C. UPARSE: Highly Accurate OTU Sequences from Microbial Amplicon Reads. *Nat. Methods* **2013**, *10*, 996–998. [CrossRef] [PubMed]

Article

Anabaenopeptins from Cyanobacteria in Freshwater Bodies of Greece

Sevasti-Kiriaki Zervou [1,*], Triantafyllos Kaloudis [1], Spyros Gkelis [2], Anastasia Hiskia [1] and Hanna Mazur-Marzec [3]

[1] Laboratory of Photo-Catalytic Processes and Environmental Chemistry, Institute of Nanoscience & Nanotechnology, National Centre for Scientific Research "Demokritos", Patriarchou Grigoriou E & 27 Neapoleos Str., 15310 Athens, Greece; t.kaloudis@inn.demokritos.gr (T.K.); a.hiskia@inn.demokritos.gr (A.H.)

[2] Department of Botany, School of Biology, Aristotle University of Thessaloniki, 54124 Thessaloniki, Greece; sgkelis@bio.auth.gr

[3] Division of Marine Biotechnology, University of Gdansk, Al. Marszałka Piłsudskiego 46, 81-378 Gdynia, Poland; hanna.mazur-marzec@ug.edu.pl

* Correspondence: s.zervou@inn.demokritos.gr

Abstract: Cyanobacteria are photosynthetic microorganisms that are able to produce a large number of secondary metabolites. In freshwaters, under favorable conditions, they can rapidly multiply, forming blooms, and can release their toxic/bioactive metabolites in water. Among them, anabaenopeptins (APs) are a less studied class of cyclic bioactive cyanopeptides. The occurrence and structural variety of APs in cyanobacterial blooms and cultured strains from Greek freshwaters were investigated. Cyanobacterial extracts were analyzed with LC–qTRAP MS/MS using information-dependent acquisition in enhanced ion product mode in order to obtain the fragmentation mass spectra of APs. Thirteen APs were detected, and their possible structures were annotated based on the elucidation of fragmentation spectra, including three novel ones. APs were present in the majority of bloom samples (91%) collected from nine Greek lakes during different time periods. A large variety of APs was observed, with up to eight congeners co-occurring in the same sample. AP F (87%), Oscillamide Y (87%) and AP B (65%) were the most frequently detected congeners. Thirty cyanobacterial strain cultures were also analyzed. APs were only detected in one strain (*Microcystis ichtyoblabe*). The results contribute to a better understanding of APs produced by freshwater cyanobacteria and expand the range of structurally characterized APs.

Keywords: anabaenopeptins; LC–qTRAP MS/MS; fragmentation spectra; structure elucidation; cyanopeptides; cyanobacterial metabolites; Greek freshwaters; cyanobacteria

Key Contribution: The first study of anabaenopeptins' occurrence and their structural variety in cyanobacterial blooms and isolated strains from freshwaters of Greece utilizing LC–qTRAP MS/MS. Possible structures for three novel anabaenopeptins are proposed, expanding the knowledge on this understudied class of cyanobacterial metabolites.

Citation: Zervou, S.-K.; Kaloudis, T.; Gkelis, S.; Hiskia, A.; Mazur-Marzec, H. Anabaenopeptins from Cyanobacteria in Freshwater Bodies of Greece. *Toxins* **2021**, *14*, 4. https://doi.org/10.3390/toxins14010004

Received: 12 October 2021
Accepted: 15 December 2021
Published: 21 December 2021

Publisher's Note: MDPI stays neutral with regard to jurisdictional claims in published maps and institutional affiliations.

1. Introduction

Anabaenopeptins (APs) are cyanobacterial metabolites with a cyclic peptide structure [1]. The presence of APs has been reported in freshwater [2–6] and marine cyanobacterial blooms [7–9], as well as in terrestrial environments, including the leaves of plants in a coastal forest [10] and the terrestrial mat in a bamboo forest [11]. APs are produced by freshwater, marine and terrestrial cyanobacteria [12] mainly belonging to the genera *Planktothrix* [13,14], *Anabaena* [15–18], *Microcystis* [19–23] and *Nostoc* [24,25]. They especially belong to the species *Planktothrix (Oscillatoria) agardhii* [3,26–32], *Planktothrix rubescens* [30,31,33,34], *Anabaena (Dolichospermum) flos-aquae* [35], *Anabaena lemmermannii* [36], *Microcystis aeruginosa* [3,37–39], *Microcystis flos-aquae* [36], *Microcystis ichthyoblabe*, *Microcystis wesenbergii* [19] and *Nostoc punctiforme* [40], as well as *Aphanizomenon flos-aquae* [41], *Nodularia spumigena* [40,42–45], *Woronichinia naegeliana* [46,47] and *Woronichinia compacta* [48]. Additionally, the cyanobacteria

Lyngbya sp. [49], *Lyngbya confervoides* [50], *Schizothrix* sp. [51], *Tychonema* sp. [52] and *Brasilonema* sp. [11] are known APs producers. APs are also produced by symbiotic cyanobacteria that have been isolated from the marine sponges *Theonella* sp. [53–55], *Theonella swinhoei* [56], *Psammocinia aff. bulbosa* [57], and from the lithistid family *Theonellidae* [58].

APs are hexapeptides with the general structure X_1-CO-[Lys-X_3-X_4-MeX_5-X_6], characterized by the presence of the amino acid lysine (Lys), which contributes to ring formation by an *N*-peptide bond with the carboxy group of amino acid X_6 and a C-peptide bond with the amino group of amino acid X_3. A side chain of one amino acid unit is attached to a five-peptide ring via an ureido bond between the a-N of Lys and the a-N of the side chain amino acid [1,35] (Figure 1). Apart from Lys, all other amino acids are variable, while the amino acid in position X_5 is usually *N*-methylated and the amino acid in position 4 is usually in homo form (Table S1).

The first characterized APs (i.e., Anabaenopeptin A and Anabaenopeptin B) were isolated in 1995 by Harada et al. from the freshwater cyanobacterial strain *Anabaena flos-aquae* NRC 525-17 [35], from which APs were named. Structural variants of APs are also known by other names such as Nodulapeptins [42], Oscillamides [27,30], Konbamide [54], Keramamides [53,55], Ferintoic acids [59], Mozamides [58], Schizopeptin [51], Brunsvicamides [52], Psymbamide A [57], Pompanopeptin B [50], Paltolides [56], Lyngbyaureidamides [49] and Nostamides [24,40]. These names were mainly based on the producing taxon or on the geographic location of discovery, complemented with suffixes describing the variety. As a consequence, nomenclature of this class is not fully systematic. Newly identified APs may also contain their molecular mass as part of their name for consistency; however, this approach may be problematic because several variants can have the same molecular mass due to the large number of possible combinations of the variable amino acid residues in the structure (Table S1).

The biosynthesis of APs is performed via nonribosomal peptide synthesis (NRPS) pathways (*aptABCD*) encoded in the genomes of a variety of cyanobacteria [40,60–62] and assembled by an NRPS enzyme complex, which has a modular structure [61]. Synthesis is performed stepwise by modules that each contain specific functional domains for the elongation of the peptide sequence through adenylation and thiolation of the activated amino acid residues, ring formation by the epimerization of Lys and *N*-methylation of amino acid X_5 [61]. The gene clusters of APs can encode either a single starter module, as in *Nostoc* and *Nodularia*, or two alternative loading modules, as in *Anabaena* sp. 90, allowing the simultaneous synthesis of multiple AP variants [40]. Possibly, due to the relaxed substrate specificity of NRPSs, numerous structural variants of cyanobacterial peptides may be generated [61]. Up to now, more than 150 AP congeners have been reported in the literature (Table S1).

Bioactivity studies have shown that APs can inhibit the enzymes responsible for the regulation of several physiological and metabolic processes [63]. Specifically, AP congeners can inhibit carboxypeptidase A [2,9,29,32,41,64], serine proteases (chymotrypsin [27,65], trypsin [51,65] and elastase [65–67]), serine/threonine protein phosphatases (PP1 and PP2A [9,30,68]) and *Mycobacterium tuberculosis* enzyme MptpB [52]. Of great pharmacological interest is their high activity against the thrombin activatable fibrinolysis inhibitor (TAFIa) (carboxypeptidase), resulting in the stimulation of fibrin clot degradation, which may help to prevent thrombosis [69]. On the other hand, APs can cause toxic effects on microorganisms such as the nematode *Caenorhabditis elegans* [70], the amoeba *Acanthamoeba castellanii* [36] and the planktic crustacean *Daphnia pulex* [71]. Furthermore, APs could possibly control cyanobacterial population density as the presence of APs (i.e., Anabaenopeptin B and Anabaenopeptin F) has been correlated with the triggering of cell lysis that ends up in the collapse of cyanobacterial blooms [72].

General structure of Anabaenopeptins

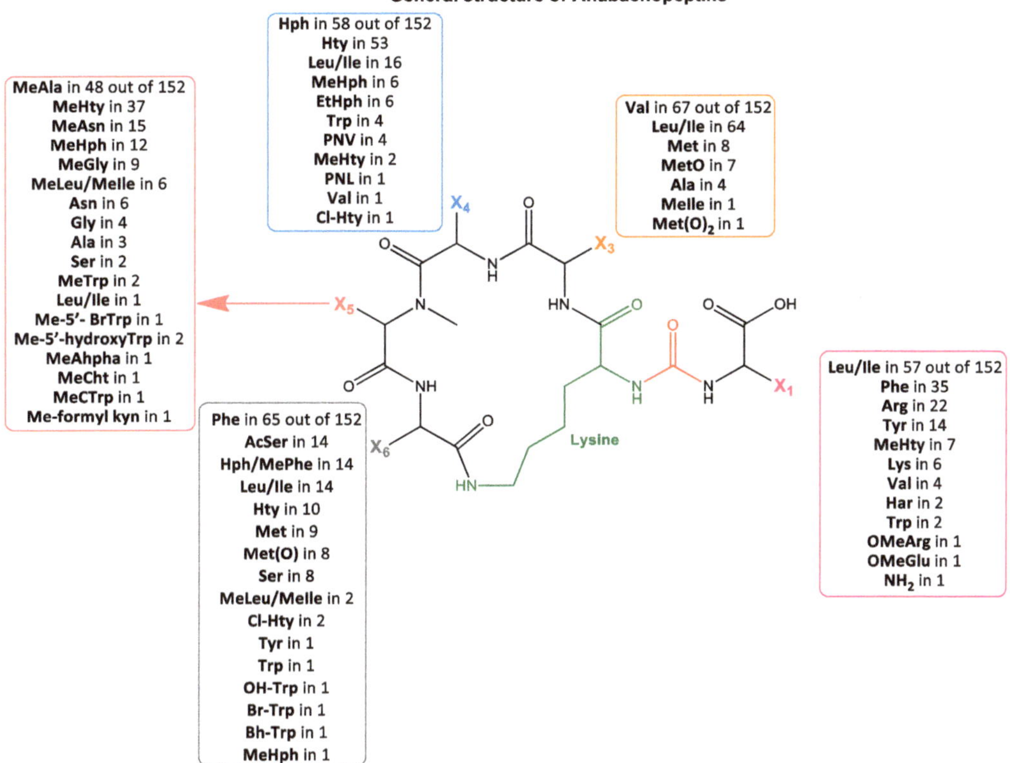

Figure 1. General structure of anabaenopeptins (APs) and an overview of their variable amino acids. Ala = alanine, AcSer = acetyl-serine, Arg = arginine, Asn = asparagine, Bh-Trp = 2-bromo-5 -hydroxy-tryptophan, Br-Trp = bromo-tryptophan, Cl-Hty = chloro-homotyrosine, EtHph = ethyl-homophenylalanine, Gly = glycine, Har = homoarginine, Hph = homophenylalanine, Hph/MePhe = homophenylalanine/methyl-phenylalanine (isobaric compounds), Hty = homotyrosine, Leu/Ile = leucine/isoleucine (isobaric compounds), Lys = lysine, Met = methionine, MetO = methionine sulfoxide, Met(O)$_2$ = methionine sulfone (S-dioxide), Me-5'-BrTrp = mehtyl-5'-bromo-tryptophan, Me-5'-hydroxyTrp = mehtyl-5'-hydroxy-tryptophan, MeAhpha = N-methyl -2-amino-6-(hydroxyl phenyl) hexanoic acid, MeAla = methyl-alanine, MeAsn = methyl-asparagine, MeCht = 6-chloro-5-hydroxy-N-methyl-tryptophan, MeCTrp = 6-chloro-N-methyl-tryptophan, Me-formyl kyn = methyl-formyl kynurenine, MeGly = methyl-glycine, MeHph = methyl-homophenylalanine, MeHty = methyl-homotyrosine, MeLeu/MeIle = methyl-leucine/methyl-isoleucine (isobaric compounds), MeTrp = methyl-tryptophan, OMeArg = arginine methyl ester, OMeGlu = glutamic acid methyl ester, OHTrp = hydroxyl-tryptophan, Phe = phenylalanine, PNV = 5-phenylnorvaline, PNL = 6-phenylnorleucine, Ser = serine, Trp = tryptophan, Tyr = tyrosine, Val = valine.

The occurrence of APs in cyanobacterial blooms and cultured strains from freshwater bodies has been reported more frequently during recent years in several countries worldwide, including Japan [2,26–28], Germany [3,13,19,20], Finland [16,18], Norway [73,74], Poland [47,75,76], Slovenia [33], Czech Republic [22,77], Austria [34], Hungary [23], Switzerland [78], Spain [4,6,79], Portugal [37,38], Italy [80–84], France [85,86], United Kingdom [85,87], Turkey [44], Israel [21,39,88], Brazil [89], Canada [59,86,90], USA [64,91–94], New Zealand [95], and India [96]. Recent studies indicate that APs could be more abundant in freshwaters than other toxic cyanobacterial metabolites such as the known cyanotoxins microcystins [6,87,90]. APs have also been detected in wild-caught fishes (fish muscles) from Pike River, Canada [97].

Structural characterization of cyanobacterial metabolites, including APs, is an emerging issue due to the great diversity of molecules, their bioactivities and possible effects on ecosystems and on human health. Nuclear magnetic resonance (NMR), after the isolation of the compound, usually from a cyanobacterial strain culture, has been used for the structural elucidation of APs e.g., [2,26,27,35]. Mass spectrometric (MS) techniques such as Matrix-Assisted Laser Desorption/Ionization Time-of-Flight (MALDI-TOF) [3,22,95] or MS coupled with liquid chromatography such as liquid chromatography–hybrid triple quadrupole/linear ion trap mass spectrometry (LC–qTRAP) [8,23,25,40,45,48] or liquid chromatography–hybrid triple quadrupole/Time-of-Flight (LC–qTOF) [24,44] are nowadays widely used as they can be applied directly to extracts of field samples or strain cultures. A significant indicator of APs' fragmentation spectrum is the characteristic fragment ion of lysine (Lys) at m/z 84 [3]. The typical fragmentation pattern of APs includes the loss of the amino acid and the CO of the side chain, resulting in the peptide ring ion [3,10,44].

Information regarding the presence of APs in Greek freshwater bodies is limited. Only three monitoring studies have been conducted so far, targeting no more than three AP congeners [5,98,99]. In the present study, an untargeted analysis approach utilizing a LC–qTRAP method was applied for the investigation of APs' presence in cyanobacteria from Greece. The main aims were (i) to report, for the first time, the structural diversity of APs in cyanobacterial bloom samples collected from lakes of Greece, (ii) to assess the ability of Greek freshwater cyanobacterial strains to produce APs and (iii) to identify the possible new structures of APs, contributing to a better understanding of the existing variety of these hexapeptide cyanobacterial metabolites.

2. Results and Discussion

2.1. Structural Elucidation of Anabaenopeptins

Thirteen APs were detected in the samples of cyanobacteria from Greek freshwater bodies. The elucidation of proposed AP structures was based on their precursor ions from full scan (MS1) (Table S1) and fragmentation (MS2) spectra, enabling annotation of the compounds [100]. Among them, the possible structures of three AP congeners were proposed for the first time in the frame of the present study. The amino acid sequences of the detected APs with their precursor ions [M + H]$^+$ and the retention time (tR) are provided in Table 1. The proposed structures, extracted ion chromatograms (EIC), full scan spectra (MS1) and fragmentation mass spectra (MS2), of the three newly annotated APs are shown in Figures 2–4, while the elucidation of their spectra are provided in the relevant captions.

Table 1. List of anabaenopeptins (APs) detected in cyanobacterial blooms and cyanobacterial strains from Greek lakes.

	m/z [M + H]$^+$	tR (min)	Name	Amino Acid Sequence							Ref.
				1 (Side Chain)	Ureido Linkage	2	3	4	5	6	
1	821.2	9.4	AP 820	Agr	CO	Lys	Val	Hph	MeAla	Phe	[22]
2	837.4	7.9	AP B	Arg	CO	Lys	Val	Hty	MeAla	Phe	[35]
3	838.3	9.2	AP 837	EtOGlu	CO	Lys	Val	Hty	MeAla	Phe	This study
4	842.3	10.6	AP 842	Tyr	CO	Lys	Ile	Hph	MeAla	Phe	[23]
5	844.2	9.9	AP A	Tyr	CO	Lys	Val	Hty	MeAla	Phe	[35]
6	851.3	8.8	AP F	Arg	CO	Lys	Ile	Hty	MeAla	Phe	[28]
7	852.2	9.4	AP 851	EtOGlu	CO	Lys	Leu/Ile	Hty	MeAla	Phe	This study
8	858.4	10.0	Osc Y	Tyr	CO	Lys	Ile	Hty	MeAla	Phe	[27]
9	870.4	10.8	AP 870	MeHty	CO	Lys	Leu/Ile	Hph	MeAla	Phe	[23]
10	872.3	10.1	AP 872	MeHty	CO	Lys	Val	Hty	MeAla	Phe	[23]
11	886.4	10.2	AP 886	MeHty	CO	Lys	Leu/Ile	Hty	MeAla	Phe	[23]
12	895.6	8.9	AP 894	Lys	CO	Lys	Leu/Ile	Hty	MeHty	Leu/Ile	This study
13	907.3	9.5	AP KB906	Arg	CO	Lys	Ile	Hph	MeHty	Ile	[101]

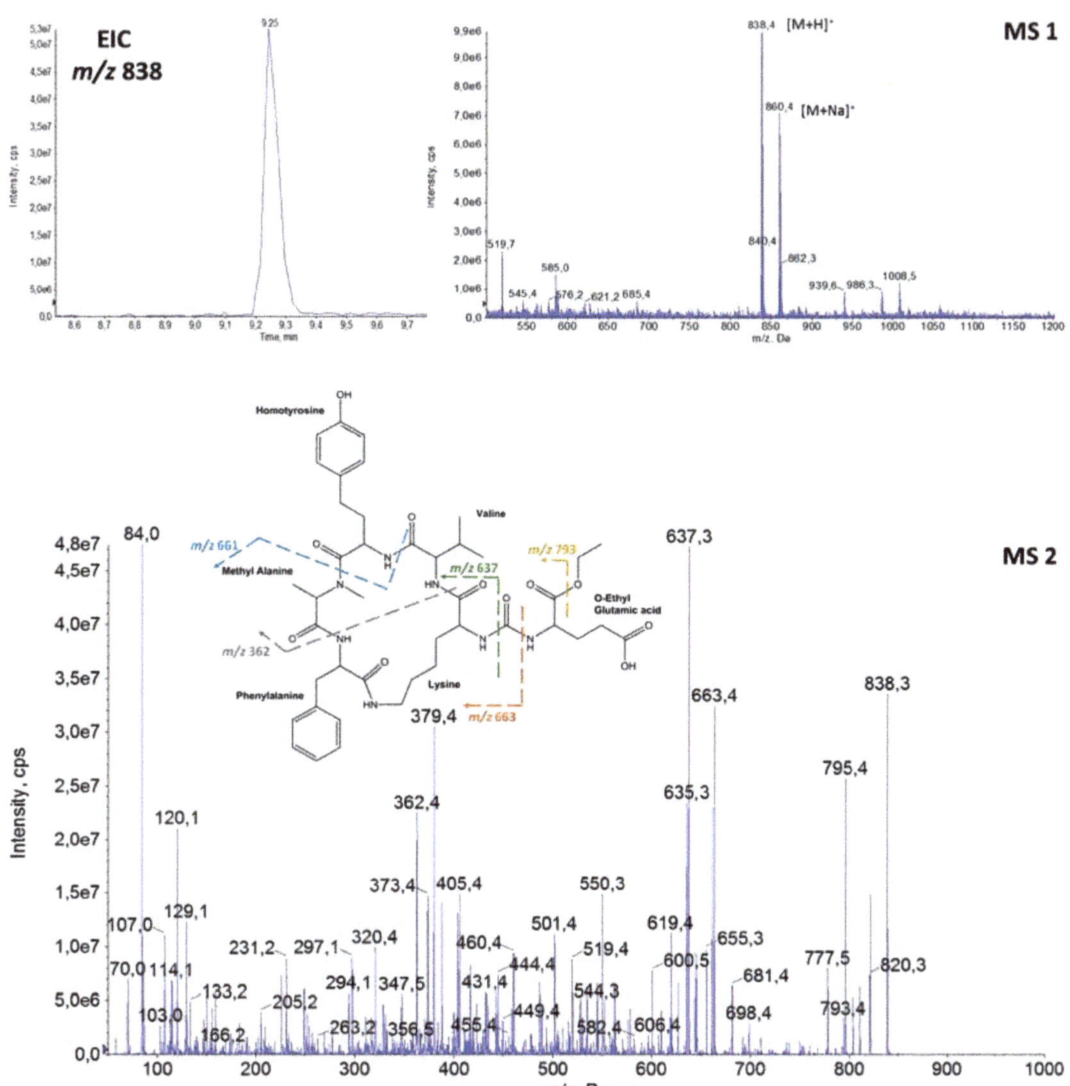

Figure 2. Extracted ion chromatogram (EIC) at *m/z* 838, full scan spectrum (MS1) at 9.25 min, fragmentation mass spectrum (MS2) and proposed structure of the new **AP 837** with [M + H]+ at *m/z* 838. (*m/z* 820 = [M + H − H$_2$O]+, *m/z* 793 = [M + H − OCH$_2$CH$_3$]+, *m/z* 663 = [M + H − OEtGlu]+, *m/z* 661 = [M + H − Hty]+, *m/z* 637 = [Lys-Val-Hty-MeAla-Phe + H]+, *m/z* 635 = [M + H − OEtGlu − CO]+, *m/z* 562 = [M + H − Val-Hty]+, *m/z* 460 = [MeAla-Phe-Lys-Val + H]+, *m/z* 405 = [MeAla-Phe-Lys-CO-NH$_2$ + H]+, *m/z* 362 = [Val-Hty-MeAla + H]+, *m/z* 320 = [Phe-Lys-CO-NH$_2$ + H]+ and/or [Lys-Val-Hty + H]+, *m/z* 150 = Hty immonium ions, *m/z* 120 = Phe immonium ion, *m/z* 107 = Hty related ion, *m/z* 102 = Glu immonium ion, *m/z* 84 = Lys immonium ion).

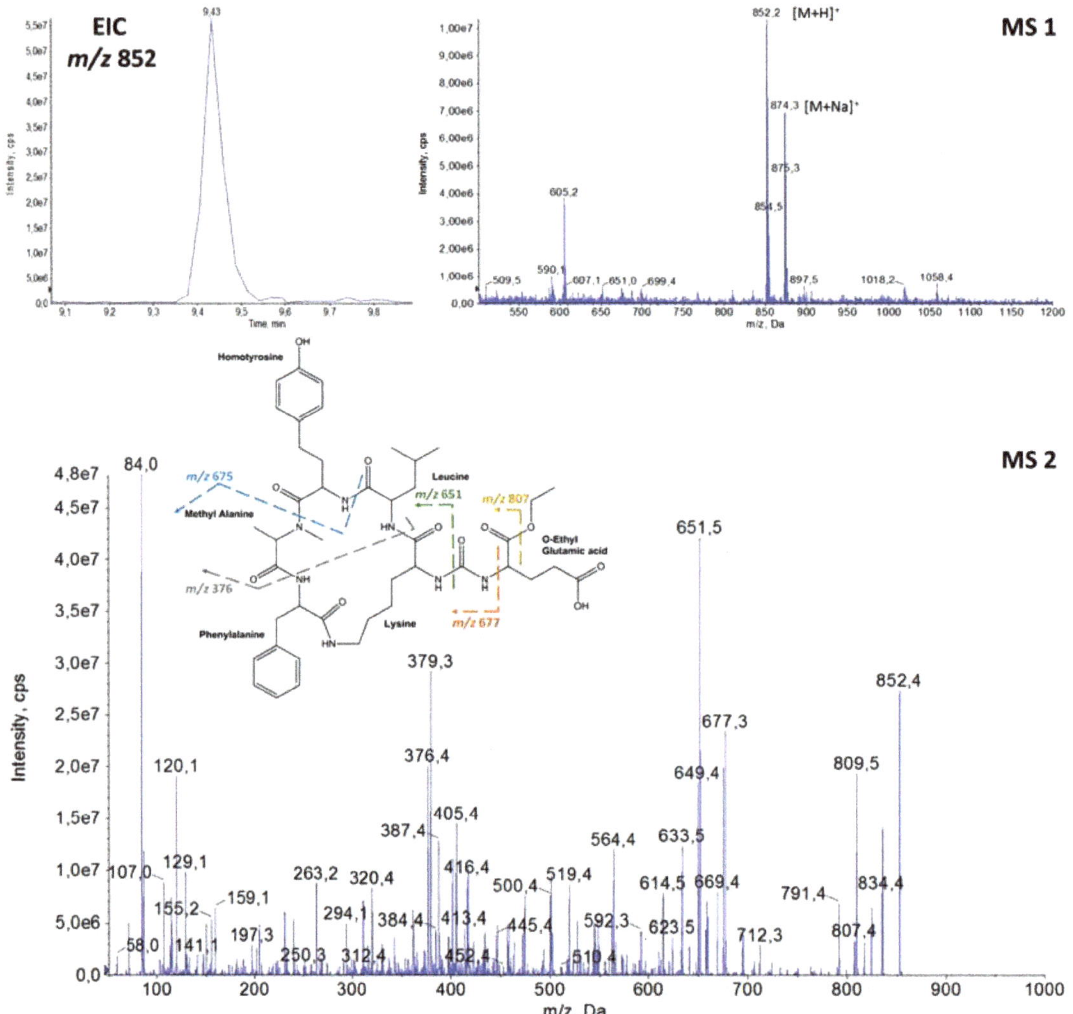

Figure 3. Extracted ion chromatogram (EIC) at m/z 852, full scan spectrum (MS1) at 9.43 min, fragmentation mass spectrum (MS2) and proposed structure of the new **AP 851** with $[M + H]^+$ at m/z 852. (m/z 834 = $[M + H - H_2O]^+$, m/z 807 = $[M + H - OCH_2CH_3]^+$, m/z 677 = $[M + H - OEtGlu]^+$, m/z 675 = $[M + H - Hty]^+$, m/z 651 = $[Lys-Leu/Ile-Hty-MeAla-Phe + H]^+$, m/z 649 = $[M + H - OetGlu - CO]^+$, m/z 564 = $[Phe-Lys-Leu/Ile-Hty + H]^+$, m/z 405 = $[MeAla-Phe-Lys-CO-NH_2 + H]^+$, m/z 376 = $[Leu/Ile-Hty-MeAla + H]^+$, m/z 320 = $[Phe-Lys-CO-NH_2 + H]^+$, m/z 263 = $[MeAla-Hty + H]^+$, m/z 150 = Hty immonium ions, m/z 120 = Phe immonium ion, m/z 107 = Hty related ion, m/z 102 = Glu immonium ion, m/z 86 = Leu/Ile immonium ion, m/z 84 = Lys immonium ion).

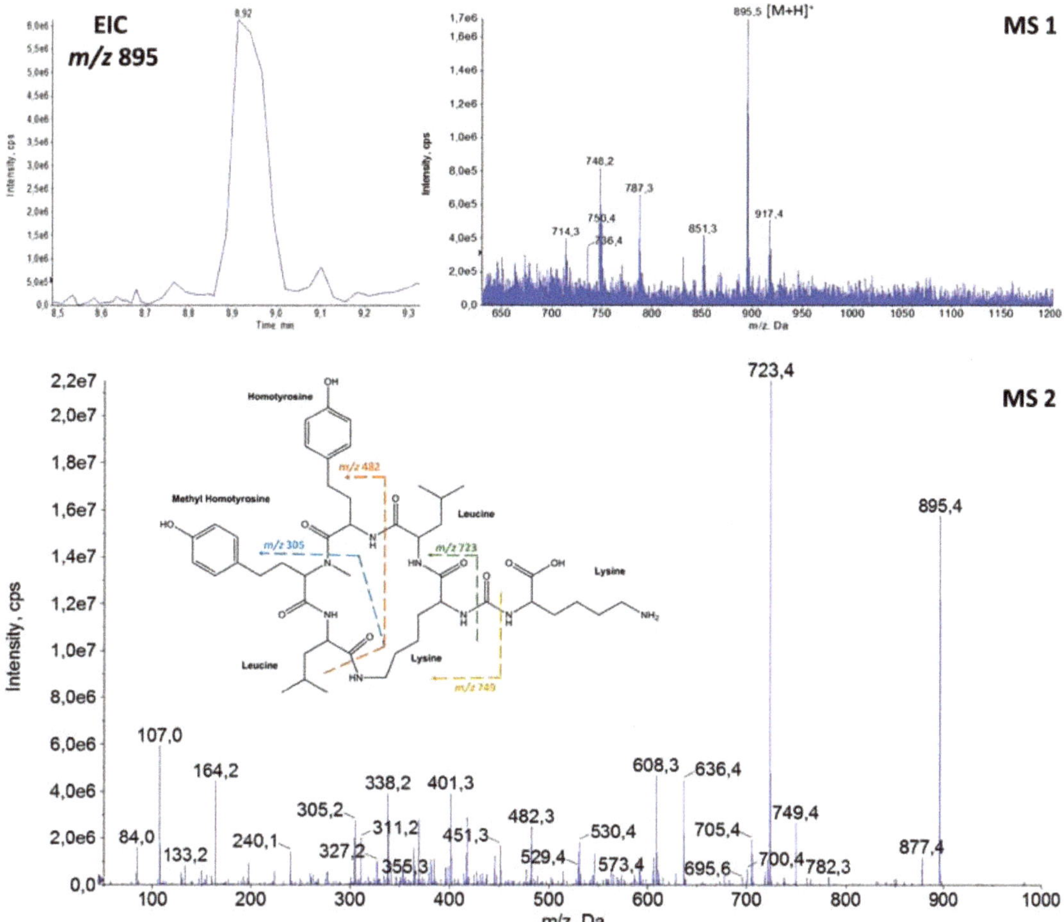

Figure 4. Extracted ion chromatogram (EIC) at *m/z* 895, full scan spectrum (MS1) at 8.92 min, fragmentation mass spectrum (MS2) and proposed structure of the new **AP 894** with [M + H]⁺ at *m/z* 895. (*m/z* 749 = [M + H − Lys]⁺, *m/z* 723 = [M + H − Lys − CO]⁺, *m/z* 705 = [Lys-Leu/Ile-Hty-MeHty-Lue/Ile − H₂O + H]⁺ or [M + H − Lys − CO − H₂O]⁺, *m/z* 636 = [CO-Lys-Leu/Ile-Hty-MeHty]⁺, *m/z* 608 = [Lys-Leu/Ile-Hty-MeHty]⁺, *m/z* 530 = [Leu/Ile-Lys-Leu/Ile-Hty]⁺, *m/z* 482 = [Leu/Ile-Hty-MeHty + H]⁺, *m/z* 369 = [Hty-MeHty + H]⁺, *m/z* 305 = [MeHty-Leu/Ile + H]⁺, *m/z* 240 = [Lys-Leu/Ile + H]⁺, *m/z* 164 = MeHty immonium ion, *m/z* 107 = Hty related ion, *m/z* 84 = Lys immonium ion).

The detection of APs was based on the diagnostic fragment ion of lysine (Lys) *m/z* 84 [3], which was present in the fragmentation spectra of all APs (Figures 2–4 and S2–S11). Structural elucidation of APs was based on fragmentation patterns described in previous studies [3,9,10,23,25,44,48] and on immonium ions of the common amino acids. In Table 1, both leucine (Leu) and isoleucine (Ile) are provided in the proposed AP sequences as these amino acids are isobaric compounds with the same chemical formula (C₆H₁₃NO₂) and they could not be distinguished.

Generally, one of the intense ions that is always present in the fragmentation spectrum of APs is the ion formed by the loss of the side chain amino acid, i.e., [M + H − X₁]⁺. Fragment ions [M + H − X₃]⁺ and [M + H − X₄]⁺ are also commonly found in the APs' spectra. Furthermore, among the most intense fragment ions of APs is the five-peptide ring ion generated after the loss of the side chain, i.e., $[Lys-X_3-X_4-MeX_5-X_6 + H]^+$.

The characteristic ions of annotated APs during this study were m/z 635 = [Lys-Leu/Ile-Hph-MeAla-Phe + H]$^+$ for AP 842 and AP 870; m/z 637 = [Lys-Val-Hty-MeAla-Phe + H]$^+$ for AP B, AP 837, AP A and AP 872; and m/z 651 = [Lys-Leu/Ile-Hty-MeAla-Phe + H]$^+$ that was present in the fragmentation spectra of AP F, AP 851, Osc Y and AP 886. Other common fragment ions of elucidated AP structures were m/z 320 = [Phe-Lys-CO-NH$_2$ + H]$^+$ and m/z 405 = [MeAla-Phe-Lys-CO-NH$_2$ + H]$^+$ (Figure 2, Figure 3, Figures S4, S5 and S7–S10) that suggested the presence of Phe amino acid in position X_6 and MeAla amino acid in position X_5.

The fragment ions m/z 550 = [MeAla-Phe-Lys-CO-Tyr $-$ H$_2$O]$^+$ and m/z 568 = [MeAla-Phe-Lys-CO-Tyr]$^+$ were characteristic of APs with Tyr as the side chain, i.e., AP 842, AP A and Osc Y (Figures S4, S5 and S7), while the fragment ions m/z 578 = [MeAla-Phe-Lys-CO-MeHty $-$ H$_2$O]$^+$ and m/z 596 = [MeAla-Phe-Lys-CO-MeHty]$^+$ indicated the presence of MeHty as the side chain, i.e., AP 870, AP 872 and AP 886 (Figures S8–S10).

Congeners of APs with arginine (Arg) as the side chain amino acid (i.e., AP B, AP F, AP 820, AP KB906) had m/z 201 = [Arg + CO + H]$^+$ as the most intense fragment ion. The other intense fragment ions of the spectra were the [M + H-Arg-CO]$^+$ and [M + H-Arg-CO-H$_2$O]$^+$. The characteristic Lys immonium ion, m/z 84, was present in the fragmentation spectra with low intensity (Figures S2, S3, S6 and S11).

Immonium ions of amino acids were also significant indicators of the peptide sequences. The presence of Phe was indicated by an intense peak at m/z 120 and Hph by m/z 134. Low intensity peaks at m/z 107 and m/z 150 suggested the presence of Hty and m/z 164 of MeHty, while m/z 136 was attributed to Tyr, m/z 58 to MeAla and m/z 86 corresponded to Leu/Ile.

2.2. Anabaenopeptins in Cyanobacterial Blooms from Greek Lakes

Samples were collected from nine different lakes of Greece during cyanobacterial bloom events, which were mainly dominated by *Microcystis* and *Dolichospermum* species, and were analyzed for the presence of APs. The detected AP congeners and the dominant cyanobacterial species of each sample are presented in Figure S1, and details are provided in Table 2. In total, thirteen different AP congeners were detected, and their amino acid sequences are shown in Table 1.

The presence of APs was confirmed in the majority of the examined samples (91%). In addition, a large within-sample structural diversity of APs was observed as at least six AP congeners were detected in each of the 11 samples (48% of total samples). Two samples contained only one AP congener. The largest diversity of APs was observed in three samples collected from lakes Kastoria (5 October 1995), Kerkini (3 August 1999) and Zazari (5 August 1999); eight APs were detected in each of them. These samples were dominated by *Microcystis* species (Table 2). A large diversity of APs was also observed in samples collected from lakes Pamvotida, Mikri Prespa, and Vistonida.

APs were not detected in two samples collected from lakes Marathonas and Karla, although cyanobacterial species that possibly produce APs were present in both lakes (i.e., *Microcystis flos-aquae* at Lake Marathonas and *Planktothrix* cf. *agardhii* at Lake Karla).

The most frequently detected APs in Greek freshwater samples were AP F (87% of samples) and Osc Y (87%), followed by AP B (65%) and AP 886 (57%). AP A and AP 872 were also common congeners among the samples. AP 820 and AP KB906 were detected in one sample from Lake Kastoria and Lake Zazari, respectively.

AP 894, whose structure is proposed for the first time in the present study, was detected in two samples collected from lakes Kerkini and Zazari. The newly proposed APs, 837 and 851, were detected in one sample collected from Lake Mikri Prespa (4 November 2014).

Table 2. Dominant cyanobacterial species and anabaenopeptins detected in samples of cyanobacterial blooms from Greek lakes.

Lake	Sampling Date	Dominant Cyanobacterial Species	Anabaenopeptins Amino Acid Sequences Are Listed in Table 1	Number of Congeners
Amvrakia	10 August 1999	*Microcystis* spp., *Dolichospermum viguieri*	AP F, Osc Y, AP 872, AP 886	4
Amvrakia *	19 August 1999	*Dolichospermum perturbatum, Microcystis* spp.	AP F	1
Amvrakia *	19 August 1999	*Dolichospermum perturbatum*	AP F, Osc Y, AP 886	3
Kastoria *	5 October 1995	*Microcystis aeruginosa, Microcystis novacekii, Microcystis wesenbergii*	AP B, AP 842, AP A, AP F, Osc Y, AP 886	6
Kastoria *	5 October 1995	*Microcystis aeruginosa, Microcystis novacekii, Microcystis wesenbergii*	AP 820, AP B, AP 842, AP A, AP F, Osc Y, AP 872, AP 886	8
Kastoria	3 July 2000	*Microcystis aeruginosa, Microcystis novacekii*	AP B, AP F, Osc Y	3
Kastoria	20 September 2000	*Microcystis aeruginosa, Microcystis flos-aquae*	AP B, AP F, Osc Y, AP 872	4
Kastoria	18 September 2014	*Microcystis aeruginosa, Microcystis flos-aquae, Microcystis* spp., *Pseudanabaena mucicola*	AP B, AP F, Osc Y, AP 886	4
Kastoria	6 October 2015	*Microcystis aeruginosa, Microcystis flos-aquae, Microcystis novacekii, Microcystis ichthyoblabe*	AP B, AP F, Osc Y	3
Kerkini	3 August 1999	*Microcystis* spp., *Microcystis wesenbergii*	AP B, AP 842, AP A, AP F, Osc Y, AP 870, AP 872, AP 886	8
Kerkini	26 August 1999	*Microcystis aeruginosa, Dolichospermum spiroides*	AP B, AP A, AP F, Osc Y, AP 872, **AP 894**	6
Mikri Prespa	5 August 1999	*Microcystis* spp., *Microcystis wesenbergii*	AP B, AP F, Osc Y, AP 870, AP 872, AP 886	6
Mikri Prespa	4 November 2014	*Microcystis aeruginosa*	AP B, **AP 837**, AP A, AP F, **AP 851**, Osc Y	6
Pamvotida	22 July 1999	*Microcystis aeruginosa*	AP B, AP 842, AP A, AP F, Osc Y, AP 886	6
Pamvotida	18 August 1999	*Dolichospermum flos-aquae, Microcystis aeruginosa*	AP F, Osc Y, AP 886	3
Pamvotida	5 August 2000	*Microcystis aeruginosa, Dolichospermum flos-aquae*	Osc Y	1
Pamvotida	17 August 2000	*Dolichospermum flos-aquae, Microcystis* spp.	AP F, Osc Y, AP 886	3
Pamvotida *	18 August 2000	*Microcystis* spp., *Dolichospermum flos-aquae*	AP B, AP F, Osc Y, AP 870, AP 872, AP 886	6
Pamvotida *	18 August 2000	*Microcystis* spp., *Dolichospermum flos-aquae*	AP B, AP 842, AP F, Osc Y, AP 872, AP 886	6
Vistonida	2 August 1999	*Microcystis aeruginosa, Microcystis* spp.	AP B, AP A, AP F, Osc Y, AP 872, AP 886	6
Zazari	5 August 1999	*Microcystis aeruginosa, Microcystis* spp.	AP B, AP 842, AP A, AP F, Osc Y, AP 872, **AP 894**, AP KB906	8
Karla	1 July 2015	*Anabaenopsis elenkinii, Raphidiopsis (Cylindrospermopsis) raciborskii, Planktothrix* cf. *agardhii, Pseudanabaena limnetica*	-	0
Marathonas	26 October 2010	*Microcystis flos-aquae, Microcystis viridis, Pseudanabaena raphidioides, Planktolyngbya limnetica*	-	0

* Samples were collected from two different sampling points for Lake Amrakia (19 August 1999), Lake Kastoria (5 October 1995) and Lake Pamvotida (18 August 2000). **BOLD:** New AP structures proposed in the frame of the present study.

In two previous monitoring studies targeting AP A and AP B by HPLC–PDA, in which cyanobacterial bloom samples were collected from up to 36 freshwater bodies of Greece, the presence of APs in lakes Zazari (AP A), Kastoria (AP A and AP B) and Pamvotis (AP A and AP B) was reported [5,98]. In the current study, both AP A and AP B were detected by mass spectrometry in lakes Kastoria, Pamvotis, Zazari, Kerkini, Mikri Prespa and Vistonida, along with several other APs congeners.

According to a three-year monitoring study of the Greek Lake Vegoritis targeting 25 cyanobacterial toxins and peptides, AP B and AP F were found to be the most frequently detected cyanobacterial metabolites; they were present in almost all the samples, followed by Osc Y [99]. These results are in agreement with the current study as AP F, Osc Y and AP B were the most commonly occurring AP congeners in the freshwaters of Greece.

The occurrence of cyanobacterial metabolites, including APs in freshwater blooms, has been investigated in a number of past studies. Analysis by MALDI-TOF MS showed the presence of AP B and AP F in samples collected from lakes in Italy [80,81,102], Germany [3], Spain [79] and Brazil [89]. In samples collected from a waterbody of Poland and analyzed by LC–qTRAP MS/MS, the most abundant AP congener was AP B, followed by AP A, AP

F, AP G, Osc Y, AP D and AP 915 [75]. The presence of AP A, AP B, AP F and Osc Y was also confirmed by LC–HRMS in samples collected from the freshwaters of Spain [6] and the Czech Republic [77], while AP B, AP A and Osc Y were identified in samples from the United Kingdom [87]. Based on the results of this study and of previous reports, it appears that AP B and AP F followed by AP A and Osc Y are the most frequently reported APs not only in Greece but also in the European continent.

AP F, Osc Y, AP B and AP A are protease inhibitors that possess activity against carboxypeptidase A and protein phosphatase 1 (PP1) [9,30,64]. AP B and AP F are also highly selective TAFIa inhibitors [69] and elastase inhibitors, with no activity towards chymotrypsin and trypsin [66], while Osc Y have presented inhibitory activity against chymotrypsin [27]. Additionally, AP A, AP B and AP F have had toxicity effects in the nematode *Caenorhabditis elegans* [70]. Even though APs' toxicity effects on animal models and microorganisms have been reported, there remains a lack of data regarding their toxicity and impact on human health [12].

APs are the 3rd class of cyanopeptides with the highest structural diversity after microcystins and cyanopeptolins [103]. In the present investigation, thirteen structures of APs from the cyanobacteria of Greek freshwaters were detected, and they had a rather low diversity of variable amino acids (Figure 5). In particular, all the moieties that composed the ring structures were represented by only two different amino acids per site. Even though the diversity was limited, it is interesting that the two amino acids that were determined in each position are among the most commonly found in known AP congeners (Figure 1).

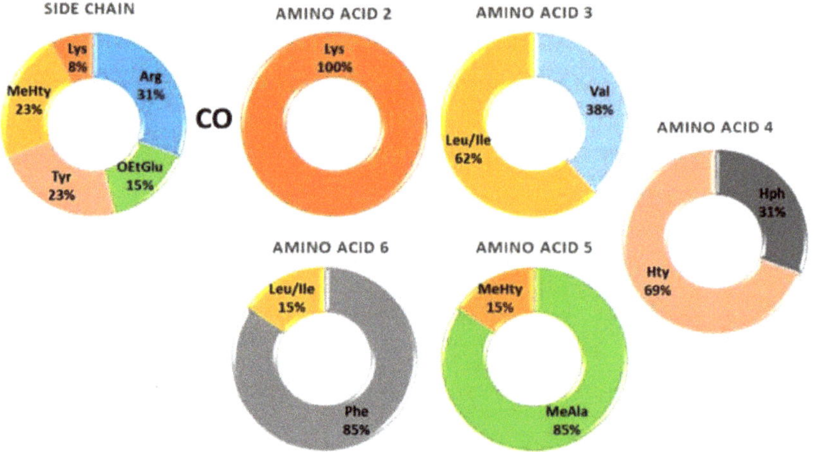

Figure 5. Diversity and frequency of variable amino acids in the structures of anabaenopeptins detected in Greek freshwaters.

Specifically, the currently known 42 APs from freshwater environments mainly consist of Val (45%) and Ile (29%) in position X_3, Hty (64%) and Hph (29%) in position X_4, MeAla (50%) and MeHty (38%) in position X_5 and Phe (45%) and Ile (24%) in position X_6 [12]. The 13 APs identified in Greek freshwaters consist of Val (38%) and Ile (62%) in position X_3, Hty (69%) and Hph (31%) in position X_4, MeAla (85%) and MeHty (15%) in position X_5 and Phe (85%) and Ile (15%) in position X_6 (Figure 5). A comparison of findings strongly supports that the variable amino acids of AP rings determined during this study are consistent with the most common ones of the known APs from freshwaters.

A higher diversity of amino acids was observed in the side chain. Arg (31%) was the most frequent, followed by Tyr (23%), MeHty (23%), OEtGlu (15%) and Lys (8%). Arg and Tyr are present in the side chains of commonly found AP congeners worldwide (i.e., AP B and AP F have Arg; AP A and Osc Y have Tyr). Contrarily, the presence of MeHty as a side chain has been reported for only seven AP congeners that were detected in cyanobacteria from Lake Balaton, Hungary [23]. The proposed side chain of the three novel APs consists of infrequent amino acids (i.e., Lys and OEtGlu). Lys (AP 894) has been determined in the side chain of six known congeners (Figure 1, Table S1), while OEtGlu (AP 837 and AP 851) is proposed for the first time. A previous study reported the presence of OMeGlu occupying the side chain amino acid position in the AP MM823 [65]. In fact, AP MM823 and the newly proposed AP 837 also share the same five-peptide ring structure. Although methylated amino acids are frequently occurring in AP structures, ethylated ones have also been reported [10,25], indicating the metabolomic potential of cyanobacteria.

2.3. Anabaenopeptins in Cyanobacterial Strains Isolated from Greek Freshwaters

Thirty cyanobacterial strains from the TAU-MAC culture collection [104], isolated from the freshwaters of Greece, were analyzed in order to evaluate their ability to produce APs (Table S2), i.e., fourteen strains of *Microcystis*, five of *Nostoc*, three of *Jaaginema*, two of *Synechococcus*, and one from the species of the genera *Anabaena, Calothrix, Chlorogloeopsis, Desmonostoc, Limnothrix* and *Nodosilinea*. APs were only detected in one strain extract out of the thirty examined. In particular, AP A and Osc Y were identified in the extract of *Microcystis ichtyoblabe* TAU-MAC 0510.

Although AP F and AP B along with Osc Y were the most frequently detected APs in cyanobacterial bloom samples in this study, they were not detected in any of the examined cyanobacterial strains. The diversity of APs in the isolated strains was limited compared to that of bloom extracts. This finding is in agreement with the results of previous studies as it was reported that *Microcystis* strains have a less diverse peptide pattern compared to that of the entire population of a bloom sample from a German lake [19], and that the *Planktothrix agardhii* samples from a Polish freshwater reservoir contained up to seven APs while the two strains isolated from the reservoir contained only one AP [75]. This was rather expected because the diversity of APs in field bloom samples reflects the high diversity of the chemotypes present in water bodies, therefore it cannot be compared with the diversity of the compounds in isolated strains [19,75]. The results of previous chemo-diversity studies of freshwater cyanobacterial strains also indicate the limited presence of AP congeners in the samples. Welker et al. reported the presence of APs in only 9% of 850 examined *Microcystis* colonies with five AP structural variants in total [22] while, in another study, 165 *Microcystis* colonies were examined and only up to four APs were detected in 21% of analyzed samples [20]. Martins et al. have also reported a limited presence of APs in *Microcystis aeruginosa* strains where one to three APs were detected in the 30% of examined strains [38]. Furthermore, in an investigation of 18 *Planktothrix* clonal strains, APs were present in 11 of them, with one, two and three APs present in seven, three and one strain, respectively [13]. The limited presence of APs in cyanobacterial strains may also be correlated with the evidence that cyanobacterial strains could lose the ability to produce cyanopeptides under laboratory conditions [105].

In a previous chemo-diversity study including 24 *Microcystis* strains isolated from the same freshwater blooms or from different populations in various geographical areas (i.e., Netherlands, Scotland, France, Senegal, Burkina Faso), it was found that AP A, AP B, AP F and Osc Y were the most commonly detected AP congeners and were mainly produced by *Microcystis aeruginosa* strains, while all the examined *Microcystis wesenbergii*/*M. viridis* strains did not produce APs. A comparison of the specific chemical footprints of the examined strains showed that the metabolite content was influenced globally by microcystin production rather than sampling locality origins [106]. In another study, it was concluded that AP B and AP E/F were among the principal cyanopeptides detected in 165 *Microcystis* sp. colonies isolated from German lakes and that APs were mostly produced by *Microcystis ichthyoblabe* colonies than by *Microcystis aeruginosa* [20]. According to Fastner et al., AP B, AP F and Osc.Y were the most prominent APs in *Microcystis ichthyoblabe* colonies isolated from a German lake, followed by AP I and AP A, while APs were rarely detected in the *Microcystis aeruginosa* colonies and not detected at all in *Microcystis wesenbergii* colonies [19]. A common conclusion of the above studies was that *Microcystis aeruginosa* colonies predominately produced microcystins; this was in contrast to *Microcystis ichthyoblabe* colonies that mainly produced APs rather than microcystins [19,20]. This is in agreement with the results of the present study where one strain belonging to cyanobacterial species *Microcystis ichthyoblabe* was found to be positive to APs while strains belonging to *Microcystis aeruginosa* and *Microcystis viridis* were negative to APs (Table S2).

In general, AP A, AP B, AP F and Osc Y are the most commonly detected APs both in *Microcystis* and *Planktothrix* strains isolated from several water bodies of European countries, such as Austria [34], the Czech Republic [22], Finland [14], Germany [13,19,20], Norway [74], Portugal [37] and Switzerland [31]. The current study constitutes the first investigation into APs' presence in several cyanobacterial strains isolated from Greek freshwaters.

3. Conclusions

The structural diversity of APs from bloom samples and cultured cyanobacteria strains of Greek freshwaters was investigated for the first time, utilizing LC–qTPAR MS/MS in IDA and EIP modes in order to structurally elucidate APs from their fragmentation spectra. Overall, thirteen APs were annotated, with three of these being reported for the first time (AP 837, AP 851 and AP 894). A variety of APs were found to occur in 21 out of 23 samples from cyanobacterial blooms from seven out of nine lakes that were mainly dominated by *Microcystis* and *Dolichospermum* species. The most frequently occurring APs in bloom samples were AP F and Osc Y, followed by AP B, AP 886 and AP A. On the other hand, in thirty samples of cultured cyanobacterial strains isolated from the freshwater bodies of Greece, APs (AP A and Osc Y) were only found in *Microcystis ichtyoblabe* TAU-MAC 0510. The results of this study are in general agreement with previous studies on the occurrence of APs in European freshwater bodies and contribute to the expansion of the range of known AP congeners by introducing three new AP structures and their mass fragmentation spectra. Considering that APs are a class of cyanobacterial bioactive metabolites that naturally occur in water bodies in high frequency and possibly in significant amounts, the results of this study highlight the need for further assessment of their environmental effects and impacts.

4. Materials and Methods

4.1. Cyanobacterial Bloom Samples

Samples were collected from nine Greek lakes (Amvrakia, Kastoria, Pamvotida, Kerkini, Zazari, Mikri Prespa, Vistonida, Karla, Marathonas) during episodes of cyanobacterial bloom (Table 2). General characteristics and location of the freshwater bodies are provided in the details of previous studies [98,107,108]. Water samples (100–1500 mL) were collected in airtight polyethylene bottles from the surface layer (0–35 cm) at the margins of the lakes where accumulation of cyanobacteria had been observed from May to October in 1995, 1999, 2000, 2010, 2014 and 2015, as previously described [5,98]. Samples were filtered through Whatman GF/C filters (Millipore, Cork, Ireland), lyophilized and stored at $-20\,^{\circ}$C until analysis. The cyanobacterial biomass of the samples ranged from 10–1000 mg/L. Dominant cyanobacterial species were characterized by microscopic analysis, as previously reported [5,98,109].

4.2. Source and Culture Conditions of Cyanobacterial Strains

Thirty cyanobacterial strains isolated from Greek freshwaters from 1999 to 2010 [109] were identified and provided by Thessaloniki Aristotle University Microalgae and Cyanobacteria (TAU-MAC) Culture Collection [104]. Strains were planktic or benthic; details of their origin and isolation are provided in [109]. Cyanobacterial strains belonging to *Chroococcales*, *Synechococcales* and *Nostocales* based on polyphasic taxonomy were classified into 10 genera (*Anabaena*, *Microcystis*, *Nostoc*, *Synechococcus*, *Limnothrix*, *Calothrix*, *Nodosilinea*, *Desmonostoc*, *Chlorogloeopsis* and *Jaaginema*) and 16 taxa, as listed in Table S2 [110]. Cyanobacterial strain cultures were grown in BG11 medium with or without nitrogen (BG11$_0$ for the nitrogen-fixing strains, see Table S2), shaken manually once per day and maintained at

25 °C (*Microcystis* strains) or 20 ± 1 °C (strains of the rest genera) under cool white light fluorescent lamps (Sylvania Standard F36W/154-T8, SLI, Sylvania, Erlangen, Germany) with a light intensity of 20–25 μmol m^{-2} s^{-1} in a 16/8 h light/dark cycle. At the end of the exponential phase of culture growth (between days 20 and 35, depending on the strain, see Table S2), the whole liquid culture (250 mL) was centrifuged and the cyanobacterial cells collected, lyophilized and stored at −20 °C until analysis. Chlorophyll-*a* was extracted from 5 mL of wet biomass with 95% (*v/v*) acetone solution and spectrophotometrically quantified, as outlined in APHA (2005) [111]. The chlorophyll-*a* concentration of the strains at the time of the collection (as an estimate of their biomass) ranged from 6.21–6.77 mg/L.

4.3. Sample Preparation and LC–MS/MS Analysis

Analysis of two different sample types, i.e., cyanobacterial blooms and cyanobacterial strain cultures, was performed. The same amount of each sample type was extracted and analyzed. Lyophilized biomass (~10 mg) of each sample was extracted with 1.5 mL of 75% methanol:25% water assisted by vortexing and sonication in an ice bath for 15 min. After centrifugation (10,000 rpm, 15 min), the supernatants were collected and further centrifuged (10,000 rpm, 5 min) prior to LC–MS/MS analysis.

Untargeted analysis was carried out with an Agilent 1200, liquid chromatography apparatus (Agilent Technologies, Waldboronn, Germany) coupled with a hybrid triple quadrupole/linear ion trap mass spectrometer (QTRAP5500, Applied Biosystems, Sciex; Concorde, ON, Canada) according to Mazur-Marzec et al., 2013 [44]. Chromatographic separation was achieved with a reversed phase column (Zorbax Eclipse XDB-C18 4.6 × 150 mm, 5 μm Agilent Technologies, Santa Clara, CA, USA) applying gradient elution. Mobile phases consisted of (A) acetonitrile and (B) 5% acetonitrile in MilliQ water, both containing 0.1% formic acid; flow rate was 0.6 mL min^{-1} and injection volume was 5 μL. Ionization was performed with electrospray (ESI) source in positive mode. For MS detection, information-dependent acquisition (IDA) mode and enhanced ion product (EIP) mode were applied. In IDA mode, a full scan from 500 to 1200 Da was acquired for detection of the compounds. EIP mode was triggered when the signal of an ion was above a threshold of 500,000 cps; the ions were fragmented in the collision cell (Q2) and fragmentation spectra were recorded from 50 to 1000 Da with a scan speed of 2000 Da s^{-1} and collision energy (CE) of 60 V with collision energy spread (CES) of 20 V. Analyst QS$^{®}$ 1.5.1 software was used for data acquisition and processing. Obtained fragmentation spectra were examined in order to elucidate the structures of occurring APs.

Supplementary Materials: The following are available online at https://www.mdpi.com/article/10.3390/toxins14010004/s1, Table S1: List of anabaenopeptins reported in the literature and their amino acid sequence, Table S2: List of the cyanobacterial strains from Greek freshwaters, examined for their ability to produce APs, Figure S1: Anabaenopeptins' presence in cyanobacterial blooms of Greek freshwaters and the dominant species, Figure S2: Fragmentation mass spectrum of Anabaenopeptin 820 with precursor ion at m/z 821 [M + H]$^+$, Figure S3: Fragmentation mass spectrum of Anabaenopeptin B with precursor ion at m/z 837 [M + H]$^+$, Figure S4: Fragmentation mass spectrum of Anabaenopeptin 842 with precursor ion at m/z 842 [M + H]$^+$, Figure S5: Fragmentation mass spectrum of Anabaenopeptin A with precursor ion at m/z 844 [M + H]$^+$, Figure S6: Fragmentation mass spectrum of Anabaenopeptin F with precursor ion at m/z 851 [M + H]$^+$, Figure S7: Fragmentation mass spectrum of Oscillamide Y with precursor ion at m/z 858 [M + H]$^+$, Figure S8: Fragmentation mass spectrum of Anabaenopeptin 870 with precursor ion at m/z 870 [M + H]$^+$, Figure S9: Fragmentation mass spectrum of Anabaenopeptin 872 with precursor ion at m/z 872 [M + H]$^+$, Figure S10: Fragmentation mass spectrum of Anabaenopeptin 886 with precursor ion at m/z 886 [M + H]$^+$, Figure S11: Fragmentation mass spectrum of Anabaenopeptin KB 906 with precursor ion at m/z 907 [M + H]$^+$.

Author Contributions: Conceptualization, S.-K.Z., T.K., S.G., A.H. and H.M.-M.; methodology, S.-K.Z., A.H. and H.M.-M.; software, S.-K.Z. and H.M.-M.; formal analysis, S.-K.Z.; investigation, S.-K.Z., T.K., S.G., A.H. and H.M.-M.; resources, T.K., S.G., A.H. and H.M.-M.; data curation, S.-K.Z. and H.M.-M.; writing—original draft preparation, S.-K.Z.; writing—review and editing, T.K., S.G., A.H.

and H.M.-M.; visualization, S.-K.Z.; supervision, A.H. and H.M.-M.; funding acquisition, T.K., A.H. and H.M.-M. All authors have read and agreed to the published version of the manuscript.

Funding: This research was co-financed by Greece and the European Union (European Social Fund—ESF) through the Operational Programme «Human Resources Development, Education and Lifelong Learning» in the context of the project "Reinforcement of Postdoctoral Researchers—2nd Cycle" (MIS-5033021), implemented by the State Scholarships Foundation (ΙΚΥ).

Institutional Review Board Statement: Not applicable.

Informed Consent Statement: Not applicable.

Data Availability Statement: Data are contained within the article or Supplementary Material.

Acknowledgments: The authors acknowledge COST Action ES 1105 "CYANOCOST—Cyanobacterial blooms and toxins in water resources: Occurrence impacts and management" www.cyanocost.net for adding value to this study through networking and knowledge sharing with European experts in the field.

Conflicts of Interest: The authors declare no conflict of interest.

References

1. Welker, M.; Von Döhren, H. Cyanobacterial peptides—Nature's own combinatorial biosynthesis. *FEMS Microbiol. Rev.* **2006**, *30*, 530–563. [CrossRef]
2. Kodani, S.; Suzuki, S.; Ishida, K.; Murakami, M. Five new cyanobacterial peptides from water bloom materials of lake Teganuma (Japan). *FEMS Microbiol. Lett.* **1999**, *178*, 343–348. [CrossRef]
3. Erhard, M.; Von Döhren, H.; Jungblut, P.R. Rapid identification of the new anabaenopeptin G from Planktothrix agardhii HUB 011 using matrix-assisted laser desorption/ionization time-of-flight mass spectrometry. *Rapid Commun. Mass Spectrom.* **1999**, *13*, 337–343. [CrossRef]
4. Barco, M.; Flores, C.; Rivera, J.; Caixach, J. Determination of microcystin variants and related peptides present in a water bloom of Planktothrix (Oscillatoria) rubescens in a Spanish drinking water reservoir by LC/ESI-MS. *Toxicon* **2004**, *44*, 881–886. [CrossRef] [PubMed]
5. Gkelis, S.; Harjunpää, V.; Lanaras, T.; Sivonen, K. Diversity of hepatotoxic microcystins and bioactive anabaenopeptins in cyanobacterial blooms from Greek freshwaters. *Environ. Toxicol.* **2005**, *20*, 249–256. [CrossRef] [PubMed]
6. Flores, C.; Caixach, J. High Levels of Anabaenopeptins Detected in a Cyanobacteria Bloom from N.E. Spanish Sau-Susqueda-El Pasteral Reservoirs System by LC–HRMS. *Toxins* **2020**, *12*, 541. [CrossRef]
7. Konkel, R.; Toruńska-Sitarz, A.; Cegłowska, M.; Ežerinskis, Ž.; Šapolaitė, J.; Mažeika, J.; Mazur-Marzec, H. Blooms of Toxic Cyanobacterium *Nodularia spumigena* in Norwegian Fjords During Holocene Warm Periods. *Toxins* **2020**, *12*, 257. [CrossRef] [PubMed]
8. Mazur-Marzec, H.; Sutryk, K.; Hebel, A.; Hohlfeld, N.; Pietrasik, A.; Błaszczyk, A. *Nodularia spumigena* Peptides—Accumulation and Effect on Aquatic Invertebrates. *Toxins* **2015**, *7*, 4404–4420. [CrossRef]
9. Spoof, L.; Błaszczyk, A.; Meriluoto, J.; Cegłowska, M.; Mazur-Marzec, H. Structures and Activity of New Anabaenopeptins Produced by Baltic Sea Cyanobacteria. *Mar. Drugs* **2016**, *14*, 8. [CrossRef] [PubMed]
10. Sanz, M.; Andreote, A.P.D.; Fiore, M.F.; Dörr, F.A.; Pinto, E. Structural Characterization of New Peptide Variants Produced by Cyanobacteria from the Brazilian Atlantic Coastal Forest Using Liquid Chromatography Coupled to Quadrupole Time-of-Flight Tandem Mass Spectrometry. *Mar. Drugs* **2015**, *13*, 3892–3919. [CrossRef]
11. Saha, S.; Esposito, G.; Urajová, P.; Mareš, J.; Ewe, D.; Caso, A.; Macho, M.; Delawská, K.; Kust, A.; Hrouzek, P.; et al. Discovery of Unusual Cyanobacterial Tryptophan-Containing Anabaenopeptins by MS/MS-Based Molecular Networking. *Molecules* **2020**, *25*, 3786. [CrossRef] [PubMed]
12. Monteiro, P.R.; do Amaral, S.C.; Siqueira, A.S.; Xavier, L.P.; Santos, A.V. Anabaenopeptins: What We Know So Far. *Toxins* **2021**, *13*, 522. [CrossRef] [PubMed]
13. Welker, M.; Christiansen, G.; Von Döhren, H. Diversity of coexisting Planktothrix (cyanobacteria) chemotypes deduced by mass spectral analysis of microystins and other oligopeptides. *Arch. Microbiol.* **2004**, *182*, 288–298. [CrossRef] [PubMed]
14. Rohrlack, T.; Skulberg, R.; Skulberg, O.M. Distribution of oligopeptide chemotypes of the cyanobacterium planktothrix and their persistence in selected lakes in fennoscandia. *J. Phycol.* **2009**, *45*, 1259–1265. [CrossRef]
15. Fujii, K.; Harada, K.-I.; Suzuki, M.; Kondo, F.; Ikai, Y.; Oka, H.; Carmichael, W.W.; Sivonen, K. Occurrence of novel cyclicpeptides together with microcystins from toxic cyanobacteria, *Anabaena* species. In *Harmful and Toxic Algal Blooms*; Yasumoto, T., Oshima, Y., Fukuyo, Y., Eds.; Intergovernmental Oceanographic Commission of UNESCO: Paris, France, 1996; pp. 559–562.
16. Fujii, K.; Sivonen, K.; Nakano, T.; Harada, K.-I. Structural elucidation of cyanobacterial peptides encoded by peptide synthetase gene in Anabaena species. *Tetrahedron* **2002**, *58*, 6863–6871. [CrossRef]
17. Grach-Pogrebinsky, O.; Carmeli, S. Three novel anabaenopeptins from the cyanobacterium *Anabaena* sp. *Tetrahedron* **2008**, *64*, 10233–10238. [CrossRef]

18. Tonk, L.; Welker, M.; Huisman, J.; Visser, P.M. Production of cyanopeptolins, anabaenopeptins, and microcystins by the harmful cyanobacteria Anabaena 90 and Microcystis PCC 7806. *Harmful Algae* **2009**, *8*, 219–224. [CrossRef]
19. Fastner, J.; Erhard, M.; Von Döhren, H. Determination of Oligopeptide Diversity within a Natural Population of Microcystis spp. (Cyanobacteria) by Typing Single Colonies by Matrix-Assisted Laser Desorption Ionization-Time of Flight Mass Spectrometry. *Appl. Environ. Microbiol.* **2001**, *67*, 5069–5076. [CrossRef] [PubMed]
20. Welker, M.; Brunke, M.; Preussel, K.; Lippert, I.; von Döhren, H. Diversity and distribution of Microcystis (cyanobacteria) oligopeptide chemotypes from natural communities studies by single-colony mass spectrometry. *Microbiology* **2004**, *150*, 1785–1796. [CrossRef]
21. Beresovsky, D.; Hadas, O.; Livne, A.; Sukenik, A.; Kaplan, A.; Carmeli, S. Toxins and biologically active secondary metabolites of Microcystis sp. isolated from Lake Kinneret. *Isr. J. Chem.* **2006**, *46*, 79–87. [CrossRef]
22. Welker, M.; Maršálek, B.; Šejnohová, L.; von Döhren, H. Detection and identification of oligopeptides in Microcystis (cyanobacteria) colonies: Toward an understanding of metabolic diversity. *Peptides* **2006**, *27*, 2090–2103. [CrossRef]
23. Riba, M.; Kiss-Szikszai, A.; Gonda, S.; Boros, G.; Vitál, Z.; Borsodi, A.K.; Krett, G.; Borics, G.; Ujvárosi, A.Z.; Vasas, G. Microcystis chemotype diversity in the alimentary tract of bigheaded carp. *Toxins* **2019**, *11*, 288. [CrossRef] [PubMed]
24. Shishido, T.K.; Jokela, J.; Fewer, D.P.; Wahlsten, M.; Fiore, M.F.; Sivonen, K. Simultaneous Production of Anabaenopeptins and Namalides by the Cyanobacterium Nostoc sp. CENA543. *ACS Chem. Biol.* **2017**, *12*, 2746–2755. [CrossRef]
25. Riba, M.; Kiss-Szikszai, A.; Gonda, S.; Parizsa, P.; Deák, B.; Török, P.; Valkó, O.; Felföldi, T.; Vasas, G. Chemotyping of terrestrial Nostoc-like isolates from alkali grassland areas by non-targeted peptide analysis. *Algal Res.* **2020**, *46*, 101798. [CrossRef]
26. Murakami, M.; Shin, H.J.; Matsuda, H.; Ishida, K.; Yamaguchi, K. A cyclic peptide, anabaenopeptin B, from the cyanobacterium Oscillatoria agardhii. *Phytochemistry* **1997**, *44*, 449–452. [CrossRef]
27. Sano, T.; Kaya, K. Oscillamide Y, a chymotrypsin inhibitor from toxic Oscillatoria agardhii. *Tetrahedron Lett.* **1995**, *36*, 5933–5936. [CrossRef]
28. Shin, H.J.; Matsuda, H.; Murakami, M.; Yamaguchi, K. Anabaenopeptins E and F, two new cyclic peptides from the cyanobacterium Oscillatoria agardhii (NIES-204). *J. Nat. Prod.* **1997**, *60*, 139–141. [CrossRef]
29. Itou, Y.; Suzuki, S.; Ishida, K.; Murakami, M. Anabaenopeptins G and H, Potent Carboxypeptidase A inhibitors from the cyanobacterium Oscillatoria agardhii (NIES-595). *Bioorganic Med. Chem. Lett.* **1999**, *9*, 1243–1246. [CrossRef]
30. Sano, T.; Usui, T.; Ueda, K.; Osada, H.; Kaya, K. Isolation of new protein phosphatase inhibitors from two cyanobacteria species, *Planktothrix* spp. *J. Nat. Prod.* **2001**, *64*, 1052–1055. [CrossRef]
31. Kosol, S.; Schmidt, J.; Kurmayer, R. Variation in peptide net production and growth among strains of the toxic cyanobacterium *Planktothrix* spp. *Eur. J. Phycol.* **2009**, *44*, 49–62. [CrossRef]
32. Okumura, H.S.; Philmus, B.; Portmann, C.; Hemscheidt, T.K. Homotyrosine-containing cyanopeptolins 880 and 960 and anabaenopeptins 908 and 915 from Planktothrix agardhii CYA 126/8. *J. Nat. Prod.* **2009**, *72*, 172–176. [CrossRef] [PubMed]
33. Grach-Pogrebinsky, O.; Sedmak, B.; Carmeli, S. Protease inhibitors from a Slovenian Lake Bled toxic waterbloom of the cyanobacterium *Planktothrix rubescens*. *Tetrahedron* **2003**, *59*, 8329–8336. [CrossRef]
34. Welker, M.; Erhard, M. Consistency between chemotyping of single filaments of *Planktothrix rubescens* (cyanobacteria) by MALDI-TOF and the peptide patterns of strains determined by HPLC-MS. *J. Mass Spectrom.* **2007**, *42*, 1062–1068. [CrossRef] [PubMed]
35. Harada, K.-I.; Fujii, K.; Shimada, T.; Suzuki, M.; Sano, H.; Adachi, K.; Carmichael, W.W. Two cyclic peptides, anabaenopeptins, a third group of bioactive compounds from the cyanobacterium Anabaena flos-aquae NRC 525-17. *Tetrahedron Lett.* **1995**, *36*, 1511–1514. [CrossRef]
36. Urrutia-Cordero, P.; Agha, R.; Cirés, S.; Lezcano, M.Á.; Sánchez-Contreras, M.; Waara, K.-O.; Utkilen, H.; Quesada, A. Effects of harmful cyanobacteria on the freshwater pathogenic free-living amoeba Acanthamoeba castellanii. *Aquat. Toxicol.* **2013**, *130–131*, 9–17. [CrossRef]
37. Saker, M.L.; Fastner, J.; Dittmann, E.; Christiansen, G.; Vasconcelos, V.M. Variation between strains of the cyanobacterium *Microcystis aeruginosa* isolated from a Portuguese river. *J. Appl. Microbiol.* **2005**, *99*, 749–757. [CrossRef] [PubMed]
38. Martins, J.; Saker, M.L.; Moreira, C.; Welker, M.; Fastner, J.; Vasconcelos, V.M. Peptide diversity in strains of the cyanobacterium *Microcystis aeruginosa* isolated from Portuguese water supplies. *Appl. Microbiol. Biotechnol.* **2009**, *82*, 951–961. [CrossRef]
39. Adiv, S.; Carmeli, S. Protease Inhibitors from *Microcystis aeruginosa* Bloom Material Collected from the Dalton Reservoir, Israel. *J. Nat. Prod.* **2013**, *76*, 2307–2315. [CrossRef]
40. Rouhiainen, L.; Jokela, J.; Fewer, D.P.; Urmann, M.; Sivonen, K. Two Alternative Starter Modules for the Non-Ribosomal Biosynthesis of Specific Anabaenopeptin Variants in Anabaena (Cyanobacteria). *Chem. Biol.* **2010**, *17*, 265–273. [CrossRef] [PubMed]
41. Murakami, M.; Suzuki, S.; Itou, Y.; Kodani, S.; Ishida, K. New anabaenopeptins, carboxypeptidaze-A inhibitors from the cyanobacterium Aphanizomenon flos-aquae. *J. Nat. Prod.* **2000**, *63*, 1280–1282. [CrossRef]
42. Fujii, K.; Sivonen, K.; Adachi, K.; Noguchi, K.; Sano, H.; Hirayama, K.; Suzuki, M.; Harada, K.-I. Comparative study of toxic and non-toxic cyanobacterial products: Novel peptides from toxic *Nodularia spumigena* AV1. *Tetrahedron Lett.* **1997**, *38*, 5525–5528. [CrossRef]
43. Schumacher, M.; Wilson, N.; Tabudravu, J.N.; Edwards, C.; Lawton, L.A.; Motti, C.; Wright, A.D.; Diederich, M.; Jaspars, M. New nodulopeptins from *Nodularia spumigena* KAC 66. *Tetrahedron* **2012**, *68*, 1622–1628. [CrossRef]

44. Mazur-Marzec, H.; Kaczkowska, M.J.; Blaszczyk, A.; Akcaalan, R.; Spoof, L.; Meriluoto, J. Diversity of Peptides Produced by *Nodularia spumigena* from Various Geographical Regions. *Mar. Drugs* **2013**, *11*, 1–19. [CrossRef] [PubMed]

45. Mazur-Marzec, H.; Bertos-Fortis, M.; Toruńska-Sitarz, A.; Fidor, A.; Legrand, C. Chemical and Genetic Diversity of *Nodularia spumigena* from the Baltic Sea. *Mar. Drugs* **2016**, *14*, 209. [CrossRef]

46. Bober, B.; Lechowski, Z.; Bialczyk, J. Determination of some cyanopeptides synthesized by Woronichinia naegeliana (Chroococcales, Cyanophyceae). *Phycol. Res.* **2011**, *59*, 286–294. [CrossRef]

47. Bober, B.; Chrapusta-Srebrny, E.; Bialczyk, J. Novel cyanobacterial metabolites, cyanopeptolin 1081 and anabaenopeptin 899, isolated from an enrichment culture dominated by *Woronichinia naegeliana* (Unger) Elenkin. *Eur. J. Phycol.* **2020**, *56*, 244–254. [CrossRef]

48. Häggqvist, K.; Toruńska-Sitarz, A.; Błaszczyk, A.; Mazur-Marzec, H.; Meriluoto, J. Morphologic, Phylogenetic and Chemical Characterization of a Brackish Colonial Picocyanobacterium (Coelosphaeriaceae) with Bioactive Properties. *Toxins* **2016**, *8*, 108. [CrossRef]

49. Zi, J.; Lantvit, D.D.; Swanson, S.M.; Orjala, J. Lyngbyaureidamides A and B, two anabaenopeptins from the cultured freshwater cyanobacterium Lyngbya sp. (SAG 36.91). *Phytochemistry* **2012**, *74*, 173–177. [CrossRef]

50. Matthew, S.; Ross, C.; Paul, V.J.; Luesch, H. Pompanopeptins A and B, new cyclic peptides from the marine cyanobacterium Lyngbya confervoides. *Tetrahedron* **2008**, *64*, 4081–4089. [CrossRef]

51. Reshef, V.; Carmeli, S. Schizopeptin 791, a New Anabeanopeptin-like Cyclic Peptide from the Cyanobacterium *Schizothrix* sp. *J. Nat. Prod.* **2002**, *65*, 1187–1189. [CrossRef]

52. Müller, D.; Krick, A.; Kehraus, S.; Mehner, C.; Hart, M.; Küpper, F.C.; Saxena, K.; Prinz, H.; Schwalbe, H.; Janning, P.; et al. Brunsvicamides A−C: Sponge-Related Cyanobacterial Peptides with Mycobacterium tuberculosis Protein Tyrosine Phosphatase Inhibitory Activity. *J. Med. Chem.* **2006**, *49*, 4871–4878. [CrossRef]

53. Kobayashi, J.I.; Sato, M.; Ishibashi, M.; Shigemori, H.; Nakamura, T.; Ohizumi, Y. Keramamide A, a novel peptide from the Okinawan marine sponge *Theonella* sp. *J. Chem. Soc. Perkin Trans. 1* **1991**, *10*, 2609–2611. [CrossRef]

54. Kobayashi, J.; Sato, M.; Murayama, T.; Ishibashi, M.; Wälchi, M.R.; Kanai, M.; Shoji, J.; Ohizumi, Y. Konbamide, a novel peptide with calmodulin antagonistic activity from the Okinawan marine sponge *Theonella* sp. *J. Chem. Soc. Chem. Commun.* **1991**, *15*, 1050–1052. [CrossRef]

55. Uemoto, H.; Yahiro, Y.; Shigemori, H.; Tsuda, M.; Takao, T.; Shimonishi, Y.; Kobayashi, J.I. Keramamides K and L, new cyclic peptides containing unusual tryptophan residue from Theonella sponge. *Tetrahedron* **1998**, *54*, 6719–6724. [CrossRef]

56. Plaza, A.; Keffer, J.L.; Lloyd, J.R.; Colin, P.L.; Bewley, C.A. Paltolides A-C, anabaenopeptin-type peptides from the Palau sponge theonella swinhoei. *J. Nat. Prod.* **2010**, *73*, 485–488. [CrossRef]

57. Robinson, S.J.; Tenney, K.; Yee, D.F.; Martinez, L.; Media, J.E.; Valeriote, F.A.; van Soest, R.W.M.; Crews, P. Probing the Bioactive Constituents from Chemotypes of the Sponge Psammocinia aff. bulbosa. *J. Nat. Prod.* **2007**, *70*, 1002–1009. [CrossRef]

58. Schmidt, E.W.; Harper, M.K.; Faulkner, D.J. Mozamides A and B, Cyclic Peptides from a Theonellid Sponge from Mozambique. *J. Nat. Prod.* **1997**, *60*, 779–782. [CrossRef]

59. Williams, D.E.; Craig, M.; Holmes, C.F.B.; Andersen, R.J. Ferintoic Acids A and B, New Cyclic Hexapeptides from the Freshwater Cyanobacterium *Microcystis aeruginosa*. *J. Nat. Prod.* **1996**, *59*, 570–575. [CrossRef]

60. Rounge, T.B.; Rohrlack, T.; Nederbragt, A.J.; Kristensen, T.; Jakobsen, K.S. A genome-wide analysis of nonribosomal peptide synthetase gene clusters and their peptides in a *Planktothrix rubescens* strain. *BMC Genom.* **2009**, *10*, 396. [CrossRef]

61. Christiansen, G.; Philmus, B.; Hemscheidt, T.; Kurmayer, R. Genetic variation of adenylation domains of the anabaenopeptin synthesis operon and evolution of substrate promiscuity. *J. Bacteriol.* **2011**, *193*, 3822–3831. [CrossRef]

62. Entfellner, E.; Frei, M.; Christiansen, G.; Deng, L.; Blom, J.; Kurmayer, R. Evolution of Anabaenopeptin Peptide Structural Variability in the Cyanobacterium Planktothrix. *Front. Microbiol.* **2017**, *8*, 219. [CrossRef]

63. Janssen, E.M.L. Cyanobacterial peptides beyond microcystins—A review on co-occurrence, toxicity, and challenges for risk assessment. *Water Res.* **2019**, *151*, 488–499. [CrossRef]

64. Harms, H.; Kurita, K.L.; Pan, L.; Wahome, P.G.; He, H.; Kinghorn, A.D.; Carter, G.T.; Linington, R.G. Discovery of anabaenopeptin 679 from freshwater algal bloom material: Insights into the structure–activity relationship of anabaenopeptin protease inhibitors. *Bioorganic Med. Chem. Lett.* **2016**, *26*, 4960–4965. [CrossRef]

65. Zafrir-Ilan, E.; Carmeli, S. Eight novel serine proteases inhibitors from a water bloom of the cyanobacterium *Microcystis* sp. *Tetrahedron* **2010**, *66*, 9194–9202. [CrossRef]

66. Bubik, A.; Sedmak, B.; Novinec, M.; Lenarčič, B.; Lah, T.T. Cytotoxic and peptidase inhibitory activities of selected non-hepatotoxic cyclic peptides from cyanobacteria. *Biol. Chem.* **2008**, *389*, 1339–1346. [CrossRef]

67. Sedmak, B.; Eleršek, T.; Grach-Pogrebinsky, O.; Carmeli, S.; Sever, N.; Lah, T.T. Ecotoxicologically relevant cyclic peptides from cyanobacterial bloom (*Planktothrix rubescens*)—A threat to human and environmental health. *Radiol. Oncol.* **2008**, *42*, 102–113. [CrossRef]

68. Gkelis, S.; Lanaras, T.; Sivonen, K. The presence of microcystins and other cyanobacterial bioactive peptides in aquatic fauna collected from Greek freshwaters. *Aquat. Toxicol.* **2006**, *78*, 32–41. [CrossRef] [PubMed]

69. Schreuder, H.; Liesum, A.; Lönze, P.; Stump, H.; Hoffmann, H.; Schiell, M.; Kurz, M.; Toti, L.; Bauer, A.; Kallus, C.; et al. Isolation, Co-Crystallization and Structure-Based Characterization of Anabaenopeptins as Highly Potent Inhibitors of Activated Thrombin Activatable Fibrinolysis Inhibitor (TAFIa). *Sci. Rep.* **2016**, *6*, 32958. [CrossRef]

70. Lenz, K.A.; Miller, T.R.; Ma, H. Anabaenopeptins and cyanopeptolins induce systemic toxicity effects in a model organism the nematode Caenorhabditis elegans. *Chemosphere* **2019**, *214*, 60–69. [CrossRef] [PubMed]
71. Pawlik-Skowrońska, B.; Toporowska, M.; Mazur-Marzec, H. Effects of secondary metabolites produced by different cyanobacterial populations on the freshwater zooplankters Brachionus calyciflorus and Daphnia pulex. *Environ. Sci. Pollut. Res.* **2019**, *26*, 11793–11804. [CrossRef] [PubMed]
72. Sedmak, B.; Carmeli, S.; Eleršek, T. "Non-toxic" cyclic peptides induce lysis of cyanobacteria—An effective cell population density control mechanism in cyanobacterial blooms. *Microb. Ecol.* **2008**, *56*, 201–209. [CrossRef]
73. Halstvedt, C.B.; Rohrlack, T.; Ptacnik, R.; Edvardsen, B. On the effect of abiotic environmental factors on production of bioactive oligopeptides in field populations of *Planktothrix* spp. (Cyanobacteria). *J. Plankton Res.* **2008**, *30*, 607–617. [CrossRef]
74. Rohrlack, T.; Edvardsen, B.; Skulberg, R.; Halstvedt, C.B.; Utkilen, H.C.; Ptacnik, R.; Skulberg, O.M. Oligopeptide chemotypes of the toxic freshwater cyanobacterium Planktothrix can form sub-populations with dissimilar ecological traits. *Limnol. Oceanogr.* **2008**, *53*, 1279–1293. [CrossRef]
75. Grabowska, M.; Kobos, J.; Toruńska-Sitarz, A.; Mazur-Marzec, H. Non-ribosomal peptides produced by Planktothrix agardhii from Siemianówka Dam Reservoir SDR (northeast Poland). *Arch. Microbiol.* **2014**, *196*, 697–707. [CrossRef]
76. Toporowska, M.; Mazur-Marzec, H.; Pawlik-Skowrońska, B. The Effects of Cyanobacterial Bloom Extracts on the Biomass, Chl-a, MC and Other Oligopeptides Contents in a Natural Planktothrix agardhii Population. *Int. J. Environ. Res. Public Health* **2020**, *17*, 2881. [CrossRef]
77. Kust, A.; Řeháková, K.; Vrba, J.; Maicher, V.; Mareš, J.; Hrouzek, P.; Chiriac, M.-C.; Benedová, Z.; Tesařová, B.; Saurav, K. Insight into Unprecedented Diversity of Cyanopeptides in Eutrophic Ponds Using an MS/MS Networking Approach. *Toxins* **2020**, *12*, 561. [CrossRef]
78. Baumann, H.I.; Jüttner, F. Inter-annual stability of oligopeptide patterns of *Planktothrix rubescens* blooms and mass mortality of Daphnia in Lake Hallwilersee. *Limnologica* **2008**, *38*, 350–359. [CrossRef]
79. Flores, C.; Caixach, J. An integrated strategy for rapid and accurate determination of free and cell-bound microcystins and related peptides in natural blooms by liquid chromatography–electrospray-high resolution mass spectrometry and matrix-assisted laser desorption/ionization time-of-flight/time-of-flight mass spectrometry using both positive and negative ionization modes. *J. Chromatogr. A* **2015**, *1407*, 76–89.
80. Ferranti, P.; Fabbrocino, S.; Cerulo, M.G.; Bruno, M.; Serpe, L.; Gallo, P. Characterisation of biotoxins produced by a cyanobacteria bloom in Lake Averno using two LC–MS-based techniques. *Food Addit. Contam.—Part A Chem. Anal. Control Expo. Risk Assess.* **2008**, *25*, 1530–1537. [CrossRef]
81. Ferranti, P.; Nasi, A.; Bruno, M.; Basile, A.; Serpe, L.; Gallo, P. A peptidomic approach for monitoring and characterising peptide cyanotoxins produced in Italian lakes by matrix-assisted laser desorption/ionisation and quadrupole time-of-flight mass spectrometry. *Rapid Commun. Mass Spectrom.* **2011**, *25*, 1173–1183. [CrossRef]
82. Bogialli, S.; Bortolini, C.; Di Gangi, I.M.; Di Gregorio, F.N.; Lucentini, L.; Favaro, G.; Pastore, P. Liquid chromatography-high resolution mass spectrometric methods for the surveillance monitoring of cyanotoxins in freshwaters. *Talanta* **2017**, *170*, 322–330. [CrossRef] [PubMed]
83. Cerasino, L.; Capelli, C.; Salmaso, N. A comparative study of the metabolic profiles of common nuisance cyanobacteria in southern perialpine lakes. *Adv. Oceanogr. Limnol.* **2017**, *8*, 22–32. [CrossRef]
84. Di Pofi, G.; Favero, G.; Nigro Di Gregorio, F.; Ferretti, E.; Viaggiu, E.; Lucentini, L. Multi-residue Ultra Performance Liquid Chromatography-High resolution mass spectrometric method for the analysis of 21 cyanotoxins in surface water for human consumption. *Talanta* **2020**, *211*, 120738. [CrossRef]
85. Roy-Lachapelle, A.; Solliec, M.; Sauvé, S.; Gagnon, C. A Data-Independent Methodology for the Structural Characterization of Microcystins and Anabaenopeptins Leading to the Identification of Four New Congeners. *Toxins* **2019**, *11*, 619. [CrossRef] [PubMed]
86. Roy-Lachapelle, A.; Solliec, M.; Sauvé, S.; Gagnon, C. Evaluation of ELISA-based method for total anabaenopeptins determination and comparative analysis with on-line SPE-UHPLC-HRMS in freshwater cyanobacterial blooms. *Talanta* **2021**, *223*, 121802. [CrossRef]
87. Filatova, D.; Jones, M.R.; Haley, J.A.; Núñez, O.; Farré, M.; Janssen, E.M.L. Cyanobacteria and their secondary metabolites in three freshwater reservoirs in the United Kingdom. *Environ. Sci. Eur.* **2021**, *33*, 29. [CrossRef]
88. Gesner-Apter, S.; Carmeli, S. Three novel metabolites from a bloom of the cyanobacterium Microcystis sp. *Tetrahedron* **2008**, *64*, 6628–6634. [CrossRef]
89. De Carvalho, L.R.; Pipole, F.; Werner, V.R.; Laughinghouse Iv, H.D.; De Camargo, A.C.M.; Rangel, M.; Konno, K.; Sant Anna, C.L. A toxic cyanobacterial bloom in an urban coastal lake, Rio Grande do Sul State, southern Brazil. *Braz. J. Microbiol.* **2008**, *39*, 761–769. [CrossRef]
90. Roy-Lachapelle, A.; Vo Duy, S.; Munoz, G.; Dinh, Q.T.; Bahl, E.; Simon, D.F.; Sauvé, S. Analysis of multiclass cyanotoxins (microcystins, anabaenopeptins, cylindrospermopsin and anatoxins) in lake waters using on-line SPE liquid chromatography high-resolution Orbitrap mass spectrometry. *Anal. Methods* **2019**, *11*, 5289–5300. [CrossRef]
91. Bartlett, S.L.; Brunner, S.L.; Klump, J.V.; Houghton, E.M.; Miller, T.R. Spatial analysis of toxic or otherwise bioactive cyanobacterial peptides in Green Bay, Lake Michigan. *J. Great Lakes Res.* **2018**, *44*, 924–933. [CrossRef]

92. Beversdorf, L.J.; Weirich, C.A.; Bartlett, S.L.; Miller, T.R. Variable cyanobacterial toxin and metabolite profiles across six eutrophic lakes of differing physiochemical characteristics. *Toxins* **2017**, *9*, 62. [CrossRef] [PubMed]

93. Beversdorf, L.J.; Rude, K.; Weirich, C.A.; Bartlett, S.L.; Seaman, M.; Kozik, C.; Biese, P.; Gosz, T.; Suha, M.; Stempa, C.; et al. Analysis of cyanobacterial metabolites in surface and raw drinking waters reveals more than microcystin. *Water Res.* **2018**, *140*, 280–290. [CrossRef]

94. Miller, T.R.; Bartlett, S.L.; Weirich, C.A.; Hernandez, J. Automated Subdaily Sampling of Cyanobacterial Toxins on a Buoy Reveals New Temporal Patterns in Toxin Dynamics. *Environ. Sci. Technol.* **2019**, *53*, 5661–5670. [CrossRef]

95. Puddick, J.; Prinsep, M.R. MALDI-TOF mass spectrometry of Cyanobacteria: A global approach to the discovery of novel secondary metabolites. *Chem. N. Z.* **2008**, *72*, 68–71.

96. Anahas, A.M.P.; Gayathri, M.; Muralitharan, G. Isolation and characterization of microcystin-producing *Microcystis aeruginosa* MBDU 626 from a freshwater bloom sample in Tamil Nadu, South India. In *Microbiological Research in Agroecosystem Management*; Springer: New Delhi, India, 2013; pp. 235–248.

97. Skafi, M.; Vo Duy, S.; Munoz, G.; Dinh, Q.T.; Simon, D.F.; Juneau, P.; Sauvé, S. Occurrence of microcystins, anabaenopeptins and other cyanotoxins in fish from a freshwater wildlife reserve impacted by harmful cyanobacterial blooms. *Toxicon* **2021**, *194*, 44–52. [CrossRef] [PubMed]

98. Gkelis, S.; Lanaras, T.; Sivonen, K. Cyanobacterial Toxic and Bioactive Peptides in Freshwater Bodies of Greece: Concentrations, Occurrence Patterns, and Implications for Human Health. *Mar. Drugs* **2015**, *13*, 6319–6335. [CrossRef]

99. Zervou, S.-K.; Moschandreou, K.; Paraskevopoulou, A.; Christophoridis, C.; Grigoriadou, E.; Kaloudis, T.; Triantis, T.M.; Tsiaoussi, V.; Hiskia, A. Cyanobacterial Toxins and Peptides in Lake Vegoritis, Greece. *Toxins* **2021**, *13*, 394. [CrossRef]

100. Sumner, L.W.; Amberg, A.; Barrett, D.; Beale, M.H.; Beger, R.; Daykin, C.A.; Fan, T.W.M.; Fiehn, O.; Goodacre, R.; Griffin, J.L.; et al. Proposed minimum reporting standards for chemical analysis. *Metabolomics* **2007**, *3*, 211–221. [CrossRef]

101. Elkobi-Peer, S.; Carmeli, S. New Prenylated Aeruginosin, Microphycin, Anabaenopeptin and Micropeptin Analogues from a *Microcystis* Bloom Material Collected in Kibbutz Kfar Blum, Israel. *Mar. Drugs* **2015**, *13*, 2347–2375. [CrossRef]

102. Ferranti, P.; Fabbrocino, S.; Chiaravalle, E.; Bruno, M.; Basile, A.; Serpe, L.; Gallo, P. Profiling microcystin contamination in a water reservoir by MALDI-TOF and liquid chromatography coupled to Q/TOF tandem mass spectrometry. *Food Res. Int.* **2013**, *54*, 1321–1330. [CrossRef]

103. Jones, M.R.; Pinto, E.; Torres, M.A.; Dörr, F.; Mazur-Marzec, H.; Szubert, K.; Tartaglione, L.; Dell'Aversano, C.; Miles, C.O.; Beach, D.G.; et al. CyanoMetDB, a comprehensive public database of secondary metabolites from cyanobacteria. *Water Res.* **2021**, *196*, 117017. [CrossRef] [PubMed]

104. Gkelis, S.; Panou, M. Capturing biodiversity: Linking a cyanobacteria culture collection to the "scratchpads" virtual research environment enhances biodiversity knowledge. *Biodivers Data J.* **2016**, *6*, e7965. [CrossRef] [PubMed]

105. Shishido, T.K.; Jokela, J.; Humisto, A.; Suurnäkki, S.; Wahlsten, M.; Alvarenga, D.O.; Sivonen, K.; Fewer, D.P. The Biosynthesis of Rare Homo-Amino Acid Containing Variants of Microcystin by a Benthic Cyanobacterium. *Mar. Drugs* **2019**, *17*, 271. [CrossRef] [PubMed]

106. Le Manach, S.; Duval, C.; Marie, A.; Djediat, C.; Catherine, A.; Edery, M.; Bernard, C.; Marie, B. Global Metabolomic Characterizations of Microcystis spp. Highlights Clonal Diversity in Natural Bloom-Forming Populations and Expands Metabolite Structural Diversity. *Front. Microbiol.* **2019**, *10*, 791. [CrossRef]

107. Gkelis, S.; Panou, M.; Chronis, I.; Zervou, S.-K.; Christophoridis, C.; Manolidi, K.; Ntislidou, C.; Triantis, T.M.; Kaloudis, T.; Hiskia, A.; et al. Monitoring a newly re-born patient: Water quality and cyanotoxin occurrence in a reconstructed shallow Mediterranean lake. *Adv. Oceanogr. Limnol.* **2017**, *8*, 33–51. [CrossRef]

108. Vardaka, E.; Moustaka-Gouni, M.; Cook, C.M.; Lanaras, T. Cyanobacterial blooms and water quality in Greek waterbodies. *J. Appl. Phycol.* **2005**, *17*, 391–401. [CrossRef]

109. Gkelis, S.; Tussy, P.F.; Zaoutsos, N. Isolation and preliminary characterization of cyanobacteria strains from freshwaters of Greece. *Open Life Sci.* **2014**, *10*, 52–60. [CrossRef]

110. Gkelis, S.; Panou, M.; Konstantinou, D.; Apostolidis, P.; Kasampali, A.; Papadimitriou, S.; Kati, D.; Di Lorenzo, G.M.; Ioakeim, S.; Zervou, S.-K.; et al. Diversity, Cyanotoxin Production, and Bioactivities of Cyanobacteria Isolated from Freshwaters of Greece. *Toxins* **2019**, *11*, 436. [CrossRef] [PubMed]

111. APHA. *Standard Methods for the Examination of Water and Wastewater*, 21st ed.; American Public Health Association, American Water Works Association, and Water Environment Federation: Washington, DC, USA, 2005.

Communication

Investigation of the Occurrence of Cyanotoxins in Lake Karaoun (Lebanon) by Mass Spectrometry, Bioassays and Molecular Methods

Noura Alice Hammoud [1,2,3], Sevasti-Kiriaki Zervou [2], Triantafyllos Kaloudis [2,4], Christophoros Christophoridis [2], Aikaterina Paraskevopoulou [2,5], Theodoros M. Triantis [2], Kamal Slim [1], Joanna Szpunar [3], Ali Fadel [1], Ryszard Lobinski [3,6] and Anastasia Hiskia [2,*]

[1] National Council for Scientific Research (CNRS), P.O. Box 11-8281, Riad El Solh, Beirut 1107 2260, Lebanon; n.hammoud@univ-pau.fr (N.A.H.); k.slim@laec-cnrs.gov.lb (K.S.); afadel@cnrs.edu.lb (A.F.)
[2] Laboratory of Photo-Catalytic Processes and Environmental Chemistry, Institute of Nanoscience and Nanotechnology, National Center for Scientific Research "Demokritos", Patr. Grigoriou E' & 27 Neapoleos Str., Agia Paraskevi, 15341 Athens, Greece; s.zervou@inn.demokritos.gr (S.-K.Z.); t.kaloudis@inn.demokritos.gr (T.K.); c.christoforidis@inn.demokritos.gr (C.C.); k.paraskevopoulou@inn.demokritos.gr (A.P.); t.triantis@inn.demokritos.gr (T.M.T.)
[3] Institut des Sciences Analytiques et de Physico-Chimie pour l'Environnement et les Matériaux, Université de Pau et des Pays de l'Adour, E2S UPPA, CNRS, IPREM UMR 5254, Hélioparc, 64053 Pau, France; joanna.szpunar@univ-pau.fr (J.S.); ryszard.lobinski@univ-pau.fr (R.L.)
[4] Department of Water Quality Control, Athens Water Supply and Sewerage Company (EYDAP SA), 156 Oropou Str., 11146 Athens, Greece
[5] Chemical Engineering Department, National Technical University, Iroon Politechniou 9, Zografou, 15780 Athens, Greece
[6] Chair of Analytical Chemistry, Warsaw University of Technology, Noakowskiego 3, 00-664 Warsaw, Poland
* Correspondence: a.hiskia@inn.demokritos.gr

Citation: Hammoud, N.A.; Zervou, S.-K.; Kaloudis, T.; Christophoridis, C.; Paraskevopoulou, A.; Triantis, T.M.; Slim, K.; Szpunar, J.; Fadel, A.; Lobinski, R.; et al. Investigation of the Occurrence of Cyanotoxins in Lake Karaoun (Lebanon) by Mass Spectrometry, Bioassays and Molecular Methods. *Toxins* **2021**, *13*, 716. https://doi.org/10.3390/toxins13100716

Received: 23 July 2021
Accepted: 5 October 2021
Published: 10 October 2021

Publisher's Note: MDPI stays neutral with regard to jurisdictional claims in published maps and institutional affiliations.

Abstract: Lake Karaoun is the largest artificial lake in Lebanon and serves multiple purposes. Recently, intensive cyanobacterial blooms have been reported in the lake, raising safety and aesthetic concerns related to the presence of cyanotoxins and cyanobacterial taste and odor (T&O) compounds, respectively. Here, we communicate for the first time results from a recent investigation by LC-MS/MS covering multiple cyanotoxins (microcystins (MCs), anatoxin-a, cylindrospermopsin, nodularin) in water and fish collected between 2019 and 2020. Eleven MCs were identified reaching concentrations of 211 and 199 µg/L for MC-LR and MC-YR, respectively. Cylindrospermopsin, anatoxin-a and nodularin were not detected. The determination of the total MCs was also carried out by ELISA and Protein Phosphatase Inhibition Assay yielding comparable results. Molecular detection of cyanobacteria (16S rRNA) and biosynthetic genes of toxins were carried out by qPCR. Untargeted screening analysis by GC-MS showed the presence of T&O compounds, such as β-cyclocitral, β-ionone, nonanal and dimethylsulfides that contribute to unpleasant odors in water. The determination of volatile organic compounds (VOCs) showed the presence of anthropogenic pollutants, mostly dichloromethane and toluene. The findings are important to develop future monitoring schemes in order to assess the risks from cyanobacterial blooms with regard to the lake's ecosystem and its uses.

Keywords: cyanotoxins (CTs); microcystins (MCs); volatile organic compounds (VOCs); taste and odor (T&O) compounds; SPE-LC-MS/MS; HS-SPME-GC/MS

Key Contribution: First report confirming the occurrence of MC congeners and cyanobacterial/algal T&O compounds in lake Karaoun; Lebanon.

1. Introduction

Cyanobacteria are photosynthetic microorganisms commonly found in surface waters. They can produce a large variety of secondary metabolites, including toxic compounds, known as cyanotoxins (CTs). CTs have various chemical structures and modes of toxicity (Figure S1). Microcystins (MCs) [1] and Nodularin (NOD) [2] are cyclic peptides, both containing the unique L-amino acid Adda ((2S,3S,8S,9S)-3-amino-9-methoxy-2,6,8-trimethyl-10-phenyldeca-4,6-dienoicacid), which is responsible for their hepatotoxic activity [3,4]. Cylindrospermopsin (CYN) is an alkaloid cyanotoxin with cytotoxic, dermatotoxic, hepatotoxic and possibly carcinogenic potency [5]. The alkaloid Anatoxin-a (ATX), also known as "Very Fast Death Factor", is a bicyclic secondary amine (2-acetyl-9-azabicyclo(1,4,2)non-2-ene) with acute neurotoxicity [6]. Among CTs, MCs is the most frequently reported class [7,8].

Under favorable environmental conditions of temperature, presence of nutrients and light, cyanobacteria can proliferate excessively and form blooms [9]. The presence of CTs in toxic cyanobacterial blooms poses a significant risk to aquatic ecosystems and drinking water sources, while it has been associated with lethal poisonings of wild animals, livestock and humans [10]. The World Health Organization (WHO) established guideline values for MCs, CYN and ATX for drinking-water (lifetime, short-term or acute) and recreational exposure and published guides to hazard identification and management of risks posed by cyanobacteria and their toxins at each step of the water-use system [11].

Lake Karaoun (Qaraoun) is the largest water body in Lebanon and has multiple uses. The reservoir was created in 1959 and is intended for irrigation, hydropower, and recreational activities, as well as an anticipated drinking water supply to the capital, Beirut [12]. However, nutrient enrichment due to the excessive use of fertilizers in the lake's watershed, untreated sewage and industrial waste water runoffs, are possibly contributing to the degradation of the reservoir's water quality [13]. In 2009, the cyanobacteria *Aphanizomenon ovalisporum* and *Microcystis aeruginosa* were first reported and were seen to dominate the lake's phytoplankton. Their proliferation was attributed to eutrophication and climate change resulting in the extension of dry periods and the rise in water temperature, as well as the pollution from multiple-industrial and agricultural sources [14]. In recent years, the blooms have become more intense, forming thick scums and emitting unpleasant odors in the area.

There is limited information on toxic cyanobacterial blooms in Lake Karaoum [14–16], with a previous study reporting the presence of CYN produced by *Aphanizomenon ovalisporum*, however, not confirmed by mass spectrometry [17]. There are no published data regarding the presence of MCs in the reservoir, especially during incidents of cyanobacterial blooms. Due to the multiple uses of the reservoir, including the supply of drinking water, there is an urgent need to definitively assess and confirm the presence of CTs in Lake Karaoun, in order to enable the design of future monitoring programs and to develop related management strategies.

In addition to CTs, cyanobacteria produce a plethora of volatile secondary metabolites, with several of them having a strong taste and odor (T&O). This can make reservoirs unacceptable as drinking water sources due to consumer complaints and can negatively impact recreational activities and tourism [18,19]. Indeed, concerning Lake Karaoun, there has been circumstantial reporting of bad smells, but the presence of T&O compounds has never been investigated and it is not clear if the origin of T&O is from cyanobacteria or from industrial pollution. T&O compounds can be present at very low concentrations and although they are generally not toxic, they are responsible for the unpleasant smell of drinking water. Their removal may require re-planning and additional investments in water treatment [20].

The primary aim of this communication is to report for the first time conclusive results of the presence of CTs (MCs, CYN, ATX) in Lake Karaoun, using mass spectrometric analytical techniques, in addition to bioassays and molecular methods. A further aim is to investigate the presence of T&O compounds, in order to understand whether they originate from cyanobacterial metabolism or from anthropogenic pollution. To meet these objectives, a set of complementary methods were used to produce reliable results for a variety of cyanotoxins and T&O compounds. Since Middle-Eastern lakes (with the exception of Lake Kinneret [21]) are under-studied in regard to toxic cyanobacteria, the results presented give important information on the presence of CTs and T&O compounds in the area, enabling future monitoring programs and management strategies for Lake Karaoun.

2. Results

2.1. Water Quality and Diversity of Cyanobacteria in Lake Karaoun

During the period of field campaigns (2019–2020), the diversity of cyanobacteria in the reservoir was limited (Figure 1). *Aphanizomenon ovalisporum* was the dominant species (>95%) from October 2019 to April 2020, while *Microcystis aeruginosa* was the most abundant on July and October 2020. In August 2019, both *Microcystis aeruginosa* and *Aphanizomenon ovalisporum* were equally present with abundances of 50% and 49%, respectively. In addition, in July of the same year, *Microcystis ichtyoblable* and *Woronichia naegeliana* were identified in very low abundances, of 1% each. A microscope image of the dominant cyanobacterial species, i.e., *Microcystis aeruginosa* and *Aphanizomenon ovalisporum* is shown in Figure 2.

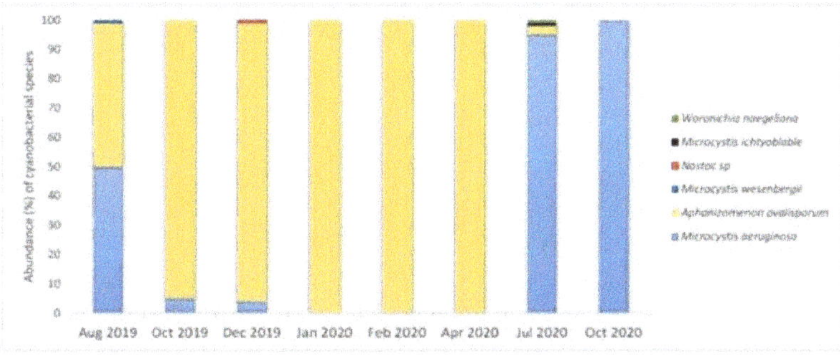

Figure 1. Abundance (%) of cyanobacterial species in Lake Karaoun per sampling month (August 2019–October 2020).

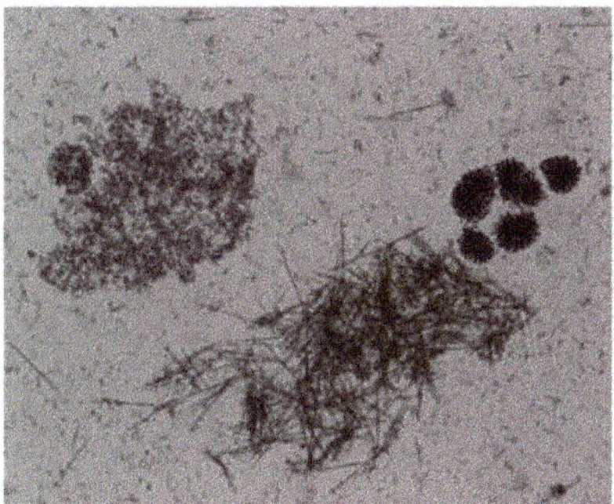

Figure 2. Dominant cyanobacteria species in Lake Karaoun: *Microcystis aeruginosa* and *Aphanizomenon ovalisporum*.

Physico-chemical parameters of the lake's water were also monitored during the sampling period by the National Litani River Authority, which is in charge of the river management (Table S5). In most of the sampling dates, concentrations of nitrate and phosphate exhibited elevated values up to 16.72, 0.28 and 0.16 expressed as mg/L of NO_3-N, NO_2-N and PO_4-P, respectively. This is in agreement with the assessment of the ecological status of the Karaoun reservoir by Fadel et al., classifying the lake as hypereutrophic, with low phytoplankton biodiversity and regular blooms of toxic cyanobacteria [15,16]. Degradation of the reservoir's water quality is considered to be mainly due to the significant loads of untreated sewage water and the discarding of agricultural and livestock remnants into the river stream.

2.2. Occurrence of Cyanotoxins (CTs) in Lake Karaoun

Occurrence of CTs in samples from Lake Karaoun was assessed by liquid chromatography tandem mass spectrometry (LC-MS/MS), complemented by ELISA, Protein Phosphatase Inhibition Assay (PPIA) and molecular detection of cyanobacteria and toxin genes by qPCR.

2.2.1. LC-MS/MS Analysis of Water Samples

Water samples were analyzed by LC-MS/MS for detection and quantitative determination of extracellular and intracellular CTs, i.e., ATX, CYN, NOD and 12 MC variants (Table 1). Cyanotoxins were found in 50% of samples (8 out of 16), extending over 2019 and 2020. CYN, ATX and NOD were not detected in any of the samples. On the other hand, 11 out of 12 MC variants i.e., dmMC-RR, MC-RR, MC-YR, dmMC-LR, MC-LR, MC-HilR, MC-WR, MC-LA, MC-LY, MC-LW and MC-LF were detected in at least one sample, while MC-HtyR was not detected in any samples. The samples with the largest diversity of MCs were sample 2 (S2-October 19, 10 variants), sample 16 (S3-October 20, 9 variants) and sample 4 (S1-December 19, 5 variants). MC-RR was the most frequently detected toxin (8 out of 16 samples). The sample with the highest concentrations of MCs was sample 16 (S3-October 20), where total MC-LR (sum of extracellular and intracellular fractions) reached 211 µg/L and MC-YR 199 µg/L. In this sample, the dissolved (extracellular) amounts of MC-LR and MC-YR were roughly 100-fold of the intracellular amounts, in contrast to samples 2 (S2-October 19) and 4 (S1-December 19), where MC-LR, MC-RR and MC-YR are found mostly as intracellular. The high proportion of extracellular MCs in sample 16 can possibly be attributed to a decaying phase of the bloom, with disruption of cyanobacteria cells and release of MCs into water. An indicative MRM chromatogram (sample 2, S2-02/10/2019—intracellular fraction) is presented in Figure S2.

Table 1. Concentrations (µg/L) of extracellular/intracellular cyanotoxins (CYN, ATX, NOD, 12 MC variants) in water samples collected from Lake Karaoun during 2019–2020, analyzed by LC-MS/MS.

Sample ID	Sample Details	CYN	ATX	dm MC-RR	MC-RR	NOD	MC-YR	MC-HtyR	dm MC-LR	MC-LR	MC-HilR	MC-WR	MC-LA	MC-LY	MC-LW	MC-LF
1	S1, 22/08/19	ND[1]/NA[2]	ND/NA	ND/NA	ND/NA	ND/NA	ND/NA	ND/NA	ND/NA	ND/NA	ND/NA	ND/NA	ND/NA	ND/NA	ND/NA	ND/NA
2	S2, 02/10/19	ND/ND	ND/ND	ND/1.60	0.339/26.4	ND/ND	ND/1.67	ND/ND	ND/1.87	0.438/91.5	ND/1.29	<LOQ[3]/ND	ND/ND	<LOQ/3.29	0.014/10.5	<LOQ/7.48
3	S4, 02/10/19	ND/ND	ND/ND	ND/ND	ND/ND	ND/ND	ND/ND	ND/ND	ND/ND	ND/ND	ND/ND	ND/ND	ND/ND	ND/ND	ND/ND	ND/ND
4	S1, 22/12/19	ND/ND	ND/ND	ND/0.53	0.050/7.42	ND/ND	0.187/8.28	ND/ND	0.050/0.70	0.158/8.43	ND/ND	ND/ND	ND/ND	ND/ND	ND/ND	ND/ND
5	S1, 15/01/20	ND/ND	ND/ND	ND/ND	0.007/ND	ND/ND	ND/ND	ND/ND	ND/ND	ND/ND	ND/ND	ND/ND	ND/ND	ND/ND	ND/ND	ND/ND
6	S4, 04/02/20	ND/ND	ND/ND	ND/ND	ND/ND	ND/ND	ND/ND	ND/ND	ND/ND	ND/ND	ND/ND	ND/ND	ND/ND	ND/ND	ND/ND	ND/ND
7	S5, 15/02/20	ND/ND	ND/ND	ND/ND	ND/ND	ND/ND	ND/ND	ND/ND	ND/ND	ND/ND	ND/ND	ND/ND	ND/ND	ND/ND	ND/ND	ND/ND
8	S5, 17/02/20	ND/ND	ND/ND	ND/ND	0.005/ND	ND/ND	ND/ND	ND/ND	ND/ND	ND/ND	ND/ND	ND/ND	ND/ND	ND/ND	ND/ND	ND/ND
9	S1, 22/02/20	ND/ND	ND/ND	ND/ND	0.007/ND	ND/ND	ND/ND	ND/ND	ND/ND	ND/ND	ND/ND	ND/ND	ND/ND	ND/ND	ND/ND	ND/ND
10	S2, 22/02/20	ND/ND	ND/ND	ND/ND	0.008/ND	ND/ND	ND/ND	ND/ND	ND/ND	ND/ND	ND/ND	ND/ND	ND/ND	ND/ND	ND/ND	ND/ND
11	S1, 15/04/20	ND/ND	ND/ND	ND/ND	ND/ND	ND/ND	ND/ND	ND/ND	ND/ND	ND/ND	ND/ND	ND/ND	ND/ND	ND/ND	ND/ND	ND/ND
12	S2, 22/04/20	ND/NA	ND/NA	ND/NA	0.090/NA	ND/NA	ND/NA	ND/NA	ND/NA	ND/NA	ND/NA	ND/NA	ND/NA	ND/NA	ND/NA	ND/NA
13	S4, 22/04/20	ND/ND	ND/ND	ND/ND	ND/ND	ND/ND	ND/ND	ND/ND	ND/ND	ND/ND	ND/ND	ND/ND	ND/ND	ND/ND	ND/ND	ND/ND
14	S1, 14/08/20	ND/ND	ND/ND	ND/ND	ND/ND	ND/ND	ND/ND	ND/ND	ND/ND	ND/ND	ND/ND	ND/ND	ND/ND	ND/ND	ND/ND	ND/ND
15	S2, 14/08/20	ND/ND	ND/ND	ND/ND	ND/ND	ND/ND	ND/ND	ND/ND	ND/ND	ND/ND	ND/ND	ND/ND	ND/ND	ND/ND	ND/ND	ND/ND
16	S3, 09/10/20	ND/ND	ND/ND	0.069/ND	7.0/0.086	ND/ND	197/1.94	ND/ND	6.9/<LOQ	209/2.18	1.6/ND	ND/ND	3.0/ND	<LOQ/ND	ND/ND	0.021/ND

[1] ND: Not detected; [2] NA: Not analyzed; [3] <LOQ: Values higher than limit of detection (LOD) and lower than limit of quantitation (LOQ); LODs and LOQs are given in Table S1.

2.2.2. ELISA and PPIA Analysis of Water Samples

Since LC-MS/MS analysis targeted only a limited number of specific MC variants (12), assessment of the presence of MCs was complemented with ELISA and Protein Phosphatase Inhibition Assay (PPIA), in order to estimate total MCs concentration based on structural (ELISA) and functional (PPIA) similarities. Samples were analyzed for extracellular (dissolved) and intracellular MCs, similarly to LC-MS/MS. Results of ELISA and PPIA, together with the sum of MCs determined by LC-MS/MS, for comparison, are presented in Table 2. In general, samples in which either MCs were not detected by LC-MS/MS or their concentrations were lower than the LODs of ELISA/PPIA (0.10/0.25 µg/L), were negative by ELISA and PPIA with the exception of samples 8 (S5–17/2/2020), 11 (S1-15/04/2020), 12 (S2-22/04/2020) and 13 (S4-22/04/2020). Especially, in the latter sample, MCs were detected only by ELISA/PPIA as extracellular, in comparable concentrations (2.30/2.32 µg/L eq. MC-LR). MCs were found mostly as extracellular in sample 16 (S3-9/10/2020) by ELISA and PPIA, in agreement to LC-MS/MS, although in this sample ELISA and PPIA gave about 55% and 32% lower concentrations than LC-MS/MS.

Table 2. Total Extracellular/total intracellular MCs concentrations in water samples from Lake Karaoun by LC-MS/MS (sum of 12 extracellular MCs/sum of 12 intracellular MCs, µg/L), ELISA (µg/L MC-LR equivalents) and PPIA (µg/L MC-LR equivalents).

ID	SAMPLE DETAILS	Total Extracellular/Intracellular MCs		
		LC-MS/MS * sum of MCs, µg/L	ELISA ** µg/L eq.MC-LR	PPIA ** µg/L eq.MC-LR
1	S1, 22/08/2019	ND/NA	ND/NA	ND/NA
2	S2, 02/10/2019	0.83/146	NA/97.2	NA/111
3	S4, 02/10/2019	ND/ND	ND/ND	ND/ND
4	S1, 22/12/2019	0.45/25.4	0.13/30.6	0.25/16.2
5	S1, 15/01/2020	0.007/ND	ND/ND	ND/ND
6	S4, 04/02/2020	ND/ND	ND/ND	ND/ND
7	S5, 15/02/2020	ND/ND	ND/ND	ND/ND
8	S5, 17/02/2020	0.005/ND	0.12/ND	ND/ND
9	S1, 22/02/2020	0.007/ND	ND/ND	ND/ND
10	S2, 22/02/2020	0.008/ND	ND/ND	ND/ND
11	S1, 15/04/2020	ND/ND	0.25/ND	0.39/ND
12	S2, 22/04/2020	0.090/NA	1.80/NA	0.87/NA
13	S4, 22/04/2020	ND/ND	2.30/ND	2.32/ND
14	S1, 14/08/2020	ND/ND	ND/ND	ND/ND
15	S2, 14/08/2020	ND/ND	ND/ND	ND/ND
16	S3, 9/10/2020	425/4.29	190/2.41	290/2.34

ND = Not detected, NA = Not Available. LODs: LC-MS/MS 0.001 µg/L; ELISA 0.10 µg/L; PPIA 0.25 µg/L. * analysis with validated methods [22,23], extracellular toxins RSD < 16%, intracellular toxins RSD < 26%; ** duplicate analysis, RSD < 25%.

2.2.3. Molecular Detection of Cyanobacteria and Cyanotoxin Genes with qPCR

Molecular detection of cyanobacteria (16S rRNA) and biosynthetic genes of MCs, NOD, CYN, and Saxitoxins (STX) was carried out by qPCR. All samples were found positive for the presence of cyanobacteria. Samples 2 (S2—02/10/2019), 4 (S1—22/12/2019) and 16 (S3—09/10/2020) were found positive for MC producing genes (*mcyE*), with sample 4 being the most abundant in both cyanobacteria and MC genes among all samples. Genes associated with production of CYN (*cyrA*) and STX (*stxA*) were not detected in any of the samples. Results (gene copies per ml of sample) are presented in Table 3.

Table 3. Detection of cyanobacteria genes (16S rRNA) and biosynthetic genes of MCs and NOD (*mcyE*), CYN (*cyrA*) and STX (*stxA*) in water samples from Lake Karaoun (gene copies/mL).

ID	SAMPLE DETAILS	Gene Copies/mL			
		16S rRNA (Cyanobacteria)	*mcyE* (MCs & NODs)	*cyrA* (CYN)	*stxA* (STX)
1	S1, 22/08/2019	NA	NA	NA	NA
2	S2, 02/10/2019	6951375	69857	ND	ND
3	S4, 02/10/2019	NA	NA	NA	NA
4	S1, 22/12/2019	9765354	127870	ND	ND
5	S1, 15/01/2020	NA	NA	NA	NA
6	S4, 04/02/2020	194446	ND	ND	ND
7	S5, 15/02/2020	9808	ND	ND	ND
8	S5, 17/02/2020	54262	ND	ND	ND
9	S1, 22/02/2020	62344	ND	ND	ND
10	S2, 22/02/2020	5093	ND	ND	ND
11	S1, 15/04/2020	40877	ND	ND	ND
12	S2, 22/04/2020	186812	ND	ND	ND
13	S4, 22/04/2020	15466	ND	ND	ND
14	S1, 14/08/2020	112948	ND	ND	ND
15	S2, 14/08/2020	34032	ND	ND	ND
16	S3, 09/10/2020	171308	6921	ND	ND

NA: not available, ND: not detected.

2.2.4. LC-MS/MS Analysis of MCs in Fish Samples

Fish (*Cyprinus caprio*) flesh and liver samples were also analyzed for 12 MC variants by LC-MS/MS, using an in-house method. MCs were not detected in the analyzed samples. An indicative LC-MS/MS MRM chromatogram of a fish liver sample along with the chromatogram of a multi-toxin standard (Figure S3), shows the absence of MCs.

2.3. Taste and Odor (T&O) and Volatile Organic Compounds (VOCs)

Untargeted screening of volatile compounds by Headspace Solid Phase Microextraction coupled to Gas Chromatography-Mass Spectrometry (HS-SPME-GC-MS) led to the detection and identification of 20 compounds belonging to the following chemical groups: terpenes/terpenoids, hydrocarbons, aldehydes/ketones, phenols, phthalates, alkyl sulfides and indoles (Table 4). The chemical structures of the identified volatile compounds are presented in Figure 3. Two typical cyanobacterial terpenoids T&O compounds, β-cyclocitral (tobacco/woody odor) and β-ionone (floral odor), were detected in samples 2, 4 and 16 and 4, 12 and 16, respectively (Table S2). Nonanal, a known cyanobacterial/algal aldehyde with a fishy odor [24], was detected in 9 samples. Dimethyl disulfide, dimethyl trisulfide and dimethyl tetrasulfide, which are alkyl sulfides associated with algae/cyanobacteria were detected in sample 4, which had a strong swampy/septic odor. Samples 2 and 4 also contained 3-methylindole (skatole), a fecal compound with a characteristic odor. A number of hydrocarbons were also detected, with 2,2,4,6,6-pentamethyl heptane being the most common, as it was found in all samples. Industrial pollutants 2,4-di-tert-butylphenol, diisobutyl phthalate and p-cresol were detected in samples 10, 4 and 2, respectively. Results of untargeted HS-SPME-GC/MS screening per sample are given in Table S2.

Table 4. Compounds detected by untargeted HS-SPME-GC/MS screening, identification level of analysis, chemical group and retention time (t_R) of detected compounds, number of samples in which they have been identified and their % peak areas.

No	Compound	Identification Level	Chemical Group	t_R (min)	Number of Samples	% Peak Area
1	α-Thujene	A	Terpenes/Terpenoids	4.82	2	0.05–0.08
2	2,2,4,6,6-pentamethyl heptane	B	Hydrocarbons	5.43	16	0.03–0.80
3	2,2,6-trimethyl-cyclohexanone	A	Aldehydes/Ketones	5.95	1	0.004
4	p-Cresol	B	Phenols	6.28	2	0.04–5.55
5	2,4-Dimethyldecane	A	Hydrocarbons	6.59	1	0.17
6	Nonanal	A	Aldehydes/Ketones	6.64	9	0.06–0.43
7	4,7-Dimethylundecane	A	Hydrocarbons	7.62	1	0.01
8	β-Cyclocitral	B,C	Terpenes/Terpenoids	7.90	3	0.05–0.16
9	2,6,6-Trimethyl-1-Cyclohexene-1-acetaldehyde	A	Terpenes/Terpenoids	8.25	1	0.007
10	6,10-Dimethyl-5,9-Undecadien-2-one	B	Aldehydes/Ketones	9.92	1	0.35
11	β-Ionone	B,C	Terpenes/Terpenoids	10.25	3	0.04–0.38
12	Pentadecane	B,C	Hydrocarbons	10.35	1	0.14
13	2,4-di-tert-butylphenol	A,C	Phenols	10.43	10	0.01–39
14	Hexadecane	B,C	Hydrocarbons	11.16	2	0.20–0.60
15	Heptadecane	B,C	Hydrocarbons	11.94	6	0.07–33
16	Diisobutyl phthalate	B	Phthalates	13.11	4	0.003–1.5
17	Dimethyl disulfide	C,D	Alkyl sulfides	5.81	1	0.02
18	Dimethyl trisulfide	C,D	Alkyl sulfides	12.35	1	0.64
19	Dimethyl tetrasulfide	D	Alkyl sulfides	17.68	1	0.78
20	3-methylindole	C,D	Indoles	20.21	2	0.005–0.02

A: Retention Index (RI) ± 20, spectral matching ≥80%; B: RI ± 20, spectral matching ≥90%; C: matching with a reference standard; D: spectral matching ≥80%.

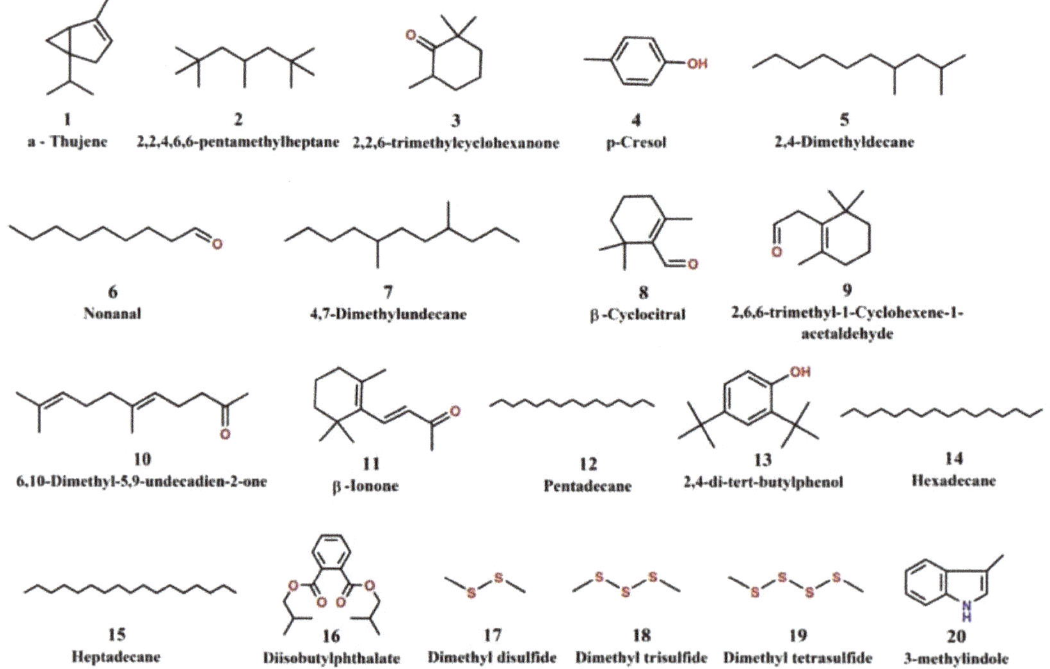

Figure 3. Chemical structures of the volatile compounds identified.

Quantitative determination of 36 VOCs (given in 5.8.2) was carried out by HS-SPME-GC/MS. The targeted compounds are typical industrial pollutants of anthropogenic origin. Concentrations of VOCs which were detected in at least one sample are presented in Table 5. All samples were found to contain dichloromethane (25.1–34.0 µg/L). Toluene was present in 12 samples at concentrations up to 0.81 µg/L. Trichloromethane (chloroform) was detected at low concentrations in samples 11 and 16 (0.40 and 0.50 µg/L, respectively). Traces (<0.25 µg/L) of trichlorobenzenes were detected in sample 6, and traces of 1,2 dichloropropane, benzene, xylenes and styrene in sample 16.

Table 5. Concentration (µg/L) of VOCs detected in at least one sample by targeted HS-SPME-GC/MS analysis

		Concentration (µg/L)								
Sample ID	Sample Details	1,2,3-Trichlorobenzene	1,2,4-Trichlorobenzene	Dichloromethane	Toluene	1,2-Dichloropropane	Benzene	Trichloromethane	o-Xylene	m+p-Xylene	Styrene
1	S1, 22/08/19	ND	ND	31.0	0.72	ND	ND	ND	ND	ND	ND
2	S2, 02/10/19	ND	ND	28.8	0.65	ND	ND	ND	ND	ND	ND
3	S4, 02/10/19	ND	ND	28.1	0.60	ND	ND	ND	ND	ND	ND
4	S1, 22/12/19	ND	ND	29.1	0.25	ND	ND	ND	ND	ND	ND
5	S1, 15/01/20	ND	ND	29.0	<0.25	ND	ND	ND	ND	ND	ND
6	S4, 04/02/20	<0.25	<0.25	27.2	<0.25	ND	ND	ND	ND	ND	ND
7	S5, 15/02/20	ND	ND	28.2	ND	ND	ND	ND	ND	ND	ND
8	S5, 17/02/20	ND	ND	26.3	<0.25	ND	ND	ND	ND	ND	ND
9	S1, 22/02/20	ND	ND	34.0	<0.25	ND	ND	ND	ND	ND	ND
10	S2, 22/02/20	ND	ND	29.4	<0.25	ND	ND	ND	ND	ND	ND
11	S1, 15/04/20	ND	ND	27.2	ND	ND	ND	0.40	ND	ND	ND
13	S4, 22/04/20	ND	ND	30.2	ND	ND	ND	ND	ND	ND	ND
14	S1, 14/08/20	ND	ND	32.3	0.46	ND	ND	ND	ND	ND	ND
15	S2, 14/08/20	ND	ND	33.6	0.35	ND	ND	ND	ND	ND	ND
16	S3, 09/10/20	ND	ND	25.1	0.81	<0.25	<0.25	0.50	<0.25	<0.25	<0.25

ND: Not Detected. Sample 12 was not available for analysis.

3. Discussion

A complementary set of analytical methods was used to investigate and confirm the occurrence of cyanotoxins in Lake Karaoun for the first time. Unambiguous determination of 12 MC congeners, NOD, ATX and CYN was carried out by LC-MS/MS, complemented by ELISA and PPIA for MCs and qPCR for MC, CYN and STX producing genes. Results confirm the presence of several congeners (variants) of microcystins in the lake at elevated concentrations during October 2019 and 2020. These early findings give important information on the occurrence, variants and levels of occurring CTs, and they can be used for the establishment of reliable future monitoring programs to support management of cyanobacteria, cyanotoxins and T&O compounds in the context of the multiple uses of the lake.

LC-MS/MS analysis confirmed the presence of dmMC-RR, MC-RR, MC-YR, dmMC-LR, MC-LR, MC-HilR, MC-WR, MC-LA, MC-LY, MC-LW, and MC-LF, with concentrations at 146.8, 25.8 and 429.3 µg/L in October 2019, December 2019 and October 2020, respectively, expressed as the sum of extra and intracellular fractions. MC-LR was the most abundant MC congener followed by MC-YR and MC-RR. CYN, ATX and NOD were not detected in any of the samples.

Results from the analysis of water samples using ELISA and PPIA for total MCs concentration were in general agreement with those obtained from LC-MS/MS (Table 2). In samples 11, 12 and 13, the higher values for extracellular fraction obtained by ELISA and

PPIA may indicate the presence of MC congeners other than those that were targeted by the LC-MS/MS method used in the study. This limitation, caused by the unavailability of commercial MC reference standards is well known in the literature, and it underscores the need for complementary quantitative analysis for total MCs (e.g., ELISA, PPIA), or for the development of non-targeted methods based on high-resolution mass spectrometry and related mass-spectral databases [25]. However, discrepancies between ELISA/PPIA and LC-MS/MS could also be attributed to the well-documented weaknesses of ELISA/PPIA. In particular, the shortcomings of ELISA are its high susceptibility to matrix effects, the variable cross-reactivity of different MC congeners and the response to degradation and/or transformation products of MCs that can result in overestimations [26]. In addition, PPIA based on the assessment of enzyme activity indicating the overall toxicity, does not have the same sensitivity to the different MCs and may also interfere with other unknown compounds present in the sample, thus resulting in under- or over-estimation of the concentration of toxins [27]. Furthermore, ELISA/PPIA lack the specificity of LC-MS/MS, as they respond to structurally/functionally similar molecules in total, and not to specific MC congeners. ELISA/PPIA are useful, quick, easy approaches for the screening of surface waters for MCs, especially where advanced laboratory infrastructure (e.g., LC-MS/MS) is not easily available. However, they must be considered as quantitative screening techniques for the detection of cyanotoxins [28,29].

Molecular analysis (qPCR) confirmed the presence of cyanobacteria (16s rRNA gene) in all the water samples collected. The gene *mcyE* which is associated with the production of MCs was found in 3 samples, i.e., samples 2 (S2, October 2019), 4 (S1, December 2019 and 16 (S3, October 2020), where the presence of MCs was also confirmed by LC-MS/MS, ELISA and PPIA. The gene responsible for the production of CYN (*cyrA*) was not detected in any of the samples (Table 3). This is in agreement with LC-MS/MS analysis where CYN was also not detected (Table 1).

Although significant concentrations of target MCs were detected in water samples, MCs were not detected in fish samples (flesh or liver). Previous studies also referred to cases where very low concentrations or total absence of MCs were reported in fish tissues collected during toxic cyanobacterial blooms [30,31]. It was shown that concentrations of MCs in fish tissues is strongly affected by local bloom conditions, fish feeding habits and ecosystem characteristics [32]. The occurrence of blooms and the presence of MCs in water are temporally and spatially variable, leading to different exposure of fish to MCs [33,34]. Inter-species differences in fish are also important factors for the intake and accumulation of MCs in their tissues. For example, zooplanktivorous fish were more likely to accumulate MCs than fish of other feeding guilds e.g., omnivorous, such as the studied *Cyprinous Carpio* [35]. The mechanisms of MC excretion and the timescales for MC elimination can also differ considerably among different fish species [36].

In the course of this study, *Microcystis aeruginosa* and *Aphanizomenon ovalisporum* were almost unique cyanobacterial species that were observed in autumn/spring and summer, respectively. These results are in agreement with a previous study reporting *Microcystis aeruginosa* and *Aphanizomenon ovalisporum* to be the most frequently encountered bloom-forming species in Lake Karaoun, either separately or together [14]. Furthermore, low phytoplankton biodiversity in Lake Karaoun was reported by Fadel and Slim who found that, since 2009, these two species had constituted >95% of the total phytoplankton biomass [16]. It was also reported that in Lake Karaoun, *Aphanizomenon ovalisporum* blooms formed both in spring and autumn, while *Microcystis aeruginosa* blooms formed at higher water temperatures observed during the summer months [15]. Although, *Aphanizomenon ovalisporum* is a well-known CYN producer, CYN was not detected during this study. On the contrary, the presence of CYN was previously reported during May to November 2012 and March to May 2013. The determination was performed with ELISA showing concentrations ranging from 0.5 to 1.7 µg/L in 2012, and from undetected to 1.7 µg/L in 2013 [17]. However, to date, the detection of CYN has never been confirmed by LC-MS/MS.

Monitoring studies of lakes and reservoirs in the Middle East are rare, with the exception of the closest natural lake, Lake Kinneret (Israel). Although Lake Kinneret is less eutrophic than the Karaoun reservoir, blooms of *Microcystis sp.* have been reported since the late 1960s [37], while *Aphanizomenon ovalisporum* and *Cylindrospermopsis raciborskii* were first detected in September 1994 and summer 1998, respectively [38,39]. The most frequently found cyanotoxins in Lake Kinneret have been CYN and MCs [21]. Despite the fact that CYN was not detected in this study, it is proposed that CYN should be included, together with MCs, in future monitoring programs to better evaluate the associated risks, since *Aphanizomenon ovalisporum* blooms continue to occur in lake Karaoun. NOD is mostly associated with brackish water cyanobacteria with a limited number of studies to report its occurrence in freshwater bodies [4,40]. Therefore, its presence is generally not expected in Lake Karaoun, while it can be determined along with MCs due to its structural similarity with them [41] in future monitoring programs. The presence of ATX should further be investigated, in cases where known cyanobacteria producers are present [4].

Screening for volatile compounds showed the presence of T&O cyanobacterial/algal terpenoids β-cyclocitral (in 3 samples) and β-ionone (in 3 samples). In particular, the presence of β-cyclocitral was reported to be associated with the occurrence of strains of *Microcystis* [42,43] which has a tobacco-woody odor with a rather high odor threshold concentration of 5 μg/L [44]. It was shown that β-cyclocitral is rapidly produced upon *Microcystis* cell rupture via a carotene oxygenase reaction and it subsequently affects grazer behavior, acting as a repellent and signal of low-quality food to grazers [43]. β-Ionone, commonly occurring in algae and higher plants, is also produced through the carotenoid cleavage of dioxygenases [45]. It has a characteristic flowery-woody odor, showing strong odor thresholds, at a level of 0.007 μg/L [46]. The release of β-ionone in lake water was positively correlated with microcystis biomass [47]. The common cyanobacteria/actimomycetes T&O compounds, geosmin and 2-methylisoborneol were not detected in any of the samples.

Nonanal, a known cyanobacterial/algal aldehyde with fishy odor [24] and an odor threshold of 1 μg/L in water [48] was detected in 9 samples. Aldehydes are mostly derivatives of polyunsaturated fatty acids and are common causes of odor in surface waters [18]. Dimethyl disulfide, dimethyl trisulfide and dimethyl tetrasulfide were detected in sample 4 that had a strong swampy/septic odor. Such undesirable odors have been a major concern for drinking water in several countries. In particular, dialkyl sulfides present strong odors described as swampy/septic, rotten, rancid and stinky, with very low odor threshold concentration levels of ng/L or less [49–51]. These and other organosulfur compounds can be produced both under oxic and anoxic conditions by a diversity of biota, biochemical pathways, enzymes and precursors [52]. 3-methylindole (skatole) was detected in samples 2 and 4. In particular, sample 4 had a strong swampy/septic odor due to the presence of dimethyl sulfides and skatole. The latter is a fecal compound with a characteristic odor that was previously reported to occur in algal cultures and field samples [18].

Several species of cyanobacteria can produce cyanotoxins and T&O compounds. For example, some strains of *Microcystis* produce microcystins together with β-cyclocitral and alkyl sulfides [42]. However, cyanobacterial T&O do not inevitably indicate the occurrence of cyanotoxins [53]. Nevertheless, since T&O can be sensed by the human nose at very low concentrations, they can serve as an early warning indicator for further investigations into the presence of toxic cyanobacteria [54]. The cyanobacterial/algal T&O detected in this study can serve as an initial list of compounds to be screened in future T&O episodes in Beirut's drinking water supplies.

The hydrocarbons detected, such as 2,2,4,6,6-pentamethyl heptane (all samples) and straight-chain hydrocarbons hexadecane (2 samples) and heptadecane (6 samples), could generally be of anthropogenic or biogenic origin. The hydrocarbon 2,2,4,6,6-Pentamethyl heptane has many industrial uses in anti-freeze products, coatings, fillers, lubricants, greases, etc. [55]. On the other hand, cyanobacteria and algae have long been known producers of alkanes, and their potential for biofuel production has been an area of increased research interest [56]. A study of volatile compounds associated with cyanobacteria and algae in freshwater by Juttner et al., [42] reported that biogenic hydrocarbons were represented exclusively by straight-chain components. Other compounds detected, such as 2,4-di-*tert*-butylphenol (antioxidant), diisobutyl phthalate (plasticizer) and p-cresol are common industrial pollutants, with uses in fuels, plastics, and production of chemicals. These were identified in samples 10, 4 and 2, respectively.

Quantitative determination of VOCs showed the presence of dichloromethane in all samples, at concentrations up to 34 µg/L. Dichloromethane is a common industrial solvent used in many chemical processes. Toluene was detected in the majority of samples at concentrations of up to 0.81 µg/L, while in sample 16 it co-occurred with traces of benzene and o,m,p-xylenes. These compounds are found in fuels and petroleum products and are also common industrial solvents. The "BTEX" volatiles (benzene, toluene, ethylbenzene, xylenes) are frequently used as an index to assess the impact of pollution caused by spills or leaks from fuel storage tanks into water bodies. Trichloromethane, detected in two samples at concentrations of up to 0.5 µg/L could indicate contamination of the lake with wastewater, since trihalomethanes, are commonly present in chlorine-treated water [57].

The above findings imply that, besides cyanobacteria and their metabolites, cyanotoxins and T&O compounds, anthropogenic pollution can also be a concern for Lake Karaoun, regarding its use as a drinking water reservoir, supporting the emerging need for studies and impact assessment of the co-occurrence of toxic cyanobacteria with other anthropogenic pollutants [58].

4. Conclusions

The results of this study demonstrate for the first time the presence of multiple MC congeners in Lake Karaoun, i.e., dmMC-RR, MC-RR, MC-YR, dmMC-LR, MC-LR, MC-HilR, MC-WR, MC-LA, MC-LY, MC-LW, and MC-LF, with total MCs reaching up to 429 µg/L (sample 16). Additionally, T&O compounds such as β-cyclocitral, β-Ionone, nonanal and dimethylsulfides were identified, while industrial pollutants of anthropogenic origin including dichloromethane and toluene were determined up to 34 (sample 9) and 0.81 (sample 16) µg/L, respectively. Complementary methods were applied aiming at the reliable determination of cyanotoxins and T&O compounds in the lake. Since blooms of *Microcystis aeruginosa* and *Aphanizomenon ovalisporum* continue to occur in Lake Karaoun, monitoring of cyanobacterial blooms is necessary in the future, for the assessment of risks related to the intended uses of the lake. With regard to the cyanotoxin analysis, the results show that MCs should be a monitored target, especially when blooms of *Microcystis aeruginosa* occur. Quantitative screening by ELISA or PPIA could be used for the monitoring of MCs if LC-MS/MS facilities and expertise are not available. It is, however, recommended to confirm the findings by LC-MS/MS which has the advantage of being compound-specific. Despite the fact that CYN was not detected in this study, CYN should be included in the ongoing monitoring schemes, especially in the presence of *Aphanizomenon ovalisporum* in the lake. Whenever incidents of unpleasant odors occur in the lake or in water supplies, further analysis of T&O and VOCs should be carried out to identify the source of T&O.

5. Materials and Methods

5.1. Study Site and Sampling

The Middle-Eastern Lake Karaoun (Qaraoun) is the largest water body in Lebanon, located in the western part of the valley Bekaa (33.34° N, 35.41° E) (Figure 4). Lake Karaoun is an artificial reservoir, created in 1959 by construction of a dam on the Litani River. It has a surface of 12 km^2 at a full capacity of 220 million m^3, a maximum depth of 60 m and a mean depth of 19 m [16]. The lake is considered to be located in a vulnerable arid to semi-arid zone, with winters being moderately cold (13 °C average temperature), the wet season extending from November to April, and summers being hot and dry, lasting from July to October [15,59].

During field campaigns that covered the wet and dry periods, 16 water samples along with a total of 9 specimens of *Cyprinus carpio* fish (weighing 215 g on average) were collected mainly by the national Litani River Authority. All water samples were collected in polyethylene bottles from 5 sampling points of Lake Karaoun (S1, S2, S3, S4 and S5). S1 was located close to the river input at the west bank, S2 was east of the dam, S3 was close to the river input at the east bank, S4 was in the middle of the lake and S5 was at the dam (Figure 4). Water samples were transported in coolers (at a temperature of 4 °C) for laboratory analysis in Beirut, Lebanon and Athens, Greece. The fish were kept on ice and were sacrificed within 24 h. Fish liver and muscle sub-samples were labeled and frozen separately at −20 °C. The frozen fish tissue samples were then lyophilized with a Labconco freeze-dryer for 48 h at −84 °C and 0.2 mbar and powdered using a pestle and a mortar.

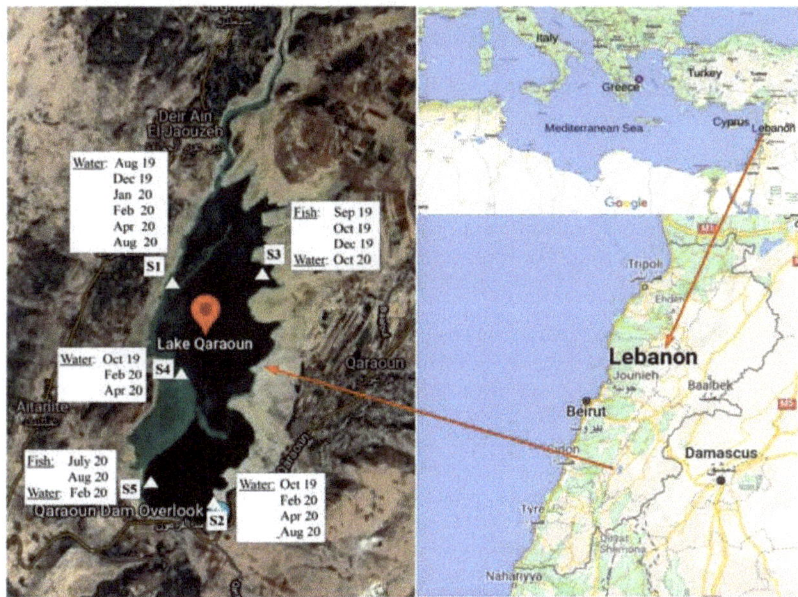

Figure 4. Map and sampling points of Lake Karaoun, Lebanon.

5.2. Chemicals and Reagents

Cyanotoxin standards of MC-RR, MC-LR, MC-YR, MC-LA and NOD were purchased from Sigma-Aldrich (Steinheim, Germany), [D-Asp3]MC-LR, [D-Asp3]MC-RR, MC-WR, MC-HtyR, MC-HilR, MC-LY, MC-LW and MC-LF from ENZO Life Science (Lausen, Switzerland), CYN from Abraxis (Warminster, PA, USA), and ATX fumarate from TOCRIS Bioscience (Bristol, UK). All toxin standards had purity >95%. A VOC 57 standard mix (44926-U) was purchased from Supelco (Darmstadt, Germany). β-Cyclocitral ($C_{10}H_{16}O$) (≥95.0%), β-ionone ($C_{13}H_{20}O$) (purity ≥ 97.0%), pentadecane ($C_{15}H_{32}$) (purity ≥ 99.8%), 2,4-di-*tert*-butylphenol ($C_{14}H_{22}O$) (99.0%), hexadecane ($C_{16}H_{34}$) (purity ≥99.8%), heptadecane ($C_{17}H_{36}$) (purity ≥ 99.5%), dimethyl disulfide (CH_3SSCH_3) (≥99.0%), dimethyl trisulfide (purity ≥ 98.5%) and 3-methylindole (C_9H_9N) (98.0%), were supplied by Sigma Aldrich (Steinheim, Germany). Acetonitrile (ACN) and methanol (MeOH) of HPLC grade (≥99.9%) as well as dichloromethane (DCM) and hexane (HXN) of analytical grade (99.9%) were supplied by Fisher Chemical (Loughborough, UK). High purity formic acid (HCOOH) (98–100%) and acetic acid (CH_3COOH) (>99.8%) were obtained from Riedel-de Haën (Seelze, Germany). Sodium chloride (NaCl) of analytical purity (99.5%) and fuming (37%) hydrochloric acid (HCl) were purchased from Merck (Darmstadt, Germany). Ethylenediaminetetraacetic acid (EDTA) of analytical grade was supplied from Serva Electrophoresis.

Sodium hydroxide (NaOH) 2M solution used for adjustment of pH was prepared from NaOH pellets (purity 98%) purchased from Sigma-Aldrich (Steinheim, Germany). Potassium carbonate (K_2CO_3) (>99.5%) was supplied from Carlo Erba Reagents. High purity water (18.2 MΩ cm at 25 °C) was produced in-house using a TEMAK TSDW10 water purification system (TEMAK, Athens, Greece).

5.3. Microscopic Examination

Microscopic examination of samples was carried out within 24 h on 20 mL water samples collected during field campaigns with a phytoplankton net and kept at low temperature (4 °C). Examination was carried out in the LAEC laboratory in Beirut, Lebanon using a phase contrast microscope (Olympus, Munster, Germany), under a ×40 objective and ×100 immersion. Identification of cyanobacteria was based on taxonomic keys as cells structures and dimensions, mucilage features and the form of colonies. Estimation of cyanobacterial abundances was performed as described elsewhere [17].

5.4. LC-MS/MS Analysis of Cyanotoxins
5.4.1. Sample Preparation of Water Samples

Preparation of water samples for LC-MS/MS analysis of cyanotoxins was based on the method of Zervou et al. [22]. For the determination of extracellular toxins, 150 mL of sample was filtered using a glass fiber filter (Millipore, Ireland). After adjusting the pH of the filtrate to 11, solid phase extraction (SPE) was performed using a dual cartridge assembly with a Supel-Select HLB (bed wt. 200 mg, volume 6 mL, Supelco) and a Supelclean ENVI-Carb (bed wt. 250 mg, volume 3 mL, Supelco) on a 12-port SPE vacuum manifold (Supelco) connected with a vacuum pump. Conditioning of cartridges was carried out with 6 mL DCM, followed by 6 mL of MeOH and 6 mL of pure water (pH 11). Sample passed through the two tandem cartridges at a 0.5 mL/min flow rate. After sample passing, cartridges were dried for 15 min (air under vacuum) and the sequence of cartridges in the assembly was reversed. Elution was done with (60:40) MeOH/DCM having 0.1% HCOOH. Eluents were evaporated to dryness under a gentle nitrogen stream, reconstituted with 150 µL of 5% (*v/v*) MeOH, and transferred into vials for LC-MS/MS analysis. For the determination of intracellular cyanotoxins, after sample filtration the filters were extracted, according to Chistophoridis et al. [23], with 9 mL 75% (*v/v*) MeOH. A 3-mL aliquot of the filtered supernatant was evaporated to dryness and the residue was re-dissolved with 500 µL of 5% (*v/v*) MeOH for LC-MS/MS analysis.

5.4.2. Sample Preparation of Fish Samples

Subsamples of 0.2 g of lyophilized powdered flesh or 0.25 g liver were extracted with 5 mL 80% MeOH containing 0.5% HCOOH by stirring for 15 min, followed by ultra-sonication for 30 min (Bandelin Sonorex Super RK106). The mixture was then transferred to a Falcon tube and centrifuged for 15 min at 4500 rpm (DuPont RMC-14 Refrigerated Micro-Centrifuge, Sorvall Instruments, Newtown, CT, USA). The supernatant was extracted three times with 1 mL of hexane. The hexane phase was discarded, and the bottom layer was transferred to a flask and diluted with 100 mL of water containing 0.3% formic acid. The extract was cleaned-up by SPE with a Supel-Select HLB cartridge (bed wt. 200 mg, volume 6 mL, Supelco) conditioned with 6 mL MeOH followed by 6 mL acidified water (0.3% HCOOH). After extraction, the cartridge was washed with 6 mL of water, dried for 15 min under vacuum and eluted with 6 mL MeOH. The eluents were dried in a water bath at 40 °C under a gentle nitrogen stream. Reconstitution was carried out with 200 µL of 5% MeOH, followed by vortexing and sonication for 3 min, prior to LC-MS/MS analysis.

5.4.3. Determination by LC-MS/MS

A Finnigan Surveyor LC system, equipped with an AS autosampler (Thermo, Waltham, MA, USA) coupled with a TSQ Quantum Discovery Max triple-stage quadrupole mass spectrometer (Thermo, Waltham, MA, USA) with electrospray ionization (ESI) source, was used for LC-MS/MS analysis. Data was acquired and processed by Xcalibur software. Targeted analysis of CYN, ATX, NOD and 12 MCs (dmMC-RR, MC-RR, MC-YR, MC-HtyR, dmMC-LR, MC-LR, MC-HilR, MC-WR, MC-LA, MC-LY, MC-LW, MC-LF) was performed as described in a previous study [22]. Detection of CTs was carried out in multiple reaction monitoring (MRM) mode, using the three most intense and characteristic precursor/product ion transitions for each toxin. Confirmation of identity was based on criteria for retention time (t_R), characteristic precursor/product ion transitions and two calculated ratios of precursor/product ion transitions. LC-MS/MS detection parameters of targeted cyanotoxins are given in Table S3. An example of MC-LR identification in a sample (S2, 2 October 2019) from Lake Karaoun is shown in Figure S4.

5.5. ELISA for Microcystins

ELISA was carried out with the Microcystins-ADDA ELISA kit (Eurofins—Abraxis, Warminster, PA, USA) in 96 well microplates, using an Infinite M200 reader (Tecan, Männedorf, Switzerland). The kit was used according to the manufacturer's instructions and concentrations of toxins were calculated using calibration curves based on the absorbance at 450 nm. For the determination of extracellular MCs, water samples were filtered through 47 mm glass fiber filters (Millipore) and analyzed without any further treatment. Dilutions with ELISA sample diluent were carried out when samples exceeded 5 µg/L MC-LR equivalents. For the analysis of intracellular MCs, after sample filtration (see 5.4.1) the filter was extracted with 9 mL of 75% MeOH. Then, 3 mL of the extract was evaporated to dryness and the residue was re-dissolved in water to avoid false results due to the high percentage of MeOH [41]. Finally, samples were analyzed in duplicate and mean values were reported when RSD < 25%.

5.6. Protein Phosphatase Inhibition Assay (PPIA) for Microcystins

PPIA was carried out using the Microcystins/Nodularins PP2A kit (Eurofins—Abraxis) in 96 well microplate using an Infinite M200 reader (Tecan, Männedorf, Switzerland) for measurements at 405 nm, according to manufacturer's instructions. Dilutions with ultrapure water were carried out when samples exceeded 2.5 µg/L MC-LR equivalents. Sample preparation prior to PPIA was the same as in ELISA. Samples were analyzed in duplicate and mean values were reported when RSD < 25%.

5.7. qPCR Assay for Total Cyanobacteria and Cyanotoxin Genes

The Phytoxigene™ CyanoDTec (Diagnostic Technology, Belrose, Australia) multiplex quantitative real-time PCR assay was applied to determine the gene copies of the 16s rRNA gene (total cyanobacteria) and the *mcyE*, *cyrA*, *sxtA* genes (microcystins, cylindrospermopsin, saxitoxins, respectively). The kit was used according to the manufacturer's instructions and PCR was carried out in a Smartcycler II system (Cepheid). In brief, a volume of water sample (1 to 15 mL, depending on visual cell density) was filtered through a Nucleopore 25 mm, 0.8 µm filter (Whatman, Little Chalfont, UK) using a syringe and filter holder. DNA extraction of filters was carried out using BioGx bead lysis tubes (Diagnostic Technology) in a BeadBug bead beater (Benchmark Scientific). Extracts were centrifuged (MicroCL 21, Thermo Fisher Scientific, Waltham, MA, USA) and proceeded to qPCR. Quantitation of gene copies was based on calibrations with the Phytoxigene™ CyanoNAS standards (100–1,000,000 copies per reaction).

5.8. GC-MS Analysis of Volatile Compounds

Analysis of volatile compounds was carried out in order to (a) screen samples to detect and identify typical cyanobacterial volatile and T&O compounds (untargeted method) and (b) to quantitatively determine a range of anthropogenic VOC pollutants (targeted method).

5.8.1. Untargeted Screening of Cyanobacterial Volatiles and T&O Compounds

Headspace Solid Phase Microextraction coupled to Gas Chromatography-Mass Spectrometry (HS-SPME-GC-MS) was used to screen water samples for volatile compounds with focus on cyanobacterial T&O. A 456 GC coupled to TQ mass spectrometer and equipped with an SPME autosampler (Bruker Daltonics, Bremen, Germany) was used. A (2–10 mL) aliquot of the sample was transferred into a 20 mL SPME glass vial containing 3 g NaCl, then adjusted to 10mL with ultrapure water and tightly sealed. HS-SPME was carried out automatically under the following conditions: 2 cm Divinylbenzene/Carboxen/Polydimethylsiloxane SPME fiber (Supelco, Bellefonte, PA, USA), equilibration 10 min at 60 °C, headspace extraction: 10 min at 60 °C, agitation 300 rpm and desorption time 2 min. GC analysis was carried out using a column RXI®- 5 Sil MS, 30 m, 0.25 mm ID, 0.25 μm df (Restek, Bellefonte, PA, USA). GC conditions were injector temp. 250 °C, splitless, constant flow 1 mL/min (He), column program: (a) 50 °C (1 min) to 250 °C at a rate of 15 °C/min, 250 °C (5 min) and (b) 35 °C (5 min) to 250 °C at a rate of 8 °C/min, 250 °C (5 min). MS conditions were: EI source (70 eV), scan 30–300 m/z, positive polarity. Fluorobenzene standard solution in methanol (Sigma – Aldrich, St. Louis, MO, USA) spiked at a concentration of 5 μg/L in the samples was used as a surrogate for evaluation of SPME efficiency.

Mass spectrometry data were processed using MSWS software (Bruker). Mass spectral deconvolution and identification of compounds was carried out with AMDIS (NIST, Gaithersburg, MD, USA) using the NIST MS library (2015) and retention index calibration with C7-C30 saturated alkanes reference standard (Sigma). Combined matching scores (retention index-RI and mass spectral matching) were used as criteria for identification (RI ± 20 of the reference RI and a spectral matching ≥80%). Identification was considered definite only if a reference standard of the suspect compound was available in the lab and retention times of the suspect and reference compounds matched within ±0.01 min, in addition to mass spectral matching.

5.8.2. Quantitative Determination of VOCs

Quantitative determination of a number of VOCs was carried out by HS-SPME-GC-MS according to EN ISO 17943:2016 (HS-SPME-GC-MS), using 5 ml samples [60]. The compounds determined were chloroform, bromoform, dibromochloromethane, bromodichloromethane, benzene, 1,2-dichloroethane, tetrachoroethene and trichloroethene, 2-ethoxy-2-methyl propane (ETBE), bromochloromethane, dibromomethane, toluene, 1,2-dibromoethane, chlorobenzene, styrene, bromobenzene, n-butyl benzene, hexachlorobutadiene, naphthalene, 1,1-dichloroethane, 1,1-dichloroethene, 1,1,2-trichloroethane, 1,1,2,2-tetrachloroethane, 1,2-dibromo-3-chloropropane, 1,2-dichloropropane, 1,2,3-trichlorobenzene, 1,2,4-trichlorobenzene, 1,3-dichloropropane, 2-methoxy-2-methyl-butane (TAME), ethylbenzene, n-propylbenzene, o-xylene, p+m-xylenes, sec-butylbenzene and tert-butylbenzene. A VOC 57 standard mix (Supelco) was used for calibration and fluorobenzene, toluene-d$_8$ and benzene-d$_6$ (Sigma) as internal standards. HS-SPME conditions were: 2 cm Divinylbenzene/Carboxen/Polydimethylsiloxane SPME fiber (Supelco), equilibration for 10 min at 40 °C, headspace extraction for 10 min at 40 °C, agitation 300 rpm and desorption time 2 min. GC conditions were: column RXI®- 624 Sil MS, 60 m, 0.32 mm ID, 1.8 μm df (Restek), injector temp. 250 °C, splitless, constant flow 1ml/min (He), column program: 35 °C (5 min) to 250 °C at a rate of 8 °C/min and at 250 °C (10min). MS conditions were: EI source (70 eV), Selected Ion Monitoring (3 ions per compound), positive polarity. LOD of each compound was 0.2 μg/L. Confirmation of determinations was carried out with the ISO criteria for retention time and ion ratios [60].

5.9. Method Validation and Quality Control Procedures

The methods applied in this study were validated either previously or in the frame of the study and method performance parameters such as specificity, linearity, precision, accuracy, and limits of detection have been assessed. Furthermore, quality control procedures including measurements of blank/negative and control/positive samples were followed in each batch of analyzed samples.

Validation data of the method applied for the determination of CTs in water by LC-MS/MS are reported in previous studies with RSD values of detected CTs being <16% and <26% for extra- and intra-cellular fractions, respectively [22,23]. This method is also accredited by ISO 17025 in the NCSR Demokritos laboratory [61].

The LC-MS/MS method for determination of CTs in fish was developed, optimized and validated in the frame of this study. For optimization, various combinations of extraction solvents, extraction/mixing times, clean-up steps and SPE cartridges were tested (Oasis HLB and HLB followed by Sep-Pak Vac, silica) as presented in Figures S5 and S6. Optimization experiments were performed with samples spiked with MC-LR and MC-RR at concentration levels of 100 and 400 ng/g dw for flesh and liver, respectively. Selection of the optimized conditions was based on maximization of % recoveries. As shown in Figure S7, method 2 provided the highest mean recoveries for the two spiked MCs in flesh (78.9% for MC-RR and 79.1% for MC-LR) and liver (77.1% for MC-RR and 75.4% for MC-LR). Subsequently, the optimized method was validated in-house. Recovery and precision were evaluated by analyzing toxins-free lyophilized flesh and liver fish samples spiked with a mixture of 12 MCs, at concentration levels of 100 and 400 ng/g dw in 3 replicates. Samples were extracted and analyzed as outlined in Section 5.4.2. Table S4 provides performance characteristics of this method. Briefly, mean recoveries of [D-Asp^3]MC-RR, MC-RR, MC-YR, [D-Asp^3]MC-LR, MC-LR, MC-HilR, MC-LA and MC-LY ranged from 68.5% to 81.6% for flesh, while liver recoveries ranged from 61.5% to 72.2%, with intra-day precision in the range of 6.8–16.5% for flesh and 5.5–15.2% for liver. LODs for fish flesh and liver samples ranged from 1.0 to 7.0 ng/g dry weight and from 0.8 to 5.6 ng/g dry weight, respectively.

In-house validation and method performance data of ELISA and PPIA for MCs in water were reported previously [41] and methods have been proven suitable for quantitative screening of MCs. Negative and positive control samples were included in each analysis batch.

The qPCR assay for total cyanobacteria and cyanotoxin genes included evaluation of negative and positive control samples in each batch of samples. The Phytoxigene™ CyanoNAS standards used for calibration and accurate quantitation were commissioned and developed by the National Measurement Institute (NMI), of the Australian Department of Industry which participates in the International Bureau of Weights and Measures (BIPM) Consultative Committee for Amount of Substance (CCQM).

The untargeted HS-SPME-GC-MS screening method for detection and identification of cyanobacterial volatile compounds has been tested with a wide range of compounds that are available in the lab of EYDAP SA and has been shown to be capable of detection/identification at concentrations generally <1µg/L. The method has also been tested successfully in interlaboratory trials for unknown odorous compounds in water. The targeted HS-SPME-GC-MS method for VOCs in water has been fully validated in-house, is accredited by ISO 17025 and has successfully been evaluated in interlaboratory tests.

Supplementary Materials: The following are available online at https://www.mdpi.com/article/10.3390/toxins13100716/s1, Figure S1. chemical structures of studied cyanotoxins, Figure S2. LC-MS/MS MRM chromatogram of sample S2, 02/10/2019 (intracellular fraction) from Lake Karaoun, Figure S3. MRM chromatogram of (a) a standard solution of 12 MCs at a concentration level corresponding to 50 ng/g dw and (b) liver sample from *Cyprinous Carpio* fish collected in September 2019, Figure S4. Example of identification of MC-LR in sample S2, 2 October 2019 (intracellular fraction) from Lake Karaoun, Figure S5. Experimental procedures tested in order to optimize the extraction of MCs in fish flesh, Figure S6. Experimental procedures tested in order to optimize the extraction of MCs in fish liver, Figure S7. Selection of method for (a) fish flesh and (b) fish liver: obtained recoveries of spiked MCs using different treatment processes; Table S1. Method LODs and LOQs for each cyanotoxin analysed in this study using LC-MS/MS, Table S2. Results of untargeted HS-SPME-GC/MS screening per sample, Table S3. LC-MS/MS detection parameters of targeted cyanotoxins, Table S4. Performance characteristics of the method for the analysis of target MCs in fish flesh and liver. Table S5. Physicochemical parameters of Lake Karaoun.

Author Contributions: Conceptualization, R.L., A.H., K.S.; Data curation, N.A.H., S.-K.Z., T.K., C.C., A.P., A.F.; Formal analysis, N.A.H., S.-K.Z., T.K., C.C., A.P., A.F.; Funding acquisition, R.L.; Investigation, R.L., J.S., T.K., T.M.T., K.S., A.H.; Methodology, T.K., S.-K.Z., A.H.; Project administration, K.S., A.H., R.L.; Resources, R.L., J.S., T.K., A.H.; Software, S.-K.Z., T.K., A.P.; Supervision, A.H., R.L.; Validation, S.-K.Z., T.K. and T.M.T.; Visualization, S.-K.Z. and T.M.T.; Writing—original draft, N.A.H., S.-K.Z., T.K. and A.H.; Writing—review & editing, S.-K.Z., T.K., T.M.T., A.F., K.S., J.S., R.L. and A.H. All authors have read and agreed to the published version of the manuscript.

Funding: Not applicable.

Institutional Review Board Statement: Not applicable.

Informed Consent Statement: Not applicable.

Data Availability Statement: Data is contained within the article or Supplementary Material.

Acknowledgments: N.A.H. thanks the National Council for Scientific Research of Lebanon (CNRS-L) and E2S UPPA (France) for granting a doctoral fellowship. The authors acknowledge COST Actions ES 1105 "CYANOCOST–Cyanobacterial blooms and toxins in water resources: Occurrence impacts and management" www.cyanocost.net and CA 18225 "WaterTOP–Taste and Odor in early diagnosis of source and drinking Water Problems" https://watertopnet.eu for adding value to this study through networking and knowledge sharing with European experts in the field. S.-K.Z. acknowledges the Action titled "National Network on Climate Change and its Impacts—Climpact", which is implemented under the sub-project 3 of the project "Infrastructure of national research networks in the fields of Precision Medicine, Quantum Technology and Climate Change", funded by the Public Investment Program of Greece, General Secretary of Research and Technology/Ministry of Development and Investments.

Conflicts of Interest: The authors declare no conflict of interest.

References

1. Bouaïcha, N.; Miles, C.; Beach, D.; Labidi, Z.; Djabri, A.; Benayache, N.; Nguyen-Quang, T. Structural diversity, characterization and toxicology of microcystins. *Toxins* **2019**, *11*, 714. [CrossRef]
2. Chen, Y.; Shen, D.; Fang, D. Nodularins in poisoning. *Clin. Chim. Acta* **2013**, *425*, 18–29. [CrossRef] [PubMed]
3. Van Apeldoorn, M.E.; Van Egmond, H.P.; Speijers, G.J.A.; Bakker, G.J.I. Toxins of cyanobacteria. *Mol. Nutr. Food Res.* **2007**, *51*, 7–60. [CrossRef] [PubMed]
4. Buratti, F.M.; Manganelli, M.; Vichi, S.; Stefanelli, M.; Scardala, S.; Testai, E.; Funari, E. Cyanotoxins: Producing organisms, occurrence, toxicity, mechanism of action and human health toxicological risk evaluation. *Arch. Toxicol.* **2017**, *91*, 1049–1130. [CrossRef]
5. De La Cruz, A.A.; Hiskia, A.; Kaloudis, T.; Chernoff, N.; Hill, D.; Antoniou, M.G.; He, X.; Loftin, K.; O'Shea, K.; Zhao, C.; et al. A review on cylindrospermopsin: The global occurrence, detection, toxicity and degradation of a potent cyanotoxin. *Environ. Sci. Process Impacts* **2013**, *15*, 1979–2003. [CrossRef]
6. Osswald, J.; Rellán, S.; Gago, A.; Vasconcelos, V. Toxicology and detection methods of the alkaloid neurotoxin produced by cyanobacteria, anatoxin-a. *Environ. Int.* **2007**, *33*, 1070–1089. [CrossRef] [PubMed]
7. Pelaez, M.; Antoniou, M.G.; He, X.; Dionysiou, D.D.; de la Cruz, A.A.; Tsimeli, K.; Triantis, T.; Hiskia, A.; Kaloudis, T.; Williams, C.; et al. Sources and Occurrence of Cyanotoxins Worldwide. In *Xenobiotics in the Urban Water Cycle: Mass Flows, Environmental Processes, Mitigation and Treatment Strategies*; Fatta-Kassinos, D., Bester, K., Kümmerer, K., Eds.; Springer: Dordrecht, The Netherlands, 2010; pp. 101–127.
8. Sivonen, K.; Jones, G. Cyanobacterial toxins. In *Toxic Cyanobacteria in Water: A Guide to Their Public Health Consequences, Monitoring, and Management*; Chorus, I., Bartram, J., Eds.; E & FN Spon: London, UK, 1999; pp. 41–111.
9. Huisman, J.; Codd, G.A.; Paerl, H.W.; Ibelings, B.W.; Verspagen, J.M.H.; Visser, P.M. Cyanobacterial blooms. *Nat. Rev. Genet.* **2018**, *16*, 471–483. [CrossRef]
10. Svirčev, Z.; Lalić, D.; Savić, G.B.; Tokodi, N.; Backović, D.D.; Chen, L.; Meriluoto, J.; Codd, G.A. Global geographical and historical overview of cyanotoxin distribution and cyanobacterial poisonings. *Arch. Toxicol.* **2019**, *93*, 2429–2481. [CrossRef] [PubMed]
11. Chorus, I.; Welker, M. Exposure to cyanotoxins: Understanding it and short-term interventions to prevent it. In *Toxic Cyanobacteria in Water: A Guide to Their Public Health Consequences, Monitoring and Management*, 2nd ed.; Chorus, I., Welker, M., Eds.; CRC Press: Boca Raton, FL, USA, 2021; pp. 295–400.
12. Osseiran, K.; Kabakian, V. Hydropower in Lebanon; History and Prospects; United Nations Development Pogramme CEDRO (UNDP/CEDRO). 2013. Available online: http://www.cedro-undp.org/content/uploads/Publication/141009092113199~{}Exchange%204.pdf (accessed on 7 July 2021).
13. Yazbek, H.; Fadel, A.; Slim, K. Facts about the degradation of Lake Qaraoun, Lebanon, and cyanobacterial harmful algal blooms (HABS). *J. Environ. Hydrol.* **2019**, *27*, 1.

14. Atoui, A.; Hafez, H.; Slim, K. Occurrence of toxic cyanobacterial blooms for the first time in Lake Karaoun, Lebanon. *Water Environ. J.* **2012**, *27*, 42–49. [CrossRef]
15. Fadel, A.; Lemaire, B.J.; Atoui, A.; Leite, B.V.; Amacha, N.; Slim, K.; Tassin, B. First assessment of the ecological status of Karaoun reservoir, Lebanon. *Lakes Reserv. Res. Manag.* **2014**, *19*, 142–157. [CrossRef]
16. Fadel, A.; Slim, K. Evaluation of the Physicochemical and Environmental Status of Qaraaoun Reservoir. In *The Litani River, Lebanon: An Assessment and Current Challenges*; Shaban, A., Hamzé, M., Eds.; Springer: Cham, Switzerland, 2018; pp. 71–86.
17. Fadel, A.; Atoui, A.; Lemaire, B.J.; Vinçon-Leite, B.; Slim, K. Dynamics of the toxin *cylindrospermopsin* and the *cyanobacterium chrysosporum (aphanizomenon)* ovalisporum in a mediterranean eutrophic reservoir. *Toxins* **2014**, *6*, 3041–3057. [CrossRef]
18. Watson, S.B. Aquatic taste and odor: A primary signal of drinking-water integrity. *J. Toxicol. Environ. Health Part A* **2004**, *67*, 1779–1795. [CrossRef]
19. McGuire, M.J. Off-flavor as the consumer's measure of drinking water safety. *Water Sci. Technol.* **1995**, *31*, 1–8. [CrossRef]
20. Stefan, M.I. (Ed.) *Advanced Oxidation Processes for Water Treatment—Fundamentals and Applications*; IWA Publishing: London, UK, 2018; pp. 1–686.
21. Sukenik, A.; Carmeli, S.; Hadas, O.; Leibovici, E.; Malinsky-Rushansky, N.; Parparov, R.; Pinkas, R.; Viner-Mozzini, Y.; Wynne, D. Water pollutants. In *Lake Kinneret: Ecology and Management*; Zohary, T., Sukenik, A., Berman, T., Nishri, A., Eds.; Springer: Dordrecht, The Netherlands, 2014; pp. 577–606.
22. Zervou, S.-K.; Christophoridis, C.; Kaloudis, T.; Triantis, T.; Hiskia, A. New SPE-LC-MS/MS method for simultaneous determination of multi-class cyanobacterial and algal toxins. *J. Hazard. Mater.* **2017**, *323*, 56–66. [CrossRef]
23. Christophoridis, C.; Zervou, S.-K.; Manolidi, K.; Katsiapi, M.; Moustaka-Gouni, M.; Kaloudis, T.; Triantis, T.M.; Hiskia, A. Occurrence and diversity of cyanotoxins in Greek lakes. *Sci. Rep.* **2018**, *8*, 17877. [CrossRef] [PubMed]
24. Lee, J.; Rai, P.K.; Jeon, Y.J.; Kim, K.-H.; Kwon, E.E. The role of algae and cyanobacteria in the production and release of odorants in water. *Environ. Pollut.* **2017**, *227*, 252–262. [CrossRef] [PubMed]
25. Jones, M.R.; Pinto, E.; Torres, M.A.; Dörr, F.; Mazur-Marzec, H.; Szubert, K.; Tartaglione, L.; Dell'Aversano, C.; Miles, C.O.; Beach, D.G.; et al. CyanoMetDB, a comprehensive public database of secondary metabolites from cyanobacteria. *Water Res.* **2021**, *196*, 117017. [CrossRef] [PubMed]
26. Birbeck, J.A.; Westrick, J.A.; O'Neill, G.M.; Spies, B.; Szlag, D.C. Comparative analysis of microcystin prevalence in Michigan lakes by online concentration LC/MS/MS and ELISA. *Toxins* **2019**, *11*, 13. [CrossRef]
27. Kumar, P.; Rautela, A.; Kesari, V.; Szlag, D.; Westrick, J.; Kumar, S. Recent developments in the methods of quantitative analysis of microcystins. *J. Biochem. Mol. Toxicol.* **2020**, *34*, e22582. [CrossRef]
28. Sanseverino, I.; Conduto António, D.; Loos, R.; Lettieri, T. Cyanotoxins: Methods and Approaches for Their Analysis and Detection; EUR 28624; Joint Research Centre (JRC), European Union. 2017. Available online: https://doi.org/10.2760/36186 (accessed on 7 July 2021).
29. Gaget, V.; Lau, M.; Sendall, B.; Froscio, S.; Humpage, A.R. Cyanotoxins: Which detection technique for an optimum risk assessment? *Water Res.* **2017**, *118*, 227–238. [CrossRef] [PubMed]
30. Hardy, F.J.; Johnson, A.; Hamel, K.; Preece, E. Cyanotoxin bioaccumulation in freshwater fish, Washington State, USA. *Environ. Monit. Assess.* **2015**, *187*. [CrossRef] [PubMed]
31. Haddad, S.P.; Bobbitt, J.M.; Taylor, R.; Lovin, L.; Conkle, J.L.; Chambliss, C.K.; Brooks, B.W. Determination of microcystins, nodularin, anatoxin-a, cylindrospermopsin, and saxitoxin in water and fish tissue using isotope dilution liquid chromatography tandem mass spectrometry. *J. Chromatogr. A* **2019**, *1599*, 66–74. [CrossRef] [PubMed]
32. Wood, J.D.; Franklin, R.; Garman, G.; McIninch, S.; Porter, A.; Bukaveckas, P. Exposure to the cyanotoxin microcystin arising from interspecific differences in feeding habits among fish and shellfish in the James River Estuary, Virginia. *Environ. Sci. Technol.* **2014**, *48*, 5194–5202. [CrossRef]
33. Wituszynski, D.M.; Hu, C.; Zhang, F.; Chaffin, J.; Lee, J.; Ludsin, S.A.; Martin, J.F. Microcystin in Lake Erie fish: Risk to human health and relationship to cyanobacterial blooms. *J. Great Lakes Res.* **2017**, *43*, 1084–1090. [CrossRef]
34. Wynne, T.T.; Stumpf, R.P. Spatial and temporal patterns in the seasonal distribution of toxic cyanobacteria in western Lake Erie from 2002–2014. *Toxins* **2015**, *7*, 1649–1663. [CrossRef]
35. Kozlowsky-Suzuki, B.; Wilson, A.E.; Ferrão-Filho, A.D.S. Biomagnification or biodilution of microcystins in aquatic foodwebs? Meta-analyses of laboratory and field studies. *Harmful Algae* **2012**, *18*, 47–55. [CrossRef]
36. Adamovský, O.; Kopp, R.; Hilscherová, K.; Babica, P.; Palíková, M.; Pašková, V.; Navrátil, S.; Maršálek, B.; Bláha, L. Microcystin kinetics (bioaccumulation and elimination) and biochemical responses in common carp (*Cyprinus carpio*) and silver carp (*hypophthalmichthys molitrix*) exposed to toxic cyanobacterial blooms. *Environ. Toxicol. Chem.* **2007**, *26*, 2687–2693. [CrossRef]
37. Pollingher, U. Phytoplankton periodicity in a subtropical lake (Lake Kinneret, Israel). *Hydrobiologia* **1986**, *138*, 127–138. [CrossRef]
38. Berman, T.; Shteinman, B. Phytoplankton development and turbulent mixing in Lake Kinneret (1992–1996). *J. Plankton Res.* **1998**, *20*, 709–726. [CrossRef]
39. Pollingher, U.; Zohary, T.; Hadas, O.; Yacobi, Y.Z.; Berman, T. *Aphanizomenon ovalisporum* (Forti) in Lake Kinneret, Israel. *J. Plankton Res.* **1998**, *20*, 1321–1339. [CrossRef]
40. Du, X.; Liu, H.; Yuan, L.; Wang, Y.; Ma, Y.; Wang, R.; Chen, X.; Losiewicz, M.D.; Guo, H.; Zhang, H. The diversity of cyanobacterial Toxins on structural characterization, distribution and identification: A systematic review. *Toxins* **2019**, *11*, 530. [CrossRef] [PubMed]

41. Triantis, T.; Tsimeli, K.; Kaloudis, T.; Thanassoulias, N.; Lytras, E.; Hiskia, A. Development of an integrated laboratory system for the monitoring of cyanotoxins in surface and drinking waters. *Toxicon* **2010**, *55*, 979–989. [CrossRef]

42. Jüttner, F. Dynamics of the volatile organic substances associated with cyanobacteria and algae in a eutrophic shallow lake. *Appl. Environ. Microbiol.* **1984**, *47*, 814–820. [CrossRef]

43. Jüttner, F.; Watson, S.B.; Von Elert, E.; Köster, O. β-Cyclocitral, a grazer defence signal unique to the cyanobacterium microcystis. *J. Chem. Ecol.* **2010**, *36*, 1387–1397. [CrossRef]

44. Buttery, R.G.; Teranishi, R.; Ling, L.C.; Turnbaugh, J.G. Quantitative and sensory studies on tomato paste volatiles. *J. Agric. Food Chem.* **1990**, *38*, 336–340. [CrossRef]

45. Simkin, A.J.; Underwood, B.A.; Auldridge, M.; Loucas, H.M.; Shibuya, K.; Schmelz, E.; Clark, D.G.; Klee, H.J. Circadian regulation of the PhCCD1 carotenoid cleavage dioxygenase controls emission of β-ionone, a fragrance volatile of petunia flowers. *Plant Physiol.* **2004**, *136*, 3504–3514. [CrossRef]

46. Paparella, A.; Shaltiel-Harpaza, L.; Ibdah, M. β-Ionone: Its occurrence and biological function and metabolic engineering. *Plants* **2021**, *10*, 754. [CrossRef] [PubMed]

47. Jiang, Y.; Cheng, B.; Liu, M.; Nie, Y. Spatial and temporal variations of taste and odor compounds in surface water, overlying water and sediment of the Western Lake Chaohu, China. *Bull. Environ. Contam. Toxicol.* **2015**, *96*, 186–191. [CrossRef]

48. Leffingwell, J.C.; Leffingwell, D. GRAS Flavor chemicals—Detection thresholds. *Perfum. Flavorist* **1991**, *16*, 1–19.

49. Wang, C.; Yu, J.; Guo, Q.; Sun, D.; Su, M.; An, W.; Zhang, Y.; Yang, M. Occurrence of swampy/septic odor and possible odorants in source and finished drinking water of major cities across China. *Environ. Pollut.* **2019**, *249*, 305–310. [CrossRef]

50. Guo, Q.; Yu, J.; Yang, K.; Wen, X.; Zhang, H.; Yu, Z.; Li, H.; Zhang, D.; Yang, M. Identification of complex septic odorants in Huangpu River source water by combining the data from gas chromatography-olfactometry and comprehensive two-dimensional gas chromatography using retention indices. *Sci. Total. Environ.* **2016**, *556*, 36–44. [CrossRef]

51. Ma, Z.; Niu, Y.; Xie, P.; Chen, J.; Tao, M.; Deng, X. Off-flavor compounds from decaying cyanobacterial blooms of Lake Taihu. *J. Environ. Sci.* **2013**, *25*, 495–501. [CrossRef]

52. Watson, S.B.; Jüttner, F. Malodorous volatile organic sulfur compounds: Sources, sinks and significance in inland waters. *Crit. Rev. Microbiol.* **2016**, *43*, 210–237. [CrossRef]

53. Khiari, D. *Managing Cyanotoxins*; Water Research Foundation: Denver, CO, USA, 2019; p. 8. Available online: https://www.waterrf.org/sites/default/files/file/2019-12/Cyanotoxins_StateOfTheScience.pdf (accessed on 7 July 2021).

54. Chorus, I.; Welker, M. (Eds.) *Toxic Cyanobacteria in Water: A Guide to Their Public Health Consequences, Monitoring and Management*, 2nd ed.; CRC Press: Boca Raton, FL, USA, 2021; pp. 1–839.

55. ECHA. European Chemicals Agency, REACH Registered Substance Portal. Available online: https://echa.europa.eu/substance-information/-/substanceinfo/100.033.401 (accessed on 8 July 2021).

56. Valentine, D.; Reddy, C.M. Latent hydrocarbons from cyanobacteria. *Proc. Natl. Acad. Sci. USA* **2015**, *112*, 13434–13435. [CrossRef] [PubMed]

57. Lim, F.Y.; Ong, S.L.; Hu, J. Recent Advances in the use of chemical markers for tracing wastewater contamination in aquatic environment: A review. *Water* **2017**, *9*, 143. [CrossRef]

58. Metcalf, J.S.; Codd, G.A. Co-occurrence of cyanobacteria and cyanotoxins with other environmental health hazards: Impacts and implications. *Toxins* **2020**, *12*, 629. [CrossRef]

59. Slim, K.; Atoui, A.; Elzein, G.; Temsah, M. Effets des facteurs environnementaux sur la qualité de l'eau et la prolifération toxique des cyanobactéries du lac KARAOUN (Liban). *LARHYSS J.* **2012**, *9*, 29–43.

60. International Organization for Standardization. *ISO 17943: Water Quality—Determination of Volatile Organic Compounds in Water—Method Using Headspace Solid-Phase Micro-Extraction (HS-SPME) Followed by gas Chromatography-Mass Spectrometry (GC-MS)*; International Organization for Standardization: Geneva, Switzerland, 2016. Available online: https://www.iso.org/standard/61076.html (accessed on 4 October 2021).

61. International Organization for Standardization. *ISO 17025:General Requirements for the Competence of Testing and Calibration Laboratories*; International Organization for Standardization: Geneva, Switzerland, 2017. Available online: https://www.iso.org/publication/PUB100424.html (accessed on 4 October 2021).

Article

Is Toxin-Producing *Planktothrix* sp. an Emerging Species in Lake Constance?

Corentin Fournier [1,†], Eva Riehle [2,†], Daniel R. Dietrich [2,*] and David Schleheck [1,3,*]

[1] Microbial Ecology and Limnic Microbiology, University of Konstanz, 78457 Konstanz, Germany;
corentin.fournier@uni-konstanz.de
[2] Human and Environmental Toxicology, University of Konstanz, 78457 Konstanz, Germany;
eva.riehle@uni-konstanz.de
[3] Limnological Institute, University of Konstanz, 78457 Konstanz, Germany
* Correspondence: daniel.dietrich@uni-konstanz.de (D.R.D.); david.schleheck@uni-konstanz.de (D.S.)
† Shared 1st authorship.

Citation: Fournier, C.; Riehle, E.; Dietrich, D.R.; Schleheck, D. Is Toxin-Producing *Planktothrix* sp. an Emerging Species in Lake Constance? *Toxins* **2021**, *13*, 666. https://doi.org/10.3390/toxins13090666

Received: 22 July 2021
Accepted: 15 September 2021
Published: 17 September 2021

Publisher's Note: MDPI stays neutral with regard to jurisdictional claims in published maps and institutional affiliations.

Abstract: Recurring blooms of filamentous, red-pigmented and toxin-producing cyanobacteria *Planktothrix rubescens* have been reported in numerous deep and stratified prealpine lakes, with the exception of Lake Constance. In a 2019 and 2020 Lake Constance field campaign, we collected samples from a distinct red-pigmented biomass maximum below the chlorophyll-a maximum, which was determined using fluorescence probe measurements at depths between 18 and 20 m. Here, we report the characterization of these deep water red pigment maxima (DRM) as cyanobacterial blooms. Using 16S rRNA gene-amplicon sequencing, we found evidence that the blooms were, indeed, contributed by *Planktothrix* spp., although phycoerythrin-rich *Synechococcus* taxa constituted most of the biomass (>96% relative read abundance) of the cyanobacterial DRM community. Through UPLC–MS/MS, we also detected toxic microcystins (MCs) in the DRM in the individual sampling days at concentrations of ≤1.5 ng/L. Subsequently, we reevaluated the fluorescence probe measurements collected over the past decade and found that, in the summer, DRM have been present in Lake Constance, at least since 2009. Our study highlights the need for a continuous monitoring program also targeting the cyanobacterial DRM in Lake Constance, and for future studies on the competition of the different cyanobacterial taxa. Future studies will address the potential community composition changes in response to the climate change driven physiochemical and biological parameters of the lake.

Keywords: *Planktothrix*; *Synechococcus*; microcystins; temperate lakes

Key Contribution: Deep water, red-pigmented cyanobacterial blooms in Lake Constance were characterized by phylogenetic community sequencing during the summers of 2019 and 2020, demonstrating the low abundance of *Planktothrix* spp., and the predominance of picoplanktic phycoerythrin-rich *Synechococcus* spp., as well as very low concentrations of cyanobacterial toxins on individual sampling days. In response to climate change, changes in the physiochemical and biological parameters of the lake may, in future, support the establishment of toxic *Planktothrix* spp. blooms and/or the mass development of potentially toxic picocyanobacteria.

1. Introduction

The formation of cyanobacterial blooms involves the complex interplay of regional and biological variables, and blooms have been reported worldwide with an increasing frequency [1,2]. Although it has been shown that climate change and eutrophication are, in many cases, major contributors to bloom formation, the mechanisms through which nutrients and temperature interact to amplify blooms varies extensively between cyanobacterial groups. Moreover, the optimal growth temperature and nutrient availability of cyanobacteria is species specific, making inferences from studies with other species more complex [2,3]. Indeed, different cyanobacterial species predominate depending on the N/P ratio [3]. The filamentous cyanobacterial genus, *Planktothrix*, occurs preferentially in prealpine and alpine lakes in temperate regions, and has been responsible for many blooms in the past (for example, in Lake Zurich [4,5], Lake Bourget [6], Lake Garda [7] and Lake Mondsee [8]), where its mass occurrence can evolve to become a major influence on the food web [9,10]. The success of *Planktothrix* spp. is attributed to its adaptability, including its ability to regulate buoyancy, and the use of phycobilins, in addition to chlorophyll-a [11,12], as chromophores. By virtue of the phycobilins, phycocyanin and phycoerythrin, *Planktothrix* spp. can absorb light in large parts of the electromagnetic spectrum; specifically, blue and green light, thus conferring its red appearance. Consequently, the free-floating *Planktothrix* spp. usually develop blooms at depths of 9–15 m [11,13]. Bloom depths within a stratified lake can be influenced by the internal waves, and can impact the growth of *Planktothrix* spp. by changing the light availability [14]. Similarly to many other cyanobacterial species, mass occurrences of *Planktothrix* spp. are supported by warmer water temperatures, mediating a more stable lake stratification in the summer [9,15]. Strikingly, *Planktothrix* spp. can, additionally, form prominent winter blooms and can thrive even beneath an ice cover, allowing for full year dominance [16,17]. Although microcystin production in winter blooms, on and under ice covers, has been reported, toxins appear to be less abundant in colder environments than in warmer environments [16,18,19].

The genus *Planktothrix* is currently distinguished in nine species, including red and green phenotypes, with the most prominent representatives being *P. rubescens* and *P. agardhii*, respectively, which mostly occur in the freshwater ecosystems of temperate regions [20]. Similarly to many other cyanobacteria, *Planktothrix* spp. are capable of producing microcystins (MCs), which are known to be toxic to humans, as well as other mammalian and non-mammalian species [15–18]. Hence, cyanobacterial blooms are often associated with the detection of increased intra- and extra-cellular toxin(s) [21]. Despite a plethora of efforts, neither the trigger for toxin production, nor the factors resulting in the development of toxic cyanobacterial blooms, have been elucidated. While toxins are suggested to be part of a defense mechanism against zooplankton or parasites [22–24], and toxin producing strains seem to have an advantage over non-toxic strains [25,26], the molecular 'switch' that turns toxin production on has not yet been discovered. Irrespective of the latter, the potential adverse impact of toxic cyanobacterial blooms on human health, society, economy and ecology highlights the importance of an improved understanding of cyanobacterial bloom formation, with or without concomitant toxin production [27,28].

Microcystins (MCs) share one common monocyclic structure with a molecular weight of approximately 1 kDa, which is composed of seven amino acids, (three *D*-amino acids, one *N*-methyldehydroalanine, two *L*-amino acids at the hypervariable positions two and four, and the unique amino acid ADDA (Figure S1) [29]). Variations in amino acid composition, and modifications such as methylations, create extreme structural diversity—at least 279 different congeners have been reported [30]. MC synthesis is encoded by 9–10 genes that constitute one *mcy* gene cluster [31]. The toxicity of MCs is induced by covalent binding, the inhibition of ser/thr protein phosphatases and the concomitant hyperphosphorylation of cellular proteins [32–34], whereby the toxicodynamically relevant biological availability of MCs highly depends on the route of exposure and the respective MC congener [35,36]. MC concentrations of <35 µg/L were reported in *Planktothrix* spp. blooms [37,38]. When considering the current World Health Organization (WHO) guideline value of 1 µg MC/L in drinking water [39], which translates to approximately 10 µg/L of raw water, largely depending on the type of water treatment used [40,41], it becomes obvious that toxin-producing *Planktothrix* spp. blooms must be taken as a serious threat to freshwater systems serving as drinking water resources.

Unlike other prealpine lakes, Lake Constance, a drinking water reservoir for more than four million inhabitants, has not yet seen prominent and recurring, well-documented blooms of *Planktothrix* spp. Although *P. rubescens* has the capability to dominate entire lake ecosystems, even at low nutrient concentrations (e.g., Lake Bourget [6]), analyses of Lake Constance samples revealed only low abundances of *Planktothrix* spp. to date [9,42]. Indeed, the first prominent appearance of *Planktothrix rubescens* was documented in 2016, when *P. rubescens* filaments were observed in various German, Austrian, and Swiss sampling sites at Lake Constance [42,43].

No molecular phylogenetic studies have been conducted, to date, to evaluate the composition of the red-pigmented cyanobacterial community in Lake Constance, in water depths that are typical for the blooms of *Planktothrix* spp. (i.e., below the chlorophyll-a maximum). In a field campaign in the Überlingen embayment of Lake Constance (47.7571° N 9.1273° E), we observed reddish colored plankton filters in the samples taken at 18 m water depth, in the summer of 2019. Subsequently, water samples were taken at the depths of maximal red pigment (phycoerythrin) concentration, as determined by a fluorescence probe, on each of the fortnightly sampling days, between June and October 2019, and between July and September 2020. The composition of the cyanobacterial community at this deep water red pigment maximum (DRM) was assessed through Illumina PCR-amplicon sequencing, using cyanobacteria-specific primers, while UPLC–MS/MS analyses were applied to determine the toxin concentrations. Potentially toxin-producing species were identified and quantified via quantitative PCR, using *Planktothrix*- and microcystin-biosynthesis-specific (*mcy*) primers, and Sanger sequencing.

2. Results

2.1. Blooms of Red-Pigmented Phytoplankton at Water Depths below the Chlorophyll-a Maximum in Lake Constance

Multiwavelength fluorometer profiles were taken along the water column, using a Moldaenke FluoroProbe, and were evaluated directly on the ship to determine the depths of the maxima of green pigment (chlorophyll-a maximum; predominantly diatoms and green algae) and red pigment concentrations (deep water red pigment maximum, DRM). An example of a depth profile for all of the recorded fluorescence channels, as well as of the interpretation of the abundance of different algae classes (as calculated by the Moldaenke FluoroProbe) is depicted in Figure 1. An increased abundance of red-pigmented cyanobacteria, such as phycoerythrin-rich *Planktothrix* spp. or *Synechococcus* spp., was suggested by the elevated fluorescence intensity at 570 nm (and 525 nm) excitation wavelength (Figure 1). According to the distinction of algae classes used by the Moldaenke FluoroProbe, these phycoerythrin-rich algae were attributed to represent 'cryptophyta'.

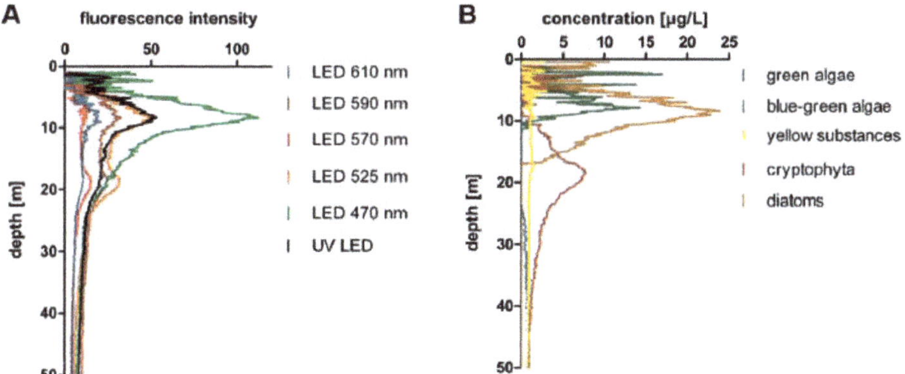

Figure 1. Representative depth profile recorded with a Moldaenke FluoroProbe multichannel fluorimeter, indicating the high abundance of red-pigmented biomass at a water depth below the chlorophyll-a maximum in Lake Constance on 1 July 2019. (**A**) Absolute fluorescence intensities recorded at the different excitation wavelengths. (**B**) Abundance of the different algae classes as attributed by FluoroProbe. Red-pigmented cyanobacteria are attributed to 'cryptophyta'. In this example, the chlorophyll-a maximum was determined at 8–10 m water depth (Figure 1A, 470 nm), and a second maximum, indicating a red-pigmented biomass, at approximately 18–20 m water depth (Figure 1A, 570 nm; 1B, cryptophyta).

2.2. Phycoerythrin-Rich Synechococcus Phylotypes Dominated the DRM Cyanobacterial Community

In order to characterize the cyanobacterial community composition that is presumably present at the DRM, the plankton biomass at the DRM was collected on Whatman GF6 glass fiber filters. The filters appeared reddish, compared to the yellow-green filters obtained from the chlorophyll-a maximum sample (see Supplementary Material, Figure S2). The total DNA was extracted from the filters, and a fragment of the 16S rRNA gene was amplified using the cyanobacteria-specific primers, CYA359F and CYA784R [44]. Illumina sequencing, with 300 bp paired-end reads, was employed. These primers amplified cyanobacteria phylotypes, which allowed for the collection of phylogenetic information at a finer resolution, and also of the low abundant cyanobacteria at the DRM. Taxonomic affiliation was carried out using two different reference databases: SILVA_138 and Greengenes. The cyanobacteria taxonomy was consistent between both databases, with the exception of the *Synechococcus* genus in Greengenes (replaced by *Cyanobium*_PCC-6307 in SILVA; *Cyanobium*_PCC-6307 is a heterotypic synonym of *Synechococcus* sp PCC-6307).

Subsequently to bioinformatic processing and the removal (filtering) of the extremely low abundant phylotypes (i.e., phylotypes represented by less than three reads in at least 20% of all samples; see Material and Methods), 35 amplicon sequence variants (ASVs) were detected in 2019, and 37 were detected in 2020 (Figure 2). Each ASV was affiliated at the level of either genus or order, depending on the last common taxonomic rank between the SILVA and Greengenes databases (note that species rank could not be affiliated by the amplicon sequencing technique that we used).

For both the 2019 and 2020 sampling campaigns, the genus *Synechococcus* clearly dominated the cyanobacterial DRM community (as examined using amplicon sequencing), occupying 96% (SILVA) or 98% (Greengenes) of the total relative read abundance, and representing 65.7% (SILVA) and 67.6% (Greengenes) of the detected ASVs in the community (see Supplementary Material, Table S2). For 2019, five *Synechococcus* ASVs represented 78% of the total relative abundance, with ASV13 being the most abundant with 21% total relative abundance (Figure 2A). For 2020, only three of the ASVs affiliated to *Synechococcus* contributed to 77% of the total relative abundance, with ASV4 contributing to almost half (45%) throughout the year (Figure 2B).

Although *Synechococcus* dominated the cyanobacterial community at the DRM in 2019 and 2020, each of the two ASVs that are affiliated to *Planktothrix* could be detected in both years. *Planktothrix* (Oscillatoriophycideae in Figure 2A,B), with 0.4% (2019) and 0.9% (2020) of the total relative abundance (Figure 2A,B), represented only low abundant taxa, together with Nostocophycideae. In addition, two *Microcystis* spp. ASVs were detected in 2019, at 0.06% of the total relative read abundance. Only one *Microcystis* spp. ASV was detected in 2020, with a very low total relative abundance of 0.008%. Although the respective relative abundances of *Planktothrix* and *Microcystis* species are low, the strong bioinformatic filtration (for details, see Section 5.6) confirms the biological significance of these amplicon sequencing results. The ASVs of the most abundant *Synechococcus* spp., as well as the *Planktothrix* spp. and *Microcystis* spp. ASVs, were used for further analyses.

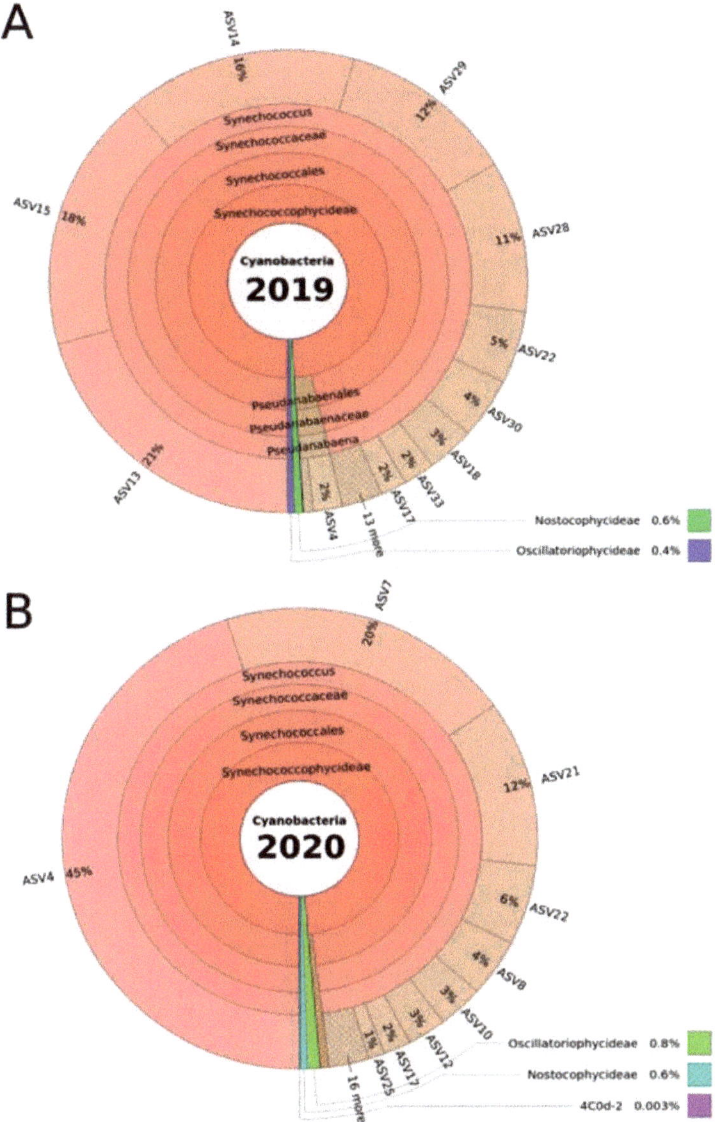

Figure 2. Krona plot of the DRM cyanobacterial community composition, as determined by 16S rRNA gene-amplicon sequencing across the sampling periods in 2019 (**A**) and 2020 (**B**). The community analysis was carried out by filtration of DRM water samples, total DNA extraction of the filters and PCR amplicon sequencing of the cyanobacteria-specific 16S rRNA gene region V3–V4 (380 bp length). For this Krona plot, and as an overview, the results shown are based on all samples combined per year. Taxonomic affiliation was carried out using the Greengenes reference database, and all taxonomic ranks are represented in the plot. Amplicon sequence variants (ASVs), as outputs of the Dada2 software package (see Section 5), distinguish sequence variations by a single nucleotide, giving ASVs a higher resolution than the operational taxonomic units (OTUs) typically used. Therefore, ASVs were used as the deepest taxonomic rank in our study. Total relative abundance was calculated by dividing the number of reads affiliated to an ASV in a sample by the total number of reads in the sample.

2.3. Synechococcus Rubescens and Cyanobium Gracile Clusters in 2019 and 2020

We examined the phylogenetic relationship of the main *Synechococcus* ASVs that were detected in 2019 and 2020, with the reference sequences of (i) all the cultivated phycoerythrin-rich *Synechococcus* spp. of Lake Constance, as established by Ernst and colleagues in 2003 [45], and (ii) *Synechococcus rubescens* and phycoerythrin-rich *Cyanobium gracile* reference sequences, as established by NCBI (Figure 3). The relationship was established using the appropriate sequence fragments, representing the PCR amplicon of 380 bp, with the IQ-TREE program [46], based on a phylogenetic inference using the maximum likelihood, coupled with ModelFinder to determine the best-fitting nucleotide substitution model [47]. Although the target sequence was shorter than the full 16S rRNA gene sequences established by Ernst et al., 2003, the phylogenetic relationship between the reference sequences remained the same, thereby confirming our analyses. The ASVs from this study were always grouped in pairs, with one ASV from 2019 (Figure 3, green font) and another from 2020 (Figure 3, blue font), as a reflection of the reoccurring *Synechococcus* phylotypes across the two years, further confirming our analysis.

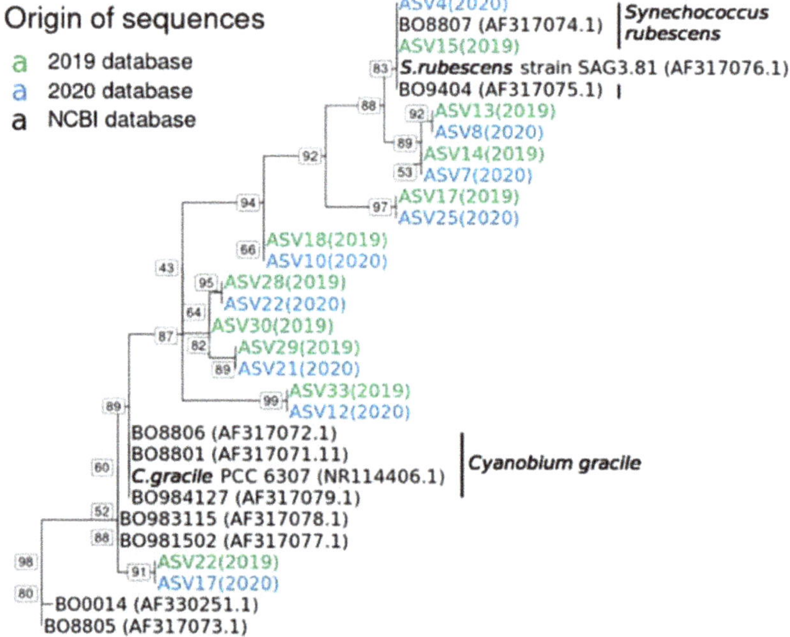

Figure 3. Illustration of the phylogenetic relationship of the *Synechococcus* spp. taxa observed in 2019 and 2020. The phylogeny is based on the cyanobacteria-specific V3–V4 (380 bp) 16S rRNA gene region. Colors correspond to the origin of the sequences, with green referring to ASVs observed in 2019 and blue referring to ASVs observed in 2020 (sampling year also in brackets). Sequences from *Synechococcus spp.* taxa, previously isolated from Lake Constance [45], and sequences of *Synechococcus rubescens* strain SAG3.81 and *Cyanobium gracile* PCC 6307 from NCBI were used as references (indicated in black). IQ-TREE was used for phylogenetic inference using maximum likelihood. The model of nucleotide substitutions used was TPM2+F + I [46], determined as the best-fit model by ModelFinder [47], based on the Bayesian information criterion scores and weights. Numbers at the internal nodes represent the percentage support of this specific node in 1000 bootstrap testing. The tree was rooted using the sequences of the cyanobacterium *Anacystis nidulans* PCC 6301 as an outgroup (not shown). The reference sequences are identified by their accession numbers in brackets.

Overall, the ASVs were grouped into two main clusters, either more closely related to the *S. rubescens* or to the *C. gracile* reference sequence (Figure 3). The top three most abundant ASVs for 2019 (ASV13, ASV14 and ASV15; see Figure 2 and below) and the top two most abundant for 2020 (ASV4 and ASV7; Figure 2), were more closely related to the *Synechococcus rubescens* NCBI reference sequence. For this group, two reference sequences of Lake Constance phycoerythrin-rich *Synechococcus* isolates [45] were available (Figure 3; BO8807 and BO9404), while for the *C. gracile* group, the reference sequences of seven Lake Constance isolates were available (Figure 3).

The megablast results were analogous to the phylogenetic tree shown, with the same ASVs in 2019 and 2020 being affiliated more closely to either *Synechococcus rubescens* or *Cyanobium gracile* (Supplementary Material, Table S3a,b), with the exception of ASV18 in 2019 and ASV10 in 2020. These ASVs showed identical percentage identities and E-values in the megablast for both *S. rubescens* and *C. gracile* (Table S3a,b) and were, therefore, grouped in between both clusters (Figure 3).

2.4. Dynamics of Synechococcus ASVs in 2019 and 2020

We examined the change in the relative abundance of the *Synechococcus* ASVs detected, over time. Figure 4 illustrates the dynamics of the *Synechococcus* ASVs as a heatmap, using relative abundance data after the log10 transformation (Figure 4A,C) and the data distribution (Figure 4B,D) as the mean and standard deviation of the relative abundances in a combined bar and jitter plot. Each dot represents the relative abundance of the taxa at a specific date (relative abundance values (%) per ASV across the sampling dates are shown in Supplementary Material, Table S4a,b).

Overall, the observed changes in the phylogenetic structure across the sampling dates suggested a high degree of successional change within the *Synechococcus* spp. community at the DRM. Some ASVs varied from being almost undetectable in the beginning to a relative abundance maximum later in the year, or showed the opposite trend (being abundant at the beginning or in the middle of the sampling campaigns), while other ASVs showed a comparatively stable (and low) relative abundance across the sampling campaigns. Indeed, while 10 ASVs were registered in 2019, only nine ASVs were registered in 2020 (Figure 4A,C). Thus, different ASVs appeared to dominate, and, therefore, provide a successional change, in the two field campaigns. For example, ASV13, as the most prominent ASV in 2019 (Figure 4B), reached its maximum in July (up to 36%; Figure 4A) and decreased thereafter in its relative abundance (to approximately 10%) at the end of September, while the two second most abundant ASVs in 2019 (ASV 14 and 15; Figure 4B) reached their maxima in mid-August and later in the year (see Figure 4A and Table S4a). Similarly, in 2020, ASV7 represented almost 50% of the total relative abundance in early July, but decreased to approximately 6% by mid-August, while the most abundant taxon in the 2020 sampling campaign (ASV4; Figure 4C,D) reached its peak at the end of August (68%) and decreased to approximately 23% by the end of the campaign (Figure 4C). Furthermore, the third most abundant ASV21 in 2020 represented only approximately 1% of the total relative abundance at the beginning of the sampling campaign, but almost 20% at its end (see Figure 4C, Table S4b).

The statistical relevance of the differences in *Synechococcus* ASV relative abundance was confirmed by a Kruskal–Wallis analysis of variance, with a *p*-value far below the threshold of 0.05 (*p*-value of 3.95×10^{-10} and 1.27×10^{-5} for 2019 and 2020, respectively). In an attempt to group the taxa according to their relative abundance differences, a post hoc Conover–Iman test was performed using a Benjamini Yekutieli *p*-value adjustment method for False Discovery Rate control (Table S5a,b). For 2019, the results indicated two groups. First, a group comprising ASVs that comparatively stably dominated the cyanobacterial community at the DRM throughout July and October (together 63% to 87% of the total relative abundance); these were ASV13, 14, 15, 28 and 29. A second group statistically differed from the first group (i.e., ASV17, 18, 22, 30 and 33), for which the abundance differences appeared to be larger and/or occurred within shorter time intervals:

For example, ASV22 reached its high relative abundance of 15% on only two sampling dates (in September and October, see Figure 4A and Figure S4a), as also discussed above. For 2020, the Conover–Iman test did not significantly separate the ASVs into two groups as for 2019 (Table S5b), although visually (Figure 4D), ASV4 dominated the cyanobacterial community with an average relative abundance of 45% throughout 2020.

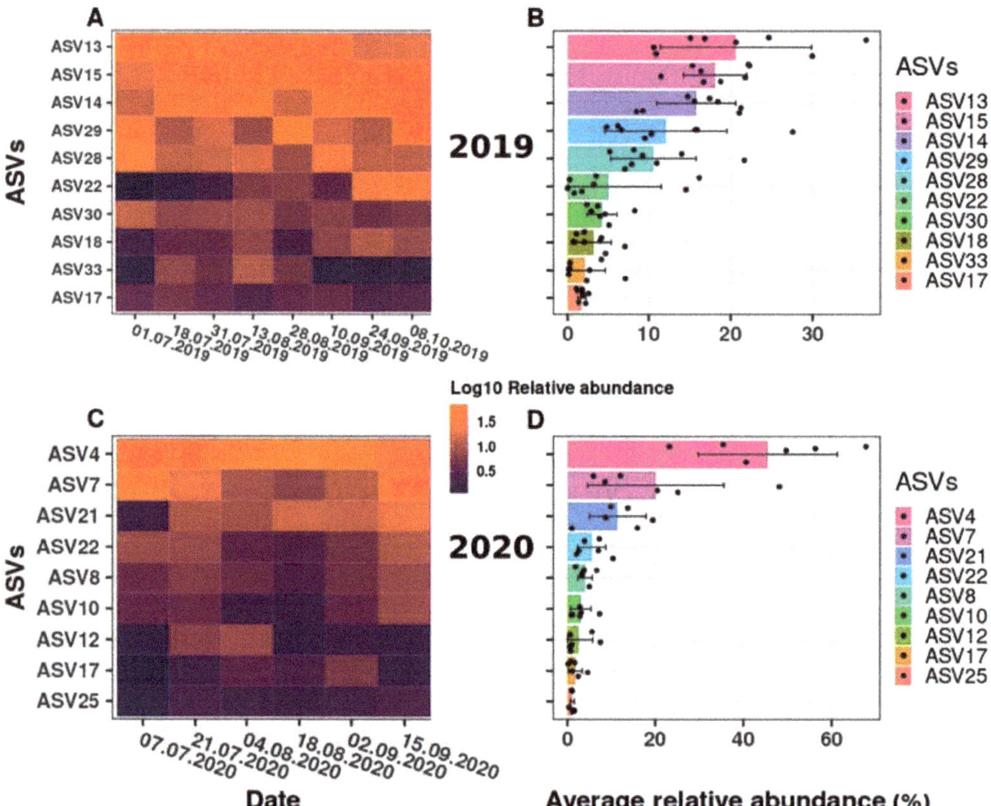

Figure 4. Dynamics of relative abundance changes for the *Synechococcus* ASVs observed during the 2019 and 2020 sampling campaigns. Shown are heatmaps across all sampling dates (**A,C**) and the ASVs' average relative abundances (**B,D**) for each year (bars) with each individual data point indicated (grey dots). For the y-axes of heatmaps (**A,C**), and average relative abundances (**B,D**), the different ASVs were ordered from the most abundant (top) to the least abundant (bottom). The heatmap data was log10 transformed from the read count matrix, and, to avoid introducing infinity for zero read counts, we added one artificial read to every cell prior to log10 transformation (log10[x + 1]). Color schemes vary between dark blue for low log10 values to bright orange for high log10 values; a higher log10 value means a higher relative abundance. Please note that the x-axes for B and D do not have the same scale. The average relative abundance across all samples of each year was calculated as it was for the Krona plot (see Figure 2). Standard deviation was also calculated and represented for each ASV; if the graphical representation of the standard deviation was below 0, the minimum error bar value was set to 0.

2.5. Dynamics of Planktothrix and Microcystis ASVs, the Abundance of Microcystin Biosynthesis Genes and the Concentration of Microcystins in Samples Taken during 2019 and 2020

In both the 2019 and the 2020 amplicon sequencing datasets, we detected two ASVs affiliated to the *Planktothrix* genus. The relative abundance of *Planktothrix* ASVs in 2019 increased from July to late September, with a peak on 31 July, where 0.75% of the total cyanobacterial community were *Planktothrix* ASVs (Figure 5A). Likewise, in 2020, *Planktothrix* ASVs were abundant from the end of July to the end of September, with a maximum of approximately 2% relative abundance on 21 July (Figure 5B). While one of the *Plank-tothrix* ASVs was affiliated to *Planktothrix rubescens* (a toxin-producing *Planktothrix* species), the amplicon sequencing of the cyanobacteria-specific 16S rRNA-gene fragments of the other ASVs did not allow for taxonomical distinction at the species level (i.e., between the mostly non-toxin-producing *P. agardhii* and the toxin-producing *P. rubescens*). For example, a megablast alignment of the *Planktothrix* ASVs suggested *Planktothrix agardhii* and *Planktothrix rubescens* were the top hits, with identical query coverage and E-values, and percentage identities varying between 99.70% and 100% for *P. agardhii*, and 99.38% to 99.74% for *P. rubescens*.

Beyond the *Planktothrix* ASVs, two ASVs that are affiliated to the *Microcystis* genus were detected in 2019; however, only one *Microcystis* ASV was detected in 2020. As both genera, *Planktothrix* and *Microcystis*, are renowned for their toxin producing species, more in depth analyses were carried out regarding the toxin producing potential of the species found in the Überlingen embayment.

To confirm the detection of the *Planktothrix* genus in Lake Constance at very low levels, particularly when compared to *Synechococcus*, quantitative PCR (qPCR) was performed to estimate the abundance of toxin-producing *Planktothrix* genotypes. Specifically, we used *Planktothrix*-specific primer pairs for the 16S rDNA gene and the *mcyBA1* gene, as described by Ostermaier and Kurmayer, 2009 (Table 1, [48]). *Planktothrix mcyBA1* encodes for the first adenylation domain in non-ribosomal peptide synthase (NRPS) gene clusters, and is present only in species that are capable of toxin production. Briefly, we created a standard curve using a dilution series of *Planktothrix* DNA (0.00001–100% *Planktothrix* DNA diluted in *Microcystis* DNA) and then calculated the relative abundance of *Planktothrix* DNA in our samples using a linear regression. The calculated relative abundance of *Planktothrix* ranged between 0.1 and 0.6% in 2019, and 0.1 and 0.01% in 2020 (Figure S3). Although the relative abundance calculated by qPCR differed from the relative abundances that were found with amplicon sequencing (Figure S3 and Figure 5), the trend aligned well between both methods. Statistical analyses of the differences between the 16S-rRNA gene and *mcyBA1* amplifications (ANOVA, see Ostermaier and Kurmayer, 2009) showed no significant difference in abundance with respect to toxin-producing or non-toxin-producing *Planktothrix* genotypes, suggesting that there was only one genotype of *Planktothrix* present in 2019 and 2020. To confirm the presence of toxin-producing cyanobacterial species in Lake Constance, we used universal *mcyE*-specific PCR primers (HEPF/R, Table 1, [49]). Being a member of the MC production gene cluster, *mcyE* is partly responsible for the synthesis of the ADDA chain in microcystins, as well as the incorporation and synthesis of D-Glu [50]. For the 2019 sampling campaign, the PCR yielded amplicons for every sample tested, suggesting the presence of potential microcystin producers throughout the year. The subsequent Sanger sequencing of the PCR products, and analysis of the consensus sequences with megablast, attributed toxin-producing capabilities to *Planktothrix* species (Figure 5A), except for the sample taken on 24 September 2019, where the *mcyE* consensus sequence had the highest alignment scores with *Microcystis*, thus matching the date with the highest relative abundance of the *Microcystis*-affiliated ASV. For 2020, *mcyE* amplicons were observed less consistently than in 2019, although, as seen in 2019, Sanger sequencing of these amplicons attributed the toxin-producing capabilities to *Planktothrix* spp. (Figure 5B).

Collectively, we provide evidence that toxin-producing *Planktothrix* spp. (and/or *Microcystis* spp.) are present in the Überlingen embayment of Lake Constance. However, our data did not allow us to conclude whether or not toxin production took place during the sampling campaigns, as many species can carry the gene cluster without actively producing the toxins [51]. Consequently, we analyzed the biomass samples that were collected independently on filters from the DRM for microcystins, using UPLC–MS/MS (Figure 5A,B). Microcystin concentrations peaked at the end of September 2019, where, in total, approximately 1.5 ng/L of intracellular microcystins were found. The microcystin variants present were MC-LR (leucine and arginine in hypervariable region) and MC-YR (leucine and tyrosine in hypervariable region), with MC-LR being almost solely responsible for the peak in toxin concentration in September 2019 (Figure 5A). Strikingly, for the samples collected during 2020, no toxins were detected under the conditions we used.

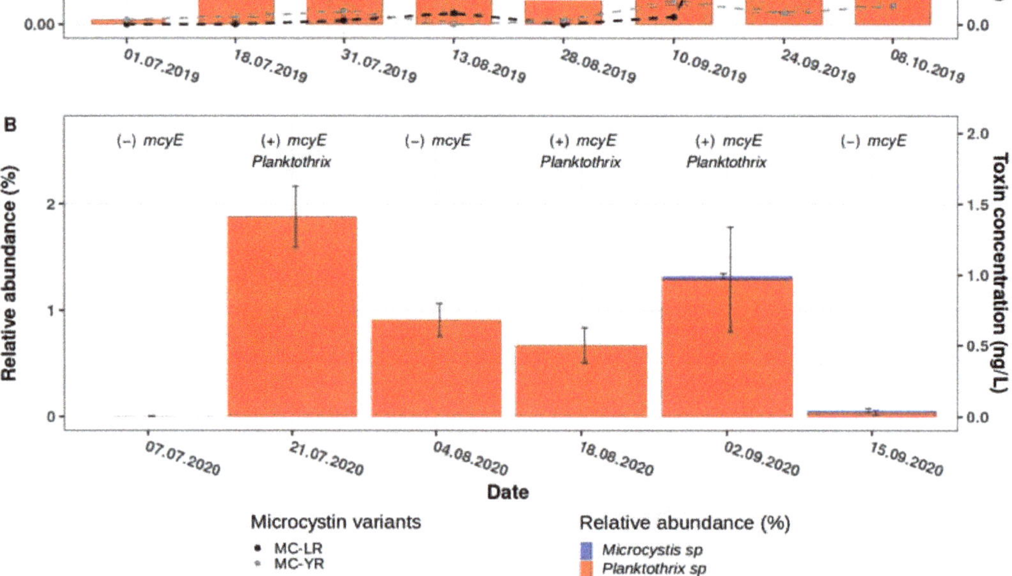

Figure 5. Relative abundance of the *Planktothrix* and *Microcystis* ASVs observed in 2019 and 2020, and of microcystin concentrations for samples taken in 2019. The bar plots represent the total relative abundance observed for the two *Planktothrix* ASVs (red bars) detected in 2019 (**A**) and 2020 (**B**), for the two *Microcystis* ASVs (blue bars) detected in 2019, and the single *Microcystis* ASV detected in 2020. The *x*-axes represent the different sampling dates and the left *y*-axis the relative abundance (%). The error bars in B represent the standard deviation of the relative abundance, as calculated from biological triplicates (n = 3); no error bars are represented for 2019, as only one sample was collected per sampling date. The right *y*-axis represents the toxin concentration (ng/L) determined for independently collected DRM filters, for which the microcystin variant concentrations are indicated by dashed lines in black (MC-LR) and grey (MC-YR) (see main text). The text on top of each bar indicates the results of the PCR amplification of the microcystin synthesis gene *mcyE* (+, detected; −, not detected) and of the phylogenetic affiliation of the *mcyE* consensus sequence to either *Planktothrix* or *Microcystis*, as established by Sanger sequencing of the PCR amplicons (see main text).

Table 1. Primers used in this study.

Name of Forward and Reverse Primers	Sequence (5′-3′)	Target	Reference
HEPF HEPR	TTTGGGGTTAACTTTTTTGGGCATAGTC AATTCTTGAGGCTGTAAATCGGGTTT	*mcyE*	Jungblut and Neilan, 2006

Table 1. *Cont.*

Name of Forward and Reverse Primers	Sequence (5′-3′)	Target	Reference
16S rDNA PTX fw 16S rDNA PTX rv 16S rDNA PTX TaqMan [1]	ATCCAAGTCTGCTGTTAAAGA CTCTGCCCCTACTACACTCTAG AAAGGCAGTGGAAACTGGAAG	16S rDNA (only *Planktothrix* spp.)	Ostermaier and Kurmayer, 2009
mcyBA1 PTX fw mcyBA1 PTX rv mcyBA1 PTX TaqMan [1]	ATTGCCGTTATCTCAAGCGAG TGCTGAAAAAACTGCTGCATTAA TTTTTGTGGAGGTGAAGCTCTTTCCTCTGA	*mcyBA1* (only *Planktothrix* spp.)	Ostermaier and Kurmayer, 2009
CYA359F CYA784R	GGGGAATYTTCCGCAATGGG ACTACWGGGGTATCTAATCCC	16S rRNA (Cyanobacteria)	Nübel et al., 1997

[1] TaqMan probes contain 5′ FAM (6-carboxyfluorescein, fluorescent reporter dye) and 3′ TAMRA (6-carboxy-tetramethylrhodamine, fluorescent quencher dye).

2.6. Retrospective Evaluation of Depth Profiles for the Lake Überlingen Routine Sampling Site

During our sampling campaigns in 2019 and 2020 (and the ongoing 2021 campaign), prominent DRM were observed from June onwards, particularly after long and stable good weather periods. This is best illustrated on 1 July 2019, when we observed a first prominent DRM at 18.4 m depth at the routine sampling site 'Wallhausen', in the Überlingen embayment of Lake Constance. This DRM appeared after the weather presented a stable window of approximately two weeks with predominant sunshine and no precipitation, low wind and elevated temperatures, as depicted in Figures S5, S7 and S8.

Three-dimensional (3D) plots of the FluoroProbe depth profiles for 'cryptophyta' content, as proxy, in the water column at 0–40 m depth (Figure 1) across the sampling campaigns in 2019 and 2020 are depicted in Figure 6A. Furthermore, we retrospectively evaluated the 'cryptophyta' depth profiles that were collected in the previous years from the routine sampling site, and transformed these into 3D plots using MATLAB (for years 2009–2018, see Supplementary Material, Figure S6). During the last twelve years, DRM have occurred multiple times (Figure S6, plots from 2011 and 2015–2020). Specifically, in 2016, we observed a prominent DRM in the summer, with maximum 'cryptophyta' concentrations of 6 µg/L on 6 September and 20 September (Figure 6B). Corresponding to the DRM, the high abundance of *Planktothrix rubescens* has been reported in various sampling sites at Lake Constance in 2016 [42,43].

Overall, prominent DRM were observed from July to October, with a maximum 'cryptophyta' content of approximately 2–4 µg/L (as estimated based on the FluoroProbe calibration), and at water depths ranging between 10 and 20 m (Figure 6A and Figure S6). The observed variation of the DRM depths likely follows the lake's internal waves, as previously found in Lake Ammer in Germany and Lake Bourget in France [14,52]. Although we could speculate that these peaks represent high abundances of *P. rubescens* in the Überlingen embayment, the absence of any appropriate samples allowing for DNA or toxin analyses available from that time preclude any corroboration.

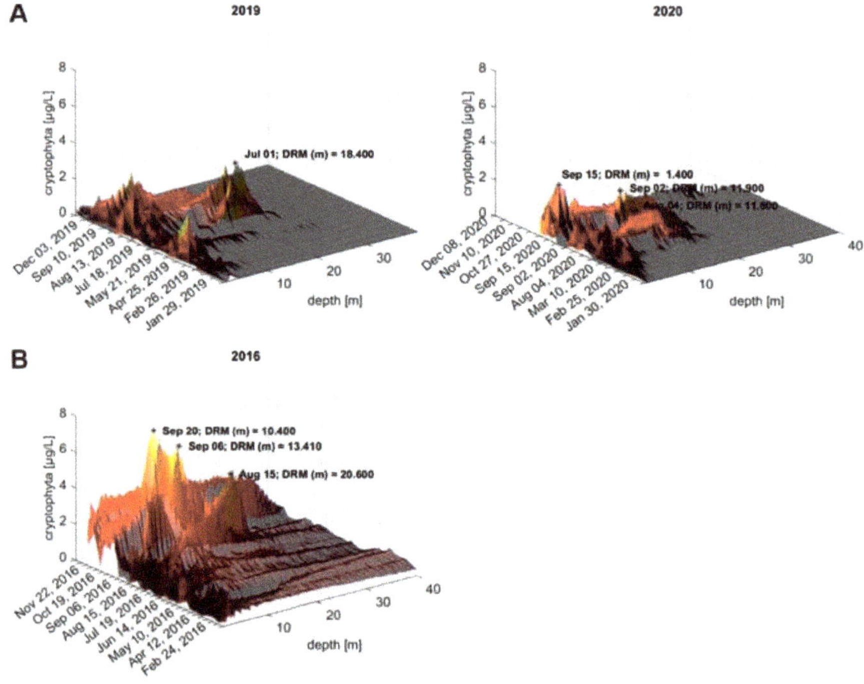

Figure 6. Depth profiles recorded with the FluoroProbe across the sampling campaigns in 2019 and 2020 (**A**), in comparison to the year 2016, in which *P. rubescens* blooms were reported in Lake Constance (**B**). Depicted are the FluoroProbe profiles for 'cryptophyta' abundance (cf. Figure 1) (expressed in µg chlorophyll-a per liter), recorded as proxy of red pigment abundance, in the water column from 0–40 m depth at the routine sampling site 'Wallhausen', in the Überlingen embayment of Lake Constance. Prominent DRM are highlighted with dates and water depths. Coordinates of the study site: 47.7571°N 9.1273°E; for an illustration, see Figure S4. In 2016, blooms of *P. rubescens* were reported for Lake Constance in September–October at various sampling sites (i.e., of the German, Austrian and Swiss sections of Lake Constance [42,43]). For 2016, FluoroProbe data is available for the Überlingen embayment of Lake Constance (**B**).

3. Discussion

The characterization of the recurring blooms of red-pigmented cyanobacteria in the Überlingen embayment of Lake Constance at water depths of 15–20 m by amplicon sequencing demonstrated the presence of *Synechococcus*, *Planktothrix* and *Microcystis*. Briefly, the relative abundance data suggested that *Synechococcus* taxa predominated the community (96–98%) at these water depths, while *Planktothrix* and *Microcystis* taxa were detectable only in very low abundances. For example, in 2020, up to 45% of the total relative abundance was represented by a single *Synechococcus* ASV (ASV4; Figures 2B and 4C,D). Moreover, the observed changes in the relative abundance of *Synechococcus* ASVs across the sampling dates suggest a high degree of successional change within the *Synechococcus* spp. community at the DRM (Figure 4).

Phycoerythrin-rich *Synechococcus* species from Lake Constance have been investigated previously [45,53,54], and, similarly to *Planktothrix* spp., they can express a large variety of phycobilins, thereby exploiting diverse light conditions [45,55]. Interestingly, the highest abundant *Synechococcus* taxa found in our studies in 2019 and 2020 are closely related to either *S. rubescens* or *C. gracile*, forming two main clusters (Figure 3, Table S3a,b). For the 2019 and 2020 sequences, a discrimination between *S. rubescens* and *C. gracile* is difficult (Figure 3), as the 380-bp PCR amplicon used in this study is cyanobacteria-specific, and allows us to affiliate taxa with high confidence only up to the genus rank. However, each *Synechococcus* taxon detected in 2019 is paired with a taxon detected in 2020, and their sequence alignment showed a 100% identity with no gap, suggesting recurring phylotypes/ecotypes, at least across the two years. The close association of our sequences with those from the early 2000s [45] suggests a stable deep water cyanobacterial community in Lake Constance. The latter interpretation is supported by the earlier finding that different lineages of *Synechococcus* spp. can adapt to, and thrive in, specific ecological niches [56].

The predominance of *Synechococcus* species over other cyanobacterial genera was reported for the experimental cocultures of *Synechococcus* and *Microcystis* strains, as well as varying phosphate and nitrogen concentrations [57,58]. Indeed, at low nutrient concentrations, which could also mirror the currently oligotrophic conditions of Lake Constance [59], *Synechococcus* outcompeted *Microcystis* in growth rate and final biomass [57,58]. The predominance of *Synechococcus* may be explained by its seemingly higher affinity for orthophosphates, more efficient nutrient uptake (due to a larger surface-to-volume ratio), as well as by the potential for competitive inhibition via the quorum sensing/quenching molecules between the two strains [57,58,60,61]. The latter laboratory findings were also corroborated in natural habitats, at least in regards to their trend [62].

Despite the seemingly stable predominance of *Synechococcus* in the deep water cyanobacterial community observed in Lake Constance over the last two decades [45], the de novo occurrence of *Planktothrix* spp. in 2016, and the reconfirmation of this earlier finding with our samples in 2019 and 2020, could suggest that *Planktothrix* spp. is in the process of establishing a stable presence in the Überlingen embayment of Lake Constance. The latter is of critical importance as mass occurrences of toxin-producing cyanobacteria at water depths of >20 m could become a threat to the water intake for the Sipplingen water treatment plant (https://www.bodensee-wasserversorgung.de, accessed on 7 May 2021 [63]) which serves >4 million people with drinking water. Indeed, we detected a low abundance of potentially MC-producing *Planktothrix* spp. using amplicon sequencing, as well as through *Planktothrix*-specific qPCR for 2019 and 2020 (Figure 5 and Figure S3). Furthermore, the presence of the microcystin biosynthesis gene cluster, *mcy*, was detected through the PCR amplification of *mcyE* (Figure 5). The subsequent Sanger sequencing of these amplicons identified *Planktothrix* spp. to be the main contributor to this gene sequence in our samples, with the exception of 24 September 2019, when the amplified *mcyE* consensus sequence identified the highest alignment scores in *Microcystis* spp. In *Planktothrix* spp., the *mcy* gene cluster can be inactivated by various mutations, including insertions or deletions, and, thus, non-toxic strains can develop [64]. Non-toxic strains are less successful in competition than their toxic relatives, and, thus, *Planktothrix* blooms are usually dominated by toxic strains [48]. Corresponding to the presence of *mcyE* in 2019, low amounts of the microcystins MC-LR and MC-YR were detected using UPLC–MS/MS. In 2020, although *mcyE* amplicons were detectable in some of our samples, concentrations of microcystins were below the detection levels of the UPLC–MS/MS that was used (LOD of UPLC–MS/MS method: 0.5 ng/mL [65]; therefore, the resulting LOD in our water samples: 62 pg/L lake water). Despite the low abundance of *Planktothrix* spp. and the low concentrations of detected toxins, the question does arise concerning whether the latter is a sign of a fundamental change in water body dynamics in the Überlingen embayment of Lake Constance (e.g., resulting from global warming). Indeed, the greater and prolonged stratification of water bodies, in conjunction with lowered nutrient levels, would promote a deep water euphotic ecosystem encompassing low-light specialized species, such as the picocyanobacterial *Synechoccocus* spp. and *Planktothrix* spp. [9,15], as was reported for other prealpine lakes (e.g., Lake Zurich [4], Lake Mondsee [8] and Lake Bourget [6]).

As *Planktothrix* spp. are stimulated by increased temperatures [15], the continuous shift toward higher overall temperatures could favor the perseverant establishment of toxin-producing *Planktothrix* spp., to the disadvantage of today's *Synechococcus*-dominated DRM ecosystems. Considering that *Synechococcus* spp. are currently markedly outcompeting other cyanobacterial species occurring at these water depths, this may suggest that the allelopathic compounds from *Synechococcus* spp. can have an adverse effect on co-occurring species. Indeed, such effects have been observed for freshwater *Synechococcus* spp., which were able to impact the growth of other freshwater cyanobacteria or green algae [66,67]. Adverse effects caused by allelopathic compounds from marine *Synechococcus* spp. were also observed on various marine invertebrates [68], as well as other bacterial species [60]. In conclusion, the described effects indicate the widespread production of allelopathic compounds by *Synechococcus* species that can even influence other bacteria, plants and invertebrates.

In most cases, the co-occurrence of species occupying the same, or similar, ecological niche leads to the dominance of one species, largely depending on the individual species' competitive advantages (e.g., nitrogen fixation, uptake of inorganic phosphorus, regulation of buoyancy, allelopathic compounds, etc.), as also shown by Weisbrod et al. [69]. This suggests that, if *Synechococcus* outcompetes *Planktothrix* in its ecological niche, allelopathic compounds (such as toxins), amongst other factors, may be used to compete against the respective other species. Indeed, allelopathic activity is one of the major competitive strategies of freshwater *Synechococcus* against coexisting phytoplankton species [66], further supporting the potential role of picocyanobacterial exudates in competition with other cyanobacteria, such as *Planktothrix*. Counterintuitively, some freshwater *Synechococcus* spp. possess a positive allelopathic activity towards *Microcystis* spp., and no effect on *Phormidium* spp., whereby *Phormidium* spp. (like *Planktothrix* spp.) is part of *Oscillatoriaceae* [66], suggesting that *Synechococcus* spp. could even promote the growth of toxin-producing species. The complex interplay of species in competition in Lake Constance emphasizes the need for further studies regarding the co-occurrence and dominance of *Synechococcus* spp., relative to *Planktothrix* and *Microcystis* spp. in this lake.

The geographical setting of the Überlingen embayment of Lake Constance, with its minor wind influence (Figures S7 and S8), is considered an additional factor that could favor the continuous development of deep water *Planktothrix* spp. populations. Indeed, light winds and convective mixing are highly important in the seasonal cycling of *P. rubescens* communities within a strongly stratified medium-sized lake [70]. In consequence, this means that toxin-producing *Planktothrix* spp. could possibly establish themselves as a dominant cyanobacteria species at deeper water levels, and may become a relevant concern for the quality of the Überlingen embayment of Lake Constance as a drinking water resource in the not too distant future.

4. Conclusions

Our study characterized the deep water red pigment maxima (DRM) in the Überlingen embayment of Lake Constance in 2019 and 2020 as being dominated by phycoerythrin-rich picocyanobacterial; namely, *Synechococcus rubescens* and *Cyanobium gracile*. Unlike other prealpine lakes, the DRM in Lake Constance is not dominated by the phycoerythrin-rich, filamentous and often toxin-producing *Planktothrix rubescens*. Indeed, the alignment of our results with the sequences from the Ernst et al. study (2003) demonstrates high sequence similarity, and suggests that the same species have been dominant at 15–20 m depth in Lake Constance for the past 20 years. However, we confirmed the reports from 2016 that *Planktothrix* spp. does occur in Lake Constance in the years 2019 and 2020, albeit at very low relative abundances. Nevertheless, microcystin concentrations of up to 1.5 ng/L were detected through UPLC–MS/MS in 2019, which appeared to be produced by *Planktothrix* and/or *Microcystis* spp. Hence, at present, Lake Constance seems to have a rather stable deep water cyanobacterial ecosystem, predominated by *Synechococcus* spp., although the geographical setting, as well as the continued climate warming, could favor the development and steady predominance of toxin-producing *Planktothrix* spp. This highlights the importance of a future monitoring program for Lake Constance, with emphasis on sequencing-based cyanobacterial community studies and microcystin monitoring, as well as the importance of competition studies regarding the different cyanobacterial taxa in relation to the physiochemical and biological parameters of the lake, particularly in respect to the ongoing climate change. Monitoring programs and hypothesis-driven competition studies may provide the required database to predict future deep water mass occurrences of toxin-producing cyanobacteria and, thus, help to secure Lake Constance as the drinking water resource for millions of people in the future.

5. Materials and Methods

5.1. Sample Collection

Samples were taken every two weeks from 1 July to 8 October in 2019, and 7 July to 15 September in 2020. The initial sampling in 2019 included only one replicate/day, while samples were collected in biological triplicates (n = 3) during 2020. A bbe Moldaenke FluoroProbe (FP) (SN: 01709; recalibrated at bbe in 2009, 2012, 2016 and 2019) was used to determine the deep red maximum (DRM). Samples were collected at that depth every other week in Upper Lake Constance (47.7571° N 9.1273° E), using a messenger-released Free Flow Water Sampler 5 L (HYDRO-BIOS, Altenholz, Germany). An overview of the samples taken during this study is illustrated in Table S1. Water was filtered through a 180 μm nylon net filter to exclude zooplankton and larger particles. Two liters of water were then filtered on one Whatman GF6 glass fiber filter while applying 2 bars pressure for the collection of plankton biomass. Filters were stored at −20 °C until analysis. This sampling method was applied to DNA extractions in 2019, and toxin extractions in 2019 and 2020. In 2020, DNA samples were collected independently (n = 3) on 0.2 μm polycarbonate filters.

5.2. DNA Extraction and PCR

DNA from the 2019 samples was extracted using the ZYMO Research Fecal/Soil/Microbe Microprep kit, following the manufacturer's instructions. DNA of 2020 samples was extracted using a phenol/chloroform/isoamylalcohol protocol, adapted from Rusch et al., [71] and the JGI protocol [72]. Standard PCR was performed with Taq polymerase, using 2X Taq MasterMix (NEB) and 30 cycles. The microcystin synthesis gene, *mcyE*, was amplified with the HEPF/R primer set [49], and DNA from cultured *Microcystis aeruginosa* strain 78 was used as a positive control for the presence of *mcyE*. Planktothrix-specific primer pairs PcPI+/− (PC-IGS) and peamso+/− (*mcyA*) [8] were used for additional PCR reactions, with *Microcystis aeruginosa* strain 78 genomic DNA as a negative control, and *Planktothrix rubescens* strain 101 genomic DNA as a positive control.

5.3. RT-PCR/TNA

Quantitative Taq Nuclease Assays (TNA or TaqMan PCR) were performed with primers specific to *Planktothrix* spp. [48]. We quantified both *Planktothrix*-specific 16S rDNA and *mcyBA1*, which encodes the first adenylation domain of *mcyB*, and is indicative of all genotypes containing the *mcy* gene cluster. Both probes contained 5′ FAM as a fluorescent marker and 3′ TAMRA as a quencher dye. Each reaction contained 50 ng template DNA, 200 nM of primers and probes, and KAPA probe fast MasterMix. Amplification and quantification were carried out in triplicates in a Bio-Rad CFX96 cycler with the following protocol: 10 min at 95 °C to activate the hot start polymerase, followed by 50 cycles of 15 s at 95 °C, 60 s at 51.5 °C for 16S, 60 s at 57 °C for *mcyBA1*, 60 s at 68 °C and a final elongation for 5 min at 68 °C. The standard curve contained *Planktothrix* DNA mixed with *Microcystis* DNA in different ratios (0.00001–100% *Planktothrix* DNA diluted in *Microcystis* DNA). Data analysis using linear regression from the standard curve was performed using BioRad CFX manager, Microsoft Excel and GraphPad Prism 5.

5.4. Sanger Sequencing

Sanger sequencing was performed with the amplicons produced by the HEPF/R primer pair to identify the main microcystin producer in the samples. PCR amplicons were purified using a QIAquick PCR cleanup kit and sent to Eurofins Genomics for analysis. The identity of the obtained sequences was determined using Nucleotide BLAST megablast [73].

5.5. Amplification and Illumina Sequencing

Amplification of the V3–V5 hypervariable regions, and the cyanobacterial-specific V3–V4 hypervariable regions of the 16S rRNA gene, was performed with 0.02 U/µL of Phusion High Fidelity DNA polymerase, 1X Phusion HF Buffer, 200 µM of dNTPs (New England Biolabs, USA) and 0.5 µM of each primer. Primers targeting the V3–V5 hypervariable regions were 357F and 926R [74,75]. Cyanobacteria-specific primers were CYA359F and CYA784R [44]. Each PCR comprised an initial denaturation step of 3 min at 98 °C, followed by 30 cycles of denaturation for 45 s at 98 °C, annealing for 20 s at 62.4 °C (V3–V5) or 60 °C (cyanobacteria-specific) and extension for 8 s at 72 °C, and a final extension step for 5 min at 72 °C. Extracted DNA was added at a final concentration of 0.12 ng/µL. No purification step was performed; the PCR products were directly sent to Eurofins Genomics for sequencing using Illumina MiSeq 2 × 300 bp with the Microbiome Profiling Indexing only package. The data presented in this study are accessible on NCBI under the bioproject number PRJNA727470.

5.6. Bioinformatics Pipeline

The analysis was carried out using the already merged dataset provided by Eurofins Genomics, Konstanz, Germany. The expected fragment sizes were 569 bp for the V3–V5 amplicon and 425 bp for the V3–V4 cyanobacterial-specific amplicon. Reads were first trimmed using Trimmomatic [76], removing all reads with a Phred quality below 3 for the start and the end of the reads, below an average quality of 10 on a window of 3 base within the reads, and below a size of 500 bp for the V3–V5 amplicons and 380 bp for the V3–V4 amplicons. FastQC was used for quality control of the reads before and after trimming [77]. The following steps were performed using QIIME2 2019.10 [78]. Filtration of chimeras using the consensus method, denoising and dereplication of the quality reads were performed using the denoise and dereplicate single-end sequences (Dada2, denoise-single) and a reads learn of 2,000,000 reads for the training error model [79]. Quality trimming had already been performed using Trimmomatic, so no trimming step was performed here. Taxonomic affiliation was achieved using the classify-consensus-vsearch program and the databases SILVA_138 and Greengenes, with a percentage identity of 80%, 90%, 97% and 100%. Taxonomic results were merged and the highest percentage identity taxonomic affiliation was retained. Reference databases were previously trained using the feature-classifier extract-reads script with the sequences of the primers used for the amplification of the hypervariable regions. After training, the databases only contained the part of interest of the 16S rRNA gene for taxonomic assignation, V3–V5 hypervariable regions for the general 16S rRNA gene amplification and V3–V4 hypervariable regions for the cyanobacteria-specific primers. Taxonomy was mostly consistent between the SILVA_138 and Greengenes databases.

5.7. Phylogenetic Analysis

A megablast (highly similar sequence) was performed on the sequences affiliated to the genera *Planktothrix*, *Microcystis* and the most abundant *Synechococcus* using the 16S rRNA sequences reference database on NCBI. The top ten hit sequences and the description table were extracted. The *Synechococcus* sequences from Lake Constance analyzed and sequenced by Anneliese Ernst [45] were downloaded from the NCBI database. Two other 16S rRNA gene sequences, belonging to *Synechococcus rubescens* and *Cyanobium gracile*, were also collected from NCBI and used as reference for the phylogenetic analysis, as previously carried out by Anneliese Ernst. The collected dataset was then compared to the most abundant *Synechococcus* sequences in 2019–2020. Sequences were merged with our sequences into a fasta file and aligned using SeaView [80]. IQ-TREE 2.1.2 [46] was used to calculate the phylogenetic tree by maximum likelihood using 1000 bootstrap parameters. The model of nucleotide substitution used was TPM2+I+F [46], determined as the best fit model by ModelFinder [47], based on the Bayesian information criterion scores and weights. The tree was visualized on R using the package ggtree [81].

5.8. Statistical and Network Analysis

Statistical analysis was performed with R software [82], using the package Phyloseq [83], vegan [84] and visualized with the package ggplot2 [85]. Features represented by less than 3 reads in below 20% of the samples were discarded. Chloroplast-affiliated features were removed from both datasets, bacteria-affiliated taxa were also removed from the cyanobacterial-specific dataset only. No rarefaction has been applied to the dataset [86]. After filtration, the lowest number of reads in a sample was 19,175 for the cyanobacteria-specific dataset and 24,329 for the general 16S rDNA dataset in 2019. The 2020 dataset consisted of only the cyanobacteria-specific amplicon and the lowest number of reads was 24,341. This number of reads instilled confidence in catching all the richness present in our dataset, as the rarefaction curves showed that we were in the stationary phase of the curve. The read count matrix was transformed in relative abundance by dividing the reads affiliated to one taxa by the total number of reads in the sample $(x/\text{sum}(x))$, and a logarithmic transformation was also applied. To avoid the presence of 0 in the matrix, one artificial read was added to each cells of the data $(\text{Log}10(x+1))$. Alpha diversity was analyzed using the Observed richness, Pielou's evenness index, Shannon diversity index and the Inverted Simpson diversity index. Community composition was observed using barplot, jitter plot and krona plot, using the relative abundance data and heatmap with the Log10 transformed data. Statistical analyses were performed using a non-parametrical statistical test, as sequencing data did not fulfill the parametric test requirements (e.g., data normality and homoscedasticity, and independence of observation) using a p-value and False discovery rate threshold of 0.05. Kruskall–Wallis one-way analysis of variance was performed on the relative abundance sub-data of the main Synechococcus taxa to test if at least one taxa dominated the community. The hypothesis that the relative abundance distribution of the tested taxa are equal was the null hypothesis (H0) and the alternative hypothesis (H1) was that the relative abundance distribution of the tested taxa are not equal. If the Kruskall–Wallis test rejected H0, a post hoc test using the Conover–Iman squared rank test, with a Benjamini Yekutieli p-value adjustment procedure, was performed to determine which taxon or group of taxa dominate the community. The Benjamini Yekutieli procedure was chosen for p value adjustment because of the dependency between the taxa's relative abundance observation. Barplots of the taxa of interest (*Planktothrix spp.*, *Microcystis spp.*, *Synechococcales*) and correlation observation of the main node from the network analysis were produced.

5.9. Toxin Extraction and Analysis

Toxins were extracted from filters using a methanolic extraction method and were, subsequently, analyzed via UPLC–MS/MS [24,65]. Briefly, 3 mL 50% (*v/v*) aqueous methanol was added to each filter and soaked for 30 min at room temperature. After vigorous mixing (vortex) for 10 min, samples were sonicated in an ultrasonic bath for 15 min, before centrifugation at $4000 \times g$ for 10 min. The supernatant was collected in a separate tube and the above steps, excluding the initial 30 min soaking time, were repeated twice with the remaining pellet. The pooled supernatant was dried overnight using a speed vacuum system (Univapo 100H) and resuspended in 250 µL 50% aqueous methanol. Toxin samples were stored in glass vials at $-20\ °C$ until analysis.

Concentrations of different MCs were measured via UPLC–MS/MS, with an internal standard containing deuterated MC-LR and MC-LF (D_5-MC-LF and D_7-MC-LR) [24,65,87]. Three different MCs, MC-RR, MC-YR and MC-LR were used as external standards for analysis, with final concentrations of 2, 10 and 100 ng/mL. An Acquity H-class liquid chromatograph and a Waters XEVO TQ-S mass spectrometer were used. For UPLC, an Acquity BEH C18 1.7 µm column (2.1 × 50 mm) with a corresponding guard column, each kept at 40 °C, was used. Solvents A and B were composed of 10% and 90% acetonitrile, respectively, 100 mM formic acid and 6 mM NH_3. Initial conditions were 25% B, held for 30 s, then 45% B within 30 s, 60% B within 180 s and 99% B within 12 s, which was held again for 30 s. The flow rate was 0.4 mL/min. Prior to application of the next sample, the column was re-equilibrated to 25% B over 78 s and held for 60 s. Injection volume was 5 µL. As described in Altaner et al., 2019, simultaneous analysis of MC congeners was carried out using five analysis windows, maximizing the scan time for each congener [65].

5.10. Evaluation of bbe Moldaenke FluoroProbe Data

For data acquisition, the Moldaenke FluoroProbe (FP) was slowly lowered (approximately 0.2 m/s) from the water surface to 100 m depth using a winch, then quickly pulled up again, while collecting data both ways. FP data was preprocessed and later plotted in 3D plots using MATLAB R2020a [88]. The datasets were chopped above 1 m depth and below 100 m depth and subsequently sorted. Data were interpolated to 0.1 m steps, and 570 nm fluorescence measurements and cryptophyta content (µg/L) data were extracted. Plotting was carried out using the MATLAB standard functions *surf* (3D) and *plot* (2D), and peaks were analyzed using *max*.

Supplementary Materials: The following are available online at https://www.mdpi.com/article/10.3390/toxins13090666/s1, Figure S1: Common microcystin structure; Figure S2: Illustration of red colored filters; Figure S3: Relative abundance as quantified with qPCR; Figure S4: Localization of study site; Figure S5: Depiction of stable weather window prior to 1 July 2019; Figure S6: Depiction of depth profiles for 2009–2020; Figures S7 and S8: Mean wind speed and directions around the study site; Table S1: Sampling dates with corresponding sampling depths; Table S2: Assignment of ASVs with SILVA and Greengenes databases; Table S3: NCBI megablast of *C. gracile* and *S. rubescens* in 2019 and 2020; Table S4: Relative abundance of *Synechococcus* ASVs per date in 2019 and 2020; Table S5: Conover–Iman test of *Synechococcus* ASVs in 2019 and 2020.

Author Contributions: Conceptualization, E.R., C.F., D.R.D. and D.S.; methodology, E.R. and C.F.; formal analysis, E.R. and C.F; data curation, E.R. and C.F.; writing—original draft preparation, E.R. and C.F; writing—review and editing, D.R.D. and D.S.; project administration, D.R.D. and D.S.; funding acquisition, D.R.D. and D.S. All authors have read and agreed to the published version of the manuscript.

Funding: This research was funded by DFG Research Training Group R3—Responses to biotic and abiotic Changes, Resilience and Reversibility of Lake Ecosystems (GRK 2272) and MTU Friedrichshafen GmbH (MTU Umweltstiftung).

Data Availability Statement: The Illumina sequence data presented in this study is accessible at NCBI under the bioproject number PRJNA727470.

Acknowledgments: We would like to highlight the excellent support received from Julia Schmidt, Microbial Ecology and Limnic Microbiology, with regard to DNA extraction from samples she helped to organize from the Überlinger embayment. The assistance with UPLC-MS/MS by Sarah Krassnig and Aswin Mangerich and the help in the laboratory by Leon Walther is thankfully acknowledged. We would also like to thank Beatrix Rosenberg for access to long-term data and Alfred Sulger, Angelika Seifried, Pia Mahler and Josef Halder at the Limnological Institute of University of Konstanz for managing the ship cruises, and Anneliese Ernst for helpful discussions. Further, we would like to thank the MTU Umweltstiftung and all members of the RTG-R3, especially Tina Romer and Frank Peeters, for their support.

Conflicts of Interest: The authors declare no conflict of interest.

References

1. Huisman, J.; Codd, G.A.; Paerl, H.W.; Ibelings, B.W.; Verspagen, J.M.H.; Visser, P.M. Cyanobacterial blooms. *Nat. Rev. Microbiol.* **2018**, *16*, 471–483. [CrossRef] [PubMed]
2. Mantzouki, E.; Campbell, J.; van Loon, E.; Visser, P.; Konstantinou, I.; Antoniou, M.; Giuliani, G.; Machado-Vieira, D.; de Oliveira, A.G.; Maronić, D.Š.; et al. A European multi lake survey dataset of environmental variables, phytoplankton pigments and cyanotoxins. *Sci. Data* **2018**, *5*, 1–13. [CrossRef]
3. Rücker, J.; Wiedner, C.; Zippel, P. Factors controlling the dominance of *Planktothrix agardhii* and *Limnothrix redekei* in eutrophic shallow lakes. *Hydrobiologia* **1997**, *342–343*, 107–115. [CrossRef]
4. Van den Wyngaert, S.; Salcher, M.M.; Pernthaler, J.; Zeder, M.; Posch, T. Quantitative dominance of seasonally persistent filamentous cyanobacteria (*Planktothrix rubescens*) in the microbial assemblages of a temperate lake. *Limnol. Oceanogr.* **2011**, *56*, 97–109. [CrossRef]

5. Walsby, A.E.; Schanz, F. Light-dependent growth rate determines changes in the population of *Planktothrix rubescens* over the annual cycle in lake Zürich, Switzerland. *New Phytol.* **2002**, *154*, 671–687. [CrossRef]
6. Jacquet, S.; Briand, J.F.; Leboulanger, C.; Avois-Jacquet, C.; Oberhaus, L.; Tassin, B.; Vinçon-Leite, B.; Paolini, G.; Druart, J.C.; Anneville, O.; et al. The proliferation of the toxic cyanobacterium *Planktothrix rubescens* following restoration of the largest natural French lake (Lac du Bourget). *Harmful Algae* **2005**, *4*, 651–672. [CrossRef]
7. Salmaso, N. Long-term phytoplankton community changes in a deep subalpine lake: Responses to nutrient availability and climatic fluctuations. *Freshw. Biol.* **2010**, *55*, 825–846. [CrossRef]
8. Kurmayer, R.; Christiansen, G.; Fastner, J.; Börner, T. Abundance of active and inactive microcystin genotypes in populations of the toxic cyanobacterium *Planktothrix* spp. *Environ. Microbiol.* **2004**, *6*, 831–841. [CrossRef]
9. Posch, T.; Köster, O.; Salcher, M.M.; Pernthaler, J. Harmful filamentous cyanobacteria favoured by reduced water turnover with lake warming. *Nat. Clim. Change* **2012**, *2*, 809–813. [CrossRef]
10. Kurmayer, R.; Jüttner, F. Strategies for the co-existence of zooplankton with the toxic cyanobacterium *Planktothrix rubescens* in Lake Zurich. *J. Plankton Res.* **1999**, *21*, 659–683. [CrossRef]
11. Kurmayer, R.; Deng, L.; Entfellner, E. Role of toxic and bioactive secondary metabolites in colonization and bloom formation by filamentous cyanobacteria Planktothrix. *Harmful Algae* **2016**, *54*, 69–86. [CrossRef]
12. Salcher, M.M.; Pernthaler, J.; Frater, N.; Posch, T. Vertical and longitudinal distribution patterns of different bacterioplankton populations in a canyon-shaped, deep prealpine lake. *Limnol. Oceanogr.* **2011**, *56*, 2027–2039. [CrossRef]
13. Walsby, A.E.; Ng, G.; Dunn, C.; Davis, P.A. Comparison of the depth where *Planktothrix rubescens* stratifies and the depth where the daily insolation supports its neutral buoyancy. *New Phytol.* **2004**, *162*, 133–145. [CrossRef]
14. Cuypers, Y.; Vinçon-Leite, B.; Groleau, A.; Tassin, B.; Humbert, J.F. Impact of internal waves on the spatial distribution of *Planktothrix rubescens* (cyanobacteria) in an alpine lake. *ISME J.* **2011**, *5*, 580–589. [CrossRef] [PubMed]
15. Paerl, H.W.; Huisman, J. Climate: Blooms like it hot. *Science* **2008**, *320*, 57–58. [CrossRef]
16. Wejnerowski, Ł.; Rzymski, P.; Kokociński, M.; Meriluoto, J. The structure and toxicity of winter cyanobacterial bloom in a eutrophic lake of the temperate zone. *Ecotoxicology* **2018**, *27*, 752–760. [CrossRef]
17. Legnani, E.; Copetti, D.; Oggioni, A.; Tartari, G.; Palumbo, M.T.; Morabito, G. *Planktothrix rubescens*' seasonal dynamics and vertical distribution in Lake Pusiano (North Italy). *J. Limnol.* **2005**, *64*, 61–73. [CrossRef]
18. Walls, J.T.; Wyatt, K.H.; Doll, J.C.; Rubenstein, E.M.; Rober, A.R. Hot and toxic: Temperature regulates microcystin release from cyanobacteria. *Sci. Total Environ.* **2018**, *610–611*, 786–795. [CrossRef]
19. Kleinteich, J.; Wood, S.A.; Küpper, F.C.; Camacho, A.; Quesada, A.; Frickey, T.; Dietrich, D.R. Temperature-related changes in polar cyanobacterial mat diversity and toxin production. *Nat. Clim. Change* **2012**, *2*, 356–360. [CrossRef]
20. Komárek, J.; Anagnostidis, K. Bd. 2/Part 2: Oscillatoriales. In *Süßwasserflora von Mitteleuropa, Bd. 19/2: Cyanoprokaryota*; Springer: Berlin/Heidelberg, Germany, 2005; p. 759, ISBN 978-3-8274-1914-9.
21. Buratti, F.M.; Manganelli, M.; Vichi, S.; Stefanelli, M.; Scardala, S.; Testai, E.; Funari, E. Cyanotoxins: Producing organisms, occurrence, toxicity, mechanism of action and human health toxicological risk evaluation. *Arch. Toxicol.* **2017**, *91*, 1049–1130. [CrossRef]
22. Jang, M.H.; Ha, K.; Joo, G.J.; Takamura, N. Toxin production of cyanobacteria is increased by exposure to zooplankton. *Freshw. Biol.* **2003**, *48*, 1540–1550. [CrossRef]
23. Jang, M.H.; Jung, J.M.; Takamura, N. Changes in microcystin production in cyanobacteria exposed to zooplankton at different population densities and infochemical concentrations. *Limnol. Oceanogr.* **2007**, *52*, 1454–1466. [CrossRef]
24. Weisbrod, B.; Riehle, E.; Helmer, M.; Martin-Creuzburg, D.; Dietrich, D.R. Can toxin warfare against fungal parasitism influence short-term Dolichospermum bloom dynamics? A field observation. *Harmful Algae* **2020**, *99*, 101915. [CrossRef] [PubMed]
25. Davis, T.W.; Berry, D.L.; Boyer, G.L.; Gobler, C.J. The effects of temperature and nutrients on the growth and dynamics of toxic and non-toxic strains of Microcystis during cyanobacteria blooms. *Harmful Algae* **2009**, *8*, 715–725. [CrossRef]
26. Dziallas, C.; Grossart, H.P. Increasing oxygen radicals and water temperature select for toxic *Microcystis* sp. *PLoS ONE* **2011**, *6*. [CrossRef]
27. Berry, M.A.; Davis, T.W.; Cory, R.M.; Duhaime, M.B.; Johengen, T.H.; Kling, G.W.; Marino, J.A.; Den Uyl, P.A.; Gossiaux, D.; Dick, G.J.; et al. Cyanobacterial harmful algal blooms are a biological disturbance to Western Lake Erie bacterial communities. *Environ. Microbiol.* **2017**, *19*, 1149–1162. [CrossRef]
28. Bingham, M.; Sinha, S.K.; Lupi, F. *Economic Benefits of Reducing Harmful Algal Blooms in Lake Erie*; Environmental Consulting & Technology Inc.: Gainesville, FL, USA, 2015.
29. Spoof, L.; Catherine, A. Appendix 3: Tables of microcystins and nodularins. In *Handbook of Cyanobacterial Monitoring and Cyanotoxin Analysis*; John Wiley & Sons, Ltd.: Chichester, UK, 2017; pp. 526–537.
30. Bouaïcha, N.; Miles, C.O.; Beach, D.G.; Labidi, Z.; Djabri, A.; Benayache, N.Y.; Nguyen-Quang, T. Structural diversity, characterization and toxicology of microcystins. *Toxins* **2019**, *11*, 714. [CrossRef]
31. Tillett, D.; Dittmann, E.; Erhard, M.; von Döhren, H.; Börner, T.; Neilan, B.A. Structural organization of microcystin biosynthesis in *Microcystis aeruginosa* PCC7806: An integrated peptide-polyketide synthetase system. *Chem. Biol.* **2000**, *7*, 753–764. [CrossRef]
32. Eriksson, J.E.; Toivola, D.; Meriluoto, J.; Karaki, H.; Han, Y.G.; Hartshorne, D. Hepatocyte deformation induced by cyanobacterial toxins reflects inhibition of protein phosphatases. *Biochem. Biophys. Res. Commun.* **1990**, *173*, 1347–1353. [CrossRef]

33. MacKintosh, C.; Beattie, K.A.; Klumpp, S.; Cohen, P.; Codd, G.A.; MacKintosh, C.; Codd, G.A.; Klumpp, S.; Beattie, K.A. Cyanobacterial microcystin-LR is a potent and specific inhibitor of protein phosphatases 1 and 2A from both mammals and higher plants. *FEBS Lett.* **2002**, *264*, 187–192. [CrossRef]

34. MacKintosh, R.W.; Dalby, K.N.; Campbell, D.G.; Cohen, P.T.; Cohen, P.; MacKintosh, C. The cyanobacterial toxin microcystin binds covalently to cysteine-273 on protein phosphatase 1. *FEBS Lett.* **1995**, *371*, 236–240. [CrossRef]

35. Feurstein, D.J.; Holst, K.; Fischer, A.; Dietrich, D.R. Oatp-associated uptake and toxicity of microcystins in primary murine whole brain cells. *Toxicol. Appl. Pharmacol.* **2009**, *234*, 247–255. [CrossRef]

36. Fischer, A.; Hoeger, S.J.; Stemmer, K.; Feurstein, D.J.; Knobeloch, D.; Nussler, A.; Dietrich, D.R. The role of organic anion transporting polypeptides (OATPs/SLCOs) in the toxicity of different microcystin congeners in vitro: A comparison of primary human hepatocytes and OATP-transfected HEK293 cells. *Toxicol. Appl. Pharmacol.* **2010**, *245*, 9–20. [CrossRef] [PubMed]

37. Catherine, A.; Quiblier, C.; Yéprémian, C.; Got, P.; Groleau, A.; Vinçon-Leite, B.; Bernard, C.; Troussellier, M. Collapse of a *Planktothrix agardhii* perennial bloom and microcystin dynamics in response to reduced phosphate concentrations in a temperate lake. *FEMS Microbiol. Ecol.* **2008**, *65*, 61–73. [CrossRef] [PubMed]

38. Yéprémian, C.; Gugger, M.F.; Briand, E.; Catherine, A.; Berger, C.; Quiblier, C.; Bernard, C. Microcystin ecotypes in a perennial *Planktothrix agardhii* bloom. *Water Res.* **2007**, *41*, 4446–4456. [CrossRef] [PubMed]

39. WHO. *Cyanobacterial Toxins: Microcystins—Background Document for Development of WHO Guidelines for Drinking-Water Quality and Guidelines for Safe Recreational Water Environments*; WHO: Geneva, Switzerland, 2020.

40. Hoeger, S.J.; Dietrich, D.R.; Hitzfeld, B.C. Effect of ozonation on the removal of cyanobacterial toxins during drinking water treatment. *Environ. Health Perspect.* **2002**, *110*, 1127–1132. [CrossRef]

41. Hoeger, S.J.; Shaw, G.; Hitzfeld, B.C.; Dietrich, D.R. Occurrence and elimination of cyanobacterial toxins in two Australian drinking water treatment plants. *Toxicon* **2004**, *43*, 639–649. [CrossRef]

42. IGKB. *Bericht Nr. 42: Limnologischer Zustand des Bodensees*; Internationale Gewässerschutzkommission für den Bodensee (IGKB): Baden, Germany, 2018.

43. Konstanz, S. Burgunderblut-Alge breitet sich im Bodensee aus. *Südkurier*, 28 March 2017; p. H1.

44. Nübel, U.; Garcia-Pichel, F.; Muyzer, G. PCR primers to amplify 16S rRNA genes from cyanobacteria. *Appl. Environ. Microbiol.* **1997**, *63*, 3327–3332. [CrossRef] [PubMed]

45. Ernst, A.; Becker, S.; Wollenzien, U.I.A.; Postius, C. Ecosystem-dependent adaptive radiations of picocyanobacteria inferred from 16S rRNA and ITS-1 sequence analysis. *Microbiology* **2003**, *149*, 217–228. [CrossRef] [PubMed]

46. Minh, B.Q.; Schmidt, H.A.; Chernomor, O.; Schrempf, D.; Woodhams, M.D.; von Haeseler, A.; Lanfear, R.; Teeling, E. IQ-TREE 2: New Models and Efficient Methods for Phylogenetic Inference in the Genomic Era. *Mol. Biol. Evol.* **2020**, *37*, 1530–1534. [CrossRef] [PubMed]

47. Kalyaanamoorthy, S.; Minh, B.Q.; Wong, T.K.F.; von Haeseler, A.; Jermiin, L.S. ModelFinder: Fast model selection for accurate phylogenetic estimates. *Nat. Methods* **2017**, *14*, 587–589. [CrossRef]

48. Ostermaier, V.; Kurmayer, R. Distribution and abundance of nontoxic mutants of cyanobacteria in lakes of the Alps. *Microb. Ecol.* **2009**, *58*, 323–333. [CrossRef] [PubMed]

49. Jungblut, A.D.; Neilan, B.A. Molecular identification and evolution of the cyclic peptide hepatotoxins, microcystin and nodularin, synthetase genes in three orders of cyanobacteria. *Arch. Microbiol.* **2006**, *185*, 107–114. [CrossRef] [PubMed]

50. Christiansen, G.; Fastner, J.; Erhard, M.; Börner, T.; Dittmann, E. Microcystin biosynthesis in *Planktothrix*: Genes, evolution, and manipulation. *J. Bacteriol.* **2003**, *185*, 564–572. [CrossRef] [PubMed]

51. Beversdorf, L.J.; Chaston, S.D.; Miller, T.R.; McMahon, K.D. Microcystin mcyA and mcyE gene abundances are not appropriate indicators of microcystin concentrations in lakes. *PLoS ONE* **2015**, *10*, e0125353. [CrossRef] [PubMed]

52. Hingsamer, P.; Peeters, F.; Hofmann, H. The consequences of internal waves for phytoplankton focusing on the distribution and production of *Planktothrix rubescens*. *PLoS ONE* **2014**, *9*. [CrossRef]

53. Weisse, T. Dynamics of autotrophic picoplankton in lake Constance. *J. Plankton Res.* **1988**, *10*, 1179–1188. [CrossRef]

54. Weisse, T.; Kenter, U. Ecological Characteristics of Autotrophic Picoplankton in a Prealpine Lake. *Int. Rev. Gesamten. Hydrobiol. Hydrogr.* **1991**, *76*, 493–504. [CrossRef]

55. Six, C.; Thomas, J.C.; Garczarek, L.; Ostrowski, M.; Dufresne, A.; Blot, N.; Scanlan, D.J.; Partensky, F. Diversity and evolution of phycobilisomes in marine *Synechococcus* spp.: A comparative genomics study. *Genome Biol.* **2007**, *8*, R259. [CrossRef]

56. Salazar, V.W.; Tschoeke, D.A.; Swings, J.; Cosenza, C.A.; Mattoso, M.; Thompson, C.C.; Thompson, F.L. A new genomic taxonomy system for the *Synechococcus* collective. *Environ. Microbiol.* **2020**, *22*, 4557–4570. [CrossRef]

57. Tan, X.; Gu, H.; Ruan, Y.; Zhong, J.; Parajuli, K.; Hu, J. Effects of nitrogen on interspecific competition between two cell-size cyanobacteria: Microcystis aeruginosa and *Synechococcus* sp. *Harmful Algae* **2019**, *89*, 101661. [CrossRef] [PubMed]

58. Tan, X.; Gu, H.; Zhang, X.; Parajuli, K.; Duan, Z. Effects of phosphorus on interspecific competition between two cell-size cyanobacteria: *Synechococcus* sp. and *Microcystis aeruginosa*. *Bull. Environ. Contam. Toxicol.* **2019**, *102*, 231–238. [CrossRef]

59. IGKB. Limnologischer Zustand des Bodensees—Jahresbericht 2018/2019. In *Jahresbericht Int. Gewässerschutzkommission für den Bodensee—Grüner Bericht*; Internationale Gewässerschutzkommission für den Bodensee (IGKB): Baden, Germany, 2020; Volume 43.

60. Santhakumari, S.; Kannappan, A.; Pandian, S.K.; Thajuddin, N.; Rajendran, R.B.; Ravi, A.V. Inhibitory effect of marine cyanobacterial extract on biofilm formation and virulence factor production of bacterial pathogens causing vibriosis in aquaculture. *J. Appl. Phycol.* **2016**, *28*, 313–324. [CrossRef]

61. Verity, P.G.; Robertson, C.Y.; Tronzo, C.R.; Andrews, M.G.; Nelson, J.R.; Sieracki, M.E. Relationships between cell volume and the carbon and nitrogen content of marine photosynthetic nanoplankton. *Limnol. Oceanogr.* **1992**, *37*, 1434–1446. [CrossRef]

62. Ruber, J.; Geist, J.; Hartmann, M.; Millard, A.; Raeder, U.; Zubkov, M.; Zwirglmaier, K. Spatio-temporal distribution pattern of the picocyanobacterium *Synechococcus* in lakes of different trophic states: A comparison of flow cytometry and sequencing approaches. *Hydrobiologia* **2018**, *811*, 77–92. [CrossRef]

63. Bodensee Wasserversorgung. Bodensee Wasserversorgung—Startseite. Available online: https://www.bodensee-wasserversorgung.de/startseite.html (accessed on 7 May 2021).

64. Christiansen, G.; Kurmayer, R.; Liu, Q.; Börner, T. Transposons inactivate biosynthesis of the nonribosomal peptide microcystin in naturally occurring *Planktothrix* spp. *Appl. Environ. Microbiol.* **2006**, *72*, 117–123. [CrossRef]

65. Altaner, S.; Puddick, J.; Fessard, V.; Feurstein, D.; Zemskov, I.; Wittmann, V.; Dietrich, D.R. Simultaneous Detection of 14 Microcystin Congeners from Tissue Samples Using UPLC-ESI-MS/MS and Two Different Deuterated Synthetic Microcystins as Internal Standards. *Toxins* **2019**, *11*, 388. [CrossRef] [PubMed]

66. Bubak, I.; Śliwińska-Wilczewska, S.; Głowacka, P.; Szczerba, A.; Możdżeń, K. The importance of allelopathic picocyanobacterium *Synechococcus* sp. on the abundance, biomass formation, and structure of phytoplankton assemblages in three freshwater lakes. *Toxins* **2020**, *12*, 259. [CrossRef]

67. Kovács, A.A.W.; Tóth, V.R.; Pálffy, K. The effects of interspecific interactions between bloom forming cyanobacteria and *Scenedesmus quadricauda* (Chlorophyta) on their photophysiology. *Acta Biol. Hung.* **2018**, *69*, 210–223. [CrossRef]

68. Costa, M.S.; Costa, M.; Ramos, V.; Leão, P.N.; Barreiro, A.; Vasconcelos, V.; Martins, R. Picocyanobacteria from a Clade of Marine *Cyanobium* Revealed Bioactive Potential Against Microalgae, Bacteria, and Marine Invertebrates. *J. Toxicol. Environ. Health* **2015**, *78*, 432–442. [CrossRef]

69. Weisbrod, B. Dynamics of Toxic Cyanobacteria in Lakes and Artificial Water Reservoirs. Ph.D. Thesis, University of Konstanz, Konstanz, Germany, 2021.

70. Fernández Castro, B.; Sepúlveda Steiner, O.; Knapp, D.; Posch, T.; Bouffard, D.; Wüest, A. Inhibited vertical mixing and seasonal persistence of a thin cyanobacterial layer in a stratified lake. *Aquat. Sci.* **2021**, *83*, 38. [CrossRef]

71. Rusch, D.B.; Halpern, A.L.; Sutton, G.; Heidelberg, K.B.; Williamson, S.; Yooseph, S.; Wu, D.; Eisen, J.A.; Hoffman, J.M.; Remington, K.; et al. The Sorcerer II Global Ocean Sampling expedition: Northwest Atlantic through eastern tropical Pacific. *PLoS Biol.* **2007**, *5*, e77. [CrossRef]

72. William, S.; Helene, F.; Copeland, A. *Bacterial DNA Isolation CTAB Protocol Bacterial Genomic DNA Isolation Using CTAB*; JGI: Walnut Creek, CA, USA, 2012; p. 4.

73. Altschul, S.F.; Gish, W.; Miller, W.; Myers, E.W.; Lipman, D.J. Basic local alignment search tool. *J. Mol. Biol.* **1990**, *215*, 403–410. [CrossRef]

74. Schuurman, T.; de Boer, R.F.; Kooistra-Smid, A.M.D.; van Zwet, A.A. Prospective Study of Use of PCR Amplification and Sequencing of 16S Ribosomal DNA from Cerebrospinal Fluid for Diagnosis of Bacterial Meningitis in a Clinical Setting. *J. Clin. Microbiol.* **2004**, *42*, 734–740. [CrossRef]

75. Walters, W.; Hyde, E.R.; Berg-Lyons, D.; Ackermann, G.; Humphrey, G.; Parada, A.; Gilbert, J.A.; Jansson, J.K.; Caporaso, J.G.; Fuhrman, J.A.; et al. Improved Bacterial 16S rRNA Gene (V4 and V4-5) and Fungal Internal Transcribed Spacer Marker Gene Primers for Microbial Community Surveys. *mSystems* **2016**, *1*. [CrossRef]

76. Bolger, A.M.; Lohse, M.; Usadel, B. Trimmomatic: A flexible trimmer for Illumina sequence data. *Bioinformatics* **2014**, *30*, 2114–2120. [CrossRef]

77. Simon, A. Babraham Bioinformatics—FastQC A Quality Control Tool for High Throughput Sequence Data. Available online: https://www.bioinformatics.babraham.ac.uk/projects/fastqc/ (accessed on 14 May 2020).

78. Bolyen, E.; Rideout, J.R.; Dillon, M.R.; Bokulich, N.A.; Abnet, C.C.; Al-Ghalith, G.A.; Alexander, H.; Alm, E.J.; Arumugam, M.; Asnicar, F.; et al. Reproducible, interactive, scalable and extensible microbiome data science using QIIME 2. *Nat. Biotechnol.* **2019**, *37*, 852–857. [CrossRef]

79. Callahan, B.; McMurdie, P.; Rosen, M.; Han, A.; Johnson, A.; Holmes, S. DADA2: High resolution sample inference from amplicon data. *bioRxiv* **2016**, 24034. [CrossRef]

80. Gouy, M.; Guindon, S.; Gascuel, O. SeaView version 4: A multiplatform graphical user interface for sequence alignment and phylogenetic tree building. *Mol. Biol. Evol.* **2010**, *27*, 221–224. [CrossRef] [PubMed]

81. Yu, G. Using ggtree to Visualize Data on Tree-Like Structures. *Curr. Protoc. Bioinform.* **2020**, *69*, e96. [CrossRef] [PubMed]

82. R Foundation for Statistical Computing. *R Core Team R: A Language and Environment for Statistical Computing*; R Foundation for Statistical Computing: Vienna, Austria, 2018.

83. McMurdie, P.J.; Holmes, S. Phyloseq: An R Package for Reproducible Interactive Analysis and Graphics of Microbiome Census Data. *PLoS ONE* **2013**, *8*, e61217. [CrossRef] [PubMed]

84. Oksanen, J.; Blanchet, F.G.; Friendly, M.; Kindt, R.; Legendre, P.; McGlinn, D.; Minchin, P.R.; O'Hara, R.B.; Simpson, G.L.; Solymos, P.; et al. *Vegan: Community Ecology Package*; R Package Version 2.5-6; R Foundation for Statistical Computing: Vienna, Austria, 2016.

85. Wickham, H. *ggplot2: Elegant Graphics for Data Analysis*; Springer: New York, NY, USA, 2016; ISBN 978-3-319-24277-4.

86. McMurdie, P.J.; Holmes, S. Waste Not, Want Not: Why Rarefying Microbiome Data Is Inadmissible. *PLoS Comput. Biol.* **2014**, *10*, e1003531. [CrossRef] [PubMed]

87. Zemskov, I.; Altaner, S.; Dietrich, D.R.; Wittmann, V. Total Synthesis of Microcystin-LF and Derivatives Thereof. *J. Org. Chem.* **2017**, *82*, 3680–3691. [CrossRef] [PubMed]
88. MathWorks. *MATLAB*; Version R2020a; MathWorks: Natick, MA, USA, 2020.

Article

Toxins and Other Bioactive Metabolites in Deep Chlorophyll Layers Containing the Cyanobacteria *Planktothrix* cf. *isothrix* in Two Georgian Bay Embayments, Lake Huron

Arthur Zastepa [1,*], Todd R. Miller [2], L. Cynthia Watson [1], Hedy Kling [3] and Susan B. Watson [4]

[1] Environment and Climate Change Canada, Canada Centre for Inland Waters, 867 Lakeshore Road, Burlington, ON L7S 1A1, Canada; linet.watson@canada.ca

[2] Joseph J. Zilber School of Public Health, University of Wisconsin-Milwaukee, Milwaukee, WI 53211, USA; millertr@uwm.edu

[3] Algal Taxonomy and Ecology Inc., P.O. Box 761, Stony Mountain, MB R0C 3A0, Canada; hedy.kling8@gmail.com

[4] School of Graduate Studies, Environmental and Life Sciences, Trent University, Peterborough, ON K9L 0G2, Canada; sboydwatson@gmail.com

* Correspondence: arthur.zastepa@canada.ca

Citation: Zastepa, A.; Miller, T.R.; Watson, L.C.; Kling, H.; Watson, S.B. Toxins and Other Bioactive Metabolites in Deep Chlorophyll Layers Containing the Cyanobacteria *Planktothrix* cf. *isothrix* in Two Georgian Bay Embayments, Lake Huron. *Toxins* **2021**, *13*, 445. https://doi.org/10.3390/toxins13070445

Received: 26 March 2021
Accepted: 23 June 2021
Published: 27 June 2021

Publisher's Note: MDPI stays neutral with regard to jurisdictional claims in published maps and institutional affiliations.

Abstract: The understanding of deep chlorophyll layers (DCLs) in the Great Lakes—largely reported as a mix of picoplankton and mixotrophic nanoflagellates—is predominantly based on studies of deep (>30 m), offshore locations. Here, we document and characterize nearshore DCLs from two meso-oligotrophic embayments, Twelve Mile Bay (TMB) and South Bay (SB), along eastern Georgian Bay, Lake Huron (Ontario, Canada) in 2014, 2015, and 2018. Both embayments showed the annual formation of DCLs, present as dense, thin, metalimnetic plates dominated by the large, potentially toxic, and bloom-forming cyanobacteria *Planktothrix* cf. *isothrix*. The contribution of *P.* cf. *isothrix* to the deep-living total biomass (TB) increased as thermal stratification progressed over the ice-free season, reaching 40% in TMB (0.6 mg/L at 9.5 m) and 65% in South Bay (3.5 mg/L at 7.5 m) in 2015. The euphotic zone in each embayment extended down past the mixed layer, into the nutrient-enriched hypoxic hypolimnia, consistent with other studies of similar systems with DCLs. The co-occurrence of the metal-oxidizing bacteria *Leptothrix* spp. and bactivorous flagellates within the metalimnetic DCLs suggests that the microbial loop plays an important role in recycling nutrients within these layers, particularly phosphate (PO_4) and iron (Fe). Samples taken through the water column in both embayments showed measurable concentrations of the cyanobacterial toxins microcystins (max. 0.4 µg/L) and the other bioactive metabolites anabaenopeptins (max. ~7 µg/L) and cyanopeptolins (max. 1 ng/L), along with the corresponding genes (max. in 2018). These oligopeptides are known to act as metabolic inhibitors (e.g., in chemical defence against grazers, parasites) and allow a competitive advantage. In TMB, the 2018 peaks in these oligopeptides and genes coincided with the *P.* cf. *isothrix* DCLs, suggesting this species as the main source. Our data indicate that intersecting physicochemical gradients of light and nutrient-enriched hypoxic hypolimnia are key factors in supporting DCLs in TMB and SB. Microbial activity and allelopathy may also influence DCL community structure and function, and require further investigation, particularly related to the dominance of potentially toxigenic species such as *P.* cf. *isothrix*.

Keywords: deep-chlorophyll layers (DCLs); cyanobacterial toxins; *Planktothrix*; allelopathy; bioactive metabolites; hypoxia; Georgian Bay

Key Contribution: Deep chlorophyll layers of *Planktothrix* cf. *isothrix* in two meso-oligotrophic embayments in Georgian Bay showed measurable concentrations of the cyanobacterial toxins microcystins (max. 0.4 µg/L) and the other bioactive metabolites anabaenopeptins (max. ~7 µg/L) and cyanopeptolins (max. 1 ng/L)—along with the corresponding genes—that can act as metabolic inhibitors (e.g., in chemical defence against grazers, parasites) and allow a competitive advantage. Supporting physicochemical data provide strongs evidence that opposing and intersecting physicochemical gradients of light and nutrients from a hypoxic hypolimnion are critical in supporting the DCLs in these embayments. Microbial processes and allelopathy may also influence the DCL community structure and function and the dominance of large, toxigenic cyanobacteria such as *P.* cf. *isothrix*.

1. Introduction

Deep chlorophyll layers (DCLs) are ecologically important but often overlooked phenomena in many deep lentic ecosystems. DCLs can account for a significant fraction of pelagic primary productivity, play a key role in biogeochemical cycling of nutrients, and influence the vertical movement of micrograzers [1–7]. DCLs are likely far more prevalent than reported, as most sampling efforts concentrate on surface water layers. Investigations of large and deep Canadian waters such as the Great Lakes, where offshore DCLs have been observed as deep as ca. 50 m, have documented assemblages dominated by diatoms, picoplankton, and cryptophytes, along with the presence of other flagellated autotrophs/mixotrophs (e.g., chrysophytes, dinoflagellates) [8–11]. Previous studies have identified the importance of light (euphotic depth, Z_{eu}) and thermal stratification in predicting the depth and thickness of deep-living phytoplankton communities, but less is known about the influence of chemical (e.g., nutrient) gradients and even less about the chemical ecology of this deep-living biota [4,7,12–14]. Several mechanisms can produce chemical gradients conducive to DCL formation and progression, including upwelling, inter-flow, groundwater, and diffusion from surficial sediments, i.e., internal loading (e.g., [15]). Internal loading often occurs in productive, stratified systems within a hypoxic hypolimnion and can provide a supply of bioavailable dissolved nutrients that are directly accessible to deep-living phytoplankton particularly in shallow waterbodies where $Z_{mix} < Z_{eu}$ [16]. Studies of DCLs in smaller and shallower (<30 m) lakes in Canada report flagellated autotrophs/mixotrophs, e.g., chrysophytes as being the most prominent [1,2,6,17]. However, a more recent study documents DCLs dominated by large cyanobacteria in two nearshore embayments of eastern Georgian Bay (Lake Huron, Ontario, Canada) (e.g., [16]).

Earlier work along the eastern coast of Georgian Bay reported DCLs in a number of sheltered embayments of intermediate size and depth (<30 m), many of which regularly undergo thermal stratification and establish pronounced chemical gradients with depth, [16]. Building on this research, we carried out two nearshore surveys (2014 and 2015) of 15 embayments along this coastline to further investigate the presence of DCLs, and selected two of these waterbodies for more focused work. In 2015, detailed measurements were done in these two embayments to characterize the seasonality and vertical structure of each documented DCL and its phytoplankton composition as well as the coincident physicochemical and nutrient gradients relevant to phytoplankton physiology. We also measured a suite of bioactive metabolites that can play a role in cyanobacterial dominance (the cyanobacterial toxins microcystins, nodularin, anatoxins, saxitoxins, and cylindrospermopsins, and several other metabolic inhibitors known to act at the food web level, anabaenopeptins, cyanopeptolins, and microginins) [13]. Samples from the DCL sites were also obtained in 2018 for additional investigations of these bioactive metabolites, including genetic screening. While the 'toxins' have been widely measured, studies of other metabolic inhibitors have been largely limited to Europe with a few recent reports from North America [13,18,19]. To our knowledge, however, none of these other bioactive metabolites have been characterized in DCLs.

2. Results and Discussion

DCLs were documented in 2 of 15 embayments, Twelve Mile Bay (TMB) and South Bay (SB), during a spatial survey conducted in 2014 (Figure 1, Table 1). In 2015, the full survey was repeated in addition to a more detailed seasonal sampling of TMB and SB (Figure 2) where profiles and discrete depth samples were collected to characterize the vertical structure of phytoplankton composition (Table S1 in Supplementary Materials), the coincident physicochemical and nutrient gradients (Figure 3, Table 2), and a suite of bioactive metabolites (Table 3). In 2018, additional samples were collected from the two DCL sites to confirm previous results on bioactive metabolites (Table 3) and construct vertical water column profiles of corresponding genes (Figure 4).

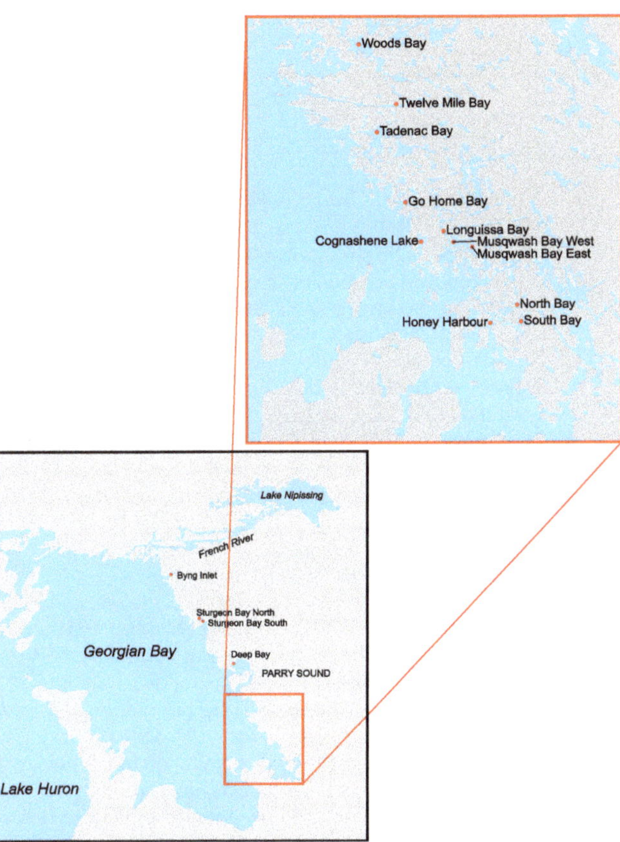

Figure 1. Locations of study sites in Georgian Bay, Lake Huron (Ontario, Canada) denoted by red circles.

Table 1. Depth of mixed (Z_{mix}) and euphotic zones (Z_{eu}) during mixed and stratified conditions and presence (Y: Yes/N: No) of deep-chlorophyll layer (DCL) in different embayments of Georgian Bay in the spring (June) and summer (August) of 2015. During mixed conditions in spring (June), Z_{mix} was equivalent to the site depth (Z_{site}).

	Mixed			Stratified		
Site	**Z_{site} (m)**	**Z_{eu} (m)**	**Z_{eu} (m)**	**Z_{mix} (m)**	**$Z_{eu}:Z_{mix}$**	**DCL (m)**
Byng Inlet	5	3.0	-	5	-	N
Sturgeon Bay North	14	2.4	4.5	8	0.6	N
Sturgeon Bay South	6	-	-	-	-	N
Deep Bay	15	2.7	3.7	8	0.5	N [a]
Woods Bay	8	-	-	-	-	N
Twelve Mile Bay	12	3.4	9.1	7	1.3	Y (~9)
Tadenac Bay	24	6.5	5.3	11	0.5	N
Go Home Bay	10	3.6	-	-	-	N
Longuissa Bay	9	3.8	4.7	8	0.6	N

Table 1. *Cont.*

Site	Z_{site} (m)	Mixed	Stratified			
		Z_{eu} (m)	Z_{eu} (m)	Z_{mix} (m)	Z_{eu}: Z_{mix}	DCL (m)
Cognashene Lake	16	3.9	9.5	7	1.4	N
Musqwash Bay West	30	-	-	-	-	N
Musqwash Bay East	20	-	-	-	-	N
North Bay	18	3.2	6.4	8	0.8	N
South Bay	10	3.9	5.7	6	1.0	Y (~7)
Honey Harbour	7	4.6	3.6	7	0.5	N

[a] Not detected during this study but reported in 2012 and 2014 by [16].

2.1. Spatiotemporal Changes in Phytoplankton Composition in Near-Surface and Deep Chlorophyll Layers in Twelve Mile Bay

Overall, TMB phytoplankton showed seasonality and a near-surface biomass, which was composed of a diverse community of flagellates, diatoms and picoplankton, typical of many meso-oligotrophic zones of the Great Lakes (e.g., [20]). While some of these near-surface taxa were also present in the deeper strata, DCLs were often composed of a distinct community, and dominated by trichomes of the cyanobacteria *P.* cf. *isothrix*, which was never observed in near-surface samples. In the summer of 2014, TMB was stratified with a distinct DCL at 10 m depth at the metalimnion, exceeding a total biomass (TB) of 1600 µg/L, almost double that in the epilimnion (Table S1). The DCL was dominated by the buoyancy-regulating cyanobacteria *Planktothrix* cf. *isothrix* (>1300 µg/L, 80% TB), with pennate diatoms (7% TB) and cryptophytes (4% TB) also abundant (Table S1). This deep-living community differed markedly from the epilimnetic assemblage, which was dominated by large flagellates—dinoflagellates (13% TB), cryptophytes (24% TB) and colonial chrysophytes (10% TB).

By autumn 2014, lake turnover brought on isothermal conditions and a more uniform vertical distribution of TB (~1000 µg/L) and composition. Both near-surface (1 m) and deep (10 m) samples had a diverse community dominated by large pennate diatoms (*Fragilaria crotonensis*, up to ~40% TB) and chrysophytes (up to ~40% TB; *Dinobryon sertularia*, *Synura* sp., ochromonads). The deep-living population of *P.* cf. *isothrix* seen during the summer was reduced significantly (<50 µg/L, 4% TB) and, again, this species was not observed near-surface. We also noted a small population of the nuisance flagellate *Gonyostomum semens* (<40 µg/L, 4% TB) near-surface. While commonplace in low numbers in softwater lakes, this raphidophyte can produce periodic blooms in boreal lakes (e.g., [21]) and outbreaks have been linked to high Fe availability [22].

During the winter ice cover of 2015, low near-surface biomass (<200 µg/L) was dominated by chrysoflagellates (55% TB), cryptophytes (24% TB) and dinoflagellates (3% TB), with low numbers of cyanobacteria also present (7% TB; *Microcystis* sp., *Planktothrix* sp.) (Figure 2, Table S1). Near-bottom (10 m) samples showed a sparse biomass (<40 µg/L), composed of mixotrophic chrysophytes (16% TB) and the large heterotrophic dinoflagellate *Gymnodinium helveticum* (14% TB), along with filaments of *Planktothrix* cf. *isothrix* (10% TB) and the diatom *Aulacoseira subarctica* (7% TB).

In spring 2015, there was a significant increase in biomass, particularly near-surface (>3300 µg/L TB) (Figure 2, Table S1). Near-surface (1 m) and deep (10 and 11.5 m) samples were all dominated by the large diatom *Fragilaria crotonensis* (up to 80% TB); large chrysoflagellates (up to 10% TB) were also important while cryptophytes (10% TB) were only present closest to the bottom (11.5 m). Low numbers of the colonial cyanobacteria *Aphanothece minutissima* and *Microcystis* sp. were present in the two deeper samples, but *Planktothrix* was not detected.

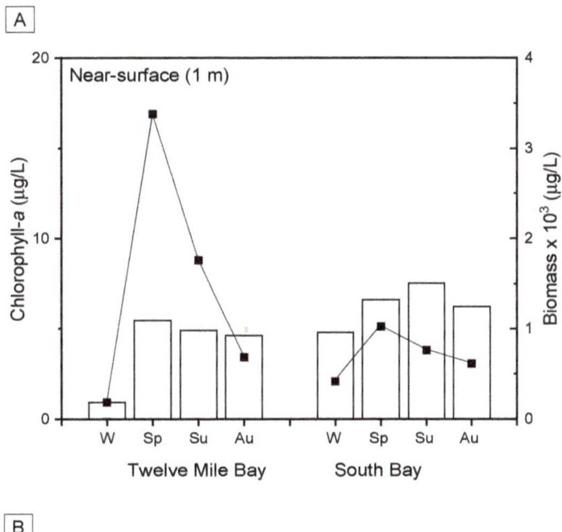

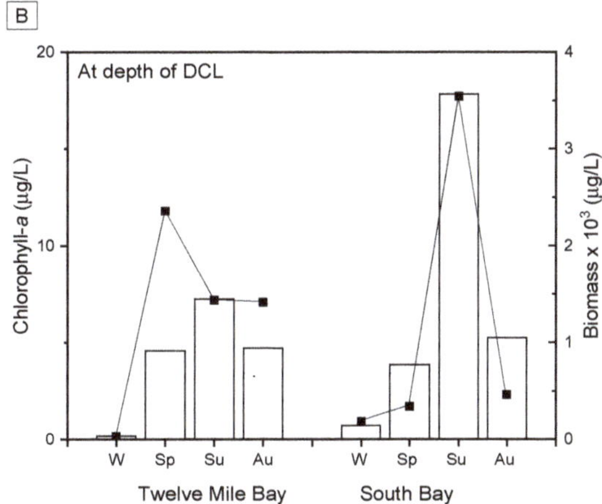

Figure 2. Seasonal (W, winter; Sp, spring; Su, summer; Au, autumn) changes in total phytoplankton biomass (line series) and extracted chlorophyll-*a* concentration (bars) in 2015. Near-surface (1 m) water column measurements are in panel (**A**) and at the depth of the recurring metalimnetic peaks, i.e., deep chlorophyll layers (DCLs) (~9 m in Twelve Mile Bay (TMB) and ~7 m in South Bay (SB), see Table 1 and Figure 3) are in panel (**B**). In the absence of DCLs during winter (W) and spring (Sp), i.e., during isothermal conditions, the sample was taken at the depth corresponding to summer DCL peak for comparison, i.e., ~9 m in TMB and ~7 m in SB. See Materials and Methods for sampling dates.

Thermal profiles in summer of 2015 revealed a stratified water column, with a mixed/epilimnetic layer (down to ~7 m) overlaying a steep thermocline to the bottom and no discernible hypolimnion (Figure 3). Epilimnetic biomass (>1700 µg/L) was again dominated by diatoms (65% TB; notably *F. crotonensis*), with large thecate dinoflagellates (13% TB) also being important (Table S1). Cyanobacteria constituted only a small fraction (14%) of this biomass, largely as colonial picocyanobacteria (*Aphanocapsa* sp., *Aphanothece minutissima*) and a few trichomes of the diazotrophic cyanobacteria *Dolichospermum* cf. *fuscum*. At depth, the phytoplankton community segregated into two distinct DCLs (~1 m apart), each showing strong peaks (>20 µg/L) in chlorophyll-*a* fluorescence (Figure 3). The more prominent DCL was concentrated within a ~1 m thick layer at the meta-hypolimnetic boundary (~9 m deep). FP mapping estimated that this DCL extended horizontally over an elliptical area of almost 0.2 km^2 until the end of the first (~9 m) contour (Figure 3,

Figure A1). This biomass peak (>1400 µg/L) was dominated by cryptophytes (50%TB; *C. reflexa*), diatoms (13% TB; possibly settling from the epilimnion) and *P.* cf. *isothrix* (8% TB) (Figure 2, Table S1). While water samples from the smaller DCL at 10 m could not be obtained, FP fluorescence indicated an assemblage dominated by "brown/golden" phytoplankton (e.g., dinoflagellates, diatoms, chrysophytes) (Figure 3).

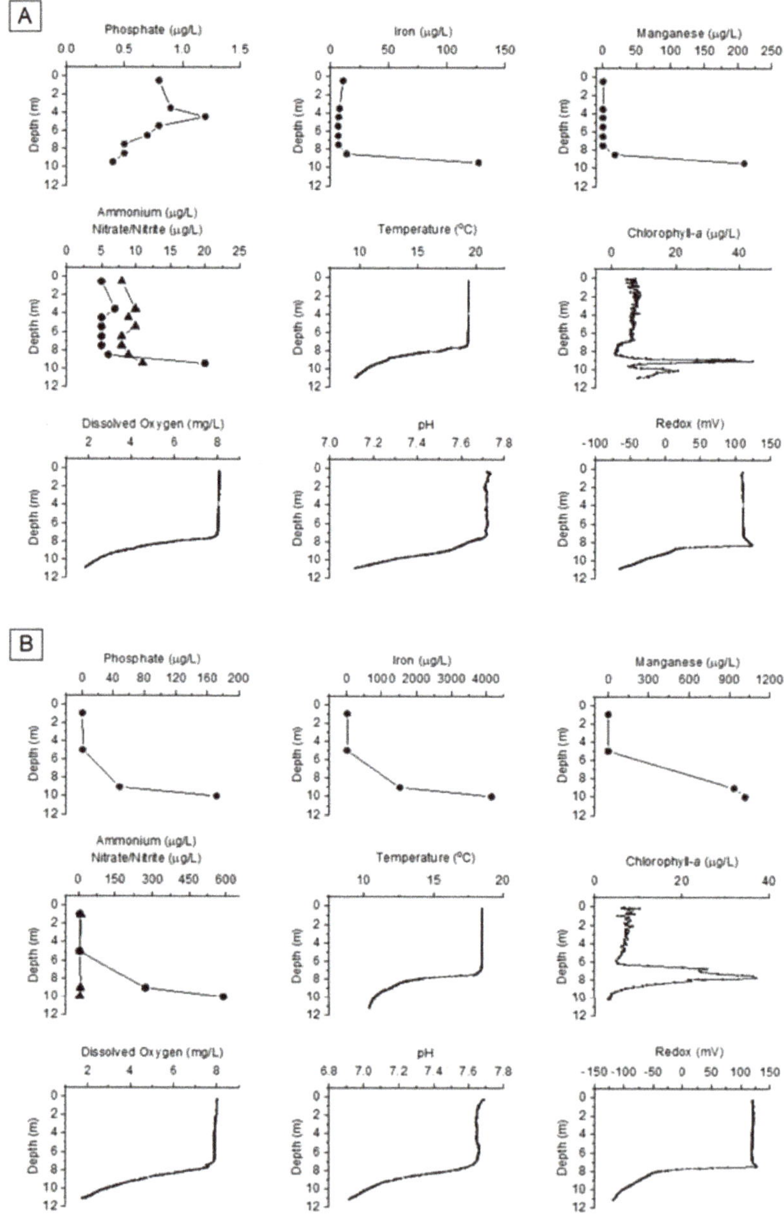

Figure 3. Water column profiles in Twelve Mile Bay (panel **A**) and South Bay (panel **B**) for: phosphate (PO$_4$), iron (Fe), manganese, ammonium (NH$_4$) (•), nitrite/nitrate (NO$_{2/3}$) (▲), temperature, chlorophyll-*a*, dissolved oxygen, pH, and redox in summer (August) of 2015. See Materials and Methods section for measurement and analytical protocols used.

In the autumn of 2015, the water column showed a deepened mixed/epilimnetic layer and a single metalimnetic DCL at ~9 m. Near-surface biomass had decreased to <700 µg/L and was dominated by large mixotrophic and non-mixotrophic chrysophytes (38% TB) and thecate dinoflagellates (14% TB) (Figure 2, Table S1). Small populations of colonial picocyanobacteria were also present but *Planktothrix* was not observed. The DCL contained double the near-surface biomass (>1400 µg/L) and was dominated by *Dinobryon* sp. (70% TB); *Peridinium* sp. (9% TB) and *P.* cf. *isothrix* (8% TB) were also important. A sample taken just above the sediments at 10 m showed a significantly lower total biomass (<700 µg/L) but an increased proportion of *P.* cf. *isothrix* (40% TB) (Figure 2, Table S1).

Although *P.* cf. *isothrix* was absent near-surface in TMB throughout all sampling, it was present in samples from both the sediment-water interface and surficial sediments throughout the seasons. Sediments—particularly those collected in winter—yielded (on qualitative inspection) viable trichomes of *P.* cf. *isothrix* under laboratory conditions simulating the summer metalimnion (Z8 media, <15 °C and <50 µmol m^{-2} s^{-1}). This suggests overwintering filaments could act as a reservoir of viable cells for the metalimnetic populations of this species. Recruitment of viable, vegetative cyanobacteria cells overwintering in sediments has been observed elsewhere with bloom-forming cyanobacteria, notably *Oscillatoria* (syn. *Planktothrix*), *Dolichospermum* (syn. *Anabaena*) and *Microcystis* [23–26]; however, in these cases, the benthic cyanobacteria were implicated in supporting surface blooms rather than DCLs. In TMB, our observations suggest that the sediments serve as a reservoir of viable *P.* cf. *isothrix* cells which seed the DCLs, and to our knowledge remain at depth and do not appear in any detectable numbers at or near the surface. While our evidence is supportive, a more direct measurement of this recruitment process is needed to confirm our hypothesis that sediment seeding may play an important role in the origins, seasonal dynamics, and fate of cyanobacteria-dominated DCLs in TMB.

2.2. Spatiotemporal Changes in Phytoplankton Composition Near-Surface and within Deep-Chlorophyll Layers in South Bay

In summer of 2014, SB was stratified with a distinct metalimnetic DCL at 6.5 m, reaching >60 µg/L chlorophyll-*a*—six times that measured near-surface (1 m)—of which ~87% was derived from cyanobacteria (based on FP fluorescence). Although a detailed taxonomic identification and enumeration of the summer 2014 DCL was not possible, results from subsequent years (reported below) suggest *P.* cf. *isothrix* as a prominent and recurring member of these deep phytoplankton communities in SB. Near-surface (1 m) summer biomass in 2014 was ~1000 µg/L, and dominated by chrysophytes (35% TB), dinoflagellates (10% TB), and picocyanobacteria (7% TB) (Table S1). Large cyanobacteria were present but in very low biomass (~3% TB; notably *Dolichospermum*).

By autumn 2014, lake turnover brought on isothermal conditions and biomass had concentrated near-surface, increasing over three times to ~3000 µg/L; dominated by large diazotrophic cyanobacteria (*Aphanizomenon flos aquae* complex, 45% TB; *Dolichospermum planctonicum*, 4% TB), with *Euglena* sp. (9% TB), *Aulacoseira ambigua* (9% TB), and *Plagioselmis nanoplanktica* (8% TB) also being abundant (Table S1).

During the winter ice cover of 2015, biomass near-surface (>400 µg/L at 1 m) was dominated by large colonial, scaled chrysoflagellates (*Synura* sp., 77% TB), with small populations (<5% each) of dinoflagellates and picocyanobacteria (Table S1). In contrast, a much lower biomass at ~8 m depth (<200 µg/L) was composed of picocyanobacteria (*Synechococcus spp.*, 45% TB), chrysophytes (37% TB), and dinoflagellates (9% TB).

By spring of 2015, under isothermal conditions, near-surface (1 m) biomass exceeded 1000 µg/L; dominated by a variety of chrysoflagellates (37% TB), cryptophytes (17% TB; mostly *C. reflexa*) and the large, horned dinoflagellate *Ceratium furcoides* (15% TB) (Figure 2, Table S1). At 5.5 m depth, a comparable biomass (>1000 µg/L) had an assemblage of scaled chrysophytes (36% TB; e.g., *Chrysosphaerella longispina*) and large dinoflagellates (*Ceratium* sp., 21% TB) with a few diazotrophic cyanobacteria (*Dolichospermum lemmermanni*, *Aphanizomenon flos-aquae*) also present. Near-bottom (10.5 m) samples had a less biomass

(<400 µg/L), composed mostly of *Cryptomonas* spp. (32% TB), *Chlamydomonas* sp. (14% TB), and *P. cf. isothrix* (9% TB).

Profiles in summer of 2015 revealed a stratified water column with near-surface (1 m) biomass (~800 µg/L) dominated by filamentous diatoms (34% TB; *Aulacoseira*), thecate dinoflagellates (15% TB), and the diazotroph *D. planktonicum* (12% TB) (Figure 2, Table S1). A distinct and sizable DCL (>3500 µg/L) was concentrated within a ~1 m thick layer at the meta-hypolimnetic boundary (~7 m). FP mapping estimated that the DCL extended horizontally over an elliptical area of almost 0.03 km^2, extending over an area bounded by the ~10 m depth contour (Figure 3, Figure A1). This deep biomass peak was dominated by *P. cf. isothrix* (65%); other cyanobacteria were also present (*Aph. flos-aquae* complex, 6% TB; *Microcystis novacekii*, 3% TB) along with dinoflagellates (7% TB) and diatoms (4% TB).

By autumn 2015, lake turnover brought on isothermal conditions and a more uniform vertical distribution of phytoplankton, i.e., no DCL was observed. A similar sized surface biomass (~600 µg/L) showed a shift in community composition from the summer assemblage towards an increased proportion of diazotrophic cyanobacteria (*D. planktonicum*, 32% TB; *Aph. flos-aquae*, 6% TB) and a decrease in the filamentous diatom *Aulacoseira* (16% TB) (Figure 2, Table S1). In the near-bottom sample (~9 m), a comparable biomass (~500 µg/L) showed similar composition (*Aulacoseira*, 17% TB; *D. planktonicum*, 15% TB). As seen in summer, *P. cf. isothrix* was again present at depth but at comparably lower proportion (9%), but other cyanobacteria (*Pseudanabaena* sp. 8% TB; *Aph. flos-aquae* complex, 7% TB) were also observed.

We note that while numerous other studies have reported DCLs that have been mixed into the epilimnion by physical processes such as turnover, storms, or artificial aeration/mixing and subsequently manifested as ephemeral surface blooms [4,18,27–32], there have been no reports of surface blooms in TMB nor SB near the locations of the DCLs, despite the high density of cottagers in the area. This suggests that the thick metalimnetic plates of *Planktothrix* remain essentially segregated from the surface community and are seeded and sustained by the physicochemical conditions in the bottom layers and sediments.

2.3. Light and Pigmentation

Light is a major factor which often controls phytoplankton vertical structure and the formation of deep-living communities (e.g., [7]), and many taxa modify or supplement their effective light regimes via vertical migration, mixotrophy, and photoadaptation. The abundance of large flagellated mixotrophs in the deep layers in TMB and SB (Table S1) suggests that motility and bactivory may provide access to alternative supplies of energy and nutrients (e.g., [17]); however, these photophagotrophs rarely predominated the DCLs in these embayments. PAR measures showed that the DCLs were positioned within the euphotic zone (Table 1), and we therefore conclude that in situ light levels were sufficient to support the development of *P. cf. isothrix*, i.e., light availability was not a primary factor limiting these DCL communities (e.g., [10,33]). Although current literature suggests phycoerythrin-rich *P. rubescens* is more often the predominant *Planktothrix* species in deep-living maxima, the predominant species *P. cf. isothrix* remained green-pigmented, suggesting minimal cellular phycoerythrin (Figure 2, Figure A2). We are aware of only one other published report of a DCL dominated by green-pigmented *Planktothrix* [3] although broader geographical surveys are needed [9,14,34–38].

2.4. Nutrients and Physicochemical Conditions

In SB, DCLs dominated by *P. cf. isothrix* were located at the metalimnion where sufficient PAR intersected with elevated levels of major nutrients and trace metals, particularly PO_4, dissolved inorganic nitrogen (DIN) and dissolved Fe (Table 1, Table 2, Figure 3). Vertical gradients of these chemical constituents increased with depth through the hypolimnion, indicative of internal loading from surficial sediments [16,39–41]. PO_4, Fe, NH_4, Mn and Co concentrations were up to three orders of magnitude higher near the sediments as conditions became hypoxic (dissolved oxygen <2 mg/L). Compared to TMB, bottom concentrations were generally higher and gradients were more pronounced in SB—which also exhibited a lower redox potential in the hypolimnion, particularly near sediments (-120 mV vs. -65 mV in TMB). Although similar gradients with depth through the hypolimnion were also evident in TMB for most measures taken, they were not as pronounced as in SB and this, along with markedly lower PO_4, is consistent with the significant difference in DCL biomass between the two embayments (Figures 2 and 3). Interestingly, the sheathed bacteria *Leptothrix* were observed in the DCL of TMB but not SB (Figure A2). These chemoorganotrophs exploit DOC-rich environments in the presence of Fe^{2+} and Mn^{2+}, which they precipitate as ferrous hydroxide and manganese oxide at the interface between oxic/anoxic zones, often co-precipitating PO_4 and making it inaccessible to phytoplankton (e.g., [42,43]). Their PO_4-precipitating activity in TMB may, to some extent, account for the minimal change in PO_4 with depth despite conditions conducive to internal loading but further study to quantify the significance of this mechanism is needed. Differences in sediment chemistry (e.g., sulphate/sulphide or aluminum) could also account for the contrasting PO_4 profiles but this has not yet been characterized nor quantified in the two embayments [39–41].

There is a considerable body of evidence to indicate that the development of opposing and intersecting vertical gradients of light and nutrients is strongly correlated with DCL formation in stratified, meso-oligotrophic systems [4,5,7,15,34,44]. Our survey showed that stratified Georgian Bay embayments lacking these opposing and intersecting vertical gradients also lacked discernible DCLs. Some, such as Sturgeon Bay, had a hypoxic hypolimnion and associated gradient in nutrients that reached the metalimnion but did not have sufficient light penetration to support a DCL, i.e., $Z_{eu} << Z_{mix}$ (Table 1). Others, such as Cognashene Lake (actually an embayment), had sufficient light at the metalimnion (Table 1) and elevated nutrients in the hypoxic hypolimnion (e.g., >1400 µg/L Fe, >600 µg/L Mn, >200 µg/L NH_4) but it was unclear if these opposing gradients intersected as we lacked sufficient resolution with depth. We speculate that the lake's greater depth and thicker hypolimnion rendered the hypolimnetic nutrients inaccessible to phytoplankton at sufficient PAR (Table 1); however, with more stable and prolonged periods of stratification, and associated hypoxia, as expected with climate change, the nutrient gradient would presumably strengthen.

Table 2. Nutrient concentrations (µg/L) and depth ratios in the hypolimnion (Hypo) and epilimnion (Epi) in each embayment (DON = dissolved organic nitrogen). Data are based on the summer 2015 sampling expedition during stratification.

Nutrient	Twelve Mile Bay			South Bay		
	Hypo	**Epi**	**Hypo:Epi**	**Hypo**	**Epi**	**Hypo:Epi**
P	15	14	1	232	13	18
Fe	127	8	16	4160	10	408
$NO_{2/3}$	11	9	1	6	8	0.8
Si	2.5	1.3	2	8	4	2
Mn	210	0.3	660	1020	0.5	2040
Co	0.2	0.01	20	0.3	0.01	26
NH_4	20	5	4	589	5.5	107
SO_4	10	11	1	2.3	7.7	0.3
DON	212	196	1	367	299	1
PO_4	0.4	0.8	0.5	172	0.7	246

Consistent with our observations, others have observed metalimnetic DCLs most commonly occuring at depths of less than 15 m, particularly those dominated by bloom-forming cyanobacteria [4,5,7,31,34,38,44–46]. Although significant DCLs dominated by potentially toxic cyanobacteria such as *Planktothrix* have been reported at depths below 15 m (e.g., [47]), this appears relatively rare [7,46]. DCLs existing deeper than 15 m (close to 30 m depth) are generally dominated by picoplankton, flagellated autotrophs/mixotrophs (e.g., large colonial chrysophytes, dinoflagellates), and diatoms [9–11,15]. The occurrence of these deeper DCLs has been documented in much larger (>19,000 km^2) and deeper systems (>60 m) including offshore Lake Ontario [10,11], Lake Michigan, Lake Superior, and Lake Huron [8,9] and in marine systems (e.g., [15]). The significant depth of these systems and corresponding thickness of the hypolimnion suggests upwelling of bottom waters [15] and recycling from biomass [1,6,48] as more likely sources of nutrients contributing to the DCLs—this, in contrast to the direct access metalimnetic DCLs in the relatively shallower TMB and SB have to the nutrient gradient generated from internal loading at the sediment-water interface. Although opposing and intersecting physicochemical gradients in nutrients and light are considered key drivers of DCL biomass formation in general; their effects on phytoplankton community composition and activity need further investigation.

2.5. Toxins Produced by Cyanobacteria

Over the course of sampling (2014, 2015, and 2018), near-surface summer samples from 5 of 15 Georgian Bay embayments showed both detectable levels of microcystins and the genetic potential to produce these hepatotoxins (positive for *mcy* gene). The highest concentration observed in any sample (0.4 µg L^{-1} in 2018) was well below the Canadian guidelines for safe drinking and recreational water (<1.5 µg/L; <20 µg/L respectively; Health Canada 2017) (Table 3, Figure 4). MC-LA was the most frequently observed microcystin variant and also present at higher concentrations than MC-LR and MC-RR. This is consistent with other reports suggesting MC-LA is more prevalent in systems with lower trophic status (e.g., [13,49]). In both TMB and SB, both the microcystin gene (*mcy*) and the microcystin toxins were present at the metalimnetic DCLs and in the hypolimnion. Based on our 2018 measurements, the *Planktothrix*-dominated DCL in TMB coincided with peaks in microcystin concentration and corresponding gene copies (*mcy*) (Figure 4, Table S1) suggesting these cyanobacteria as the primary source. Other species of *Planktothrix* are known for the production of a wide range of bioactive metabolites including microcystins [18]; however, production by *P.* cf. *isothrix* has yet to be directly confirmed.

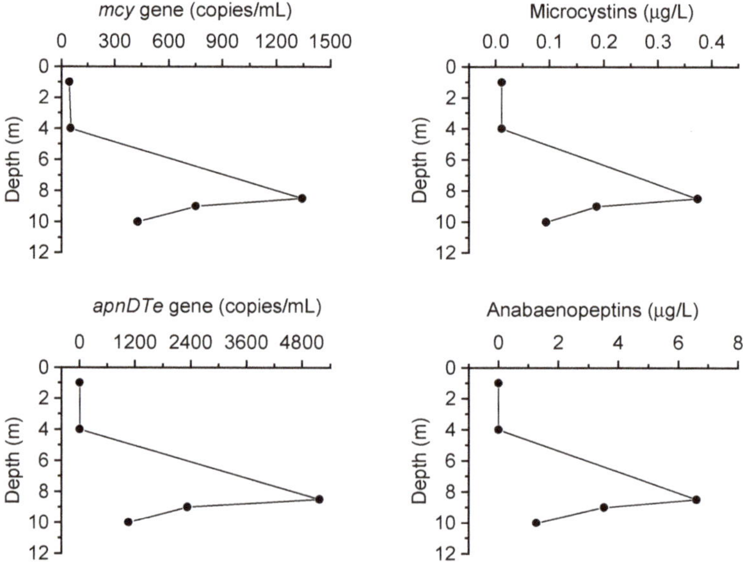

Figure 4. Vertical water column gradients of bioactive metabolites (microcystins and anabaenopeptins) produced by cyanobacteria and corresponding genes (*mcy* and *apnDTe*, respectively) in Twelve Mile Bay in summer (August) of 2018. Note that higher concentrations of these bioactive metabolites (and their genes) co-occur with biomass peaks, notably of *Planktothrix isothrix*, at depth (Table S1, 2018 data).

The presence of the *mcy* gene and trace amounts of microcystin in the hypolimnion of Tadenac Bay is also noteworthy (Table 3) as this embayment is relatively unimpacted (TP < 10 μg/L, TDP < 5 μg/L) with minimal shoreline development. The presence and low-level expression of the *mcy* gene in Tadenac Bay suggests an innate potential for toxin production despite the oligotrophic nature of this system (Table 3). None of the other toxins for which the samples were screened were detected, i.e., anatoxins, saxitoxins, cylindrospermopsins, nodularins (Table 3). In North Bay, a single sample showed the presence of low copy numbers of the saxitoxin gene (*sxtA*), but not the toxin.

Table 3. Major bioactive metabolites (right column) [1] and corresponding genes (left column) [2] (detect "+"; non-detect "-"; not measured "NM"; or maximum concentration (μg/L), any depth) in water samples from embayments along eastern Georgian Bay during 2014, 2015, and 2018. MC: Microcystins, NOD: Nodularin, CYN: Cylindrospermopsins, STX: Saxitoxins, ATX: Anatoxins, APT: Anabaenopeptins, CPT: Cyanopeptolins, MG: Microginins. NOD, CYN, and ATX were not detected in any of the samples.

Sites	MCs		STXs		APTs		CPTs		MGs	
Byng Inlet	-	-	-	-	-	-	-	-	-	-
Sturgeon Bay North	+	0.01	-	-	-	-	-	-	-	-
Sturgeon Bay South	NM	NM	NM	NM	NM	NM	NM	NM	NM	NM
Deep Bay	-	-	-	-	-	-	-	-	-	-
Woods Bay	NM	NM	NM	NM	NM	NM	NM	NM	NM	NM
Twelve Mile Bay [3]	+	0.4	-	-	+	6.6	+	0.001	-	-
Tadenac Bay	+	0.01	-	-	-	-	-	-	-	-
Go Home Bay	-	-	-	-	-	-	-	-	-	-
Longuissa Bay	-	-	-	-	-	-	-	-	-	-
Cognashene Lake	-	-	-	-	-	-	-	-	-	-
Musqwash Bay West	NM	NM	NM	NM	NM	NM	NM	NM	NM	NM
Musqwash Bay East	NM	NM	NM	NM	NM	NM	NM	NM	NM	NM
North Bay	-	-	+	-	-	-	-	-	-	-
South Bay	+	0.03	-	-	+	0.01	+	0.001	+	-
Honey Harbour	+	0.04	-	-	-	-	-	-	-	-

[1] ELISA and PPIA results verified by LC-MS/MS. [2] PCR primers listed in Table A2 (adapted from [18,50]. [3] Peaks in Twelve Mile Bay were measured in 2018 and the highest concentrations of these bioactive metabolites (and their genes) co-occurred with increasing biomass, notably of *Planktothrix* cf. *isothrix*, at the depth of the deep chlorophyll layer (Figure 4, Table S1, 2018 data).

2.6. Other Bioactive Metabolites Produced by Cyanobacteria

When associated with cyanobacteria, the term 'toxins' is generally applied to the small fraction of bioactive metabolites that affect humans and other large vertebrates; however, many bioactive metabolites produced by cyanobacteria have no known effect on human, pet, or livestock health, but are active towards more proximal elements of the food web including competing phytoplankton, bacteria, pathogens (e.g., viruses, chytrids), and grazers (e.g., [13]). These bioactive metabolites, which include a variety of peptides, have been reported previously from some European and US lakes and are known to inhibit metabolic processes (e.g., protease inhibitors) and function in chemical defence [13,19]. Our study represents a first-time analysis of samples from Canadian waters for both cyanobacterial toxins and these other bioactive metabolites. As seen with microcystins, several anabaenopeptins and one cyanopeptolin (1007) as well as their corresponding genes (*apn, oci*) were detected in both TMB and SB (Table 3); however, clear peaks coincident with the *Planktothrix*-dominated DCL were only evident in TMB (Figure 4). Anabaenopeptins (A, B, and F combined) peaked in 2018 at 6.6 μg L^{-1} in the DCL of the meso-oligotrophic TMB, which was comparable to concentrations found in eutrophic surface waters in the United States and Europe (e.g., [51–53]). Anabaenopeptins are thought to play a role in chemical defence via their ability to inhibit proteases used by parasites of cyanobacteria, such as chytrids, to digest their hosts [54]. This anti-parasitic activity could afford cyanobacteria in TMB an additional competitive advantage to establish dominance within the DCL.

Our results indicate that these (and potentially other) bioactive metabolites—typically reported from eutrophic systems (e.g., [51])—are also produced in meso-oligotrophic waters. Furthermore, while most studies have concentrated efforts on surface waters where visible blooms of cyanobacteria are most evident, our data demonstrate the potential importance of these semiochemicals in the often-overlooked communities in DCLs.

3. Conclusions

We observed the annual formation of persistent DCLs in TMB and SB during the stratification period, with a predominance of the large cyanobacteria *Planktothrix* cf *isothrix*. These deep-living layers were formed near the hypoxic hypolimnion at the intersection of opposing vertical gradients of PAR and nutrients, notably Fe, Mn and PO_4. Similar DCLs were not observed in the other (13) embayments surveyed along the same coastline, where analogous gradients of PAR and water chemistry were not detected. We also observed the Fe- and PO_4-precipitating bacteria *Leptothrix* within the DCL of TMB but not SB, suggesting that microbial processes may contribute to differences in their metalimnetic chemical gradients and DCL composition and biomass. In both embayments, the *Planktothrix*-dominated DCLs coincided with measurable levels of cyanobacterial toxins and other bioactive metabolites (most notably microcystins and anabaenopeptins), along with the associated genes. This is the first report of these compounds in DCLs from meso-oligotrophic Canadian lakes, and it merits more focused work to understand if and how these compounds function in the establishment and maintenance of these deep-living communities. The presence of significant numbers of live *Planktothrix* in surficial sediments and overlying water suggest these act as important seed populations to the DCLs, meriting further investigation. Overall, the prevalence of these significant DCLs dominated by large, potentially harmful cyanobacteria across different waterbodies is unknown, and their formation in apparently oligotrophic systems requires more research to understand the driving factors and track their change with changing environmental conditions and anthropogenic development in respective regions.

4. Materials and Methods

4.1. Study Sites

Fifteen embayments along the eastern shore of Georgian Bay, Lake Huron (Ontario, Canada) were surveyed, with a focus on sheltered embayments of intermediate size and depth (<30 m), particularly those that are reported to regularly undergo stratification and establish pronounced chemical gradients facilitated by hypolimnetic hypoxia (Figure 1) [16,55]. All 15 embayments were sampled during summer (July/August) and autumn (September/October) stratification of 2014 and 2015 (Figures 1 and A1, Table A1). Twelve Mile Bay (TMB) and South Bay (SB) were selected, based on the presence of DCLs, for additional and more detailed sampling and analysis in 2015 in winter (February/March), spring (June), summer (August), and autumn (September/October). Based on the earlier results, TMB and SB were sampled again during stratification in the summer of 2018 (July/August) and analyzed for a broader suite of bioactive metabolites beyond the cyanobacterial toxins commonly measured (Table 3).

4.2. Physicochemical Profiles of Water Column

At each sampling, photosynthetically active radiation (PAR; 400–700 nm) was measured with a LI-193 spherical underwater quantum sensor (LI-COR Biosciences, NE, USA) at 0.5 m depth intervals. Measurements were corrected for variance in incident irradiance using a LI-190R quantum sensor (LI-COR Biosciences, NE, USA), and euphotic zone depth calculated (i.e., <1% sub-surface irradiance). A YSI multi-parameter sonde (Xylem Inc., New York, NY, USA) was used to obtain depth profiles of temperature, dissolved oxygen, pH, specific conductivity, redox, and turbidity at each site, while a FluoroProbe (FP) (bbe Moldaenke GmbH, Schwentinental, Germany) was used to measure depth profiles of in situ, fluorescence-based, phytoplankton pigment class-specific chlorophyll ('green algae', 'cyanobacteria', 'brown' algae' (diatoms, chrysophytes and dinoflagellates), and 'crypto-phytes').

4.3. Water Sampling

At each site and date, whole water samples were collected from near-surface (1 m) and bottom (1 m from sediments) using a 10 L Niskin water sampler, subsampled into acid washed polyethylene bottles and kept in the dark at 4 °C until processed (within 24 h). During stratification, samples were also collected from the middle of the thermocline (metalimnion), the epilimnion (1 m), and hypolimnion (1 m from sediments). Samples were also taken at additional depths where DCLs were detected in the FP profiles.

4.4. Sediment Sampling

Surficial sediments were sampled using a modified gravity corer (Uwitech, Austria). Overlying water was siphoned and surficial 1.0 cm of the core was extruded on-site, placed in pre-labeled Whirl-Pak® bags, and immediately transported to the laboratory on ice in a dark cooler for microscopic imaging before being frozen in the dark at −20 °C.

4.5. Water Quality Analysis

Dissolved and particulate components were analyzed using filtrate from a cellulose acetate filter (0.45 μm pore size, 47 mm diameter) and material collected on a Whatman GF/C filter (1.2 μm nominal pore size, 47 mm diameter), respectively. These included major nutrients (dissolved inorganic and organic carbon (DIC, DOC), particulate organic carbon (POC), nitrate/nitrite ($NO_{2/3}$), ammonium (NH_4), total dissolved Kjeldahl nitrogen (TKN), particulate organic nitrogen (PON), phosphorus (P), dissolved phosphorus (DP), phosphate (PO_4)), dissolved silica, and extracted chlorophyll-*a*, all of which were analyzed at the National Laboratory for Environmental Testing (NLET, Burlington, Ontario) using standard methods [56]. Dissolved organic nitrogen (DON) was calculated by subtracting NH_4^{1+} from TKN.

4.6. Extraction and Analysis of Cyanobacterial Bioactive Metabolites

Whole water samples were concentrated onto Whatman GF/C filters (1.2 μm nominal pore size, 47 mm diameter) and stored at −80 °C in the dark until extraction. Cyanobacterial toxins and other bioactive metabolites were extracted in 10 mL of analytical grade aqueous methanol (1:1 *v/v*) amended with analytical grade formic acid (0.1%) using probe sonication (three 30 sec pulses) (Fisher Scientific Co. Qsonica Sonicator Q500, Ontario, Canada). Samples were then centrifuged at 3000 rpm for 15 min to pellet debris. A PTFE syringe filter (1.0 μm pore size, 30 mm diameter) attached to a 10 mL gas-tight glass syringe was used to filter the resulting supernatant into a glass vial, which was then evaporated to dryness using nitrogen gas-flow and heat (30 °C). The resulting residue was reconstituted with 1 mL of analytical grade aqueous methanol (1:1 *v/v*) and vortexed. A final filtration was done using a PTFE syringe filter (0.45 μm pore size, 15 mm diameter) attached to a gas-tight glass syringe into a 1.5 mL HPLC amber glass vial and stored at −80 °C in the dark until analysis.

Nineteen cyanobacterial peptides and five alkaloids were analyzed by high performance liquid chromatography tandem mass spectrometry (HPLC-MS/MS), as described previously [51]. Certified reference standards of microcystin-LR, -[Dha⁷]LR and nodularin (NOD-R) were purchased from the National Research Council (NRC) of Canada Biotoxins program (Nova Scotia, Canada). Microcystin-RR, -LA, -LF, -YR, -WR, -LY, -LW, -HtyR, and -HilR were purchased from Enzo Life Sciences (New York, NY, USA). Anabaenopeptin-A, -B, and -F; cyanopeptolin-1007, -1021, and -1040; as well as microginin-690 were purchased from MARBIONC (NC, USA). Anatoxin-a fumarate was purchased from Tocris Bioscience (MN, USA) as a racemic mixture. Homo-anatoxin-a was purchased from Abraxis. Cylindrospermopsin was purchased from Enzo Life Sciences. Saxitoxin and neosaxtoxin were purchased from the NRC. Each analyte exceeded 95% purity as per the manufacturer's certification.

HPLC-MS/MS analysis was performed on a Sciex 4000 QTRAP (AB Sciex, USA) tandem mass spectrometer equipped with a Shimadzu Prominence HPLC. Peptides were chromatographically separated in 20 μL injections of extracts on a Luna C8 column (Phenomenex, CA, USA) using gradient elution where the mobile phase consisted of A (0.1% formic acid and 5 mM ammonium acetate in HPLC grade water) and B (0.1% formic acid and 5 mM ammonium acetate in 95% acetonitrile). The gradient began at 30% B for 3 min, increasing over a linear gradient to 95% B at 9 min, and held at 95% B until 15 min at which point B was returned to the starting condition for 5 min. The mass spectrometer was operated in positive mode using a scheduled multiple reaction monitoring method. Alkaloids were separated by hydrophilic interaction liquid chromatography (HILIC) (SeQuant®, 5 μm, 150 × 2.1 mm I.D., EMD Millipore Corporation, MA, USA) with mobile phases of (A) HPLC water with 60 mM formic acid and (B) 100% acetonitrile. Isocratic elution (60% B) was used for the HILIC method. The concentration of target analytes was determined based on a linear regression model of peak area relative to the known concentration of an 8-point calibration curve prepared in 1:1 methanol:water. Regression coefficients of R > 0.98 were accepted at an accuracy of >90% at each calibration level.

4.7. Taxonomic Identification and Enumeration

Unfiltered sample aliquots of 100 mL were preserved in Lugol's iodine solution (2% *v/v*) for later taxonomic identification, abundance, and biomass using the Utermöhl technique [57].

4.8. DNA Extraction and PCR Amplification

Water samples were filtered onto 0.22 μm pore size, 47 mm polycarbonate filters and stored at −80 °C for DNA extraction. Samples were extracted using DNeasy PowerWater kit (Qiagen, Canada) following the manufacturer's protocol.

The availability of a commercial multiplex, quantitative PCR (qPCR) kit for genes of cylindrospermopsin (*cyrA*), microcystin/nodularin (*mcyE/ndA*), and saxitoxin (*sxtA*) facilitated the simultaneous measurement of respective gene copy numbers. The qPCR assays were performed using the Phytoxigene™ kit (Diagnostic Technology, Australia) according to the manufacturer's protocol. Briefly, the lyophilized master mix was first spun down and then reconstituted in PCR-grade water. Each qPCR reaction consisted of 20 μL of the reconstituted master mix and 5 μL of template DNA (10–50 ng) and was carried out in a Bio-Rad CFX96 cycler (Bio-Rad, USA). The cycling conditions consisted of an initial denaturation at 95 °C for 2 min, followed by 40 cycles of denaturation at 95 °C for 15 sec and annealing-extension at 60 °C for 30 s. Gene copies per sample were calculated using a standard curve (target gene copy number vs. Ct) determined for each target gene. Standard curves for all target genes were constructed using standards purchased from the same manufacturer (correlation coefficient and efficiency: *mcyE/ndA*: R^2 = 0.999; E = 100.7%; *cyrA*: R^2 = 0.999; E = 100.7%; *sxtA*: R^2 = 1.000; E = 102.1%).

All samples were screened by conventional endpoint PCR for the presence of anatoxin-a synthesis gene and several synthesis genes of oligopeptides know to be bioactive metabolites: aeruginoside, anabaenopeptin, cyanopeptolin, microcystin, microginin, and prenylagaramide. PCR was performed in a total volume of 25 μL composed of 5 μL of GoTaq 10X buffer (Promega), 0.5 μL of dNTPs (10 mM each), 0.5 μL each forward and reverse primers (10 μM), 0.125 μL GoTaq Polymerase (5 units μL^{-1}), 2 μL of template DNA (10–50 ng), and 16 μL of nuclease-free water (see Table A2 for primer sequences). The thermal cycling condition consisted of an initial denaturation step at 95° C for 3 min followed by 35 cycles of denaturation at 95° C for 1 min, annealing for 30 sec (variable annealing temperature; see Table A2), and extension at 72° for 30 s. PCR products were evaluated on a 2% agarose gel.

Anabaenopeptin gene copies were quantified using digital PCR (dPCR). Each dPCR mixture consisted of 7.5 μL of QuantStudio 3D digital PCR master mix v.2 (Thermo Fisher

Scientific, Canada), 1.4 µL of each 900 nM forward and reverse primer, 0.75 µL of 250 nM probe, 2 µL of DNA extract, and 195 µL nuclease-free water. The reaction mixtures were loaded into a QuantStudio 3D digital PCR 20K chip v2 (Thermo Fisher Scientific, Canada) using a QuantStudio 3D chip loader (Thermo Fisher Scientific, Canada). The reactions were carried out in a ProFlex thermocycler (Thermo Fisher Scientific, Canada) using the following cycling conditions: 96 °C for 10 min, 40 cycles of 60 °C for 2 min and 98 °C for 30 s, and a final step of 60 °C for 2 min. The chips were read on a QuantStudio 3D digital PCR instrument and the results analyzed using the QuantStudio AnalysisSuite software (V 3.2, 2019) (Thermo Fisher Scientific, Canada).

Supplementary Materials: The following are available at https://www.mdpi.com/article/10.3390/toxins13070445/s1, Table S1: Taxonomic identification and enumeration of phytoplankton in Twelve Mile Bay and South Bay embayments of Georgian Bay, Lake Huron (Ontario, Canada).

Author Contributions: Conceptualization, A.Z. and S.B.W.; methodology, T.R.M., L.C.W. and H.K.; validation, A.Z.; formal analysis, A.Z. and S.B.W.; investigation, A.Z. and S.B.W.; resources, A.Z. and S.B.W.; data curation, A.Z., S.B.W., T.R.M., L.C.W. and H.K.; writing—original draft preparation, A.Z.; writing—review and editing, A.Z., S.B.W., T.R.M., L.C.W. and H.K; visualization, A.Z.; supervision, A.Z. and S.B.W.; project administration, A.Z. and S.B.W.; funding acquisition, A.Z. and S.B.W. All authors have read and agreed to the published version of the manuscript.

Funding: This study was supported by Environment and Climate Change Canada's Lake Simcoe/Southeastern Georgian Bay Clean-Up Fund.

Institutional Review Board Statement: Not applicable.

Informed Consent Statement: Not applicable.

Data Availability Statement: Data is contained within the article or supplementary material.

Acknowledgments: We would like to extend our sincerest gratitude to the hardworking students and staff, Kaitlin Burek, Shelby Grassick, Flora (Zi Wan) Dong, Jay Guo, Joseph Gzik, Anqi Liang, and the skilled technical team of the Research Support Section, Adam Morden, Daniel Abbey, and Corey Treen, within the Watershed Hydrology and Ecology Research Division at Environment and Climate Change Canada for their assistance in the field and laboratory. Thank you also to Patricia Chow-Fraser and Stuart Campbell at McMaster University as well as Aisha Chiandet at the Severn Sound Environmental Association for their support in the field as well as for their knowledge, guidance, and navigation in planning and execution of our science in this complex watershed. A thank you also to Bob Rowsell, David Depew and Jacqui Milne for their support, encouragement, and engagement in the early stages of the work.

Conflicts of Interest: The authors declare no conflict of interest.

Appendix A

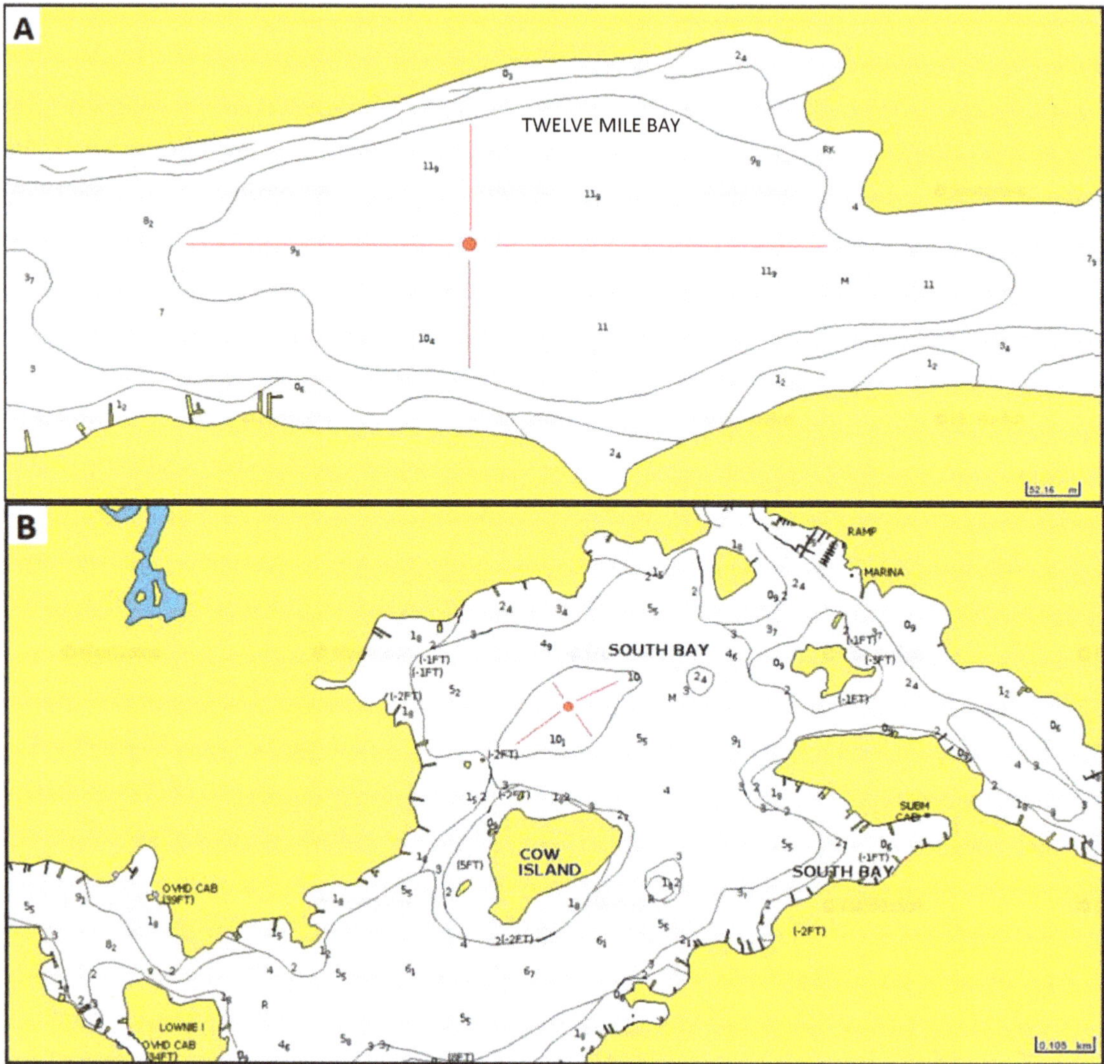

Figure A1. Location within (**A**) Twelve Mile Bay (TMB), scale 52.16 m and (**B**) South Bay (SB), scale 0.105 km, in Georgian Bay, Lake Huron, Ontario, Canada where each of the deep-chlorophyll layers were sampled, denoted by a filled, red circle. Three-dimensional mapping by Fluoroprobe© revealed that each DCL extended horizontally to the end of the first contour demarcated on the map (~9 m in TMB and ~10 m in SB, approximate extent indicated by red lines) and therefore each of these contours were used to estimate the elliptical area over which each of the DCLs existed. Maps reproduced from Navionics Web API v2, www.navionics.com accessed on January 2021.

Table A1. Coordinates and depths of sites under study in Georgian Bay, Lake Huron (Ontario, Canada). All 15 embayments were sampled during summer (July/August) and autumn (September/October) stratification of 2014 and 2015. Twelve Mile Bay and South Bay were selected (based on presence of DCLs) for additional and more detailed sampling and analysis in 2015 in winter (February/March), spring (June), summer (August), and autumn (September/October). TMB and SB were again sampled during stratification in the summer of 2018 (July/August) and analyzed for a broader suite of bioactive metabolites beyond the cyanobacterial toxins commonly measured (Table 3).

Site	Latitude	Longitude	Depth (m)
Byng Inlet	45.77056	−80.56836	3
Sturgeon Bay North	45.6133	−80.4325	14
Sturgeon Bay South	45.6055	−80.4096	6
Deep Bay	45.3953	−80.2236	15
Woods Bay	45.13785	−79.9894	8
Twelve Mile Bay	45.0838	−79.946	12
Tadenac Bay	45.0588	−79.9775	24
Go Home Bay	44.9913	−79.9376	10
Longuissa Bay	44.9613	−79.8884	8
Cognashene Lake	44.9510	−79.9186	16
Musqwash Bay West	44.9520	−79.8787	30
Musqwash Bay East	44.9483	−79.85006	20
North Bay	44.8919	−79.7925	18
South Bay	44.8762	−79.7857	10
Honey Harbour	44.8770	−79.8263	7

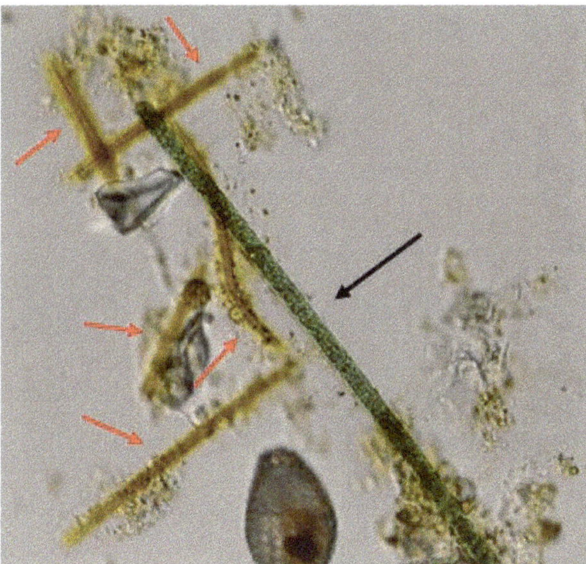

Figure A2. Microscopic imaging of water sample at the DCL (~9 m) in Twelve Mile Bay showing *Planktothrix* cf. *isothrix* (black arrow) and *Leptothrix* with Fe oxide precipitates (five red arrows). Images are not to scale and are only for representative purpose.

Table A2. Primers used to amplify regions of genes coding for cyanobacterial toxins and other bioactive metabolites (oligopeptides).

Gene Loci	Primer [1]	Sequence (5'-3') [2]	Amplicon Length (bp)	Annealing Temperature (°C)
Aeruginoside aerD	aerD-F	GAAACCAGTAGTGAACAGACCTTAAATTATC	493	60
	aerD-R	GACCAACTCATCTCAATTTTCCC		
Anabaenopeptin apnD	apnDTe-F	GACACGCCTTCTATTTTCGAGA	581	60
	apnDTe-R	CGCGAAATAAAACAATAGGGG		
apnC	apnNMT-F	CGTGCAGATGATGACCTATCCA	470	55
	apnNMT-R	AAGGTTCGCAATACTTCAGGGTT		
Anatoxin [3] anaC	anaC-gen-F2	TCTGGTATTCAGTMCCCTCYAT	366	58
	anaC-gen-R2	CCCAATARCCTGTCATCAA		
Cyanopeptolin ociC	cptDTe-F	GATCTCTATCAACAGTTTGGAGCAA	690	61
	cptDTe-R	ACTGTTCGGCTAACACTTGAACAT		
ociB	ociB-F	TGGTTTTTAGATCAATTTGAGTCCG	512	66
	ociB-R	CCACTGTTTTTGCCAAAGAGTG		
Microcystin mcyC	mcyCTe-F	TTACAAGCGATGAATCTCATGG	503	60
	mcyCTe-R	GGGATTTAATAAGAAACCATCAACC		
mcyE	Hep-F	TTTGGGGGTTAACTTTTTTGGGCATAGTC	466	60
	Hep-R	AATTCTTGAGGCTGTAAATCGGGTTT		
Microginin micD	mgnTe-F	TGGTCAATGGGAGGAGTGATAG	514	60
	mgnTe-R	CTGTAGTGATCTCCACTAATCCATTG		
micA	mcnA-F	AAACCCTTTGATTTGAGCCAG	428	60
	mcnA-R	CAGCAAGGTACAGCCCTGTT		
Prenylagaramide pagA	acyA-F	CCCTGGAAAAGATATTTTAGGAGC	489	60
	acyA-R	ATTTGAAGTACGGTTTGACATGG		

[1] Primers for cylindrospermopsin, nodularin, and saxitoxin are not listed here since these toxins were only measured using the Phytoxigene™ kit (Diagnostic Technology, Australia) and the primers are included in the kit's master mix. [2] [18]. [3] [50].

References

1. Pick, F.R.; Nalewajko, C.; Lean, D.R.S. The origin of a metalimnetic chrysophyte peak. *Limnol. Oceanogr.* **1984**, *29*, 125–134. [CrossRef]
2. Fee, E.J. The vertical and seasonal distribution of chlorophyll in lakes of the Experimental Lakes Area, northwestern Ontario: Implications for primary production estimates. *Limnol. Oceanogr.* **1976**, *21*, 767–783. [CrossRef]
3. Pannard, A.; Planas, D.; Noac'h, P.; Bormans, M.; Jourdain, M.; Beisner, B. Contribution of the deep chlorophyll maximum to primary production, phytoplankton assemblages and diversity in a small stratified lake. *J. Plankton Res.* **2020**, *42*, 630–649. [CrossRef]
4. Pannard, A.; Planas, D.; Beisner, B. Macrozooplankton and the persistence of the deep chlorophyll maximum in a stratified lake. *Freshw. Biol.* **2015**, *60*, 1717–1733. [CrossRef]
5. Lofton, M.; Leach, T.; Beisner, B.; Carey, C. Relative importance of top-down vs. bottom-up control of lake phytoplankton vertical distributions varies among fluorescence-based spectral groups. *Limnol. Oceanogr.* **2020**, *65*, 2485–2501. [CrossRef]
6. Pick, F.R.; Lean, D.R.S.; Nalewajko, C. Nutrient status of metalimnetic phytoplankton peaks. *Limnol. Oceanogr.* **1984**, *29*, 960–971. [CrossRef]
7. Leach, T.H.; Beisner, B.E.; Carey, C.C.; Pernica, P.; Rose, K.C.; Huot, Y.; Brentrup, J.A.; Domaizon, I.; Grossart, H.-P.; Ibelings, B.W.; et al. Patterns and drivers of deep chlorophyll maxima structure in 100 lakes: The relative importance of light and thermal stratification. *Limnol. Oceanogr.* **2018**, *63*, 628–646. [CrossRef]
8. Barbiero, R.P.; Tuchman, M.L. The Deep chlorophyll maximum in Lake Superior. *J. Great Lakes Res.* **2004**, *30*, 256–268. [CrossRef]
9. Bramburger, A.J.; Reavie, E.D. A comparison of phytoplankton communities of the deep chlorophyll layers and epilimnia of the Laurentian Great Lakes. *J. Great Lakes Res.* **2016**, *42*, 1016–1025. [CrossRef]

10. Twiss, M.; Ulrich, C.; Zastepa, A.; Pick, F. On phytoplankton growth and loss rates to microzooplankton in the epilimnion and metalimnion of Lake Ontario in mid-summer. *J. Great Lakes Res.* **2012**, *38*, 146–153. [CrossRef]
11. Scofield, A.E.; Watkins, J.M.; Weidel, B.C.; Luckey, F.J.; Rudstam, L.G. The deep chlorophyll layer in Lake Ontario: Extent, mechanisms of formation, and abiotic predictors. *J. Great Lakes Res.* **2017**, *43*, 782–794. [CrossRef]
12. Legrand, C.; Rengefors, K.; Fistarol, G.; Granéli, E. Allelopathy in phytoplankton—Biochemical, ecological and evolutionary aspects. *Phycologia* **2003**, *42*, 406–419. [CrossRef]
13. Janssen, E.M.L. Cyanobacterial peptides beyond microcystins—A review on co-occurrence, toxicity, and challenges for risk assessment. *Water Res.* **2019**, *151*, 488–499. [CrossRef]
14. Camacho, A. On the occurrence and ecological features of deep chlorophyll maxima (DCM) in Spanish stratified lakes. *Limnetica* **2006**, *25*, 453–478. [CrossRef]
15. Silsbe, G.; Malkin, S. Where light and nutrients collide: The global distribution and activity of subsurface chlorophyll maximum layers. In *Aquatic Microbial Ecology and Biogeochemistry: A Dual Perspective*; Glibert, P.M., Kana, T.M., Eds.; Springer: Cham, Switzerland, 2016; pp. 141–152.
16. Verschoor, M.; Powe, C.; McQuay, E.; Schiff, S.; Venkiteswaran, J.; Li, J.; Molot, L. Internal iron loading and warm temperatures are pre-conditions for cyanobacterial dominance in embayments along Georgian Bay, Great Lakes. *Can. J. Fish. Aquat. Sci.* **2017**, *74*, 1439–1453. [CrossRef]
17. Bird, D.F.; Kalff, J. Phagotrophic sustenance of a metalimnetic phytoplankton peak. *Limnol. Oceanogr.* **1989**, *34*, 155–162. [CrossRef]
18. Kurmayer, R.; Deng, L.; Entfellner, E. Role of toxic and bioactive secondary metabolites in colonization and bloom formation by filamentous cyanobacteria Planktothrix. *Harmful Algae* **2016**, *54*, 69–86. [CrossRef]
19. Śliwińska-Wilczewska, S.; Wisniewska, K.; Konarzewska, Z.; Cieszyńska, A.; Felpeto, A.B.; Lewandowska, A.; Latała, A. The current state of knowledge on taxonomy, modulating factors, ecological roles, and mode of action of phytoplankton allelochemicals. *Sci. Total. Environ.* **2021**, *773*, 145681. [CrossRef]
20. Reavie, E.D.; Barbiero, R.P.; Allinger, L.E.; Warren, G.J. Phytoplankton trends in the Great Lakes, 2001–2011. *J. Great Lakes Res.* **2014**, *40*, 618–639. [CrossRef]
21. Findlay, D.L.; Paterson, J.J.; Hendzel, L.L.; Kling, H.J. Factors influencing *Gonyostomum semen* blooms in a small boreal reservoir lake. *Hydrobiologia* **2005**, *533*, 243–252. [CrossRef]
22. Lebret, K.; Östman, Ö.; Langenheder, S.; Drakare, S.; Guillemette, F.; Lindström, E.S. High abundances of the nuisance raphidophyte *Gonyostomum semen* in brown water lakes are associated with high concentrations of iron. *Sci. Rep.* **2018**, *8*, 13463. [CrossRef]
23. Head, R.M.; Jones, R.I.; Bailey-Watts, A.E. An assessment of the influence of recruitment from the sediment on the development of planktonic populations of cyanobacteria in a temperate mesotrophic lake. *Freshw. Biol.* **1999**, *41*, 759–769. [CrossRef]
24. Brunberg, A.-K.; Blomqvist, P. Recruitment of Microcystis (Cyanophyceae) from lake sediments: The importance of littoral inocula. *J. Phycol.* **2003**, *39*, 58–63. [CrossRef]
25. Holland, D.P.; Walsby, A.E. Viability of the cyanobacterium Planktothrix rubescens in the cold and dark, related to over-winter survival and summer recruitment in Lake Zürich. *Eur. J. Phycol.* **2008**, *43*, 179–184. [CrossRef]
26. Kravchuk, E.; Ivanova, E.; Gladyshev, M. Spatial distribution of resting stages (akinetes) of the cyanobacteria *Anabaena flos-aquae* in sediments and its influence on pelagic populations. *Mar. Freshw. Res.* **2011**, *62*, 450–461. [CrossRef]
27. Nürnberg, G.K.; LaZerte, B.D.; Olding, D.D. An Artificially induced Planktothrix rubescens surface bloom in a small kettle lake in Southern Ontario compared to blooms world-wide. *Lake Reserv. Manag.* **2003**, *19*, 307–322. [CrossRef]
28. Grach-Pogrebinsky, O.; Sedmak, B.; Carmeli, S. Seco[D-Asp3]microcystin-RR and [D-Asp3,D-Glu(OMe)6]microcystin-RR, two new microcystins from a toxic water bloom of the cyanobacterium Planktothrix rubescens. *J. Nat. Prod.* **2004**, *67*, 337–342. [CrossRef]
29. Cantin, A.; Beisner, B.E.; Gunn, J.M.; Prairie, Y.T.; Winter, J.G. Effects of thermocline deepening on lake plankton communities. *Can. J. Fish. Aquat. Sci.* **2011**, *68*, 260–276. [CrossRef]
30. Salmaso, N.; Boscaini, A.; Shams, S.; Cerasino, L. Strict coupling between the development of *Planktothrix rubescens* and microcystin content in two nearby lakes south of the Alps (lakes Garda and Ledro). *Ann. Limnol. Int. J. Limnol.* **2013**, *49*, 309–318. [CrossRef]
31. Kasprzak, P.; Shatwell, T.; Gessner, M.; Gonsiorczyk, T.; Kirillin, G.; Selmeczy, G.; Padisak, J.; Engelhardt, C. Extreme weatherevent triggers cascade towards extreme turbidity in a clear-water Lake. *Ecosystems* **2017**, *20*, 1407–1420. [CrossRef]
32. Giling, D.P.; Nejstgaard, J.C.; Berger, S.A.; Grossart, H.-P.; Kirillin, G.; Penske, A.; Lentz, M.; Casper, P.; Sareyka, J.; Gessner, M.O. Thermocline deepening boosts ecosystem metabolism: Evidence from a large-scale lake enclosure experiment simulating a summer storm. *Glob. Chang. Biol.* **2017**, *23*, 1448–1462. [CrossRef]
33. de Marsac, N.T. Occurrence and nature of chromatic adaptation in cyanobacteria. *J. Bacteriol.* **1977**, *130*, 82. [CrossRef]
34. Micheletti, S.; Schanz, F.; Walsby, A.E. The daily integral of photosynthesis by *Planktothrix rubescens* during summer stratification and autumnal mixing in Lake Zürich. *New Phytol.* **1998**, *139*, 233–246. [CrossRef]
35. Davis, P.A.; Beard, S.J.; Walsby, A.E. Variation in filament width and gas vesicles of red and green isolates of Planktothrix spp. *Arch. Hydrobiol. Suppl.* **2003**, *146*, 15–29.
36. Legnani, E.; Copetti, D.; Oggioni, A.; Tartari, G.; Maria, T.; Palumbo, G.; Morabito, G. Planktothrix rubescens' seasonal dynamics and vertical distribution in Lake Pusiano (North Italy). *J. Limnol.* **2005**, *64*, 61–73. [CrossRef]

37. Copetti, D.; Tartari, G.; Morabito, G.; Oggioni, A.; Legnani, E.; Imberger, J. A biogeochemical model of Lake Pusiano (North Italy) and its use in the predictability of phytoplankton blooms: First preliminary results. *J. Limnol.* **2006**, *65*, 59–64. [CrossRef]

38. Halstvedt, C.B.; Rohrlack, T.; Andersen, T.; Skulberg, O.; Edvardsen, B. Seasonal dynamics and depth distribution of *Planktothrix* spp. in Lake Steinsfjorden (Norway) related to environmental factors. *J. Plankton Res.* **2007**, *29*, 471–482. [CrossRef]

39. Boström, B.; Andersen, J.M.; Fleischer, S.; Jansson, M. Exchange of phosphorus across the sediment-water interface. *Hydrobiologia* **1988**, *170*, 229–244. [CrossRef]

40. Katsev, S.; Dittrich, M. Modeling of decadal scale phosphorus retention in lake sediment under varying redox conditions. *Ecol. Model.* **2013**, *251*, 246–259. [CrossRef]

41. Loh, P.S.; Molot, L.A.; Nowak, E.; Nürnberg, G.K.; Watson, S.B.; Ginn, B. Evaluating relationships between sediment chemistry and anoxic phosphorus and iron release across three different water bodies. *Inland Waters* **2013**, *3*, 105–118. [CrossRef]

42. Rentz, J.; Turner, I.; Ullman, J. Removal of phosphorus from solution using biogenic iron oxides. *Water Res.* **2009**, *43*, 2029–2035. [CrossRef] [PubMed]

43. Buliauskaitė, R.; Wilfert, P.; Kumar, P.S.; de Vet, W.W.J.M.; Witkamp, G.-J.; Korving, L.; van Loosdrecht, M.C.M. Biogenic iron oxides for phosphate removal. *Environ. Technol.* **2020**, *41*, 260–266. [CrossRef]

44. Hamre, K.D.; Lofton, M.E.; McClure, R.P.; Munger, Z.W.; Doubek, J.P.; Gerling, A.B.; Schreiber, M.E.; Carey, C.C. In situ fluorometry reveals a persistent, perennial hypolimnetic cyanobacterial bloom in a seasonally anoxic reservoir. *Freshw. Sci.* **2018**, *37*, 483–495. [CrossRef]

45. Padisak, J.; Barbosa, F.; Koschel, R.; Krienitz, L. Deep layer cyanoprokaryota maxima in temperate and tropical lakes. *Arch. Hydrobiol.* **2003**, *58*, 175–199.

46. Molot, L.A.; Watson, S.B.; Creed, I.F.; Trick, C.G.; McCabe, S.K.; Verschoor, M.J.; Sorichetti, R.J.; Powe, C.; Venkiteswaran, J.J.; Schiff, S.L. A novel model for cyanobacteria bloom formation: The critical role of anoxia and ferrous iron. *Freshw. Biol.* **2014**, *59*, 1323–1340. [CrossRef]

47. Selmeczy, G.; Tapolczai, K.; Casper, P.; Krienitz, L.; Padisak, J. Spatial- and niche segregation of DCM-forming cyanobacteria in Lake Stechlin (Germany). *Hydrobiologia* **2015**, *764*, 229–240. [CrossRef]

48. Padisak, J.; Krienitz, L.; Koschel, R.; Nedoma, J. Deep-layer autotrophic picoplankton maximum in the oligotrophic Lake Stechlin, Germany: Origin, activity, development and erosion. *Eur. J. Phycol.* **1997**, *32*, 403–416. [CrossRef]

49. Taranu, Z.E.; Zastepa, A.; Creed, I.; Pick, F.R. Beyond total microcystins, the importance of examining different microcystin variants. *LakeLine* **2017**, *37*, 14–18.

50. Rantala-Ylinen, A.; Känä, S.; Wang, H.; Rouhiainen, L.; Wahlsten, M.; Rizzi, E.; Berg, K.; Gugger, M.; Sivonen, K. Anatoxin-a synthetase gene cluster of the cyanobacterium *Anabaena* sp. Strain 37 and molecular methods to detect potential producers. *Appl. Environ. Microbiol.* **2011**, *77*, 7271–7278. [CrossRef]

51. Beversdorf, L.J.; Weirich, C.A.; Bartlett, S.L.; Miller, T.R. Variable cyanobacterial toxin and metabolite profiles across six eutrophic lakes of differing physiochemical characteristics. *Toxins* **2017**, *9*, 62. [CrossRef] [PubMed]

52. Bartlett, S.; Brunner, S.; Klump, J.; Houghton, E.M.; Miller, T.R. Spatial analysis of toxic or otherwise bioactive cyanobacterial peptides in Green Bay, Lake Michigan. *J. Great Lakes Res.* **2018**, *44*, 924–933. [CrossRef] [PubMed]

53. Chorus, I.; Sivonen, K.; Codd, G.A.; Börner, T.; Von Doehren, H.; Welker, M.; Dittmann, E.; Claussner, Y.; Christopffersen, K.; Schober, E.; et al. Toxic and Bioactive Peptides in Cyanobacteria. 2006. Available online: https://ci.nii.ac.jp/naid/10030356501/ (accessed on 26 June 2021).

54. Rohrlack, T.; Christiansen, G.; Kurmayer, R. Putative antiparasite defensive system involving ribosomal and nonribosomal oligopeptides in cyanobacteria of the genus Planktothrix. *Appl. Environ. Microbiol.* **2013**, *79*, 2642–2647. [CrossRef] [PubMed]

55. Charlton, M.; Mayne, G. Science and Monitoring Synthesis for South-Eastern Georgian Bay; Report prepared for Environment Canada. 2013. Available online: https://www.stateofthebay.ca/wp-content/uploads/2019/03/Science-and-Monitoring-Synthesis-for-South-Eastern-Georgian-Bay_Charlton-and-Mayne_2013.pdf (accessed on 26 June 2021).

56. NLET (National Laboratory for Environmental Testing). *Schedule of Services. Environment and Climate Change Canada, C.C.f.I.W.*; NLET (National Laboratory for Environmental Testing): Burlington, ON, Canada, 1997.

57. Findlay, D.; Kling, H. Protocols for Measuring Biodiversity: Phytoplankton in Freshwater. 2003. Available online: https://www.researchgate.net/publication/264881321_Protocols_for_measuring_biodiversity_Phytoplankton_in_freshwater (accessed on 26 June 2021).

Article

Cyanobacterial Toxins and Peptides in Lake Vegoritis, Greece

Sevasti-Kiriaki Zervou [1], Kimon Moschandreou [2], Aikaterina Paraskevopoulou [1], Christophoros Christophoridis [1], Elpida Grigoriadou [3], Triantafyllos Kaloudis [1], Theodoros M. Triantis [1], Vasiliki Tsiaoussi [2] and Anastasia Hiskia [1,*]

[1] Laboratory of Photo-Catalytic Processes and Environmental Chemistry, Institute of Nanoscience & Nanotechnology, National Center for Scientific Research "Demokritos", Patriarchou Grigoriou E & 27 Neapoleos Str, 15310 Agia Paraskevi, Athens, Greece; s.zervou@inn.demokritos.gr (S.-K.Z.); k.paraskevopoulou@inn.demokritos.gr (A.P.); c.christoforidis@inn.demokritos.gr (C.C.);t.kaloudis@inn.demokritos.gr (T.K.); t.triantis@inn.demokritos.gr (T.M.T.)

[2] The Goulandris Natural History Museum—Greek Biotope/Wetland Centre, 14th km Thessaloniki-Mihaniona, Thermi P.O. Box 60394, 57001 Thessaloniki, Greece; kmosch@ekby.gr (K.M.); vasso@ekby.gr (V.T.)

[3] Water Resources Management Agency of West Macedonia, 50100 Kozani, Decentralized Administration of Epirus—Western Macedonia, Greece; grig.elpida@gmail.com

* Correspondence: a.hiskia@inn.demokritos.gr

Citation: Zervou, S.-K.; Moschandreou, K.; Paraskevopoulou, A.; Christophoridis, C.; Grigoriadou, E.; Kaloudis, T.; Triantis, T.M.; Tsiaoussi, V.; Hiskia, A. Cyanobacterial Toxins and Peptides in Lake Vegoritis, Greece. *Toxins* **2021**, *13*, 394. https://doi.org/10.3390/toxins13060394

Received: 28 April 2021
Accepted: 27 May 2021
Published: 1 June 2021

Publisher's Note: MDPI stays neutral with regard to jurisdictional claims in published maps and institutional affiliations.

Abstract: Cyanotoxins (CTs) produced by cyanobacteria in surface freshwater are a major threat for public health and aquatic ecosystems. Cyanobacteria can also produce a wide variety of other understudied bioactive metabolites such as oligopeptides microginins (MGs), aeruginosins (AERs), aeruginosamides (AEGs) and anabaenopeptins (APs). This study reports on the co-occurrence of CTs and cyanopeptides (CPs) in Lake Vegoritis, Greece and presents their variant-specific profiles obtained during 3-years of monitoring (2018–2020). Fifteen CTs (cylindrospermopsin (CYN), anatoxin (ATX), nodularin (NOD), and 12 microcystins (MCs)) and ten CPs (3 APs, 4 MGs, 2 AERs and aeruginosamide (AEG A)) were targeted using an extended and validated LC-MS/MS protocol for the simultaneous determination of multi-class CTs and CPs. Results showed the presence of MCs (MC-LR, MC-RR, MC-YR, dmMC-LR, dmMC-RR, MC-HtyR, and MC-HilR) and CYN at concentrations of <1 µg/L, with MC-LR (79%) and CYN (71%) being the most frequently occurring. Anabaenopeptins B (AP B) and F (AP F) were detected in almost all samples and microginin T1 (MG T1) was the most abundant CP, reaching 47.0 µg/L. This is the first report of the co-occurrence of CTs and CPs in Lake Vegoritis, which is used for irrigation, fishing and recreational activities. The findings support the need for further investigations of the occurrence of CTs and the less studied cyanobacterial metabolites in lakes, to promote risk assessment with relevance to human exposure.

Keywords: cyanotoxins; microcystins; cylindrospermopsin; cyanopeptides; anabaenopeptins; microginins; aeruginosins; aeruginosamide; SPE; LC-MS/MS; Lake Vegoritis

Key Contribution: First report of the co-occurrence of multi-class cyanotoxins and cyanopeptides in Lake Vegoritis. Simultaneous determination of cyanotoxins and cyanopeptides in intra- and extracellular fractions with an extended and validated LC-MS/MS protocol. Three-year monitoring study that revealed the co-occurrence of microcystins and cylindrospermopsin with anabaenopeptins, microginins, aeruginosins, and aeruginosamide in Lake Vegoritis, Greece

1. Introduction

Cyanobacteria are common photosynthetic microorganisms found in lakes and surface water reservoirs, which can, under favorable conditions, grow massively to form blooms [1]. Several cyanobacteria species produce potent toxic compounds as secondary metabolites, called cyanotoxins (CTs) [2,3]. Several incidents of wild and domestic animal poisoning as well as human health effects due to toxic cyanobacterial blooms have been reported [4–7].

Cyanotoxins comprise a large number of compounds presenting a variety of chemical structures (Figure S1). Microcystins (MCs) [8] and nodularins (NODs) [9] are cyclic peptides typically characterized by the presence of the unique amino acid Adda ((2S,3S,8S,9S)-3-amino-9-methoxy-2,6,8-trimethyl-10-phenyl deca-4,6-dienoic acid) in their structure, which is associated with their hepatotoxicity [2,10,11]. The alkaloid cylindrospermopsin (CYN) is cytotoxic, dermatotoxic, hepatotoxic, and possibly carcinogenic [12,13]. Anatoxin-a (ATX) is a bicyclic secondary amine (2-acetyl-9-azabicyclo (1,2,4) non-2-ene) with acute neurotoxicity [14].

In addition to CTs, cyanobacteria can also produce a wide variety of other metabolites, including compounds of peptide structure such as microginins (MGs), aeruginosins (AERs), aeruginosamides (AEGs), and anabaenopeptins (APs) [15] (Figure S2). MGs are a group of linear oligopeptides characterized by the presence of a decanoic acid derivative, 3-amino-2-hydroxy-decanoic acid (Ahda) at the *N*-terminus [16]. Although not fully investigated, MGs were shown to present strong protease inhibition with MG variants displaying ecotoxicological effects [17]. AERs are linear peptides that include both a derivative of hydroxyl-phenyl lactic acid (Hpla) at the *N*-terminus, the amino acid 2-carboxy-6-hydroxyoctahydroindole (Choi) and an arginine derivative at the C-terminus [18]. Studies on their bioactivity revealed that they inhibit serine proteases trypsin and thrombin, while AER 828A was found to be toxic to *Thamnocephalus platyurus* [19]. The linear peptides AEGs, characterized by the presence of prenyl and thiazole groups, are an understudied group of cyanobacterial metabolites for which limited knowledge is available with regards to their occurrence in cyanobacterial blooms and their bioactivity [20,21]. APs are cyclic peptides with the general structure of X1-CO-[Lys-X3-X4-MeX5-X6]. Lysine (Lys) is present in all variants while X1, X3, X4, X5 and X6 are variable amino acids. A side chain of one amino acid is attached to the ring through an ureido bond with Lys [15]. Recently, it has been reported that APs can be very abundant in nature [22,23]. To date, little is known about the potential health effects of APs on animals and humans [17]. Anabaenopeptin F (AP F) is considered a protease inhibitor and it was shown to inhibit protein phosphatases similarly to MCs [24]. Additionally, Anabaenopeptin B (AP B) and AP F were shown to induce lysis of the cyanobacteria *Microcystis aeruginosa* that can drastically influence cyanobacterial community dynamics and trigger the release of toxins into surface waters [25].

Toxic cyanobacteria blooms occur worldwide [26], with climate predictions suggesting their increase in the future in terms of frequency and severity [27–29]. Therefore, there is an urgent need to monitor toxic cyanobacteria and their toxins, especially in water bodies intended to be used as drinking water supplies or for recreational activities, particularly by children [30]. At the same time, there is need to better assess the occurrence, bioactivity, and effects of other cyanobacterial peptides (CPs) in order to improve risk assessment and the development of management strategies for cyanobacterial blooms.

Lake Vegoritis is a large natural lake covering an area of 60 km^2 in the region of Western Macedonia, in north-western Greece. The banks of the lake are an ideal refuge for many wild birds and it has remarkable fish fauna, which includes a large variety of species. The lake's sensitive ecosystem belongs to the European Network of Protected Areas (NATURA 2000), due to its important habitats and rich biodiversity. A part of its littoral zone was also designated as a bathing area, according to Directive 2006/7/EC. The ecological and historical background of toxic cyanobacterial blooms of Lake Vegoritis, as well as the recreational activities that it offers, create a growing concern about the possible effect of CTs and CPs to its ecosystem and human health. Besides its ecological importance, the lake is used for irrigation, fishing, and recreational activities.

Recently, a multi-lake survey covering 14 lakes in Greece was conducted with the aim to assess the presence of a wide range of CTs from different classes including MCs, NODs, CYN and ATX, using liquid chromatography coupled to tandem mass spectrometry (LC-MS/MS) [31]. In the frame of that study, it was found that water from Lake Vegoritis contained MCs and traces of CYN. Although the occurrence of CTs in Lake Vegoritis was confirmed and documented, it was based only on individual samples and the study was not designed to provide information concerning the spatial and temporal variation of CTs. Furthermore, there is complete lack of information regarding the co-occurrence of other cyanobacterial peptides such as MGs, AERs, AEGs, and APs.

In response to the above study, a 3-year monitoring program of Lake Vegoritis was initiated in 2018, with the aim to characterize the cyanobacterial species present and to assess the occurrence of various classes of CTs and CPs in the lake. Fifteen CTs (i.e., CYN, ATX, NOD and dmMC-RR, MC-RR, MC-YR, MC-HtyR, dmMC-LR, MC-LR, MC-HilR, MC-WR, MC-LA, MC-LY, MC-LW, and MC-LF) and ten CPs (i.e., MG FR1, MG FR3 MG T1, MG T2, AER 602/K139, AER 298A, AEG A, AP B, AP F and oscillamide (OSC Y)) were targeted. Selection of the targeted cyanobacterial metabolites was based on their frequency of detection in other Greek lakes [31,32]. To implement this monitoring program, a new method for simultaneous determination of various classes of CPs in addition to CTs was developed and validated, extending a previously validated LC-MS/MS analytical protocol [31,33] to include MGs, AERs, AEGs and APs. Using this new protocol, the detection of several classes of CTs and CPs would be possible in a single analytical run.

Results obtained by this 3-year study enable risk assessment and management of toxic cyanobacterial blooms by the lake's authorities. The study's findings also facilitate the reliable and effective communication of the risks to the general public and stakeholders, with regards to the uses of the lake, such as irrigation, fishing, and recreational activities.

2. Results and Discussion

2.1. Physico-Chemical Parameters of Lake Vegoritis

Lake Vegoritis is under pressure from point source and diffuse pollution. As reported by monitoring results from 2012–2015, the lake was in moderate ecological and good chemical status [34]. The concentrations of most of the physicochemical quality parameters, with the exception of nitrates and sulfates, did not fluctuate considerably during the study period (Table S1). The F^-, NO_2^-, Br^- and PO_4^{3-} ions were measured below the method's limit of quantification (LOQ) during the whole study period. The Cl^- ranged between 32–40 mg/L, with a single high measurement (58 mg/L) in July 2020. The NO_3^- ions ranged from below LOQ to 1.0 mg/L, with the highest values detected in the winter of 2018 and 2020 and in August 2020. The greatest variability was noticed for SO_4^{2-} levels that ranged between 84 and 200 mg/L, with no seasonal or other temporal pattern observed. The cation (Na^+, K^+, Mg^{2+}, Ca^{2+}) concentrations measured showed relative stability throughout the study period and no temporal distribution pattern.

In 2018, the highest levels of total phosphorus (TP) were measured. Concentrations ranged from 22 to 60 µg/L and displayed the highest values in February and April (60 and 50 µg/L, respectively). In June 2018, TP decreased to 43 µg/L, while in July 2018 the mean value of the TP concentration was 38 µg/L. Concentrations declined in the following months (<29 µg/L). The following years, 2019 and 2020, TP concentrations were measured at slightly lower levels, 28–50 µg/L and 16–38 µg/L, respectively.

The transparency of the water was measured using the Secchi disc and in June 2018 there was a significant reduction to 0.3 m, from 6.5 m and 6.0 m measured in February 2018 and April 2018, respectively. Such a low value was measured for the first time in the lake throughout the operation of the National Monitoring Water Network (2012–2018). Since then, the transparency of the water in the lake, including bathing area sampling points, presented a noteworthy increase (2.4 m on 4 July 2018 and 3.0 m on 12 July 2017 and 17 July 2018). The Secchi disc was visible to the bottom of the bathing area stations until the autumn. The following years 2019–2020, transparency of the water at the NMWN (National Monitoring Water Network) sampling point ranged from 1.6 m to 4.8 m.

The total suspended solids (TSS) exhibited a high value of 8.46 mg/L in July 2018. During the summer, their concentrations in the lake decreased both at the NMWN sampling point and the bathing area (Site 1). In the next two years, TSS did not exceed 2.54 mg/L, except for two cases in June 2019 (3.75 mg/L) and in September 2020 (5.67 mg/L).

2.2. Chlorophyll α

The concentration of chlorophyll α in Lake Vegoritis in June 2018 was high (15.9 µg/L), with a decreasing trend during the following summer months at both the NMWN sampling point and the bathing area (Site 1) (3.0–8.0 µg/L). This reduction was in line with the improvement in water transparency values observed during the same period. Similar values of chlorophyll α concentration (4.2–7.7 µg/L) were measured during the following years, 2019 and 2020. In two samples collected on 10 September 2019 and 22 September 2020 much higher values (19.4 and 31.8 µg/L, respectively) were measured, but no discoloration of water was observed (Table S1).

2.3. Phytoplankton

During June 2018, the total phytoplankton biomass was estimated at 2.3 mg/L. The most abundant taxa were the green alga *Sphaerocystis schroeteri* (39,889,429 cells/L), cyanobacteria of the genus *Dolichospermum* (14,864,850 cells/L) and the species *Aphanizomenon flos-aquae* (3,383,107 cells/L). Chlorophyta and cyanobacteria comprised 88% of the biomass. In the subsequent samplings, no bloom, mat or scum (as defined in Directive 2006/7/EC) [35] were visually observed. Furthermore, phytoplankton biomass in samples from NMWN gradually decreased, mainly due to the decrease in the biomass of chlorophytes and dinophytes. The *Dolichospermum* biomass also declined sharply (from 0.75 mg/L in June 2018 to 0.02 mg/L in July 2018 and 0.04 mg/L in August 2018).

At the same period, summer 2018, in the samples from the bathing area the cyanobacteria *Microcystis* spp., *Aphanocapsa* spp., *Dolichospermum* spp., and *Aphanizomenon flos-aquae* were dominant. However, their biomass did not exceed 0.7 mg/L (*Aphanizomenon flos-aquae* in July 2018). The biomass values of the dominant cyanobacteria were measured at higher levels in July (1.6 mg/L), followed by sharp decline in the next months. Regarding the chlorophyte *Sphaerocystis schroeteri*, its biomass was measured at lower levels compared to the measurement of June 2018 at the NMWN sampling point and showed a gradual further decrease during August and September.

During the warm period of 2019 the cyanobacteria biomass (four samples from NMWN sampling point) displayed the opposite trend. Biomass gradually increased during summer until the maximum value of 5.4 mg/L (September 2019), where *Aphanizomenon* spp. biomass was estimated at 2.1 mg/L, *Lemmermanniella* spp. at 2.1 mg/L, and *Raphidiopsis raciborskii* at 0.5 mg/L. In all the other three samples of 2019, *Aphanizomenon* spp. and *Dolichospermum* spp. were the main representatives, though with lower biomass values (up to 0.82 mg/L).

No specific trend was observed during the warm period of 2020. In June 2020 the biomass of cyanobacteria was minimal (0.1 mg/L). Over the next three months it increased up to 1.8 mg/L, as measured in August 2020. In July a *Dolichospermum* species dominated (1.1 mg/L), but low biomass values were measured for *Aphanocapsa* cf. *holsatica* and *Cyanodictyon* species (up to 0.2 mg/L). In August and September, species of the genus *Aphanizomenon* prevailed (1.6 and 0.9 mg/L, respectively). Low biomass values of *Microcystis* species (0.1 and 0.3 mg/L) were also measured. Total phytoplankton and cyanobacteria biomass concentrations measured during the study period are given in Table S1.

2.4. Occurrence of Cyanotoxins (CTs) in Lake Vegoritis

A range of CTs (extracellular and intracellular fractions), including CYN, ATX, dmMC-RR, MC-RR, NOD, MC-YR, MC-HtyR, dmMC-LR, MC-LR, MC-HilR, MC-WR, MC-LA, MC-LY, MC-LW, and MC-LF were determined by LC-MS/MS in samples taken during the study (2018–2020). Results are presented in Tables S2 and S3.

In 2018, 14 samples were analyzed from July to November, seven from each one of the sampling sites. None of these samples were found to contain detectable amounts of CYN, ATX and NOD. The analysis of filtered water showed the presence of extracellular MCs, i.e., MC-LR, MC-RR and MC-YR, at concentrations up to 0.029, 0.023 and 0.014 µg/L, respectively. Intracellular (cell-bound) MCs were also detected, including MC-RR, MC-LR,

MC-YR, and dmMC-RR at concentrations of up to 0.074, 0.055, 0.026 and 0.003 μg/L, respectively.

During 2019, a total of 24 samples from the two sampling sites were analyzed from April to October. ATX and NOD were not detected in any of the samples. Contrary to findings of 2018 monitoring, CYN was detected mainly in the intracellular (cell-bound) fraction, at concentrations ranging from 0.032 to 0.685 μg/L, and at a lower level in the extracellular (dissolved) fraction. MCs were also detected, with MC-LR and dmMC-LR up to 0.233 and 0.080 μg/L (extracellular fraction) and MC-RR, dmMC-LR, MC-LR, and MC-HilR up to 0.029, 0.055, 0.241, and 0.027 μg/L, respectively (intracellular fraction).

In 2020, 20 samples were analyzed from May to October. Co-occurrence of CYN and MCs was again observed with CYN present during this period at concentrations reaching 0.128 and 0.075 μg/L in extra- and intracellular fractions, respectively. MC-RR, MC-HtyR, dmMC-LR, and MC-HilR (extracellular fraction) were also present at concentrations of up to 0.315 μg/L (22 June 2020, Site 2), while dmMC-RR, MC-RR, MC-YR, MC-HtyR, dmMC-LR, MC-LR, and MC-HilR (intracellular fraction) were up to 0.674 μg/L (7 September 2020, Site 1).

The intracellular and extracellular fractions of MCs and CYN, as well as the total concentrations (sum of intracellular and extracellular) per sampling date and site, are presented in Figures 1 and 2. The occurrence of individual CTs in Lake Vegoritis are shown in Figure 3.

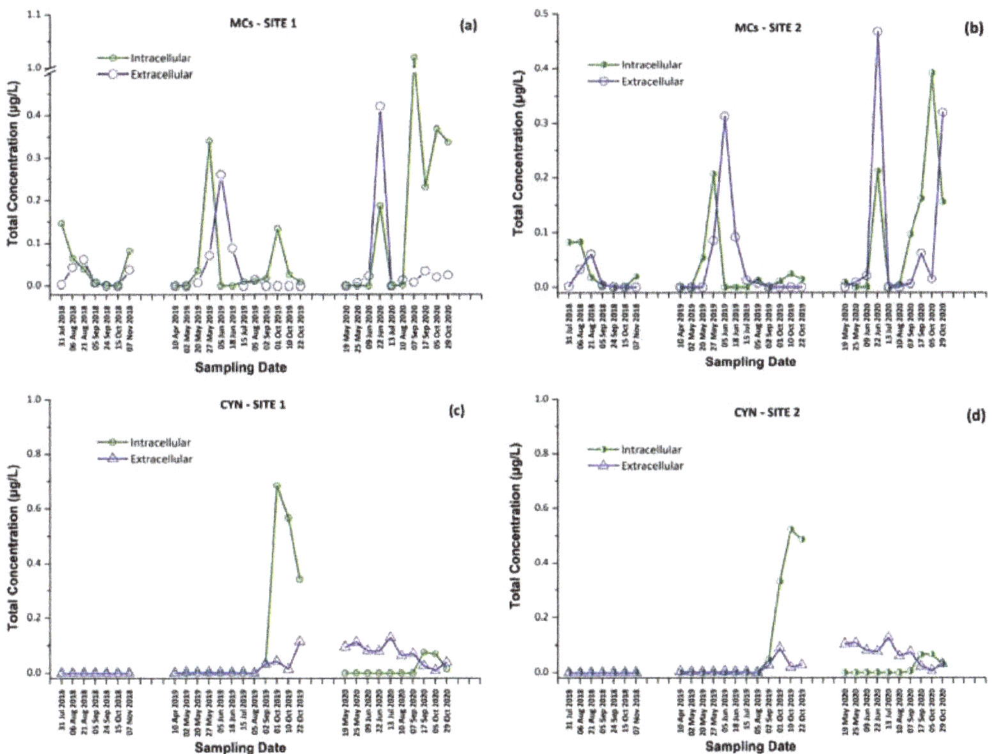

Figure 1. Intracellular and extracellular fractions of microcystins (MCs) and cylindrospermopsin (CYN) per sampling date; MCs at (**a**) Site 1 and (**b**) Site 2, and CYN at (**c**) Site 1 and (**d**) Site 2.

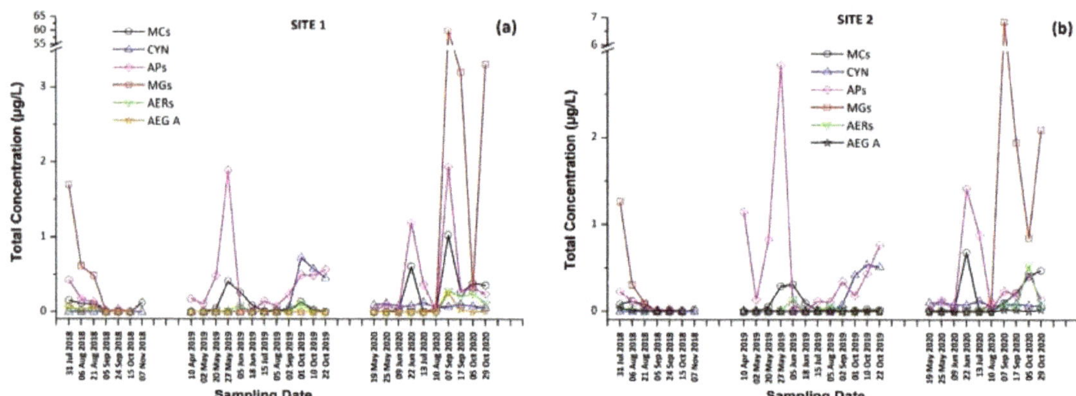

Figure 2. Total concentration (sum of intracellular and extracellular) of cyanotoxins (CTs) and cyanopeptides (CPs) detected per sampling date at (**a**) Site 1 and (**b**) Site 2.

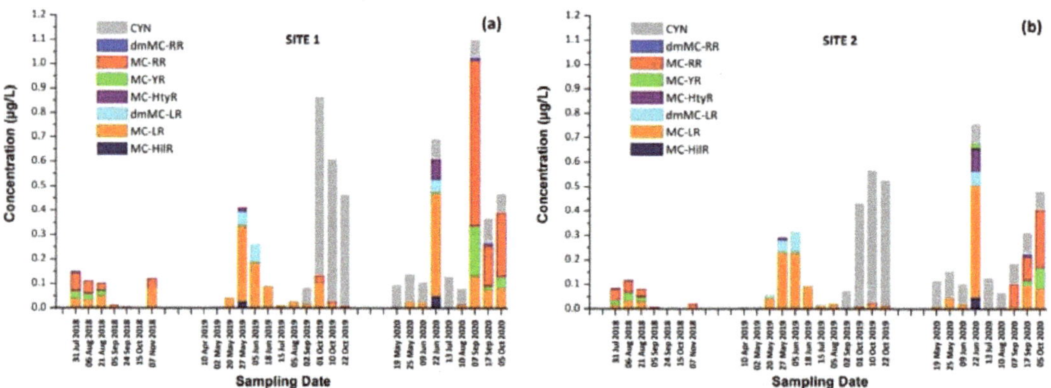

Figure 3. Occurrence of CTs in Lake Vegoritis at (**a**) Site 1 and (**b**) Site 2.

Results for both sampling sites presented a similar seasonal trend in CT concentrations, with an increase during the summer followed by a second outbreak in the autumn. As expected, intracellular MCs tended to appear first, followed by extracellular MCs, due to the release of cell-bound MCs into water after the lysis of cyanobacterial cells, with the exception of June 2020 (Figure 1a,b). The maximum of total MCs, 1.02 µg/L, consisted of dmMC-RR, MC-RR, MC-YR, and MC-LR, and was measured on 7 September 2020 at Site 1, with MC-RR being the most abundant MC variant. The percentage of samples in which each CT was detected is shown in Table 1. MC-LR was detected in 79% of samples through the entire monitoring program, while MC-RR was detected in 50% of samples and was largely absent during 2019.

CYN was also detected in 71% of samples of Lake Vegoritis, while detections occurred only during 2019 and 2020, and not in samples taken in 2018. CYN was found in both extra- and intracellular fractions. The maximum concentration of total CYN was 0.727 µg/L at Site 1 on 1 September 2019 (Figure 1c).

Table 1. Percentage of samples where cyanotoxins (CTs) and cyanopeptides (CPs) were detected during the monitoring period (2018–2020).

	CTs							
	CYN	dm MC-RR	MC-RR	MC-YR	MC-HtyR	dm MC-LR	MC-LR	MC-HilR
% Presence	71	12	50	24	7	17	79	5

	CPs								
	AP B	AP F	Osc Y	MG FR1	MG FR3	MG T1	MG T2	AER 602/K139	AER 298A	AEG A
% Presence	100	98	68	27	27	29	29	45	9	23

The presence of MCs could be related to the dominant cyanobacteria *Dolichospermum* spp. (*Anabaena* spp.), *Aphanizomenon* spp., and *Microcystis* spp. that were identified in Lake Vegorits during the study period [36]. The presence of CYN may also be attributed to the cyanobacteria *Aphanizomenon* spp. and *Dolichospermum* spp. as well as to *Raphidiopsis raciborskii* [12,37]. These results concur with a previous study where Lake Vegoritis was found to be mainly dominated by *Dolichospermum* spp., *Aphanizomenon* spp., and *Microcystis* spp. [31]. In the same study, CYN, MC-RR and MC-LR were identified in one sample of biomass from Lake Vegoritis (September 2008) at concentrations of <LOQ, 0.118 and 0.049 µg/L, respectively. In a water sample (July 2014), extracellular dmMC-RR, MC-RR, MC-YR, dmMC-LR, MC-LR, and MC-LY (MC-RR: 104 µg/L and MC-LR 96.3 µg/L) were also detected [31]. Although the findings were based on only two samples, the toxin profile (MC-RR, MC-LR) is in agreement with the present study. However, the concentrations reported in the previous study were far higher, possibly because the sampling was targeted rather than systematic, aiming at localized bloom formations.

While there are no recreational beach monitoring programs for toxins in Greece, this is the first study to investigate the presence, the concentration, and the diversity of CTs in a popular lake beach of North Greece devoted to recreational activities. In all cases, the measured concentration of CTs did not exceed the provisional guideline values proposed for recreational water by the World Health Organization (WHO) that have been recently updated and were set at 24 and 6 µg/L for MC-LR and CYN, respectively [38,39].

2.5. Detection, Identification and Occurrence of Cyanobacterial Peptides (CPs) in Lake Vegoritis

Although the occurrence of MCs in fresh water bodies is well documented due to the development of analytical protocols [1,31,33,40], data bases [41,42], and a number of commercially available standards, less is known regarding the presence of other CPs. Recent studies showed the presence of CPs in water bodies [17,43], but analysis was mainly done in cyanobacterial biomass, not in water samples. Analytical protocols have not been developed and validated for water samples (extracellular–intracellular fractions), to include cleanup and pre-concentration steps [22]. In this study, we present method performance and validation results for targeted LC-MS/MS analysis of water samples for CPs, based on an analytical workflow previously used for analysis of CTs [33]. The targeted CPs were MG FR1, MG FR3 MG T1, MG T2, AER 602/K139, AER 298A, AEG A, AP B, AP F and OSC Y. The validated method was then used to analyze water samples from Lake Vegoritis.

2.5.1. Chromatographic Separation and MS/MS Identification of CPs

A sample of cyanobacterial mass from a bloom in Lake Kastoria, Greece (September 2014) containing all target CPs was used as a reference sample. Efficient chromatographic separation of target CPs was achieved with a reversed-phase C18 column (Atlantis T3, Waters), previously applied for MCs, NODs, CYN and ATX [33].

Identification of CPs was performed using tandem mass spectrometry in a multiple reaction monitoring (MRM) mode (Table 2). The total ion chromatogram (TIC) and MRM chromatograms of the selected quantifier transitions obtained from cyanobacterial mass extract from Lake Vegoritis (7 September 2020, Site1) are presented in Figure 4. MGs presented a characteristic fragment ion, at m/z 128.2, attributed to a part of Ahda and, at m/z 162.1, to a part of chlorinated Ahda [44,45]. The most intense common ions of MGs, which share the amino acid sequence proline (Pro)-tyrosine (Tyr)-tyrosine (Tyr) at the C-terminus in their structure (i.e., MG FR3, MG T1 and MG T2), were at m/z 233.0 [Pro-Tyr-CO + H]$^+$ and m/z 442.2 [Pro-Tyr-Tyr + H]$^+$ [45,46]. AERs were characterized by m/z 140.0 and m/z 122.0, which are the Choi immonium ion and dehydrated Choi immonium ion, respectively [47]. Additionally, m/z 221.2 was attributed to Leucine (Leu)-Choi fragment (or Isoleucine (Ile)-Choi fragment), while m/z 311.0 was indicative of the presence of the arginine derivative—argininol in the structure of AERs [45,48]. APs were characterized by m/z 84, which corresponds to the lysine (Lys) immonium ion [49]. The fragment ion m/z 201.0 is characteristic of APs that contain arginine (Arg) as a side chain [49]. Fragment ions from the loss of the side chain amino acid with the CO linkage (i.e., m/z 637.3 for AP B and m/z 651.4 for AP F) or amino acid from the ring (i.e., m/z 681.4 for OSC Y [50]) were also considered. AEGs were characterized by m/z 112.0 and was annotated in a previous study as TzlCO in case of AEG A, since it is a common ion in the fragmentation spectra of some other AEGs [20]. The structure of fragment ions m/z 86.0 and m/z 154.2 was proposed in the frame of this study as [PreNH$_3$]$^+$ and [(Pre)$_2$NH$_2$]$^+$, respectively (Figure 5). The fragmentation pathways of AEG A, involving the m/z 154.2 and 86.0 fragment ions based on in silico fragmentation (Mass Frontier 8.0, Thermo Scientific), are presented in Figure S3.

Table 2. LC–MS/MS detection parameters of CPs.

Cyanopeptide	t_R (min)	Precursor Ion	Product Ions	Collision Energy (eV)	Product Ion Assignment	Ref.
MG FR1	18.9	728.0 [M+H]$^+$	100.0	40	MeLeu immonium ion	[46]
			128.2 ^{Q}	40	Ahda fragment (C$_8$H$_{18}$N)	[45]
			384.2	40	[M + H-Tyr-Tyr]$^+$	[46]
MG FR3	15.6	728.0 [M+H]$^+$	128.2	40	Ahda fragment (C$_8$H$_{18}$N)	[45]
			233.0 ^{Q}	40	[Pro-Tyr-CO + H]$^+$	[45]
			442.0	40	[Pro-Tyr-Tyr + H]$^+$	[46]
MG T1	15.5	732.0 [M+H]$^+$	162.1	40	Cl-Ahda fragment (C$_8$H$_{17}$NCl)	[44]
			233.0 ^{Q}	40	[Pro-Tyr-CO + H]$^+$	[45]
			442.2	40	[Pro-Tyr-Tyr + H]$^+$	[46]
MG T2	15.7	698.0 [M+H]$^+$	128.2	40	Ahda fragment (C$_8$H$_{18}$N)	[45]
			233.0 ^{Q}	40	[Pro-Tyr-CO + H]$^+$	[45]
			442.2	40	[Pro-Tyr-Tyr + H]$^+$	[46]
AER 602/K139	13.8	603.2 [M+H]$^+$	122.0	40	[Choi immonium-H$_2$O]$^+$	[47]
			140.0	40	Choi immonium ion	[47]
			221.2 ^{Q}	40	Leu-Choi fragment	[45]
AER 298A	13.6	605.3 [M+H]$^+$	122.0	40	[Choi immonium-H$_2$O]$^+$	[47]
			140.0	40	Choi immonium ion	[47]
			311.0 ^{Q}	40	[Choi-Argininol-NH$_2$ + H]$^+$	[48]
AEG A	24.6	561.4 [M+H]$^+$	86.0	40	[PreNH$_3$]$^+$	This study
			112.0	40	TzlCO	[20]
			154.2 ^{Q}	40	[(Pre)$_2$NH$_2$]$^+$	This study
AP B	14.8	837.4 [M+H]$^+$	84.0	40	Lys immonium ion	[49]
			201.1 ^{Q}	40	CO-Arg (side chain)	[49]
			637.3	40	[Lys-Phe-MeAla-HTyr-Val + 2H]$^+$	[49]
AP F	15.2	851.3 [M+H]$^+$	84.0	40	Lys immonium ion	[49]
			201.0 ^{Q}	40	CO-Arg (side chain)	[49]
			651.4	40	[Lys-Phe-MeAla-HTyr-Ile + 2H]$^+$	[49]
OSC Y	19.9	858.4 [M+H]$^+$	84.0	40	Lys immonium ion	[50]
			405.0 ^{Q}	40	[M + H-Tyr-(Htyr-Ile)]$^+$	[50]
			681.4	40	[M + H-Htyr]$^+$	[50]

^{Q} quantifier ion.

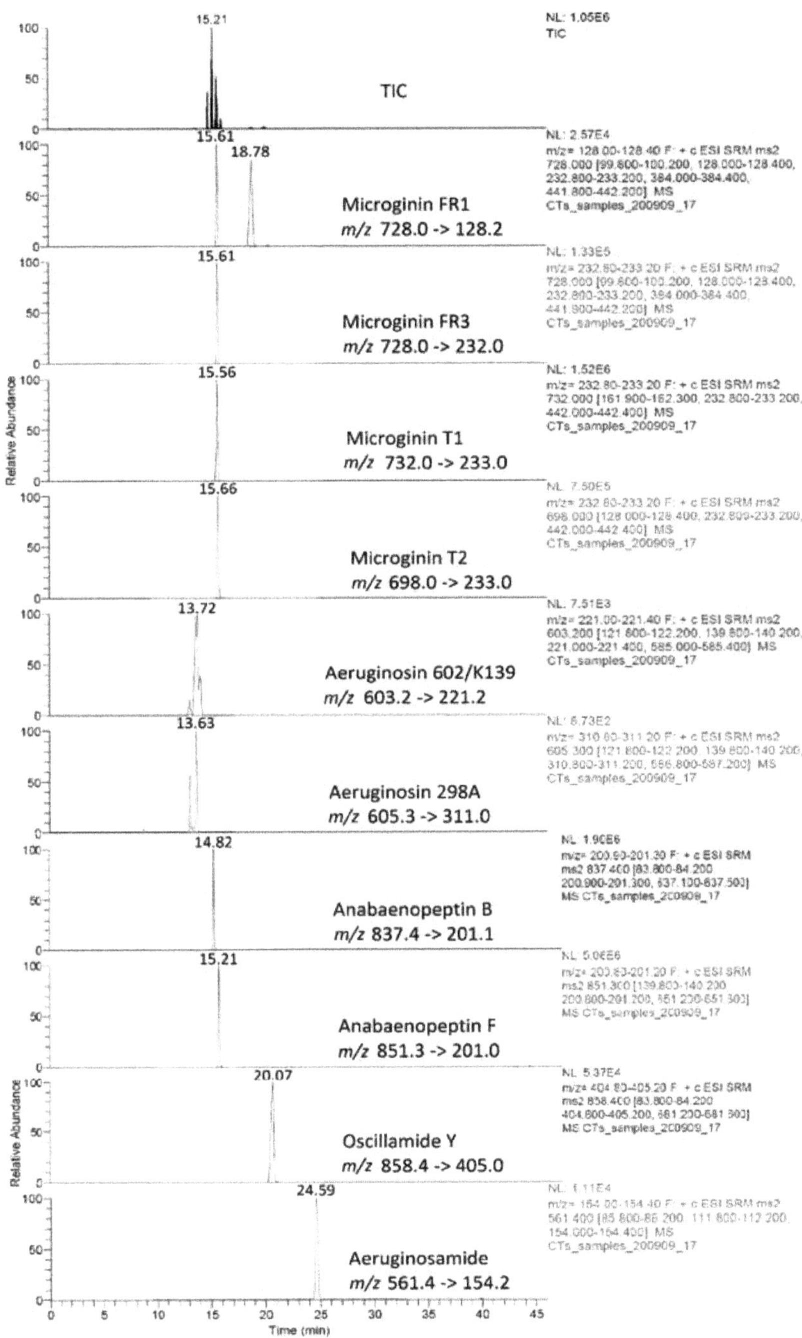

Figure 4. Example of TIC and MRM chromatograms of quantifier transitions for the intracellular fraction of CPs (sample taken on 7 September 2020, Site 1).

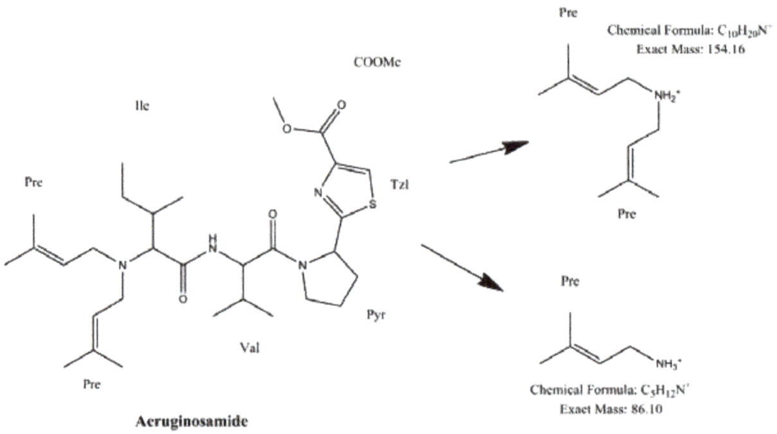

Figure 5. Structure of proposed fragment ions of AEG A.

2.5.2. Method Performance and Validation Results

The ability of the method for accurate quantification was assessed for AP B, for which an analytical reference standard was available. The response (quantification ion peak area) over the range 5–100 μg L^{-1} was linear (r$^2 \geq$ 0.999). Precision, expressed as relative standard deviation (%RSD), was 8.6% under repeatability (n = 3) conditions and 15.9% under reproducibility (different days, n = 15) conditions. The limit of detection (LOD) of AP B was 0.001 μg/L and the LOQ was 0.003 μg/L. The LOD was estimated from measurements (n = 8) of standard solution (5 μg/L) using the formula: LOD = t ($n-1$, 0.95) × SD, where t ($n-1$, 0.95) was the t-test value for n—1 degrees of freedom at 95% confidence level, (1.895 for n = 8) and SD was the standard deviation of measurements. Limit of quantification (LOQ) was estimated as 3 × LOD.

Quantification of APs was carried out using the class equivalent approach with concentrations expressed as AP B equivalents, while for the rest CPs concentrations were expressed as MC-LR equivalents.

Recoveries of CPs (extracellular and intracellular fractions) were evaluated by analyzing spiked samples using CP-free water and cyanobacterial biomass as matrices and the reference sample from the Lake Kastoria bloom for spiking. Results are presented in Table 3. Recovery experiments were carried out in triplicate and mean recoveries in the extracellular fraction ranged from 77.0–129.2% for all target CPs, except for AEG A which was poorly recovered (17.1%) and AER 602/K139 that showed a recovery of 163.5%. Mean recoveries in the intracellular fraction were in the range of 73.4–98.3%, except for AEG A which had a low recovery (7.5%) (Table 3). In all recovery estimations, %RSD was <28.4%.

The validated analytical protocol can be used for detection and identification of target CTs and CPs of various chemical classes using a single analytical method. It could further serve as a basic template for analysis of cyanobacterial metabolites, expanding to more CTs and CPs in the future as they become commercially available as standards or included in mass spectral databases.

Table 3. Recoveries of target CPs from water (extracellular and intracellular).

	MG FR1	MG FR3	MG T1	MG T2	AER 602/K139	AER 298A	AEG A	AP B	AP F	Osc Y
Extracellular Recovery (%RSD, $n = 3$)	103.7% (9.5)	79.7% (9.2)	86.5% (6.7)	77.0% (6.3)	163.5% (7.4)	129.2% (8.8)	17.1% (25.2)	102.7 (8.6)	108.6% (6.5)	95.8% (6.5)
Intracellular Recovery (%RSD, $n = 3$)	74.0% (16.9)	75.5% (2.2)	75.1% (9.9)	75.6% (13.8)	88.8% (5.6)	98.3% (13.7)	7.5% (28.4)	87.2% (3.3)	96.5% (9.1)	73.4% (10.0)

2.5.3. Occurrence of CPs in Lake Vegoritis

Concentrations of extracellular and intracellular target CPs (MG FR1, MG FR3, MG T1, MG T2, AER 602/K139, AER 298A, AEG A, AP B, AP F, and OSC Y) during the monitoring period are presented in Tables S4 and S5, respectively.

In 2018, MG FR1, MG FR3, MG T1, MG T2, AER 602/K139, and OSC Y were found only in intracellular fraction, with MG T1 to be the most abundant one, reaching 1.16 µg/L. AEG A, AP B, and AP F were found mostly in the intracellular fraction. AER 298A was not detected in any sample. During 2019, none of the samples analyzed were found to contain detectable amounts of MG FR1, MG FR3, MG T1, MG T2, and AEG A. AER 602/K139 and AER 298A were found only in the intracellular fraction. Target APs were all present in both extracellular and intracellular fractions, with the intracellular being at higher concentrations than the extracellular. In 2020, MG FR1, MG FR3, MG T1, MG T2, AER 298A, and AEG A were present only in intracellular form with MG T1 found to be the most abundant, reaching 47.0 µg/L. AER 602/K139, AP B, AP F, and OSC Y were found in both extracellular and intracellular form. AP F was the most abundant, up to 1.382 µg/L in the intracellular fraction, while AER 602/K139 was the most abundant (0.154 µg/L) in the extracellular fraction. The results present a similar trend in the CPs profile in both sampling sites (Figure 6).

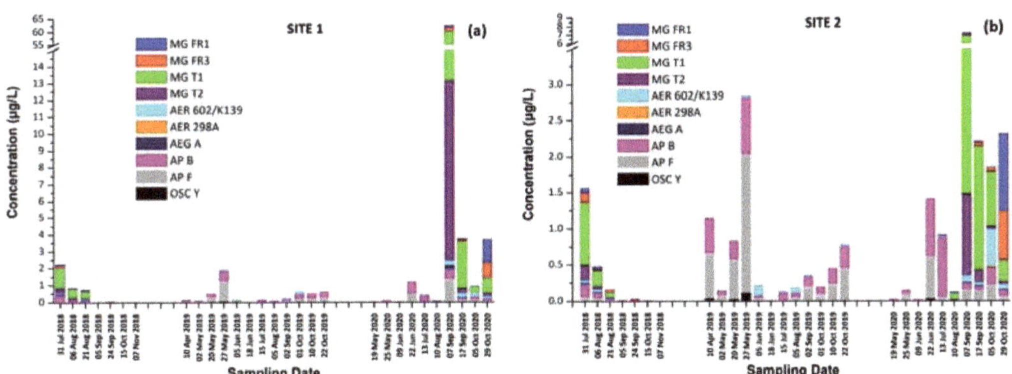

Figure 6. Occurrence of CPs detected in Lake Vegoritis at (**a**) Site 1 and (**b**) Site 2.

These findings consist of the first report of the occurrence of a variety of CPs, in addition to MCs and CYN, in Lake Vegoritis (Figure S4), showing that all 10 target CPs were detected, with MGs and Aps found to be the most abundant classes of CPs compared to other classes (Figure 7). The highest concentration of total MGs (sum of intra- and extra-cellular), 60.0 µg/L, was measured on 7 September 2020 (Site 1), and consisted of MG FR1, MG FR3, MG T1, and MG T2, where MG T1 was the most abundant. The maximum of total APs (AP B, AP F, and OSC Y), 2.83 µg/L (Figure 2), was measured on 27 May 2019 (Site 2), with AP F presenting the highest concentration (Figure 6).

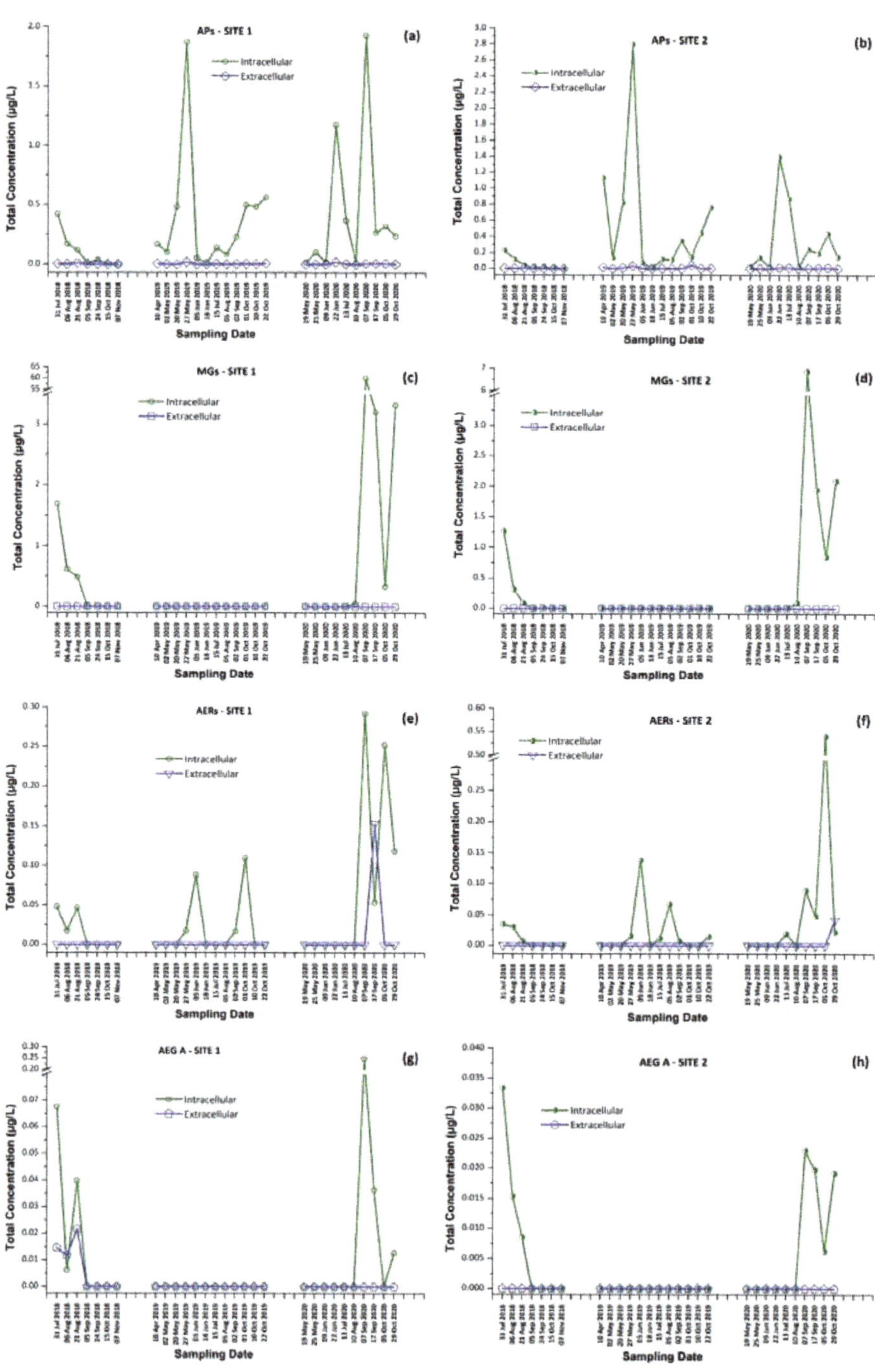

Figure 7. Intracellular and extracellular fractions of APs, MGs, AERs and AEG A per sampling date; APs at (**a**) Site 1 and (**b**) Site 2; MGs at (**c**) Site 1 and (**d**) Site 2; AERs at (**e**) Site 1 and (**f**) Site 2; and AEG A at (**g**) Site 1 and (**h**) Site 2.

To date, there is only one report in the literature related to the presence of MGs in cyanobacterial bloom samples collected from Greek lakes other than Vegoritis, providing only qualitative data with no information on the presence of MGs in the water phase [32]. According to that study, the most frequently detected MGs were MG FR1 (70% of samples) followed by MG T1 (52%). In another study on the occurrence of APs in the freshwater bodies of Greece, the presence of AP B and AP A was reported, however, analysis was carried out using HPLC-UV without confirmation by mass spectrometry [51].

There is still lack of information concerning the occurrence of AERs and AEG A in Greek lakes. Furthermore, there is a gap in the knowledge related to the co-existence of CPs with CTs which may cause adverse health effects to humans and animals. In the frame of the present study, it was found that in water samples of Lake Vegoritis, AP B and AP F, which were detected in almost all samples (100% and 98%, respectively), co-existed with the frequently found MC-LR, CYN, and MC-RR (79%, 71%, and 50%, respectively), (Table 1). OSC Y was also present in 68% of the samples. The frequency of occurrence of the rest of the CPs ranged from 9% to 45%, while the detected CTs ranged from 5% to 24%, respectively.

Cyanobacteria possess a great metabolic potential and are able to co-produce several peptides from different classes [15]. MGs and AEG A exhibited a similar trend in Lake Vegoritis and their production could be attributed to *Microcystis* spp. [21,47,52], since both CP classes were detected in 2018 and 2020, but not in 2019, similarly to *Microcystis* spp. Low concentrations of AERs were determined in all sampling periods and their presence in Lake Vegoritis might be related to *Aphanizomenon* spp. and *Dolichospermum* spp. [53,54] as well as *Microcystis* spp. [18]. *Dolichospermum* spp. (*Anabaena* spp.), the dominant cyanobacterial species during all sampling periods, are possibly related to the high frequency of APs detection, especially AP B, which was present in all samples from Lake Vegoritis [55].

Our results are in agreement with other studies investigating the presence of CPs in inland water bodies. MCs, Aps, and AERs were the main cyanobacterial metabolites identified in biomass during a bloom episode in a dam for drinking water on Lake Occhito, near the town of Foggia in Southern Italy [56]. Similarly, MCs and APs were identified in the biomass from Siemianówka Dam Reservoir (northeast Poland) during a study from 2009 to 2012 [57]. In a more recent study of six eutrophic lakes in USA by Beversdorf et al., APs were detected in all lakes together with MGs at concentrations of the same order of magnitude found in Lake Vegoritis [22]. Similarly, both MCs and CPs (APs, MGs, and cyanopeptolins) were detected in surface and raw drinking waters from the eutrophic Lake Winnebago, Wisconsin [58]. In another study on the diversity and spatial distribution of MCs, NODs, Aps, and MGs in Green Bay, Lake Michigan, an important recreational resource, the presence of MCs (mainly MC-RR and -LR) and CPs (mainly APs and MGs) was reported with the mean of total MCs and APs as 1.28 and 0.20 µg/L, respectively [23].

The occurrence of cyanobacterial metabolites was also reported for lake water samples (mostly recreational) from Canada, in the frame of a collaborative citizen-science project. CTs were present in 75% of the samples, from ng/L up to µg/L, and AP A and AP B were in 38% of the samples at concentrations up to 10 µg/L [59]. High levels of APs (µg mg/L) were also detected in water samples from the Sau-Susqueda-El Pasteral reservoir system in Spain in the autumn of 2015, although MCs were <0.3 µg/L [60].

In most of the published studies, the intracellular (cell-bound) or total concentration (sum of extracellular and intracellular fractions) of CPs was reported. There is a general lack of knowledge on whether CPs are released from cyanobacterial cells into water or remain as cell-bound in the intracellular fraction [58]. In the frame of the current study, as intra- and extracellular fractions were quantified separately, it seems that the CPs studied are mostly intracellular, with the extracellular fraction concentrations being generally lower by an order of magnitude (Tables S4, S5 and Figure 7).

From Figure 2, it was also observed that the maximum concentrations of CPs occurred at the same time period with CTs, following the same trend. Taking into consideration the co-occurrence of CTs with CPs of different classes with unknown behavior, bioactivity and environmental levels in lake water, more research and monitoring programs are urgently needed for assessing possible threats to humans and the environment. This study was the first of this kind in Greece and in Lake Vegoritis, where both recreational and fishing activities take place.

3. Conclusions

This study reports, for the first time, the co-occurrence of CTs and CPs in Lake Vegoritis, situated in the north of Greece. Furthermore, the study describes the variant-specific changes of CT and CP profiles over a 3-year monitoring period as well as the basic water quality parameters of the lake and phytoplankton-cyanobacteria composition.

In order to realize the study, a previously validated LC-MS/MS analytical protocol for the simultaneous determination of multi-class cyanotoxins in water was further extended and validated to be applied for the detection of multiple CPs, i.e., MGs (MG FR1, MG FR3 MG T1, and MG T2), AERs (AER 602/K139 and AER 298A), AEG A and APs (AP B, AP F, and OSC Y). Samples of Lake Vegoritis were analyzed for CTs and CPs in both extracellular and intracellular fractions.

Using the above protocol, CTs were detected in two sampling sites of Lake Vegoritis, during 2018–2020, consisting of MC-LR, MC-RR, MC-YR, dmMC-LR, dmMC-RR, MC-HtyR, MC-HilR and CYN, with MC-LR (79%) being the most frequently detected, followed by CYN (71%). The concentrations of MCs and CYN were generally low (<1 µg/L) and they did not exceed the guideline values proposed for recreational water by WHO (24 and 6 µg/L for MC-LR and CYN, respectively). CPs belonging to the classes of APs, AERs, MGs, and AEG A were also detected, with AP B and AP F present in almost all water samples. The co-occurrence of two potent cyanotoxin classes (MCs and CYN) with multiple CPs in the lake's water supports the need for future studies on the interactions between multiple cyanobacteria metabolites with regards to possible effects on human health.

4. Materials and Methods

4.1. Study Area Description and Sample Collection

Lake Vegoritis, one of the largest lakes in Greece, is located in Western Macedonia region in North-Western Greece and occupies the lowest area of the Ptolemaida basin (Figure 8a). It is considered as one of the most important water resources of Western Macedonia for its multiple uses and benefits for humans [34]. The investigative monitoring of the condition of Lake Vegoritis was designed and implemented from 2018 to 2020. In compliance with the WFD provisions [35], integrated samples (from the euphotic zone, 2.5× Secchi Depth) were taken from the pelagic zone of the lake (40°44′40.70″N, 21°47′3.90″E, National Monitoring Water Network sampling point, NMWN point), with a Nansen water sampler (Hydro-Bios, Germany). Physicochemical features were measured seasonally and each month during the growing season (May to October). For phytoplankton and chlorophyll α analysis, 2–4 samples were obtained during the growing season. Grab samples (1.5 L) for toxins analysis were collected in disposable plastic bottles from the surface layer of the lake (0–30 cm) at least monthly, from July to November, April to October, and May to October during 2018, 2019 and 2020, respectively. Sampling took place at two points: Site 1, 40°43′29.5″N, 21°45′13.6″E from the designated bathing area of the lake, and Site 2, 40°43′12.1″N, 21°45′07.2″E from the pier (Figure 8b). The samples were transported to the laboratory within 24 h of collection, in dark containers and at low temperature (≈4 °C).

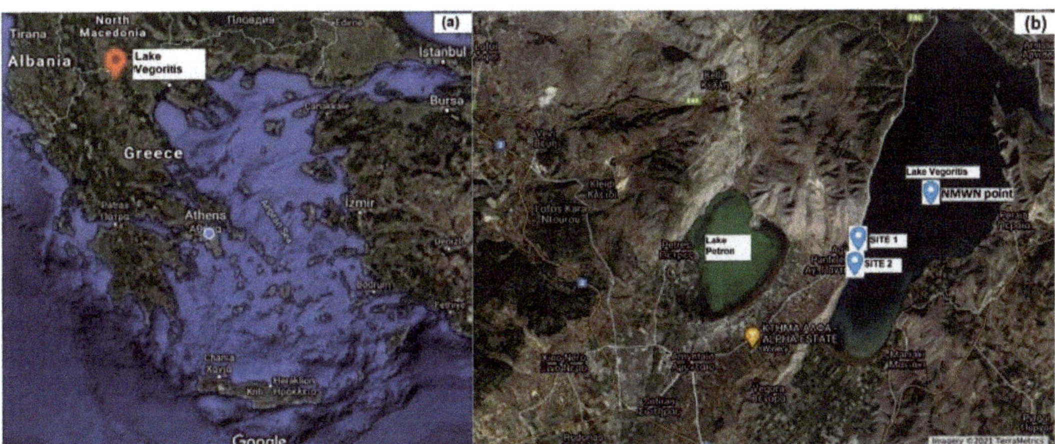

Figure 8. The (**a**) map and (**b**) sampling points of Lake Vegoritis.

4.2. Chemicals and Instrumentation

[D-Asp^3]MC-LR, [D-Asp^3]MC-RR, MC-WR, MC-HtyR, MC-HilR, MC-LY, MC-LW, MC-LF, and AP B standards were supplied by ENZO Life Science. MC-RR, MC-LR, MC-YR, MC-LA, and NOD standards were supplied by Sigma-Aldrich. CYN was purchased from Abraxis and ($\pm$) Anatoxin-a fumarate from TOCRIS Bioscience. All substances had a purity of >95%. Organic solvents i.e., acetonitrile (ACN) and methanol (MeOH) of HPLC grade (99.9%) and dichloromethane (DCM) of analytical reagent grade (99.9%), were supplied by Fisher Chemical. Formic acid (FA) was obtained from Riedel de Haen. Ultra-pure water (18.2 MΩ cm at 25 °C) was produced in the lab using a Temak TSDW10 water purification system (TEMAK, Athens, Greece).

Lake water samples were filtered through glass fiber filters with a 47 mm diameter and 0.7 μm pore size (Millipore, Cork, Ireland) and a glass vacuum filtration device. Solid-phase extraction (SPE) was carried out using a 12-port SPE vacuum manifold with large volume samplers connected through PTFE tubes (Supelco, Bellefonte, PA, USA) and a diaphragm vacuum pump (KNF Laboport, Freiburg, Germany). SPE cartridges used for cleanup and pre-concentration purposes were Supel-Select HLB (bed wt. 200 mg, volume 6 mL, Supelco, St. Louis, MO, USA) and Supelclean ENVI-Carb (bed wt. 250 mg, volume 3 mL, Supelco, St. Louis, MO, USA).

The analysis of target analytes was carried out using a TSQ Quantum Discovery Max triple-stage quadrupole mass spectrometer (Thermo Electron Corporation, San Jose, PA, USA), with an electrospray ionization (ESI) source coupled to a Finnigan Surveyor LC system, equipped with a Surveyor AS autosampler (Thermo Electron Corporation, San Jose, PA, USA). Xcalibur software 2.0 was used to control the MS parameters for data acquisition and data analysis. The chromatographic column used was an Atlantis T3 (2.1 mm × 100 mm, 3 μm, Waters, Wexford, Ireland).

4.3. Physico-Chemical Parameters

Ion chromatography (IC) was applied for separation, analysis, and quantification of both anions (F$^-$, Cl$^-$, NO$_2$$^-$, Br$^-$, NO$_3$$^-$, PO$_4$$^{3-}$, and SO$_4$$^{2-}$) and cations (Na$^+$, K$^+$, Mg^{2+}, and Ca^{2+}) in water samples. Subsamples were filtered through 0.45 μm glass fiber filters within 48 h of sampling and kept at 2–6 °C for up to 4 days. For cation analysis, samples were acidified to a pH of 3 $\pm$ 0.5 with nitric acid. Dissolved anions and cations were determined using a Dionex Aquion IC System (Thermo Fisher Scientific, Waltham, MA, USA) according to EN ISO 10,304 and 14,911 guidelines, respectively [61,62].

Determination of total phosphorus (TP) was carried out in unfiltered water subsamples (about 250 mL, stored at $-18\ ^\circ$C for up to six months), using the ascorbic acid method following persulfate digestion [63]. Orthophosphates were determined based on spectrophotometry (Hitachi U-5100 UV/VIS, Hitachi, Tokyo, Japan).

For determination of total suspended solids (TSS), a measured volume of a water subsample (about 500 mL) was filtered through a pre-weighed glass microfiber filter (Whatman Grade 934-AH® RTU circles). The filter was heated at 103–105 $^\circ$C for an hour and then weighed. The TSS were calculated as the mass increase divided by the water volume filtered [63]. Analysis was conducted within seven days from sampling and the subsamples were kept at 4 $^\circ$C in the dark.

4.4. Chlorophyll α Analysis

Subsamples of 1 L were filtered through Whatman GF/F glass fiber filters within 48 h of sampling. Chlorophyll α was measured using 90% acetone and application of the trichromatic equation [63,64]. Absorbance was measured using a Cary 60 UV-Vis spectrophotometer (Agilent Technologies, Santa Clara, CA, USA).

4.5. Microscopic Analysis

For phytoplankton analysis, subsamples (about 500 mL) were transferred to plastic bottles and preserved with acid Lugol's solution [65]. Additional samples were obtained by vertical plankton net hauls (20 µm mesh, Hydro-Bios, Altenholz, Germany) through the euphotic zone, and preserved with formaldehyde. Phytoplankton identification and enumeration were based on the Utermöhl settling technique [66] as described in EN 15,204 guidelines [65]. The analysis was carried out using sedimentation chambers (Hydro-Bios) and an inverted microscope Zeiss AxioObserver.A1 equipped with an AxioCam (Carl Zeiss, Oberkochen, Germany). The phytoplankton biomass was estimated following EN 16,695 guidelines [67].

4.6. Analysis of CTs and CPs

4.6.1. Sample Preparation

Analysis of CTs (CYN, ATX, NOD, dmMC-RR, MC-RR, MC-YR, MC-HtyR, dmMC-LR, MC-LR, MC-HilR, MC-WR, MC-LA, MC-LY, MC-LW, and MC-LF) and CPs (MG FR1, MG FR3, MG T1, MG T2, AER 602/K139, AER 298A, AEG A, AP B, AP F, and OSC Y) in water samples was carried out by liquid chromatography coupled with tandem mass spectrometry (LC-MS/MS). For the determination of intra- and extracellular CTs and CPs, water samples were first filtered through GF/F filters and then the filters and filtered water were analyzed. Intracellular CTs and CPs were extracted from the filters' biomass by an extraction mixture containing 75% MeOH:25% H_2O. After evaporation of the extract and reconstitution with MeOH: H_2O (5:95 *v/v*), the final solution was injected into the LC-MS/MS for analysis [31]. Filtered water samples were pre-treated using the dual cartridge (HLB and Envi-Carb) SPE process [33]. Briefly, water samples, after adjustment to pH 11, were passed through a dual cartridge assembly of HLB and ENVI-Carb. Recovery of extracellular CTs and CPs was achieved by reversing the cartridges and eluting with a mixture of 10 mL DCM:MeOH (40:60, *v/v*), containing 0.5% FA. The extract was dried and the residue was re-dissolved with 400 µL MeOH: H_2O (5:95, *v/v*) prior to LC-MS/MS analysis.

4.6.2. LC-MS/MS Analysis

A gradient elution program was applied for chromatographic separation with solvents (A) ACN and (B) water, both containing 0.5% FA. The gradient started at 5% A (held for 3 min), which increased to 20% A in 1 min (held for 2 min), further to 35% A in 1 min (held for 7 min), 70% A in 14 min, and finally 90% in 1 min (held for 3 min). An equilibration time of 10 min was kept after each sample run. The flow rate was set at 0.2 mL/min with 20 µL injection volume and the column temperature was set at 30 $^\circ$C.

Electrospray ionization (ESI) in positive mode was used and the three most intense and characteristic precursor–product ion transitions were selected for detection and identification of each compound in MRM mode. LC-MS/MS detection parameters for the target CTs was set according to Zervou et al. [33]. Parameters for the CPs detection are given in Table 2. The selection of MRM transitions was based on fragmentation spectra from previous studies [46,49,50,57,68] or in the frame of this study. In all cases, single protonated $[M+H]^+$ ions were set as the precursor ions. The most intense product ion was chosen to be the quantifier ion, while quantification was performed using external standards at concentrations of 5, 20, and 100 µg/L. For APs, since only AP B was commercially available as a standard, quantification was carried out using the class equivalent approach with AP F and OSC Y concentrations expressed as AP B equivalents. All other CPs for which no class equivalent standard was available, were expressed as MC-LR equivalents [69].

4.6.3. Validation of Methods for the Determination of CPs

The analytical work flow used for the target CTs [31] was also validated for its performance for CPs. Recovery studies were carried out by spiking samples with an extract of cyanobacterial mass (Lake Kastoria, September 2014) containing all target CPs. The TIC and MRM chromatograms of the selected quantification ions obtained from cyanobacterial mass extract from Lake Kastoria (September 2014) are shown in Figure S5.

To obtain the extract of cyanobacterial mass used for spiking, 10 mg of lyophilized biomass (Lake Kastoria, September 2014) was extracted two times with 1.5 mL of 75% MeOH: 25% H_2O and a third time with 1.5 mL *n*-butanol. Each time the mixture was vortexed, sonicated for 15 min in a sonication bath (Bandelin Sonorex Super RK106) and then centrifuged at 4000 rpm for 10 min at room temperature (DuPont RMC-14 Refrigerated Micro-Centrifuge, Sorvall Instruments, Newtown, CT, USA) and the supernatant was separated from the pellet. All supernatants were pooled together. One milliliter of the extract was evaporated to dryness, the residue was re-dissolved in 400 µL of MeOH: H_2O (5:95 *v/v*) and analyzed by LC-MS/MS. The rest of the extract was used for spiking CP-free biomass (retained on filter after passing 200 mL of lake water) and filtered water samples (400 mL) to carry out recovery experiments. Additionally, recovery experiments were carried out by spiking filtered CP-free water samples (400 mL) with AP B at the concentration level of 100 ng/L.

Supplementary Materials: The following are available online at https://www.mdpi.com/article/10.3390/toxins13060394/s1, Figure S1: structures of several classes of cyanotoxins. For the classes of *Microcystins* and *Nodularins*, R_x stands for variable L-amino acids, Figure S2: structures of several classes of cyanopeptides. For the classes of *Anabaenopeptins* and *Microginins*, A_x stands for variable L-amino acids, Figure S3: fragmentation pathways of AEG A involving the *m/z* 154.2 and 86.0 fragment ions. Fragmentation pathways predicted by Mass Frontier 8.0 software, Figure S4: heat maps showing all CP and CT concentrations detected in Lake Vegoritis throughout the two sampling points (Site 1 and Site 2), Figure S5: TIC and MRM chromatograms of quantifier ions for the cyanopeptides obtained from the biomass extract of Lake Kastoria (sampling September 2014), and Table S1: physico-chemical parameters, total phytoplankton biomass and cyanobacteria biomass concentrations measured during the study period. Table S2: concentrations of extracellular CTs from Lake Vegoritis (µg/L), Table S3: concentrations of intracellular CTs from Lake Vegoritis (µg/L), Table S4: concentrations of extracellular CPs from Lake Vegoritis (µg/L), and Table S5: concentrations of intracellular CPs from Lake Vegoritis (µg/L).

Author Contributions: Conceptualization, E.G. and A.H.; Data curation, S.-K.Z., K.M. and V.T.; Formal analysis, S.-K.Z., K.M., A.P., C.C. and V.T.; Funding acquisition, A.H.; Investigation, K.M., E.G., V.T. and A.H.; Methodology, S.-K.Z., K.M., E.G., V.T. and A.H.; Project administration, Theodoros M. Triantis and A.H.; Resources, K.M., E.G., V.T. and A.H.; Software, S.-K.Z.; Supervision, A.H.; Validation, S.-K.Z., T.K. and T.M.T.; Visualization, S.-K.Z. and T.M.T.; Writing—original draft, K.M., V.T. and A.H.; Writing—review & editing, S.-K.Z., T.K. and T.M.T. All authors have read and agreed to the published version of the manuscript.

Funding: The cyanotoxins data presented in this research were financed by the Decentralized Administration of Epirus—Western Macedonia through the program "Determination of cyanotoxins in water samples of Lake Vegoritida" (Duration: 4/2019-4/2021). The phytoplankton, chlorophyll α, and physicochemical data used in this research come from Act MIS 5001204 financed by the European Union Cohesion Fund (Partnership Agreement 2014–2020).

Institutional Review Board Statement: Not applicable.

Informed Consent Statement: Not applicable.

Data Availability Statement: Data is contained within the article or Supplementary Material.

Acknowledgments: The authors acknowledge Vasileios Mixelakis, Coordinator of Decentralized Administration of Epirus-West Macedonia, for his administrative support and encouragement. S.-K.Z. acknowledges the Action titled "National Network on Climate Change and its Impacts—Climpact", which is implemented under the sub-project 3 of the project "Infrastructure of national research networks in the fields of Precision Medicine, Quantum Technology and Climate Change", funded by the Public Investment Program of Greece, General Secretary of Research and Technology/Ministry of Development and Investments.

Conflicts of Interest: The authors declare no conflict of interest.

References

1. Meriluoto, J.; Spoof, L.; Codd, G.A. *Handbook of Cyanobacterial Monitoring and Cyanotoxin Analysis*; Wiley: Chichester, West Sussex, UK, 2017; pp. 1–548.
2. Van Apeldoorn, M.E.; Van Egmond, H.P.; Speijers, G.J.A.; Bakker, G.J.I. Toxins of cyanobacteria. *Mol. Nutr. Food Res.* **2007**, *51*, 7–60. [CrossRef] [PubMed]
3. Buratti, F.M.; Manganelli, M.; Vichi, S.; Stefanelli, M.; Scardala, S.; Testai, E.; Funari, E. Cyanotoxins: Producing organisms, occurrence, toxicity, mechanism of action and human health toxicological risk evaluation. *Arch. Toxicol.* **2017**, *91*, 1049–1130. [CrossRef]
4. Carmichael, W.W.; Azevedo, S.M.F.O.; An, J.S.; Molica, R.J.R.; Jochimsen, E.M.; Lau, S.; Rinehart, K.L.; Shaw, G.R.; Eaglesham, G.K. Human fatalities form cyanobacteria: Chemical and biological evidence for cyanotoxins. *Environ. Health Perspect.* **2001**, *109*, 663–668. [CrossRef]
5. Jochimsen, E.M.; Carmichael, W.W.; An, J.; Cardo, D.M.; Cookson, S.T.; Holmes, C.E.M.; De Antunes, M.B.C.; De Melo Filho, D.A.; Lyra, T.M.; Barreto, V.S.T.; et al. Liver failure and death after exposure to microcystins at a hemodialysis center in Brazil. *N. Engl. J. Med.* **1998**, *338*, 873–878. [CrossRef]
6. Krienitz, L.; Ballot, A.; Kotut, K.; Wiegand, C.; Pütz, S.; Metcalf, J.S.; Codd, G.A.; Pflugmacher, S. Contribution of hot spring cyanobacteria to the mysterious deaths of Lesser Flamingos at Lake Bogoria, Kenya. *FEMS Microbiol. Ecol.* **2003**, *43*, 141–148. [CrossRef]
7. Briand, J.F.; Jacquet, S.; Bernard, C.; Humbert, J.F. Health hazards for terrestrial vertebrates from toxic cyanobacteria in surface water ecosystems. *Vet. Res.* **2003**, *34*, 361–377. [CrossRef] [PubMed]
8. Bouaïcha, N.; Miles, C.O.; Beach, D.G.; Labidi, Z.; Djabri, A.; Benayache, N.Y.; Nguyen-Quang, T. Structural Diversity, Characterization and Toxicology of Microcystins. *Toxins* **2019**, *11*, 714. [CrossRef]
9. Chen, Y.; Shen, D.; Fang, D. Nodularins in poisoning. *Clin. Chim. Acta* **2013**, *425*, 18–29. [CrossRef]
10. Carmichael, W.W. The Cyanotoxins. In *Advances in Botanical Research*; Callow, J.A., Ed.; Academic Press: Cambridge, MA, USA, 1997; Volume 27, pp. 211–256.
11. Dawson, R.M. The toxicology of microcystins. *Toxicon* **1998**, *36*, 953–962. [CrossRef]
12. De La Cruz, A.A.; Hiskia, A.; Kaloudis, T.; Chernoff, N.; Hill, D.; Antoniou, M.G.; He, X.; Loftin, K.; O'Shea, K.; Zhao, C.; et al. A review on cylindrospermopsin: The global occurrence, detection, toxicity and degradation of a potent cyanotoxin. *Environ. Sci. Process. Impacts* **2013**, *15*, 1979–2003. [CrossRef]
13. Poniedziałek, B.; Rzymski, P.; Kokociński, M. Cylindrospermopsin: Water-linked potential threat to human health in Europe. *Environ. Toxicol. Pharmacol.* **2012**, *34*, 651–660. [CrossRef]
14. Osswald, J.; Rellán, S.; Gago, A.; Vasconcelos, V. Toxicology and detection methods of the alkaloid neurotoxin produced by cyanobacteria, anatoxin-a. *Environ. Int.* **2007**, *33*, 1070–1089. [CrossRef]
15. Welker, M.; Von Döhren, H. Cyanobacterial peptides-Nature's own combinatorial biosynthesis. *FEMS Microbiol. Rev.* **2006**, *30*, 530–563. [CrossRef]
16. Okino, T.; Matsuda, H.; Murakami, M.; Yamaguchi, K. Microginin, an angiotensin-converting enzyme inhibitor from the blue-green alga Microcystis aeruginosa. *Tetrahedron Lett.* **1993**, *34*, 501–504. [CrossRef]
17. Janssen, E.M.L. Cyanobacterial peptides beyond microcystins–A review on co-occurrence, toxicity, and challenges for risk assessment. *Water Res.* **2019**, *151*, 488–499. [CrossRef]

18. Ersmark, K.; Del Valle, J.R.; Hanessian, S. Chemistry and biology of the aeruginosin family of serine protease inhibitors. *Angew. Chem. Int. Ed.* **2008**, *47*, 1202–1223. [CrossRef]

19. Kohler, E.; Grundler, V.; Häussinger, D.; Kurmayer, R.; Gademann, K.; Pernthaler, J.; Blom, J.F. The toxicity and enzyme activity of a chlorine and sulfate containing aeruginosin isolated from a non-microcystin-producing Planktothrix strain. *Harmful Algae* **2014**, *39*, 154–160. [CrossRef]

20. Cegłowska, M.; Szubert, K.; Wieczerzak, E.; Kosakowska, A.; Mazur-Marzec, H. Eighteen New Aeruginosamide Variants Produced by the Baltic Cyanobacterium Limnoraphis CCNP1324. *Mar. Drugs* **2020**, *18*, 446. [CrossRef] [PubMed]

21. Lawton, L.A.; Morris, L.A.; Jaspars, M. A bioactive modified peptide, aeruginosamide, isolated from the cyanobacterium Microcystis aeruginosa. *J. Org. Chem.* **1999**, *64*, 5329–5332. [CrossRef]

22. Beversdorf, L.J.; Weirich, C.A.; Bartlett, S.L.; Miller, T.R. Variable cyanobacterial toxin and metabolite profiles across six eutrophic lakes of differing physiochemical characteristics. *Toxins* **2017**, *9*, 62. [CrossRef]

23. Bartlett, S.L.; Brunner, S.L.; Klump, J.V.; Houghton, E.M.; Miller, T.R. Spatial analysis of toxic or otherwise bioactive cyanobacterial peptides in Green Bay, Lake Michigan. *J. Great Lakes Res.* **2018**, *44*, 924–933. [CrossRef] [PubMed]

24. Sano, T.; Usui, T.; Ueda, K.; Osada, H.; Kaya, K. Isolation of new protein phosphatase inhibitors from two cyanobacteria species, Planktothrix spp. *J. Nat. Prod.* **2001**, *64*, 1052–1055. [CrossRef]

25. Sedmak, B.; Carmeli, S.; Eleršek, T. "Non-toxic" cyclic peptides induce lysis of cyanobacteria-An effective cell population density control mechanism in cyanobacterial blooms. *Microb. Ecol.* **2008**, *56*, 201–209. [CrossRef]

26. Pelaez, M.; Antoniou, M.G.; He, X.; Dionysiou, D.D.; de la Cruz, A.A.; Tsimeli, K.; Triantis, T.; Hiskia, A.; Kaloudis, T.; Williams, C.; et al. Sources and Occurrence of Cyanotoxins Worldwide. In *Xenobiotics in the Urban Water Cycle: Mass Flows, Environmental Processes, Mitigation and Treatment Strategies*; Fatta-Kassinos, D., Bester, K., Kümmerer, K., Eds.; Springer Netherlands: Dordrecht, The Netherlands, 2010; pp. 101–127. [CrossRef]

27. Balbus, J.M.; Boxall, A.B.A.; Fenske, R.A.; McKone, T.E.; Zeise, L. Implications of global climate change for the assessment and management of human health risks of chemicals in the natural environment. *Environ. Toxicol. Chem.* **2013**, *32*, 62–78. [CrossRef]

28. Paerl, H.W.; Huisman, J. Climate change: A catalyst for global expansion of harmful cyanobacterial blooms. *Environ. Microbiol. Rep.* **2009**, *1*, 27–37. [CrossRef]

29. Mantzouki, E.; Lürling, M.; Fastner, J.; de Senerpont Domis, L.; Wilk-Woźniak, E.; Koreivienė, J.; Seelen, L.; Teurlincx, S.; Verstijnen, Y.; Krztoń, W.; et al. Temperature effects explain continental scale distribution of cyanobacterial toxins. *Toxins* **2018**, *10*, 156. [CrossRef] [PubMed]

30. Weirich, C.A.; Miller, T.R. Freshwater harmful algal blooms: Toxins and children's health. *Curr. Probl. Pediatric Adolesc. Health Care* **2014**, *44*, 2–24. [CrossRef]

31. Christophoridis, C.; Zervou, S.K.; Manolidi, K.; Katsiapi, M.; Moustaka-Gouni, M.; Kaloudis, T.; Triantis, T.M.; Hiskia, A. Occurrence and diversity of cyanotoxins in Greek lakes. *Sci. Rep.* **2018**, *8*, 17877. [CrossRef]

32. Zervou, S.K.; Gkelis, S.; Kaloudis, T.; Hiskia, A.; Mazur-Marzec, H. New microginins from cyanobacteria of Greek freshwaters. *Chemosphere* **2020**, *248*, 125961. [CrossRef]

33. Zervou, S.K.; Christophoridis, C.; Kaloudis, T.; Triantis, T.M.; Hiskia, A. New SPE-LC-MS/MS method for simultaneous determination of multi-class cyanobacterial and algal toxins. *J. Hazard. Mater.* **2017**, *323*, 56–66. [CrossRef]

34. Ministry of Environment and Energy. *First Revision of the River Basin Management Plan of West Makedonia (EL09)*; Ministry of Environment and Energy: Athens, Greece, 2017; p. 296, Special Secretariat for Water.

35. European Union. Directive 2006/7/EC of the European parliament and of the council of 15 February 2006 concerning the management of bathing water quality and repealing Directive 76/160/EEC. *Off. J. Eur. Union* **2006**, *64*, 37–51.

36. Rastogi, R.P.; Sinha, R.P.; Incharoensakdi, A. The cyanotoxin-microcystins: Current overview. *Rev. Environ. Sci. Bio/Technol.* **2014**, *13*, 215–249. [CrossRef]

37. Huang, I.S.; Zimba, P.V. Cyanobacterial bioactive metabolites—A review of their chemistry and biology. *Harmful Algae* **2019**, *83*, 42–94. [CrossRef] [PubMed]

38. Cyanobacterial Toxins: Microcystins. *Background Document for Development of WHO Guidelines for Drinking-Water Quality and Guidelines for Safe Recreational Water Environments*; World Health Organization: Geneva, Switzerland, 2020.

39. Cyanobacterial Toxins: Cylindrospermopsins. *Background Document for Development of WHO Guidelines for Drinking-Water Quality and Guidelines for Safe Recreational Water Environments*; World Health Organization: Geneva, Switzerland, 2020.

40. EN 16698:2015. *Water Quality. Guidance on Quantitative and Qualitative Sampling of Phytoplankton from Inland Waters*; Comité European de Normalization (CEN): Brussels, Belgium, 2015.

41. Jones, M.R.; Pinto, E.; Torres, M.A.; Dörr, F.; Mazur-Marzec, H.; Szubert, K.; Tartaglione, L.; Dell'Aversano, C.; Miles, C.O.; Beach, D.G.; et al. CyanoMetDB, a comprehensive public database of secondary metabolites from cyanobacteria. *Water Res.* **2021**, *196*, 117017. [CrossRef]

42. Spoof, L.; Catherine, A. Appendix 3: Tables of Microcystins and Nodularins. In *Handbook of Cyanobacterial Monitoring and Cyanotoxin Analysis*; Meriluoto, J., Spoof, L., Codd, G.A., Eds.; Wiley: Chichester, West Sussex, UK, 2017; pp. 526–537. [CrossRef]

43. Kust, A.; Reháková, K.; Vrba, J.; Maicher, V.; Mareš, J.; Hrouzek, P.; Chiriac, M.-C.; Benedová, Z.; Tesarová, B.; Saurav, K. Insight into Unprecedented Diversity of Cyanopeptides in Eutrophic Ponds Using an MS/MS Networking Approach. *Toxins* **2020**, *12*, 561. [CrossRef] [PubMed]

44. Ishida, K.; Matsuda, H.; Murakami, M.; Yamaguchi, K. Microginins 299-A and -B, leucine aminopeptidase inhibitors from the cyanobacterium *Microcystis aeruginosa* (NIES-299). *Tetrahedron* **1997**, *53*, 10281–10288. [CrossRef]

45. Welker, M.; Maršálek, B.; Šejnohová, L.; von Döhren, H. Detection and identification of oligopeptides in Microcystis (cyanobacteria) colonies: Toward an understanding of metabolic diversity. *Peptides* **2006**, *27*, 2090–2103. [CrossRef] [PubMed]

46. Zervou, S.K.; Kaloudis, T.; Hiskia, A.; Mazur-Marzec, H. Fragmentation mass spectra dataset of linear cyanopeptides-microginins. *Data Brief* **2020**, *31*, 105825. [CrossRef]

47. Welker, M.; Brunke, M.; Preussel, K.; Lippert, I.; von Döhren, H. Diversity and distribution of Microcystis (cyanobacteria) oligopeptide chemotypes from natural communities studies by single-colony mass spectrometry. *Microbiology* **2004**, *150*, 1785–1796. [CrossRef] [PubMed]

48. Liu, L.; Budnjo, A.; Jokela, J.; Haug, B.E.; Fewer, D.P.; Wahlsten, M.; Rouhiainen, L.; Permi, P.; Fossen, T.; Sivonen, K. Pseudoaeruginosins, Nonribosomal Peptides in Nodularia spumigena. *ACS Chem. Biol.* **2015**, *10*, 725–733. [CrossRef] [PubMed]

49. Erhard, M.; Von Döhren, H.; Jungblut, P.R. Rapid identification of the new anabaenopeptin G from Planktothrix agardhii HUB 011 using matrix-assisted laser desorption/ionization time-of-flight mass spectrometry. *Rapid Commun. Mass Spectrom.* **1999**, *13*, 337–343. [CrossRef]

50. Häggqvist, K.; Toruńska-Sitarz, A.; Błaszczyk, A.; Mazur-Marzec, H.; Meriluoto, J. Morphologic, Phylogenetic and Chemical Characterization of a Brackish Colonial Picocyanobacterium (Coelosphaeriaceae) with Bioactive Properties. *Toxins* **2016**, *8*, 108. [CrossRef] [PubMed]

51. Gkelis, S.; Lanaras, T.; Sivonen, K.; Taglialatela-Scafati, O. Cyanobacterial toxic and bioactive peptides in freshwater bodies of Greece: Concentrations, occurrence patterns, and implications for human health. *Mar. Drugs* **2015**, *13*, 6319–6335. [CrossRef] [PubMed]

52. Martins, J.; Vasconcelos, V. Cyanobactins from Cyanobacteria: Current Genetic and Chemical State of Knowledge. *Mar. Drugs* **2015**, *13*, 6910–6946. [CrossRef]

53. Österholm, J.; Popin, R.V.; Fewer, D.P.; Sivonen, K. Phylogenomic analysis of secondary metabolism in the toxic cyanobacterial genera Anabaena, Dolichospermum and aphanizomenon. *Toxins* **2020**, *12*, 248. [CrossRef] [PubMed]

54. Anas, A.R.J.; Kisugi, T.; Umezawa, T.; Matsuda, F.; Campitelli, M.R.; Quinn, R.J.; Okino, T. Thrombin inhibitors from the freshwater cyanobacterium Anabaena compacta. *J. Nat. Prod.* **2012**, *75*, 1546–1552. [CrossRef]

55. Harada, K.-i.; Fujii, K.; Shimada, T.; Suzuki, M.; Sano, H.; Adachi, K.; Carmichael, W.W. Two cyclic peptides, anabaenopeptins, a third group of bioactive compounds from the cyanobacteriumAnabaena flos-aquae NRC 525-17. *Tetrahedron Lett.* **1995**, *36*, 1511–1514. [CrossRef]

56. Ferranti, P.; Fabbrocino, S.; Chiaravalle, E.; Bruno, M.; Basile, A.; Serpe, L.; Gallo, P. Profiling microcystin contamination in a water reservoir by MALDI-TOF and liquid chromatography coupled to Q/TOF tandem mass spectrometry. *Food Res. Int.* **2013**, *54*, 1321–1330. [CrossRef]

57. Grabowska, M.; Kobos, J.; Toruńska-Sitarz, A.; Mazur-Marzec, H. Non-ribosomal peptides produced by Planktothrix agardhii from Siemianówka Dam Reservoir SDR (northeast Poland). *Arch. Microbiol.* **2014**, *196*, 697–707. [CrossRef] [PubMed]

58. Beversdorf, L.J.; Rude, K.; Weirich, C.A.; Bartlett, S.L.; Seaman, M.; Kozik, C.; Biese, P.; Gosz, T.; Suha, M.; Stempa, C.; et al. Analysis of cyanobacterial metabolites in surface and raw drinking waters reveals more than microcystin. *Water Res.* **2018**, *140*, 280–290. [CrossRef] [PubMed]

59. Roy-Lachapelle, A.; Vo Duy, S.; Munoz, G.; Dinh, Q.T.; Bahl, E.; Simon, D.F.; Sauvé, S. Analysis of multiclass cyanotoxins (microcystins, anabaenopeptins, cylindrospermopsin and anatoxins) in lake waters using on-line SPE liquid chromatography high-resolution Orbitrap mass spectrometry. *Anal. Methods* **2019**, *11*, 5289–5300. [CrossRef]

60. Flores, C.; Caixach, J. High Levels of Anabaenopeptins Detected in a Cyanobacteria Bloom from N.E. Spanish Sau-Susqueda-El Pasteral Reservoirs System by LC–HRMS. *Toxins* **2020**, *12*, 541. [CrossRef]

61. EN ISO 14911:1998. *Water Quality. Determination of Dissolved Li+, Na+, NH4+, K+, Mn2+, Ca2+, Mg2+, Sr2+ and Ba2+ Using Ion Chromatography. Method for Water and Waste Water*; Comité Europeen de Normalization (CEN): Brussels, Belgium, 1998.

62. EN ISO 10304-1:2007. *Water Quality–Determination of Dissolved Anions by Liquid Chromatography of Ions–Part 1: Determination of Bromide, Chloride, Fluoride, Nitrate, Nitrite, Phosphate and Sulfate*; Comité Europeen de Normalization (CEN): Brussels, Belgium, 2007.

63. Rice, E.W.; Bridgewater, L.; American Public Health Association; American Water Works Association; Water Environment Federation. *Standard Methods for the Examination of Water and Wastewater*, 22nd ed.; American Public Health Association: Washington, DC, USA, 2012.

64. Jeffrey, S.W.; Humphrey, G.F. New spectrophotometric equations for determining chlorophylls a, b, c1 and c2 in higher plants, algae and natural phytoplankton. *Biochem. Physiol. Pflanz.* **1975**, *167*, 191–194. [CrossRef]

65. EN 15204:2006. *Water Quality—Guidance Standard on the Enumeration of Phytoplankton Using Inverted Microscopy (Utermöhl Technique)*; Comité Europeen de Normalization (CEN): Brussels, Belgium, 2006.

66. Utermohl, H. Zur Vervollkommung der quantitativen phytoplankton-methodik. *Mitt. Int. Limnol.* **1958**, *9*, 38.

67. EN16695:2015. *Water Quality–Guidance on the Estimation of Phytoplankton Biovolume*; Comité Europeen de Normalization (CEN): Brussels, Belgium, 2015.

68. Lombardo, M.; Pinto, F.C.R.; Vieira, J.M.S.; Honda, R.Y.; Pimenta, A.M.C.; Bemquerer, M.P.; Carvalho, L.R.; Kiyota, S. Isolation and structural characterization of microcystin-LR and three minor oligopeptides simultaneously produced by Radiocystis feernandoi (Chroococcales, Cyanobacteriae): A Brazilian toxic cyanobacterium. *Toxicon* **2006**, *47*, 560–566. [CrossRef] [PubMed]
69. Natumi, R.; Janssen, E.M.L. Cyanopeptide Co-Production Dynamics beyond Mirocystins and Effects of Growth Stages and Nutrient Availability. *Environ. Sci. Technol.* **2020**, *54*, 6063–6072. [CrossRef] [PubMed]

toxins

MDPI

Article

Is the Cyanobacterial Bloom Composition Shifting Due to Climate Forcing or Nutrient Changes? Example of a Shallow Eutrophic Reservoir

Morgane Le Moal [1], Alexandrine Pannard [1], Luc Brient [1], Benjamin Richard [2], Marion Chorin [1], Emilien Mineaud [1,3] and Claudia Wiegand [1,*]

[1] ECOBIO (Ecosystems, Biodiversity, Evolution) UMR 6553, University Rennes 1, CNRS, 263 av. du général Leclerc, 35700 Rennes, France; morgane.lemoal@hotmail.fr (M.L.M.); alexandrine.pannard@univ-rennes1.fr (A.P.); luc.brient@univ-rennes1.fr (L.B.); marion.chorin@univ-rennes1.fr (M.C.); emilien.mineaud@gmail.com (E.M.)

[2] Département Santé-Environnement, Agence Régionale de Santé de Bretagne, 32 bd de la Résistance, 56000 Vannes, France; Benjamin.RICHARD@ars.sante.fr

[3] Department of Biological Sciences, University of Quebec, C. P. 8888, Succ. Centre-Ville, Montréal, QC H3C 3P8, Canada

* Correspondence: claudia.wiegand@univ-rennes1.fr

Citation: Le Moal, M.; Pannard, A.; Brient, L.; Richard, B.; Chorin, M.; Mineaud, E.; Wiegand, C. Is the Cyanobacterial Bloom Composition Shifting Due to Climate Forcing or Nutrient Changes? Example of a Shallow Eutrophic Reservoir. *Toxins* **2021**, *13*, 351. https://doi.org/10.3390/toxins13050351

Received: 30 March 2021
Accepted: 11 May 2021
Published: 13 May 2021

Publisher's Note: MDPI stays neutral with regard to jurisdictional claims in published maps and institutional affiliations.

Abstract: Cyanobacterial blooms in eutrophic freshwater is a global threat to the functioning of ecosystems, human health and the economy. Parties responsible for the ecosystems and human health increasingly demand reliable predictions of cyanobacterial development to support necessary decisions. Long-term data series help with identifying environmental drivers of cyanobacterial developments in the context of climatic and anthropogenic pressure. Here, we analyzed 13 years of eutrophication and climatic data of a shallow temperate reservoir showing a high interannual variability of cyanobacterial development and composition, which is a less occurring and/or less described phenomenon compared to recurrant monospecific blooms. While between 2007–2012 *Planktothrix agardhii* dominated the cyanobacterial community, it shifted towards *Microcystis* sp. and then *Dolichospermum* sp. afterwards (2013–2019). The shift to *Microcystis* sp. dominance was mainly influenced by generally calmer and warmer conditions. The later shift to *Dolichospermum* sp. was driven by droughts influencing, amongst others, the N-load, as P remained unchanged over the time period. Both, climatic pressure and N-limitation contributed to the high variability of cyanobacterial blooms and may lead to a new equilibrium. The further reduction of P-load in parallel to the decreasing N-load is important to suppress cyanobacterial blooms and ameliorate ecosystem health.

Keywords: cyanobacteria; eutrophication; long term monitoring; water quality

Key Contribution: Long term (13 years) data were used to explain high variability in cyanobacterial bloom development and composition in a lowland reservoir; N-reduction and climatic forces (draughts increasing residence time) were identified as the main drivers. The phytoplankton community may be shifting from being dominated by *Planktothrix* sp. in the past to *Microcystis* sp. and *Dolichospermum* sp. in future, with the potential of increased production of cyanobacterial toxins.

1. Introduction

Recurrent and persistent mass developments (blooms) of cyanobacteria are one of the main outcomes of eutrophication of freshwater ecosystems, that is the natural increase in organic matter production and accumulation in an aquatic ecosystem, accelerated during the last decades by human activities [1–4]. Natural control of cyanobacterial blooms, e.g., by zooplankton, is limited by their low palatability owed to long filaments or colony formation and their low nutritional value, besides their toxicity [5]. Cyanobacterial dominance thus causes a misbalance and malfunctioning of the aquatic ecosystem by supressing the biodiversity of zooplankton, phytoplankton (via competition) and submerged macropytes (by shading), thereby disturbing trophic chains and fluxes [5–7]. The decay of phytoplankton including cyanobacteria consumes oxygen, affecting organisms in the water column and ultimately at the sediment. Bottom oxygen depletion limits phosphorus (P) fixation and enhances its release, thus accelerates eutrophication [8].

As various cyanobacteria produce bioactive or toxic metabolites, their mass development threatens organisms in the water body or depending on it [9–11]. To ensure human safety, many countries use thresholds of cyanobacterial cell densities combined with toxin concentration to restrict access for recreational activities, and the WHO has established guidelines of 1 μg L^{-1} of total microcystins in drinking water [12,13]. Managers and actors responsible for ecosystem and human health increasingly demand reliable predictions of seasonal cyanobacterial development as those regulations may cause financial consequences, including losses in income via recreational activities or increasing costs for drinking water purification. The different cyanobacteria genera comprise distinctive capabilities to form blooms or to produce specific toxins [6], which underlines the necessity to predict bloom formation, composition, duration and heterogeneity in a given water body.

Despite the urge to return or shift towards the dominance of eukaryotic phytoplankton, this remains difficult, as several physiological mechanisms enable cyanobacteria to outperform eukaryotic phytoplankton. Cyanobacteria grow better in low-light conditions, which occur either during mixing events or due to high phytoplankton biomass, moreover some cyanobacterial species possess gas-vesicles enabling them to adjust their light requirements via buoyancy in the water column [14]. They often benefit from faster uptake kinetics for CO_2, a higher temperature optimum and lower N concentrations [14]. Additionally, several species are able to directly use dinitrogen as an N-source [8]. All of these features lead to a higher growth rate during summer, by which cyanobacteria can outcompete eukaryotic phytoplankton, leading to recurrent bloom situations.

Eutrophication is the main driver influencing cyanobacteria blooms, but temperature intensifies bloom frequency, duration and intensity [15]. Shallow lowland lakes suffer more from cyanobacteria because here nutrient loadings meet higher temperatures, and P is easily resuspended from the sediment due to their often polymictic character [6,16]. This applies even more to shallow reservoirs, which continuously receive nutrients (P, N) from the inflowing rivers, charging their sediments as they act as nutrient trap. If not reduced, eutrophication in combination with climate change will thus in future increase cyanobacterial blooms [6,17].

Due to the interaction of the driving factors and the differences from one water body to another, it remains difficult to foresee the development and density of a phytoplankton bloom, and particularly its species composition. Long-term data series help identify environmental drivers of cyanobacterial blooms' occurrence and composition in the context of climatic and anthropogenic pressure and improve predictions for managers. Long term studies previously illustrated the trajectories of eutrophication. This knowledge was applied to reduce point source P pollution that led to a gradual decrease of phytoplanktonic biomass and/or changes in its composition towards eukaryotic species, as demonstrated for eight lakes or reservoirs of many examples by Fastner et al. [18]. The intensification of agricultural practices, however, increased again N and P flux towards adjacent waters, causing numerous lakes and reservoirs to experience or re-experience cyanobacterial blooms. Nevertheless, where effort has been made to reduce agricultural P and N emissions to a lake, it was followed by the amelioration of its water quality [19,20]. Long term (37 years) surveys showed changes in the cyanobacterial community composition towards diazotrophic species as consequence of N decrease while P remained constant, [21], whereas, in another lake, reduction in both N and P decreased also the proportion of N_2 fixers in the phytoplankton community [20].

In accordance with their habitat characteristics in terms of depth, nutrient load, temperature, etc, the same species will dominate recurrent blooms in most lakes. *Microcystis sp.*

continuously dominates for example Lake Taihu, China [22,23] or the Aguieira reservoir in Portugal [24], while *Planktothrix agardhii* perennially dominates shallow eutrophic water bodies in lowland areas of the Netherlands and Northern Germany [25,26]. Compared to large monospecific blooms, interannual variation amongst cyanobacterial dominance needs better understanding as it can be a good indicator for changes in climatic and anthropogenic pressure. Interannual variations may indicate a shift towards a new state of the composition of the bloom.

The investigated shallow reservoir has received its eutrophication with the incoming rivers during the past decades. It shows a high variability of cyanobacterial development between years, for which the driving factors are not yet identified. Aims were therefore 1) to characterize the seasonal dynamic of the nutrient pattern as well as the phytoplankton and zooplankton succession over a one-year period, and 2) to identify the driving factors of cyanobacterial development over 13 years of monitoring, testing the hypothesis that nutrients, phyto- and zooplankton in the reservoir differ along the transect from entrance to outlet (H1), and that even in this short time series, climatic forcing changes the dynamics of cyanobacterial blooms (H2).

2. Results

2.1. Seasonal Cycle

2.1.1. Nutrients' Concentration

Nutrients' concentrations were very similar at the entrance, in the middle and lower basin of the reservoir during the 2018–2019 seasonal cycle (Figure 1). No significant difference was observed between the three stations in terms of nutrient concentrations but nitrogen and phosphorus concentrations revealed different patterns of seasonal variability. Total dissolved nitrogen and nitrate–nitrite were highly correlated (Spearman correlations, $r_S > 0.92$, $p < 0.001$) and reached maximum concentrations in winter. Particulate N was low, but still corresponded to the particulate P concentrations during that period. Using the theoretical composition of the phytoplankton with the Redfield ratio and the ratios given by Reynolds (2006), we find that 150 µg of P corresponds to 1.05 mg N/L (mass ratio of 47:7:1 for C:N:P) and are therefore consistent with particulate N concentrations of 1.25 mg N/L. During periods of high concentrations in N, dissolved nitrogen reached almost 100% of the total N from January to May, as particulate N remained below LOQ for several sampling dates. During that period, nitrate and nitrite accounted for 65% of the total dissolved nitrogen.

Particulate phosphorus, total dissolved phosphorus and phosphate maximum concentrations were recorded during summer, and both particulate P and TDP were correlated (Spearman correlations, $r_S = 0.65$, $p < 0.001$). Phosphorus concentrations of all fractions were very low from January to May, but reached almost 180 µg L^{-1} for particulate P and 125 µg L^{-1} for TDP at the end of summer, in September 2018. From June to December, total phosphorus was mainly composed of particulate phosphorus for two thirds and TDP for one third. The dissolved fraction accounted on average for 49 ± 24%, 40 ± 10% and 40 ± 13% at the entrance, middle and in the lower basin of the reservoir, respectively. Among this dissolved fraction, more than half was composed of phosphate (mean = 53 ± 10%, 59 ± 23% and 53 ± 12% at the entrance, middle and in the lower basin of the reservoir, respectively). N/P Redfield ratios were low at the entrance of the reservoir (<10 with mean = 6 ± 4), while there were closer to 16 in the middle (>10 with mean = 12.6 ± 1.6) and the lower basin (>10 with mean = 14.9 ± 3.5; Table 1).

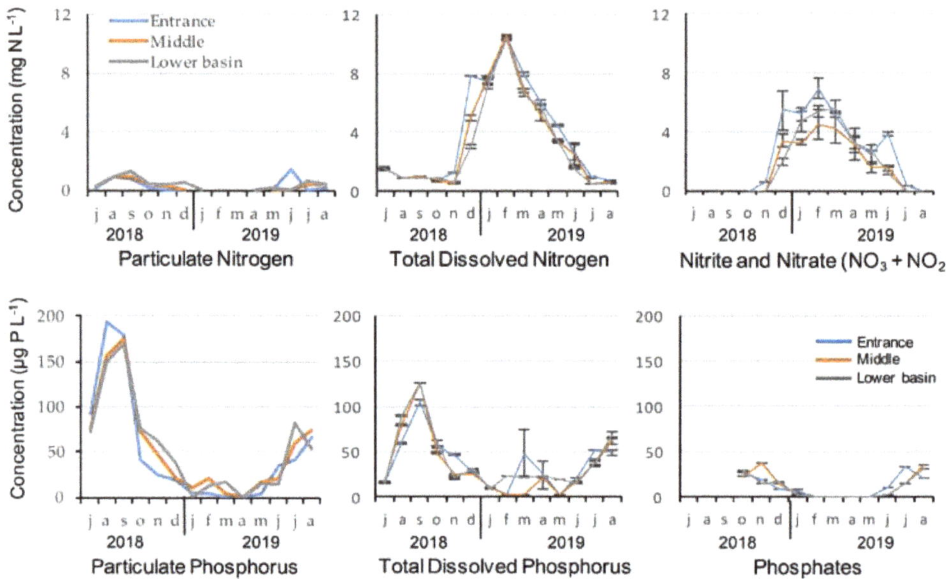

Figure 1. Nutrients concentration at the entrance, middle and in the lower basin of the reservoir during the 2018–2019 seasonal cycle.

Table 1. N/P Redfield ratios of the particulate matter at the entrance, middle and in the lower basin of the reservoir, respectively. N/P Redfield ratios were calculated only for the July–November period, as during the rest of the year, there was no particulate nitrogen and/or phosphorus concentrations were below limit of quantification.

		N/P Ratios of the Particulate Matter		
		Entrance	**Middle**	**Lower Basin**
	July	5.45	10.16	10.34
	August	10.67	12.99	14.62
2018	September	10.00	12.43	16.41
	October	7.61	11.62	11.18
	November	1.15	12.23	13.83
2019	July	0.28	15.45	17.80
	August	6.78	13.16	19.92
	Mean ± SD	5.99 ± 4.03	12.58 ± 1.62	14.87 ± 3.46

2.1.2. Phytoplankton

The relative proportions of the different groups of phytoplankton and the total abundance of cyanobacteria were very similar at the entrance, middle and in the lower basin, showing no significant difference (Figure 2A). The relative contribution of phytoplankton is calculated from cell number, biovolume being unfortunately unknown, as counting was performed at the genus level. Cyanobacteria dominated planktonic community from June to December 2018 and from June to at least August 2019. There were no cyanobacteria in winter and spring, and their concentrations reached 1,500,000 cells mL^{-1} in late summer (September 2018). A second bloom of cyanobacteria was observed during autumn 2018 in the middle and the lower basin, reaching 600,000 and 1,200,000 cells mL^{-1}, respectively, in November. Cyanobacterial density was one order of magnitude lower during summer 2019. Phytoplankton eukaryotic community ranged from 200 cells mL^{-1} in winter to 72,000 cells mL^{-1} in early summer (July 2018). Chlorophytes were the most abundant in this community almost all yearlong, except in spring (March–April) during which diatoms or in february, where euglenophytes were dominating. In January, there was a co-domnance of chlorophytes with Chrysophytes at the entrance but with diatoms in the middle and lower basin.

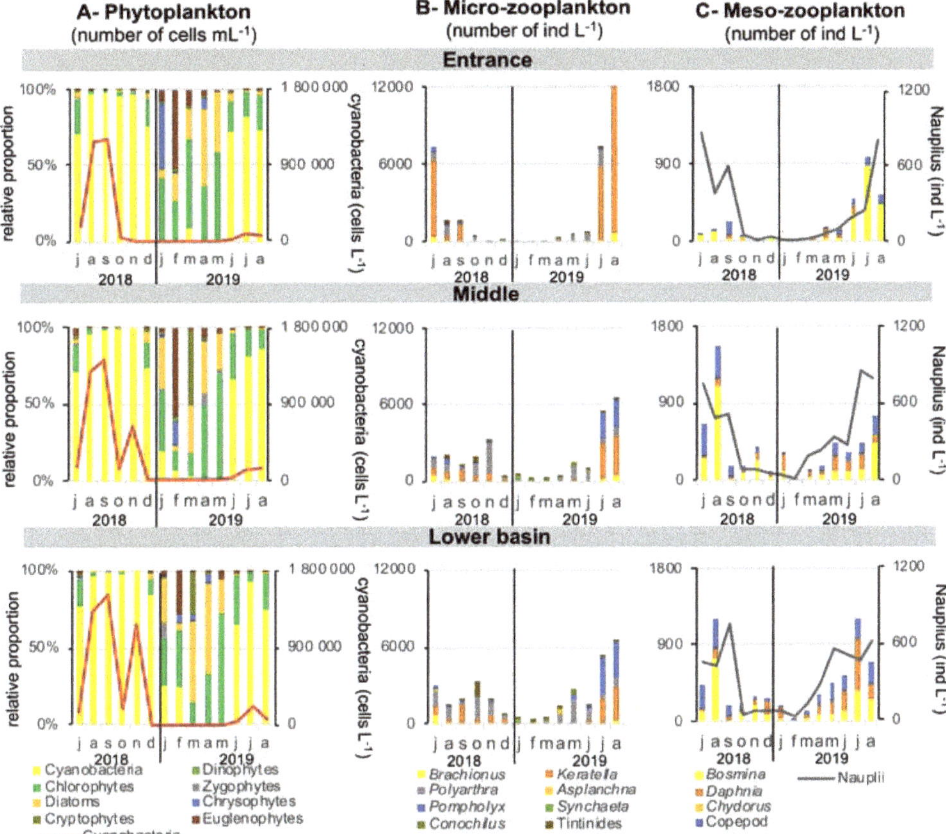

Figure 2. Phytoplankton and zooplankton dynamics over 2018–2019 seasonal cycle at the entrance, middle and in the lower basin of the reservoir. No statistical differences were observed between abundances of the three stations, except for *Polyarthra* (rotifer), copepods and *Daphnia*, for which the entrance present a lower abundance than the lower basin ($p < 0.05$).

2.1.3. Zooplankton

Zooplanktonic community abundance and composition was different between the entrance of the reservoir and middle–lower basin, apart from a common pattern of higher density during summers than winter (Figure 2B). Abundances of the rotifer *Polyathra* were significantly higher in the lower basin compared with the entrance ($p < 0.05$). The Cladocera *Daphnia* and copepods had also higher abundances in both the middle and lower basins compared with the entrance ($p < 0.05$). At the seasonal scale, the rotifer species *Keratella* strongly dominated the micro-zooplankton community during both summers at the entrance of the reservoir, reaching 6000 and 12,000 individuals L^{-1} in July 2018 and August 2019, respectively. In the middle and in the lower basin, *Pompholyx* and *Polyarthra* species accompanied *Keratella*, reaching a total maximum concentration of 3000 and 6000 individuals L^{-1} in July 2018 and August 2019, respectively. Concerning meso-zooplankton, the Cladoceran *Bosmina* genus dominated the community in the entire reservoir, together with copepods in a smaller proportion (Figure 2C). The total concentration of meso-zooplankton reached 1200 and 1500 individuals L^{-1} in the middle and the lower basin during summer 2018, while they were 4 to 5 times lower at the entrance of the reservoir (300 individuals L^{-1}). During summer 2019, *Daphnia* developed in the middle and lower basin, dominating at 51% the community in July 2019 in the lower basin. Concentration of Nauplii was quite homogenous in the whole reservoir, ranging between few individuals in winter to 800 individuals L^{-1} in summer.

2.2. Inter-Annual Variability of Summer Periods

2.2.1. Abiotic Parameters

With the exception of 2010, summers from 2013-2019 were globally warmer, sunnier and dryer compared to 2007–2012 (Table 2). GAMs were applied successfully to temperature, nitrate concentrations, residence time and flow showing their significant changes over the tested time period (Figure 3, Figure S1 and Table 3). Even during this relatively short time period with respect to long term data, a slight increase of temperature was observed (Figure 3A, Figure 4A) During summers 2010 and 2013–2019, either warm temperature ($\geq$20 days above 20 °C) or high light intensity ($\geq$10 days above 2800 J m^{-2}) or dry months (less than 1.5 million m^3 water entering into the reservoir) were recorded (Table 2). Before 2013 (except 2009), a strong and frequent wind was measured in the summers, with at least 19 days of wind with an average daily speed greater than 4 m s^{-1} (Figure 4E, Table 2). After 2013, only 2019 was classified as a windy summer. GAM could not be adjusted on wind due to the high daily variability and explained less than 4% of the deviance (not shown).

The biggest differences between the years of the study concerned the amount of water entering the reservoir during the hydrological year, ranging from 19 to 155 million m^3 per year (Table 2). 2012, 2017 and 2019 were the driest hydrological years, having received less than 40 million m^3. This increased the residence time and lowered the level, and water outflow stopped from July onwards (Figure 3B,C, Figure 4C,D). On the contrary, 2007, 2008 and 2014 presented high flow (Figure 3B) all year round (Table 2).

Table 2. Hydrological and meteorological characterisation of summer periods between 2007 and 2019. "Annual river inflow" was measured between November of the previous year and October, while "Summer river inflow" was measured between June and September. Dates of downstream flow stop (out of the lake) are indicated and depends on the water level (overflow). The date of the autumn increase in river inflow is also indicated in the last column.

	Number of Days from June to September			River Inflow in the Reservoir:		Date at Which Downstream Flow is Closed to 0	Date of Autumn Discharge Beginning
	With Water Temperature > 20 °C	With Light > 2800 J cm^{-2}	With Daily Wind > 4m s^{-1}	Annual River Inflow (m^3)	Summer River Inflow (m^3)		
2007	2 days	1 days	25 days	104,466,499	14,288,573	outflow all year round	
2008	4 days	6 days	29 days	82,705,190	6,988,378	outflow all year round	
2009	10 days	5 days	11 days	72,962,554	3,228,854	26/08/2009	17/11/2009
2010	19 days	15 days	27 days	74,001,686	900,547	27/07/2010	07/10/2010
2011	11 days	9 days	33 days	53,371,440	879,984	24/09/2011	15/12/2011
2012	14 days	3 days	22 days	30,433,277	3,377,462	14/09/2012	05/10/2012
2013	25 days	10 days	18 days	105,320,477	3,085,603	12/09/2013	06/11/2013
2014	18 days	16 days	7 days	155,305,037	2,709,418	outflow all year	
2015	12 days	18 days	18 days	68,434,762	1,434,931	05/07/2015	23/11/2015
2016	20 days	1 days	7 days	62,350,906	2,336,688	25/07/2016	26/01/2017
2017	23 days	3 days	11 days	19,428,682	1,200,269	03/07/2017	15/12/2017
2018	30 days	8 days	10 days	71,447,098	7,710,250	23/07/2018	07/12/2018
2019	25 days	13 days	29 days	38,914,906	2,677,622	12/07/2019	26/10/2019

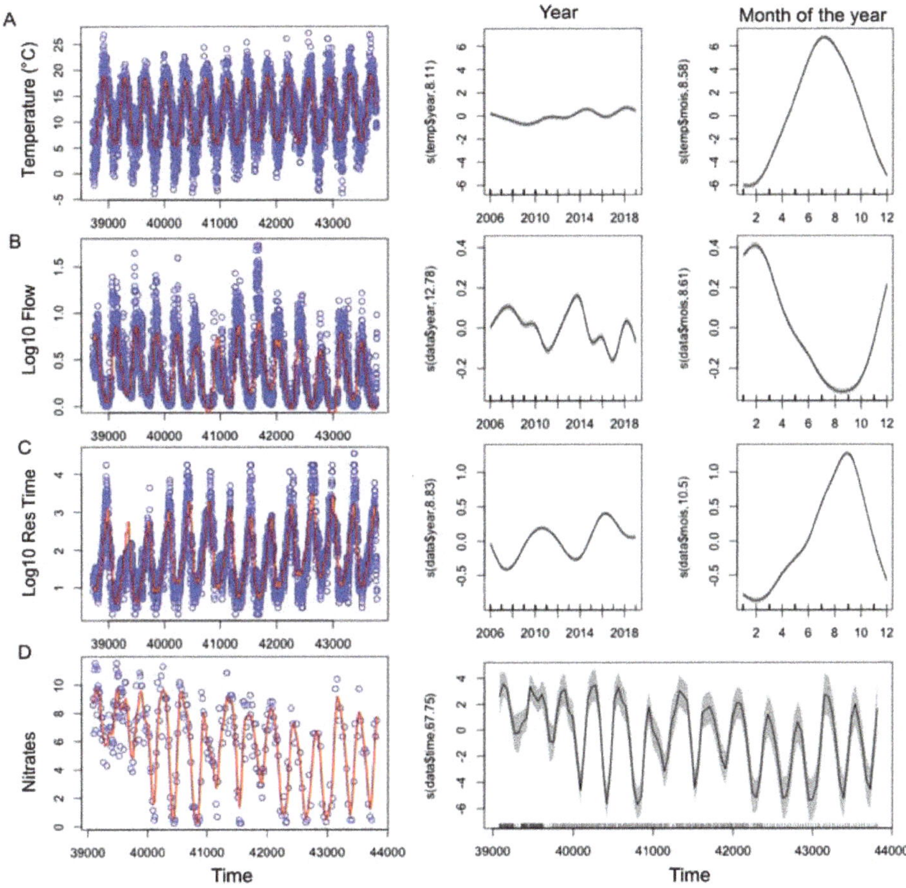

Figure 3. (**left**) Time series of the abiotic factors (blue points), showing the adjusted GAM (Table 3) in red. (**right**) Smooth fit with confidence bands performed with mgcv's gam function, showing the effect of year and month (or time for Nitrates) on the abiotic factors.

Table 3. Summary of the GAM results performed on the abiotic parameters, with their smoothing functions. Plots are shown in Figure S1. Adj., adjusted.

Dependant Variable (Number of Data Points)	Adj. R^2	GAMs Model DE (%)	REML	Smoothing Functions	Estimated Degrees of Freedom	Fisher Test	*p*-Value
Temperature (n = 5019)	0.7	70.1%	12598	S (year)	8.105	12.04	4.92×10^{-16}
				S (month)	8.584	1292.95	$<2 \times 10^{-16}$
Log10 Res Time (n = 5052)	0.719	72%	3360.3	S (year)	8.829	129.8	$<2 \times 10^{-16}$
				S (month, k = 12)	10.497	1073.5	$<2 \times 10^{-16}$
Nitrates (n = 283)	0.817	86.1%	612.37	S (time)	67.75	15.15	$<2 \times 10^{-16}$
Flow (n = 5027)	0.651	65.2%	−970.35	S (year)	12.78	71.01	$<2 \times 10^{-16}$
				S (month)	8.61	8.61	$<2 \times 10^{-16}$

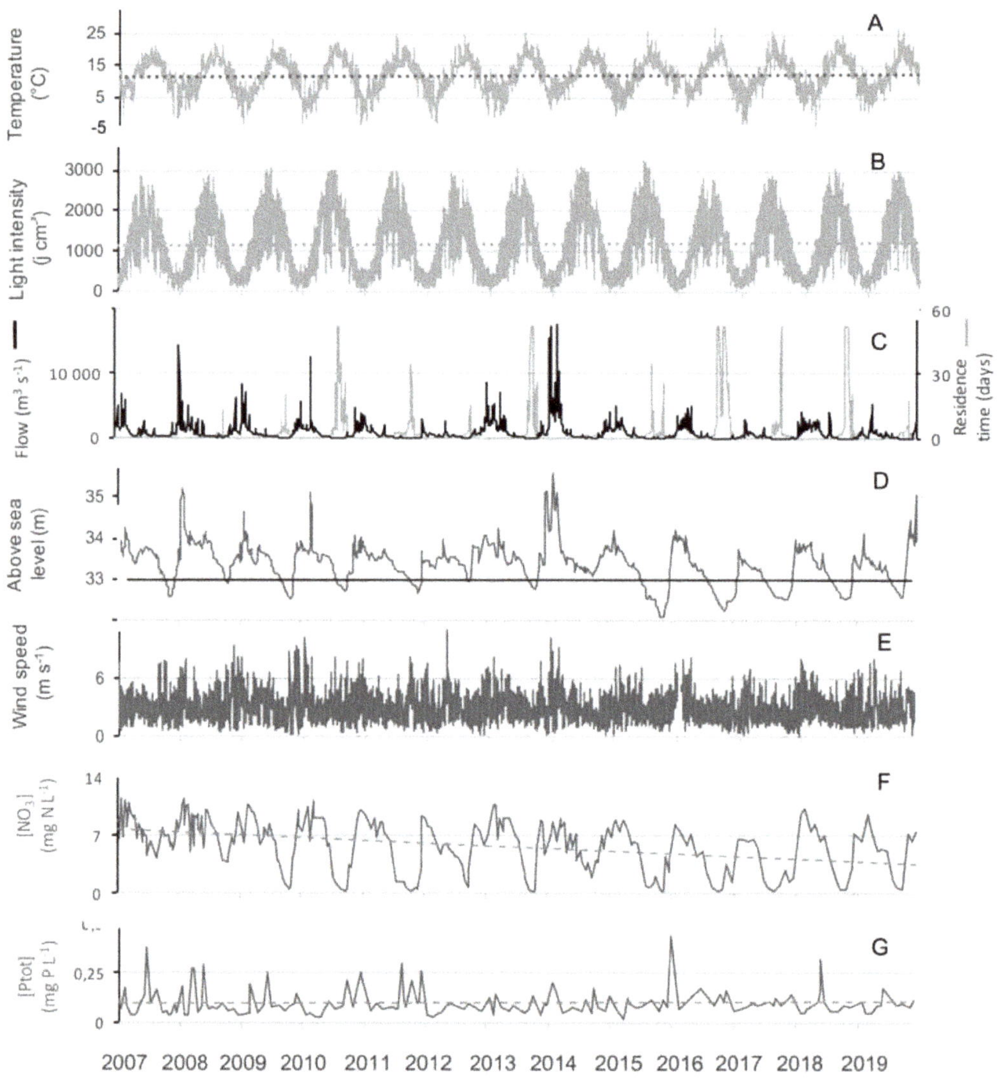

Figure 4. Time series of (**A**) the air temperature, (**B**) the light intensity, (**C**) the residence time and the river inflow, (**D**) the lake water level, (**E**) the mean daily wind speed, (**F**) the nitrates and (**G**) the total phosphorus concentration over 13 years (2007–2019). The light grey areas mark the summer periods. The different dotted lines show tendencies over the 13 years period. The black line in plot D indicates the hight above sea level of the outflow.

2.2.2. Nutrients

A strong seasonal pattern was observed for nitrates (Figures 3E and 4F), with maximum concentrations in winter (11.5 mg N-NO$_3^-$ L^{-1}) and minimum ones in summer. Owing to the low data points (n = 283), GAM was not able to discouple the seasonal pattern from the interannual one, and we smooth only one independent variable, the time (Figure 3D). Nitrate concentrations slightly decreased over the studied period, and since 2009 concentrations were at the LOD at the end of summer periods, except in 2014 (Figures 3D and 4F). Contrastingly, no seasonal pattern nor long-term tendency was observed for total phosphorus for the

2007–2019 period (Figure 4G). Total phosphorus concentration remained high with on average 100 ± 67 μg P-PO$_4$$^{3-}$ L^{-1} over the period studied.

2.2.3. Cyanobacterial Blooms

The cyanobacterial community varied strongly interannually, even including years without blooms, 2010, 2014 and 2015 (Figure 5). *Planktothrix agardhii* was the most abundant species, dominating 6 of the 13 monitored years and reaching a biomass of at least 35–55 mm^3 L^{-1} in 2007, 2011, 2012 and 2018. *Microcystis* (mainly *M. aeruginosa*) dominated in low densities in 2013, 2015 and early 2019, with associated detection of microcystin in low concentration in 2013 and 2015 (Figure 5, Table S1). *Microcystis* and microcystin (-LR, -RR, -YR, -LA, but not the demethylated congeners) were also detected in low density in 2018, a *Planktothrix* dominated year. *Dolichospermum* (mainly *spiroides*) dominated in huge densities in 2017 (75 mm^3 L^{-1}), and was present to a lesser extent in 2012, 2018 and 2019. Saxitoxin was detected in low concentration in 2017 and 2018. Anatoxin-a and cylindrospermopsin remained below LOD. *Aphanizomenon* dominated the cyanobacterial community in 2009. *Aphanothece* and *Aphanocapsa* (picocyanobacteria) dominated the cyanobacterial community in 2010 despite their very small biovolume per cell. All these species were present during summer 2014, but none dominated or bloomed.

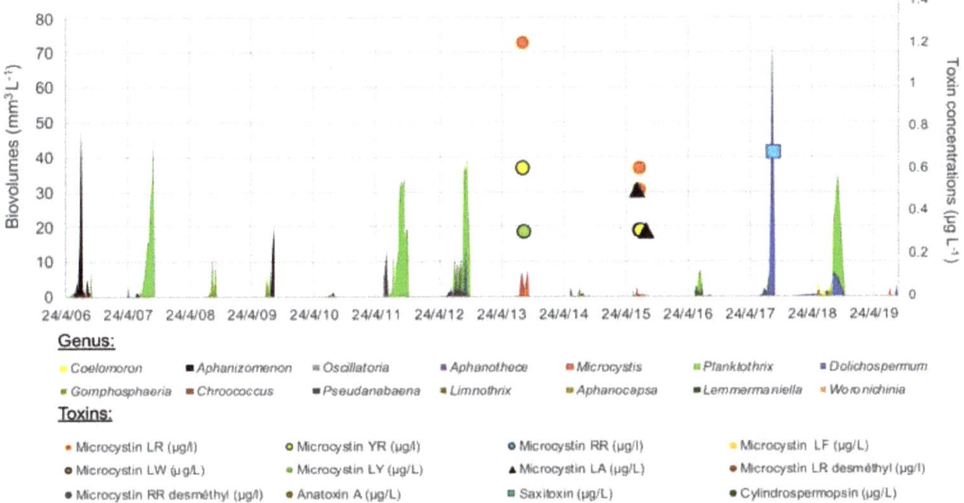

Figure 5. Cyanobacterial genus (in biovolumes, areas) and toxins' concentrations (points) over 2007–2019 summer periods in the lower basin of the reservoir.

2.2.4. Coupling Blooming Species with Environmental Parameters

To link the cyanobacteria species composition depending on time and environmental parameters, a canonical correspondence analysis has been performed on 195 sampling dates performed during summer. The final CCA included four environmental variables and time, which explain 10.2% of the total variability in the species composition (Figure 6). The CCA discriminated the 12 cyanobacteria species of which 6 formed a bloom. The two first axes of the analysis presented here represent 7.5% of the total variability in species blooming (total inertia) and 73.9% of the constrained inertia. The first axis correlated with time ($r = 0.92$), hence when samples are grouped by sampling dates, old dates appear on the left while recent ones are on the right (Figure 6a). The first axis correlated moreover to air temperature ($r = 0.35$), residence time ($r = 0.3$) and light ($r = 0.25$) (all towards the positive side).

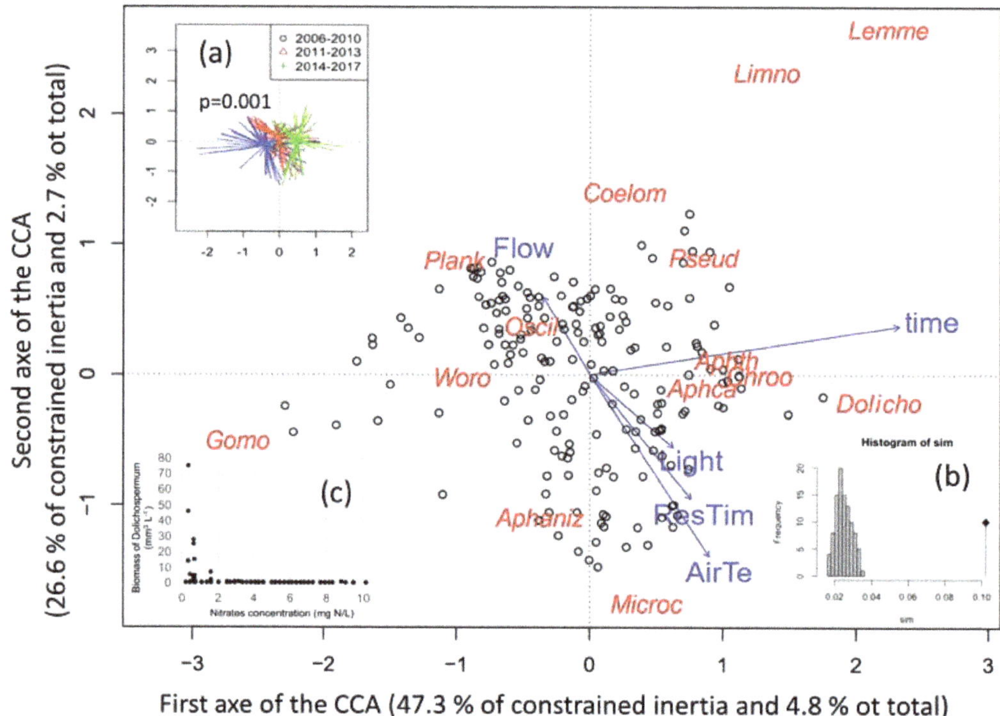

Figure 6. Distance biplot of the Canonical Correspondence Analysis (CCA) linking cyanobacterial biovolumes to environmental parameters measured during the six previous days. Data from summer periods between 2006 and 2017 were used. (**a**) Significant change of cyanobacteria composition with time was tested by permutation test (function envfit). (**b**) Histogramm of the permutation test for the ordination model. (**c**) Biomass of *Dolichospermum* depending on nitrates concentration, showing a threshold effect. ResTim: water residence time, AirTe: mean air temperature, Light: mean daily global radiation, Flow: mean river inflow, and time: sampling dates as open circles. The less significant environmental variables (rainfall, wind in intensity and variability) were eliminated from the analysis based on Akaike Information Criterion (AIC). Cyanobacteria genus by alphabetic order are: Aphaniz: *Aphanizomenon*; Aphca: *Aphanocpasa*, Aphth: *Ahanothece*, Chroo: *Chroococcus*, Coelom: *Coelomoron*, Dolicho: *Dolichospermum*; Gomo: *Gomphosphaeria*, Lemne: *Lemmermaniella*, Limno: *Limnothrix*, Microc: *Microcystis*; Oscill: *Oscillatoria*, Plank: *Planktothrix agardhii*, Pseud: *Pseudanabaena*, Woro: *Woronichinia*.

The first axis is also explained by *Planktothrix* (23.7%) and *Gomphosphaeria* (20.3%) on the negative side and *Dolichospermum* (14.2%), *Aphanothece* (13.9%), *Lemmermaniella* (8.3%) and *Aphanocapsa* (5.6%) on the positive side.

The second axis negatively correlates with air temperature (r = −0.73), residence time (r = −0.50) and light (r = −0.29), and positively with flow (r = 0.31) (Figure 6). Sampling dates with low flows thus coincide with high temperature, residence time and light. Air temperature and residence time thus contribute to both axes. The second axis is explained by *Microcystis* (28.2%) and *Aphanizomenon* (24.3%) on the negative side (bottom part — low flow), and by *Planktothrix* (18.2%), *Lemmermaniella* (11.7%), *Coelomoron* (7.0%) and *Limnothrix* (3.5%) on the positive side (upper part–high flow). All blooming species thus contributed at least to one of the two axes, while *Planktothrix* contributes negatively to both of them, in opposition with air temperature and residence time.

It should be noticed that nutrients were not included in the CCA analyses because they were measured less frequently and not necessarily at the same times than the phytoplankton community, moreover P concentrations did not change during the period of the data series. *Dolichospermum* sp. however seems to benefit below concentrations of 2 mg L^{-1} of nitrogen (Figure 6c): blooms of *Dolichospermum* were indeed only observed at very low nitrate concentrations.

3. Discussion

A low spatial variability but a high interannual variability in cyanobacteria biomass and composition have been revealed by this study. The seasonal cycle characterized this shallow reservoir as relatively homogeneous, with a similar evolution of nutrient concentrations and phytoplankton abundances, but different patterns for zooplankton along the increasing distance and depth from the entrance to the lower basin. These findings reject our first hypothesis assuming differeces concerning most parameters, but it could be accepted for the zooplankton dynamics after further investigations.

Maximum concentrations of phosphorus and nitrogen were both high, but with strong opposite seasonal patterns: the highest phosphorus concentrations occurred in summer when nitrogen was the lowest. Phosphorus concentrations of the lake were 3 times higher in the reservoir in summer 2018 compared to stations measured in headwaters of the incoming river Yvel [27]. As both nutrients usually enter during the wet winter months [28], their opposing seasonal pattern could indicate either enhanced uptake or denitrification for N, whereas the P concentrations exceeding those of the incoming water could have been released from the sediments during summer, as known from other lakes [20,29,30].

In summer, the Redfield N/P ratios in the total fraction were always below 20, confirming a deficiency of bioavailable N in the water column during this period [31]. This result is in line with an analyse of 369 German lakes concluding that N limitation seems to predominate during summer in shallow polymictic lakes [29].

Phytoplankton and zooplankton succession and densities were similar to many eutrophic water bodies. Total abundances of meso-zooplankton appeared high, but remained in agreement with other shallow eutrophic lakes [32–34]. Small zooplanktonic taxa seemed to dominate at the detriment of larger cladocerans, e.g., *Daphnia* sp., as known from eutrophic temperate lakes of both Europe [34] and the USA [35]. As sampling size was low in this study, more research focusing on zooplankton dynamics in that reservoir would be beneficial.

An interesting element in the seasonal cycle was the interannual variability between the two summers: while the bloom of cyanobacteria reached an exceptional high value of 1.5 million cells mL^{-1} in 2018, exceeding the limit allowing bathing by a factor of 15 in France and other countries [36], it was one order of magnitude inferior in 2019. At the same time, in the lower basin the zooplanktonic community switched from a dominance of *Polyarthra* and *Bosmina* during 2018 to a dominance of *Keratella, Pompholyx* and *Daphnia* in 2019, which could have been provoked by decreasing cyanobacteria in the phytoplankton community. Some zooplankton taxa such as the raptorial rotifer *Polyarthra*, or to a lesser extent the cladocerans *Bosmina*, are highly selective feeders avoiding cyanobacteria compared to filter feeders like *Daphnia* that thrive better in the absence of cyanobacteria [37–40]. A zooplankton community can also be top-down controlled by zooplanktivorous fish preferring large zooplankton such as *Daphnia* [41]. Despite the Lac-au-Duc reservoir being a frequented fishing site, the lack of published data on the fish compartment does not allow us to discuss their potential role in structuring the planktonic community.

The contrast of cyanobacteria abundance between the summer of 2018 and 2019 is in the range of the interannual variability of blooms' intensity and species composition in the reservoir Lac-au-Duc observed during 13 years. Within the 2007–2019 period, densities of 40 to 80 mm^3 L^{-1} were reached during five summers, while in the other eight summers it ranged at maximum from 1 to 20 mm^3 L^{-1}.

Towards the end of the time series (2018–2019), a dominance or co-dominance of *Dolichospermum* replaced the dominance of *Planktothrix agardhii* until 2013 together with a co-occurance of *Microcystis* (mainly *aeruginosa*) in low densities in 2013, 2015 and early 2019. This shift in 2013 of cyanobacterial dominance seems to correlate to dryer and warmer summers connected to an increase of water residence time, a slight decrease of wind and a decrease of N sources, which in total confirms our second hypothesis (climate forces as main drivers). Before 2013, with the exception of 2009, summers were indeed characterized by more windy days. Blooms of *Planktothrix agardhii* dominated during that first period, as this species tolerates low average insolation in turbid waters of polymictic lakes [26,41]. *P. agardhii* also grows best at temperatures between 10–20 °C [42], and may become disadvantaged by warmer temperatures. Additionally, at the regional scale in Brittany, *P. agardhii* was the dominant taxa of the freshwater cyanobacterial community between 2004 and 2011 [43].

Surprisingly, none of the measured microcystin congeners was detected during *Planktothrix agardhii* blooms despite *Planktothrix* sp. being able to form toxic blooms in temperate freshwater ecosystems [44,45]. Variation of toxin production can be explained by (i) presence/absence of *mcy* genes necessary for their synthesis and (ii) individual variation within *mcy* genotypes with inactivation or regulation at the level of genes expression [46,47]. Natural *Planktothrix agardhii* blooming populations can be composed of microcystin producing (*mcy*) and non-producing (non-*mcy*) genotypes, and their proportion can vary considerably [47–49]. Moreover, it was demonstrated on non-mcy strains of *P. agardhii* isolated from nine European freshwater bodies to have lost more than 90% of their *mcy* genes during evolutionary processes [50]. Based on the analyse of 138 *Planktothrix* strains from three continents, the variable spatial distribution of *mcy* and non-*mcy* genotypes was suggested to depend on ecophysiological adaptation [51]. In laboratory experiments, non-mcy strains of *P. agardhii* seem to have better fitness than *mcy* strains under non-limiting conditions [52]. Similar results were obtained for *Microcystis*: non-*mcy* strains dominated under optimal growth conditions [53]. The authors of these studies then hypothesized that when cyanobacteria grow under favourable environmental conditions, the cost of producing microcystins becomes too high compared to the advantages it can bring [52]. Based on these results, it is tempting to hypothesize that non-limiting growth conditions concerning P and N may have favoured the selection of non-*mcy Planktothrix agardhii* strains in the Lac-au-Duc reservoir. It would be interesting to verify the presence of *mcy* genes, especially since microcystin detections were demonstrated to negatively correlate to *Planktothrix* biovolume in several Brittany lakes [43], suggesting that non-*mcy P. agardhii* strains are prevalent at the regional scale. From the composition of the cyanobacterial bloom, also other toxins, such as anatoxin-a and cylindrospermopsin could have been expected [44,54], they were however never detected during the monitoring between 2007–2019.

Since summer 2013, *Microcystis* and *Dolichospermum* became dominant or codominant, related to generally calmer conditions. Both genera are known to benefit greatly from water column stability as they can regulate their buoyancy according to their requirements in the illuminated area of stable lakes [55–57]. *Microcystis* occurrence was moreover strongly related to light intensity and temperature, which enhances the stabilization of the water column and favoured the development and persistence of *Microcystis* blooms in other lakes as well [16,58]. Temperature also benefits directly cyanobacterial development through their growth rate [59], and has been the most important factor driving the development of *Microcystis* sp. as analysed in more than 1000 lakes [60]. At the regional scale, light intensity was identified as the main climatic driver for *Microcystis* sp. [43]. Despite dominant in 2013 and 2015, *Microcystis* abundances and associated toxin production remained low in Lac-au-Duc compared to other lakes (e.g., [16,24], suggesting that the favourable conditions for its development were not fully met. The optimal growth rate for *Microcystis* is well above 25 °C [59], a temperature rarely reached in Lac-au-Duc or in the surrounding region [43]. We can also hypothesize that *Microcystis* was limited by nitrogen in 2018 and 2019, when the species dominated in early summer but disappeared thereafter. Although *Microcystis* is able to use diverse forms of nitrogen, N availability appears to be an essential element controlling the development of this cyanobacteria and its toxin production [61,62].

This context of N limitation in Lac au Duc could also explain the emerging occurrence of the diazotrophic *Dolichospermum* in recent years, as underlined by the threshold above which its occurance decreases. Nitrate concentrations progressively declined during the last decade in the river entering the reservoir, as in other rivers at the regional scale [63]. Thus, N-fixation ability provides an ecological advantage for this cyanobacterium [41].

Indeed, the first occurrence of *Dolichospermum* in 2012 and its largest bloom in 2017 corresponded to the driest hydrological years of the time series, that is the years with the lowest nitrogen recharge. Drought also seems to have an impact on the hydrological functioning of the reservoir: from 2015 onwards, due to the succession of rather dry hydrological years, the water level remained low, when outflow stopped from beginning of July until autumn. Water movements were doubtless strongly reduced and this phenomenon was probably exacerbated during dry summers such as in 2017 and 2018, when the inflow of water ceased already in early summer. Air temperature, residence time and light contributed to both axis of the CCA, therefore jointly with time on the first axis, indicating that both parameters tend to increase over time. The occurrence of *Dolichospermum* was seems partly related to a long water residence time producing a favourable environment for these buoyant cyanobacteria. These results are in accordance with those of Hayes et al. (2015) who found evidence in 42 lakes from agricultural watersheds that droughts strongly influence the system towards N limitation and induce the development of diazotrophic cyanobacteria.

We hypothesize that the lake is in the progress to shift to a new equilibrium, dominated by N_2 fixing cyanobacteria. This is potentially connected to toxin production, but evidence for a causal link between reduced N loading and diazotrophic cyanobacteria such as *Dolichospermum* abundance or biovolume is mixed [57]. In addition, evidence is increasing that N_2 fixation cannot always compensate significantly for the N deficiency, underlying the need to continue reducing emission of both nitrogen and phosphorus in the catchment [20,64–67]. Thus, a further monitoring of the cyanobacterial community is recommended.

This study provides another proof for the necessity to apply a catchment wide approach to limit P and N entrance into lakes and reservoirs. While reducing N was successful, it remains difficult to reduce non-point source P, despite efforts undertaken by farmers in changing agricultural practices. Moreover, depending on the water-bodies' bathymetry and the P stocked in the sediment, the time to reach and restore less eutrophe conditions needs to be taken into consideration. Many environmental parameters can be intercorrelated (nutrient availability, residence time, surface water temperature, stratification, etc.), making it impossible to separate the individual effect of these factors on a community. The complementary approach to such environmental surveys is to use very long term data, a multiple sites approach or mesocosms experiments. The intensity (frequency, stations and parameters) of long term sampling campaigns may present shortcomings, as in our case due to the restauration attempts with applications of $CuSO_4$ amongst others, thus we were forced to eliminate several of the years, which may have weakened the data set.

4. Conclusions

To conclude, over only 13 years, two major shifts in the cyanobacterial community have been recorded: a first shift in 2013 from a mixing tolerant population of *Planktothrix* to a buoyant species, *Microcystis*, preferring more stable water conditions, influenced by generally calmer and warmer conditions. A second shift from 2017 onwards was driven simultaneously by droughts (representing the biggest change) and reduced N loadings favouring the diazothrophic *Dolichospermum*. Our study also points out that if P reduction is not successful, dominance of *Microcystis* sp. and *Dolichospermum* sp. potentially increase. It is tempting to predict that these buoyant cyanobacteria will replace the *Planktothrix agardhii* population permanently in the context of climate forcing favouring warmer and dryer conditions at the global scale if N and in particular P cannot be reduced below their requirements [45,68]. This could imply many changes in terms of potential toxin production and increased persistance of cyanobacterial populations from year to year with the growing stock of cells in sediment as both *Microcystis* and *Dolichospermum* have dormant cells, while *Dolichospermum* possesses in addition akinetes.

Changes can also concern curative actions as buoyant cyanobacteria are sensitive to artificial mixing for example, in contrast to *Planktothrix* [69]. Thus, it will be necessary to follow the evolution of future years to confirm or deny the persistent installation of *Microcystis* or *Dolichospermum* in relation to climatic variation.

5. Materials and Methods

5.1. Study Site

The study site, called "Lac au Duc", is one of the largest shallow water bodies (250 ha, 2.6-m depth in mean) in Brittany (western France, Figure 7). This 3-million m^3 recreational and drinking water reservoir drains an agricultural catchment (37,000 ha) with the Yvel river as the main tributary. The lake is used for bathing, fishing and nautical activities.

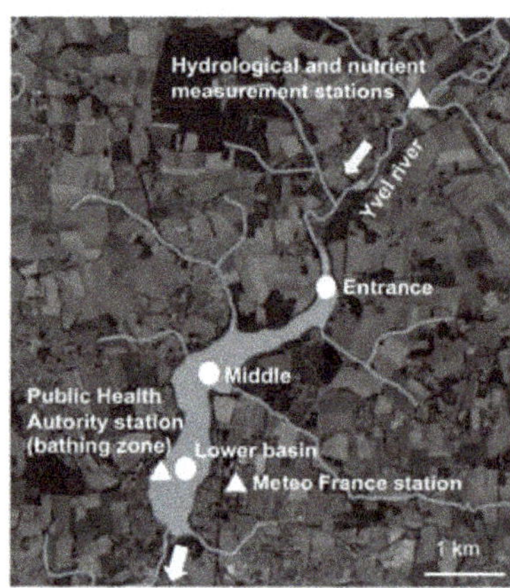

Figure 7. Localization of the Lac-au-Duc in France and stations monitored during the 2018–2019 seasonal cycle (round) for the 13 summer period analyses (triangle). Maps extracted from Geoportail.

Two types of monitoring data were used in this study: (i) data acquired monthly over a 2018–2019 seasonal cycle at three points in the lake and (ii) data acquired almost weekly over 13 summers, from 2007 to 2019 in a bathing zone. Additional environmental data were collected from governmental and meteorological data bases.

5.2. Seasonal Cycle 2018–2019

5.2.1. Sampling Sites and Sampling

To examine if nutrients, phyto- and zooplankton were evenly distributed in the reservoir we realized a seasonal cycle. Samples were collected from July 2018 to August 2019 at the entrance, in the middle and in the lower basin of the reservoir (Figure 7) from a boat. Duplicates of five liters of sub-surface water were collected with a 1 m vertical tube sampler at each point for nutrient, phytoplankton and zooplankton analyses. Duplicate samples for dissolved nutrient analyses were filtered on board on sterile Minisart CA 0.45 μm. Nutrient samples were kept at 4 °C until return to the lab, where they were stocked at −20 °C until analysis. Temperature and oxygen were measured along depth profiles with an Idronaut Ocean Seven 316 Plus CTD.

5.2.2. Nutrients

Particulate phosphorus, total dissolved phosphorus (TDP), particulate nitrogen and total dissolved nitrogen (TDN) were measured by colorimetry after digestion with persulfate according to [70], with a limit of detection of 6 μg P L^{-1} and 50 μg N L^{-1}. Orthophosphate ($PO_4{}^{3-}$) was analysed by the ammonium molybdate method [71] with a limit of detection of 3 µg P L^{-1}. After nitrate ($NO_3{}^-$) reduction into nitrite ($NO_2{}^-$) with vanadium chloride, $NO_2{}^-$ (originally present and reduced nitrates) was measured by colorimetry using sulphanilamide and N-1-naphthylethylenediamine dihydrochloride [72], with a limit of detection of 50 µg N L^{-1}. Colorimetric measurements were realised using a Gallery Photometric Analyser Gallery Plus (Thermo Fisher, Saint Herblain, France).

5.2.3. Phytoplankton

For the seasonal cycle, microalgae and cyanobacteria composition was determined directly in the fresh sample when possible or were preserved in Lugols solution and stored at 4 °C. Identification were realized to the genus level according to Komárek. Phytoplankton cell counts were carried out with a Nageotte cell according to [73]. For large colonies, the number of cells was estimated using photos. Microscopic photos were used to assess the number of cells in colonies of picocyanobacterial, using Pegasus software. At least, 400 individuals were counted per sample. For both, a light microscope (Olympus BX 50, Rungis, France) was used. Prior to cell counts and identification, if necessary, microalgae and cyanobacteria were concentrated on a 1 µm Poretics polycarbonate membrane filter, thanks to a filtration pump but with low vacuum pressure. Transfer and resuspension of cells in 1 ml of water was done while the membrane was still wet. Intact colonial chlorophytes and cyanobacteria were checked on the microscope to ensure that damage due to the concentration of cells was low.

5.2.4. Zooplankton

Zooplankton was concentrated from two liters of water by filtering through a 50-µm net, then narcotized with soda water and preserved in 70% ethanol at 4 °C until counting. For counting, 1 mL of the concentrated sample was distributed on a Sedgewick-Rafter counting chamber. For each sample, a minimum of 400 rotifers were counted and identified at the genus or species level, according to USEPA protocol [74] with a Zeiss microscope, based on the identification manuals for rotifers [75] and cladocerans [76]. In low-abundance samples, the totality of the 1 mL was observed. In addition, crustaceans and cladocerans were counted.

5.3. Long Term Series 2007 to 2019
5.3.1. Phytoplankton and Toxins

Each summer since 2004 the regional public health authorities (French Agence Régionale de Santé, ARS) of Brittany monitor water from a bathing zone located in the lower basin of the reservoir (Figure 7). Depending on the bloom density, samples were collected weekly from June to September, except in 2009 (20/7 to 31/8) and 2010 (24/6 to 30/8). Monitoring stopped in mid-September in 2015–2019, while it continued until end of October in 2011 and 2012. Cyanobacteria abundance and species composition as well as toxin concentrations were analysed by local laboratories. Toxins were analysed with LC-MS/MS [77] after extraction with methanol. Microcystins analysed comprised the congeners microcystin—LR, YR, RR, LF, LW, LY, LA, desmethyl-LR and desmethyl RR with a limit of quantification (LOQ) of 0.3 µg L^{-1}. MC-LR-eqivalents were calculated according to Wolf and Frank [78] and Fastner and Humpage [79] with MC-LR (1 per definition), MC-LA, MC-YR, MC-YM (1.0) and MC-RR (0.1). Saxitoxin was determined with an LOQ of 0.6 µg L^{-1}, anatoxin LOQ < 0.3 µg L^{-1} and cylindrospermopsin LOQ < 0.6 µg L^{-1}. All cyanobacterial cell densities data were converted to biovolume with the use of relevant geometric formulas and data literature [80,81], to account for differences in contribution from larger species and smaller species.

5.3.2. Abiotic Parameters

Daily meteorological data (air temperature, global radiation intensity and wind speed) were collected from Météo-France database from 2007 to 2019. Measurements were realised at the Météo-France station of Ploërmel, located 0.5 km from the reservoir (Figure 7). The light energy sensor broke in July 2017, from this date measurements from the Rennes-St-Jacques Météo-France station, located 60 km from the reservoir, were used. The number of days with an air temperature above 20 °C, or with solar radiations greater than 2800 J cm^{-2} were calculated. The water flow as well as nutrient concentrations were measured in the Yvel river, 3 km upstream the entrance of the reservoir (Figure 7). The water flow data were provided for the 2007–2019 period from the national database Hydro France [82] and corresponded to the daily mean values calculated from continuous stage records. From these data, we calculated the residence time of the water in the reservoir, knowing that its volume is approximately 3 million m^3. To take into account the time needed between climate variation and phytoplankton response, hydrological and meteorological data were averaged over 6 days preceding the sampling date of cyanobacteria. Wind variation completed the average data. Windy summers were identified by the number of days of wind above a threshold of 4 m s^{-1} during a 24 h average following [82]. Monthly data for nitrate and total phosphorus concentrations were provided from the database of the Yvel-Yvet watershed manager (the Syndicat Mixte du Grand Bassin de l'Oust, SMGBO) from 2007 to 2019. By combining these nutrient data with flow data, we calculated the quantities of NO$_3{}^-$ and Ptot entering the reservoir over the summer period or the complete hydrological year (November to November). The city of Ploërmel also provided reservoir water level data from the drinking water purification station.

5.3.3. Statistical Analyses

Despite being available, we discarded data from 2002 to 2005 as some copper sulphate treatments had been applied in the whole reservoir up to summer 2005. Treatments with calcium carbonate and hydrogen peroxide also took place, but locally in the bathing area, between 2013 and 2015 and in July 2018, respectively. During these years, except 2018, no significant difference was observed between the bathing area and the rest of the lake. We then chose to keep the data from the long-term monitoring bathing area for 2013–2015, while we used data from outside of the enclosed area only in 2018. Monitoring of nutrient concentrations started in 2007, thus data from 2006 were not used in analyses, leaving an extra year after the copper sulphate treatments.

All statistical analyses were performed in R Studio version 1.2.5019 [83]. In order to link cyanobacterial blooming genera with environmental parameters and characterize the sampling dates, a canonical correspondence analysis (CCA) was performed on Hellinger transformed genera biovolume (response matrix) and centred-reduced environmental parameters (explanatory variables). To reduce the number of explanatory variables, the ordistep function of the vegan package has been used to find the most parsimonious model based on Akaike Information Criterion (AIC). The less significant explanatory variables were eliminated from the CCA, in order to identify only the significant ones. The significance of the final CCA has been tested through a permutation test with the function envfit. A reference distribution under H0 from the data themselves is generated by permutations of rows and columns and by calculating the new percentage of constrained variance (sum of all canonical eigenvalues). The collinearity between explanatory parameters has also been checked.

We used generalized additive models (GAM) for time series of abiotic parameters, to analyze their temporal dynamics according to the season and the year. GAM takes into account the nonlinear response between the dependent variable (the abiotic parameter) and the explanatory variables (time, month of the year and year), using smooth functions called splines. Finally, the final shape of the relationship is determined by the data themselves. The independant variables in the initial model were the month of the year and the year, except for nitrates, for which it was only time (date) due to little data (Table 3). The best model was selected based on the gain in deviance explained (DE) relative to the initial model, while minimizing the Akaike's information criterion (AIC). Models were adjusted using the R package 'mgcv' and the function gam() with the selection of the Restricted

Maximum Likelihood (REML) method. The family was Gaussian and the link function was identity. A Log 10 transformation was done on the abiotic parameter when needed.

Supplementary Materials: The following are available online at https://www.mdpi.com/article/10.3390/toxins13050351/s1, Figure S1. Plots from the function gam.check () for each abiotic parameters: At the top left, the Q—Q plot compares the model residuals to a normal distribution., Table S1. Microcystin types meas-ured over the 2007–2019 summer periods.

Author Contributions: Conceptualization, M.L.M., A.P. and C.W.; methodology, M.L.M., A.P. and C.W.; formal analysis, M.L.M. and A.P.; investigation, M.L.M., M.C., L.B. and E.M.; resources, B.R.; data curation, A.P.; writing—original draft preparation, M.L.M.; writing—review and editing, A.P., C.W., L.B. and B.R.; visualization, M.L.M. and A.P.; supervision, C.W. and L.B.; project administration, C.W.; funding acquisition, C.W. All authors have read and agreed to the published version of the manuscript.

Funding: This research was funded by the Interreg CPES program 2017–2020 (https://www.cpes-interreg.eu; last accessed date 12 May 2021).

Institutional Review Board Statement: Not applicable.

Informed Consent Statement: Not applicable.

Data Availability Statement: Long term data series were obtained from the following sources: Cyanobacteria: ARS Bretagne, France; Temperature, radiation and wind speed: Meteo France, Ploermel, respectively, Rennes, St Jaques, nutrients: Syndicat Mixte du Grand Bassin de l'Oust, SMGBO, Bretagne, France.

Acknowledgments: We thank the French regional public health agency of Brittany for providing the data on cyanobacteria, and Meteo France and DREAL Bretagne-Hydro France for meteorological and hydrological data, respectively. We also thank the Syndicat Mixte du Grand Bassin de l'Oust for providing data on nutrients, as well as the city of Ploërmel for the data on the water level in the reservoir. The nautical base of Ploërmel is thanked for its help collecting samples. We thank also André-Jean Francez for his kind help in zooplankton identification, as well as Enora Briand for her useful comments on the discussion on microcystin production. Nutrient analysis were conducted at the Plateforme EcoChimie (EcoChim, UMS OSUR 3343).

Conflicts of Interest: The authors declare no conflict of interest. The funders had no role in the design of the study; in the collection, analyses, or interpretation of data; in the writing of the manuscript, or in the decision to publish the results.

References

1. Smith, V.H.; Tilman, G.D.; Nekola, J.C. Eutrophication: Impacts of Excess Nutrient Inputs on Freshwater, Marine, and Terrestrial Ecosystems. *Environ. Pollut.* **1999**, *100*, 179–196. [CrossRef]
2. Le Moal, M.; Gascuel-Odoux, C.; Menesguen, A.; Souchon, Y.; Etrillard, C.; Levain, A.; Moatar, F.; Pannard, A.; Souchu, P.; Lefebvre, A.; et al. Eutrophication: A New Wine in an Old Bottle? *Sci. Total Environ.* **2019**, *651*, 1–11. [CrossRef]
3. O'Neil, J.M.; Davis, T.W.; Burford, M.A.; Gobler, C.J. The Rise of Harmful Cyanobacteria Blooms: The Potential Roles of Eutrophication and Climate Change. *Harmful Algae* **2012**, *14*, 313–334. [CrossRef]
4. Paerl, H.W.; Otten, T.G. Blooms Bite the Hand That Feeds Them. *Science* **2013**, *342*, 433–434. [CrossRef]
5. Ger, K.A.; Hansson, L.A.; Lurling, M. Understanding Cyanobacteria-Zooplankton Interactions in a More Eutrophic World. *Freshw. Biol.* **2014**, *59*, 1783–1798. [CrossRef]
6. Huisman, J.; Codd, G.A.; Paerl, H.W.; Ibelings, B.W.; Verspagen, J.M.H.; Visser, P.M. Cyanobacterial Blooms. *Nat. Rev. Microbiol.* **2018**, *16*, 471–483. [CrossRef] [PubMed]
7. Bockwoldt, K.A.; Nodine, E.R.; Mihuc, T.B.; Shambaugh, A.D.; Stockwell, J.D. Reduced Phytoplankton and Zooplankton Diversity Associated with Increased Cyanobacteria in Lake Champlain, USA. *J. Contemp. Water Res. Educ.* **2017**, *160*, 100–118. [CrossRef]
8. Nurnberg, G.K.; LaZerte, B.D. More than 20 Years of Estimated Internal Phosphorus Loading in Polymictic, Eutrophic Lake Winnipeg, Manitoba. *J. Great Lakes Res.* **2016**, *42*, 18–27. [CrossRef]
9. Carmichael, W.W. The Cyanotoxins. *Adv. Bot. Res.* **1997**, 211–256. [CrossRef]
10. Wiegand, C.; Pflugmacher, S. Ecotoxicological Effects of Selected Cyanobacterial Secondary Metabolites a Short Review. *Toxicol. Appl. Pharmacol.* **2005**, *203*, 201–218. [CrossRef]
11. Metcalf, J.S.; Codd, G.A. Cyanotoxins. In *Ecology of Cyanobacteria II*; Springer: Berlin/Heidelberg, Germany, 2012; pp. 651–675.
12. *World Health Organization Guidelines for Drinking Water Quality, Addendum to Volume 2: Health Criteria and Other Supporting Information*; WHO Publications: Geneva, Switzerland, 1998.

13. Ibelings, B.W.; Backer, L.C.; Kardinaal, W.E.A.; Chorus, I. Current Approaches to Cyanotoxin Risk Assessment and Risk Management around the Globe. *Harmful Algae* **2014**, *40*, 63–74. [CrossRef]

14. Dokulil, M.T.; Teubner, K. Cyanobacterial Dominance in Lakes. *Hydrobiologia* **2000**, *438*, 1–12. [CrossRef]

15. Taranu, Z.E.; Gregory-Eaves, I.; Leavitt, P.R.; Bunting, L.; Buchaca, T.; Catalan, J.; Domaizon, I.; Guilizzoni, P.; Lami, A.; McGowan, S.; et al. Acceleration of Cyanobacterial Dominance in North Temperate-Subarctic Lakes during the Anthropocene. *Ecol. Lett.* **2015**, *18*, 375–384. [CrossRef]

16. Jöhnk, K.D.; Huisman, J.; Sharples, J.; Sommeijer, B.; Visser, P.M.; Stroom, J.M. Summer Heatwaves Promote Blooms of Harmful Cyanobacteria: Heatwaves Promote Harmful Cyanobacteria. *Glob. Chang. Biol.* **2008**, *14*, 495–512. [CrossRef]

17. Kosten, S.; Huszar, V.L.M.; Becares, E.; Costa, L.S.; van Donk, E.; Hansson, L.A.; Jeppesenk, E.; Kruk, C.; Lacerot, G.; Mazzeo, N.; et al. Warmer Climates Boost Cyanobacterial Dominance in Shallow Lakes. *Glob. Chang. Biol.* **2012**, *18*, 118–126. [CrossRef]

18. Fastner, J.; Abella, S.; Litt, A.; Morabito, G.; Vörös, L.; Pálffy, K.; Straile, D.; Kümmerlin, R.; Matthews, D.; Phillips, M.G.; et al. Combating Cyanobacterial Proliferation by Avoiding or Treating Inflows with High P Load—Experiences from Eight Case Studies. *Aquat. Ecol.* **2016**, *50*, 367–383. [CrossRef]

19. Jeppesen, E.; Sondergaard, M.; Jensen, J.P.; Havens, K.E.; Anneville, O.; Carvalho, L.; Coveney, M.F.; Deneke, R.; Dokulil, M.T.; Foy, B.; et al. Lake Responses to Reduced Nutrient Loading—An Analysis of Contemporary Long-Term Data from 35 Case Studies. *Freshw. Biol.* **2005**, *50*, 1747–1771. [CrossRef]

20. Shatwell, T.; Köhler, J. Decreased Nitrogen Loading Controls Summer Cyanobacterial Blooms without Promoting Nitrogen-fixing Taxa: Long-term Response of a Shallow Lake. *Limnol. Oceanogr.* **2019**, *64*, S166–S178. [CrossRef]

21. Schindler, D.W.; Hecky, R.E.; Findlay, D.; Stainton, M.; Parker, B.; Paterson, M.; Beaty, K.; Lyng, M.; Kasian, S. Eutrophication of Lakes Cannot Be Controlled by Reducing Nitrogen Input: Results of a 37-Year Whole-Ecosystem Experiment. *Proc. Natl. Acad. Sci.* **2008**, *105*, 11254–11258. [CrossRef]

22. Liu, X.; Lu, X.H.; Chen, Y.W. The Effects of Temperature and Nutrient Ratios on *Microcystis* Blooms in Lake Taihu, China: An 11-Year Investigation. *Harmful Algae* **2011**, *10*, 337–343. [CrossRef]

23. Deng, J.; Salmaso, N.; Jeppesen, E.; Qin, B.; Zhang, Y. The Relative Importance of Weather and Nutrients Determining Phytoplankton Assemblages Differs between Seasons in Large Lake Taihu, China. *Aquat. Sci.* **2019**, *81*, 48. [CrossRef]

24. Vasconcelos, V.; Morais, J.; Vale, M. Microcystins and Cyanobacteria Trends in a 14 Year Monitoring of a Temperate Eutrophic Reservoir (Aguieira, Portugal). *J. Environ. Monit.* **2011**, *13*, 668–672. [CrossRef] [PubMed]

25. Rücker, J.; Wiedner, C.; Zippel, P. Factors controlling the dominance of *Planktothrix agardhii* and Limnothrix redekei in eutrophic shallow lakes. In *Shallow Lakes' 95*; Springer: Dordrecht, The Netherlands, 1997; pp. 107–115.

26. Kurmayer, R.; Deng, L.; Entfellner, E. Role of Toxic and Bioactive Secondary Metabolites in Colonization and Bloom Formation by Filamentous Cyanobacteria Planktothrix. *Harmful Algae* **2016**, *54*, 69–86. [CrossRef] [PubMed]

27. Casquin, A.; Gu, S.; Dupas, R.; Petitjean, P.; Gruau, G.; Durand, P. River Network Alteration of C-N-P Dynamics in a Mesoscale Agricultural Catchment. *Sci. Total Environ.* **2020**, *749*. [CrossRef]

28. Sondergaard, M.; Jensen, J.P.; Jeppesen, E. Seasonal Response of Nutrients to Reduced Phosphorus Loading in 12 Danish Lakes. *Freshw. Biol.* **2005**, *50*, 1605–1615. [CrossRef]

29. Dolman, A.M.; Mischke, U.; Wiedner, C. Lake-Type-Specific Seasonal Patterns of Nutrient Limitation in German Lakes, with Target Nitrogen and Phosphorus Concentrations for Good Ecological Status. *Freshw. Biol.* **2016**, *61*, 444–456. [CrossRef]

30. Sondergaard, M.; Jensen, J.P.; Jeppesen, E.; Møller, P.H. Seasonal Dynamics in the Concentrations and Retention of Phosphorus in Shallow Danish Lakes after Reduced Loading. *Aquat. Ecosyst. Health Manag.* **2002**, *5*, 19–29. [CrossRef]

31. Guildford, S.J.; Hecky, R.E. Total Nitrogen, Total Phosphorus, and Nutrient Limitation in Lakes and Oceans: Is There a Common Relationship? *Limnol. Oceanogr.* **2000**, *45*, 1213–1223. [CrossRef]

32. Abrantes, N.; Antunes, S.; Pereira, M.; Gonçalves, F. Seasonal Succession of Cladocerans and Phytoplankton and Their Interactions in a Shallow Eutrophic Lake (Lake Vela, Portugal). *Acta Oecologica* **2006**, *29*, 54–64. [CrossRef]

33. Cazzanelli, M.; Warming, T.P.; Christoffersen, K.S. Emergent and Floating-Leaved Macrophytes as Refuge for Zooplankton in a Eutrophic Temperate Lake without Submerged Vegetation. *Hydrobiologia* **2008**, *605*, 113–122. [CrossRef]

34. Zingel, P.; Haberman, J. A comparison of zooplankton densities and biomass in Lakes Peipsi and Võrtsjärv (Estonia): Rotifers and crustaceans versus ciliates. In *European Large Lakes Ecosystem Changes and Their Ecological and Socioeconomic Impacts*; Springer: Dordrecht, The Netherlands, 2007; pp. 153–159.

35. Moody, E.K.; Wilkinson, G.M. Functional Shifts in Lake Zooplankton Communities with Hypereutrophication. *Freshw. Biol.* **2019**, *64*, 608–616. [CrossRef]

36. Chorus, I.; Bartram, J. *Toxic Cyanobacteria in Water: A Guide to Their Public Health Consequences, Monitoring and Management*; CRC Press: Boca Raton, FL, USA, 1999; ISBN 1-4822-9506-7.

37. Barnett, A.J.; Finlay, K.; Beisner, B.E. Functional Diversity of Crustacean Zooplankton Communities: Towards a Trait-Based Classification. *Freshw. Biol.* **2007**, *52*, 796–813. [CrossRef]

38. Obertegger, U.; Flaim, G. Community Assembly of Rotifers Based on Morphological Traits. *Hydrobiologia* **2015**, *753*, 31–45. [CrossRef]

39. Burian, A.; Kainz, M.J.; Schagerl, M.; Yasindi, A. Species-Specific Separation of Lake Plankton Reveals Divergent Food Assimilation Patterns in Rotifers. *Freshw. Biol.* **2014**, *59*, 1257–1265. [CrossRef]

40. Jeppesen, E.; Jensen, J.P.; Sondergaard, M.; Lauridsen, T.; Landkildehus, F. Trophic Structure, Species Richness and Biodiversity in Danish Lakes: Changes along a Phosphorus Gradient. *Freshw. Biol.* **2000**, *45*, 201–218. [CrossRef]
41. Reynolds, C.S. *The Ecology of Phytoplankton*; Cambridge University Press: Cambridge, UK, 2006; ISBN 1-139-45489-7.
42. Suda, S. Taxonomic Revision of Water-Bloom-Forming Species of Oscillatorioid Cyanobacteria. *Int. J. Syst. Evol. Microbiol.* **2002**, *52*, 1577–1595. [CrossRef]
43. Pitois, F.; Thoraval, I.; Baures, E.; Thomas, O. Geographical Patterns in Cyanobacteria Distribution: Climate Influence at Regional Scale. *Toxins* **2014**, *6*, 509–522. [CrossRef]
44. Sivonen, K.; Jones, G. Cyanobacterial Toxins. *Toxic Cyanobacteria Water A Guide Public Health Conseq. Monit. Manag.* **1999**, *1*, 43–112.
45. Paerl, H.W.; Huisman, J. Climate—Blooms like It Hot. *Science* **2008**, *320*, 57–58. [CrossRef]
46. Dittmann, E.; Neilan, B.A.; Erhard, M.; vonDohren, H.; Borner, T. Insertional Mutagenesis of a Peptide Synthetase Gene That Is Responsible for Hepatotoxin Production in the Cyanobacterium *Microcystis Aeruginosa* PCC 7806. *Mol. Microbiol.* **1997**, *26*, 779–787. [CrossRef]
47. Kurmayer, R.; Christiansen, G.; Fastner, J.; Borner, T. Abundance of Active and Inactive Microcystin Genotypes in Populations of the Toxic Cyanobacterium Planktothrix Spp. *Environ. Microbiol.* **2004**, *6*, 831–841. [CrossRef] [PubMed]
48. Yepremian, C.; Gugger, M.F.; Briand, E.; Catherine, A.; Berger, C.; Quiblier, C.; Bernard, C. Microcystin Ecotypes in a Perennial Planktothrix Agardhii Bloom. *Water Res.* **2007**, *41*, 4446–4456. [CrossRef]
49. Briand, E.; Gugger, M.; Francois, J.C.; Bernard, C.; Humbert, J.F.; Quiblier, C. Temporal Variations in the Dynamics of Potentially Microcystin-Producing Strains in a Bloom-Forming Planktothrix Agardhii (Cyanobacterium) Population. *Appl. Environ. Microbiol.* **2008**, *74*, 3839–3848. [CrossRef] [PubMed]
50. Christiansen, G.; Molitor, C.; Philmus, B.; Kurmayer, R. Nontoxic Strains of Cyanobacteria Are the Result of Major Gene Deletion Events Induced by a Transposable Element. *Mol. Biol. Evol.* **2008**, *25*, 1695–1704. [CrossRef] [PubMed]
51. Kurmayer, R.; Blom, J.F.; Deng, L.; Pernthaler, J. Integrating Phylogeny, Geographic Niche Partitioning and Secondary Metabolite Synthesis in Bloom-Forming Planktothrix. *ISME J.* **2015**, *9*, 909–921. [CrossRef] [PubMed]
52. Briand, E.; Yéprémian, C.; Humbert, J.; Quiblier, C. Competition between Microcystin-and Non-microcystin-producing Planktothrix Agardhii (Cyanobacteria) Strains under Different Environmental Conditions. *Environ. Microbiol.* **2008**, *10*, 3337–3348. [CrossRef] [PubMed]
53. Briand, E.; Bormans, M.; Quiblier, C.; Salencon, M.J.; Humbert, J.F. Evidence of the Cost of the Production of Microcystins by *Microcystis Aeruginosa* under Differing Light and Nitrate Environmental Conditions. *PLoS ONE* **2012**, *7*. [CrossRef]
54. De la Cruz, A.A.; Hiskia, A.; Kaloudis, T.; Chernoff, N.; Hill, D.; Antoniou, M.G.; He, X.; Loftin, K.; O'Shea, K.; Zhao, C.; et al. A Review on Cylindrospermopsin: The Global Occurrence, Detection, Toxicity and Degradation of a Potent Cyanotoxin. *Environ. Sci. Process. Impacts* **2013**, *15*, 1979. [CrossRef]
55. Ibelings, B.W.; Mur, L.R.; Walsby, A.E. Diurnal Changes in Buoyancy and Vertical-Distribution in Populations of Microcystins in 2 Shallow Lakes. *J. Plankton Res.* **1991**, *13*, 419–436. [CrossRef]
56. Huisman, J.; Sharples, J.; Stroom, J.M.; Visser, P.M.; Kardinaal, W.E.A.; Verspagen, J.M.H.; Sommeijer, B. Changes in Turbulent Mixing Shift Competition for Light between Phytoplankton Species. *Ecology* **2004**, *85*, 2960–2970. [CrossRef]
57. Li, X.C.; Dreher, T.W.; Li, R.H. An Overview of Diversity, Occurrence, Genetics and Toxin Production of Bloom-Forming Dolichospermum (Anabaena) Species. *Harmful Algae* **2016**, *54*, 54–68. [CrossRef]
58. Harke, M.J.; Steffen, M.M.; Gobler, C.J.; Otten, T.G.; Wilhelm, S.W.; Wood, S.A.; Paerl, H.W. A Review of the Global Ecology, Genomics, and Biogeography of the Toxic Cyanobacterium, Microcystis Spp. *Harmful Algae* **2016**, *54*, 4–20. [CrossRef]
59. Thomas, M.K.; Litchman, E. Effects of Temperature and Nitrogen Availability on the Growth of Invasive and Native Cyanobacteria. *Hydrobiologia* **2016**, *763*, 357–369. [CrossRef]
60. Rigosi, A.; Carey, C.C.; Ibelings, B.W.; Brookes, J.D. The Interaction between Climate Warming and Eutrophication to Promote Cyanobacteria Is Dependent on Trophic State and Varies among Taxa. *Limnol. Oceanogr.* **2014**, *59*, 99–114. [CrossRef]
61. Gobler, C.J.; Burkholder, J.M.; Davis, T.W.; Harke, M.J.; Johengen, T.; Stow, C.A.; Van de Waal, D.B. The Dual Role of Nitrogen Supply in Controlling the Growth and Toxicity of Cyanobacterial Blooms. *Harmful Algae* **2016**, *54*, 87–97. [CrossRef]
62. Chaffin, J.D.; Davis, T.W.; Smith, D.J.; Baer, M.M.; Dick, G.J. Interactions between Nitrogen Form, Loading Rate, and Light Intensity on Microcystis and Planktothrix Growth and Microcystin Production. *Harmful Algae* **2018**, *73*, 84–97. [CrossRef]
63. Dupas, R.; Minaudo, C.; Gruau, G.; Ruiz, L.; Gascuel-Odoux, C. Multidecadal Trajectory of Riverine Nitrogen and Phosphorus Dynamics in Rural Catchments. *Water Resour. Res.* **2018**, *54*, 5327–5340. [CrossRef]
64. Lewis, W.M.; Wurtsbaugh, W.A. Control of Lacustrine Phytoplankton by Nutrients: Erosion of the Phosphorus Paradigm. *Int. Rev. Hydrobiol.* **2008**, *93*, 446–465. [CrossRef]
65. Moss, B.; Jeppesen, E.; Sondergaard, M.; Lauridsen, T.L.; Liu, Z.W. Nitrogen, Macrophytes, Shallow Lakes and Nutrient Limitation: Resolution of a Current Controversy? *Hydrobiologia* **2013**, *710*, 3–21. [CrossRef]
66. Scott, J.T.; Grantz, E.M. N-2 Fixation Exceeds Internal Nitrogen Loading as a Phytoplankton Nutrient Source in Perpetually Nitrogen-Limited Reservoirs. *Freshw. Sci.* **2013**, *32*, 849–861. [CrossRef]
67. Kolzau, S.; Dolman, A.M.; Voss, M.; Wiedner, C. The Response of Nitrogen Fixing Cyanobacteria to a Reduction in Nitrogen Loading. *Int. Rev. Hydrobiol.* **2018**, *103*, 5–14. [CrossRef]

68. Paerl, H.W.; Scott, J.T.; McCarthy, M.J.; Newell, S.E.; Gardner, W.S.; Havens, K.E.; Hoffman, D.K.; Wilhelm, S.W.; Wurtsbaugh, W.A. It Takes Two to Tango: When and Where Dual Nutrient (N & P) Reductions Are Needed to Protect Lakes and Downstream Ecosystems. *Environ. Sci. Technol.* **2016**, *50*, 10805–10813. [CrossRef] [PubMed]
69. Mantzouki, E.; Visser, P.M.; Bormans, M.; Ibelings, B.W. Understanding the Key Ecological Traits of Cyanobacteria as a Basis for Their Management and Control in Changing Lakes. *Aquat. Ecol.* **2016**, *50*, 333–350. [CrossRef]
70. Grasshoff, P. Methods of Seawater Analysis. Verlag Chemie. *FRG* **1983**, *419*, 61–72.
71. US EPA United States Environmental Protection Agency. *Method 365.1: Determination of Phosphorus by Semi-Automated Colorimetry*; O'Dell, J.W., Ed.; Office of Research and Development: Cincinnati, OH, USA, 1993.
72. US EPA United States Environmental Protection Agency. *Method 353.2. Determination of Nitrate-Nitrite Nitrogen by Automated Colorimetry*; O'Dell, J.W., Ed.; Office of Research and Development: Cincinnati, OH, USA, 1993.
73. Brient, L.; Lengronne, M.; Bertrand, E.; Rolland, D.; Sipel, A.; Steinmann, D.; Baudin, I.; Legeas, M.; Le Rouzic, B.; Bormans, M. A Phycocyanin Probe as a Tool for Monitoring Cyanobacteria in Freshwater Bodies. *J. Environ. Monit.* **2008**, *10*, 248–255. [CrossRef]
74. US EPA United States Environmental Protection Agency. Standard Operating Procedure for Zooplankton Analysis (Method LG403). *Great Lakes Natl. Off.* **2016**. Available online: https://www.epa.gov/sites/production/files/2017-01/documents/sop-for-zooplankton-analysis-201607-22pp.pdf (accessed on 1 May 2018).
75. Pourriot, R.; Francez, A.-J. Introduction Pratique à La Systématique Des Organismes Des Eaux Continentales Françaises.-8: Rotifères. *Publ. De La Société Linnéenne De Lyon* **1986**, *55*, 148–176. [CrossRef]
76. Amoros, C. Introduction Pratique à La Systématique Des Organismes Des Eaux Continentales Françaises-5. Crustacés Cladocères. *Publ. De La Société Linnéenne De Lyon* **1984**, *53*, 72–107.
77. *Méthode Interne ME 230*; Laboratoire Départemental d'analyses Du Morbihan: Brittany, France, 2016.
78. Wolf, H.U.; Frank, C. Toxicity Assessment of Cyanobacterial Toxin Mixtures. *Environ. Toxicol.* **2002**, *17*, 395–399. [CrossRef]
79. Fastner, J.; Humpage, A. Chapter 2.1.; Hepatotoxic cyclic peptides—Microcystins and Nodularins. In *Toxic Cyanobacteria in Water: A Guide to Their Public Health Consequences, Monitoring and Management*; CRC Press: Boca Raton, FL, USA, 2021; pp. 21–52.
80. US EPA Method LG 401, Standard Operating Procedure for Phytoplankton Analysis. *Environ. Monit. Syst. Lab.* **2003**.
81. Rimet, F.; Druart, J.C. A Trait Database for Phytoplankton of Temperate Lakes. *Ann. De Limnol. Int. J. Limnol.* **2018**, *54*. [CrossRef]
82. HydroFrance. Available online: Http://Www.Hydro.Eaufrance.Fr/ (accessed on 10 January 2020).
83. R Core Team. *R: A Language and Environment for Statistical Computing*; Software Version R Version 3.6.1 (2019-07-05); R Foundation for Statistical Computing: Vienna, Austria, 2019.

Article

Cyanotoxin Screening in BACA Culture Collection: Identification of New Cylindrospermopsin Producing Cyanobacteria

Rita Cordeiro [1,2,*], Joana Azevedo [3], Rúben Luz [1,2], Vitor Vasconcelos [3,4], Vítor Gonçalves [1,2] and Amélia Fonseca [1,2]

[1] CIBIO, Centro de Investigação em Biodiversidade e Recursos Genéticos, InBIO Laboratório Associado, Pólo dos Açores, Universidade dos Açores, 9500-321 Ponta Delgada, Portugal; ruben.fs.luz@uac.pt (R.L.); vitor.mc.goncalves@uac.pt (V.G.); maria.ao.fonseca@uac.pt (A.F.)
[2] Faculdade de Ciências e Tecnologia, Universidade dos Açores, 9500-321 Ponta Delgada, Portugal
[3] Interdisciplinary Centre of Marine and Environmental Research—CIIMAR/CIMAR, University of Porto, Terminal de Cruzeiros do Porto de Leixões, Av. General Norton de Matos s/n, 4450-208 Matosinhos, Portugal; joana.azevedo@ciimar.up.pt (J.A.); vmvascon@fc.up.pt (V.V.)
[4] Department of Biology, Faculty of Sciences, University of Porto, 4069-007 Porto, Portugal
[*] Correspondence: rita.ip.cordeiro@uac.pt

Abstract: Microcystins (MCs), Saxitoxins (STXs), and Cylindrospermopsins (CYNs) are some of the more well-known cyanotoxins. Taking into consideration the impacts of cyanotoxins, many studies have focused on the identification of unknown cyanotoxin(s)-producing strains. This study aimed to screen strains from the Azorean Bank of Algae and Cyanobacteria (BACA) for MCs, STX, and CYN production. A total of 157 strains were searched for *mcy*, *sxt*, and *cyr* producing genes by PCR, toxin identification by ESI-LC-MS/MS, and cyanotoxin-producing strains morphological identification and confirmation by 16S rRNA phylogenetic analysis. Cyanotoxin-producing genes were amplified in 13 strains and four were confirmed as toxin producers by ESI-LC-MS/MS. As expected *Aphanizomenon gracile* BACA0041 was confirmed as an STX producer, with amplification of genes *sxt*A, *sxt*G, *sxt*H, and *sxt*I, and *Microcystis aeruginosa* BACA0148 as an MC-LR producer, with amplification of genes *mcy*C, *mcy*D, *mcy*E, and *mcy*G. Two nostocalean strains, BACA0025 and BACA0031, were positive for both *cyr*B and *cyr*C genes and ESI-LC-MS/MS confirmed CYN production. Although these strains morphologically resemble *Sphaerospermopsis*, the 16S rRNA phylogenetic analysis reveals that they probably belong to a new genus.

Keywords: microcystin; saxitoxin; cylindrospermopsin; ESI-LC-MS/MS; 16S rRNA phylogeny; Azores

Key Contribution: 16s rRNA phylogenetic analysis revealed evidence for new toxic cyanobacteri-ataxa, in two identified CYN producing nostocalean strains BACA0025 and BACA0031.

Citation: Cordeiro, R.; Azevedo, J.; Luz, R.; Vasconcelos, V.; Gonçalves, V.; Fonseca, A. Cyanotoxin Screening in BACA Culture Collection: Identification of New Cylindrospermopsin Producing Cyanobacteria. *Toxins* **2021**, *13*, 258. https://doi.org/10.3390/toxins 13040258

Received: 8 February 2021
Accepted: 1 April 2021
Published: 3 April 2021

Publisher's Note: MDPI stays neutral with regard to jurisdictional claims in published maps and institutional affiliations.

1. Introduction

Due to their impacts on ecosystem degradation, public health risk, and associated economical losses, cyanotoxins are one of the most studied primary and secondary metabolites from cyanobacteria [1–4]. Microcystins (MCs), Saxitoxins (STXs), and Cylindrospermopsins (CYNs) are some of the more well-known cyanotoxins, for which a high number of studies were done covering different aspects such as their taxonomic distribution, genetic organization, biosynthesis pathways, and mechanisms of action [5–10].

Microcystins are hepatotoxins produced by non-ribosomal pathways, and the most common and more well-studied cyanotoxins, with over 240 variants (e.g., MC-LR, MC-LY, MC-RR). Mostly produced by *Microcystis aeruginosa* [11], MCs are, however, currently known to be produced by over 40 species [12]. The gene cluster responsible for the biosynthesis of MCs (*mcy*) contains 10 encoding genes organized in two operons (*mcy*A-C and *mcy*D-J) in which each gene has a specific function for the toxin synthesis, regulation, and release [13,14]. This hydrophilic peptide has an Adda group in its structure, responsible for its toxicity [15,16]. With over 240 analogs, other groups as ADMAdda have been identified and characterized as responsible for MCs more toxic variants [17].

The alkaloid STX is one of the most potent natural neurotoxin [18] identified in both cyanobacteria and dinoflagellates [19]. In cyanobacteria, STXs have been identified in several species as *Aphanizomenon gracile*, *Cylindrospermum stagnale* [20], *Dolichospermum circinale* [21], and *Raphidiopsis raciborskii* [22]. The gene cluster *sxt* encodes 26 genes responsible for the biosynthesis of STXs [19] and its gene organization varies between species, which may indicate gene loss or transfer between species [6,23,24]. This scenario has been observed in the STX gene cluster where the *sxt*A gene, recognized as the encoding gene for the polyketide synthase-like structure involved in the first step of STXs production [6], was found in both STXs-producing and non-producing *A. gracile* strains, but all non-producing strains lost at least one of the other genes of the cluster (e.g. *sxt*G, *sxt*H, *sxt*I, or *sxt*X) [25].

Cylindrospermopsins are cytotoxic alkaloids initially identified in *Raphidiopsis raciborskii* [26] but also reported to be produced by other species as *Aphanizomenon gracile* [27] or *Chrysosporum ovalisporum* [28]. As MCs, CYNs biosynthesis is done by a non-ribosomal pathway, where the gene cluster *cyr* (15 encoding genes) has been identified as responsible for CYNs synthesis, regulation, and exportation in *R. raciborskii* [29].

Cyanobacteria have high morphological and ecological variability [30], and due to this, its taxonomical classification is relatively intricate and is in constant change [31]. New cyanobacteria families, genera, and species, as Oculatellaceae and Trichocoleaceae [32], *Aliinostoc* [33], *Halotia* [34], and *Lusitaniella* [35], and *Hyella patelloides* [35] and *Compactonostoc shennongjiaensis* [36] have been reported in recent years. These taxonomic redefinitions are challenging for the identification and distribution of toxins in many of the new genera [37], thus increasing the lack of knowledge on toxic strains. Although cyanobacteria are widely distributed through many environments [7,38], most cyanobacteria and cyanotoxins studies are focused on planktonic species from inland freshwater systems [7,20], with less information in extreme environments (e.g., hot springs, caves, hypersaline lakes, polar deserts), especially regarding their toxicity [7,39].

Taking into consideration cyanotoxins impacts, either in environmental or public health, many scientific studies have focused on the identification of unknown cyanotoxin(s)-producing species, usually by a combination of methods as immunoassays (e.g., ELISA—enzyme-linked immunosorbent assay), by targeting biosynthesis encoding genes (PCR—polymerase chain reaction) and/or by analytical methods, as liquid chromatography [40,41]. The liquid chromatography-mass spectrometry (LC-MS) is a highly sensitive and selective method for cyanotoxin detection and identification [40,41]. It can be used for cyanotoxins confirmation even in very low concentrations, as well as analog identification, even without available standards [40].

The main aim of this study was to screen for cyanotoxin-producing cyanobacteria in strains isolated from several environments (freshwater, terrestrial and thermal) from the Azores islands using a polyphasic approach, based on the detection of MCs, STX, and CYN encoding genes, the identification of their production by LC-MS, and the phylogenetic distribution of toxic strains through 16S rRNA molecular identification.

2. Results

2.1. Detection of Cyanotoxin(s)-Producing Cyanobacteria

From the 157 screened cyanobacteria strains, 13 were identified as cyanotoxin(s) potential producers by PCR amplification of the STX, CYN, and/or MC encoding genes, and four were confirmed as toxin producers by ESI-LC-MS/MS (Figures S1–S6; Table 1; Table S2).

Table 1. Detection of genes involved in microcystin, cylindrospermopsin, and saxitoxin production, and ESI-LC-MS/MS analysis on BACA strains. Only strains with positive detection of either the genes or the toxins are shown.

Strains	Habitat	Species	PCR			ESI-LC-MS/MS		
			SXT	CYN	MC	SXT	CYN	MC-LR
BACA0007	Lake	*Kamptonema* sp.	G	-	-	-	-	-
BACA0025	Lake	unidentified species	-	B, CC	-	-	+	-
BACA0031	Lake	unidentified species	-	B, CC	-	-	+	-
BACA0041	Lake	*Aphanizomenon gracile*	A, G, H, I	-	-	+	-	-
BACA0081	Lake	*Cylindrospermum* sp.	H	-	-	-	-	-
BACA0091	Lake	*Nostoc* sp.	-	-	E	-	-	-
BACA0109	Lake	*Nostoc* sp.	-	B	-	-	-	-
BACA0142	Thermal	*Leptolyngbya* sp.	-	B	-	-	-	-
BACA0146	Thermal	*Leptolyngbya* sp.	-	B	-	-	-	-
BACA0148	Lake	*Microcystis aeruginosa*	-	-	C, D, E, G	-	-	+
BACA0203	Lake	*Leptodesmis* sp.	G, H	-	-	-	-	-
BACA0204	Lake	*Leptolyngbya* sp.	G	-	-	-	-	-
BACA0223	Lake	*Anathece minutissima*	G	-	-	-	-	-

"-": absence of biosynthesis-encoding genes amplification or absence of toxin in the ESI-LC-MS/MS analysis; C: *mcy*C, D: *mcy*D, E: *mcy*E, G: *mcy*G, A: *sxt*A, G: *sxt*G, H: *sxt*H, I: *sxt*I, B: *cyr*B, CC: *cyr*C, amplification of respective biosynthesis-encoding genes.

STX producing genes were detected in six strains, however, the only strain with amplification of all the searched STX-encoding genes, *Aphanizomenon gracile* BACA0041, was the only one confirmed as saxitoxin producer by ESI-LC-MS/MS (Table 1).

CYN producing genes were detected in five strains, although only BACA0025 and BACA0031 had amplification of both searched *cyr*B and *cyr*C genes. These two strains were also the only ones with CYN detection by ESI-LC-MS/MS. The remaining three strains with positive results for the detection of CYN encoding genes only amplified the *cyr*B gene (Table 1).

MC-encoding genes were only detected in *Microcystis aeruginosa* BACA0148 (*mcy*C, *mcy*D, *mcy*E, and *mcy*G) and *Nostoc* sp. BACA0091 (*mcy*E). However, in the ESI-LC-MS/MS analysis, MC-LR ions were identified only in *M. aeruginosa* BACA0148.

2.2. Phylogenetic Characterization

Strains BACA0025 and BACA0031 had similarities with sequences deposited in the GenBank NCBI between 96 and 97%, which shows the distinctiveness of these strains, while strains BACA0041 and BACA0148 had similarities greater than 98% with *Aphanizomenon gracile* and *Microcystis aeruginosa* strains, respectively (Table S3). In the 16S rRNA phylogenetic tree strains BACA0041 and BACA0148 are positioned in clades with strains close to their initial identification, while strains BACA0025 and BACA0031 had high phylogenetic distances from the closest morphological identified strains.

The strain BACA0041, morphologically identified as *Aphanizomenon gracile* (Figure 1F,G), was positioned near other *A. gracile* and *Aphanizomenon flos-aquae* strains, and closer to know STX producing strains as *A. gracile* PMC638.10 or *A. flos-aquae* NIES 81 (Figure 2).

Strains BACA0025 (Figure 1A–C) and BACA0031 (Figure 1E), morphologically identified as *Sphaerospermopsis* sp. are positioned in cluster II between *Nostoc* spp., *Fortiea* spp., *Desikacharya* spp., *Trichormus* spp., *Minunostoc cylindricum*, and *Desmonostoc* spp. strains, however, with significant phylogenetic distance from all these strains (Figure 2).

The strain BACA0148, morphologically identified as *Microcystis aeruginosa* (Figure 1D), was positioned near other *M. aeruginosa* strains, and closer to know MC producing strains as *M. aeruginosa* VN481 (AB666064) or *M. bengalensis* VN486 (AB666082) (Figure 3).

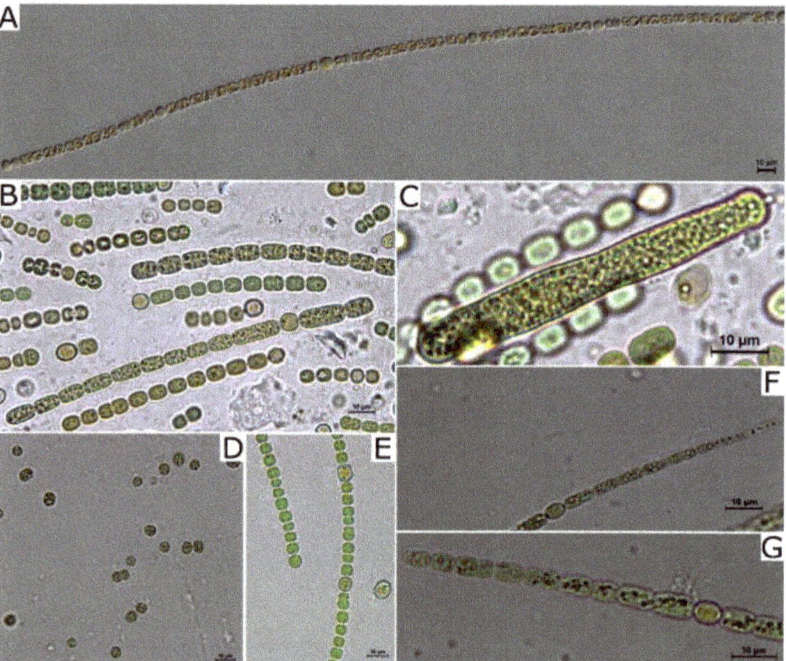

Figure 1. Cyanotoxin-producing strains included in the phylogenetic analysis: BACA0025 (**A–C**), *Microcystis aeruginosa* BACA0148 (**D**), BACA0031 (**E**), and *Aphanizomenon gracile* BACA0041 (**F,G**). Scale bars –10 μm.

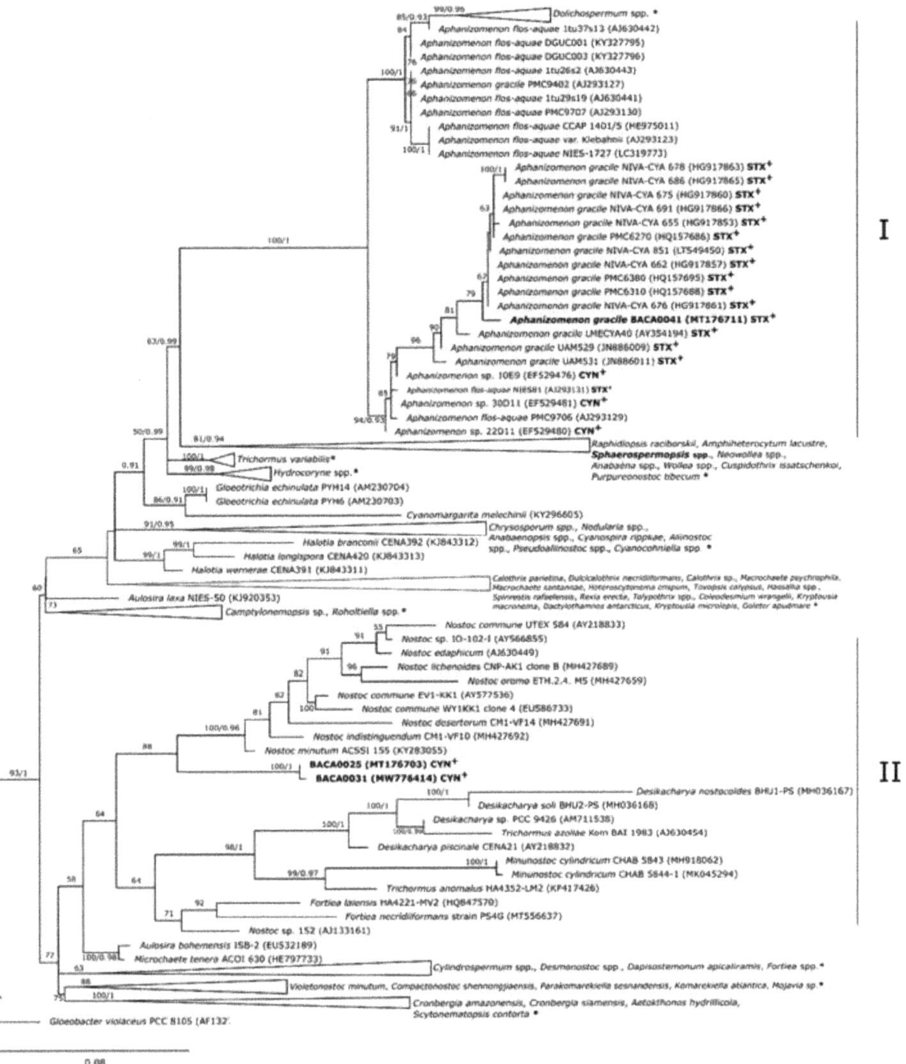

Figure 2. Maximum likelihood tree cluster I of partial 16S rRNA gene sequences (1070 bp). Bootstrap (greater than 50%) and probabilities values (greater than 0.9) are presented in front of the pertinent nodes. **CYN⁺**: Cylindrospermopsin producing strains; STX+: Saxitoxin producing strains. * Collapsed sequences (NCBI accession numbers): AJ630425, FN691925, FN691926, KC297496, FN691923, FN691907, FM242088, FN691908, AB551474, FN691909, KR154316, KR154314, AB551455, KR154298, FN691922, KR154304, AJ630413, AF092504, HQ407325, LR590627, JQ237770, AY763117, AJ582102, LR590629, LR746263, MG921181, MG921182, HQ730086, KT583658, AY701557, FM161349, FM161350, LC474825, LC474824, KT290381, AJ293110, AJ630428, KT290378, KM019920, KT290326, GU434226, AJ630446, MT294032, MT294033, AY196087, KM245026, MN381942, MN381943, DQ234830, AJ630457, AJ630456, KC346266, KC346265, KC346267, JN385287, AF160256, JF768744, EU076457, FJ234895, EF529489, EF529482, EF529488, EU076457, FJ234884, FJ234885, FJ234890, FJ234897, AB608023, JQ237772, LN997860, AJ133177, AJ781131, AJ781149, AJ781145, KM199731, FM177481, AY038033, AM773306, FR774773, MK503791, KY403996, MK503792, MK503793, MH497064, KJ737427, MN243143, AF334695, KY863521, AP018280, KT336439.2, KT336441, MG970541, MG970549, MG970536, MG970538, JN695681, JN695682, FR822753, AF334690, AF334692, JQ083655, KF017617,

KF934181, JX088105, HG970655, AF334701, AF334703, KY508609, KY508610, HG970652, MN626663, MN626664, KM199732, KY508607, KY508612, KY508611, KY508608, HG970653, KF417425, JN385289, JN385292, HQ847564, KM268886, KM268884, KY098849, KM268889, KF934180, KF052614, 125687, KF052599, 114701, GQ287650, KF052617, KF052610, KF052603, KF052605, KU161661, AM711524, HG004586, KT166436, MH497066, MH291266, MF642332, KX787933, KJ566945, KJ566947, KF052607, 125685, KF052616, HE797731, KB235930, MN400069, MN400070, MH598843, MT044191, MT044192, KX638483, KX638487, KX638489, KU161674, KU161676, KU161650, AY577534, MF002129, KM019950, AY785313, HQ847558, HQ847561.

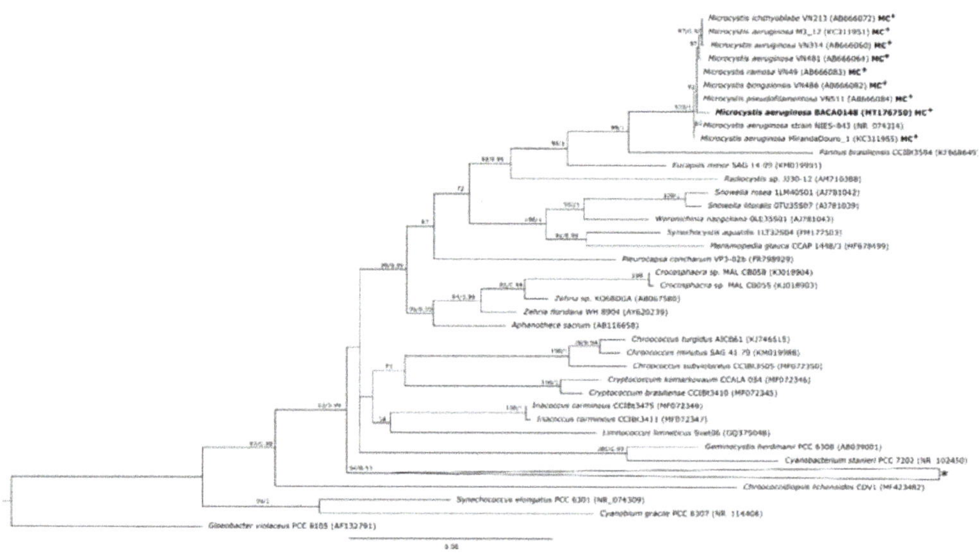

Figure 3. Maximum likelihood tree of the partial 16S rRNA gene sequences (1324 bp). Bootstrap (greater than 50%) and probabilities values (greater than 0.9) are presented in front of the pertinent nodes. MC$^+$: Microcystin-producing strains. * Collapsed sequences (NCBI accession numbers): *Myxosarcina* sp. (AJ344562; AJ344561), *Dermocarpella incrassata* (AJ344559), *Synechocystis pevalekii* (KM350249), *Gloeocapsopsis crepidinum* (KF498710), *Foliisarcina bertiogensis* (KT731151), *Hyella patelloides* (HQ832901), *Chroococcidiopsis* sp. (AJ344557), *Xenococcus* sp. (AF132783), *Stanieria cyanosphaera* (112109), *Chroococcopsis gigantea* (KM019987).

3. Discussion

The ability to produce cyanotoxins depends on the simultaneous existence of several genes involved in their biosynthesis pathways [42]. As seen in previous studies, the *sxt* cluster has suffered several modifications and its gene composition and organization varies between taxa [6,43–45]. Although the amplification of the *sxt*A gene and the absence of toxin production in *Aphanizomenon* strains was seen in other studies [10,25,46,47], in the studied strains only strains with *sxt*A amplification were able to produce STX (Table 1). Cirés et al. [47] states that the *sxt*A gene does not allow the distinction between STX-producing and non-producing *Anabaena*/*Aphanizomenon* strains, nonetheless the presence of *sxt*A in non-producing strains could have been due to gene loss/inactivation within the *sxt* cluster [46]. The presence of *sxt*G or *sxt*H amplifications in several non-STX producing genera (Table 1; Table S2; Figure S3) in this study are interesting new results that require further investigation to fully understand *sxt* gene distribution among cyanobacteria taxa. Nonetheless, to our knowledge, the presence of *sxt*G, and/or *sxt*H genes in *Kamptonema*, *Leptodesmis*, *Anathece minutissima*, or *Leptolyngbya* strains, have not been reported before.

Aphanizomenon gracile BACA0041 was confirmed as an STX producer with amplification of the *sxt*A, *sxt*G, *sxt*H, and *sxt*I genes. The ESI-LC-MS/MS spectra of *A. gracile* BACA0041 matched the fragmented pattern of STX standard spectra (Figure S4), with identification of the precursor ion 300 m/z and product ions 282 m/z, 265 m/z, 241 m/z,

240.25 m/z, 204 m/z, and 186 m/z [10]. Phylogenetic analysis confirms *Aphanizomenon gracile* BACA0041 identification's, this strain is positioned in a well-supported clade of several STX-producing *A. gracile* strains (Figure 2), as *A. gracile* UAM531 [47] and *A. gracile* PMC 638.10 [10]. In our study, we report another freshwater STX-producing *A. gracile* strain, the first STX producer identified in the Azores islands.

Strains BACA0025 and BACA0031 are quite similar, these were both initially identified as *Sphaerospermopsis* sp. (isolated from similar freshwater lakes from the same island; Table S1), with similar BlastN results (Table S3). Strains BACA0025 and BACA0031 were positioned in cluster II in the phylogenetic tree (Figure 2) close to Nostocaceae genera, however with significant phylogenetic distance to conclude that these strains might belong to a new genus. Both were confirmed as CYN producers with detection of both *cyr*B and *cyr*C genes (Table 1). The ESI-LC-MS/MS spectra of BACA0025 and BACA0031 both matched the fragmented pattern of CYN standard spectra (Figure S5), with identification of the precursor ion 416 m/z and product ions 336 m/z, 318 m/z, 274 m/z, and 194 m/z [48], confirming these two strains as CYN producers.

The *cyr*B gene was also amplified in *Nostoc* sp. BACA0109, and in two thermal *Leptolyngbya* sp. strains BACA0142 and BACA0146, however without CYN identification in the ESI-LC-MS/MS. Amplification of *cyr* genes without CYN production confirmation has been reported previously, as is the case of non-CYN producing *Chrysosporum bergii* and *Chrysosporum ovalisporum* strains with amplification of *cyr*A, *cyr*B, and *cyr*C genes [49]. *Nostoc* and *Leptolyngbya* strains are known to produce cyanotoxins, however, MCs [12,17,50] and not CYN, and as far as we know, the presence of the *cyr*B and *cyr*C genes has not been previously reported in *Nostoc* or *Leptolyngbya* strains.

Microcystins are the most common and more well-studied cyanotoxins, being MC-LR one of the most prevalent and toxic congeners [51]. *Microcystis aeruginosa* was the first cyanobacteria species identified as MCs producer and is the most studied species regarding MCs [11]. Our results show that the Azorean strain *M. aeruginosa* BACA0148 is also an MC-LR producer, with the detection of *mcy*C, *mcy*D, *mcy*E, and *mcy*G genes. Characteristic MC-LR fragmentation pattern was observed in *M. aeruginosa* BACA0148 (Figure S6), with identification of precursor ion 995 m/z and product ions 977 m/z, 866 m/z, 599 m/z, and 553 m/z [52]. The *mcy*E gene amplification in *Nostoc* sp. BACA0091, despite the absence of MC-LR ions in the ESI-LC-MS/MS, or the absence of the other searched *mcy* genes (*mcy*C, *mcy*D, and *mcy*G), can be explained due to gene(s) recombination or loss [6,14,53]. As stated by Dittmann et al. [6], the *mcy* gene cluster has high repetitive sequences, enabling recombination events that ultimately cause changes in the final product, which can be confirmed by the growing reported number of MCs congeners [54,55].

All strains identified as toxin producers (BACA0025, BACA0031, *A. gracile* BACA0041, and *M. aeruginosa* BACA0148) were isolated from lakes in Pico and São Miguel islands (Table S1, S2). The presence of toxic strains in these lakes represents environmental and public health hazards. Contrarily, none of the strains isolated from thermal and terrestrial habitats were identified as cyanotoxin producers, although in some of them, cyanotoxin-encoding genes were detected, as in the thermal *Leptolyngbya* strains BACA0112, BACA0123, BACA0142, BACA0144 and BACA0146, and *Coleospermum* sp. BACA0119.

4. Conclusions

Within the BACA collection, we identified and reported another MC-LR-producing *M. aeruginosa* strain (BACA0148) and another STX-producing *A. gracile* strain (BACA0041). Phylogenetic analysis revealed evidence for new cyanobacteria taxa BACA0025 and BACA0031, confirmed as CYN producers. Further studies are necessary to confirm and describe these new taxa, with morphological characterization and 16S rRNA and ITS analysis.

The identification of new cyanotoxin-producing strains, and unreported toxins, in the Azores, confirms the risk of toxicity and threat to environmental and public health; thus, an appropriate monitoring program should be implemented/updated to search MCs,

STX, and CYN. Future efforts should also be made to avoid cyanobacteria blooms and consequently cyanotoxins released in high concentrations.

5. Materials and Methods

5.1. BACA Strains and Growth Conditions

A total of 157 strains, isolated from various environments (Table S1), were retrieved from the Azorean Bank of Algae and Cyanobacteria (BACA) created in the framework of the REBECA project (MAC/1.1a/060). For genetic analysis, 50 mL cultures were prepared without agitation, whereas for toxin extraction, the cultures were scaled up to 1 L, with filtered aeration. All strains were grown in liquid BG-11 media (with or without combined nitrogen) [56], in a climate-controlled room with a 14:10 h light: dark (170 µmol photons $m^{-2}\,s^{-1}$) photoperiod at 25 °C [56,57]. Cyanobacterial cells were harvested by centrifugation (4000× g for 15 min), after 3–5 weeks, and lyophilized. The lyophilized cyanobacteria biomass were stored at −20 °C.

5.2. Cyanotoxins Analysis

5.2.1. Toxin extraction

The lyophilized cyanobacteria, 157 samples in total, were weighed (80–100 mg) to a glass vial and extracted with a 5% methanolic solution (2–10 mL). Solutions were then submitted to ultrasounds for 1–5 min, 60 Hz, in an ice bath, and transferred to falcons to be centrifuged (5000× g, 5 min, 4°C). Pellet was then submitted to a second extraction and left in the dark at 4 °C overnight. Supernatants were pooled together and lyophilized (extraction solvent completely freeze-dried).

Residues were finally dissolved in 200–500 µL 50% methanol LC-MS grade acidified with 0.1% Formic Acid and filtered with a nylon membrane 0.2 µm before analysis (or centrifuged at 10,000× g for 5 min). Samples were protected from light through all the processes and stored at −80 °C until analysis.

5.2.2. ESI-LC-MS/MS analysis

Saxitoxin (CRM-00-STX, Lot 16-001, 99% purity), microcystin-LR (CRM-00-MC-LR, Lot 19-001, 96% purity), and cylindrospermopsin (CRM-03-CYN, Lot 16-001, 99% purity) standards were all supplied by Cifga (Lugo, Spain). Although MC-RR, MC-YR, and MC-LA standards were not used, their mass was searched in the spectra.

All the standards were injected individually and then as a standard mixture with a concentration interval from 10 ppb to 300 ppb. ESI-LC-MS/MS analysis was performed to confirm the presence or absence of these three cyanotoxins in the selected cyanobacteria strains.

Samples were injected in a Liquid Chromatograph Thermo Finnigan Surveyor HPLC System (Thermo Scientific, Waltham, MA, USA), coupled with a Mass Spectrometry LCQ Fleet™ Ion Trap Mass Spectrometer (Thermo Scientific, Waltham, MA, USA), with a column TSKgel® Amide-80 Phase carbamoyl (250 mm× 2 mm i.d., 5 µm) (TOSOH Bioscience-Lot082B, Tokyo, Japan).

The eluents used were methanol (A) and water (B) both acidified with formic acid at 0.1% (v/v). The gradient program started at 10% B (held for 10 min), increasing to 50% B in 5 min, turning back to initial conditions in 5 min, equilibrating more 10 min with 10% B. The injection volume was 10 µL with a flow of 0.2 mL min^{-1} and column kept at 30 °C.

Mass spectrometry analysis acquisition parameters were as follows: ESI source, positive ionization using collision-induced dissociation (CID). Table S4 resumes analysis parameters for each searched toxin.

5.3. DNA Extraction, PCR Amplification, and Sequencing

Total genomic DNA was extracted with the PureLinkTM Genomic DNA Mini Kit (Invitrogen, Carlsbad, CA, USA), as previously described by Cordeiro et al. [58]. DNA samples were stored at −20 °C.

Genes *mcy*C, *mcy*D, *mcy*E, and *mcy*G were targeted for MCs production potential, *sxt*A, *sxt*G, *sxt*I, and *sxt*H for STX, and *cyr*B and *cyr*C for CYN, using specific primer pairs available in the literature (Table 2). For the 16S rRNA gene amplification primers 27F [59], CYA359F [60], and 1494R [59] were used, whereas for sequencing it was also used primers CYA781F [60] and CYA781R (Table 2).

Table 2. Primers used to amplify and/ or sequence cyanotoxins biosynthesis genes and 16S rRNA.

Gene	Primer	Fragment length (bp)	Sequence (5'–3')	References
*mcy*C	PSCF1	674	GCAACATCCCAAGAGCAAAG	
	PSCR1		CCGACAACATCACAAAGGC	
*mcy*D	PKDF1	647	GACGCTCAAATGATGAAAC	
	PKDR1		GCAACCGATAAAAACTCCC	Ouahid et al. [61]
*mcy*E	PKEF1	755	CGCAAACCCGATTTACAG	
	PKER1		CCCCTACCATCTTCATCTTC	
*mcy*G	PKGF1	425	ACTCTCAAGTTATCCTCCCTC	
	PKGR1		AATCGCTAAAACGCCACC	
*sxt*A	sxtA F	602	AGGTCTTTGACTTGCATCCAA	Ledreux et al. [10]
	sxtA R		AACCGGCGACATAGATGATA	
*sxt*G	sxtGf	893	AGGAATTCCCTATCCACCGGAG	
	sxtGr		CGGCGAACATCTAACGTTGCAC	
*sxt*H	sxtHf	812	AAGACCACTGTCCCCACCGAGG	Casero et al. [25]
	sxtHr		CTGTGCAGCGATCTGATGGCAC	
*sxt*I	sxtIf	910	AGCGCTGCCGCTATGGTTGTCG	
	sxtIr		ACGCAATTGAGGGCGACACCAC	
*cyr*B	M13	597	GGCAAATTGTGATAGCCACGAGC	
	M14		GATGGAACATCGCTCACTGGTG	Schembri et al. [62]
*cyr*C	M4	650	GAAGCTCTGGAATCCGGTAA	
	M5		AATCCTTACGGGATCCGGTGC	
16S rRNA	27F		AGAGTTTGATCCTGGCTCAG	Neilan et al. [59]
	CYA359F		GGGGAATYTTCCGCAATG GG	Nubel et al. [60]
	1494R		TACGGCTACCTTGTTACGAC	Neilan et al. [59]
	CYA781R		GACTACTGGGGTATCTAATCCCATT	Nubel et al. [60]
	CYA781F		AATGGGATTAGATACCCCAGTAGTC	This study

PCRs were carried out in a ProFlex™ 3 × 32-well PCR System (Thermo Fischer, Waltham, MA, USA), according to the literature [9,10,25,58,61,62]. The PCR products were visualized by electrophoresis on 1.5% agarose gels stained with SYBR™ SAFE (0.2 g mL^{-1}) and visualized using the transilluminator Molecular Imager® Gel Doc™ XR^{+} (BioRad, Hercules, CA, USA).

16S rRNA amplification products were purified using the EXTRACTME® DNA clean-up kit (Blirt, Gdańsk, Poland), following the manufacturer's protocol. Sequencing was done by Macrogen Ltd. (Madrid, Spain). Nucleotide sequences were deposited in the NCBI Genbank under the accession numbers MT176703, MW776414, MT176711, and MT176750.

5.4. Phylogenetic Analysis

Partial 16S sequences were amplified for the four strains with toxin identification by ESI-LC-MS/MS. All the sequences obtained in this study were compared with sequences deposited in the GenBank NCBI by BlastN tool.

The databases were constructed with the 16S rRNA sequences from this study and identified strains with MCs, STX, and CYN production retrieved from the literature [10,17,25,43,63–67] and GenBank. A final database of 267 OTUs (operational taxonomic units) were aligned for Nostocales and 50 OTUs for Chroococcales, using MAFFT v7.475 [68]. The sequence data matrixes with a final length of 1070 bp (Nostocales) and 1324 bp (Chroococcales) were used to infer phylogenetic distances.

The 16S rRNA gene phylogenetic relations were calculated using maximum likelihood (ML) and Bayesian inference (BI). jModelTest 2.1.10 [69] was used to select the best-fit nucleotide model for our database, on which the general time-reversible evolutionary model of substitution with gamma-distributed evolutionary rates and with an estimated proportion of invariable sites (GTR+G+I) was selected. ML was calculated using the IQ-Tree online version v1.6.12 [70] with 1000 ultrafast bootstrap and BI was calculated using MrBayes v3.2.7a [71], applying two separate runs with four chains each and 50,000,000 Markov chain Monte Carlo generations (sampling every 100 generations with a 0.25 burn-in). The tree was drawn with FigTree 1.4.4 (http://tree.bio.ed.ac.uk/software/figtree, accessed on 8 February 2021) and Inkscape 1.0.1 (https://inkscape.org/pt/, accessed on 8 February 2021). Only the ML tree is presented, with bootstrap percentages (ML) and BI probabilities for branch support, since ML and BI methods resulted in similar trees. Only probabilities above 0.9 and bootstrap percentages above 50 are shown at the branch nodes of the phylogenetic distance trees. *Gloeobacter violaceus* PCC 8105 (AF132791) was used as the out-group.

Supplementary Materials: The following are available online at https://www.mdpi.com/article/10.3390/toxins13040258/s1, Figure S1: Electrophoresis gel photos of *mcy*C (674 bp), *mcy*D (647 bp), *mcy*E (755 bp), and *mcy*G (425 bp) biosynthesis genes amplifications, Figure S2: Electrophoresis gel photos of *cyr*B (650 bp) and *cyr*C (597 bp) biosynthesis genes amplifications, Figure S3: Electrophoresis gel photos of *sxt*A (602 bp), *sxt*G (893 bp), *sxt*H (812 bp), and *sxt*I (910 bp) biosynthesis genes amplifications, Figure S4: Total ion chromatograms and spectra of a STX standard solution (**A**) and sample *Aphanizomenon gracile* BACA0041 (**B**), Figure S5: Total ion chromatograms and spectra of a CYN standard solution (**A**), sample BACA0025 (**B**) and sample BACA0031 (**C**), Figure S6. Total ion chromatograms and spectra of an MC-LR standard solution (**A**) and sample *Microcystis aeruginosa* BACA0148 (**B**), Table S1: Strains information, Table S2: PCR amplifications of MC (*mcy*C, *mcy*D, *mcy*E, *mcy*G), STX (*sxt*A, *sxt*G, *sxt*H, *sxt*I) and CYN (*cyr*B, *cyr*C) biosynthesis-encoding genes and ESI-LC-MS/MS toxicity confirmation, Table S3: Sequence identity (%) of 16S rRNA gene fragment between BACA strains and other cyanobacterial sequences available in GenBank (NCBI), Table S4: ESI-LC-MS/MS analysis parameters for the identification of CYN, MC-LR, and STX.

Author Contributions: Conceptualization, R.C., V.V., V.G., and A.F.; methodology, R.C., J.A., and R.L.; formal analysis, R.C.; investigation, R.C.; resources, V.V., V.G., and A.F.; writing—original draft preparation, R.C.; writing—review and editing, R.C., J.A., R.L., V.V., V.G., and A.F.; visu-alization, R.C.; supervision, V.V., V.G., and A.F.; project administration, V.G.; funding acquisition, V.V., V.G., and A.F. All authors have read and agreed to the published version of the manuscript.

Funding: This research was funded by FEDER funds through the Interreg-MAC 2014-2020 program under the projects REBECA—Red de Excelencia en Biotecnología Azul (algas) de la Región Macaronesia (MAC/1.1a/060) and REBECA-CCT—Red de Excelencia en Biotecnología Azul de la Región Macaronésica. Consolidación, Certificación y Transferencia (MAC2/1.1b/269). Rita Cordeiro was supported by a Ph.D. grant (M3.1.a/F/017/2015) from the Fundo Regional da Ciência e Tecnologia (FRCT). CIIMAR acknowledges the project H2020 RISE project EMERTOX—Emergent Marine Toxins in the North Atlantic and the Mediterranean: New Approaches to Assess their Occurrence and Future Scenarios in the Framework of Global Environmental Changes (grant agreement no. 778069), and FCT Projects UIDB/04423/2020 and UIDP/04423/2020. This work was also funded by FEDER funds through the Operational Programme for Competitiveness Factors - COMPETE and by National Funds through FCT - Foundation for Science and Technology under the UID/BIA/50027/2020 and POCI-01-0145-FEDER-006821. The APC was funded by REBECA-CCT (MAC2/1.1b/269).

Institutional Review Board Statement: Not applicable.

Informed Consent Statement: Not applicable.

Data Availability Statement: Data is contained within the article or Supplementary Material. The data presented in this study are available in https://www.mdpi.com/article/10.3390/toxins13040258/s1.

Acknowledgments: We would like to thank Carmo Barreto from the University of the Azores, for the use of Biochemistry Lab facilities, without it, this work would not be possible.

Conflicts of Interest: The authors declare no conflict of interest. The funders had no role in the design of the study; in the collection, analyses, or interpretation of data; in the writing of the manuscript, or in the decision to publish the results.

References

1. Hamilton, D.P.; Salmaso, N.; Paerl, H.W. Mitigating harmful cyanobacterial blooms: Strategies for control of nitrogen and phosphorus loads. *Aquat. Ecol.* **2016**, *50*, 351–366. [CrossRef]
2. Brooks, B.W.; Lazorchak, J.M.; Howard, M.D.A.; Johnson, M.-V.V.; Morton, S.L.; Perkins, D.A.K.; Reavie, E.D.; Scott, G.I.; Smith, S.A.; Steevens, J.A. Are harmful algal blooms becoming the greatest inland water quality threat to public health and aquatic ecosystems? *Environ. Toxicol. Chem.* **2016**, *35*, 6–13. [CrossRef] [PubMed]
3. Holland, A.; Kinnear, S. Interpreting the Possible Ecological Role(s) of Cyanotoxins: Compounds for Competitive Advantage and/or Physiological Aide? *Mar. Drugs* **2013**, *11*, 2239–2258. [CrossRef] [PubMed]
4. Chorus, I.; Welker, M. *Toxic Cyanobacteria in Water*, 2nd ed.; CRC Press, on behalf of the World Health Organization: Boca Raton, CH, Geneva, 2021.
5. Van Apeldoorn, M.E.; Van Egmond, H.P.; Speijers, G.J.A.; Bakker, G.J.I. Toxins of cyanobacteria. *Mol. Nutr. Food Res.* **2007**, *51*, 7–60. [CrossRef] [PubMed]
6. Dittmann, E.; Fewer, D.P.; Neilan, B. Cyanobacterial toxins: Biosynthetic routes and evolutionary roots. *FEMS Microbiol. Rev.* **2013**, *37*, 23–43. [CrossRef] [PubMed]
7. Cirés, S.; Casero, M.C.; Quesada, A. Toxicity at the edge of life: A review on cyanobacterial toxins from extreme environments. *Mar. Drugs* **2017**, *15*, 233. [CrossRef] [PubMed]
8. Quiblier, C.; Wood, S.; Echenique-Subiabre, I.; Heath, M.; Villeneuve, A.; Humbert, J.-F. A review of current knowledge on toxic benthic freshwater cyanobacteria–Ecology, toxin production and risk management. *Water Res.* **2013**, *47*, 5464–5479. [CrossRef]
9. Rantala-Ylinen, A.; Känä, S.; Wang, H.; Rouhiainen, L.; Wahlsten, M.; Rizzi, E.; Berg, K.; Gugger, M.; Sivonen, K. Anatoxin-a synthetase gene cluster of the cyanobacterium Anabaena sp. strain 37 and molecular methods to detect potential producers. *Appl. Environ. Microbiol.* **2011**, *77*, 7271–7278. [CrossRef] [PubMed]
10. Ledreux, A.; Thomazeau, S.; Catherine, A.; Duval, C.; Yéprémian, C.; Marie, A.; Bernard, C. Evidence for saxitoxins production by the cyanobacterium Aphanizomenon gracile in a French recreational water body. *Harmful Algae* **2010**, *10*, 88–97. [CrossRef]
11. Codd, G.A.; Meriluoto, J.; Metcalf, J.S. Introduction: Cyanobacteria, Cyanotoxins, Their Human Impact, and Risk Management. In *Handbook of Cyanobacterial Monitoring and Cyanotoxin Analysis*; Codd, G.A., Meriluoto, J., Metcalf, J.S., Eds.; John Wiley and Sons, Ltd: Chichester, UK, 2017; pp. 1–8.
12. Bernard, C.; Ballot, A.; Thomazeau, S.; Maloufi, S.; Furey, A.; Mankiewicz-Boczek, J.; Pawlik-Skowrońska, B.; Capelli, C.; Salmaso, N. Appendix 2: Cyanobacteria associated with the production of cyanotoxins. In *Handbook of Cyanobacterial Monitoring and Cyanotoxin Analysis*; Meriluoto, J., Spoof, L., Codd, G.A., Eds.; John Wiley and Sons, Ltd: Chichester, UK, 2017.
13. Tillett, D.; Dittmann, E.; Erhard, M.; Von Döhren, H.; Börner, T.; Neilan, B.A. Structural organization of microcystin biosynthesis in *Microcystis aeruginosa* PCC7806: An integrated peptide-polyketide synthetase system. *Chem. Biol.* **2000**, *7*, 753–764. [CrossRef]
14. Heck, K.; Alvarenga, D.O.; Shishido, T.K.; Varani, A.M.; Dörr, F.A.; Pinto, E.; Rouhiainen, L.; Jokela, J.; Sivonen, K.; Fiore, M.F. Biosynthesis of microcystin hepatotoxins in the cyanobacterial genus *Fischerella*. *Toxicon* **2018**, *141*, 43–50. [CrossRef] [PubMed]
15. Bláha, L.; Babica, P.; Maršálek, B. Toxins produced in cyanobacterial water blooms–toxicity and risks. *Interdiscip. Toxicol.* **2009**, *2*, 10102–10109. [CrossRef] [PubMed]
16. Metcalf, J.S.; Codd, G.A. Cyanotoxins. In *Ecology of Cyanobacteria II*; Whitton, B., Ed.; Springer: Dordrecht, The Netherlands, 2012; pp. 651–675.
17. Oksanen, I.; Jokela, J.; Fewer, D.P.; Wahlsten, M.; Rikkinen, J.; Sivonen, K. Discovery of rare and highly toxic microcystins from lichen-associated cyanobacterium *Nostoc* sp. Strain IO-102-I. *Appl. Environ. Microbiol.* **2004**, *70*, 5756–5763. [CrossRef]
18. Wiese, M.; D'Agostino, P.M.; Mihali, T.K.; Moffitt, M.C.; Neilan, B. a Neurotoxic alkaloids: Saxitoxin and its analogs. *Mar. Drugs* **2010**, *8*, 2185–2211. [CrossRef] [PubMed]
19. Pearson, L.A.; Dittmann, E.; Mazmouz, R.; Ongley, S.E.; D'Agostino, P.M.; Neilan, B.A. The genetics, biosynthesis and regulation of toxic specialized metabolites of cyanobacteria. *Harmful Algae* **2016**, *54*, 98–111. [CrossRef] [PubMed]
20. Borges, H.L.F.; Branco, L.H.Z.; Martins, M.D.; Lima, C.S.; Barbosa, P.T.; Lira, G.A.S.; Bittencourt-Oliveira, M.C.; Molica, R.J.R. Cyanotoxin production and phylogeny of benthic cyanobacterial strains isolated from the northeast of Brazil. *Harmful Algae* **2015**, *43*, 46–57. [CrossRef]
21. Humpage, A.; Rositano, J.; Bretag, A.; Brown, R.; Baker, P.; Nicholson, B.; Steffensen, D. Paralytic shellfish poisons from Australian cyanobacterial blooms. *Mar. Freshw. Res.* **1994**, *45*, 761. [CrossRef]
22. Sivonen, K.; Himberg, K.; Luukkainen, R.; Niemelä, S.I.; Poon, G.K.; Codd, G.A. Preliminary characterization of neurotoxic cyanobacteria blooms and strains from Finland. *Toxic. Assess.* **1989**, *4*, 339–352. [CrossRef]
23. Kellmann, R.; Mihali, T.K.; Jeon, Y.J.; Pickford, R.; Pomati, F.; Neilan, B. a Biosynthetic Intermediate Analysis and Functional Homology Reveal a Saxitoxin Gene Cluster in Cyanobacteria. *Appl. Environ. Microbiol.* **2008**, *74*, 4044–4053. [CrossRef]
24. Mihali, T.K.; Kellmann, R.; Neilan, B. a Characterisation of the paralytic shellfish toxin biosynthesis gene clusters in Anabaena circinalis AWQC131C and Aphanizomenon sp. NH-5. *BMC Biochem.* **2009**, *10*, 8. [CrossRef]

25. Casero, M.C.; Ballot, A.; Agha, R.; Quesada, A.; Cirés, S. Characterization of saxitoxin production and release and phylogeny of *sxt* genes in paralytic shellfish poisoning toxin-producing Aphanizomenon Gracile. *Harmful Algae* **2014**, *37*, 28–37. [CrossRef]

26. Hawkins, P.R.; Runnegar, M.T.C.; Jackson, A.R.B.; Falconer, I.R. Severe hepatotoxicity caused by the tropical cyanobacterium (blue-green alga) Cylindrospermopsis raciborskii (Woloszynska) Seenaya and Subba Raju isolated from a domestic water supply reservoir. *Appl. Environ. Microbiol.* **1985**, *50*, 1292–1295. [CrossRef] [PubMed]

27. Kokociński, M.; Mankiewicz-Boczek, J.; Jurczak, T.; Spoof, L.; Meriluoto, J.; Rejmonczyk, E.; Hautala, H.; Vehniäinen, M.; Pawełczyk, J.; Soininen, J. Aphanizomenon gracile (Nostocales), a cylindrospermopsin-producing cyanobacterium in Polish lakes. *Environ. Sci. Pollut. Res.* **2013**, *20*, 5243–5264. [CrossRef] [PubMed]

28. Banker, R.; Carmeli, S.; Hadas, O.; Teltsch, B.; Porat, R.; Sukenik, A. Identification of cylindrospermopsin in *Aphanizomenon ovalisporum* (Cyanophyceae) isolated from lake Kinneret, Israel. *J. Phycol.* **1997**, *33*, 613–616. [CrossRef]

29. Mihali, T.K.; Kellmann, R.; Muenchhoff, J.; Barrow, K.D.; Neilan, B.A. Characterization of the gene cluster responsible for cylindrospermopsin biosynthesis. *Appl. Environ. Microbiol.* **2008**, *74*, 716–722. [CrossRef]

30. Whitton, B.; Potts, M. Introduction to the Cyanobacteria. In *The Ecology of Cyanobacteria II–Their Diversity in Time and Space*; Whitton, B.A., Ed.; Springer: Dordrecht, The Netherlands, 2012; pp. 1–13. ISBN 978-94-007-3854-6.

31. Komárek, J.; Kastovsky, J.; Mares, J.; Johansen, J.R. Taxonomic classification of cyanoprokaryotes (cyanobacterial genera) 2014, using a polyphasic approach. *Preslia* **2014**, *86*, 295–335.

32. Mai, T.; Johansen, J.R.; Pietrasiak, N.; Bohunická, M.; Martin, M.P. Revision of the Synechococcales (Cyanobacteria) through recognition of four families including Oculatellaceae fam. nov. and Trichocoleaceae fam. nov. and six new genera containing 14 species. *Phytotaxa* **2018**, *365*, 1–59. [CrossRef]

33. Bagchi, S.N.; Dubey, N.; Singh, P. Phylogenetically distant clade of Nostoc-like taxa with the description of *Aliinostoc* gen. nov. and Aliinostoc morphoplasticum sp. nov. *Int. J. Syst. Evol. Microbiol.* **2017**, *67*, 3329–3338. [CrossRef]

34. Genuário, D.B.; Vaz, M.G.M.V.; Hentschke, G.S.; Sant'Anna, C.L.; Fiore, M.F. Halotia gen. nov., a phylogenetically and physiologically coherent cyanobacterial genus isolated from marine coastal environments. *Int. J. Syst. Evol. Microbiol.* **2015**, *65*, 663–675. [CrossRef]

35. Brito, Â.; Ramos, V.; Mota, R.; Lima, S.; Santos, A.; Vieira, J.; Vieira, C.P.; Kaštovský, J.; Vasconcelos, V.M.; Tamagnini, P. Description of new genera and species of marine cyanobacteria from the Portuguese Atlantic coast. *Mol. Phylogenet. Evol.* **2017**, *111*, 18–34. [CrossRef]

36. Cai, F.; Li, X.; Yang, Y.; Jia, N.; Huo, D.; Li, R. Compactonostoc shennongjiaensis gen. & sp. nov. (Nostocales, Cyanobacteria) from a wet rocky wall in China. *Phycologia* **2019**, *58*, 200–210. [CrossRef]

37. Österholm, J.; Popin, R.V.; Fewer, D.P.; Sivonen, K. Phylogenomic Analysis of Secondary Metabolism in the Toxic Cyanobacterial Genera Anabaena, Dolichospermum and Aphanizomenon. *Toxins* **2020**, *12*, 248. [CrossRef] [PubMed]

38. Humbert, J.-F.; Fastner, J. Ecology of Cyanobacteria. In *Handbook of Cyanobacterial Monitoring and Cyanotoxin Analysis*; Meriluoto, J., Spoof, L., Codd, G.A., Eds.; John Wiley and Sons, Ltd: Chichester, UK, 2017; pp. 9–18.

39. Cordeiro, R.; Luz, R.; Vasconcelos, V.; Fonseca, A.; Gonçalves, V. A Critical Review of Cyanobacteria Distribution and Cyanotoxins Occurrence in Atlantic Ocean Islands. *Cryptogam. Algol.* **2020**, *41*, 73–89. [CrossRef]

40. Moreira, C.; Ramos, V.; Azevedo, J.; Vasconcelos, V. Methods to detect cyanobacteria and their toxins in the environment. *Appl. Microbiol. Biotechnol.* **2014**, *98*, 8073–8082. [CrossRef] [PubMed]

41. Sanseverino, I.; António, D.C.; Loos, R.; Lettieri, T. *Cyanotoxins: Methods and Approaches for Their Analysis and Detection*; Publications Office of the European Union: Luxembourg, 2017.

42. Merel, S.; Walker, D.; Chicana, R.; Snyder, S.; Baurès, E.; Thomas, O. State of knowledge and concerns on cyanobacterial blooms and cyanotoxins. *Environ. Int.* **2013**, *59*, 303–327. [CrossRef]

43. Ballot, A.; Cerasino, L.; Hostyeva, V.; Cirés, S. Variability in the *sxt* Gene Clusters of PSP Toxin Producing *Aphanizomenon gracile* Strains from Norway, Spain, Germany and North America. *PLoS ONE* **2016**, *11*, 0167552. [CrossRef]

44. Murray, S.A.; Mihali, T.K.; Neilan, B.A. Extraordinary Conservation, Gene Loss, and Positive Selection in the Evolution of an Ancient Neurotoxin. *Mol. Biol. Evol.* **2011**, *28*, 1173–1182. [CrossRef] [PubMed]

45. Moustafa, A.; Loram, J.E.; Hackett, J.D.; Anderson, D.M.; Plumley, F.G.; Bhattacharya, D. Origin of saxitoxin biosynthetic genes in cyanobacteria. *PLoS ONE* **2009**, *4*, 5758. [CrossRef]

46. Ballot, A.; Fastner, J.; Wiedner, C. Paralytic shellfish poisoning toxin-producing cyanobacterium *Aphanizomenon gracile* in Northeast Germany. *Appl. Environ. Microbiol.* **2010**, *76*, 1173–1180. [CrossRef]

47. Cirés, S.; Wörmer, L.; Ballot, A.; Agha, R.; Wiedner, C.; Velázquez, D.; Casero, M.C.; Quesada, A. Phylogeography of cylindrospermopsin and paralytic shellfish toxin-producing nostocales cyanobacteria from mediterranean europe (Spain). *Appl. Environ. Microbiol.* **2014**, *80*, 1359–1370. [CrossRef]

48. Akcaalan, R.; Köker, L.; Oğuz, A.; Spoof, L.; Meriluoto, J.; Albay, M. First report of cylindrospermopsin production by two cyanobacteria (Dolichospermum mendotae and Chrysosporum ovalisporum) in Lake Iznik, Turkey. *Toxins* **2014**, *6*, 3173–3186. [CrossRef] [PubMed]

49. Ballot, A.; Ramm, J.; Rundberget, T.; Kaplan-Levy, R.N.; Hadas, O.; Sukenik, A.; Wiedner, C. Occurrence of non-cylindrospermopsin-producing Aphanizomenon ovalisporum and Anabaena bergii in Lake Kinneret (Israel). *J. Plankton Res.* **2011**, *33*, 1736–1746. [CrossRef]

50. Gantar, M.; Sekar, R.; Richardson, L.L. Cyanotoxins from black band disease of corals and from other coral reef environments. *Microb. Ecol.* **2009**, *58*, 856–864. [CrossRef] [PubMed]
51. Du, X.; Liu, H.; Yuan, L.; Wang, Y.; Ma, Y.; Wang, R.; Chen, X.; Losiewicz, M.D.; Guo, H.; Zhang, H. The diversity of cyanobacterial toxins on structural characterization, distribution and identification: A systematic review. *Toxins* **2019**, *11*, 530. [CrossRef] [PubMed]
52. Mayumi, T.; Kato, H.; Imanishi, S.; Kawasaki, Y.; Hasegawa, M.; Harada, K. Structural Characterization of Microcystins by LC/MS/MS under Ion Trap Conditions. *J. Antibiot.* **2006**, *59*, 710–719. [CrossRef]
53. Rantala, A.; Fewer, D.P.; Hisbergues, M.; Rouhiainen, L.; Vaitomaa, J.; Börner, T.; Sivonen, K. Phylogenetic evidence for the early evolution of microcystin synthesis. *Proc. Natl. Acad. Sci. USA* **2004**, *101*, 568–573. [CrossRef] [PubMed]
54. Meriluoto, J.; Spoof, L.; Codd, G.A. Appendix 3: Tables of Microcystins and Nodularins. In *Handbook of Cyanobacterial Monitoring and Cyanotoxin Analysis*; Meriluoto, J., Spoof, L., Codd, G.A., Eds.; John Wiley and Sons, Ltd: Chichester, UK, 2017; pp. 526–537.
55. Galetović, A.; Azevedo, J.; Castelo-Branco, R.; Oliveira, F.; Gómez-Silva, B.; Vasconcelos, V. Absence of Cyanotoxins in Llayta, Edible Nostocaceae Colonies from the Andes Highlands. *Toxins* **2020**, *12*, 382. [CrossRef] [PubMed]
56. Rippka, R. Isolation and purification of cyanobacteria. In *Methods in Enzymology*; Elsvier: Amsterdam, The Netherlands, 1988; Volume 167, pp. 3–27.
57. Rippka, R.; Waterbury, J.; Stanier, R. Isolation and Purification of Cyanobacteria: Some General Principles. In *The Prokaryotes: A Handbook on Habitats, Isolation and Identification of Bacteria.*; Starr, M.P., Stolp, H., Trüper, H.G., Balows, A., Schlegel, H.G., Eds.; Springer: Berlin/Heidelberg, Germany, 1981.
58. Cordeiro, R.; Luz, R.; Vasconcelos, V.; Gonçalves, V.; Fonseca, A. Cyanobacteria Phylogenetic Studies Reveal Evidence for Polyphyletic Genera from Thermal and Freshwater Habitats. *Diversity* **2020**, *12*, 298. [CrossRef]
59. Neilan, B.A.; Jacobs, D.; Del Dot, T.; Blackall, L.L.; Hawkins, P.R.; Cox, P.T.; Goodman, A.E. rRNA sequences and evolutionary relationships among toxic and nontoxic cyanobacteria of the genus *Microcystis*. *Int. J. Syst. Bacteriol.* **1997**, *47*, 693–697. [CrossRef]
60. Nübel, U.; Garcia-Pichel, F.; Muyzer, G. PCR primers to amplify 16S rRNA genes from cyanobacteria. *Appl. Environ. Microbiol.* **1997**, *63*, 3327–3332. [CrossRef]
61. Ouahid, Y.; Pérez-Silva, G.; Del Campo, F.F. Identification of potentially toxic environmental *Microcystis* by individual and multiple PCR amplification of specific microcystin synthetase gene regions. *Environ. Toxicol.* **2005**, *20*, 235–242. [CrossRef] [PubMed]
62. Schembri, M.A.; Neilan, B.A.; Saint, C.P. Identification of genes implicated in toxin production in the cyanobacterium Cylindrospermosis raciborskii. *Environ. Toxicol.* **2001**, *16*, 413–421. [CrossRef] [PubMed]
63. Yilmaz, M.; Foss, A.J.; Selwood, A.I.; Özen, M.; Boundy, M. Paralytic shellfish toxin producing *Aphanizomenon gracile* strains isolated from Lake Iznik, Turkey. *Toxicon* **2018**, *148*, 132–142. [CrossRef] [PubMed]
64. Ballot, A.; Swe, T.; Mjelde, M.; Cerasino, L.; Hostyeva, V.; Miles, C.O. Cylindrospermopsin- and Deoxycylindrospermopsin-Producing Raphidiopsis raciborskii and Microcystin-Producing Microcystis spp. in Meiktila Lake, Myanmar. *Toxins* **2020**, *12*, 232. [CrossRef] [PubMed]
65. Stuken, A.; Campbell, R.J.; Quesada, A.; Sukenik, A.; Dadheech, P.K.; Wiedner, C. Genetic and morphologic characterization of four putative cylindrospermopsin producing species of the cyanobacterial genera Anabaena and Aphanizomenon. *J. Plankton Res.* **2009**, *31*, 465–480. [CrossRef]
66. Liu, Y.; Chen, W.; Li, D.; Shen, Y.; Liu, Y.; Song, L. Analysis of paralytic shellfish toxins in Aphanizomenon DC-1 from Lake Dianchi, China. *Environ. Toxicol.* **2006**, *21*, 289–295. [CrossRef] [PubMed]
67. Lyra, C.; Suomalainen, S.; Gugger, M.; Vezie, C.; Sundman, P.; Paulin, L.; Sivonen, K. Molecular characterization of planktic cyanobacteria of Anabaena, Aphanizomenon, Microcystis and Planktothrix genera. *Int. J. Syst. Evol. Microbiol.* **2001**, *51*, 513–526. [CrossRef]
68. Katoh, K.; Standley, D.M. MAFFT Multiple Sequence Alignment Software Version 7: Improvements in Performance and Usability. *Mol. Biol. Evol.* **2013**, *30*, 772–780. [CrossRef]
69. Darriba, D.; Taboada, G.L.; Doallo, R.; Posada, D. jModelTest 2: More models, new heuristics and parallel computing. *Nat. Methods* **2012**, *9*, 772. [CrossRef]
70. Trifinopoulos, J.; Nguyen, L.-T.; von Haeseler, A.; Minh, B.Q. W-IQ-TREE: A fast online phylogenetic tool for maximum likelihood analysis. *Nucleic Acids Res.* **2016**, *44*, 232–235. [CrossRef]
71. Ronquist, F.; Teslenko, M.; van der Mark, P.; Ayres, D.L.; Darling, A.; Höhna, S.; Larget, B.; Liu, L.; Suchard, M.A.; Huelsenbeck, J.P. MrBayes 3.2: Efficient Bayesian Phylogenetic Inference and Model Choice Across a Large Model Space. *Syst. Biol.* **2012**, *61*, 539–542. [CrossRef] [PubMed]

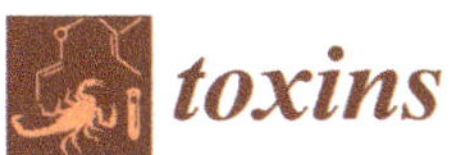 

Article

Potentially Toxic Planktic and Benthic Cyanobacteria in Slovenian Freshwater Bodies: Detection by Quantitative PCR

Maša Zupančič [1,2,*], Polona Kogovšek [3], Tadeja Šter [4], Špela Remec Rekar [4], Leonardo Cerasino [5], Špela Baebler [3], Aleksandra Krivograd Klemenčič [4] and Tina Eleršek [1]

[1] Department of Genetic Toxicology and Cancer Biology, National Institute of Biology, 1000 Ljubljana, Slovenia; tina.elersek@nib.si

[2] Jozef Stefan International Postgraduate School, 1000 Ljubljana, Slovenia

[3] Department of Biotechnology and Systems Biology, National Institute of Biology, 1000 Ljubljana, Slovenia; polona.kogovsek@nib.si (P.K.); spela.baebler@nib.si (Š.B.)

[4] Slovenian Environment Agency, 1000 Ljubljana, Slovenia; tadeja.ster@gov.si (T.Š.); spela.remec-rekar@gov.si (Š.R.R.); aleksandra.krivograd-klemencic@gov.si (A.K.K.)

[5] Department of Sustainable Agro-Ecosystems and Bioresources, Research and Innovation Centre, Fondazione Edmund Mach, 38010 San Michele all'Adige, Italy; leonardo.cerasino@fmach.it

[*] Correspondence: masa.zupancic@nib.si

Abstract: Due to increased frequency of cyanobacterial blooms and emerging evidence of cyanotoxicity in biofilm, reliable methods for early cyanotoxin threat detection are of major importance for protection of human, animal and environmental health. To complement the current methods of risk assessment, this study aimed to evaluate selected qPCR assays for detection of potentially toxic cyanobacteria in environmental samples. In the course of one year, 25 plankton and 23 biofilm samples were collected from 15 water bodies in Slovenia. Three different analyses were performed and compared to each other; qPCR targeting *mcyE*, *cyrJ* and *sxtA* genes involved in cyanotoxin production, LC-MS/MS quantifying microcystin, cylindrospermopsin and saxitoxin concentration, and microscopic analyses identifying potentially toxic cyanobacterial taxa. qPCR analyses detected potentially toxic *Microcystis* in 10 lake plankton samples, and potentially toxic *Planktothrix* cells in 12 lake plankton and one lake biofilm sample. A positive correlation was observed between numbers of *mcyE* gene copies and microcystin concentrations. Potential cylindrospermopsin- and saxitoxin-producers were detected in three and seven lake biofilm samples, respectively. The study demonstrated a potential for cyanotoxin production that was left undetected by traditional methods in both plankton and biofilm samples. Thus, the qPCR method could be useful in regular monitoring of water bodies to improve risk assessment and enable timely measures.

Keywords: cyanotoxin detection; harmful cyanobacterial blooms; next-generation biomonitoring; real-time PCR; qPCR; LC-MS/MS; microcystin; cylindrospermopsin; saxitoxin

Key Contribution: Currently used biomonitoring methods are not sufficient for detection of cyano-toxic potential. In addition to planktic cyanobacteria, benthic species in biofilm can be a potential source of cyanotoxins and therefore both groups should be included in biomonitoring for risk assessment.

Citation: Zupančič, M.; Kogovšek, P.; Šter, T.; Remec Rekar, Š.; Cerasino, L.; Baebler, Š.; Krivograd Klemenčič, A.; Eleršek, T. Potentially Toxic Planktic and Benthic Cyanobacteria in Slovenian Freshwater Bodies: Detection by Quantitative PCR. *Toxins* **2021**, *13*, 133. https://doi.org/10.3390/toxins13020133

Received: 18 January 2021
Accepted: 9 February 2021
Published: 11 February 2021

Publisher's Note: MDPI stays neutral with regard to jurisdictional claims in published maps and institutional affiliations.

1. Introduction

Cyanobacterial blooms and a subsequent release of cyanotoxins into the environment are becoming more frequent due to eutrophication, global warming and other anthropogenic pressures. They can have negative effects on all ecosystem services as well as human and animal health and can cause economical damage by affecting tourism, recreation, industry, agriculture and drinking water supply. On the European Union level, there is no legislation prescribing regular monitoring of cyanotoxin concentration in surface waters. The most frequently used guideline is the one for drinking water from the World Health Organisation, setting the upper limit of 1 µg/L of microcystin-LR equivalents [1].

Early detection of cyanotoxin threat could help water resources managers take timely and appropriate measures. Current approaches for identification and quantification of cyanobacterial cells and cyanotoxins—microscopic count and analytical methods, such as high-performance liquid chromatography (HPLC), liquid chromatography–mass spectrometry (LC-MS), enzyme-linked immunosorbent assay (ELISA) or protein phosphatase 1A (PP1A) analyses—each have their advantages, but they can often be time-consuming, costly or technically demanding. Emerging molecular methods, such as quantitative PCR (qPCR), could enable fast, highly sensitive and cost-effective detection of potentially toxic cyanobacteria [2]. This approach is based on the extraction of community DNA from environmental samples and can therefore provide a full picture of the cyanobacterial diversity. However, for its use in regular monitoring programs, these methods have to be thoroughly tested and optimised in order to achieve comparability with current methods and thus applicability in monitoring schemes.

Analytical methods for detection of cyanotoxins can determine their concentration only at a certain point in time, whereas cyanotoxin content can vary significantly throughout the day (depending on hydrological conditions and the presence of bacterial decomposers). Besides, due to a high number of variants of cyanotoxins (e.g., over 248 variants of microcystins [3]) and a lack of standards for each of these variants, not all of them can be measured. These methods are also too expensive to monitor the concentration daily. On the other hand, detection and quantification of genes involved in cyanotoxin synthesis could enable cost-effective monitoring of the potential for cyanotoxin production on a daily basis at various locations. Moreover, all of the cyanotoxins with known genetic basis can be analysed. This can give us comprehensive information on the toxigenic potential of cyanobacterial communities in the environment.

The qPCR method has been applied in various studies, which have been summed up by Pacheco et al. [2]. Majority of the studies target genes involved in microcystin (MC) synthesis (e.g., [4], followed by cylindrospermopsin (CYN) (e.g., [5] and saxitoxin (SXT) synthesis (e.g., [6]). There have been attempts to optimise the method for its use directly in the field ([7] or to target various genes at once in multiplex reactions (e.g., [8]). However, there is still no consensus on the applicability of qPCR in regular monitoring since gene copy numbers reveal only the potential for cyanotoxin production, which is not always in correlation with actual cyanotoxin concentrations in the environment. Precisely this contrast between the methods indicates the advantage of qPCR over analytical methods focused on cyanotoxin measurement, as the former one could predict also future risk rather than assessing only the current situation.

The majority of the cyanobacterial qPCR studies are directed at microcystins, while detection of cylindrospermopsins and especially saxitoxins is still relatively rare [2]. Moreover, most of the studies focus on plankton samples, while cyanobacteria in biofilm are still underrepresented in molecular studies despite increasing evidence of their toxicity with potential acute effects on animals [9–13]. Furthermore, few of these studies include comparison of qPCR method and microscopy [2]. As microscopy is the preferred method of biomonitoring in many countries, it is important to investigate its effectiveness in detecting potential risk of cyanotoxin production. Additionally, all such studies are geographically limited with little or no focus on the central European region [2]. Taking into account the high genetic variability in naturally occurring cyanobacterial strains throughout the world (e.g., [14]), the assays should be tested in different regions and different water bodies to assure their wide applicability.

Therefore, the aim of this study was to expand the evaluation of qPCR assays for detection of cyanotoxin threat to understudied benthic cyanobacteria in biofilm samples, with the emphasis on comparison of results with microscopy as well as LC-MS/MS. We focused on water bodies in different regions of Slovenia (central Europe) and on three groups of cyanotoxins: microcystins, cylindrospermopsins and saxitoxins. We employed five previously published qPCR assays for detection of the microcystin- [15–17], cylindrospermopsin- [18] and saxitoxin-producing cyanobacteria [6]. Although the cyanotoxin potential is not necessarily linked to cyanotoxin concentration, we evaluated the correlation between the number of gene copies, microscopically determined cell number of potentially toxic species and cyanotoxin concentration. This is the first study in Slovenia aiming to detect cyanobacterial toxic potential with qPCR, and one of the few studies employing this method in environmental biofilm samples. Only thorough understanding of strengths and weaknesses of the qPCR method can enable its implementation into existing environmental monitoring strategies.

2. Results

2.1. Evaluation of the qPCR Assays

For our study, we have chosen previously published assays mcyE-Ana, mcyE-Mic, mcyE-Pla, cyrJ and sxtA, targeting microcystin-producers from genera *Dolichospermum* (ex *Anabaena*), *Microcystis* and *Planktothrix*, cylindrospermopsin-producers and saxitoxin-producers, respectively ([6,15–18]; Table 1). First, we evaluated the selected qPCR assays in terms of their specificity, sensitivity and robustness.

Table 1. Primers used for qPCR amplification of selected target regions. bp—base pair.

Target Cyanotoxins/Organisms	Assay	Target Gene	Primer Label	Nucleotide Sequence $(5' \rightarrow 3')$	Fragment Length [bp]	Reference
Microcystins (genus *Dolichospermum*)	mcyE-Ana	*mcyE*	mcyE-F2	GAA ATT TGT GTA GAA GGT GC	247	[15]
			AnamcyE-12R	CAA TCT CGG TAT AGC GGC		[16]
Microcystins (genus *Microcystis*)	mcyE-Mic	*mcyE*	mcyE-F2	GAA ATT TGT GTA GAA GGT GC	247	[15]
			MicmcyE-R8	CAA TGG GAG CAT AAC GAG		[16]
Microcystins (genus *Planktothrix*)	mcyE-Pla	*mcyE*	mcyE-F2	GAA ATT TGT GTA GAA GGT GC	249	[15]
			PlamcyE-R3	CTC AAT CTG AGG ATA ACG AT		[17]
Cylindrospermopsins	cyrJ	*cyrJ*	cyrJ207-F	CCC CTA CAA CCT GAC AAA GCT T	77	[18]
			cyrJ207-R	CCC GCC TGT CAT AGA TGC A		
Saxitoxins	sxtA	*sxtA*	sxtA-F	GAT GAC GGA GTA TTT GAA GC	125	[6]
			sxtA-R	CTG CAT CTT CTG GAC GGT AA		
(Cyano-)bacteria and plant chloroplasts	16S-cyano	16S rRNA	cyano-real16S-F	AGC CAC ACT GGG ACT GAG ACA	73	[6]
			cyano-real16S-R	TCG CCC ATT GCG GAA A		

2.1.1. Selection and Specificity of the qPCR Assays

Based on a literature review (Supplementary file S1), we selected nine published assays for specificity evaluation. In addition to the assays shown in Table 1, two assays targeting all microcystin-producers (McyE-F2b/R4 [15,19] and DQmcy [20]), and assay anaC-gen targeting anatoxin-producers [21,22] were evaluated. Assay anaC-gen was excluded due to too high specificity indicated in the original paper [21] and too long amplicon, originally designed for end-point PCR application. The other eight assays were evaluated in vitro on test environmental samples. Assays mcyE-F2b/R4 and DQmcy were excluded based on suboptimal performance with Slovenian environmental samples (dissociation curves indicating non-target amplification (multiple peaks) and certain results inconsistent with microscopic observations, which could be due to regional differences in cyanobacterial genotypes; data not shown)).

The remaining six assays were further characterised in silico and in vitro. Specificity evaluation done in the original papers demonstrated appropriate specificity of all assays, while BLAST analysis in this study revealed that four assays (namely mcyE-Ana, mcyE-Pla, cyrJ and sxtA) are specific for desired target organisms and genes. Assays mcyE-Mic and 16S-cyano showed non-target alignment; the former with *Pseudanabaena* sp. CCM-UFV065, and the latter with plant chloroplasts as well as some heterotrophic bacteria, such as *Chryseobacterium* or *Actinobacterium*, which can be found in freshwater habitats and could thus be amplified in environmental samples. In original studies, specificity of assays mcyE-Mic [16] and 16S-cyano [6] was in vitro tested with selected non-target genera, namely *Planktothrix*, *Dolichospermum* and *Nostoc*, and with several target cyanobacterial cultures, respectively. None of the assays showed any cross-reactivity, however, unspecific reaction predicted with in silico analysis was not evaluated.

In cyanobacterial cultures, specific amplification occurred only in strains producing target cyanotoxins and not in other strains (Table 2). In assay 16S-cyano, strong amplification was observed in selected plant samples, confirming cross-reactivity with plant 16S rRNA genes (Supplementary file S2). This could lead to false positive results and overestimation of cyanobacterial abundance in environmental samples; thus, this assay was used only for evaluation of DNA extraction and control of inhibition in qPCR reactions.

Table 2. Specificity of selected qPCR assays, with positive results shaded in grey. Average quantification cycle (Cq) values between three technical replicates of DNA in 10^{-2} dilution and reference melting temperatures (Tm, in °C), calculated as an average of all DNA dilutions within quantification range are shown. Raw data is available in Supplementary file S4. MC—microcystins, CYN—cylindrospermopsins, SXT—saxitoxins, N—no specific amplification.

Cyanobacterial Cultures		qPCR Assays											
Strain	Toxicity	16S-cyano		mcyE-Ana		mcyE-Mic		mcyE-Pla		cyrJ		sxtA	
		Cq	Tm	Cq	Tm	Cq	Tm	Cq	Tm	Cq	Tm	Cq	Tm
Anabaena sp. UHCC 0315	MC-producer	23.13	81.1	25.86	75.7	N	N	N	N	N	N	N	N
Microcystis aeruginosa PCC 7806	MC-producer	20.14	81.0	N	N	19.82	78.0	N	N	N	N	N	N
Planktothrix sp. NIVA-CYA126/8	MC-producer	23.55	81.1	N	N	N	N	24.37	77.6	N	N	N	N
Aphanizomenon ovalisporum ILC-164	CYN-producer	23.87	81.2	N	N	N	N	N	N	24.79	80.0	N	N
Aphanizomenon gracile NIVA-CYA 851	SXT-producer	18.79	81.3	N	N	N	N	N	N	N	N	18.72	79.6
	Reference Tm	81.1		75.9		78.2		77.6		79.9		79.5	

In some of the environmental samples, melting temperatures of the amplified products (Tm) obtained via dissociation curve analysis indicated the presence of non-target amplicons. This was mostly observed with assays mcyE-Ana and mcyE-Mic, which showed up to 12.3 °C and up to 10.0 °C higher Tm then the reference Tm (Table 2, Supplementary file S4), respectively. Additionally, primer dimers were detected in some samples with assays mcyE-Ana, mcyE-Mic and mcyE-Pla (Tm < 70 °C). Therefore, results for all cyanotoxin-specific assays were considered positive if their Tm values were within the expected range (±0.5 °C). For the assay 16S-cyano, however, a wider range of obtained Tm values (80.7–82.7 °C; Supplementary files S2–S5) were considered positive, as they were consistent between technical replicates and the Cq values were below 30. Reference Tm values from pure cultures corresponded closely to Tm of synthetic DNA fragments for all assays (± ≤0.5 °C, data not shown). Gel electrophoresis of the samples with multiple Tm peaks (18 samples for mcyE-Ana, 3 samples for mcyE-Pla and 2 samples for cyrJ) produced multiple bands of different lengths, in contrast to environmental samples with one distinctive Tm peak that produced a single band (data not shown), confirming non-specific amplification of various DNA fragments in the former group. Therefore, such samples were considered negative.

2.1.2. Sensitivity of the qPCR Assays.

Dilution series of the cyanobacterial culture DNA was prepared to evaluate the sensitivity of the assays (Table 3; calibration curves in Supplementary file S6). All six qPCR assays showed high sensitivity, ranging from 10 to 30 cells/mL for all assays, except sxtA, which showed the highest sensitivity, detecting less than 1 cell/mL (Table 3). Amplification efficiency determined from the dilution series of all assays was between 63% and 98%.

Table 3. Limit of detection (LOD), limit of quantification (LOQ), amplification efficiency and correlation coefficient for selected qPCR assays based on calibration curves of reference cyanobacterial strains DNA. LOQ and LOD are expressed in cells/μL DNA and in cells/mL sample, which depends on the individual sample volume. Individual calibration curves are shown in Supplementary file S6.

Assay	LOD (Cells/μL DNA, Cells/mL Sample)	LOQ (Cells/μL DNA, Cells/mL Sample)	Amplification Efficiency	Correlation Coefficient (R^2) of Linear Range of Curve
mcyE-Ana	3.3, 30.1	33, 301	0.63	1.0000
mcyE-Mic	16.5, 205.9	165, 2059	0.75	0.9973
mcyE-Pla	2.4, 12.5	238, 1252	0.68	0.9969
cyrJ	5.7, 114.6	6, 115	0.73	0.9991
sxtA	0.3, 1.5	3, 15	0.98	0.9988
16S-cyano	1.6, 20.6	165, 2059	0.79	0.9968

2.1.3. Robustness of the qPCR Assays

Performance of the assays was evaluated also on typical environmental samples with known cyanobacterial taxa composition (as determined by microscopy). Up to 14 samples of plankton community DNA (2 from lakes, 6 from urban ponds, 1 from urban stream and 5 cyanobacterial bloom samples) were tested with specific assays. The presence of the target organisms was mostly confirmed in samples where it was expected (Supplementary file S7), thus proving the suitability of the method for target gene detection in environmental samples sampled in our region.

For intra-assay variability, the absolute difference between Cq values from three technical replicates in positive samples was determined. Relatively high intra-assay variability was observed in samples with target gene concentration close to LOD of the assays, which is probably due to a stochastic effect, while lower variability was observed in the rest of the samples (Supplementary file S5). For the assay mcyE-Ana, we could not assess intra-assay variability as we did not get any positive qPCR results from the environmental samples. When using cyanobacterial monocultures as a template, the variability was in general lower than with environmental samples (Supplementary files S3 and S5). This is expected, as the presence of inhibitory compounds in environmental samples can interfere with qPCR amplification [23], which can result in higher intra-assay variability. Inter-assay variability, evaluated for assays mcyE-Pla, cyrJ and sxtA in two separate runs, differed between assays but showed reasonably good reproducibility, taking into account degradation of DNA due to freeze/thaw cycle and different inhibitory substances in environmental samples (data not shown).

Additionally, possible inhibition of qPCR reactions was checked by testing two subsequent dilutions of 10 randomly selected samples. Only one sample showed a potential for inhibition of a qPCR reaction (BL1.5, Supplementary file S8), therefore analysis of this sample was repeated with more diluted DNA. Robustness of the assays was demonstrated by testing DNA extracted from different matrices (cyanobacterial monocultures, synthetic DNA fragments, frozen or lyophilised bloom samples, environmental samples of plankton or biofilm), of variable purity and of variable DNA concentration (data not shown), where adequate performance was observed in all cases.

2.2. Presence of Target Genes in Environmental Samples

For the analysis of the environmental samples, negative and positive controls were included in every qPCR run, where the former showed no amplification and the latter showed specific amplification in all cases. Additionally, negative controls were included in every DNA extraction and in every sampling, and their NanoDrop measurements showed no presence of DNA. This confirmed appropriate assay performance and absence of contamination during field sampling, DNA extraction and preparation of qPCR reaction mixtures. Successful DNA extraction and qPCR amplification were additionally confirmed by positive results of the assay 16S-cyano amplifying 16S rRNA genes in all samples with a Cq range between 14 and 24 (Supplementary file S5).

Detailed results of qPCR, LC-MS/MS and microscopic analyses are depicted in Supplementary file S8, raw data is available in Supplementary file S5. Potentially toxic *Microcystis* cells (assay mcyE-Mic) were detected in 10 lake plankton samples (all samples from Lake Vogrscek, Slivnica and Pernica). Potentially toxic *Planktothrix* cells (assay mcyE-Pla) were detected in 12 lake plankton samples (all samples from Lake Bled, in low amounts also in Lake Bohinj) and in low amounts in one biofilm sample (Lake Bled). Potentially toxic *Dolichospermum* species (assay mcyE-Ana) were not detected in any plankton nor biofilm sample. Potential cylindrospermopsin producers (assay cyrJ) were detected in three lake biofilm samples (Lake Bled, Sava River), while potential saxitoxin producers (assay sxtA) were detected in seven lake biofilm samples (Lake Bled, Koseze Pond) (Table 4, Supplementary file S8). In some samples, only 1/3 or 2/3 technical replicates were positive; all of these samples were close to LOD of the assay. This means that the target genes were present in low quantities and thus not amplified in every subsample (stochastic effect).

Table 4. Abundance of target gene copies in environmental samples. Plankton [gc/mL]: * 1–10^2, ** 10^2–10^4, *** 10^4–10^6; biofilm [gc/g dry weight]: * 10^3–10^5, ** 10^5–10^7. Due to low amplification efficiency, the gc values are not reliable and should be used as qualitative observation. Sample description and quantification data is available in Supplementary file S8. Sample BL1.4 has only been analysed for cyanotoxin content and not by qPCR and is thus not represented in this table. gc—gene copies, empty—below LOD.

	Sample	mcyE-Ana	mcyE-Mic	mcyE-Pla	cyrJ	sxtA
				qPCR Assays		
PLANKTON [gc/mL]	BL1.1			***		
	BL1.2			***		
	BL1.3			***		
	BL1.5			***		
	BL1.6			**		
	BL1.7			***		
	BL1.8			***		
	BL1.9			**		
	BL1.10			***		
	BL1.11			**		
	BL1.12			***		
	BO1					
	BO2					
	BO3			*		
	PE1		***			
	PE2		***			
	PE3		***			
	PE4		**			
	SL1		**			
	SL2		***			
	SL3		**			
	VO1		**			
	VO2		**			
	VO3		**			

Table 4. *Cont.*

		qPCR Assays				
	Sample	**mcyE-Ana**	**mcyE-Mic**	**mcyE-Pla**	**cyrJ**	**sxtA**
BIOFILM [gc/g ry weight]	BL2.1					**
	BL2.2				*	*
	BL2.3					**
	BL2.4					*
	BL2.5					
	BL2.6				*	
	BL2.7			*		*
	BL2.8					
	BL2.9					
	BL2.10					**
	SO1					
	SO2					
	SO3					
	LU					
	BI					
	PS					
	SA1				*	
	SA2					
	RI					
	LJ					
	KO					**
	TI					
	GL					

2.3. Temporal Variability of Microcystin Abundance

Cylindrospermopsins or saxitoxins were not detected with LC-MS/MS in any of the samples. Microcystins were detected in 16 out of 23 plankton samples (out of which one was uncertain as it was too close to the background noise) and 4 out of 22 biofilm samples (out of which 3 were uncertain) with 5 different variants observed (MC-RR, MC-RRdm, MC-HtyRdm, MC-LRdm, MC-LR). The highest concentrations of microcystins were measured in plankton samples from Lake Bled during winter months (up to 660 ng/L in February, BL1.2, Supplementary file S8), which complies with the highest cell concentration of the microcystin-producing species *Planktothrix rubescens* (See Section 2.5. Correlation between Parametres). Moreover, all three microcystin variants found in these samples are typically produced by *Planktothrix rubescens*. Figure 1 shows the microcystin diversity in plankton samples of Lake Bled, where a temporal trend can be observed. In all other samples, microcystins were either not detected or their concentrations were low; up to 6.9 ng/L in plankton samples (SL2, Supplementary file S8) and up to 11.9 ng/g dry weight in biofilm samples (BL2.6, Supplementary file S8) and thus their diversity is not represented in the Figure 1.

2.4. Microscopic Analyses

With microscopic analyses, we found 17 potentially toxic taxa in plankton samples (Aphanizomenon sp., Aphanizomenon flos-aquae, Aphanizomenon issatschenkoi, Cylindrospermopsis raciborskii, Dolichospermum crassum, Dolichospermum flos-aquae, Dolichospermum lemmermannii, Dolichospermum planctonicum, Microcystis aeruginosa, Microcystis flos-aquae, Phormidium sp., Phormidium amoenum, Planktothrix agardhii, Planktothrix rubescens, Pseudanabaena sp., Pseudanabaena catenata, Pseudanabaena limnetica) and 4 in biofilm samples (Oscillatoria sp., Phormidium sp., Phormidium autumnale, Pseudanabaena catenata) (Figure 2, Supplementary file S9). The highest diversity of potentially toxic planktic cyanobacteria was found in Lake Pernica (13 taxa), while in Lake Bohinj none was detected. In the majority of the lakes, there was a higher diversity of potentially toxic taxa observed in summer months (June, July, August) than in the rest of the year. Out of the 23 biofilm samples, in the majority of them there was one potentially toxic taxon detected under microscope, while six of them contained none and two of them contained two potentially toxic taxa (Koseze Pond, Lake Bled). The most common potentially toxic genus was Phormidium, which was detected in 17 samples.

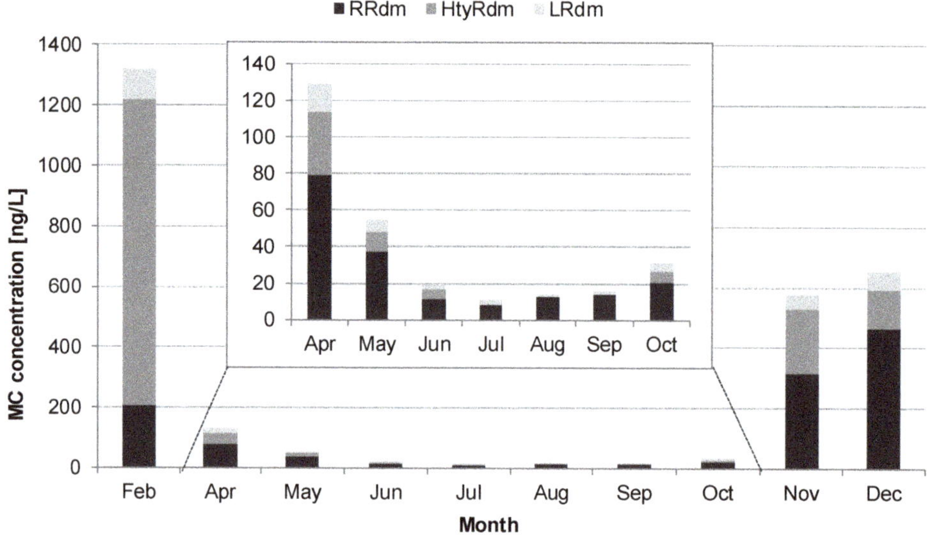

Figure 1. Microcystin temporal variability and diversity in plankton samples from Lake Bled. The three congeners (RRdm, HtyRdm and LRdm) are all demethylated variants, typical of *Planktothrix rubescens*. The temporal changes of proportions can be explained by the succeeding of different chemotypes of the same species. Samples Feb–Dec correspond to samples BL1.2–BL1.12 (Supplementary file S8).

2.5. Correlation between Parametres

To further evaluate qPCR as a method to detect cyanotoxin production potential, correlations (presented as Pearson correlation coefficient) between qPCR, LC-MS/MS and microscopy results were determined. For plankton samples (N = 25), the numbers of *mcyE* gene copies (sum of values produced by assays mcyE-Ana, mcyE-Mic and mcyE-Pla) were positively correlated with microcystin concentrations measured by LC-MS/MS (r = 0.8375, Figure 3A), while there was no correlation with cell number or biovolume of all potential microcystin-producing taxa. There was also no correlation found between *Microcystis*- or *Planktothrix*-specific *mcyE* gene copies and cell numbers or biovolumes of *Microcystis* or *Planktothrix* cells, respectively. However, when results from Lake Bled were analysed separately, there was a positive correlation between *Planktothrix*-specific *mcyE* gene copies and both cell numbers and biovolumes of *Planktothrix* cells (r = 0.8831 in both cases, the former one is presented on Figure 3B). More detailed graphical representation of results from Lake Bled produced with different methods (Figure 4) shows similar temporal trend observed with qPCR, microscopy and LC-MS/MS. Elevated abundances of *Planktothrix* cells and microcystin concentrations in winter months correspond to a scarlet-coloured blooms of *Planktothrix rubescens* (Figure 2), which were observed on Lake Bled in February 2019 and January 2020.

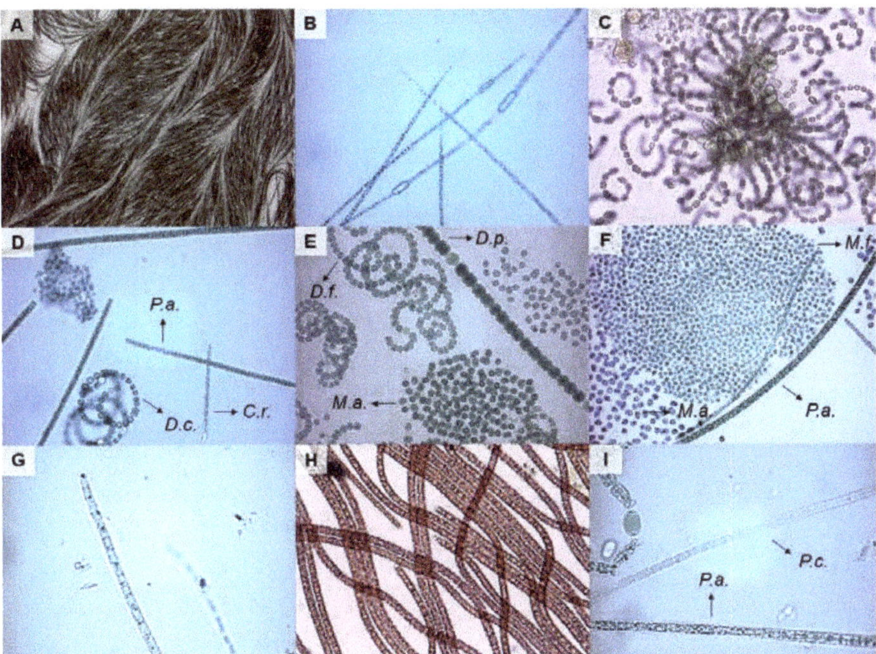

Figure 2. Potentially toxic cyanobacterial taxa found in environmental samples. (**A**)—Aphanizomenon flos-aquae, (**B**)—Aphanizomenon issatschenkoi, (**C**)—Dolichospermum lemmermanii, (**D**)—C.r. Cylindrospermpsis raciborskii, D.c. Dolichospermum crassum, P.a. Planktothrix agardhii, (**E**)—D.f. Dolichorpermum flos-aquae, D.p. Dolichospermum planctonicum, M.a. Microcystis aeruginosa, (**F**)—M.a. Microcystis aeruginosa, M.f. Microcystis flos-aqaue, P.a. Planktothrix agardhii, (**G**)—Phormidium amoenum, (**H**)—Planktothrix rubescens, (**I**)—P.a. Planktothrix agardhii, P.c. Pseudoanabaena catenata. The photos were taken under light microscope with 160× (**A**), 400× (**C,H**), 640× (**B,D,E,F**) or 1600× magnification (**G,I**).

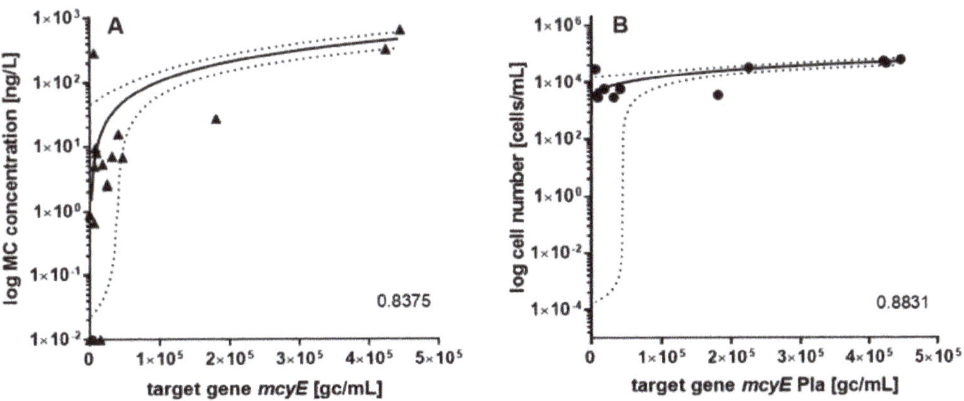

Figure 3. Scatterplots showing the correlations between different parameters for plankton samples. Solid lines represent linear regression curves, dotted lines represent 95% confidence band, and Pearson correlation coefficient (r) is given in the bottom right-hand corners. (**A**) correlation between *mcyE* copies concentration (the sum of concentrations obtained by assays mcyE-Ana, mcyE-Mic and mcyE-Pla) and MC concentration. (**B**) correlation between *Planktothrix*-specific *mcyE* gene copies concentration and cell concentration of all *Planktothrix* species (only samples from Lake Bled, N = 11).

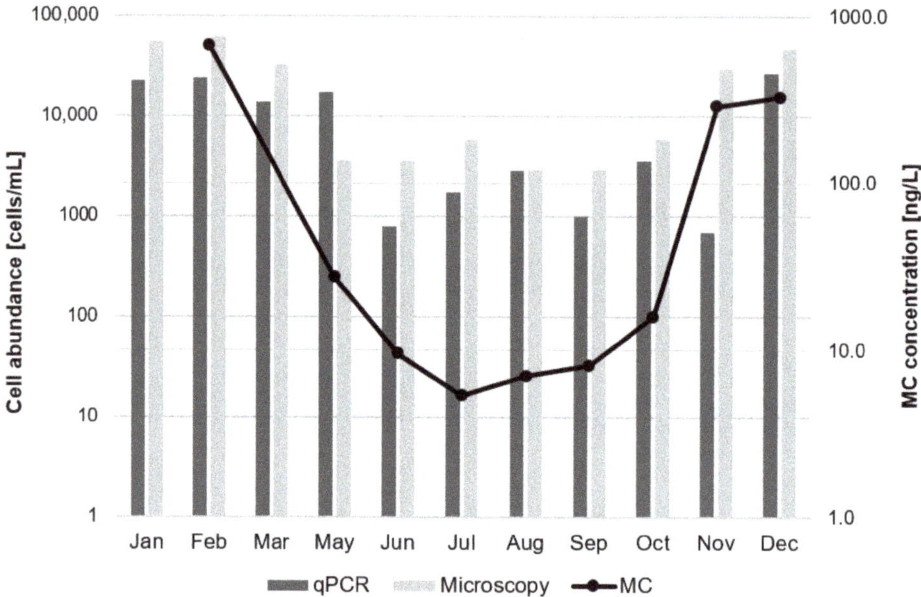

Figure 4. *Planktothrix* cell abundance in Lake Bled in 2019 determined by qPCR (assay mcyE-Pla) and microscopic analyses. Due to low amplification efficiency and possible variability of gene copy numbers per cell, the calculated cell abundance values are not reliable and should be used as qualitative observation. Total MC concentration measured by LC-MS/MS is included for comparison. In microscopy, 10% variation is expected, which is the average standard deviation between technical replicates with Bürker Türk counting chamber in our laboratory.

For plankton samples, correlations could not be determined for assays mcyE-Ana, cyrJ and sxtA as we did not detect target genes in any of the plankton samples. For biofilm samples (N = 23), correlations between target gene copies and relative species abundance were determined and there was no correlation found for any of the assays. Species abundance was evaluated semi-quantitatively (values 1–5), which might have impacted the results.

3. Discussion

The study aimed to evaluate the suitability of a qPCR method for early detection of potentially toxic cyanobacteria in surface water bodies in the central European region. Systematic search for publications describing molecular assays (Supplementary file S1) revealed several qPCR assays that were applied for detection of cyanobacteria-specific target genes. We performed a selection of the amplicons, where we took into consideration their specificity, sensitivity and suitability for qPCR reaction conditions. Furthermore, we tested the performance of selected assays (Table 1) in vitro. Even though all assays showed good performance in pure cultures (Table 2) and synthetic DNA fragments, some of them showed unspecific amplification in environmental samples. This is a consequence of heterogenous samples originating from water bodies and presents a high risk in application of SYBR Green chemistry detection in environmental samples, especially when the genome of the target organism is unknown. Nevertheless, we were able to filter the true positive samples from the unspecific samples with reference Tm values. Based on these results, all five selected cyanotoxin-specific assays are suitable for detection of cyanotoxin potential in water bodies. However, as we did not detect cylindrospermopsins or saxitoxins in any of the environmental samples, further research is needed to confirm the suitability of the assays for potential producers of these cyanotoxins. On the other hand, assay 16S-cyano was shown to be inappropriate for detection of cyanobacteria, since it also amplifies plant chloroplast DNA (Supplementary file S2). Thus, we used it as a DNA extraction and qPCR inhibition control.

The qPCR results revealed some new information that was unknown up to date and could not be obtained by microscopy alone. The most novel finding is the potential for cylindrospermopsin- and saxitoxin-production in biofilm in certain water bodies (Table 4), where it has never been reported before. In Slovenia, cylindrospermopsins have been detected once in low amounts in a planktic sample (data not published), while saxitoxins have never been detected. Our study indicates that despite these cyanotoxins not being detected in the moment of sampling (Supplementary file S8), a potential for their production exists, which might be important information for the future monitoring schemes and research studies.

Another important finding is the discovery of cyanotoxic potential in biofilm (Table 4). Even though the first observation of potentially toxic cyanobacteria in biofilm samples was reported in 1997 [24] and cyanotoxins from benthic cyanobacteria are believed to have caused animal death on a few occasions [9–13,24], such studies are still scarce. These findings suggest anatoxin-a and microcytins as the prevalent cyanotoxins in cyanobacterial mats. However, our results indicate such microbial mats might also possess cylindrospermopsin- and saxitoxin-producing potential (found in over a third of the biofilm samples, Table 4), which could not be detected by either microscopy or LC-MS/MS. Although there have been some prior publications about cylindrospermopsin- [25–27] and saxitoxin-production [28–30] by benthic cyanobacteria, the issue is still poorly investigated. This information, together with some novel findings regarding microcystins in Slovenia (presence of potentially toxic *Microcystis* in Lake Slivnica and Vogrscek throughout the whole year, which has not been reported by regular monitoring before) might be a valuable guideline for future water management.

In addition to applying the qPCR method to environmental samples, our aim was also to compare its performance to traditionally used methods. While many studies have compared qPCR results with cyanotoxins measurements (e.g., [5,6,31]), comparisons with microscopic counts are harder to find, so one of our goals was to evaluate qPCR in comparison with microscopy-based biomonitoring methods. There was no correlation found between gene copy numbers and cell numbers or biovolumes for microcystin-producing cyanobacterial species, neither when analysed separately by genus nor as a whole group. Regarding *Microcystis* genus, there were four samples where *Microcystis*-specific *mcyE* genes were detected, while *Microcystis* cells were not observed under microscope (Supplementary file S8). This could mean that the numbers of *Microcystis* cells in these lakes were below LOD of microscopy, but could be detected by qPCR, which is expected due to much higher sensitivity of qPCR. Alternatively, discrepancy between results could be caused by cross-contamination of field equipment while transferring it between lakes or by low specificity of the assay mcyE-Mic (detecting *mcyE* genes in other genera or other non-target products). The possibility of non-target detection was confirmed also by BLAST analysis, revealing *Pseudanabaena* sp. as one of the assay's potential targets. However, the majority of the results cannot be explained by this, as *Pseudanabaena* was microscopically observed only in two of these samples. This could be further investigated by DNA sequencing of obtained qPCR products. On the other hand, there were also two samples where *Microcystis* cells were microscopically identified, while *Microcystis*-specific *mcyE* genes were not detected (Supplementary file S8). This might be due to the fact that toxic and non-toxic *Microcystis* cells cannot be distinguished morphologically [32].

For the *Planktothrix* species, elevated cell and microcystin concentrations in Lake Bled in the beginning of the year (Figure 4) correspond to a moderate *Planktothrix rubescens* bloom in 2019. Elevated concentrations at the end of the year contributed to a massive bloom formation that occurred in the end of January 2020 (field observations), which was influenced also by other nutritional factors taking place that month, so it cannot be explained only by our measurement in 2019. Despite the lack of correlation between microscopic and qPCR-based abundance of *Planktothrix* cells in the whole dataset, there was a positive correlation when the analysis was performed only with samples from Lake Bled (Figure 3). Those 11 samples represent a majority of qPCR-positive samples for this assay (mcyE-Pla; Table 4), which indicates that the lack of correlation in other samples is primarily caused by samples with negative qPCR and positive microscopy results. In most of such samples, the dominant *Planktothrix* species was *P. agardhii* (Lake Pernica, Supplementary file S9), which might suggest that the assay does not amplify target genes in the whole genus equally. Alternatively, the discrepancy might be caused by the presence of non-toxic *P. agardhii* strains and the inability of microscopic analyses to differentiate between them, which has been shown in previous studies [33]. Regarding *Dolichospermum* genus, the complete lack of genus-specific *mcyE* genes in all analysed samples (assay mcyE-Ana) despite some microscopic observations (Lake Pernica, Lake Bled; Supplementary file S9) might indicate that the assay does not amplify target genes in all *Dolichospermum* strains, or that the present taxa were in fact not possessing *mcyE* genes.

Furthermore, qPCR was also compared to LC-MS/MS results. A positive correlation was found between *mcyE* gene copy numbers (sum of all analysed genera) and microcystin concentrations (Figure 3), which corroborate numerous prior studies (e.g., [16,31,34]). However, there were also some discrepancies between qPCR and LC-MS/MS results. In some samples, target genes were detected (mostly below LOQ), while cyanotoxins were not (Supplementary file S8). Similar inconsistencies have been observed in other studies as well (e.g., [6,35].) Authors' potential explanations include low concentration of cyanotoxins (below LOD of analytical method), degradation of cyanotoxins in the samples, lack of gene expression or mutations leading to non-toxicity. It has to be noted that these results are not always expected to match, as analytical methods (such as LC-MS/MS) measure the actual cyanotoxin concentration in a particular moment of sampling, while qPCR detects only the potential for cyanotoxin production. It has been indicated that despite the presence of *mcy* genes, their expression can vary in time significantly [36]. The toxin production depends on physical parameters (e.g., temperature), growth phase [37] and nutrient content [38]. Therefore, quantification of gene copies alone cannot always predict toxin concentration.

Regarding the methodology itself, our study confirmed that qPCR has significantly higher sensitivity (LOD = 1.5–205.9 cells/mL, Table 3) than microscopic cell count (Bürker Türk counting chamber, LOD = 10.000 cells/mL) of Slovenian samples, which are not pre-concentrated. Detection of less than 1 cell/mL (assay sxtA) could be explained by multiple gene copies of the target gene per cell [6] and possible free DNA in the sample. Specificity of all assays was tested and proved appropriate in the original publications. On top of that, our experiments showed that the assays are highly specific in cyanobacterial monocultures (Table 2), while there was some non-specific amplification observed in environmental samples—especially with assays mcyE-Ana and mcyE-Mic. Some of these amplicons are probably primer dimers (Tm < 70 °C), which were observed also in the original study [16]. In order to eliminate false positives, the authors measured fluorescence at a temperature higher than Tm of primer dimers (77 °C). On the other hand, we also observed non-target amplicons with higher Tm (mostly > 80 °C), which indicates amplification of non-target regions. These samples did not show distinctive Tm peaks, but rather multiple smaller peaks, and we confirmed the presence of various non-specific amplicons also by gel electrophoresis. Cq values of such samples were therefore a product of amplification of various templates and could not be considered positive. These false positive signals were excluded from further analyses. The results suggest that the SYBR Green chemistry might not be the most suitable for environmental samples. Specificity and quantification could be improved by using TaqMan chemistry (Roche Molecular Systems Inc., USA) with fluorescent probes or by complementing qPCR results with sequencing of the products.

In some of the assays, amplification efficiency was relatively low (Table 3). While in the original studies amplification efficiency exceeded 90% [6,16,18] for all evaluated assays, in our study that was the case only for assay sxtA. This difference might be caused by sequence variability of uncharacterised cyanobacterial cultures, which is even more significant between different geographical regions of sampling. Possible mismatches in the target regions due to high sequence variability between cyanobacterial strains can lead to low amplification efficiency. Moreover, the effect of the sample impurities and secondary structure of genomic DNA has to be considered as well. Therefore, the LOD and LOQ values, as well as calculated gene copy numbers and cell numbers, might not be fully reliable and should be taken only as a qualitative observation. The reliability of the assay mcyE-Pla might be additionally decreased by a narrow linear dynamic range (Supplementary file S6), which should be addressed in future studies. Moreover, the efficiency of DNA extraction from environmental samples should be evaluated for a more accurate quantification of cells and comparability of results.

For implementation in existing monitoring programs, it is important to quantify cells of the target organisms, not merely gene copies. This might be uncertain in the phylum of cyanobacteria due to unknown number of target gene copies per cell. Genetic cluster *mcy* is thought to appear in only one copy per genome [39,40], while gene *sxtA* appears on average in 3.58 copies per cell in strain *A. circinalis* AWQC131C [6]. Besides, cyanobacteria can contain up to 10 or even more copies of genome per cell [41]; ploidy level differs between species and strains, while it depends also on the growth phase and environmental parameters [42,43]. Therefore, it has to be noted that the calculated cell numbers represent an average for all the genotypes containing target genes from environmental samples, estimated based on assumption that they contain the same number of gene copies as the reference strains. This issue could be avoided if the risk assessment guidelines were adapted for operation with number of gene copies instead of cells.

To enable a thorough cyanotoxin risk assessment in regular monitoring, an assay for detection of anatoxin-a production potential should be designed and optimised. In this study, anatoxin-a was excluded, because literature search did not reveal any appropriate qPCR assays for detection of all anatoxin-producing cyanobacteria. What is more, it would be beneficial to optimise a single assay for detection of *mcyE* genes in all potential microcystin-producers instead of genus-specific assays in order to simplify the test and lower the costs.

4. Conclusions

This is the first study in Slovenia aiming to detect cyanotoxic potential with qPCR, as well as one of the few studies employing this method in environmental biofilm samples. We conclude that the method is appropriate for detection of potentially toxic cyanobacteria in water bodies for the purpose of rapid screening and early warning, which could improve risk assessment and protection of human and ecosystem health. Its advantages are early risk detection, short time of analysis and cost effectiveness, while the main downside of the tested assays is suboptimal specificity in environmental samples as a result of SYBR Green chemistry used.

In particular, we aimed to expand the evaluation of the qPCR method also to understudied benthic cyanobacteria in biofilm samples, with the emphasis on comparison with microscopy and LC-MS/MS. The study demonstrated that in both plankton and biofilm samples there might be a potential for cyanotoxin production which is left undetected by traditional methods. This might be especially important in urban water bodies with regular human and animal visitors. In such water bodies, qPCR could provide additional information if implemented in biomonitoring programs, ensuring appropriate precautions to avoid negative effects of acute and chronic exposure to cyanotoxins.

Furthermore, the study indicated that microscopy as the preferred and often the only method of regular biomonitoring is not sufficient for detecting cyanotoxic potential. Similarly, LC-MS/MS did not detect cyanotoxins in all the samples with observed potential for their production. Implementation of the qPCR method with monitoring strategies could serve for assessing potential toxicity of cyanobacterial blooms or microbial mats and forming recommendations for visitors, as well as in evaluating the efficiency of implemented measures for removal or destruction of cyanobacterial cells or cyanotoxins.

5. Methods and Materials

5.1. Cyanobacterial Cultures and Synthetic DNA Fragments

For specificity testing and cell quantification, reference cyanobacterial cultures were used: microcystin-producing *Anabaena* sp. UHCC 0315 (University of Helsinki, Finland) [16], *Microcystis aeruginosa* PCC 7806 (Pasteur Institute, France) [16] and *Planktothrix agardhii* NIVA-CYA 126 (Norwegian Institute for Water Research, Norway) [17], cylindrospermopsin-producing *Aphanizomenon ovalisporum* ILC-164 (Israel Oceanographic and Limnological Research, Israel) [44] and saxitoxin-producing *Aphanizomenon gracile* NIVA-CYA 851 (Norwegian Institute for Water Research, Norway) [45]. The cultures were grown in standard medium BG11 (Gibco, ThermoFisher Scientific, Waltham, MA, USA) under natural light conditions at room temperature. For DNA extraction (see Section 5.3. DNA Extraction and Quality Control), the cultures were filtered through Sterivex columns (Milipore Sterivex-GP Pressure Filter Unit, Merck KGaA, Germany) with 0.22 μm pore size; the volume filtered was between 5 and 19 mL. We estimated cell concentration using Bürker Türk counting chamber and light microscopy.

For positive controls of qPCR reactions and gene copy quantification, synthetic DNA fragments (gBlocks, Integrated DNA Technologies, Coralville, IA, USA) specific to the selected target regions were used. Their nucleotide sequences were determined based on reference sequences from NCBI GenBank [46]; *Anabaena* sp. 90 (AJ536156.1) for mcyE-Ana, *Microcystis aeruginosa* PCC 7806 (AF183408.1) for mcyE-Mic, *Planktothrix agardhii* 213 (EU151891.1) for mcyE-Pla, *Aphanizomenon* sp. 10E6 (GQ385961.1) for cyrJ, *Anabaena circinalis* AWQC131C (DQ787201.1) for sxtA and *Anabaena circinalis* AWQC131C (AF247589.1) for 16S-cyano (Supplementary file S10).

5.2. Environmental Sampling

In total, 25 plankton and 23 biofilm samples were collected from 7 lakes or reservoirs and 8 rivers or streams in Slovenia in 2019 (Supplementary file S8); one of them (plankton, BL1.4) was included only in cyanotoxin analysis. Plankton samples were collected in lakes with integrating water sampler (Hydro-Bios, IWS III, Germany) 11 times in the pilot area, Lake Bled, and 3–4 times elsewhere. Biofilm samples were collected once in rivers and selected lakes, by brushing biofilm off stones or—if stones were not available—macrophytes, wooden substrate or bricks. Sampling was performed according to national guidelines for monitoring of ecological state of water bodies intercalibrated in the frame of Water Framework Directive implementation. All field equipment used for community DNA samples was treated beforehand with 10% H_2O_2 solution and rinsed with distilled water. The sampling procedure was controlled in the field by using blank controls with Milli-Q water (Merck KGaA).

Each sample was collected in three aliquots for community DNA extraction, microscopic cell count and cyanotoxin analysis. For DNA extraction from plankton samples, 60–1000 mL of water was filtered through Sterivex columns (Milipore Sterivex-GP Pressure Filter Unit, Merck KGaA) with pore size 0.22 μm. The columns were stored on –20 °C for up to 161 days. For biofilm, 10 mL of biofilm-Mili-Q water mixture was mixed with 40 mL of absolute ethanol and stored on 4 °C for up to 147 days. For cyanotoxin analysis, plankton (200–2000 mL) or biofilm (5 or 10 mL) samples were filtered through GF/C filters (Whatman, GE Healthcare, Chicago, IL, USA) with pore size 1.2 μm and dry weight was determined (for biofilm). Filters were stored on –20 °C for up to 14 months before analysed with LC-MS/MS. Cyanobacterial cell count was performed under light microscope with counting chambers (e.g., Hydro-Bios, Germany). For plankton samples, cell numbers were counted, and volumes were calculated for each species (i.e., biovolumes). For filamentous cyanobacterial species, where cell numbers cannot be directly counted, first their biovolumes were calculated based on the measurements of filament length and width, and then this was converted to cell numbers based on species-specific cell biovolumes known from literature or from our previous measurements. For biofilm samples, species abundance was evaluated semi-quantitatively by assigning each species a value of 1–5 based on their abundance (0—not detected, 1—very rare, 2—rare, 3present, 4—frequent, 5—dominant).

5.3. DNA Extraction and Quality Control

DNA extraction was performed using commercially available kits following manufacturers' instructions; for cyanobacterial cultures and plankton samples DNeasy PowerWater Sterivex kit (Qiagen, Germany) and for biofilm samples NucleoSpin Soil kits (Macherey-Nagel, Germany) were used. DNA was stored at –20 °C for further analyses. In each extraction, a blank control using sterile water (B. Braun, Germany) was included. DNA concentration and purity of the samples and blank controls were evaluated using spectrophotometer NanoDrop (Thermo Scientific, ThermoFisher Scientific) with 1.5–2 µL of DNA sample and elution buffers from DNA extraction kits as a background.

5.4. qPCR Setup, Assay Validation and Quantification

After a literature review of previously designed qPCR assays for the detection of potentially toxic cyanobacteria, nine published assays were selected and evaluated based on their performance described in the published papers and in silico and in vitro characterisation in this study. In addition to the specificity evaluation done in the original papers (in silico and in vitro for all primers), in silico specificity check was performed with NCBI Primer-BLAST [47] using nr/nt database. Furthermore, specificity was tested in vitro on five different cyanobacterial cultures (see below) for all assays, and for assay 16S-cyano additionally with selected heterotrophic bacterial strains that could be present in freshwater habitats (*Salmonella enterica*, *Escherichia coli*, *Pseudomonas fluorescens*, *Brevundimonas* sp., *Arcobacter butzleri*) and with selected plant samples to check for amplification of plant chloroplasts (*Solanum lycopersicum*, *Vitis vinifera*, *Alnus glutinosa*, *Clematis* sp.). The final set of assays included five qPCR assays to target cyanotoxin-producing cyanobacteria and an additional assay to target 16S rRNA gene, which served as a DNA extraction and qPCR inhibition control (Table 1). Amplification was performed on qPCR cycler Applied Biosystems 7900HT (ThermoFisher Scientific). First, different primer concentrations (0.3 µM and 0.9 µM) and reaction volumes (10 µL and 20 µL) were tested and optimised for each assay. Final reaction volume was 10 µL, consisting of 5 µL of SYBR Green PCR Master Mix (Applied Biosystems, ThermoFisher Scientific), 0.9 µM (assays mcyE-Ana, mcyE-Pla, cyrJ, sxtA and 16S-cyano) or 0.3 µM (assay mcyE-Mic) of each primer and 2 µL of DNA template in 10^{-1} dilution. Potential for qPCR inhibition was evaluated on up to five selected samples for assays mcyE-Pla, cyrJ and sxtA, by analysing two subsequent dilutions (10^{-1} and 10^{-2}) for each sample. Reactions were performed in 384-well clear PCR plates (Thermo Scientific, ThermoFisher Scientific), covered by MicroAmp™ optical adhesive film (Applied Biosystems, ThermoFisher Scientific). Temperature profile was as follows: 2 min on 50 °C, 10 min on 95 °C, followed by 45 cycles of 15 s on 95 °C and 1 min on 60 °C. Dissociation stage with initial denaturation for 15 s on 95 °C, followed by 15 s on 60 °C and a gradual increase up to 95 °C, was added at the end. Every reaction was performed in three technical replicates. Positive (specific synthetic DNA fragments) and negative controls (nuclease-free water, Sigma-Aldrich, St. Louis, MS, USA) were included in every experiment. qPCR amplification conditions were the same for synthetic DNA and DNA isolated from cultures or environmental samples.

qPCR results were analysed with software SDS (version 2.4.1, Applied Biosystems, ThermoFisher Scientific). Threshold values were set manually for each assay within linear part of exponential curve, allowing for the comparison of Cq values between runs. For each sample, the amplification curve (giving Cq value) and dissociation curve of the amplified product (giving Tm value) was checked. To exclude false positives, results were considered positive if the following two criteria were met: Cq was within detection range and there was a distinctive peak with appropriate Tm. Cut-off values for Cq (values at the highest dilution within detection range) and reference values for Tm (average Tm values of all DNA dilutions within quantification range) were determined based on results from cyanobacterial cultures. Tm values served to control the specificity of amplification, therefore only values in the expected range were considered positive.

Samples that resulted in multiple peaks with one of them showing appropriate Tm were further analysed with agarose gel electrophoresis. qPCR products were run on 2% agarose gel with $1\times$ Tris Acetate-EDTA (TAE) buffer stained with ethidium bromide at 100 V for 90 min. Samples contained 4 µL of each product and 1 µL of $6\times$ Mass Ruler DNA Loading Dye (Thermo Scientific, ThermoFisher Scientific), and GeneRuler 100 bp DNA Ladder (Thermo Scientific, ThermoFisher Scientific) was used to determine the product length. Visualisation of amplicons was performed under UV light using UVP ChemStudio PLUS Imaging System (Analytik Jena, Germany).

Assay sensitivity (LOD, LOQ, linear dynamic range) of the assays was evaluated from the cyanobacterial culture calibration curves. LOD was defined as a number of cells where at least 2/3 technical replicates produced a positive result. LOQ was defined as number of cells in sample that gave Cq value at the lower end of the linear curve and the CV of technical replicates did not exceed 2%. Amplification efficiency (e) was calculated by the equation $e = 10^{-1/S}-1$, where S represents the slope of the linear dynamic range of the calibration curve. Potential for qPCR inhibition was evaluated by comparing differences between average Cq values of two subsequent dilutions for the same sample.

For results with 3/3 positive technical replicates, average Cq value (which was the basis for quantification of cells and gene copies) and absolute difference between Cq values of technical replicates (which served for assessing intra-assay variability) were calculated. Quantification was performed using calibration curves approach. The calibration curves were generated from eight subsequent dilutions of cyanobacterial cultures DNA for quantification of target cells, or of synthetic DNA fragments for quantification of target gene copies. DNA concentration of stock solution of DNA fragments was 10 ng/µL (according to the manufacturer), from which gene copy numbers were calculated based on the following equation:

$$gene\ copy\ number = \frac{DNA\ amount\,[ng] \times 6.022 \times 10^{23}}{DNA\ fragment\ lenght\,[bp] \times 650 \times 10^{9}}$$

Results with calculated values below LOQ were given a value of LOQ/2. Results with 1/3 or 2/3 positive technical replicates were given a value of LOQ/10. These values were used for all following analyses. To ensure comparability of results despite variable volumes of plankton samples filtered prior to DNA extraction and variable densities of biofilm samples, gene copies per microliter of DNA were converted into values per millilitre of water (plankton) or gram of dry weight (biofilm).

Intra-assay variability was tested by using three technical replicates within each experiment. Repeatability (inter-assay variability) of qPCR reactions was tested by repeating the experiment two times with the same template DNA and under the same conditions on up to five selected samples for assays mcyE-Pla, cyrJ and sxtA, and comparing average Cq values of the two technical repetitions. Robustness of the assays was evaluated by testing them on diverse set of samples; cyanobacterial monocultures, synthetic DNA fragments, frozen or lyophilised cyanobacterial bloom samples and environmental samples (plankton, biofilm).

5.5. Cyanotoxin Analysis with LC-MS/MS

Intracellular cyanotoxins were extracted from filters applying the protocol described by Cerasino and Salmaso [48] and quantified with LC-MS/MS. The extraction was carried out by using a mixture of acetonitrile in water (60/40 v/v), containing 0.1% formic acid. Extracted toxins were injected into a LC-MS/MS system, composed of a Waters Acquity UPLC system (Waters, Milford, MA, USA) coupled to a SCIEX 4000 QTRAP mass spectrometer (AB Sciex Pte. Ltd., Singapore). The mass detector was operated in scheduled MRM (Multiple Reaction Monitor) mode, using positive electrospray ionisation (ESI+). Quantification of microcystins was performed following the protocol from Cerasino and Salmaso [48], which has the capability of determining the 11 most common microcystin variants, namely

RR, [D-Asp3]-RR (RRdm), [D-Asp3]-HtyrR (HtyRdm), YR, LR, [D-Asp3]-LR (LRdm), WR, LA, LY, LW, LF. Analysis of cylindrospermopsins and saxitoxins was performed following the protocol from Ballot et al. [49], targeting CYN, STX, dcSTX, NeoSTX, GTX1, GTX4, GTX5, C1 and C2.

5.6. Data Analysis

Calibration curves and their Pearson correlation coefficients were prepared in Microsoft Excel (2007). Correlations and linear regression curves between gene copy numbers and microcystin concentrations, cell numbers and biovolumes were determined with Pearson correlation coefficient, using Prism 6 (GraphPad Inc., San Diego, CA, USA) with 95% confidence interval. For biofilm samples, relative abundance was calculated by summing up the values of individual target species and normalising the summed value to 100%. In order to include negative values (below LOD) in correlation analysis, they were replaced with a minimum detected value for each assay divided by 100 (for microscopy and LC-MS/MS results) or by 10 (for qPCR results).

Supplementary Materials: The following are available online at https://www.mdpi.com/2072-665 1/13/2/133/s1.

Author Contributions: Š.R.R., T.Š., A.K.K. and M.Z. carried out the sampling campaign. Š.R.R., T.Š., A.K.K. and T.E. performed microscopic analyses. L.C. performed LC-MS/MS measurements. P.K., M.Z. and T.E. planned the qPCR experiments with the help of Š.B. M.Z. performed molecular laboratory work and associated data analyses. T.E. performed statistical analyses. P.K. and Š.B. contributed to interpretation of qPCR results. All authors have read and agreed to the published version of the manuscript.

Funding: This research was funded by ARRS research programs P1-0245 and P4-0165, project Interreg Eco-AlpsWater (569) and MORS contract no. 848-23 and annex no. 1/2019.

Data Availability Statement: Majority of the data produced in this study is available in supplementary files. Additional data is available upon request to the corresponding author.

Acknowledgments: We are grateful to Rainer Kurmayer (UIBK) for contributing comments to the manuscript, Assaf Sukenik (IOLR) for providing us with cyanobacterial culture *Aphanizomenon ovalisporum* ILC-164, Zala Kogej (NIB) for technical help with laboratory work, Janja Zabukovnik for providing us with test environmental samples, Nataša Mehle (NIB) for providing us with plant samples, and Valentina Turk (NIB) and Olga Zorman Rojs (UL, Veterinary faculty) for providing us with heterotrophic bacteria samples.

Conflicts of Interest: The authors declare no conflict of interest.

References

1. Sanseverino, I.; António, D.C.; Loos, R.; Lettieri, T. *Cyanotoxins: Methods and Approaches for Their Analysis and Detection*; EUR 28624; Publications Office of the European Union: Luxembourg, 2017. [CrossRef]
2. Pacheco, A.B.F.; Guedes, I.A.; Azevedo, S.M.F.O. Is qPCR a Reliable Indicator of Cyanotoxin Risk in Freshwater? *Toxins* **2016**, *8*, 172. [CrossRef]
3. Spoof, L.; Catherine, A. Tables of Microcystins and Nodularins. In *Handbook of Cyanobacterial Monitoring and Cyanotoxin Analysis*; Meriluoto, J., Spoof, L., Codd, G.A., Eds.; John Wiley and Sons Ltd.: Hoboken, NJ, USA, 2017; pp. 526–537. [CrossRef]
4. Ngwa, F.F.; Madramootoo, C.A.; Jabaji, S. Comparison of Cyanobacterial Microcystin Synthetase (Mcy) E Gene Transcript Levels, Mcy E Gene Copies, and Biomass as Indicators of Microcystin Risk under Laboratory and Field Conditions. *MicrobiologyOpen* **2014**, *3*, 411–425. [CrossRef]
5. Orr, P.T.; Rasmussen, J.P.; Burford, M.A.; Eaglesham, G.K.; Lennox, S.M. Evaluation of Quantitative Real-Time PCR to Characterise Spatial and Temporal Variations in Cyanobacteria, *Cylindrospermopsis raciborskii* (Woloszynska) Seenaya et Subba Raju and Cylindrospermopsin Concentrations in Three Subtropical Australian Reservoirs. *Harmful Algae* **2010**, *9*, 243–254. [CrossRef]
6. Al-Tebrineh, J.; Mihali, T.K.; Pomati, F.; Neilan, B.A. Detection of Saxitoxin-Producing Cyanobacteria and *Anabaena circinalis* in Environmental Water Blooms by Quantitative PCR. *Appl. Environ. Microbiol.* **2010**, *76*, 7836–7842. [CrossRef] [PubMed]
7. Marbun, Y.R.; Yen, H.-K.; Lin, T.-F.; Ramos, Y.; Lin, H.-L.; Michinaka, A. Rapid On-Site Monitoring of Cylindrospermopsin-Producers in Reservoirs Using Quantitative PCR. *Sustain. Environ. Res.* **2012**, *22*, 143–151.
8. Al-Tebrineh, J.; Pearson, L.A.; Yasar, S.A.; Neilan, B.A. A Multiplex qPCR Targeting Hepato- and Neurotoxigenic Cyanobacteria of Global Significance. *Harmful Algae* **2012**, *15*, 19–25. [CrossRef]

9. Gugger, M.; Lenoir, S.; Berger, C.; Ledreux, A.; Druart, J.C.; Humbert, J.F.; Guette, C.; Bernard, C. First Report in a River in France of the Benthic Cyanobacterium *Phormidium favosum* Producing Anatoxin-a Associated with Dog Neurotoxicosis. *Toxicon* **2005**, *45*, 919–928. [CrossRef]

10. Wood, S.A.; Heath, M.W.; Holland, P.T.; Munday, R.; McGregor, G.B.; Ryan, K.G. Identification of a Benthic Microcystin-Producing Filamentous Cyanobacterium (Oscillatoriales) Associated with a Dog Poisoning in New Zealand. *Toxicon* **2010**, *55*, 897–903. [CrossRef]

11. Faassen, E.J.; Harkema, L.; Begeman, L.; Lurling, M. First Report of (Homo)Anatoxin-a and Dog Neurotoxicosis after Ingestion of Benthic Cyanobacteria in The Netherlands. *Toxicon* **2012**, *60*, 378–384. [CrossRef] [PubMed]

12. Fastner, J.; Beulker, C.; Geiser, B.; Hoffmann, A.; Kröger, R.; Teske, K.; Hoppe, J.; Mundhenk, L.; Neurath, H.; Sagebiel, D.; et al. Fatal Neurotoxicosis in Dogs Associated with Tychoplanktic, Anatoxin-a Producing *Tychonema* sp. in Mesotrophic Lake Tegel, Berlin. *Toxins* **2018**, *10*, 60. [CrossRef]

13. Bauer, F.; Fastner, J.; Bartha-Dima, B.; Breuer, W.; Falkenau, A.; Mayer, C.; Raeder, U. Mass Occurrence of Anatoxin-a- and Dihydroanatoxin-a-Producing *Tychonema* sp. in Mesotrophic Reservoir Mandichosee (River Lech, Germany) as a Cause of Neurotoxicosis in Dogs. *Toxins* **2020**, *12*, 726. [CrossRef]

14. Kurmayer, R.; Gumpenberger, M. Diversity of Microcystin Genotypes among Populations of the Filamentous Cyanobacteria *Planktothrix rubescens* and *Planktothrix agardhii*. *Mol. Ecol.* **2006**, *15*, 3849–3861. [CrossRef] [PubMed]

15. Rantala, A.; Fewer, D.P.; Hisbergues, M.; Rouhiainen, L.; Vaitomaa, J.; Börner, T.; Sivonen, K. Phylogenetic Evidence for the Early Evolution of Microcystin Synthesis. *Proc. Natl. Acad. Sci. USA* **2004**, *101*, 568–573. [CrossRef] [PubMed]

16. Vaitomaa, J.; Rantala, A.; Halinen, K.; Rouhiainen, L.; Tallberg, P.; Mokelke, L.; Sivonen, K. Quantitative Real-Time PCR for Determination of Microcystin Synthetase E Copy Numbers for *Microcystis* and *Anabaena* in Lakes. *Appl. Environ. Microbiol.* **2003**, *69*, 7289–7297. [CrossRef]

17. Rantala, A.; Rajaniemi-Wacklin, P.; Lyra, C.; Lepistö, L.; Rintala, J.; Mankiewicz-Boczek, J.; Sivonen, K. Detection of Microcystin-Producing Cyanobacteria in Finnish Lakes with Genus-Specific Microcystin Synthetase Gene E (McyE) PCR and Associations with Environmental Factors. *Appl. Environ. Microbiol.* **2006**, *72*, 6101–6110. [CrossRef] [PubMed]

18. Campo, E.; Lezcano, M.-Á.; Agha, R.; Cirés, S.; Quesada, A.; El-Shehawy, R. First TaqMan Assay to Identify and Quantify the Cylindrospermopsin-Producing Cyanobacterium *Aphanizomenon ovalisporum* in Water. *Adv. Microbiol.* **2013**, *3*, 430–437. [CrossRef]

19. Rantala, A.; Rizzi, E.; Castiglioni, B.; De Bellis, G.; Sivonen, K. Identification of Hepatotoxin-Producing Cyanobacteria by DNA-Chip. *Environ. Microbiol.* **2008**, *10*, 653–664. [CrossRef]

20. Al-Tebrineh, J.; Gehringer, M.M.; Akcaalan, R.; Neilan, B.A. A New Quantitative PCR Assay for the Detection of Hepatotoxigenic Cyanobacteria. *Toxicon* **2011**, *57*, 546–554. [CrossRef]

21. Sabart, M.; Crenn, K.; Perrière, F.; Abila, A.; Leremboure, M.; Colombet, J.; Jousse, C.; Latour, D. Co-Occurrence of Microcystin and Anatoxin-a in the Freshwater Lake Aydat (France): Analytical and Molecular Approaches during a Three-Year Survey. *Harmful Algae* **2015**, *48*, 12–20. [CrossRef] [PubMed]

22. Rantala-Ylinen, A.; Känä, S.; Wang, H.; Rouhiainen, L.; Wahlsten, M.; Rizzi, E.; Berg, K.; Gugger, M.; Sivonen, K. Anatoxin-a Synthetase Gene Cluster of the Cyanobacterium Anabaena sp. Strain 37 and Molecular Methods to Detect Potential Producers. *Appl. Environ. Microbiol.* **2011**, *77*, 7271–7278. [CrossRef]

23. Rački, N.; Dreo, T.; Gutierrez-Aguirre, I.; Blejec, A.; Ravnikar, M. Reverse Transcriptase Droplet Digital PCR Shows High Resilience to PCR Inhibitors from Plant, Soil and Water Samples. *Plant. Methods* **2014**, *10*, 1–10. [CrossRef] [PubMed]

24. Mez, K.; Hanselmann, K.; Beattie, K.; Codd, G.; Hauser, B.; Naegeli, H.; Preisig, H. Identification of a Microcystin in Benthic Cyanobacteria Linked to Cattle Deaths on Alpine Pastures in Switzerland. *Eur. J. Phycol.* **1997**, *32*, 111–117. [CrossRef]

25. Seifert, M.; McGregor, G.; Eaglesham, G.; Wickramasinghe, W.; Shaw, G. First Evidence for the Production of Cylindrospermopsin and Deoxy-Cylindrospermopsin by the Freshwater Benthic Cyanobacterium, *Lyngbya wollei* (Farlow Ex Gomont) Speziale and Dyck. *Harmful Algae* **2007**, *6*, 73–80. [CrossRef]

26. Mazmouz, R.; Chapuis-Hugon, F.; Mann, S.; Pichon, V.; Méjean, A.; Ploux, O. Biosynthesis of Cylindrospermopsin and 7-Epicylindrospermopsin in *Oscillatoria* sp. strain PCC 6506: Identification of the Cyr Gene Cluster and Toxin Analysis. *Appl. Environ. Microbiol.* **2010**, *76*, 4943–4949. [CrossRef] [PubMed]

27. Gaget, V.; Humpage, A.R.; Huang, Q.; Monis, P.; Brookes, J.D. Benthic Cyanobacteria: A Source of Cylindrospermopsin and Microcystin in Australian Drinking Water Reservoirs. *Water Res.* **2017**, *124*, 454–464. [CrossRef]

28. Smith, F.M.J.; Wood, S.A.; van Ginkel, R.; Broady, P.A.; Gaw, S. First Report of Saxitoxin Production by a Species of the Freshwater Benthic Cyanobacterium, *Scytonema* Agardh. *Toxicon* **2011**, *57*, 566–573. [CrossRef]

29. Lajeunesse, A.; Segura, P.A.; Gélinas, M.; Hudon, C.; Thomas, K.; Quilliam, M.A.; Gagnon, C. Detection and Confirmation of Saxitoxin Analogues in Freshwater Benthic *Lyngbya wollei* Algae Collected in the St. Lawrence River (Canada) by Liquid Chromatography-Tandem Mass Spectrometry. *J. Chromatogr. A* **2012**, *1219*, 93–103. [CrossRef]

30. Belykh, O.I.; Tikhonova, I.V.; Kuzmin, A.V.; Sorokovikova, E.G.; Fedorova, G.A.; Khanaev, I.V.; Sherbakova, T.A.; Timoshkin, O.A. First Detection of Benthic Cyanobacteria in Lake Baikal Producing Paralytic Shellfish Toxins. *Toxicon* **2016**, *121*, 36–40. [CrossRef]

31. Otten, T.G.; Xu, H.; Qin, B.; Zhu, G.; Paerl, H.W. Spatiotemporal Patterns and Ecophysiology of Toxigenic Microcystis Blooms in Lake Taihu, China: Implications for Water Quality Management. *Environ. Sci. Technol.* **2012**, *46*, 3480–3488. [CrossRef]

32. Ouahid, Y.; Pérez-Silva, G.; Del Campo, F.F. Identification of Potentially Toxic Environmental Microcystis by Individual and Multiple PCR Amplification of Specific Microcystin Synthetase Gene Regions. *Environ. Toxicol.* **2005**, *20*, 235–242. [CrossRef]

33. Christiansen, G.; Molitor, C.; Philmus, B.; Kurmayer, R. Nontoxic Strains of Cyanobacteria Are the Result of Major Gene Deletion Events Induced by a Transposable Element. *Mol. Biol. Evol.* **2008**, *25*, 1695–1704. [CrossRef] [PubMed]
34. Ostermaier, V.; Kurmayer, R. Application of Real-Time PCR to Estimate Toxin Production by the Cyanobacterium *Planktothrix* sp. *Appl. Environ. Microbiol.* **2010**, *76*, 3495–3502. [CrossRef]
35. Al-Tebrineh, J.; Merrick, C.; Ryan, D.; Humpage, A.; Bowling, L.; Neilan, B.A. Community Composition, Toxigenicity, and Environmental Conditions during a Cyanobacterial Bloom Occurring along 1100 Kilometers of the Murray River. *Appl. Environ. Microbiol.* **2012**, *78*, 263–272. [CrossRef]
36. Wood, S.A.; Rueckert, A.; Hamilton, D.P.; Cary, S.C.; Dietrich, D.R. Switching Toxin Production on and off: Intermittent Microcystin Synthesis in a *Microcystis* Bloom. *Environ. Microbiol. Rep.* **2011**, *3*, 118–124. [CrossRef] [PubMed]
37. El Semary, N.A. Investigating Factors Affecting Growth and Cellular McyB Transcripts of *Microcystis aeruginosa* PCC 7806 Using Real-Time PCR. *Ann. Microbiol.* **2010**, *60*, 181–188. [CrossRef]
38. Pimentel, J.S.M.; Giani, A. Microcystin Production and Regulation under Nutrient Stress Conditions in Toxic *Microcystis* Strains. *Appl. Environ. Microbiol.* **2014**, *80*, 5836–5843. [CrossRef] [PubMed]
39. Kaebernick, M.; Dittmann, E.; Börner, T.; Neilan, B.A. Multiple Alternate Transcripts Direct the Biosynthesis of Microcystin, a Cyanobacterial Nonribosomal Peptide. *Appl. Environ. Microbiol.* **2002**, *68*, 449–455. [CrossRef] [PubMed]
40. Tanabe, Y.; Kaya, K.; Watanabe, M.M. Evidence for Recombination in the Microcystin Synthetase (Mcy) Genes of Toxic Cyanobacteria *Microcystis* spp. *J. Mol. Evol.* **2004**, *58*, 633–641. [CrossRef]
41. Labarre, J.; Chauvat, F.; Thuriaux, P. Insertional Mutagenesis by Random Cloning of Antibiotic Resistance Genes into the Genome of the Cyanobacterium *Synechocystis* Strain PCC 6803. *J. Bacteriol.* **1989**, *171*, 3449–3457. [CrossRef]
42. Zerulla, K.; Ludt, K.; Soppa, J. The Ploidy Level of *Synechocystis* sp. PCC 6803 is Highly Variable and is Influenced by Growth Phase and by Chemical and Physical External Parameters. *Microbiology* **2016**, *162*, 730–739. [CrossRef]
43. Becker, S.; Fahrbach, M.; Bo, P.; Ernst, A. Quantitative Tracing, by Taq Nuclease Assays, of a *Synechococcus* Ecotype in a Highly Diversified Natural Population. *Appl. Environ. Microbiol.* **2002**, *68*, 4486–4494. [CrossRef] [PubMed]
44. Jiang, Y.; Xiao, P.; Yu, G.; Shao, J.; Liu, D.; Azevedo, S.M.F.O.; Li, R. Sporadic Distribution and Distinctive Variations of Cylindrospermopsin Genes in Cyanobacterial Strains and Environmental Samples from Chinese Freshwater Bodies. *Appl. Environ. Microbiol.* **2014**, *80*, 5219–5230. [CrossRef] [PubMed]
45. Ballot, A.; Cerasino, L.; Hostyeva, V.; Cirés, S. Variability in the Sxt Gene Clusters of PSP Toxin Producing *Aphanizomenon gracile* Strains from Norway, Spain, Germany and North America. *PLoS ONE* **2016**, *11*. [CrossRef]
46. Clark, K.; Karsch-Mizrachi, I.; Lipman, D.J.; Ostell, J.; Sayers, E.W. GenBank. *Nucleic Acids Res.* **2016**, *44*, 67–72. [CrossRef] [PubMed]
47. Madden, T. The BLAST Sequence Analysis Tool. In *The NCBI Handbook*, 2nd ed.; National Center for Biotechnology Information: Bethesda, MD, USA, 2013. Available online: https://www.ncbi.nlm.nih.gov/books/NBK153387/ (accessed on 23 June 2020).
48. Cerasino, L.; Salmaso, N. Co-Occurrence of Anatoxin-a and Microcystins in Lake Garda and Other Deep Perialpine Lakes. *Adv. Oceanogr. Limnol.* **2020**, *11*, 11–21. [CrossRef]
49. Ballot, A.; Swe, T.; Mjelde, M.; Cerasino, L.; Hostyeva, V.; Miles, C.O. Cylindrospermopsin- And Deoxycylindrospermopsin-Producing Raphidiopsis Raciborskii and Microcystin-Producing Microcystis Spp- And Meiktila Lake, Myanmar. *Toxins* **2020**, *12*, 232. [CrossRef] [PubMed]

Article

Roles of Nutrient Limitation on Western Lake Erie CyanoHAB Toxin Production

Malcolm A. Barnard [1,*], Justin D. Chaffin [2], Haley E. Plaas [1], Gregory L. Boyer [3], Bofan Wei [3], Steven W. Wilhelm [4], Karen L. Rossignol [1], Jeremy S. Braddy [1], George S. Bullerjahn [5], Thomas B. Bridgeman [6], Timothy W. Davis [5], Jin Wei [7], Minsheng Bu [7] and Hans W. Paerl [1,*]

[1] Institute of Marine Sciences, University of North Carolina at Chapel Hill, Morehead City, NC 28557, USA; hplaas@live.unc.edu (H.E.P.); krossign@email.unc.edu (K.L.R.); jbraddy@email.unc.edu (J.S.B.)
[2] Stone Laboratory and Ohio Sea Grant, The Ohio State University, Put-In-Bay, OH 43456, USA; chaffin.46@osu.edu
[3] Department of Chemistry, State University of New York College of Environmental Science and Forestry Campus, Syracuse, NY 13210, USA; glboyer@esf.edu (G.L.B.); bwei101@syr.edu (B.W.)
[4] Department of Microbiology, University of Tennessee at Knoxville, Knoxville, TN 37996, USA; wilhelm@utk.edu
[5] Department of Biological Sciences, Bowling Green State University, Bowling Green, OH 43403, USA; bullerj@bgsu.edu (G.S.B.); timdavi@bgsu.edu (T.W.D.)
[6] Lake Erie Center, University of Toledo, Oregon, OH 43616, USA; Thomas.Bridgeman@utoledo.edu
[7] Key Laboratory of Integrated Regulation and Resource Development on Shallow Lakes, Ministry of Education, College of Environment, Hohai University, Nanjing 210098, China; weijin@hhu.edu.cn (J.W.); hjxyym@hhu.edu.cn (M.B.)
* Correspondence: malcolm.barnard@unc.edu (M.A.B.); hans_paerl@unc.edu (H.W.P.); Tel.: +1-252-726-6841 (ext. 254) (M.A.B.); +1-252-726-6841 (ext. 133) (H.W.P.); Fax: +1-252-726-2426 (M.A.B. & H.W.P.)

Citation: Barnard, M.A.; Chaffin, J.D.; Plaas, H.E.; Boyer, G.L.; Wei, B.; Wilhelm, S.W.; Rossignol, K.L.; Braddy, J.S.; Bullerjahn, G.S.; Bridgeman, T.B.; et al. Roles of Nutrient Limitation on Western Lake Erie CyanoHAB Toxin Production. *Toxins* 2021, 13, 47. https://doi.org/10.3390/toxins13010047

Received: 10 December 2020
Accepted: 6 January 2021
Published: 9 January 2021

Publisher's Note: MDPI stays neutral with regard to jurisdictional claims in published maps and institutional affiliations.

Abstract: Cyanobacterial harmful algal bloom (CyanoHAB) proliferation is a global problem impacting ecosystem and human health. Western Lake Erie (WLE) typically endures two highly toxic CyanoHABs during summer: a *Microcystis* spp. bloom in Maumee Bay that extends throughout the western basin, and a *Planktothrix* spp. bloom in Sandusky Bay. Recently, the USA and Canada agreed to a 40% phosphorus (P) load reduction to lessen the severity of the WLE blooms. To investigate phosphorus and nitrogen (N) limitation of biomass and toxin production in WLE CyanoHABs, we conducted in situ nutrient addition and 40% dilution microcosm bioassays in June and August 2019. During the June Sandusky Bay bloom, biomass production as well as hepatotoxic microcystin and neurotoxic anatoxin production were N and P co-limited with microcystin production becoming nutrient deplete under 40% dilution. During August, the Maumee Bay bloom produced microcystin under nutrient repletion with slight induced P limitation under 40% dilution, and the Sandusky Bay bloom produced anatoxin under N limitation in both dilution treatments. The results demonstrate the importance of nutrient limitation effects on microcystin and anatoxin production. To properly combat cyanotoxin and cyanobacterial biomass production in WLE, both N and P reduction efforts should be implemented in its watershed.

Keywords: cyanotoxins; Maumee Bay; Sandusky Bay; *Microcystis*; *Planktothrix*; microcystin; anatoxin

Key Contribution: Nutrient limitation of cyanobacterial harmful algal blooms (CyanoHABs) was investigated with respect to the production of the cyanotoxins microcystin and anatoxin in Maumee Bay and Sandusky Bay in Western Lake Erie. This is one of the first studies investigating nutrient limitation effects on anatoxin production in Lake Erie and one of the first studies to evaluate the effects of nutrient reduction on Western Lake Erie CyanoHABs using nutrient dilution assays. To reduce CyanoHABs and their toxicity, both N and P reductions are needed in the Western Lake Erie watershed.

1. Introduction

Freshwater ecosystems are critical for sustaining life and supporting civilizations throughout history [1]. As the global human population grows, increased urbanization, agricultural and industrial productions, combined with insufficient wastewater treatment practices, have led to a widespread increase in nutrient pollution of these ecosystems, threatening clean and safe water supplies [2]. Excessive inputs of nitrogen (N) and phosphorus (P) have accelerated eutrophication, the process of increasing organic enrichment, which is largely attributable to increased microalgal and aquatic macrophyte growth [3]. The major detrimental impacts of eutrophication include harmful algal bloom (HAB) formation, decreased water transparency (increased turbidity), O_2 depletion, and reduced biodiversity [3,4]. HAB formation has been a major water quality issue in the U.S. since the 1960s, as noted in a 1965 White House Report indicating HABs as a major source of environmental degradation [5]. Furthermore, nutrient-driven eutrophication of lakes and rivers is one the most significant causes of water quality decline globally [3,6–8]. In particular, there are growing concerns about the proliferation and diversification of N- and P-based fertilizers, as they are potent stimulants of aquatic primary production along the freshwater to marine continuum [9,10]. Additionally, climate change (e.g., warming and changing precipitation patterns) is increasing the likelihood of more expansive blooms, exposing human and animal populations (e.g., pets, wildlife, cattle, fish, birds) to waterborne and aerosolized toxins [7,11–14]. Despite CyanoHAB toxicity being a major human and ecosystem health hazard, the causes and controls of underlying toxicity mechanisms remain poorly understood [15].

Blooms of cyanobacteria in Lake Erie, largely dominated by filamentous heterocystous (N_2-fixing) forms (*Anabaena/Dolichospermum, Aphanizomenon*), were common in the late 1950s through to the 1970s. These blooms dissipated following the signing in of the *Great Lakes Water Quality Agreement* of 1972, which was updated in 2012. However, the blooms returned as non-N_2 fixing *Microcystis* blooms in the early 2000s, which have continued and perhaps worsened [16,17], leading to major environmental degradation and increased human health risks [7]. In August 2014, a toxic *Microcystis* spp. bloom in Western Lake Erie (WLE) created a water crisis, forcing public water supplies to be shut down for over 400,000 people in Toledo, OH, USA [7,18]. Nutrient runoff from agricultural nonpoint sources has been a major factor promoting CyanoHABs in WLE [7]. Primary production in Maumee Bay of Lake Erie (largely dominated by *Microcystis* spp. in the summer) shifts from P-limitation to N-limitation with spatial nutrient limitation heterogeneity with N- and P-limitation occurring several km apart [19–21]. Prior studies revealed that during the summer months, N was often depleted in embayments such as Sandusky and Maumee Bay [22–26], where summertime molar N:P ratios for Sandusky Bay remained below the canonical Redfield ratio (16:1) [26–28]. This suggests the presence of strong N sinks, mediated by denitrification and/or active N cycling and N uptake by high amounts of algal biomass [28–30]. The primary fertilizers used in the agriculturally dominated drainage basin of Lake Erie are inorganic fertilizers (ammonium nitrate, urea, and phosphate) and manure, which has low N:P ratios (~5:1), is about 20% [31–34]. There is an urgent need to determine the linkage between different bioreactive forms of N and P and the promotion of toxic CyanoHABs, to establish the necessary reduction in these nutrient forms to ensure the security of surface potable water. Nutrient reduction will likely need to be even greater as climate change increases the N and P reduction thresholds required for CyanoHAB mitigation [35,36]. The *in situ* bioassay-based study reported here is among the first to use an experimental approach to investigate the response of a natural CyanoHAB community dominated by either *Microcystis* (Maumee Bay) or *Planktothrix* (Sandusky Bay) to actual reductions in N, P or both, under natural conditions in Lake Erie. Satellite and field images of the 2019 WLE blooms can be seen in Figure 1.

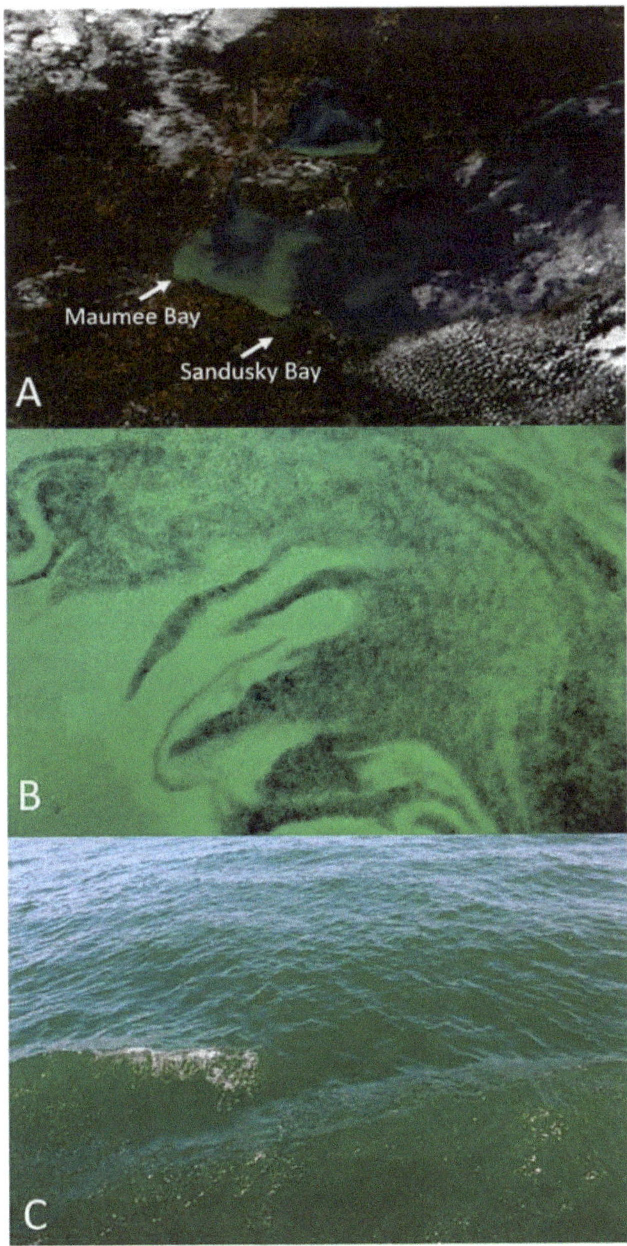

Figure 1. Images of the 2019 WLE CyanoHABs. (**A**) Satellite imagery from the NASA Terra satellite of the WLE CyanoHAB on 19 August 2019 as provided by NOAA MODIS [37]; (**B**) Maumee Bay *Microcystis*-dominated cyanobacterial harmful algal bloom (CyanoHAB) on 4 August 2019 during sampling for the August 2019 bioassays. Photo by H. Plaas; (**C**) Sandusky Bay *Planktothrix*-dominated bloom on 4 August 2019 during sampling for the August 2019 bioassays. Photo by H. Plaas.

A recent review suggested that management efforts to reduce P pollution without controlling N inputs have caused nutrient imbalances in eutrophic systems, which may favor toxic CyanoHABs incapable of fixing atmospheric N_2 gas, i.e., requiring combined N sources [24]. Prior to P load reductions in the 1970s, CyanoHABs in Lake Erie were mostly the N_2-fixers *Aphanizomenon* and *Dolichospermum*, formerly *Anabaena* [38]; now, CyanoHABs are primarily non-N_2-fixing *Microcystis* and *Planktothrix* [16]. In WLE, molecular analysis of the *Microcystis* community indicates a shift from toxic to non-toxic strains that correlates with NO_3 availability [39], although there appears to be a temporal disconnect as a multiyear analysis found no correlation between the proportion of microcystin-producing genotypes of *Microcystis* and the concentration of microcystin [40]. Recent work has strengthened links between N availability, dominant strain shifts, and toxicity by showing seasonal trends in these patterns [24]. The inability of these cyanobacteria to fix atmospheric N_2, and their strong affinity for reduced N forms (e.g., NH_4 and urea), suggests that N delivered through agricultural runoff and internal N recycling mechanisms play critical roles in modulating total phytoplankton biomass, CyanoHAB community composition, and toxicity [39,41].

The prominent cyanotoxins, microcystin and anatoxin, have molecular structures containing N, suggesting that their syntheses may be linked to N availability; hence, there is a need to investigate the potential roles N fertilizers (i.e., NH_4, NO_3, and urea) play in bloom dynamics and toxin production in Lake Erie [23,36,42]. A recent study showed that there are N concentration reduction thresholds at which bloom microcystin levels will decrease, leading to further evidence that N limitation may play a role in controlling cyanotoxin production in the WLE blooms [41]. Due to the shift to non-N_2-fixing CyanoHABs, a major unknown concerning this shift in nutrient limitation is how specific microcystin and anatoxin production potentials are linked to nutrient input reductions.

The US Environmental Protection Agency (EPA) and Environment and Climate Change Canada have recommended a 40% reduction in springtime P loading into WLE to help control the blooms [43–46]. The 40% P load reduction was the result of a multiple modeling exercise included in the Great Lakes Water Quality Agreement between the US and Canada [47]. As both N and P have been shown to influence the WLE CyanoHABs, it is crucial to investigate the effects of both 40% reductions in both N and P in addition to the investigations of the effects of N and P addition. Here, we addressed the following questions: (1) how do nutrients influence WLE microcystin and anatoxin production? (2) Do the same nutrients limit toxin production and CyanoHAB biomass? (3) Will the 40% P reduction as recommended by the US EPA be effective in reducing CyanoHAB microcystin and anatoxin and biomass production in WLE? (4) Is P reduction alone enough to decrease WLE CyanoHAB biomass and microcystin and anatoxin production, or is a combined N and P reduction strategy needed? Given the relatively high content of N in the cyanotoxins microcystin and anatoxin, we predicted that cyanotoxin production is N-limited and that excessive N inputs promote toxicity of these non-N_2-fixing CyanoHABs.

2. Results

2.1. June 2019 Experiement

June 2019 bioassay experiments were characterized by a late spring diatom bloom shortly before the onset of a summer *Microcystis* bloom in Maumee Bay and the very early *Planktothrix* bloom in Sandusky Bay. In both Maumee and Sandusky Bays, there were high N concentrations—over 200 μmol L^{-1} nitrate plus nitrite in Maumee Bay and over 100 μmol L^{-1} nitrate plus nitrite in Sandusky Bay and relatively low P concentrations of 1–2 μmol L^{-1} dissolved reactive phosphorus (DRP) (Table 1).

Table 1. Initial nutrient concentrations in the June 2019 bioassay water collected from control Cubitainers. All data are $n = 3$.

Nutrient Parameter	Maumee Bay		Sandusky Bay	
	No Dilution	**40% Dilution**	**No Dilution**	**40% Dilution**
$NO_3 + NO_2$ ($\mu mol\ L^{-1}$)	223.67 ± 25.43	137.64 ± 35.00	101.45 ± 5.95	58.46 ± 8.46
NH_4 ($\mu mol\ L^{-1}$)	1.34 ± 1.01	3.67 ± 0.60	24.28 ± 0.66	17.14 ± 0.85
DRP ($\mu mol\ L^{-1}$)	2.24 ± 0.23	1.50 ± 0.07	1.20 ± 0.14	0.85 ± 0.05
Silicate ($\mu mol\ L^{-1}$)	139.14 ± 12.23	100.46 ± 2.53	130.42 ± 19.50	78.88 ± 16.20

In the June Maumee Bay experiment, growth rates significantly differed ($p < 0.001$) among nutrient treatments, but there was no difference between the undiluted and diluted treatments ($p = 0.76$). The +P and +P&N treatments resulted in a higher growth rate than the control and +N treatments, indicating P-limited growth, in both the undiluted and 40% dilution treatments, likely due to the high concentrations of N in the bay (Figure 2; Table 1, Tables S1 and S2). Cyanotoxins were not detected in the June Maumee Bay experiment.

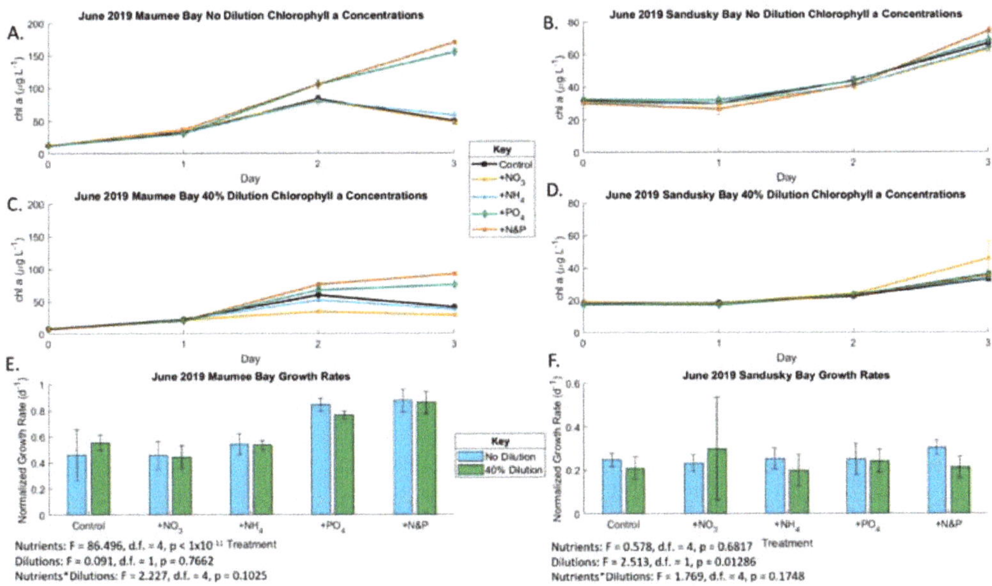

Figure 2. Growth rates of phytoplankton, as determined by chlorophyll *a* accumulation during the course of incubation in the June 2019 bioassays. (**A**) Undiluted Maumee Bay water (also see Table S1); (**B**) undiluted Sandusky Bay water (also see Table S1); (**C**) 40% dilution Maumee Bay water (also see Table S1); (**D**) 40% dilution Sandusky Bay water (also see Table S1); (**E**) Maumee Bay growth rates under the various nutrient addition treatments at the two locations of T3 compared to T0 (also see Table S2). Error bars are standard error; (**F**) Maumee Bay growth rates under the various nutrient addition treatments at the two locations of T3 compared to T0 (also see Table S2). Error bars are standard error. Significances between treatments for (**E,F**) are from two-factor ANOVAs.

In the June Sandusky Bay experiment, nutrient enrichment did not impact growth rates ($p = 0.68$), but growth rates were lower in the 40% diluted treatments ($p = 0.013$; Figure 2; Tables S1 and S2), which indicates nutrient-replete conditions. The initial undiluted total microcystin concentration was 0.136 µg/L and total anatoxin concentration was 0.053 µg/L (Tables S3 and S5). Microcystin concentrations increased throughout the experiment in the no dilution treatments but not the 40% dilution (Figure 3). The Sandusky Bay microcystin production rate was slightly yet not significantly affected by nutrient enrichment ($p = 0.067$), becoming significant in the biomass-normalized analyses ($p < 0.001$). However, the production rate was lower in the 40% diluted treatments (Figure 3 and Tables S3—S6). The June 2019 Sandusky Bay anatoxin production rate was not affected by dilution treatment with a slight nutrient effect in the biomass-normalized analyses ($p < 0.01$) (Figure 4).

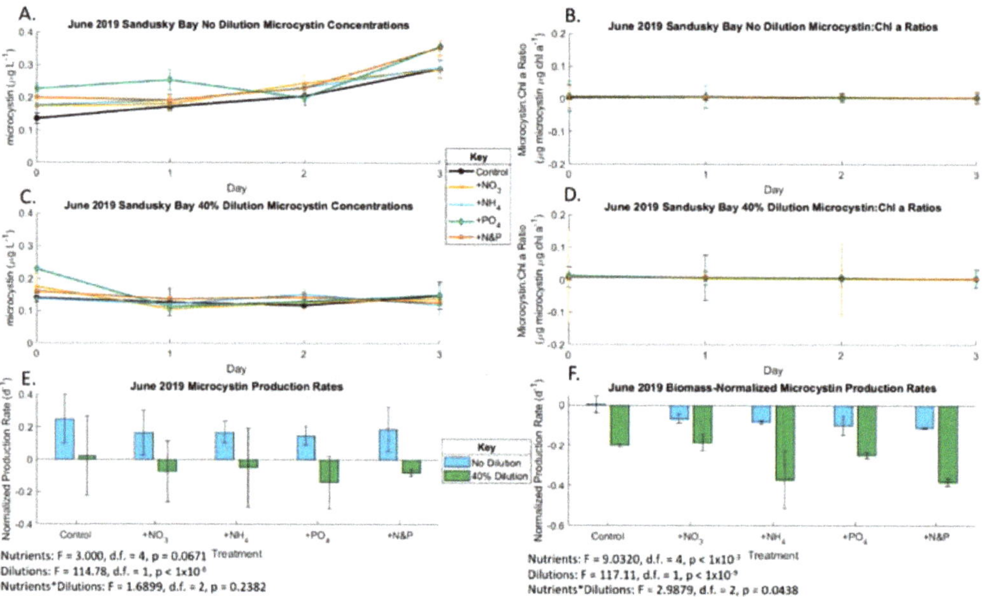

Figure 3. Production rates of microcystin during the June 2019 bioassays. Only Sandusky Bay produced microcystin in June. (**A**) Undiluted Sandusky Bay microcystin concentrations (also see Table S3); (**B**) undiluted Sandusky Bay biomass-normalized microcystin concentrations (also see Table S4); (**C**) 40% dilution Sandusky Bay microcystin concentrations (also see Table S3); (**D**) 40% dilution Sandusky Bay biomass-normalized microcystin concentrations (also see Table S4); (**E**) Maumee Bay microcystin production rates under the various nutrient addition treatments at the two locations of T3 compared to T0 (also see Table S5); (**F**) Maumee Bay biomass-normalized microcystin production rates under the various nutrient addition treatments at the two locations of T3 compared to T0 (also see Table S6). Error bars are standard error. Significance for (**E**,**F**) is from *n*-factor ANOVA analysis due to unbalanced data sets.

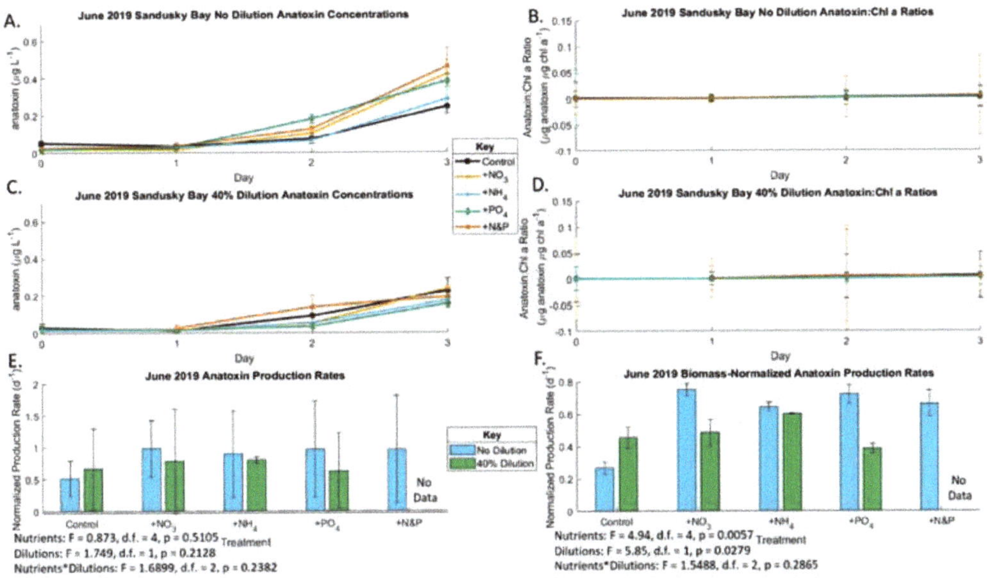

Figure 4. Chlorophyll *a*-based production rates of anatoxin during the June 2019 bioassays. Only Sandusky Bay produced anatoxin in June. (**A**) Undiluted Sandusky Bay anatoxin concentrations (also see Table S7); (**B**) undiluted Sandusky Bay biomass-normalized anatoxin concentrations (also see Table S8); (**C**) 40% dilution Sandusky Bay anatoxin concentrations (also see Table S7); (**D**) 40% dilution Sandusky Bay biomass-normalized anatoxin concentrations (also see Table S8); (**E**) Maumee Bay anatoxin production rates under the various nutrient addition treatments at the two locations of T3 compared to T0 (also see Table S9); (**F**) Maumee Bay biomass-normalized anatoxin production rates under the various nutrient addition treatments at the two locations of T3 compared to T0 (also see Table S10). Error bars are standard error. Significance for (**E,F**) is from *n*-factor ANOVA analysis due to unbalanced data sets.

2.2. August 2019 Experiment

August experiments were characterized by dense blooms of *Microcystis* in Maumee Bay and *Planktothrix* in Sandusky Bay. Maumee Bay had high N concentrations—over 100 μmol L^{-1} nitrate plus nitrite—but Sandusky Bay had low N concentrations with 6.5 μmol L^{-1} nitrate plus nitrite (Table 2). Both Maumee and Sandusky Bay had low P concentrations of 0.03 to 0.20 μmol L^{-1} DRP (Table 2).

Table 2. Initial concentrations of nutrients in the August 2019 bioassay water taken from T0 control Cubitainers. All data are *n* = 3.

Nutrient Parameter	Maumee Bay		Sandusky Bay	
	No Dilution	40% Dilution	No Dilution	40% Dilution
$NO_3 + NO_2$ (μmol L^{-1})	127.12 $\pm$ 10.82	60.10 $\pm$ 12.94	6.59 $\pm$ 0.29	6.61 $\pm$ 0.05
NH_4 (μmol L^{-1})	0.70 $\pm$ 0.42	1.74 $\pm$ 1.58	1.05 $\pm$ 0.69	1.04 $\pm$ 0.06
Urea (μmol L^{-1})	3.45 $\pm$ 0.61	1.59 $\pm$ 1.15	2.91 $\pm$ 1.46	3.99 $\pm$ 1.09
DRP (μmol L^{-1})	0.20 $\pm$ 0.20	0.05 $\pm$ 0.01	0.03 $\pm$ 0.01	0.10 $\pm$ 0.07
Silicate (μmol L^{-1})	124.33 $\pm$ 7.13	95.44 $\pm$ 6.24	51.12 $\pm$ 12.23	64.91 $\pm$ 18.52

Chlorophyll *a* concentrations decreased throughout incubation of the very dense bloom in the August Maumee Bay experiment, coinciding with negative growth rates (Figure 5). The diluted treatments had reduced algal mortality compared to the undiluted treatments likely due to lower initial starting biomass ($p < 0.001$). The growth rate was significantly affected by nutrients ($p < 0.001$) and the interaction between nutrients and dilution ($p = 0.012$), but there was no discernable pattern, leading to a lack of ecological significance. The initial undiluted total microcystin concentration in the August Maumee Bay experiment was 18.06 µg/L. Microcystin concentration and production rates (Figure 6) followed a similar pattern to chlorophyll with a non-significant nutrient effect ($p = 0.14$) and significant dilution effect without biomass-normalization ($p < 0.001$) and a non-significant effect in biomass-normalized analysis ($p = 0.5452$). Anatoxin was not detected in the Maumee Bay August experiment.

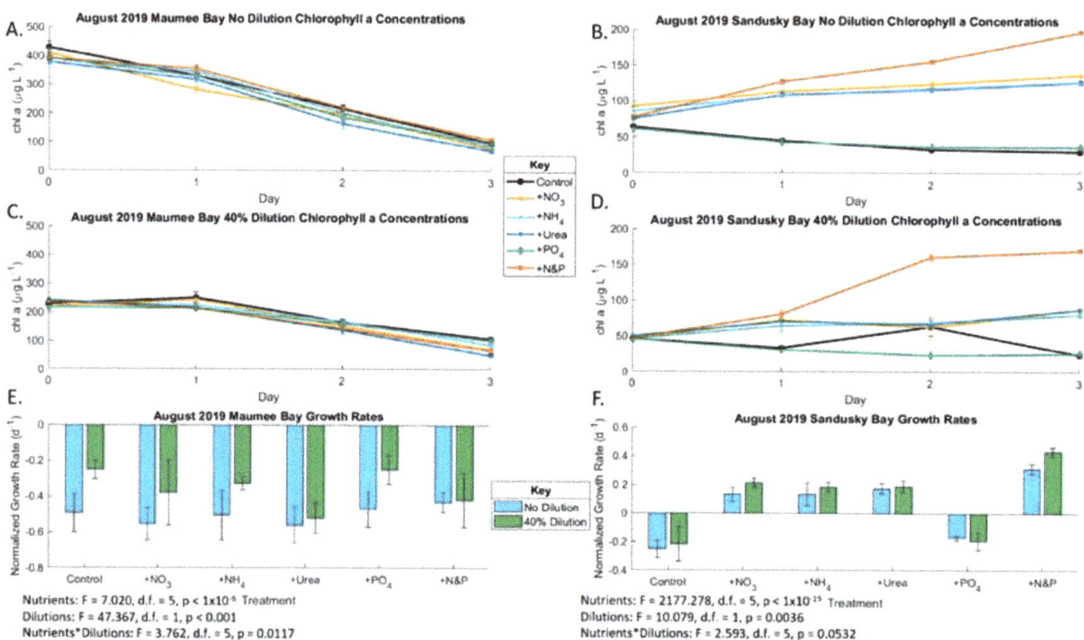

Figure 5. Growth rates of phytoplankton in the August 2019 bioassays. (**A**) Undiluted Maumee Bay Chlorophyll *a* (also see Table S1); (**B**) undiluted Sandusky Bay Chlorophyll *a* (also see Table S1); (**C**) undiluted Maumee Bay Chlorophyll *a* (also see Table S1); (**D**) 40% dilution Sandusky Bay Chlorophyll *a* (also see Table S1); (**E**) Maumee Bay growth rates under the various nutrient addition treatments at the two locations at T3 compared to T0 (also see Table S2). Error bars are standard error. Significance for (**E**) is from two-factor ANOVA analysis; (**F**) Maumee Bay growth rates under the various nutrient addition treatments at the two locations at T3 compared to T0 (Table S2). Error bars are standard error. Significance for (**F**) is from *n*-factor ANOVA analysis due to unbalanced data sets.

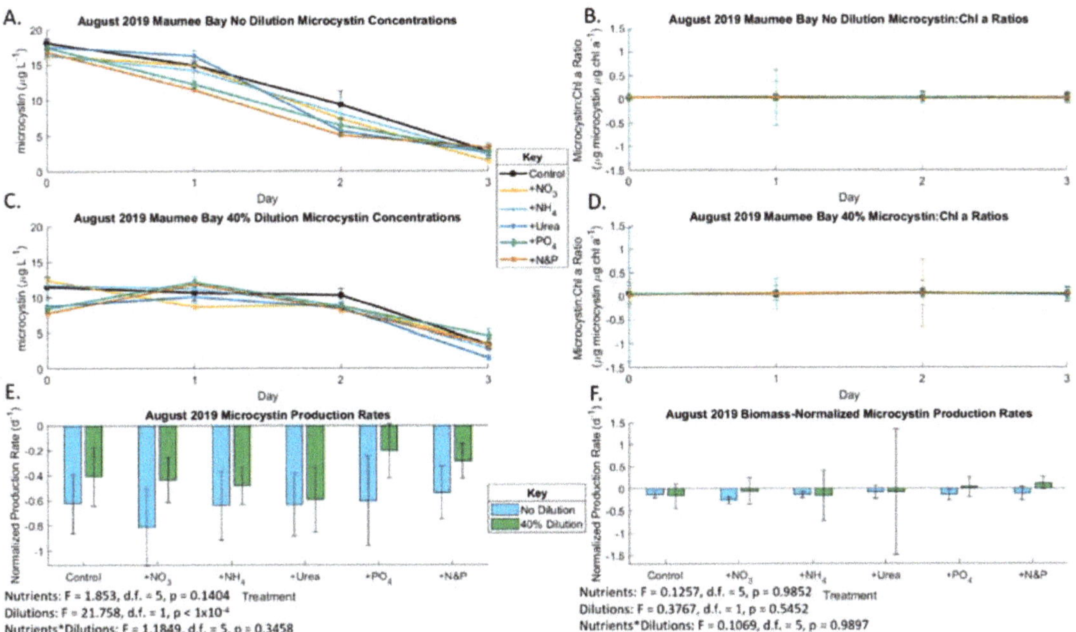

Figure 6. Production rates of microcystin during the August 2019 bioassays. Only Maumee Bay produced microcystin in all samples. (**A**) Undiluted Maumee Bay microcystin concentrations (also see Table S3); (**B**) undiluted Maumee Bay biomass-normalized microcystin concentrations (also see Table S4); (**C**) 40% dilution Maumee Bay microcystin concentrations (also see Table S3); (**D**) 40% dilution Maumee Bay biomass-normalized microcystin concentrations (also see Table S4); (**E**) Maumee Bay microcystin production rates under the various nutrient addition treatments at the two locations of T3 compared to T0 (also see Table S5); (**F**) Maumee Bay biomass-normalized microcystin production rates under the various nutrient addition treatments at the two locations of T3 compared to T0 (also see Table S6). Error bars are standard error. Significance for (**E,F**) is from 2-factor ANOVA analysis.

In the August Sandusky Bay experiment, chlorophyll concentration increased throughout the incubation in the three N-only treatments and the + N&P treatment, while it declined in the control and P-only treatment in both the diluted and non-diluted treatments (Figure 5), which indicates N was the primary limiting nutrient. The various forms of N did not exert a discernable difference on growth rates. The highest growth rates were measured in the +N&P treatments, which indicates a secondary P limitation. The dilution effect was also significant ($p = 0.004$). The initial undiluted anatoxin concentration was 0.596 µg/L (Figure 7). Anatoxin production was primarily N-limited both with and without biomass normalization ($p < 0.001$), like growth rates, but P was not secondarily limiting. Unlike chlorophyll, which decreased throughout incubation in the control and P-only treatment, anatoxin concentrations in the control and P-only treatment remained constant throughout the incubation due to production rates of anatoxin increasing throughout the incubation.

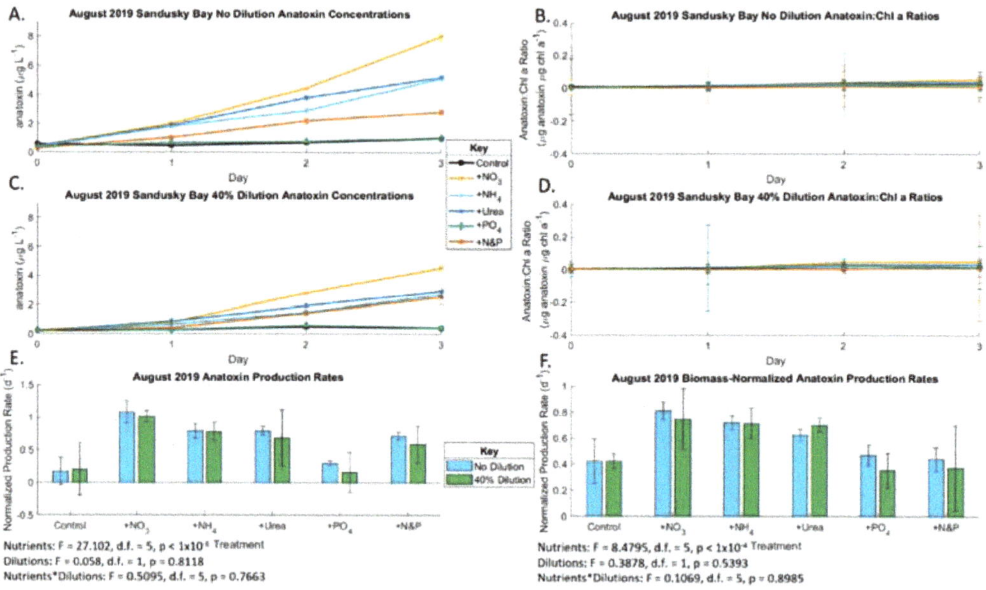

Figure 7. Production rates of anatoxin during the August 2019 bioassays. Only Sandusky Bay produced anatoxin in August. (**A**) Undiluted Sandusky Bay anatoxin concentrations (also see Table S7); (**B**) undiluted Sandusky Bay biomass-normalized anatoxin concentrations (also see Table S8); (**C**) 40% dilution Sandusky Bay anatoxin concentrations (also see Table S7); (**D**) 40% dilution Sandusky Bay biomass-normalized anatoxin concentrations (also see Table S8); (**E**) Maumee Bay anatoxin production rates under the various nutrient addition treatments at the two locations of T3 compared to T0 (also see Table S9); (**F**) Maumee Bay biomass-normalized anatoxin production rates under the various nutrient addition treatments at the two locations of T3 compared to T0 (also see Table S10). Error bars are standard error. Significance for (**E,F**) is from 2-factor ANOVA analysis.

3. Discussion

Given that CyanoHABs and their associated cyanotoxins have led to adverse human and ecosystem health outcomes in WLE [18], it is important to clarify the major driver(s) of CyanoHAB toxicity. This study investigated nutrient limitation on biomass production and cyanotoxin production, focusing on microcystin and anatoxin. We found that high concentrations of both major nutrients, P and N, drove CyanoHAB growth and microcystin and anatoxin production in WLE. We also found times when the 40% reduction in nutrients could slow microcystin production during nutrient replete conditions (Figure 3E,F).

We found that the June 2019 late spring diatom bloom in Maumee Bay was P-limited, which was induced in both the undiluted and 40% dilution samples due to high ambient N concentrations (>100 µmol/L), while the June 2019 Sandusky Bay *Planktothrix* bloom was not affected by nutrient addition, but growth was slowed following a 40% reduction in nutrients. This is possibly explained by the rapid growth associated with the early bloom, with the 40% reduction in nutrients dropping below the threshold needed to support this bloom [48]. During the bloom maxima in August 2019, the Maumee Bay *Microcystis* bloom was nutrient replete under both undiluted and 40% dilution treatments, with less of a decline in the biomass due to the 40% lower starting biomass following dilution. Additionally, ammonium concentration was higher in the initial 40% dilution than the undiluted sample in both the June and August 2019 Maumee Bay, likely due to an initial die off in the subsample, leading to increased regenerated N as ammonium. These results are likely due to bottle effects attributable to the very high biomass; restricted exchange of gases and nutrients [49–51]. The August Sandusky Bay *Planktothrix* bloom was N-limited in both the 40% reduction and the undiluted samples. All nutrient concentrations in the August 2019 Sandusky Bay 40% dilution were higher than concentrations in the undiluted treatment, likely due to the rapid growth of the *Planktothrix* bloom using up more nutrients in the undiluted control group prior to sample filtration, when compared to the reduced biomass in the 40% dilution. Differences between the effects of the different N species were not significant at either location during either experimental period, which has been seen previously in strongly N-limited blooms in WLE [52], but differs from past findings in WLE during periods of weaker N-limitation [22,53–55]. This could be due to the high ambient concentrations of NO_3 paired with low NH_4 (Tables 1 and 2). Our findings of N limitation contradict the previous assumption that P availability exclusively controls CyanoHABs [56–59]. Instead, these findings support the paradigm shift to also consider N input reductions to mitigate CyanoHABs [19,29,60,61].

During the early Sandusky Bay *Planktothrix* bloom (June 2019), microcystin production shifted from between N and P co-limitation in the undiluted samples to nutrient deplete conditions in the 40% dilution samples. This is likely due to the bloom's use of nutrient resources early on to support biomass production rather than produce secondary metabolites, e.g., cyanotoxins, possibly due to the genetic inability of the June populations to produce the microcystin as seen in prior years [62,63]. Alternatively, the cells could have lysed due to viral or other processes and the dissolved microcystin was not captured on the 0.7 μm porosity GF/F filters or degraded [64,65]. At its peak in August 2019, the *Microcystis* bloom in Maumee Bay was the only bloom that produced microcystin. This production of microcystin occurred under nutrient replete conditions, with less of a decline in microcystin concentrations with slight P limitation in the diluted samples and no apparent nutrient limitation in the undiluted samples. Neither experiments showed significant effects of the various forms of N.

Even though cyanobacteria require N to produce N-rich microcystin, P is also required for cellular growth to allow for higher microcystin concentrations. As the ratio of microcystin to chlorophyll *a* in both June and August was nearly linear (Figures 3 and 6), we conclude that the primary bloomers—*Planktothrix* in Sandusky Bay and *Microcystis* in Maumee Bay—were the primary producers of microcystin. The P requirement for microcystin production has been observed in prior studies in Lake Erie, and in several German lakes [66]. This deviates from previous studies that clearly demonstrated links between N availability and higher N:P and bloom toxicity in microcystin-producing blooms [7,67–69]. This could be due to microcystin being an "N bargain" with a C:N ratio of 4.9:1 compared to the average of 3.6:1 in a survey of 2000 proteins [69]. However, P-limitation of microcystin production has been shown to occur in chemostat experiments [70] and in a transcriptome experiment on Lake Erie blooms [40]. The microcystin congener pattern observed in these experiments followed what was expected for North American lakes, including Lake Erie, with microcystin LR, YR, RR being the dominant congeners [18].

We observed anatoxin production in the Sandusky Bay *Planktothrix* bloom during both early and peak blooms. This is the first study showing anatoxin production in Lake Erie, although it has been shown that anatoxin production can occur during *Planktothrix* blooms accompanied by other cyanobacteria, including *Cuspidothrix issatschenkoi*, which has previously been identified in Sandusky Bay [23,71–75]. This was likely the case, as the biomass normalized anatoxin production mirrors the anatoxin production in the non-normalized analysis (Figures 4 and 7), meaning that secondary cyanobacterial species may be driving the anatoxin production in Sandusky Bay. During the early *Planktothrix* bloom in June 2019, there was no apparent nutrient limitation in the undiluted treatments. However, there was co-limitation by both N and P in the diluted treatments. During the peak bloom in August 2019, anatoxin production was N-limited in both the undiluted and 40% diluted samples. While no differences were found between forms of N added in the June bioassay, during the peak bloom in August, NO_3 additions led to higher concentrations of anatoxin compared to NH_4 and urea additions. Additionally, N limitation of anatoxin production has been shown previously [76]. As observed in this experiment, higher overall

N concentrations lead to higher anatoxin concentrations, with NO_3 enrichment leading to the largest increase in anatoxin production, which parallels results from other limnetic anatoxin-producing CyanoHABs [77–81]. Anatoxin production in Sandusky Bay and other *Planktothrix*-dominated bodies of water needs further examination, given the neurotoxicity and potential developmental toxicity of anatoxin [82,83] as well as its multiple deleterious environmental effects [84,85].

Nutrient concentrations were very high during both the early and peak 2019 bloom in Maumee Bay with 223.67 ± 25.43 µg L^{-1} NO_3 and 2.224 ± 1.008 µg L^{-1} DRP in June and 127.12 ± 10.82 µg L^{-1} combined NO_3 and NO_2 in August and 0.203 ± 0.199 µg L^{-1} DRP. Similar to Maumee Bay, Sandusky Bay exhibited high nutrient concentrations in June with 101.45 ± 5.95 µg L^{-1} NO_3 and 0.203 ± 0.138 µg L^{-1} DRP in June, but had lower nutrient concentrations in August with 127.12 ± 10.82 µg L^{-1} NO_3 in August and 0.032 ± 0.012 µg L^{-1} DRP. This is likely due to larger nutrient loads from the Maumee River than from the Sandusky River, as seen previously in 2007 [86]. The high nutrient loads were exacerbated by elevated precipitation associated with a very wet winter in 2019 [87], which will likely continue to be an issue as high precipitation events are predicted to continue in the future [88–90]. Denitrification and assimilation draw down nitrate to concentrations below the threshold of detection (<0.5 µmol/L) throughout summer and fall in western Lake Erie and Sandusky Bay [28,91], which is a pattern that occurs independent of tributary nutrient loads [19]. Our Maumee Bay experiments occurred before nitrate depletion, and therefore, we would expect to have observed N-limited growth and microcystin production following the N depletion [25]. However, it remains to be seen how a 40% dilution in nutrients (N and P) would affect N-limited *Microcystis* in late summer. Therefore, nutrient input reductions need to target both N and P rather than just P as recommended by the US EPA and Environment and Climate Change Canada [43–45,92]. While P reduction is actively pursued [93], N management strategies are required as well [35,94,95].

4. Conclusions

Our results suggest that nutrient dynamics play a crucial role in the WLE CyanoHABs for both biomass production as well as microcystin and anatoxin production in the eutrophic Sandusky and Maumee Bays. During the peak bloom periods when microcystin and anatoxin concentrations are highest, microcystin production was nutrient deplete and anatoxin production was N-limited. Maumee Bay biomass shifted from P-limited immediately prior to the *Microcystis* bloom to nutrient deplete during peak bloom, while the Sandusky Bay *Planktothrix* bloom shifted from nutrient deplete to N-limited from early bloom to peak bloom. A 40% reduction in N and P led to a slight reduction in biomass and microcystin and anatoxin production. However, further studies are needed to investigate the long-term nutrient reduction thresholds needed to control CyanoHABs. With N and P enrichment stimulating the WLE CyanoHABs, there is a need to constrain external loads of both N and P, and impose stricter nutrient-limited conditions in order to help mitigate the CyanoHAB problem in WLE [52,96–98]. Our study took place only in eutrophic bays and we showed that a 40% reduction might not be enough in Maumee and Sandusky Bay because growth and toxin production could still be nutrient-saturated. Future studies are needed to determine if a 40% reduction is adequate for the open waters of WLE. Furthermore, an adaptive management approach is needed to determine if the 40% reduction goal needs to be adjusted with changes in land use practices and climate change [99]. Additionally, future studies should focus on drawing direct functional links between nutrient enrichment and cyanotoxin production, e.g., Krausfeldt et al. [36]. Lastly, anatoxin should be more closely monitored in WLE, as it is a potent neurotoxin with human health-associated implications [100].

5. Materials and Methods

5.1. Bioassay Methods

We performed experimental manipulations of natural Maumee Bay (Oregon, OH, USA) and Sandusky Bay (Sandusky, OH, USA) phytoplankton communities that were collected from nearshore docks (Figure 8; Table S11). Water was pumped from 1 m below the surface into pre-cleaned (flushed with lake water) 20 L carboys using a non-destructive diaphragm pump and was transported to The Ohio State University Stone Laboratory on South Bass Island (Put-in-Bay, OH, USA) (Figure 8).

This experiment deployed in situ bioassays, using 4 L pre-cleaned polyethylene Cubitainers to which natural lake water was added from Maumee and Sandusky Bays using the methodology described in Paerl et al. [101] and Xu et al. [102]. Microcosm treatments were individually amended with either 100 µM N of NO_3 (as KNO_3), 100 µM N of NH_4 (as NH_4Cl), 6 µM PO_4 (as KH_2PO_4), 100 µM N and 6 µMP added as a combined addition of 50 µM NO_3, 50 µM NH_4, and 6 µM PO_4, and, in August 2019, urea (50 µM urea to achieve 100 µM N), yielding similar total dissolved nutrient concentrations (for each treatment) and falling within a range matching riverine dissolved inorganic nutrient discharge into Lake Erie nearshore waters. To avoid silica or dissolved inorganic carbon limitation in Cubitainers during the incubation period, we added 50 µM Si as Na_2SiO_3 and 10 mg L^{-1} (83.25 µM) DIC as $NaHCO_3$ based on previous Si and DIC values from Hanson et al. [103] and Rockwell et al. [104]. We used a major ion solution (MIS) specific to WLE to provide 40% dilutions to mimic the EPA-recommended reductions in P inputs to WLE as well as a parallel 40% reduction in N, as both N and P have been shown to influence WLE CyanoHAB bloom dynamics [22,39,43]. The 40% dilution control investigated a 40% reduction in both N and P. Incubations were run for 72 h at a lake site near the Stone Laboratory at ambient lake water temperatures and light conditions [23,101,102]. Based on previous work on eutrophic Lake Taihu, China [95], a 72 h maximum incubation period was chosen to minimize "bottle effects", while having ample time to examine phytoplankton growth, microcystin, and anatoxin production responses.

To perform nutrient dilutions, we developed a major ion solution (MIS) for WLE, which provided a N- and P-free dilution media to minimize hypertonic and hypotonic effects on the organisms in the samples by balancing major dissolved ions in the system (Table 3). As an example, artificial seawater is the MIS for the open ocean. For WLE, we based the ambient ion concentrations on a past study by Chapra et al. [105]. As there is substantial natural variability due to rainfall and evaporative effects and the ions in the MIS are in micromolar concentrations and pulse events change the ions in WLE, these deviations are considered reasonable. The compounds used in the MIS are found in Table S12.

Table 3. Concentrations of major ions in the ambient Lake Erie water and the major ion solution (MIS) used for the dilutions in the bioassays.

Ion [1]	Average Ambient Concentration (mg/L) [105]	MIS [1] Concentration (mg/L)	MIS [1] Concentration (µM)	Percent Difference between Chapra et al. [105] and MIS Concentrations
Ca^{2+}	32.11	32	800	− 0.34%
Mg^{2+}	8.89	8.88	370	− 0.11%
Na^+	8.58	4.6	200	− 46.39% [2]
K^+	1.431	1.56	40	9.01% [3]
Cl^-	14.58	16.33	460	12.00% [3]
SO_4^{2-}	22.81	43.2	450	89.39% [3]

[1] Constituents of MIS can be found in Table S12; [2] lower concentration compared to ambient concentration; [3] higher concentration compared to average ambient concentrations.

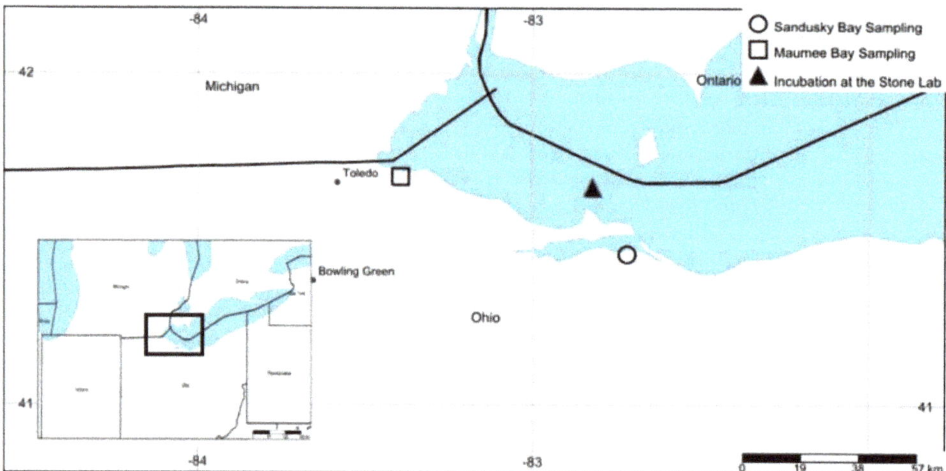

Figure 8. Map of the sampling sites and the location of the incubation. Maumee Bay water was collected off a bulkhead dock near the University of Toledo Lake Erie Center in Oregon, OH, USA. Sandusky Bay sampling took place at a dock outside the Paper District Marina in Sandusky, OH, USA. Incubation took place at The Ohio State Stone Laboratory on South Bass Island (Put-In-Bay, OH, USA). GPS coordinates for the sampling and incubation sites can be found in Table S11. This figure was created with www.simplemappr.net [106].

5.2. Phytoplankton Biomass Determination

Chlorophyll *a*, as an indicator of phytoplankton biomass, was measured on subsampled samples by filtering 50 mL of sample water onto Whatman glass fiber filters (GF/F). Filters were frozen at -20 °C and subsequently extracted using a tissue grinder in 90% acetone [107,108]. Chlorophyll *a* in extracts was measured using the non-acidification method of Welschmeyer [109] on a Turner Designs Trilogy fluorometer calibrated with pure Chlorophyll *a* standards (Turner Designs, Sunnyvale, CA, USA).

5.3. Nutrient Concentration Determination

Nutrient samples were collected in 50 mL Falcon tubes by collecting the GF/F filtered water from the chlorophyll *a* sample collection and frozen at -20 °C. A continuous segmented flow auto-analyzer (QuAAtro SEAL Analytical Inc., Mequon, WI, USA) was used to quantify nitrate, nitrite, ammonium, dissolved reactive P, and silicate using standard U.S. EPA methods [110]. Urea concentration (as urea-N) was determined spectrophotometrically [52,111,112].

5.4. Anatoxin and Microcystin Determinations

Cyanotoxins were measured on subsampled samples by filtering 50 mL of the sample water onto Whatman GF/F. Filters were frozen at -20 °C until extraction with ultrasonic sonication in 5 mL of 50% methanol and 1% acetic acid. Samples were centrifuged at $14,000 \times g$ for 10 min at 4 °C. The supernatants were filtered through 0.45 µm pore-size nylon syringe filters (Corning, CLS431225) and stored at -20 °C until analysis. Microcystin was quantified via coupled liquid chromatography/mass spectrometry using methods modified from Boyer [113] and Peng et al. [114]. Reverse-phase liquid chromatography using a Waters 2695 solvent delivery system (Waters, Milford, MA, USA) coupled to a Waters ZQ4000 mass spectrometer (Waters, Milford, MA, USA) (m/z 500–1250 amu) and a 2996 photodiode array detector (Waters, Milford, MA, USA) (210 to 400 nm wavelength) was used to screen for molecular ions of 22 common microcystin congeners (RR, dRR, mRR, H4YR, hYR, YR, LR, mLR, zLR. dLR, meLR, AR, FR, WR, LA, dLA, mLA, LL, LY, LW,

LF, WR). Separation conditions used an ACE 5 C18, 150 × 3.0 mm column and a 30–70% aqueous acetonitrile gradient containing 0.1% formic acid at a flow rate of 0.3 mL min. Individual congener concentrations were quantified using the peak area of the extracted ion relative to standards of microcystin-LR (Enzo Life Sciences, Ann Arbor, MI, USA). This allows quantification of congeners where standards are not available. Detection of congeners was validated by co-occurring presence of the diagnostic UV signature from the ADDA group. Full methodological details and the standard operating protocols are available from Protocols.io [115].

Anatoxin-a, dihydro-anatoxin-a and homoanatoxin-a were determined by LC-MS/MS using one quantification ion and two confirmation ions for each compound. Separation was achieved with an ACE 5 4.6 × 150 mm column (MacMod Analytical, Chadds Ford, PA, USA) assembly with solvent flow of 0.5 mL/min from a Waters Alliance 2695 solvent system (Waters, Milford, MA, USA). The solvent system was: A, 0.1% formic acid in water; B, 0.1% formic acid in acetonitrile. The separation gradient was: 0 to 20% B from 0 to 10 min, 20% to 80% B from 10 to 20 min, and 80% to 100% B from 20 to 23 min, followed by equilibration back to 0% B for 7 min. Toxins were identified using a Waters Acquity TQD mass spectrometer (Waters, Milford, MA, USA) operated in positive mode with capillary voltage 3.5 kV, desolvation and cone gasses at 30 and 800 $\mathrm{Lh^{-1}}$, respectively, desolvation and source temperatures of 400 and 150 °C, respectively. Retention times and fragmentation patterns were determined using anatoxin-a (BioMOL International, Farmingdale, NY, USA), homoanatoxin-a isolated from natural sources and α and β dihydroanatoxin synthesized by catalytic hydrogenation/reduction of anatoxin-a [116]. Calibration was performed with anatoxin-a; dihydro-anatoxin-a and homoanatoxin-a concentrations were estimated using the anatoxin-a standard curve. A phenylalanine standard was run with each set to confirm the baseline resolution between anatoxin-a and phenylalanine. Multiple reaction monitoring quantitation transitions were: anatoxin-a (166.09 > 131.00, collision energy (CE) 15 eV), dihydro-anatoxin-a (168.20 > 43.10, CE 23 eV), homoanatoxin-a (180.10 > 163.10, CE 15 eV). Confirmation transitions were: anatoxin-a (166.09 > 148.90, CE 15 eV; 166.09 > 90.90, CE 17 eV), dihydro-anatoxin-a (168.20 > 55.90, CE 22 eV; 168.20 > 67.00, CE 26 eV), homoanatoxin-a (180.10 > 145.10, CE 15 eV; 180.10 > 105.00, CE 17 eV).

5.5. Data Transformation and Analysis

To remove biomass effects on toxin to better measure nutrient effects on microcystin and anatoxin production, microcystin and anatoxin concentrations are normalized to biomass as proxied by chlorophyll *a*. Microcystin:chl *a* and anatoxin:chl *a* ratios are calculated using Equation (1):

$$toxin : chl\ a\ ratio\left(\mu g\ microcystin\ or\ anatoxin\ \mu g\ chlorophyll\ a^{-1}\right) = \frac{[toxin]}{[biomass]} \quad (1)$$

where [toxin] is the concentration of either microcystin or anatoxin (in μg $\mathrm{L^{-1}}$) and [biomass] is the concentration of chlorophyll *a* (in μg $\mathrm{L^{-1}}$).

For comparison between dilution treatments, we calculated production rates from the chlorophyll *a* and biomass-normalized microcystin and biomass-normalized anatoxin concentrations. Production rate ($\mathrm{d^{-1}}$) is a method to ln normalize the changes in concentrations, where a production of 0.693 $\mathrm{d^{-1}}$ is a doubling of the concentration per day, a production of 0.0 $\mathrm{d^{-1}}$ indicates no change, and a production of −0.693 $\mathrm{d^{-1}}$ represents a halving of the concentration. Production is calculated using Equation (2):

$$Production\left(d^{-1}\right) = \ln\left(\frac{\mu_{T3}}{\mu_{T0}}\right) * \frac{1}{t} \quad (2)$$

where μ_{T0} is the average value of the measurement for the initial time point (T0), μ_{T3} is the average value of the measurement for the time point of 3 days (T3), and t is the time difference between the samplings (in days), which in this case is t = 3 days. To calculate the standard deviation for the production, propagated standard deviation is used, as calculated by Equation (3):

$$Propagated\ Standard\ Deviation = \sqrt{\left(\frac{\sigma_{T0}}{\mu_{T0}}\right)^2 + \left(\frac{\sigma_{T3}}{\mu_{T3}}\right)^2} \tag{3}$$

where μ_{T0} is the average value of the measurement for T0, ς_{T0} is the standard deviation for the measurement at T0, μ_{T3} is the average value of the measurement for T3, and ς_{T3} is the standard deviation for the measurement at T3. For error bars, standard error is used, which is calculated using Equation (4):

$$Standard\ Error = \frac{\sigma}{\sqrt{n}} \tag{4}$$

where ς is the standard deviation for Figure 2a–d, Figure 3a–d, Figure 4a–d, Figure 5a–d, Figure 6a–d, and Figure 7a–d, ς is the propagated standard deviation for Figure 2e–f, Figure 3e–f, Figure 4e–f, Figure 5e–f, Figure 6e–f, and Figure 7e–f, and n is the number of data points. The standard errors are available in the WLE_Barnard_et_al_Toxins GitHub repository [117].

5.6. Statistical Analysis

To evaluate the source of the variation between the treatments, ANOVA analyses were performed. For this experiment, two-factor ANOVA analyses were run on balanced data sets (all data $n = 3$), and n-factor ANOVA analyses were run on unbalanced data sets (one or more treatments were characterized as $n = 1$ or $n = 2$) using MATLAB ver. R2018b [118]. Both the two-factor and n-factor ANOVA analyses calculate degrees of freedom (d.f.) as the number of treatments (n) minus one (d.f. $= n-1$). The homogeneity of variances was tested for with Levene's Absolute test using MATLAB ver. R2018b [118]. All data and corresponding n-values are in Tables S1, S3, S4, S7, and S8.

Supplementary Materials: The following are available online at https://www.mdpi.com/2072-6651/13/1/47/s1, Table S1: Chlorophyll *a* data, Table S2: Chlorophyll *a* production rates, Table S3: Microcystin data, Table S4: Biomass-normalized microcystin data, Table S5: Microcystin production rates, Table S6: Biomass-normalized microcystin production rates, Table S7: Anatoxin data, Table S8: Biomass-normalized anatoxin data, Table S9: Anatoxin production rates, Table S10: Biomass-normalized anatoxin production rates, Table S11: GPS coordinates of the Western Lake Erie sampling sites, Table S12: Compounds comprising the major ion solution. The following are available online at www.doi.org/10.5281/zenodo.4281127, Code used to produce Figures 2–7, importable data file formatted for the code, Key to the importable data file.

Author Contributions: Conceptualization, M.A.B. and H.W.P.; data curation, M.A.B.; formal analysis, M.A.B., G.L.B. and B.W.; funding acquisition, M.A.B., J.D.C., G.L.B., S.W.W., G.S.B., T.B.B. and T.W.D.; investigation, M.A.B., J.D.C., H.E.P., G.L.B., B.W., S.W.W., K.L.R., J.S.B., G.S.B., T.B.B., T.W.D., J.W., M.B. and H.W.P.; methodology, M.A.B., J.D.C., H.E.P., G.L.B., B.W., K.L.R., J.S.B., G.S.B., T.B.B., T.W.D., J.W., M.B. and H.W.P.; project administration, G.S.B.; supervision, J.D.C., G.L.B., S.W.W., G.S.B., T.B.B., T.W.D. and H.W.P.; writing-original draft, M.A.B.; writing-review and editing, J.D.C., H.E.P., G.L.B., B.W., S.W.W., K.L.R., J.S.B., G.S.B., T.B.B., T.W.D., J.W., M.B. and H.W.P.; all authors have read and agreed to the published version of the manuscript. All authors have read and agreed to the published version of the manuscript.

Funding: This research was funded by the United States National Science Foundation (OCE 0812913, OCE 0825466, OCE 1840715, CBET 0826819, IOS 1451528 and DEB 1831096), the United States National Institutes of Health (NIEHS P01ES028939), a Grant-in-Aid of Research from Sigma Xi, The Scientific Research Society (G201903158412545) [MAB], a Kenan Graduate Student Award from the University of North Carolina at Chapel Hill Department of Marine Sciences [MAB], and the NOAA/North Carolina Sea Grant Program R/MER-43, R/MER-47 [MAB, HEP, KLR, JSB, HWP].

Institutional Review Board Statement: Not applicable.

Informed Consent Statement: Not applicable.

Data Availability Statement: The data formatted for analysis and executable MATLAB code used to produce Figures 2–7 can be found on GitHub at www.doi.org/10.5281/zenodo.4281127 [117]. The data presented in this study are also available in table form in the accompanying supplementary material: https://www.mdpi.com/2072-6651/13/1/47/s1.

Acknowledgments: We thank R. Sloup, N. Hall, and B. Abare of the UNC Institute of Marine Sciences, as well as laboratory technicians and students from The Ohio State University and University of Toledo Lake Erie Center at the for their help with experimental work.

Conflicts of Interest: The authors declare no conflict of interest. The funders had no role in the design of the study; in the collection, analyses, or interpretation of data; in the writing of the manuscript, or in the decision to publish the results.

References

1. Haas, M.; Baumann, F.; Castella, D.; Haghipour, N.; Reusch, A.; Strasser, M.; Eglinton, T.I.; Dubois, N. Roman-driven cultural eutrophication of Lake Murten, Switzerland. *Earth Planet. Sci. Lett.* **2019**, *505*, 110–117. [CrossRef]
2. Mohamed, M.N.; Wellen, C.; Parsons, C.T.; Taylor, W.D.; Arhonditsis, G.; Chomicki, K.M.; Boyd, D.; Weidman, P.; Mundle, S.O.C.; Van Cappellen, P.; et al. Understanding and managing the re-eutrophication of Lake Erie: Knowledge gaps and research priorities. *Freshw. Sci.* **2019**, *38*, 675–691. [CrossRef]
3. Smith, V.H.; Tilman, G.D.; Nekola, J.C. Eutrophication: Impacts of excess nutrient inputs on freshwater, marine, and terrestrial ecosystems. *Environ. Pollut.* **1999**, *100*, 179–196. [CrossRef]
4. Conley, D.J.; Paerl, H.W.; Howarth, R.W.; Boesch, D.F.; Seitzinger, S.P.; Havens, K.E.; Lancelot, C.; Likens, G.E. Controlling eutrophication: Nitrogen and phosphorus. *Science* **2009**, *323*, 1014–1015. [CrossRef] [PubMed]
5. *Environmental Pollution Panel President's Science Advisory Committee in Restoring the Quality of Our Environment*; White House Publishing: Cottage Grove, OR, USA, 1965.
6. Sharpley, A.N.; Herron, S.; Daniel, T. Overcoming the challenges of phosphorus-based management in poultry farming. *J. Soil Water Conserv.* **2007**, *62*, 375–389.
7. Bullerjahn, G.S.; McKay, R.M.; Davis, T.W.; Baker, D.B.; Boyer, G.L.; D'Anglada, L.V.; Doucette, G.J.; Ho, J.C.; Irwin, E.G.; Kling, C.L.; et al. Global solutions to regional problems: Collecting global expertise to address the problem of harmful cyanobacterial blooms. A Lake Erie case study. *Harmful Algae* **2016**, *54*, 223–238. [CrossRef] [PubMed]
8. Harke, M.J.; Steffen, M.M.; Gobler, C.J.; Otten, T.G.; Wilhelm, S.W.; Wood, S.A.; Paerl, H.W. A review of the global ecology, genomics, and biogeography of the toxic cyanobacterium, *Microcystis* spp. *Harmful Algae* **2016**, *54*, 4–20. [CrossRef]
9. Paerl, H.W.; Scott, J.T.; McCarthy, M.J.; Newell, S.E.; Gardner, W.S.; Havens, K.E.; Hoffman, D.K.; Wilhelm, S.W.; Wurtsbaugh, W.A. It takes two to tango: when and where dual nutrient (N & P) reductions are needed to protect lakes and downstream ecosystems. *Environ. Sci. Technol.* **2016**, *50*, 10805–10813. [CrossRef]
10. Paerl, H.W.; Otten, T.G.; Kudela, R. Mitigating the expansion of harmful algal blooms across the freshwater-to-marine continuum. *Environ. Sci. Technol.* **2018**, *52*, 5519–5529. [CrossRef]
11. Paerl, H.W.; Huisman, J. Climate change: A catalyst for global expansion of harmful cyanobacterial blooms. *Environ. Microbiol. Rep.* **2009**, *1*, 27–37. [CrossRef]
12. Chapra, S.C.; Boehlert, B.; Fant, C.; Bierman, V.J.; Henderson, J.; Mills, D.; Mas, D.M.L.; Rennels, L.; Jantarasami, L.; Martinich, J.; et al. Climate change impacts on harmful algal blooms in U.S. freshwaters: a screening-level assessment. *Environ. Sci. Technol.* **2017**, *51*, 8933–8943. [CrossRef] [PubMed]
13. Backer, L.C.; McNeel, S.V.; Barber, T.; Kirkpatrick, B.; Williams, C.; Irvin, M.; Zhou, Y.; Johnson, T.B.; Nierenberg, K.; Aubel, M.; et al. Recreational exposure to microcystins during algal blooms in two California lakes. *Toxicon* **2010**, *55*, 909–921. [CrossRef] [PubMed]
14. Plaas, H.E.; Paerl, H.W. Toxic cyanobacteria: a growing threat to water and air quality. *Environ. Sci. Technol.* **2021**, *55*, 44–64. [CrossRef] [PubMed]
15. Smith, Z.; Martin, R.; Wei, B.; Wilhelm, S.; Boyer, G. Spatial and temporal variation in paralytic shellfish toxin production by benthic *Microseira (Lyngbya) wollei* in a freshwater New York lake. *Toxins (Basel)* **2019**, *11*, 44. [CrossRef]
16. Steffen, M.M.; Belisle, B.S.; Watson, S.B.; Boyer, G.L.; Wilhelm, S.W. Status, causes and controls of cyanobacterial blooms in Lake Erie. *J. Great Lakes Res.* **2014**, *40*, 215–225. [CrossRef]
17. Pauer, J.J.; Anstead, A.M.; Melendez, W.; Taunt, K.W.; Kreis, R.G. Revisiting the Great Lakes Water Quality Agreement phosphorus targets and predicting the trophic status of Lake Michigan. *J. Great Lakes Res.* **2011**, *37*, 26–32. [CrossRef]
18. Steffen, M.M.; Davis, T.W.; McKay, R.M.L.; Bullerjahn, G.S.; Krausfeldt, L.E.; Stough, J.M.A.; Neitzey, M.L.; Gilbert, N.E.; Boyer, G.L.; Johengen, T.H.; et al. Ecophysiological examination of the Lake Erie *Microcystis* bloom in 2014: linkages between biology and the water supply shutdown of Toledo, OH. *Environ. Sci. Technol.* **2017**, *51*, 6745–6755. [CrossRef]
19. Chaffin, J.D.; Bridgeman, T.B.; Bade, D.L.; Mobilian, C.N. Summer phytoplankton nutrient limitation in Maumee Bay of Lake Erie during high-flow and low-flow years. *J. Great Lakes Res.* **2014**, *40*, 524–531. [CrossRef]

20. Belisle, B.S.; Steffen, M.M.; Pound, H.L.; Watson, S.B.; DeBruyn, J.M.; Bourbonniere, R.A.; Boyer, G.L.; Wilhelm, S.W. Urea in Lake Erie: Organic nutrient sources as potentially important drivers of phytoplankton biomass. *J. Great Lakes Res.* **2016**, *42*, 599–607. [CrossRef]

21. Wilhelm, S.; DeBruyn, J.; Gillor, O.; Twiss, M.; Livingston, K.; Bourbonniere, R.; Pickell, L.; Trick, C.; Dean, A.; McKay, R. Effect of phosphorus amendments on present day plankton communities in pelagic Lake Erie. *Aquat. Microb. Ecol.* **2003**, *32*, 275–285. [CrossRef]

22. Chaffin, J.D.; Bridgeman, T.B.; Bade, D.L. Nitrogen constrains the growth of late summer cyanobacterial blooms in Lake Erie. *Adv. Microbiol.* **2013**, *3*, 16–26. [CrossRef]

23. Davis, T.W.; Bullerjahn, G.S.; Tuttle, T.; McKay, R.M.; Watson, S.B. Effects of increasing nitrogen and phosphorus concentrations on phytoplankton community growth and toxicity during *Planktothrix* Blooms in Sandusky Bay, Lake Erie. *Environ. Sci. Technol.* **2015**, *49*, 7197–7207. [CrossRef] [PubMed]

24. Gobler, C.J.; Burkholder, J.A.M.; Davis, T.W.; Harke, M.J.; Johengen, T.; Stow, C.A.; Van de Waal, D.B. The dual role of nitrogen supply in controlling the growth and toxicity of cyanobacterial blooms. *Harmful Algae* **2016**, *54*, 87–97. [CrossRef] [PubMed]

25. Chaffin, J.D.; Davis, T.W.; Smith, D.J.; Baer, M.M.; Dick, G.J. Interactions between nitrogen form, loading rate, and light intensity on *Microcystis* and *Planktothrix* growth and microcystin production. *Harmful Algae* **2018**, *73*, 84–97. [CrossRef] [PubMed]

26. Conroy, J.D.; Kane, D.D.; Quinlan, E.L.; Edwards, W.J.; Culver, D.A. Abiotic and biotic controls of phytoplankton biomass dynamics in a freshwater tributary, estuary, and large lake ecosystem: Sandusky Bay (lake Erie) chemostat. *Inland Waters* **2017**, *7*, 473–492. [CrossRef]

27. Redfield, A.C.; Ketchum, B.H.; Richards, F.A. The influence of organisms on the composition of sea-water. In *The Sea: Ideas and Observations on Progress in the Study of the Seas*; Interscienc Publishers: New York, NY, USA, 1963.

28. Salk, K.R.; Bullerjahn, G.S.; McKay, R.M.L.; Chaffin, J.D.; Ostrom, N.E. Nitrogen cycling in Sandusky Bay, Lake Erie: Oscillations between strong and weak export and implications for harmful algal blooms. *Biogeosciences* **2018**, *15*, 2891–2907. [CrossRef]

29. Hampel, J.J.; McCarthy, M.J.; Neudeck, M.; Bullerjahn, G.S.; McKay, R.M.L.; Newell, S.E. Ammonium recycling supports toxic *Planktothrix* blooms in Sandusky Bay, Lake Erie: Evidence from stable isotope and metatranscriptome data. *Harmful Algae* **2019**, *81*, 42–52. [CrossRef]

30. Shen, Y.; Huang, Y.; Hu, J.; Li, P.; Zhang, C.; Li, L.; Xu, P.; Zhang, J.; Chen, X. The nitrogen reduction in eutrophic water column driven by Microcystis blooms. *J. Hazard. Mater.* **2020**, *385*, 121578. [CrossRef]

31. Muenich, R.L.; Kalcic, M.; Scavia, D. Evaluating the impact of legacy p and agricultural conservation practices on nutrient loads from the Maumee River watershed. *Environ. Sci. Technol.* **2016**, *50*, 8146–8154. [CrossRef]

32. Cousino, L.K.; Becker, R.H.; Zmijewski, K.A. Modeling the effects of climate change on water, sediment, and nutrient yields from the Maumee River watershed. *J. Hydrol. Reg. Stud.* **2015**, *4*, 762–775. [CrossRef]

33. Michalak, A.M.; Anderson, E.J.; Beletsky, D.; Boland, S.; Bosch, N.S.; Bridgeman, T.B.; Chaffin, J.D.; Cho, K.; Confesor, R.; Daloglu, I.; et al. Record-setting algal bloom in Lake Erie caused by agricultural and meteorological trends consistent with expected future conditions. *Proc. Natl. Acad. Sci. USA* **2013**, *110*, 6448–6452. [CrossRef] [PubMed]

34. Smith, D.R.; Owens, P.R.; Leytem, A.B.; Warnemuende, E.A. Nutrient losses from manure and fertilizer applications as impacted by time to first runoff event. *Environ. Pollut.* **2007**, *147*, 131–137. [CrossRef] [PubMed]

35. Paerl, H.W.; Barnard, M.A. Mitigating the global expansion of harmful cyanobacterial blooms: Moving targets in a human- and climatically-altered world. *Harmful Algae* **2020**, *96*, 101845. [CrossRef] [PubMed]

36. Krausfeldt, L.E.; Farmer, A.T.; Castro, H.F.; Boyer, G.L.; Campagna, S.R.; Wilhelm, S.W. Nitrogen flux into metabolites and microcystins changes in response to different nitrogen sources in *Microcystis aeruginosa* NIES-843. *Environ. Microbiol.* **2020**, 1462–2920. [CrossRef] [PubMed]

37. NOAA MODIS. Algal Bloom in Lake Erie. Available online: https://modis.gsfc.nasa.gov/gallery/individual.php?db_date=2019-08-19 (accessed on 19 August 2019).

38. Wacklin, P.; Hoffmann, L.; Komarek, J. Nomenclatural validation of the genetically revised cyanobacterial genus *Dolichospermum* (RALFS ex BORNET et FLAHAULT) comb. nova. *Fottea* **2009**, *9*, 59–64. [CrossRef]

39. Davis, T.W.; Berry, D.L.; Boyer, G.L.; Gobler, C.J. The effects of temperature and nutrients on the growth and dynamics of toxic and non-toxic strains of *Microcystis* during cyanobacteria blooms. *Harmful Algae* **2009**, *8*, 715–725. [CrossRef]

40. Rinta-Kanto, J.M.; Konopko, E.A.; DeBruyn, J.M.; Bourbonniere, R.A.; Boyer, G.L.; Wilhelm, S.W. Lake Erie Microcystis: Relationship between microcystin production, dynamics of genotypes and environmental parameters in a large lake. *Harmful Algae* **2009**, *8*, 665–673. [CrossRef]

41. Newell, S.E.; Davis, T.W.; Johengen, T.H.; Gossiaux, D.; Burtner, A.; Palladino, D.; McCarthy, M.J. Reduced forms of nitrogen are a driver of non-nitrogen-fixing harmful cyanobacterial blooms and toxicity in Lake Erie. *Harmful Algae* **2019**, *81*, 86–93. [CrossRef]

42. Osswald, J.; Rellán, S.; Gago, A.; Vasconcelos, V. Toxicology and detection methods of the alkaloid neurotoxin produced by cyanobacteria, anatoxin-a. *Environ. Int.* **2007**, *33*, 1070–1089. [CrossRef]

43. *U.S. Action Plan for Lake Erie*; United States Environmental Protection Agency: Washington, DC, USA, 2018.

44. *Phosphorus Loading Targets For Lake Erie*; Annex 4 Objectives and Targets Task Team Final Report to the Nutrients Annex Subcommittee; United States Environmental Protection Agency: Washington, DC, USA, 2015.

45. Environment and Climate Change Canada. *Canada-Ontario Lake Erie Action Plan Partnering on Achieving Phosphorus Loading Reductions to Lake Erie from Canadian Sources*; Ontario Ministry of the Environment and Climate Change: Toronto, ON, Canada, 2018.

46. Maccoux, M.J.; Dove, A.; Backus, S.M.; Dolan, D.M. Total and soluble reactive phosphorus loadings to Lake Erie: A detailed accounting by year, basin, country, and tributary. *J. Great Lakes Res.* **2016**, *42*, 1151–1165. [CrossRef]

47. Scavia, D.; DePinto, J.V.; Bertani, I. A multi-model approach to evaluating target phosphorus loads for Lake Erie. *J. Great Lakes Res.* **2016**, *42*, 1139–1150. [CrossRef]

48. Yan, P.; Guo, J.S.; Zhang, P.; Xiao, Y.; Li, Z.; Zhang, S.Q.; Zhang, Y.X.; He, S.X. The role of morphological changes in algae adaptation to nutrient stress at the single-cell level. *Sci. Total Environ.* **2021**, *754*, 142076. [CrossRef] [PubMed]

49. Spivak, A.C.; Vanni, M.J.; Mette, E.M. Moving on up: Can results from simple aquatic mesocosm experiments be applied across broad spatial scales? *Freshw. Biol.* **2011**, *56*, 279–291. [CrossRef]

50. Del Giorgio, P.; Williams, P. *Respiration in Aquatic Ecosystems*; Oxford University Press: Oxford, UK, 2005; ISBN 9780191713347.

51. Nogueira, P.; Domingues, R.B.; Barbosa, A.B. Are microcosm volume and sample pre-filtration relevant to evaluate phytoplankton growth? *J. Exp. Mar. Bio. Ecol.* **2014**, *461*, 323–330. [CrossRef]

52. Chaffin, J.D.; Bridgeman, T.B. Organic and inorganic nitrogen utilization by nitrogen-stressed cyanobacteria during bloom conditions. *J. Appl. Phycol.* **2014**, *26*, 299–309. [CrossRef]

53. Chaffin, J.D.; Bridgeman, T.B.; Heckathorn, S.A.; Mishra, S. Assessment of *Microcystis* growth rate potential and nutrient status across a trophic gradient in western Lake Erie. *J. Great Lakes Res.* **2011**, *37*, 92–100. [CrossRef]

54. North, R.L.; Guildford, S.J.; Smith, R.E.H.; Havens, S.M.; Twiss, M.R. Evidence for phosphorus, nitrogen, and iron colimitation of phytoplankton communities in Lake Erie. *Limnol. Oceanogr.* **2007**, *52*, 315–328. [CrossRef]

55. Moon, J.; Carrick, H. Seasonal variation of phytoplankton nutrient limitation in Lake Erie. *Aquat. Microb. Ecol.* **2007**, *48*, 61–71. [CrossRef]

56. Likens, G.E. *Nutrients and Eutrophication: The Limiting-Nutrient Controversy: Proceedings*; American Society of Limnology and Oceanography: Waco, TX, USA, 1972.

57. Motew, M.; Chen, X.; Booth, E.G.; Carpenter, S.R.; Pinkas, P.; Zipper, S.C.; Loheide, S.P.; Donner, S.D.; Tsuruta, K.; Vadas, P.A.; et al. The Influence of Legacy P on Lake Water Quality in a Midwestern Agricultural Watershed. *Ecosystems* **2017**, *20*, 1468–1482. [CrossRef]

58. Schindler, D.W. The dilemma of controlling cultural eutrophication of lakes. *Proc. R. Soc. B Biol. Sci.* **2012**, *279*, 4322–4333. [CrossRef]

59. Schindler, D.W.; Hecky, R.E.; Findlay, D.L.; Stainton, M.P.; Parker, B.R.; Paterson, M.J.; Beaty, K.G.; Lyng, M.; Kasian, S.E.M. Eutrophication of lakes cannot be controlled by reducing nitrogen input: Results of a 37-year whole-ecosystem experiment. *Proc. Natl. Acad. Sci. USA* **2008**, *105*, 11254–11258. [CrossRef] [PubMed]

60. Scott, J.T.; McCarthy, M.J.; Paerl, H.W. Nitrogen transformations differentially affect nutrient-limited primary production in lakes of varying trophic state. *Limnol. Oceanogr. Lett.* **2019**. [CrossRef]

61. Davis, T.; Harke, M.; Marcoval, M.; Goleski, J.; Orano-Dawson, C.; Berry, D.; Gobler, C. Effects of nitrogenous compounds and phosphorus on the growth of toxic and non-toxic strains of *Microcystis* during cyanobacterial blooms. *Aquat. Microb. Ecol.* **2010**, *61*, 149–162. [CrossRef]

62. Jankowiak, J.; Hattenrath-Lehmann, T.; Kramer, B.J.; Ladds, M.; Gobler, C.J. Deciphering the effects of nitrogen, phosphorus, and temperature on cyanobacterial bloom intensification, diversity, and toxicity in western Lake Erie. *Limnol. Oceanogr.* **2019**, *64*, 1347–1370. [CrossRef]

63. Stucken, K.; John, U.; Cembella, A.; Murillo, A.A.; Soto-Liebe, K.; Fuentes-Valdés, J.J.; Friedel, M.; Plominsky, A.M.; Vásquez, M.; Glöckner, G. the smallest known genomes of multicellular and toxic cyanobacteria: comparison, minimal gene sets for linked traits and the evolutionary implications. *PLoS ONE* **2010**, *5*, e9235. [CrossRef]

64. McKindles, K.M.; Manes, M.A.; DeMarco, J.R.; McClure, A.; McKay, R.M.; Davis, T.W.; Bullerjahn, G.S. Dissolved microcystin release coincident with lysis of a *Microcystis*-dominated bloom in western Lake Erie attributed to a novel cyanophage. *Appl. Environ. Microbiol.* **2020**. [CrossRef]

65. Thees, A.; Atari, E.; Birbeck, J.; Westrick, J.A.; Huntley, J.F. Isolation and characterization of Lake Erie bacteria that degrade the cyanobacterial microcystin toxin MC-LR. *J. Great Lakes Res.* **2019**, *45*, 138–149. [CrossRef]

66. Dolman, A.M.; Rücker, J.; Pick, F.R.; Fastner, J.; Rohrlack, T.; Mischke, U.; Wiedner, C. Cyanobacteria and cyanotoxins: the influence of nitrogen versus phosphorus. *PLoS ONE* **2012**, *7*, e38757. [CrossRef]

67. Beversdorf, L.J.; Miller, T.R.; McMahon, K.D. Long-term monitoring reveals carbon-nitrogen metabolism key to microcystin production in eutrophic lakes. *Front. Microbiol.* **2015**, *6*, 456. [CrossRef]

68. Harris, T.D.; Smith, V.H.; Graham, J.L.; de Waa, D.B.V.; Tedesco, L.P.; Clercin, N. Combined effects of nitrogen to phosphorus and nitrate to ammonia ratios on cyanobacterial metabolite concentrations in eutrophic Midwestern USA reservoirs. *Inl. Waters* **2016**, *6*, 199–210. [CrossRef]

69. Jover, L.F.; Effler, T.C.; Buchan, A.; Wilhelm, S.W.; Weitz, J.S. The elemental composition of virus particles: Implications for marine biogeochemical cycles. *Nat. Rev. Microbiol.* **2014**, *12*, 519–528. [CrossRef] [PubMed]

70. Oh, H.M.; Lee, S.J.; Jang, M.H.; Yoon, B.D. Microcystin production by *Microcystis aeruginosa* in a phosphorus-limited chemostat. *Appl. Environ. Microbiol.* **2000**, *66*, 176–179. [CrossRef] [PubMed]

71. Viaggiu, E.; Melchiorre, S.; Volpi, F.; Di Corcia, A.; Mancini, R.; Garibaldi, L.; Crichigno, G.; Bruno, M. Anatoxin-a toxin in the cyanobacterium *Planktothrix rubescens* from a fishing pond in northern Italy. *Environ. Toxicol.* **2004**, *19*, 191–197. [CrossRef] [PubMed]

72. Carrasco, D.; Moreno, E.; Paniagua, T.; de Hoyos, C.; Wormer, L.; Sanchis, D.; Cirés, S.; Martín-del-Pozo, D.; Codd, G.A.; Quesada, A. Anatoxin-a occurrence and potential cyanobacterial anatoxin-a producers in Spanish reservoirs. *J. Phycol.* **2007**, *43*, 1120–1125. [CrossRef]

73. Toporowska, M.; Pawlik-Skowrońska, B.; Kalinowska, R. Mass development of diazotrophic cyanobacteria (*Nostocales*) and production of neurotoxic anatoxin-a in a *Planktothrix (Oscillatoriales)* dominated temperate lake. *Water Air Soil Pollut.* **2016**, *227*, 1–13. [CrossRef]

74. Smith, Z.J.; Conroe, D.E.; Schulz, K.L.; Boyer, G.L. Limnological differences in a two-basin lake help to explain the occurrence of anatoxin-a, paralytic shellfish poisoning toxins, and microcystins. *Toxins (Basel)* **2020**, *12*, 559. [CrossRef]

75. Mahmood, N.A.; Carmichael, W.W. The pharmacology of anatoxin-a(s), a neurotoxin produced by the freshwater cyanobacterium Anabaena flos-aquae NRC 525-17. *Toxicon* **1986**, *24*, 425–434. [CrossRef]

76. Gagnon, A.; Pick, F.R. Effect of nitrogen on cellular production and release of the neurotoxin anatoxin-a in a nitrogen-fixing cyanobacterium. *Front. Microbiol.* **2012**, *3*, 211. [CrossRef]

77. Boopathi, T.; Ki, J.-S. Impact of environmental factors on the regulation of cyanotoxin production. *Toxins (Basel)* **2014**, *6*, 1951–1978. [CrossRef]

78. Osswald, J.; Rellán, S.; Gago-Martinez, A.; Vasconcelos, V. Production of anatoxin-a by cyanobacterial strains isolated from Portuguese fresh water systems. *Ecotoxicology* **2009**, *18*, 1110–1115. [CrossRef]

79. Neilan, B.A.; Pearson, L.A.; Muenchhoff, J.; Moffitt, M.C.; Dittmann, E. Environmental conditions that influence toxin biosynthesis in cyanobacteria. *Environ. Microbiol.* **2013**, *15*, 1239–1253. [CrossRef] [PubMed]

80. Heath, M.; Wood, S.A.; Young, R.G.; Ryan, K.G. The role of nitrogen and phosphorus in regulating *Phormidium* sp. (cyanobacteria) growth and anatoxin production. *FEMS Microbiol. Ecol.* **2016**, *92*, 21. [CrossRef] [PubMed]

81. Rapala, J.; Sivonen, K.; Luukkainen, R.; Niemelä, S.I. Anatoxin-a concentration in *Anabaena* and *Aphanizomenon* under different environmental conditions and comparison of growth by toxic and non-toxic *Anabaena*-strains—A laboratory study. *J. Appl. Phycol.* **1993**, *5*, 581–591. [CrossRef]

82. Stolerman, I.P.; Albuquerque, E.X.; Garcha, H.S. Behavioural effects of anatoxin, a potent nicotinic agonist, in rats. *Neuropharmacology* **1992**, *31*, 311–314. [CrossRef]

83. Rogers, E.H.; Hunter, E.S.; Moser, V.C.; Phillips, P.M.; Herkovits, J.; Muñoz, L.; Hall, L.L.; Chernoff, N. Potential developmental toxicity of anatoxin-a, a cyanobacterial toxin. *J. Appl. Toxicol.* **2005**, *25*, 527–534. [CrossRef]

84. Anderson, B.; Voorhees, J.; Phillips, B.; Fadness, R.; Stancheva, R.; Nichols, J.; Orr, D.; Wood, S.A. Extracts from benthic anatoxin-producing *Phormidium* are toxic to 3 macroinvertebrate taxa at environmentally relevant concentrations. *Environ. Toxicol. Chem.* **2018**, *37*, 2851–2859. [CrossRef]

85. Lovin, L.M.; Brooks, B.W. Global scanning of anatoxins in aquatic systems: Environment and health hazards, and research needs. *Mar. Freshw. Res.* **2020**, *71*, 689. [CrossRef]

86. Richards, R.P.; Baker, D.B.; Crumrine, J.P.; Stearns, A.M. Unusually large loads in 2007 from the Maumee and Sandusky Rivers, tributaries to Lake Erie. *J. Soil Water Conserv.* **2010**, *65*, 450–462. [CrossRef]

87. Ma, C.G.; Chang, E.K.M.; Wong, S.; Zhang, R.; Zhang, M.; Del Genio, A. Impacts of storm track variations on wintertime extreme precipitation and moisture budgets over the Ohio Valley and northwestern United States. *J. Clim.* **2020**, *33*, 5371–5391. [CrossRef]

88. Paerl, H.W.; Hall, N.S.; Hounshell, A.G.; Rossignol, K.L.; Barnard, M.A.; Luettich, R.A.; Rudolph, J.C.; Osburn, C.L.; Bales, J.; Harding, L.W. Recent increases of rainfall and flooding from tropical cyclones (TCs) in North Carolina (USA): Implications for organic matter and nutrient cycling in coastal watersheds. *Biogeochemistry* **2020**, *4*. [CrossRef]

89. Paerl, H.W.; Havens, K.E.; Hall, N.S.; Otten, T.G.; Zhu, M.; Xu, H.; Zhu, G.; Qin, B. Mitigating a global expansion of toxic cyanobacterial blooms: Confounding effects and challenges posed by climate change. *Mar. Freshw. Res.* **2019**. [CrossRef]

90. Trenberth, K.E. The impact of climate change and variability on heavy precipitation, floods, and droughts. In *Encyclopedia of Hydrological Sciences*; American Cancer Society: Atlanta, GA, USA, 2008; ISBN 978-0-470-84894-4.

91. Boedecker, A.R.; Niewinski, D.N.; Newell, S.E.; Chaffin, J.D.; McCarthy, M.J. Evaluating sediments as an ecosystem service in western Lake Erie *via* quantification of nutrient cycling pathways and selected gene abundances. *J. Great Lakes Res.* **2020**. [CrossRef]

92. *Preventing Eutrophication: Scientific Support for Dual Nutrient Criteria Fact sheet*; US EPA Office of Water: Washington, DC, USA, 2015.

93. Baker, D.B.; Johnson, L.T.; Confesor, R.B.; Crumrine, J.P.; Guo, T.; Manning, N.F. Needed: Early-term adjustments for Lake Erie phosphorus target loads to address western basin cyanobacterial blooms. *J. Great Lakes Res.* **2019**, *45*, 203–211. [CrossRef]

94. Levy, S. Microcystis Rising: Why Phosphorus Reduction Isn't Enough to Stop CyanoHABs. *Environ. Health Perspect.* **2017**, *125*, A34–A39. [CrossRef] [PubMed]

95. Xu, H.; Paerl, H.W.; Qin, B.; Zhu, G.; Hall, N.S.; Wu, Y. Determining critical nutrient thresholds needed to control harmful cyanobacterial blooms in eutrophic Lake Taihu, China. *Environ. Sci. Technol.* **2015**, *49*, 1051–1059. [CrossRef] [PubMed]

96. Müller, S.; Mitrovic, S.M. Phytoplankton co-limitation by nitrogen and phosphorus in a shallow reservoir: Progressing from the phosphorus limitation paradigm. *Hydrobiologia* **2015**, *744*, 255–269. [CrossRef]

97. Zohary, T.; Herut, B.; Krom, M.D.; Fauzi, C.; Mantoura, R.; Pitta, P.; Psarra, S.; Rassoulzadegan, F.; Stambler, N.; Tanaka, T.; et al. P-limited bacteria but N and P co-limited phytoplankton in the Eastern Mediterranean—A microcosm experiment. *Deep Sea Res. Part II Top. Stud. Oceanogr.* **2005**, *52*, 3011–3023. [CrossRef]

98. Wilhelm, S.W.; Bullerjahn, G.S.; McKay, R.M.L. The complicated and confusing ecology of *Microcystis* blooms. *mBio* **2020**, *11*, e00529-20. [CrossRef]
99. Stow, C.A.; Glassner-Shwayder, K.; Lee, D.; Wang, L.; Arhonditsis, G.; DePinto, J.V.; Twiss, M.R. Lake Erie phosphorus targets: An imperative for active adaptive management. *J. Great Lakes Res.* **2020**, *46*, 672–676. [CrossRef]
100. Testai, E.; Scardala, S.; Vichi, S.; Buratti, F.M.; Funari, E. Risk to human health associated with the environmental occurrence of cyanobacterial neurotoxic alkaloids anatoxins and saxitoxins. *Crit. Rev. Toxicol.* **2016**, *46*, 385–419. [CrossRef]
101. Paerl, H.W.; Hall, N.S.; Calandrino, E.S. Controlling harmful cyanobacterial blooms in a world experiencing anthropogenic and climatic-induced change. *Sci. Total Environ.* **2011**, *409*, 1739–1745. [CrossRef]
102. Xu, H.; Paerl, H.W.; Qin, B.; Zhu, G.; Gaoa, G. Nitrogen and phosphorus inputs control phytoplankton growth in eutrophic Lake Taihu, China. *Limnol. Oceanogr.* **2010**, *55*, 420–432. [CrossRef]
103. Hanson, P.C.; Carpenter, S.R.; Armstrong, D.E.; Stanley, E.H.; Kratz, T.K. Lake dissolved inorganic carbon and dissolved oxygen: Changing drivers from days to decades. *Ecol. Monogr.* **2006**, *76*, 343–363. [CrossRef]
104. Rockwell, D.C.; Warren, G.J.; Bertram, P.E.; Salisbury, D.K.; Burns, N.M. The U.S. EPA Lake Erie indicators monitoring program 1983-2002: Trends in phosphorus, silica, and chlorophyll a in the central basin. *J. Great Lakes Res.* **2005**, *31*, 23–34. [CrossRef]
105. Chapra, S.C.; Dove, A.; Warren, G.J. Long-term trends of Great Lakes major ion chemistry. *J. Great Lakes Res.* **2012**, *38*, 550–560. [CrossRef]
106. Shorthouse, D.P. SimpleMappr. Available online: https://www.simplemappr.net (accessed on 14 September 2020).
107. Arar, E.J.; Collins, G.B. *Method 445.0 In Vitro Determination of Chlorophyll a and Pheophytin in Marine and Freshwater Algae by Fluorescence*; United States Environmental Protection Agency: Washington, DC, USA, 1997.
108. Arar, E.J.; Collins, G.B. *Using the Turner Designs Model 10 Analog, The 10AU Field, or the TD-700 Laboratory Fluorometer with EPA Method 445.0*; United States Environmental Protection Agency: Washington, DC, USA, 2001.
109. Welschmeyer, N.A. Fluorometric analysis of chlorophyll a in the presence of chlorophyll b and pheopigments. *Limnol. Oceanogr.* **1994**, *39*, 1985–1992. [CrossRef]
110. Chaffin, J.D.; Mishra, S.; Kane, D.D.; Bade, D.L.; Stanislawczyk, K.; Slodysko, K.N.; Jones, K.W.; Parker, E.M.; Fox, E.L. Cyanobacterial blooms in the central basin of Lake Erie: Potentials for cyanotoxins and environmental drivers. *J. Great Lakes Res.* **2019**, *45*, 277–289. [CrossRef]
111. Goeyens, L.; Kindermans, N.; Abu Yusuf, M.; Elskens, M. A room temperature procedure for the manual determination of urea in seawater. *Estuar. Coast. Shelf Sci.* **1998**, *47*, 415–418. [CrossRef]
112. Mulvenna, P.F.; Savidge, G. A modified manual method for the determination of urea in seawater using diacetylmonoxime reagent. *Estuar. Coast. Shelf Sci.* **1992**, *34*, 429–438. [CrossRef]
113. Boyer, G.L. The occurrence of cyanobacterial toxins in New York lakes: Lessons from the MERHAB-Lower Great Lakes program. In *Lake and Reservoir Management*; Taylor & Francis Group: Abingdon, UK, 2007; Volume 23, pp. 153–160.
114. Peng, G.; Martin, R.M.; Dearth, S.P.; Sun, X.; Boyer, G.L.; Campagna, S.R.; Lin, S.; Wilhelm, S.W. Seasonally relevant cool temperatures interact with N chemistry to increase microcystins produced in lab cultures of *Microcystis aeruginosa* NIES-843. *Environ. Sci. Technol.* **2018**, *52*, 4127–4136. [CrossRef]
115. Boyer, G.L. LCMS-SOP Determination of microcystins in water samples by high performance liquid chromatography (HPLC) with single quadrupole mass spectrometry (MS). *Protocols* **2020**. [CrossRef]
116. Yang, X. *Occurrence of the Cyanobacterial Neurotoxin, Anatoxin-A, in New York State Waters*; State University of New York: Syracuse, NY, USA, 2007.
117. Barnard, M.A. GitHub: malcolmbarnard/WLE_Barnard_et_al_Toxins. *Zenodo* **2021**. [CrossRef]
118. *Mathworks MATLAB*, version 9.5.0.1049112 (R2018b); The MathWorks: Natick, MA, USA, 2018.

Article

Mass Occurrence of Anatoxin-a- and Dihydroanatoxin-a-Producing *Tychonema* sp. in Mesotrophic Reservoir Mandichosee (River Lech, Germany) as a Cause of Neurotoxicosis in Dogs

Franziska Bauer [1,*], Jutta Fastner [2], Bernadett Bartha-Dima [3], Wolfram Breuer [3], Almuth Falkenau [4], Christian Mayer [4] and Uta Raeder [1]

[1] Limnological Research Station Iffeldorf, Aquatic Systems Biology Unit, Technical University of Munich, Hofmark 1-3, 82393 Iffeldorf, Germany; uta.raeder@tum.de

[2] German Environment Agency, Schichauweg 58, 12307 Berlin, Germany; jutta.fastner@uba.de

[3] Bavarian Health and Food Safety Authority, Veterinärstraße 2, 85764 Oberschleißheim, Germany; Bernadett.Bartha-Dima@lgl.bayern.de (B.B.-D.); Wolfram.Breuer@lgl.bayern.de (W.B.)

[4] Center for Clinical Veterinary Medicine, Institute for Veterinary Pathology, Ludwig-Maximilians-University, Veterinärstraße 13, 80539 Munich, Germany; falkenau@patho.vetmed.uni-muenchen.de (A.F.); mayer@patho.vetmed.uni-muenchen.de (C.M.)

* Correspondence: Franziska.bauer@tum.de

Received: 28 October 2020; Accepted: 17 November 2020; Published: 20 November 2020

Abstract: In August 2019, three dogs died after bathing in or drinking from Mandichosee, a mesotrophic reservoir of the River Lech (Germany). The dogs showed symptoms of neurotoxic poisoning and intoxication with cyanotoxins was considered. Surface blooms were not visible at the time of the incidents. Benthic *Tychonema* sp., a potential anatoxin-a (ATX)-producing cyanobacterium, was detected in mats growing on the banks, as biofilm on macrophytes and later as aggregations floating on the lake surface. The dogs' pathological examinations showed lung and liver lesions. ATX and dihydroanatoxin-a (dhATX) were detected by LC-MS/MS in the stomachs of two dogs and reached concentrations of 563 and 1207 µg/L, respectively. Anatoxins (sum of ATX and dhATX, ATXs) concentrations in field samples from Mandichosee ranged from 0.1 µg/L in the open water to 68,000 µg/L in samples containing a large amount of mat material. Other (neuro)toxic substances were not found. A molecular approach was used to detect toxin genes by PCR and to reveal the cyanobacterial community composition by sequencing. Upstream of Mandichosee, random samples were taken from other Lech reservoirs, uncovering *Tychonema* and ATXs at several sampling sites. Similar recent findings emphasize the importance of focusing on the investigation of benthic toxic cyanobacteria and applying appropriate monitoring strategies in the future.

Keywords: cyanobacteria; anatoxin-a; dihydroanatoxin-a; *Tychonema*; neurotoxicosis; cyanotoxins; macrophytes; benthic; tychoplanktic; reservoir

Key Contribution: Occurrence of anatoxin-a- and dihydroanatoxin-a-producing *Tychonema* sp. in the reservoir Mandichosee of the river Lech with resulting intoxication of dogs.

---

1. Introduction

Cyanobacteria are common components in aquatic systems. Phototrophic cyanobacteria are ubiquitously distributed, can form mass populations under eutrophic conditions and certain strains of many taxa can produce different toxins [1–5]. Therefore, cyanobacteria can cause major problems in drinking water supplies, fish farming and bathing waters. Cyanotoxins are among the strongest biogenic toxins found in nature [6,7]. The toxicity of cyanotoxins can be ranked between the tetrodotoxin of marine poisonous animals and strychnine from *Strychnos nux-vomica*, known as arrow poison [6]. According to their mode of action, cyanotoxins can be divided into hepatotoxins, neurotoxins and cytotoxins, leading to liver or respiratory failure after ingestion of hazardous amounts of toxins [6]. Beyond this, effects reported after contact with cyanobacteria include, e.g., odor impairments, skin reactions and gastrointestinal problems. These are, however, rather attributed to other metabolites from cyanobacteria or associated bacteria than to cyanotoxins.

In the past, it was assumed that cyanobacterial blooms occur mainly in eutrophic waters [8]. Numerous findings demonstrate that planktic cyanobacteria dominate in lakes with higher trophic levels and that blooms are more frequent there [9–11]. However, it is known that cyanobacteria colonizing the bottom of rivers and lakes can also be found in waters with lower trophic levels and that they can produce the same toxins as planktic cyanobacteria [12]. Nevertheless, the knowledge about benthic cyanobacteria in lakes and rivers is still very limited. To date, studies on toxic benthic cyanobacteria are often descriptive and based on case studies. These were often performed in the context of animal poisonings, which have been increasingly reported in recent years [12]. Until now, the European Bathing Water Directive (Directive 76/160/EEC concerning the quality of bathing water) regulates only the surveillance of planktic cyanobacteria and their threat to humans.

For example, repeated incidents of dog poisoning occurred from toxic cyanobacteria in mesotrophic Lake Tegel (Berlin, Germany) not detected by standard surveillance protocols for bathing water. No planktic blooms were present at the time of the incidents. However, there was a mass occurrence of detached water moss (*Fontinalis antipyretica*), which was densely populated by the anatoxin-a (ATX)-producing cyanobacterium *Tychonema* [13]. This incident showed for the first time a problematic occurrence of tychoplanktic toxic cyanobacteria in a German lake. Therefore, Fastner et al. [13] requested a revision of the bathing water surveillance strategies at that time.

Since the first incident in May 2017, ATX-producing cyanobacteria associated with macrophytes have occurred regularly in Lake Tegel and, most probably, other dogs died in 2019 due to ATX (Fastner et al., unpublished data). The question arises as to whether this problem is limited to Lake Tegel or whether toxic benthic/tychoplanktic cyanobacteria are to date simply undiscovered and underestimated in German inland waters. In general, however, the incidents at Lake Tegel have increased attention on such cases. Apart from *Tychonema*, other cyanobacterial genera known for their capability of producing ATX are *Anabaena*, *Aphanizomenon*, *Cylindrospermum*, *Microcoleus*, *Oscillatoria*, *Phormidium* and *Raphidiopsis* [14–22].

In August 2019, three dogs died after bathing in or drinking from Mandichosee, a reservoir of the River Lech (Bavaria, South Germany). Intoxication was suspected and, in particular, poisoned baits were presumed, but poisoning by toxic cyanobacteria has also been taken into consideration. Aware of the incidents at Lake Tegel, the aim of the present study was to reveal the causes of the dog deaths by means of a sampling campaign initiated immediately after the incidents at Mandichosee.

Surface algal blooms were not visible in Mandichosee at the time of the incidents. Instead, there were red-brown mats on the shore areas as well as macrophytes in the shallow water densely covered with red-brown biofilms. Later, red-brown aggregations were also floating at the lake surface in the shore areas. Microscopy revealed mass abundances of benthic *Tychonema* sp. Toxin analyses were conducted from the dogs' stomachs, and different environmental samples. Furthermore, a molecular approach was used to detect toxin genes by PCR and to reveal the cyanobacterial community composition by Illumina MiSeq sequencing. In order to better estimate the extent of the distribution of benthic *Tychonema* sp., the Lech reservoirs were also sampled upstream.

2. Results

2.1. Macroscopic and Microscopic Appearance of Tychonema

Red-brown biofilms were detected on dead and living plant material of the macrophytes *Elodea*, *Potamogeton*, *Myriophyllum* and unidentified filamentous algae (Figure 1A) at the shorelines of the reservoir Mandichosee (Figure 2, sites 23.1–23.3). Furthermore, red-brown aggregations were found growing on the sediment (Figure 1A) along the shorelines (Figure 2, site 23.2). Near the bathing area and at the shores in the southern reservoir (Figure 2, site 23.1), red-brown slimy aggregates were floating at the surface from the second sampling date onwards (Figure 1C).

Figure 1. Appearance of benthic *Tychonema* sp. in Mandichosee (**A–C,E**) and River Lech upstream (**D,F**). (A) Red-brown biofilm at the shorelines and on dead plant material in Mandichosee. (**B**) Red-brown clumps floating on the water surface in Lechaue. (**C**) *Tychonema* aggregates floating on the water surface in Mandichosee. (**D**) Mass occurrence of *Tychonema* aggregates in the reservoir Schwabstadl. (**E**) Microscopy of red-brown biofilm taken from Mandichosee (×1000). (**F**) Microscopy of red-brown biofilm taken from Lechaue.

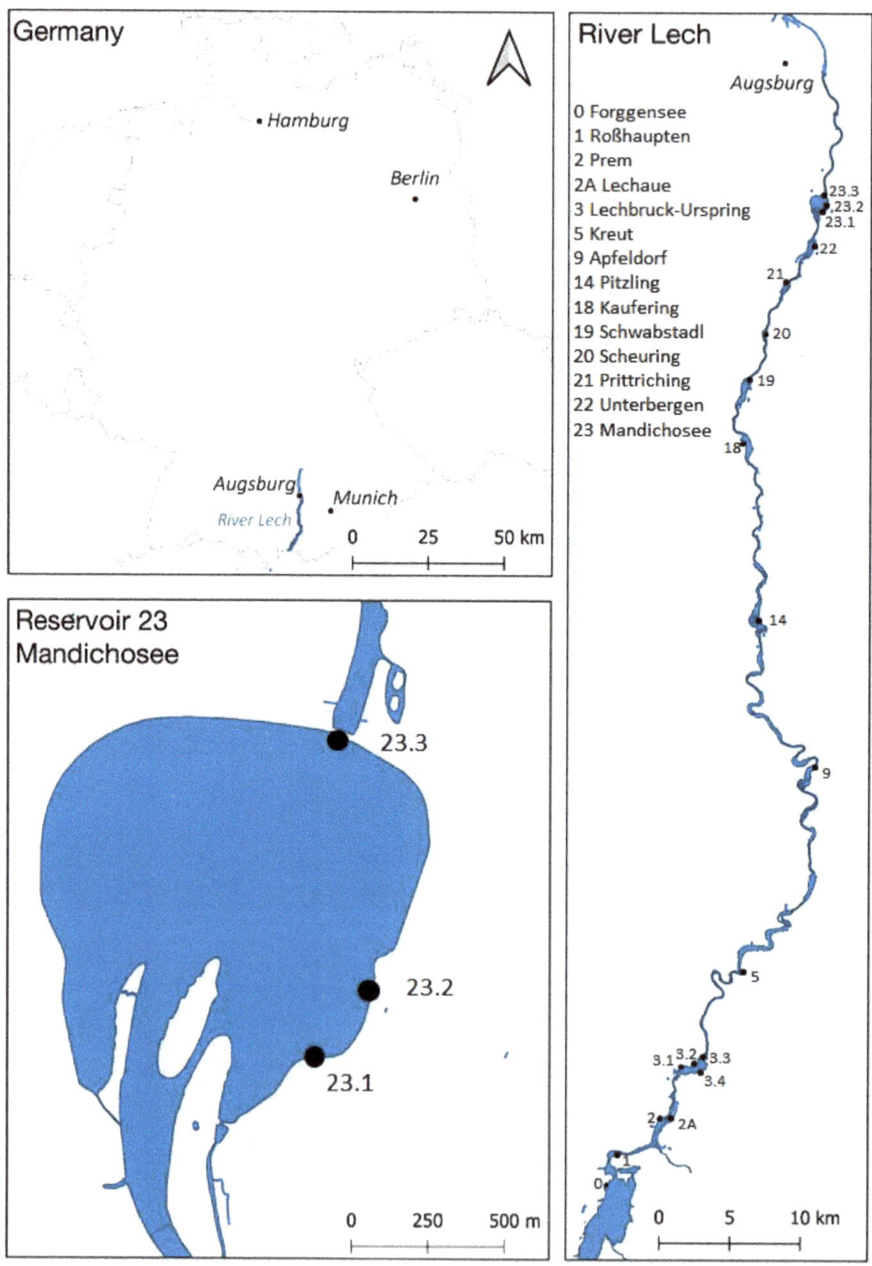

Figure 2. Map of the study area and the sampling sites (dots) in Mandichosee and River Lech. Site 23.1 is the bathing area where the dog casualties occurred. Built with QGIS (http://www.qgis.org), data/maps copyright: Geofabrik GmbH and OpenStreetMap Contributors (https://download.geofabrik.de/europe/germany/bayern.html).

Upstream of the River Lech, red-brown scums were also detected in different abundances in reservoirs Lechaue (2A), Lechbruck-Urspring (3.2 and 3.4), Kaufering (18), Schwabstadl (19), Scheuring (20) and Prittriching (21) (Figure 2). The shorelines of the reservoirs Schwabstadl (19) and Prittriching (21) were heavily covered with these scums (Figure 1D). In the reservoir Lechaue (2A), red-brown biofilm has also been found growing on dead wood.

The microscopic analyses of the biofilms of sampled plants, sediments and of the scums revealed filamentous cyanobacteria, which could be identified as *Tychonema* sp. based on their morphological traits (Figure 1E,F). Trichome widths varied between 5.5 and 14.0 μm. The trichomes of the *Tychonema* samples taken in August and early September 2019 in the northern waters between the reservoir Kaufering (18) and the reservoir Mandichosee (23) differed from those collected later in September 2019 from the southern reservoirs Lechaue (2A) and Lechbruck-Urspring (3): centripetal cross walls were clearly visible in northern samples (Figure 1E), whereas filaments were less structured in the southern samples and rather showed granulated cross walls (Figure 1F). An overview of the microscopic results is shown in Table S1 (Supplementary Material).

2.2. Animal Necroscopy and Histopathology

The symptoms of two dogs (D1 and D2) had been retching, salivation and dyspnoea. The third dog (D3) had shown acute seizures and oral discharge. Post-mortem examination of D1 and D2 revealed congested lungs and livers, accompanied by an edema of the wall of the gallbladder in the case of D1. In D3, the lung was diffusely solidified and the liver was moderately swollen. Histologically, the lungs of D1 and D2 showed slight congestion and edema, and in D3, severe pulmonary congestion, edema and mild diffuse interstitial pneumonia with type II pneumocyte hyperplasia, infiltration of interalveolar septa with macrophages, lymphocytes, plasma cells and single neutrophils and similar mild multifocal perivascular infiltration. The livers of D1 and D2 were mildly hyperemic. Sudan red stains were inconspicuous. The liver of D3 displayed some randomly scattered foci with minimal hepatocellular degeneration and resorptive infiltration. In D1, multifocal hemorrhage was seen in the brainstem. Furthermore, the stomach and intestine of D3 were affected by a mild multisegmental mucosal lymphoplasmacytic infiltration (mild gastroenteritis). Other organs showed no structural anomalies of inflammatory, degenerative, malformative or neoplastic origin.

2.3. Toxin Analyses

DhATX and ATX were detected in nearly all samples containing *Tychonema* mats, macrophytes covered with *Tychonema* and sediments, as well as in the stomachs of the dogs (Figure 3, Table 1). In contrast, they could hardly be found in the open water, even near the mats (Table 1). Cylindrospermopsin (CYN), deoxycylindrospermopsin (Deoxy-CYN), homoanatoxin (HATX) and dihydrohomoanatoxin (DhHATX) were not detected in any sample.

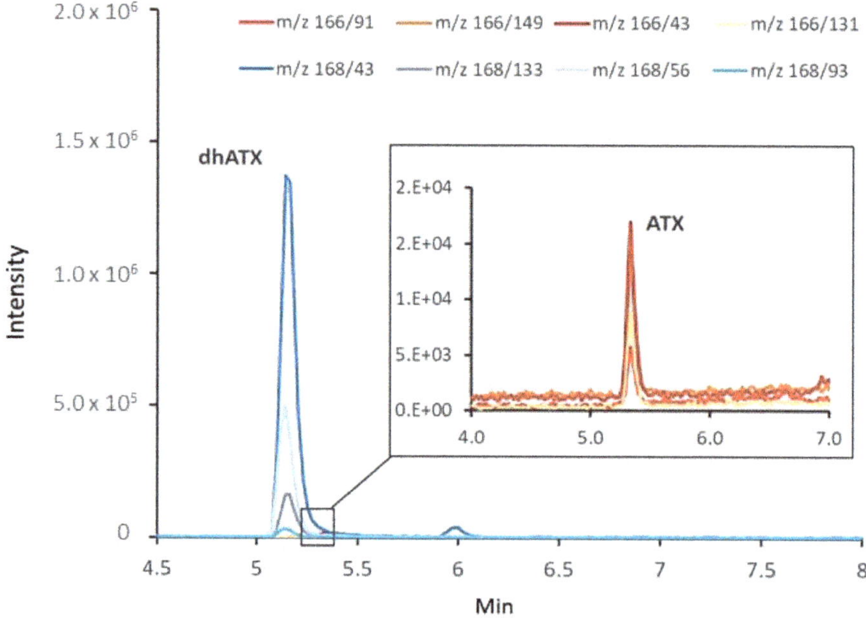

Figure 3. Reconstructed LC-MS/MS chromatogram from a sample from Mandichosee, 23 August 2019. Bluish lines represent transitions indicative for dihydroanatoxin-a (DhATX) and reddish ones, those for anatoxin-a (ATX).

Table 1. ATX and DhATX concentrations in samples from Mandichosee, other Lech reservoirs and the investigated dogs.

Sample Origin (Site)	Sample	Date (Day Month Year)	ATX (µg/L)	DhATX (µg/L)
Dog (D3)	Stomach content	14 August 19	n.d.	563
Dog (D1)	Stomach content	14 August 19	25.2	1182
Reservoir Mandicho (23.1)	Mat from shoreline	21 August 19	24.4–320	1917–39,528
Reservoir Mandicho (23.1)	Floating mat	21 August 19	241–453	16,954–67,622
Reservoir Mandicho (23.2)	Mat from shoreline	21 August 19	28.2–214	2375–23,488
Reservoir Mandicho (23.2)	Water and macrophytes	14 August 19	9.9–40.9	607–3012
Reservoir Mandicho (23.2)	*Elodea* with water	21 August 19	10.3–12.9	320–326
Reservoir Mandicho (23.2)	Macrophytes with water	21 August 19	n.d.	0.7
Reservoir Mandicho (23.2)	*Myriophyllum* with water	21 August 19	3.5	157
Reservoir Mandicho (23.1)	Macrophytes/Periphyton	23 August 19	7.7–272	3113–29,335

Table 1. Cont.

Sample Origin (Site)	Sample	Date (Day Month Year)	ATX (µg/L)	DhATX (µg/L)
Reservoir Mandicho (23.1)	Macrophytes and water	17 September 19	2.1	310
Reservoir Mandicho (23.1)	Macrophytes and water	17 September 19	0.1	0.2
Reservoir Mandicho (23.2)	Water surface above mat	21 August 19	<LOQ	2.7
Reservoir Mandicho (23.2)	Water surface 20 cm above mat	21 August 19	<LOQ	1.0
Reservoir Mandicho (23.2)	Water surface 50 cm above mat	21 August 19	<LOQ	1.4
Reservoir Mandicho (23.2)	Water surface 100 cm above mat	21 August 19	<LOQ	5.0
Reservoir Mandicho (23.2)	Water	14 August 19	<LOQ	3.8
Reservoir Mandicho (23.1)	Water	21 August 19	n.d.	0.6
Reservoir Prittriching (21)	Water	17 September 19	n.d.	n.d.
Reservoir Scheuring (20)	Water	17 September 19	n.d.	n.d.
Reservoir Schwabstadl (19)	Water	17 September 19	n.d.	0.7
Reservoir Kaufering (18)	Water	17 September 19	n.d.	n.d.

LOQ: Limit of Quantification.

The sum of ATX and dhATX in samples containing *Tychonema* (mats, macrophytes, sediments) ranged from 0.1 µg/L in the open water up to almost 68,000 µg/L in samples containing a very large amount of mat material (Table 1). Concentrations were overall higher in Mandichosee in August than in other reservoirs of the river Lech sampled in September.

Contents of ATXs in *Tychonema* mat biomass ranged between 86 and 353 µg/g fresh weight (FW) (Table S2, Supplementary Material). Assuming a dry weight (DW) of 10% of fresh biomass, contents per DW were between 860 and 3537 µg/g.

The concentration in the dogs' stomachs was 563 and 1207 µg/L, respectively (Table 1).

2.4. Detection of Toxin Genes

None of the samples contained the target genes for microcystin or saxitoxin. The target gene for ATX *anaC* could be detected in the samples taken from Mandichosee next to the bathing area (site 23.1, on 21 August 2019), next to the parking area (site 23.2, on 19 and 21 August and 10 September) and next to the power plant (site 23.3, on 19, 21 August and on 10 September 2019). Furthermore, *anaC* genes were detected in scum samples taken on 12 September 2019 upstream of Mandichosee in the reservoirs Schwabstadl (19) and Prittriching (21) (Figure 2, sites 19 and 21). The results are summarized in Table S1 (Supplementary Material).

2.5. Illumina MiSeq Sequencing

The first sequencing run revealed a total of about 600,000 sequences, clustered into 407 bacterial operational taxonomic units (OTUs). Following the SILVA classification [23], the most frequently occurring OTU was assigned to the cyanobacterial genus *Tychonema*. Besides the *Tychonema* OTU, there were five other cyanobacterial OTUs: three OTUs classified as *Cyanobium*, and one OTU each attributed to the genera *Limnothrix* and *Pseudanabaena*. Furthermore, two cyanobacterial OTUs were not classified down to genus level but were assigned to the order Oxyphotobacteria and Caenarcaniphilales (Melainabacteria). The run of the second part of the samples revealed a comparable result.

The samples from the reservoirs in Kaufering (18), Schwabstadl (19) and Scheuring (20) exclusively contained *Tychonema* as cyanobacterial OTUs. The *Tychonema* OTU was present in all samples taken from Mandichosee except the open water sample taken next to the power plant. Furthermore, the *Tychonema* OTU was present in the bank sample taken downstream of the power plant at dam 23, which forms the reservoir Mandichosee. In addition, *16S rRNA gene* sequences of *Tychonema* were found in several samples taken in the reservoirs of River Lech upstream. In all these samples, the presence of *Tychonema* was already suspected based on macroscopic examination and detected microscopically: Lechaue (2A), Lechbruck-Urspring (3.2, 3.4), Kaufering (18), Schwabstadl (19), Scheuring (20) and Prittriching (21). In the samples of reservoirs Prem (2), Kreut (5), Apfeldorf (9) and Unterbergen (22), *Tychonema* could only be identified by sequencing. Reservoirs Forggensee (0), Roßhaupten (1) and sampling site Pitzling (14) were tested negatively for *Tychonema* genes. An overview of *Tychonema*-positive and -negative sampling sites is given in Table S1 (Supplementary Material).

2.6. Hydro-Chemical and Hydro-Physical Characterization

Hydro-chemical and hydro-physical parameters are summarized in Table S1 (Supplementary Material). The total phosphorus contents of 10 samples, which were taken from the site below the reservoir Pitzling (14) downstream to the reservoir Mandichosee (23) varied between 11.1 and 20.3 µg/L. These few phosphorus values give a first indication of the trophic status of water bodies, which can be classified between oligo-mesotrophic and mesotrophic according to Schneider et al. [24]. In all samples, the soluble reactive phosphorus was below the detection limit of 5 µg/L. The concentrations of inorganic nitrogen were only measured in the reservoir Mandichosee. The nitrate nitrogen content was 0.5 mg/L and the ammonia nitrogen content was 0.02 mg/L.

The hydro-physical parameters measured at different times along the shoreline of the reservoir Mandichosee were almost identical, except for oxygen saturation in regions with and without benthic *Tychonema* occurrence. The surface temperature varied between 14.3 and 19.8 °C, pH values between 8.4 and 8.7, and conductivity between 307 and 341 µS/cm. Oxygen saturation varied between 115 and 142%, with supersaturations indicating photosynthetic activity of *Tychonema* aggregations.

Additionally, the investigated sites in the other reservoirs of the River Lech where *Tychonema* was found were described hydro-physically. The temperature ranged from 11.8 °C to 19.3 °C. The pH values ranged between 8.1 and 9.4. The conductivity was between 324 and 494 µS/cm and thus tended to be slightly higher than in the reservoir Mandichosee. The oxygen saturation was between 96 and 177%. The high supersaturations were determined directly in the *Tychonema* aggregates. At these locations, an optically recognizable bubble formation already indicated the oxygen production by the primary producers.

3. Discussion

The filamentous cyanobacteria detected in mats from Mandichosee and the River Lech were identified as *Tychonema* due to their morphological traits [25]. This presumption was confirmed by molecular biological methods. The cyanobacteria filaments occurred as a benthic species and could also be found in the form of floating scums. Although the cell widths overlap with other *Tychonema* species, such as *Tychonema bourellyi* (4.0–6.0 µm) or *T. tenue* (5.5–8 µm), *Tychonema* in the River Lech is probably *Tychonema bornetii* due to its benthic occurrence. Trichome widths of *Tychonema bornetii* usually vary between 7.0 and 16 µm [25]. The DNA sequencing confirmed *Tychonema* to the genus level and indicated that it was only a single species and not several different *Tychonema* species. Nevertheless, morphological and molecular differentiation is difficult and sometimes ambiguous. The cells at the sampling sites in the southern part of the River Lech sometimes also showed morphological characteristics of *Phormidium*. Based on *16S rRNA gene* analyses, *Tychonema* and *Phormidium* are very similar and are assigned to a common lineage together with *Microcoleus* [26].

Literature on the occurrence of *Tychonema* is limited, though, in part, this may be due to the taxonomic separation of the genus *Tychonema* from *Oscillatoria* in 1988 by Anagnostidis and Komárek [27]. However, in recent years, the occurrence of *Tychonema* has been increasingly reported. A massive occurrence of tychoplanktic, ATX-producing *Tychonema* sp. has recently been described for Lake Tegel, Berlin, Germany [13]. As was the case with Mandichosee, the neurotoxicosis of several dogs attracted attention and led to further investigations. In Lake Tegel, *Tychonema* sp. was found primarily associated with water moss (*Fontinalis antipyretica*). Trichome widths of *Tychonema* sp. originating from Lake Tegel were narrower than those found in Mandichosee and ranged only between 6.8 and 8.9 µm. In contrast to Lake Tegel, the *Tychonema* found in Mandichosee and the River Lech was observed growing in the form of benthic mats also overgrowing various macrophytes (*Elodea*, *Myriophyllum* and *Potamogeton* species).

Tychonema has also been found in the large lakes of the Southern Alps [28–30]. However, in contrast to the findings in Lake Tegel and Mandichosee, the pelagic form *Tychonema bourellyi* is present in these lakes [28].

Tychonema is known as cold-stenotherm genus with prevalence in clear water bodies in northern temperate regions [31]. The optimum growth temperature of a *T. bourellyi* strain from Cumbrian lakes was determined between 17 °C and 25 °C [32]. Comparably, *Tychonema* in Lake Tegel and Lake Garda show their growth maximum in spring at around 17 °C and disappear in summer with higher temperatures [13,29,30]. The water temperature in the River Lech and reservoir Mandichosee at *Tychonema*-positive sampling sites ranged between 11.8 and 19.8 °C. As the measurements started only after the dog casualties had occurred, the intensive growth phase may have already occurred when temperatures were higher. Monthly monitoring during bathing water surveillance revealed a maximum surface temperature of 24 °C in July 2019. It has already been presumed earlier that, in addition to temperature, the oligotrophication of lakes may play an important role, leading to shifts in cyanobacterial communities in favor of *Tychonema* species [15]. This is true for the lakes south of the Alps or Lake Tegel, but the water quality of mesotrophic Mandichosee did not change in recent years. It seems that *Tychonema* meets its requirement for cold, clear water bodies in middle Europe in spring in mesotrophic lakes and rivers or in rivers having their origin in mountains, but can also tolerate higher temperatures.

High concentrations of ATXs (sum of dhATX and ATX) were found in the reservoir Mandichosee, with the proportion of dhATX usually accounting for more than 95%. Also in other producers such as *Oscillatoria*, *Cylindrospermum stagnale* and *Phormidium/Microcoleus*, the concentration of dhATX was shown to be much higher than that of ATX [26,33–36].

Toxins from benthic mats are usually related to biomass, however, risk assessment often requires information on toxin concentrations per liter of water. As benthic mats are not homogeneously distributed in the water body, such as planktic cyanobacteria, the determined toxin concentrations depend on the biomass of mats in a certain water volume. We have tried to simulate a possible up-take scenario and analyzed some parts of mat samples in a small volume of water (1 mL). Expressing the toxin concentration per liter then lead to extremely high concentrations of up to 68,000 ATXs µg/L, showing that large amounts of toxins can be taken up with a small volume of water when it contains a large mat fraction. ATX concentrations in the range of mg per liter have also been reported in tychoplanktic *Tychonema* and benthic *Phormidium* species [13,36].

ATX content in mat samples from Mandichosee were in the same mg per DW range as observed for *Cylindrospermum stagnale* (1200 µg dhATX/g DW) [35] and *Kamptonema* (*Phorm.*) *formosum* (8000 µg/g DW) [37].

The presence of anatoxins in mats and in the dogs' stomachs indicates that they are the most likely reason for the neurotoxicosis of dogs at Mandichosee. The large amount of dhATX in the samples suggests this toxin to be the primary toxic agent, however, the data on the toxicity of dhATX in relation to ATX are contradictory. While earlier studies suggest that dhATX is about ten times less toxic than ATX [38,39], a more recent study shows that an intraperitoneal (i.p.) injection resulted in a lower toxicity of dhATX compared to ATX, while oral consumption caused greater toxicity [40]. Even if dhATX is less toxic than ATX, its high contents in *Tychonema* mats in Mandichosee can easily lead to the intake of lethal concentrations. High concentrations of ATXs in bathing waters such as in the reservoir Mandichosee are not only hazardous for dogs, but also for small children taking up mat material while playing with them or during bathing accidents. Adults are of lower risk as they would avoid swallowing mat material unless during near-drowning situations. In contrast to mammals, macrozoobenthos organisms such as *Chironomus* and *Deleatidium* larvae proved to be very resistant to high ATX concentrations (300–600 µg/L) and showed almost no mortality and no or only low accumulation of ATX [41,42]. Also the rainbow trout (*Oncorhynchus mykiss*) was not affected at similar concentrations but accumulated up to 23 µg ATX/g BW [43]. The consumption of such contaminated fish does not result in an acute dose but may exceed the tolerable daily intake concentration value (for lifetime consumption) of ATX [44].

The incidents at Mandichosee have shown that mass occurrences of benthic cyanobacteria cannot be detected with the current surveillance strategies of bathing waters. There are several reasons for this: When assessing the threat from cyanobacteria, the focus of monitoring is mainly on nutrient-rich stagnant and slow-flowing waters where primarily planktonic cyanobacteria may be present. The turbidity of the water body, determined as Secchi transparency, and cyanobacterial abundance (biovolume or cyano-chlorophyll-a) are the decisive parameters for more detailed analyses and, if indicated, for the decision to prohibit bathing. Special attention is given to surface accumulations of cyanobacteria. Thus, benthic mass occurrences of cyanobacteria are not surveyed and are usually only discovered after animal fatalities. To our knowledge, only New Zealand currently includes surveying benthic cyanobacteria in their guidelines for recreational fresh waters [45].

However, observations in recent years indicate an increase in potential toxic benthic cyanobacteria, even in German lakes. The reason for this phenomenon is not clear. Possible reasons are higher transparency due to improved water quality, as has also been described for Lake Tegel in recent years [46] or due to sinking water levels in dry summers. The fact that the incidents have occurred mainly in recent years suggest that they might be directly or indirectly connected to climate change. The results of the present study indicate that not only Mandichosee is affected by the toxic benthic *Tychonema* occurrence, but also other reservoirs upstream and downstream, since it is a system connected by the River Lech.

4. Conclusions

Based on the preliminary findings, a systematic monitoring of the *Tychonema* occurrence at the Lech reservoirs is proposed. If dog deaths had not occurred at Lake Tegel and the reservoir Mandichosee, the occurrence of the toxin producer *Tychonema* would almost certainly not have been detected. It cannot be excluded that *Tychonema* is already present in more lakes than is currently known and has not yet been detected due to lack of knowledge and insufficient monitoring strategies. Literature on the conditions for *Tychonema* occurrence is very limited to date. Therefore, it is important to investigate the distribution and growth conditions of the toxin-producing cyanobacterium *Tychonema* in detail, to focus on the investigation of toxic benthic cyanobacteria and to develop appropriate surveillance strategies.

5. Materials and Methods

5.1. Study Site

The investigations have focused on the reservoir Mandichosee. This reservoir is situated in Southern Bavaria, Germany, near the city of Augsburg (48°15′45″ N. 10°56′0″ E). With a surface area of 1.6 km^2, it is the second largest of the numerous reservoirs of the River Lech. Due to its proximity to a metropolitan area, Mandichosee is a very popular lake, which is used intensively for leisure activities and recreation. According to the local district authorities, Mandichosee has been known for excellent bathing water quality.

With a length of 256 km, the River Lech is one of the largest and most important rivers in Bavaria. The River Lech has its source in Vorarlberg, Austria, and streams through South Bavaria before it flows into the River Danube (Bavaria, Germany). As a typical alpine river, the River Lech was originally characterized by early summer floods due to snow melting and since precipitation in the Alpine region is very high in summer, flood events could theoretically occur at any time. As in all alpine rivers, the water temperature was mostly low, oxygen content and flow velocity high. To protect against flooding by regulating the water level and to generate electricity by hydro power, 23 dams and a cascade of reservoirs were built along the River Lech over a distance of about 90 km from 1954 to 1983. In the south, the largest and highest situated reservoir, the Forggensee, was first put into operation, and successively the other ones, of which the Mandichosee is the most northerly. These reservoirs have a great influence on the temperature, the flow velocity and the bed load of the River Lech. Because of this, it is considered to be the most modified river in Bavaria today. Along the River Lech, there are several protected areas for the use of drinking water supply. Various reservoirs serve for leisure and recreational purposes.

5.2. Animal Necroscopy and Histopathology

Two of the accidentally intoxicated dogs found along the reservoir Mandichosee (D1: Husky, adult, spayed; D2: Yorkshire terrier, juvenile, male) were submitted to the Bavarian Health and Food Safety Authority for necropsy. Formalin-fixed specimens (brain, bone marrow, heart, lung, kidney, liver, intestine) were processed routinely for histological investigation (paraffin embedded, cut in 4 µm thick sections, mounted on glass slides and stained with hematoxylin and eosin). In addition, formalin-fixed frozen tissue (liver) was stained with Sudan red (14 µm thick sections, mounted on glass slides).

In the third case (D3: Jack Russell terrier, juvenile, male), necropsy was conducted at the Institute of Veterinary Pathology of the University of Munich according to a standardized post-mortem examination protocol. For histopathological examination, tissue samples of several organs were taken considering a defined collection protocol. With reference to the preliminary report (acute seizures, oral discharge), a special neuropathological examination was performed according to directives of the International Veterinary Epilepsy Task Force (IVETF). The formalin-fixed (7% neutral-buffered formaldehyde) and paraffin-embedded samples were routinely processed and stained with hematoxylin and eosin as well as Giemsa stain.

After the histopathological examination, the stomach contents of D1 and D3 were sent from the Bavarian Health and Food Safety Authority (LGL) to the German Environment Agency (UBA) for toxin analyses by LC-MS/MS.

5.3. Sampling and in Situ Measurements of Hydro-physical Parameters

Sampling and in situ measurements at the reservoir Mandichosee were carried out on 19 August, 21 August and 10 September, at the reservoirs Kaufering (18), Schwabstadl (19), Scheuring (20), Prittriching (21) and Unterbergen (22) on 12 September, and at the reservoirs Forggensee (0), Roßhaupten (1), Prem (2), Lechaue (2a), Lechbruck-Ursprung (3), Kreut (5), Dornau (6), Apfeldorf (9) and Pitzling (14) on 19 September 2019.

Samples were taken several times at different locations along the shoreline of the reservoir Mandichosee, specifically in the bathing area (23.1) where the dog casualties occurred (see introduction), and once at each location of the other reservoirs (Figure 2).

At each sampling, benthos from different substrates (sediment, stones, wood), detached benthos floating as tychoplanktic material and macrophytes, e.g., *Elodea*, *Myriophyllum*, were collected for microscopic and molecular analysis. For further molecular analyses, 0.5 L of water were taken from the benthic or floating material, from the plant stock and from the adjacent free water. In addition, 0.5 L water were taken at selected sampling sites for hydrochemical analyses. During each sampling campaign, the hydro-physical parameters, i.e., temperature, pH value, conductivity and oxygen saturation, were measured directly on site with a multi-parameter probe (MPP 930 IDS, WTW, Weilheim, Germany).

Additional samples for toxin measurements were taken on 14, 21, 23 August and 17 September from Mandichosee and on 17 September from reservoirs 18–22. The samples originate from different sample types (water, algal mat, macrophytes). The exact sample origins can be found in Table 1.

5.4. Post-Processing of the Samples

The water samples for the molecular analysis were filtered through 0.2 μm filters (Sartorius, Göttingen, Germany), and stored at −20 °C until DNA extraction.

The red-brown biofilms of benthic mats and macrophytes were isolated with a sterile inoculation loop, transferred into 1.5 mL Eppendorf tubes and were stored at −20 °C until DNA extraction.

For toxin analysis, sub-samples were mostly taken from well-mixed water, water/mat or water/macrophyte samples. In order to relate ATX contents to fresh weight, a few sub-samples were prepared by weighing around 100 mg fresh mat material and adding 1 mL of 0.1% formic acid. All sub-samples were frozen prior to analysis.

5.5. Toxin Analyses

Analysis of CYN, deoxy-CYN, ATX and HATX as well as dhATX and dhHATX, was carried out as detailed previously [13]. Briefly, water samples with or without parts of *Tychonema* mats or macrophytes covered with *Tychonema* and the stomach contents of the dogs were acidified with formic acid to a final concentration of 0.1% formic acid and frozen/thawed twice. The samples were ultrasonicated for 10 min, shaken for 1 h, centrifuged and filtered (0.2 μm, PVDF, Whatman, Maidstone, UK) before analysis by LC-MS/MS.

LC-MS analysis was carried out on an Agilent 2900 series HPLC system (Agilent Technologies, Waldbronn, Germany) coupled to an API 5500 QTrap mass spectrometer (AB Sciex, Framingham, MA, USA) equipped with a turbo-ionspray interface. Chromatographic separation of 10 μl crude extracts was performed on an Atlantis C18 column (2.1 mm, 150 mm, Waters, Eschborn, Germany) at 30 °C. Compounds were eluted at a flow rate of 0.25 mL min^{-1} using a linear gradient of 0.1% formic acid in water (A) and 0.1% formic acid in methanol (B) from 1% to 25% B within 5 min.

Identification of CYN, deoxy-CYN, ATX and HATX was performed in the positive multiple reaction monitoring (MRM) mode with the following transitions: CYN *m/z* 416.1 [M + H]+ to 194 and 416.1/176, deoxy-CYN *m/z* 400.1 [M + H]+ to 320 and 400.1/194, ATX *m/z* 166.1 [M + H]+ to 149, 166.1/131, 166.1/91, 166.1/43 and HATX *m/z* 180.0 [M + H]+ to 163, 180.0/145 and 180.0/91. CYN and ATX were from the National Research Council (Canada) and deoxy-CYN and HATX were from Novakits (Nantes, France). The detection limit for CYN and deoxy-CYN was 0.02 μg L^{-1}, for ATX 0.02 μg L^{-1} and for HATX 0.04 μg L^{-1}.

In addition, dihydroanatoxin-a (DhATX) and dihydrohomoanatoxin-a (DhHATX) were analyzed using fragment ions described in the literature as no reference material was available [47]. DhATX was quantified with ATX using the fragment *m/z* 43 present in both substances as a quantifier ion.

5.6. Microscopy

Algae or cyanobacteria trichomes were isolated immediately after sampling in the laboratory with a sterile inoculation loop from the benthic mats, from the floating material or from the macrophytes and transferred to microscopic slides. The subsequent analysis was performed on a microscope (Leica DMRBE, Leica, Wetzlar, Germany) with photographic equipment (Zelos 285 GV, Kappa, Norderstedt, Germany). Potential *Tychonema* trichomes were measured, morphologically characterized and photographed.

5.7. Molecular Approach

5.7.1. DNA Extraction and Purification

DNA was extracted from the filters or the Eppendorf tubes using the phenol/chloroform-based method described by Zwirglmaier et al. [48]. Subsequently, DNA was purified using Sure Clean Plus (Bioline, London, UK) and eluted in sterile ultrapure water. The DNA was then stored frozen until further analysis.

5.7.2. Polymerase Chain Reaction

Polymerase chain reaction was performed to detect the target genes for ATX, microcystin (MCY) and saxitoxin (SXT). Primer pairs for the target genes *anaC*, *mcyE* and *sxtA* were anxgen-F and anxgen-R [21], mcyE-F2 and mcyE-R4 [49] and sxtA-F and sxtA R [50]. The strain PCC6506 (*Oscillatoria* sp., obtained from the Pasteur Culture Collection of Cyanobacteria, Paris, France) served as positive control for *anxC*, strain SAG14.85 (*Microcystis aeruginosa*, obtained from the Culture Collection of Algae at Göttingen University, Göttingen, Germany) served as positive control for *mcyE* and strain NIVA-CYA655 (*Aphanizomenon gracile*, obtained from the Norwegian Culture Collection of Algae, Oslo, Norway) as positive control for *sxtA*. The cycling conditions in the CFX thermocycler (CFX Connect, Biorad, Germany) were as follows: 94 °C for 10 min, 30 cycles of 94 °C for 30 s, annealing temperature for 30 s and 72 °C for 30 s, followed by a final elongation step of 72 °C for 7 min. Annealing temperatures were 55.8 °C for *anaC*, 56 °C for *mcyE* and 57.5 °C for *sxtA*.

5.7.3. Sequencing

Samples were bidirectionally sequenced at the Core Facility Microbiome, ZIEL, TUM, Freising using Illumina MiSeq v3 2 × 300 paired-end sequencing. Polymerase chain reaction (PCR) primers used for the first step were S-DBact-0341-b-S-17 (5′ *TCGTCGGCAGCGTCAGATGTGTATAAGAG ACAGCCTACGGGNGGCWGCAG* 3′) and S-D-Bact-0785-a-A-21 (5′ *GTCTCGTGGGCTCGGAGATGTG TATAAGAGACAGG*ACTACHVGGGTATCTAATCC 3′) (Illumina overhang adapter in italics). These primers cover the *16S rRNA gene* variable regions V3-V4. These hypervariable regions combined with a paired-end sequence configuration are recommended as the most effective study design [51]. Processing of Illumina MiSeq sequence data was done within IMNGS [52]. The operational taxonomic units (OTUs) were clustered at 97% sequence identity and subsequently classified with SINA online [53]. For classification, SILVA taxonomy [23] was used, which was implemented in SINA online. Sequencing was carried out in two runs, which were analyzed separately. Sequence data have been submitted to NCBI's Sequence Reads Archive (BioProject IDs PRJNA671879 and PRJNA672047).

5.8. Hydrochemical Analysis

The water samples for hydrochemical analysis were filtered through a 0.2 μm pore size membrane filter (GE Healthcare, Amersham, UK) prior to measurement of nitrate–nitrogen (NO_3^--N) and ammonium–nitrogen (NH_4^+-N) concentrations. Total phosphorus (TP) concentration was calculated from unfiltered water samples. Values of TP and NH_4^+-N were determined following established methods by the German Chemists' Association [54]. NO_3^--N values were determined using the method described by [55].

Supplementary Materials: The following are available online at http://www.mdpi.com/2072-6651/12/11/726/s1, Table S1: Sample overview and results of microscopy, hydro-physical and hydro-chemical measurements as well as molecular analyses, Table S2: ATX and DhATX contents in mat material from Mandichosee.

Author Contributions: Conceptualization, F.B., J.F. and U.R.; Data curation, F.B., J.F., W.B., A.F. and C.M.; Formal analysis, F.B.; Funding acquisition, F.B., J.F. and U.R.; Investigation, F.B., J.F., B.B.-D. and U.R.; Methodology, F.B., J.F., W.B., A.F. and C.M.; Project administration, F.B., J.F., B.B.-D. and U.R.; Resources, F.B., J.F., B.B.-D., W.B., A.F., C.M. and U.R.; Software, F.B. and J.F.; Supervision, J.F. and U.R.; Validation, F.B., J.F., B.B.-D., W.B., A.F., C.M.

and U.R.; Visualization, F.B. and J.F.; Writing—original draft, F.B., J.F., W.B., A.F. and C.M.; Writing—review and editing, B.B.-D. and U.R. All authors have read and agreed to the published version of the manuscript.

Funding: This research was conducted within the framework of the CYTOXKLIMA study (grant number K3-8503-PN 17-12-D17105/2017) as part of the network initiative "Climate Change and Health" (https://www.vkg.bayern.de) funded by the Bavarian State Ministry of the Environment and Consumer Protection and the Bavarian State Ministry of Health and Care.

Acknowledgments: We sincerely thank Katrin Zink (LfU) and Veronika Weilnhammer (LGL) for constructive discussion and valuable suggestions and David Konkol (UBA, Berlin) for skilled technical assistance in the laboratory. Furthermore, we would also like to thank Markus Motzke, chairman of the water rescue service Mering, for the information on the incidents with the dead dogs and for his observations at the reservoir Mandichosee.

Conflicts of Interest: The authors declare no conflict of interest.

References

1. Carmichael, W.W. (Ed.) Freshwater blue-green algae (cyanobacteria) toxins—A review. In *The Water Environment*; Springer: Boston, MA, USA, 1981; pp. 1–13. [CrossRef]
2. Sivonen, K. Cyanobacterial toxins. In *Encyclopedia of Microbiology*; Schaechter, M., Ed.; Elsevier: Oxford, UK, 2009; pp. 290–307.
3. Van Apeldoorn, M.E.; van Egmond, H.P.; Speijers, G.J.A.; Bakker, G.J.I. Toxins of cyanobacteria. *Mol. Nutr. Food Res.* **2007**, *51*, 7–60. [CrossRef]
4. Jakubowska, N.; Szeląg-Wasielewska, E. Toxic picoplanktonic cyanobacteria. *Mar. Drugs* **2015**, *13*, 1497–1518. [CrossRef] [PubMed]
5. Leão, P.N.; Niclas, E.; Antunes, A.; Gerwick, W.H.; Vasconcelos, V. The chemical ecology of cyanobacteria. *Nat. Prod. Rep.* **2012**, *29*, 372–391. [CrossRef] [PubMed]
6. Morscheid, H.; Fromme, H.; Krause, D.; Kurmayer, R.; Morscheid, H.; Teuber, K. *Toxinbildende Cyanobakterien (Blaualgen) in Bayerischen Gewässern*; Bayerisches Landesamt für Umwelt: Augsburg, Germany, 2006; Volume 125, pp. 1–145.
7. Buratti, F.M.; Manganelli, M.; Vichi, S.; Stefanelli, M.; Scardala, S.; Testai, E.; Funari, E. Cyanotoxins: Producing organisms, occurrence, toxicity, mechanism of action and human health toxicological risk evaluation. *Arch. Toxicol.* **2017**, *91*, 1049–1130. [CrossRef] [PubMed]
8. Merel, S.; Walker, D.; Chicana, R.; Snyder, S.; Baurès, E.; Thomas, O. State of knowledge and concerns on cyanobacterial blooms and cyanotoxins. *Environ. Int.* **2013**, *59*, 303–327. [CrossRef] [PubMed]
9. O'Neil, J.M.; Davis, T.W.; Burford, M.A.; Gobler, C.J. The rise of harmful cyanobacteria blooms: The potential roles of eutrophication and climate change. *Harmful Algae* **2012**, *14*, 313–334. [CrossRef]
10. Xu, H.; Paerl, H.W.; Qin, B.; Zhu, G.; Hall, N.S.; Wu, Y. Determining critical nutrient thresholds needed to control harmful cyanobacterial blooms in eutrophic Lake Taihu, China. *Environ. Sci. Technol.* **2015**, *49*, 1051–1059. [CrossRef]
11. Huisman, J.; Codd, G.A.; Paerl, H.W.; Ibelings, B.W.; Verspagen, J.M.H.; Visser, P.M. Cyanobacterial blooms. *Nat. Rev. Microbiol.* **2018**, *16*, 471–483. [CrossRef]
12. Quiblier, C.; Wood, S.; Echenique-Subiabre, I.; Heath, M.; Villeneuve, A.; Humbert, J.F. A review of current knowledge on toxic benthic freshwater cyanobacteria–ecology, toxin production and risk management. *Water Res.* **2013**, *47*, 5464–5479. [CrossRef]
13. Fastner, J.; Beulker, C.; Geiser, B.; Hoffmann, A.; Kröger, R.; Teske, K.; Hoppe, J.; Mundhenk, L.; Neurath, H.; Sagebiel, D.; et al. Fatal neurotoxicosis in dogs associated with tychoplanktic, anatoxin-a producing *Tychonema* sp. in mesotrophic Lake Tegel, Berlin. *Toxins* **2018**, *10*, 60. [CrossRef]
14. Namikoshi, M.; Murakami, T.; Watanabe, M.; Oda, T.; Yamada, J.; Tsujimura, S.; Nagai, H.; Oishi, S. Simultaneous production of homoanatoxin-a, anatoxin-a, and a new non-toxic 4-hydroxyhomoanatoxin-a by the cyanobacterium *Raphidiopsis mediterranea* Skuja. *Toxicon* **2003**, *42*, 533–538. [CrossRef]
15. Aráoz, R.; Nghiem, H.; Rippka, R.; Palibroda, N.; Tandeau de Marsac, N.; Herdman, M. Neurotoxins in axenic oscillatorian cyanobacteria: Coexistence of anatoxin-a and homoanatoxin-a determined by ligand-binding assay and GC/MS. *Microbiology* **2005**, *151*, 1263–1273. [CrossRef]

16. Méjean, A.; Mann, S.; Maldiney, T.; Vassiliadis, G.; Lequin, O.; Ploux, O. Evidence that Biosynthesis of the Neurotoxic Alkaloids Anatoxin-a and Homoanatoxin-a in the Cyanobacterium *Oscillatoria* PCC 6506 Occurs on a Modular Polyketide Synthase Initiated by l-Proline. *J. Am. Chem. Soc.* **2009**, *131*, 7512–7513. [CrossRef] [PubMed]

17. Méjean, A.; Paci, G.; Gautier, V.; Ploux, O. Biosynthesis of anatoxin-a and analogues (anatoxins) in cyanobacteria. *Toxicon* **2014**, *91*, 15–22. [CrossRef] [PubMed]

18. Wood, S.A.; Heath, M.W.; Kuhajek, J.; Ryan, K.G. Fine-scale spatial variability in anatoxin-a and homoanatoxin-a concentrations in benthic cyanobacterial mats: Implication for monitoring and management. *J. Appl. Microbiol.* **2010**, *109*, 2011–2018. [CrossRef] [PubMed]

19. Wood, S.A.; Smith, F.M.J.; Heath, M.W.; Palfroy, T.; Gaw, S.; Young, R.G.; Ryan, K.G. Within-Mat Variability in Anatoxin-a and Homoanatoxin-a Production among Benthic *Phormidium* (Cyanobacteria) Strains. *Toxins* **2012**, *4*, 900–912. [CrossRef] [PubMed]

20. Wood, S.A.; Biessy, L.; Puddick, J. Anatoxins are consistently released into the water of streams with *Microcoleus autumnalis*-dominated (cyanobacteria) proliferations. *Harmful Algae* **2018**, *80*, 88–95. [CrossRef]

21. Rantala-Ylinen, A.; Känä, S.; Wang, H.; Rouhiainen, L.; Wahlsten, M.; Rizzi, E.; Berg, K.; Gugger, M.; Sivonen, K. Anatoxin-a synthetase gene cluster of the cyanobacterium *Anabaena* sp. strain 37 and molecular methods to detect potential producers. *Appl. Environ. Microbiol.* **2011**, *77*, 7271–7278. [CrossRef]

22. Faassen, E.; Harkema, L.; Begeman, L.; Lurling, M. First report of (homo)anatoxin-a and dog neurotoxicosis after ingestion of benthic cyanobacteria in The Netherlands. *Toxicon* **2012**, *60*, 378–384. [CrossRef]

23. Quast, C.; Pruesse, E.; Yilmaz, P.; Gerken, J.; Schweer, T.; Yarza, P.; Peplies, J.; Glöckner, F.O. The SILVA ribosomal RNA gene database project: Improved data processing and web-based tools. *Nucleic Acids Res.* **2012**, *41*, 590–596. [CrossRef]

24. Schneider, S.; Melzer, A. The Trophic Index of Macrophytes (TIM)–a new tool for indicating the trophic state of running waters. *Int. Rev. Hydrobiol.* **2003**, *88*, 49–67. [CrossRef]

25. Komárek, J.; Anagnostidis, K. Cyanoprokaryota-Oscillatoriales. In *Freshwater Flora of Central Europe*; Büdel, B., Gärtner, G., Krienitz, L., Schagerl, M., Eds.; Springer Spektrum: Heidelberg, Germany, 2005; pp. 369–375.

26. Conklin, K.Y.; Stancheva, R.; Otten, T.G.; Fadness, R.; Boyer, G.L.; Read, R.; Zhang, X.; Sheath, R.G. Molecular and morphological characterization of a novel dihydroanatoxin-a producing *Microcoleus* species (cyanobacteria) from the Russian River, California, USA. *Harmful Algae* **2020**, *93*, 101767. [CrossRef] [PubMed]

27. Castenholtz, R.W.; Rippka, R.; Herdmann, M. Form-*Tychonema*. In *Bergey's Manual of Systematics of Archaea and Bacteria*; Whitman, W.B., Ed.; Wiley: Hoboken, NJ, USA, 2015; p. 561. [CrossRef]

28. Shams, S.; Capelli, C.; Cerasino, L.; Ballot, A.; Dietrich, D.R.; Sivonen, K.; Salmaso, N. Anatoxin-a producing *Tychonema* (Cyanobacteria) in European waterbodies. *Water Res.* **2015**, *69*, 68–79. [CrossRef] [PubMed]

29. Salmaso, N.; Cerasino, L.; Boscaini, A.; Capelli, C. Planktic *Tychonema* (Cyanobacteria) in the large lakes south of the Alps: Phylogenetic assessment and toxigenic potential. *FEMS Microbiol. Ecol.* **2016**, *92*, 155. [CrossRef]

30. Capelli, C.; Cerasino, L.; Boscaini, A.; Salmaso, N. Molecular tools for the quantitative evaluation of potentially toxigenic *Tychonema bourrellyi* (Cyanobacteria, Oscillatoriales) in large lakes. *Hydrobiologia* **2018**, *824*, 109–119. [CrossRef]

31. Komárek, J.; Komárkova, J.; Kling, H. Filamentous Cyanobacteria. In *Freshwater Algae of North America*; Wehr, J.D., Sheat, R.G., Eds.; Elsevier: Waltham, MA, USA, 2003; pp. 117–196.

32. Butterwick, C.; Heaney, S.I.; Talling, J.F. Diversity in the influence of temperature on the growth rates of freshwater algae, and its ecological relevance. *Freshw. Biol.* **2005**, *50*, 291–300. [CrossRef]

33. James, K.J.; Furey, A.; Sherlock, I.R.; Stack, M.A.; Twohig, M.; Caudwell, F.B.; Skulberg, O.M. Sensitive determination of anatoxin-a, homoanatoxin-a and their degradation products by liquid chromatography with fluorimetric detection. *J. Chromatogr.* **1998**, *798*, 147–157. [CrossRef]

34. Mann, S.; Cohen, M.; Chapuis-Hugon, F.; Pichon, V.; Mazmouz, R.; Méjean, A. Synthesis, configuration assignment, and simultaneous quantification by liquid chromatography coupled to tandem mass spectrometry, of dihydroanatoxin-a and dihydrohomoanatoxin-a together with the parent toxins, in axenic cyanobacterial strains and in environmental samples. *Toxicon* **2012**, *60*, 1404–1414. [CrossRef]

35. Méjean, A.; Dalle, K.; Paci, G.; Bouchonnet, S.; Mann, S.; Pichon, V.; Ploux, O. Dihydroanatoxin-a is biosynthesized from Proline in *Cylindrospermum stagnale* PCC 7417: Isotopic incorporation experiments and mass spectrometry analysis. *J. Nat. Prod.* **2016**, *79*, 1775–1782. [CrossRef]

36. Wood, S.A.; Puddick, J.; Fleming, R.; Heussner, A.H. Detection of anatoxin-producing *Phormidium* in a New Zealand farm pond and an associated dog death. *N. Z. J. Bot.* **2017**, *55*, 36–46. [CrossRef]

37. Gugger, M.; Lenoir, S.; Berger, C.; Ledreux, A.; Druart, J.C.; Humbert, J.F.; Guette, C.; Bernard, C. First report in a river in France of the benthic cyanobacterium *Phormidium favosum* producing anatoxin-a associated with dog neurotoxicosis. *Toxicon* **2005**, *45*, 919–928. [CrossRef] [PubMed]

38. Bates, H.A.; Rapoport, H. Synthesis of anatoxin-a via intramolecular cyclization of iminium salts. *J. Am. Chem. Soc.* **1979**, *101*, 1259–1265. [CrossRef]

39. Wonnacott, S.; Jackman, S.; Swanson, K.L.; Rapoport, H.; Albuquerque, E.X. Nicotinic pharmacology of anatoxin analogs. Side chain structure-activity relationships at neuronal nicotinic ligand binding sites. *J. Pharmacol. Exp. Ther.* **1991**, *259*, 387–391. [PubMed]

40. Puddick, J.; van Ginkel, R.; Page, C.D.; Murray, J.S.; Greenhough, H.E.; Bowater, J.; Selwood, A.I.; Wood, S.A.; Prinsep, M.R.; Truman, P.; et al. Acute toxicity of dihydroanatoxin-a from *Microcoleus autumnalis* in comparison to anatoxin-a. *Chemosphere* **2020**, *263*, 127937. [CrossRef] [PubMed]

41. Toporowska, M.; Pawlik-Skowrońska, B.; Kalinowska, R. Accumulation and effects of cyanobacterial microcystins and anatoxin-a on benthic larvae of *Chironomus* spp. (Diptera: Chironomidae). *Eur. J. Entomol.* **2014**, *111*, 83–90. [CrossRef]

42. Kelly, L.T.; Puddick, J.; Ryan, K.G.; Champeau, O.; Wood, S.A. An ecotoxicological assessment of the acute toxicity of anatoxin congeners on New Zealand *Deleatidium* species (mayflies). *Inland Waters* **2019**, *10*, 101–108. [CrossRef]

43. Osswald, J.; Azevedo, J.; Vasconcelos, V.; Guilhermino, L. Experimental determination of the bioconcentration factors for anatoxin-a in juvenile rainbow trout (*Oncorhynchus mykiss*). *Proc. Int. Acad. Ecol. Environ. Sci.* **2017**, *1*, 77.

44. Testai, E.; Scardala, S.; Vichi, S.; Buratti, F.M.; Funari, E. Risk to human health associated with the environmental occurrence of cyanobacterial neurotoxic alkaloids anatoxins and saxitoxins. *Crit. Rev. Toxicol.* **2016**, *46*, 385–419. [CrossRef]

45. Wood, S.A.; Hamilton, D.P.; Paul, W.J.; Safi, K.A.; Williamson, W.M. *New Zealand Guidelines for Cyanobacteria in Recreational Fresh Waters: INTERIM Guidelines (Commissioned Report for External Body)*; Ministry for the Environment: Wellington, New Zealand, 2009; pp. 1–82.

46. Chorus, I.; Schauser, I. *Oligotrophication of Lake Tegel and Schlachtensee, Berlin—Analysis of System Components, Causalities and Response Thresholds Compared to Responses of Other Waterbodies*; Umweltbundesamt: Dessau-Roßlau, Germany, 2011; pp. 1–157.

47. James, K.J.; Crowley, J.; Hamilton, B.; Lehane, M.; Skulberg, O.; Furey, A. Anatoxins and degradation products, determined using hybrid quadrupole time-of-flight and quadrupole ion-trap mass spectrometry: Forensic investigations of cyanobacterial neurotoxin poisoning. *Rapid Commun. Mass Spectrom.* **2005**, *19*, 1167–1175. [CrossRef]

48. Zwirglmaier, K.; Keiz, K.; Engel, M.; Geist, J.; Raeder, U. Seasonal and spatial patterns of microbial diversity along a trophic gradient in the interconnected lakes of the Osterseen Lake District, Bavaria. *Front. Microbiol.* **2015**, *6*, 1168. [CrossRef]

49. Rantala, A.; Rajaniemi-Wacklin, P.; Lyra, C.; Lepistö, L.; Rintala, J.; Mankiewicz-Boczek, J.; Sivonen, K. Detection of microcystin-producing cyanobacteria in Finnish lakes with genus-specific microcystin synthetase gene E (*mcyE*) PCR and associations with environmental factors. *Appl. Environ. Microbiol.* **2006**, *72*, 6101–6110. [CrossRef] [PubMed]

50. Al-Tebrineh, J.; Mihali, T.K.; Pomati, F.; Neilan, B.A. Detection of saxitoxin-producing cyanobacteria and *Anabaena circinalis* in environmental water blooms by quantitative PCR. *Appl. Environ. Microbiol.* **2010**, *76*, 7836–7842. [CrossRef] [PubMed]

51. Mizrahi-Man, O.; Davenport, E.R.; Gilad, Y. Taxonomic classification of bacterial 16S rRNA genes using short sequencing reads: Evaluation of effective study designs. *PLoS ONE* **2013**, *8*, 53608. [CrossRef] [PubMed]

52. Lagkouvardos, I.; Joseph, D.; Kapfhammer, M.; Giritli, S.; Horn, M.; Haller, D.; Clavel, T. IMNGS: A comprehensive open resource of processed 16S rRNA microbial profiles for ecology and diversity studies. *Sci. Rep* **2016**, *6*, 1–9. [CrossRef] [PubMed]

53. Pruesse, E.; Peplies, J.; Glöckner, F.O. SINA: Accurate high-throughput multiple sequence alignment of ribosomal RNA genes. *Bioinformatics* **2012**, *28*, 1823–1829. [CrossRef]

54. *Deutsche Einheitsverfahren zur Wasser-, Abwasser-und Schlammuntersuchung*; Wasserchemische Gesellschaft in der Gesellschaft Deutscher Chemiker; Wiley-VCH: Weinheim, Germany, 2013. [CrossRef]
55. Navone, R. Proposed method for nitrate in potable waters. *J. Am. Water Works Assoc.* **1964**, *56*, 781–784. [CrossRef]

Publisher's Note: MDPI stays neutral with regard to jurisdictional claims in published maps and institutional affiliations.

Article

Is a Central Sediment Sample Sufficient? Exploring Spatial and Temporal Microbial Diversity in a Small Lake

Barbara Weisbrod [1,*], **Susanna A. Wood** [2], **Konstanze Steiner** [2], **Ruby Whyte-Wilding** [2], **Jonathan Puddick** [2], **Olivier Laroche** [2] **and Daniel R. Dietrich** [1,*]

1   Human and Environmental Toxicology, Department of Biology, University of Konstanz, Universitätsstrasse 10, 78457 Konstanz, Germany
2   Cawthron Institute, 98 Halifax Street East, Nelson 7010, New Zealand; Susie.Wood@cawthron.org.nz (S.A.W.); Konstanze.Steiner@cawthron.org.nz (K.S.); rubyw.wilding@gmail.com (R.W.-W.); Jonathan.Puddick@cawthron.org.nz (J.P.); Olivier.Laroche@cawthron.org.nz (O.L.)
*   Correspondence: barbara.weisbrod@uni-konstanz.de (B.W.); daniel.dietrich@uni-konstanz.de (D.R.D.)

Received: 5 August 2020; Accepted: 7 September 2020; Published: 9 September 2020

Abstract: (1) Background: Paleolimnological studies use sediment cores to explore long-term changes in lake ecology, including occurrences of harmful cyanobacterial blooms. Most studies are based on single cores, assuming this is representative of the whole lake, but data on small-scale spatial variability of microbial communities in lake sediment are scarce. (2) Methods: Surface sediments (top 0.5 cm) from 12 sites ($n = 36$) and two sediment cores were collected in Lake Rotorua (New Zealand). Bacterial community (16S rRNA metabarcoding), *Microcystis* specific 16S rRNA, microcystin synthetase gene E (*mcyE*) and microcystins (MCs) were assessed. Radionuclide measurements (^{210}Pb, ^{137}Cs) were used to date sediments. (3) Results: Bacterial community, based on relative abundances, differed significantly between surface sediment sites ($p < 0.001$) but the majority of bacterial amplicon sequence variants (88.8%) were shared. Despite intense MC producing *Microcystis* blooms in the past, no *Microcystis* specific 16S rRNA, *mcyE* and MCs were found in surface sediments but occurred deeper in sediment cores (approximately 1950's). ^{210}Pb measurements showed a disturbed profile, similar to patterns previously observed, as a result of earthquakes. (4) Conclusions: A single sediment core can capture dominant microbial communities. Toxin producing *Microcystis* blooms are a recent phenomenon in Lake Rotorua. We posit that the absence of *Microcystis* from the surface sediments is a consequence of the Kaikoura earthquake two years prior to our sampling.

Keywords: cyanobacteria; earthquakes; harmful algal blooms; microcystin; sediment; sediment cores

Key Contribution: High spatial variability of microbial communities between sampling sites in lake sediment. Historical analyses of toxin producing cyanobacterial blooms. Possible contribution of seismic events for bloom recurrence. Single central sediment samples can represent whole lake microbial community.

---

1. Introduction

Lake ecosystems worldwide are experiencing dramatic changes in water quality, nutrient concentrations and phytoplankton community composition due to global warming, eutrophication and introductions of non-native species [1,2]. Consequently, the occurrence of harmful cyanobacterial blooms has increased significantly [3,4]. Many bloom-forming species produce cyanotoxins, which have neurotoxic, cytotoxic, and hepatotoxic effects on organisms [5]. Contact with or ingestion of water contaminated with cyanotoxins can lead to poisoning, or death of humans, pets, stock, and wildlife [5]. Amongst the cyanobacterial toxins, the cyclic heptapeptide microcystins (MCs) are the most frequently found [6–10].

MCs are potent inhibitors of eukaryotic protein phosphatases resulting in a wide range of organ toxicities, but primarily causing liver damage [11]. Cyanobacteria genera known to produce MCs include *Planktothrix, Oscillatoria, Dolichospermum* and most prominently *Microcystis* [5]. At least 271 structural variants of MCs have been described [12]. MCs are synthesized non-ribosomally by a large multifunctional enzyme complex which contains both non-ribosomal peptide synthetase and polyketide synthase domains [13]. The *mcy* gene cluster (*mcyA-J*) encoding these biosynthetic enzymes has been employed as the target region for the molecular detection of potentially toxin producing cyanobacteria strains in water and sediment [6,14,15].

Considerable effort has been invested in developing management and restoration plans for water bodies experiencing cyanobacterial blooms [16–18]. Beyond the need for effective bloom mitigation strategies, there is an increasing demand for understanding how lake ecosystems and their catchments have changed historically and how this might correlate with increasing prevalence of cyanobacterial blooms. Unfortunately, long-term monitoring data are scarce. An emerging technique that can contribute historic data is the application of molecular biological methods to sediment core analysis [19]. Employing molecular techniques, e.g., metabarcoding and droplet digital PCR (ddPCR), allows reconstruction of historical cyanobacteria communities using phylogenetic marker genes like the 16S ribosomal RNA (16S rRNA) but also functional genes (such as the *mcy* gene cluster) identified in sediment samples [6,7]. Accumulation of cyanobacterial DNA and toxins in sediments occurs via natural cell sedimentation during and after blooms [20]. This is especially true for *Microcystis* sp. that overwinter in surface sediments [21,22].

Both cyanobacterial DNA and MCs are thought to be well preserved in sediments, and both have been detected in sediment layers dating back over 200 years [6–10]. Most paleolimnological studies on cyanobacteria [6] but also eukaryotes [23,24], collect sediment cores from a single central site or deepest point in a lake, assuming a relatively even distribution across the lake. However, during a bloom cell densities of buoyant cyanobacteria, e.g., *Microcystis* spp., are highly dependent on wind and current drift, hence there is often high spatial variability across a lake surface [25–27]. Cell and toxin concentrations at one site can change dramatically within short time periods. Given the spatial patchiness of cyanobacterial blooms, and the accumulation of buoyant scums along shorelines, it is likely that cyanobacterial cells and toxins may also be unevenly distributed on lake sediment. In addition, MC sediment concentration resulting from surface blooms are likely influenced by a complex interplay of microbial degradation, thermal decomposition and photolysis via UV radiation which can all vary within different sites of the same lake [28,29]. Therefore, there is an uncertainty as to whether single core samples are representative of the whole lake and provide an accurate record of bloom history. Most research regarding spatial distribution of sediment microbial communities has been undertaken on intermediate (different lakes) [30] to global (thousands of kilometers) scales [31–33]. Hence, a better understanding of small-scale (tens to hundreds of meters) distribution patterns within a lake is required to ensure a robust interpretation of paleolimnological results.

This study focuses on Lake Rotorua, a small hypertrophic lake in the northeast of the South Island of New Zealand [34]. The lake has experienced cyanobacterial blooms since at least the 1970s with *Dolichospermum, Aphanizomenon* and MC-producing *Microcystis* present in high abundances. However, there is uncertainty as to whether these species have always been present in the lake and when blooms begun. The aims of this study were to: (i) determine the spatial variability in bacterial and cyanobacterial species in the surface sediment of Lake Rotorua, with a focus on the abundance of *Microcystis*, MC synthetase gene E (*mcyE*) copy numbers and MCs; and (ii) explore the historical bacterial (in particular cyanobacteria) composition in two sediment cores.

2. Results

2.1. Physicochemical Parameters

Temperature, pH and salinity did not differ markedly between sampling sites (Table S1). Dissolved oxygen concentration varied between 0.4 and 7.6 mg L^{-1} and was highest in the embayment (S01, 7.6 and S02, 6.3 mg L^{-1}) and at one littoral site (S08, 6.6 mg L^{-1}; Table S1).

2.2. Surface Sediments

2.2.1. Detection of *Microcystis* Sp. Specific 16S rRNA and *mcyE*

All samples, besides the littoral sample S06, contained low levels of the *Microcystis* sp. 16S rRNA gene (Figure 1a). The littoral samples S07, S09 and S12 contained more than 10^6 gene copy numbers g^{-1} sediment (Figure 1a). The *mcyE* gene copy numbers were relatively high in the three mid-lake samples (S03, S04, S05) (69,443–1.3 × 10^6 gene copy numbers g^{-1} sediment), low (≤15,565 gene copy numbers g^{-1} sediment) in samples S01–02, S06–07, S10 and S12, and no copies were detected in samples S08, S09 and S11 (Figure 1b). As a single copy gene, *mcyE* gene copy numbers should be present at a lower concentration compared to the *Microcystis* 16S rRNA copy numbers. This was observed for samples S01-03, and S06-12, while higher *mcyE* gene copy numbers than corresponding 16S rRNA copy numbers were observed for samples S04 and S05.

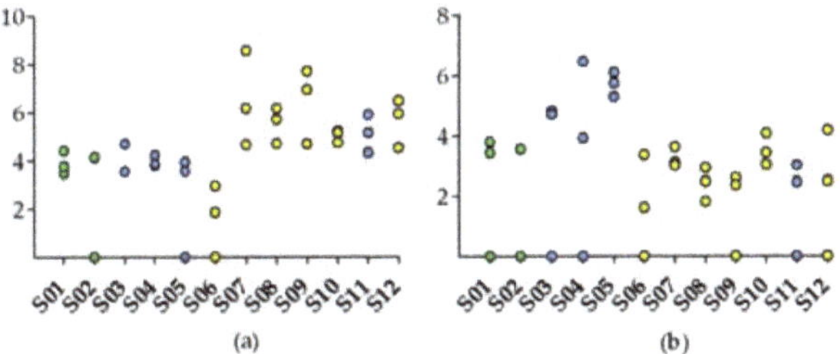

Figure 1. Abundance of *Microcystis* specific, (**a**) 16S rRNA gene, and (**b**) *mcyE* gene in surface sediment samples (S01–S12). Sample sites are indicated by color (embayment = green, mid-lake = blue, littoral = yellow). Data are log(x + 1)-transformed.

2.2.2. Bacterial Community

The total number of unique 16S rRNA amplicon sequence variants (ASVs) after quality-filtering and pre-processing was 1011. Negative controls contained 372 reads assigned to 12 ASVs. These were excluded when ASV that accounted for >0.01% of reads were removed. A total of 546 ASVs (54%) were shared between embayment, littoral and mid-lake sampling sites (Figure 2) and 65 ASVs (36%) were shared between mid-lake and littoral sites (in total 90.1% shared ASVs). Mid-lake and embayment sites shared three ASVs, and littoral and embayment sites shared 55 ASVs (Figure 2). The most common seven bacterial phyla were Proteobacteria, Bacteroidetes, Acidobacteria, Fibrobacteres, Verrucomicrobia and Planctomycetes. The relative abundance of cyanobacteria in surface sediments was low, accounting for 0.14–0.59% of the total bacterial community (Figure 3a). Nine unique cyanobacteria ASVs were detected, representing four genera; *Snowella*, *Planktothrix*, *Dolichospermum* and *Cyanobium* (Figure 3b). Contrary to expectations based on *Microcystis* sp. specific 16S rRNA gene analyses, no *Microcystis* ASVs were detected (Figure 3b). Further integration of the raw data showed that *Microcystis* ASVs were present in very low numbers, but because they represented <0.01% of all ASVs, they were discarded in the bioinformatic pipeline.

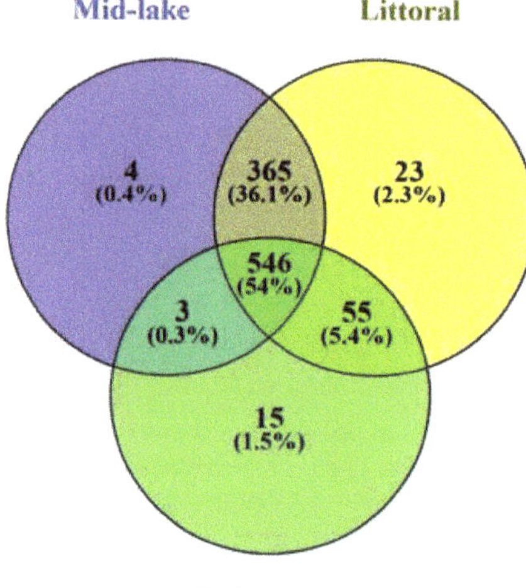

Figure 2. Venn diagram illustrating the unique and shared amplicon sequence variants (ASVs) between the three sampling sites of surface sediments.

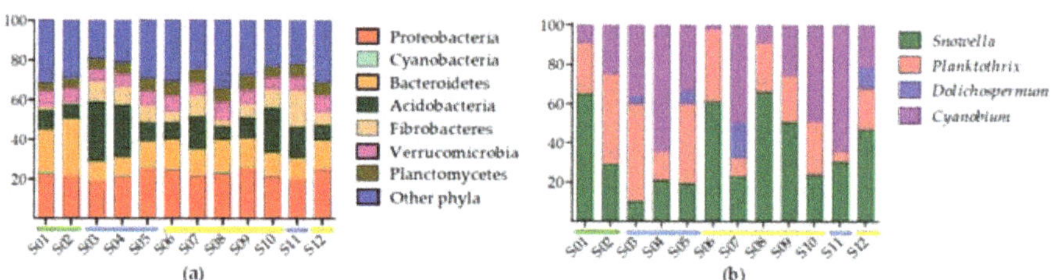

Figure 3. Bacterial community of surface sediment samples (S01–S12) based on 16S rRNA sequences. Relative abundance of; (**a**) cyanobacteria and the seven most abundant bacteria phyla the in the total bacterial community, and (**b**) cyanobacteria genera in the cyanobacteria portion of the data. Sample sites are indicated by colour (embayment = green, mid-lake = blue, littoral = yellow).

The NMDS ordination showed clear grouping of the total bacterial community by sampling site (Figure 4). The communities in the embayment sites (S01 and S02) were significantly different from littoral and mid-lake sites (S03–S12), (one-way distance-based permutational analysis of variance (PERMANOVA), emb.-littoral: $t = 3.142$, $p = 0.0001$; emb.-mid.: $t = 4.011$ $p = 0.0007$; mid.-littoral: $t = 2.285$, $p = 0.0001$; Figure 4a). Cyanobacterial communities were also significantly different between sites (PERMANOVA, emb.-littoral: $t = 2.153$, $p = 0.001$; emb.-mid.: $t = 5.577$, $p = 0.0006$; mid.-littoral: $t = 3.026$, $p = 0.0001$; Figure 4a).

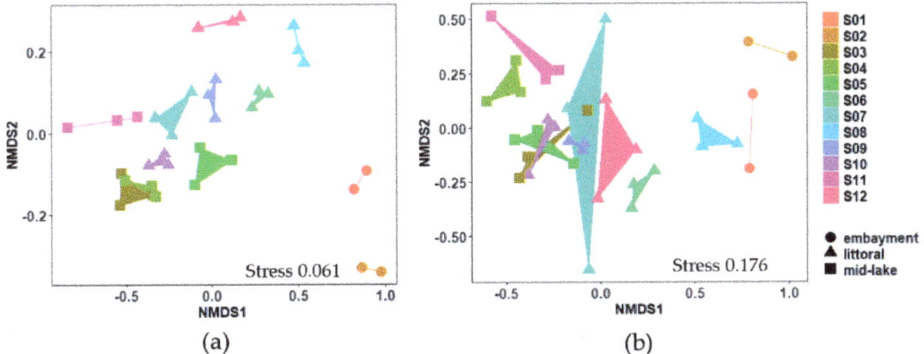

Figure 4. Two-dimensional non-metric multidimensional scaling (NMDS) ordination of microbial communities in the surface sediment samples (S01–S12). Plots are based on Bray–Curtis dissimilarities of 16S rRNA sequences of the; (**a**) total bacterial community, and (**b**) cyanobacterial community. Sampling site is indicated by shape (circle, triangle, square). Each point represents a sample. Replicates form a polygon. Stress value indicates fit of ordination, values ≤0.1 are considered fair, values ≤0.05 indicate good fit.

2.2.3. Toxin Analysis

MCs were not detected (<0.02 ng mL^{-1} or <2 µg g^{-1} wet weight in a 10 g sample) in any of the surface sediments collected. Recovery rates of the spiked MC congeners ranged between 25% for MC-RR, 73% for MC-LR and 126% for MC-LA (average result for the four spiked samples), indicating absorption to the sample matrix for certain congeners (Table S2). Recovery rates also differed between the four samples fortified with MCs, reinforcing this observation.

2.3. Sediment Cores

2.3.1. Bacterial Community and Detection of *Microcystis* Sp. Specific 16S rRNA

High throughput sequencing of 16S rRNA amplicons of sediment core samples resulted in a total of 1712 unique ASVs of which 1521 (89%) were shared among both cores (Figure S1). An additional 159 ASVs occurred only in the mid-lake core, and 32 ASVs were unique to the littoral core (Figure S1). NMDS ordination of the total bacterial community shifted with increasing depth in both cores, indicating an increasing dissimilarity between communities with increasing depth (Figure 5). In total, nine ASVs were assigned to cyanobacteria representing four genera *Microcystis*, *Aphanizomenon*, *Dolichospermum* and *Cyanobium* (Figure 6b). In the mid-lake core, the cyanobacterial ASVs occurred mainly in the upper 17 cm (0.4–16.5% relative abundance) and in the littoral core, in the upper 12 cm (0.3–10.4% relative abundance), showing in both cores a decreasing trend in relative abundance with increasing sediment depth (Figure 6a and Figure S1a). In deeper sediment core layers, cyanobacterial ASV abundance was below 0.1% of total bacterial community (Figure 6a and Figure S1a).

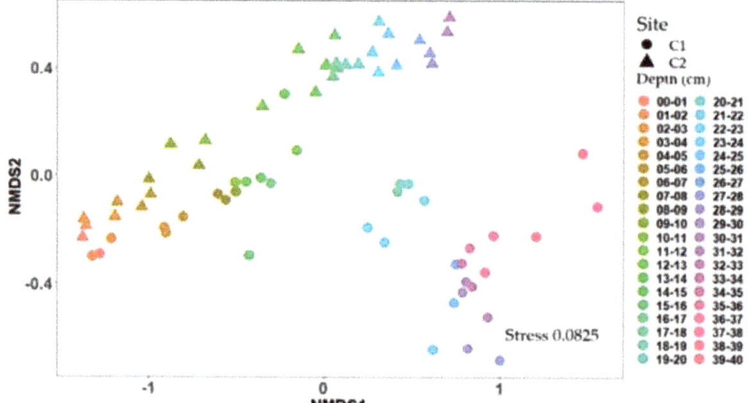

Figure 5. Two-dimensional non-metric multidimensional scaling (NMDS) ordination of bacterial community of sediment cores. Plots are based on Bray–Curtis similarities of 16S rRNA sequences of the total bacterial community. Color indicates depth layer, sampling site is indicated by shape (circles = C1 (mid-lake), triangles = C2 (littoral)). Each point represents a sample.

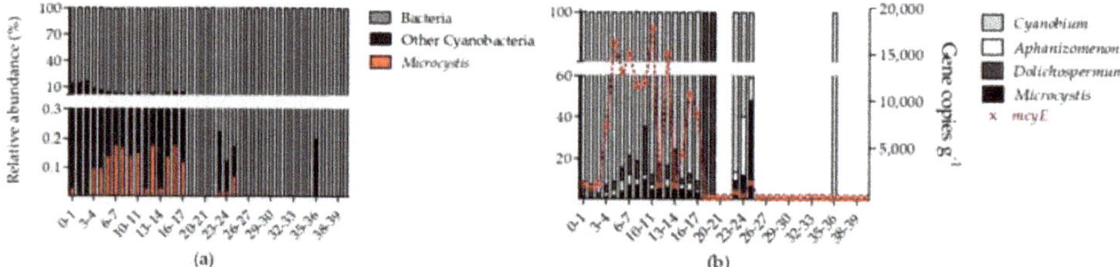

Figure 6. Bacterial community of the sediment core layers (cm) from the mid-lake (C1). Relative abundance of; (**a**) *Microcystis* and other cyanobacteria in the total bacterial community, and (**b**) Cyanobacteria genera in the total cyanobacteria community based on 16S rRNA sequences. *mcyE* gene copies (g^{-1} sediment) were detected with droplet digital PCR. Missing bars indicate that cyanobacterial taxa accounted for <0.1% of the total bacterial community in these samples.

In the littoral core *Microcystis* was detected only in traces in the uppermost layer (0–1 cm, 0.03% Figure 5a), but was not detected in the upper 1–3 cm. No *Microcystis* was detected in the upper 0–3 cm of littoral core (Figure S1). Cyanobacterial taxa found in these upper layers belonged to the genera of *Dolichospermum*, *Cyanobium* and *Aphanizomenon* (Figure 5b and Figure S2). In the mid-lake core, *Microcystis* was first detected at 4 cm depth and was present to a depth of 25 cm (Figure 5a). *Microcystis* abundance ranged between <1% to 48% of the cyanobacterial community accounting for 0–0.18% of the total bacterial community in the sediment layers (Figure 5). A peak in the relative abundance for *Microcystis* was observed at 22–25 cm depth, accounting for 9%, 12% and 48% of the cyanobacterial community, respectively. In the littoral core, *Microcystis* was first detected at 5 cm. The relative abundance of *Microcystis* in the cyanobacterial community was much lower compared to the mid-lake core accounting for ≤14% (Figure S2).

2.3.2. Detection of *Microcystis* Specific *mcyE*

The vertical profile of *mcyE* copy numbers in the mid-lake core showed a similar pattern to the sequencing results (Figure 5b). Copy numbers of *mcyE* were detected in lower numbers in the upper 3 cm of the core (1089–1361 copies g^{-1} sediment) but increased from 4 to 17 cm. A peak concentration of 18,217 *mcyE* copies g^{-1} sediment was observed at 10–11 cm. Beyond 17 cm, *mcyE* copy numbers were not detected (<100 copies g^{-1} sediment).

2.3.3. Sediment Chronology

The ^{137}Cs chronomarker profile showed a typical increase in activity over sediment depth. The increase was from 4 cm sediment depth down to a depth of 14–15 cm and was below background level for the two deeper layers (24–25 cm and 32–33 cm; Figure 7). The 1963–1964 (nuclear weapon testing) ^{137}Cs activity peak was not directly measured. However, the peak was assumed to be located between the highest ^{137}Cs activity (14 cm depth) and the first below background measurement (24 cm; Figure 7). Due to the low temporal resolution, no dating model was applicable to this ^{137}Cs profile. ^{210}Pb activity showed a non-linear activity pattern with no clear increasing or decreasing trend. Therefore, the pattern could not be interpreted as a function of radioactive decay and no ^{210}Pb dating model could be applied.

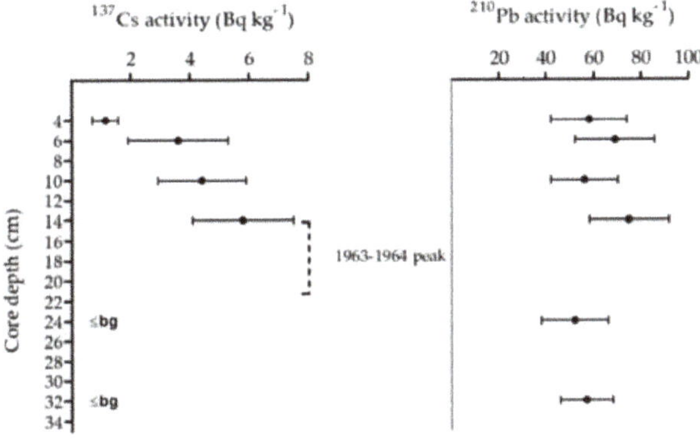

Figure 7. Radionuclide profile of ^{137}Cs and ^{210}Pb of the sediment core from the mid-lake (C1). Supported ^{210}Pb (^{226}Rd) has been subtracted from plotted ^{210}Pb data. Measured value consistent with the background measurement are indicated with (≤bg). Error bars indicate uncertainty based on the combined standard uncertainty multiplied by a coverage factor providing a level of confidence of 95%. "1963-1964 peak" refers to the presumed marker position of the ^{137}Cs maximum from 1963–1964.

3. Discussion

3.1. Bacterial Community and Microcystis Abundance in Surface Lake Sediment

In this study, we initially set out to explore the spatial variability of *Microcystis* and other cyanobacteria in surface sediments to improve knowledge of fine-scale distribution patterns needed for paleolimnological research. Initially, we planned to explore only the cyanobacterial component of the metabarcoding data but given the unexpectedly low abundance of cyanobacteria, we expanded our analysis to the entire bacterial community. This also gives these results greater relevance to paleolimnological studies which are now incorporating molecular microbial analysis into their suite of techniques [23,24,35].

There are only a few studies available which focus on spatial variability of sediment DNA on a fine scale. O'Donnell et al., [36] assessed a near-shore marine habitat and observed a decrease in similarity of sediment microbial communities with increasing distance between sampling sites. To our knowledge there is just one other paleolimnological study focusing on historic cyanobacteria communities which sampled several cores in the same lake [37]. However, the latter study focused on comparing the quality and integrity of a preserved DNA fragments between cores, rather than characterizing the cyanobacterial community.

The results of the present study demonstrate that surface sediment samples taken from a central point in the lake capture approximately 90% of the microbial diversity in the mid-lake and littoral zones. This is especially important as most paleolimnological studies use single sediment cores from the deepest point in a lake. In contrast, the bacterial communities in the embayment sites were significantly different to the littoral and mid-lake sites and varied markedly among each other. This is most likely due to differences between abiotic parameters compared to littoral and mid-lake sites as reflected by lower dissolved oxygen and water depth. In addition, the different physical and chemical sediment characteristics might have an impact on DNA preservation between the embayment sites [19]. However, correlating differences in environmental parameters to differences of bacterial community structure was not the focus of this study and requires a more comprehensive sampling approach. It is important to recognize that Lake Rotorua is a small lake. In larger lakes, a higher spatial diversity of abiotic parameters and bacterial communities might be expected.

In the present study we expected to observe a high number of *Microcystis* 16S rRNA and *mcyE* gene copies and MCs in the surface sediment samples. This assumption was based on the occurrence of intense *Microcystis* blooms recorded within the last decade [21,38] and the occurrence of high numbers of *Microcystis* cells on surface sediments during a study undertaken in 2014 [21]. *Microcystis* can overwinter on the sediment surface [22,39,40]. The overwintering cells can resist harsh environmental conditions and form an inoculum for new spring blooms [41,42]. The abundance of overwintering cells has been shown to vary spatially, with higher abundance generally occurring at greater depths [21,41]. However, in the present study, only traces of *Microcystis* 16S rRNA gene copies and none of the eleven MCs analyzed were detected in the sediment surface layers. The absence of *Microcystis* and MCs in the surface sediment, despite the recent reports of MC-producing *Microcystis* in the lake, prompted us to take sediment cores to ensure that our detection methods were viable and that DNA and toxins were preserved in the sediment of Lake Rotorua.

3.2. Historic Microcystis Populations

In agreement with the results from surface sediments, *Microcystis* 16S rRNA and *mcyE* gene copies were absent in the upper 3 cm of the sediment cores. However, *Microcystis* 16S rRNA and *mcyE* gene copies were detected in greater depths indicating that toxin producing strains of this species have been present in the lake since around the 1950s. The reasons why there was no *Microcystis* and a generally low abundance of cyanobacteria before the 1950s cannot be readily explained with the data presented here. It might be linked to changes in land use of the lake catchment. The land around the lake was once podocarp forest, but this was burnt and replaced by pasture. Together with the rising usage of fertilizers, this has caused significant nutrient enrichments in the region's waterways [43]. An increase of cyanobacteria during the 1900s was also observed in other lake sediment studies of North America [4] and Europe [6] and was linked to nutrient enrichment as a result of intensified agriculture.

Differences in historic bacterial communities between the littoral and mid-lake cores are probably due to water level fluctuations of Lake Rotorua. We assume that the current littoral area for the lake may have been swamp or wetland for some time or even for reoccurring time periods. This was supported by the fact that we were unable to obtain a longer core as we hit what we presume was thick historic vegetation and root. The mid-lake core is therefore more representative for looking at the historic bacterial communities of Lake Rotorua.

3.3. Why Is Microcystis Absent in the Surface Sediments?

The absence of *Microcystis* 16S rRNA and *mcyE* gene copies and MCs in the uppermost layers of all sediment samples of Lake Rotorua, corresponds with the observation that since 2016 (July and October 2017, January 2018) no *Microcystis* cells have been recorded in water column samples (Environment Canterbury, unpublished monitoring data). In November 2016, an earthquake of magnitude 7.8 struck the Kaikoura region. This was the second biggest earthquake recorded in New Zealand since European settlement. The earthquake triggered an estimated 100,000 landslides in the region, although no major ones were visible directly around the lake [44,45]. It is possible that this event resulted in a significant flux of sediment into the lake. Lake sediments have previously been used to reconstruct the chronology of natural disasters including floods [46] and earthquakes [47,48]. Earthquakes are linked to increased sediment fluxes to rivers and basins as a consequence of seismically induced landslides which often leads to a short-term increase of local and distant erosion rates [49]. These sediment fluxes can be identified as lake deposits (turbidites), which differ in structure from sediments originating from regular sedimentation processes, and can help to reconstruct long-term earthquake occurrences as previously shown for the New Zealand alpine fault [50].

While the ^{137}Cs profile of the sediment core was as expected, ^{210}Pb did not display its typical linear decay profile known from undisturbed sediments. Non-linear ^{210}Pb profiles where previously shown for Lake Antern in the French Alps [51] and the Chilean lake district [52] as a result of earthquake-induced sediment changes. These changes can result from direct seismological mixing of old sediment layers containing ^{210}Pb depleted material [51], or from landslides triggered by earthquakes, which increase erosion rates and sediment fluxes to rivers and basins [49], thereby overlaying lake sediments. The thickness of the earthquake-induced mass movement deposits varies between lakes and depends on the magnitude of the seismic event, site and size of triggered landslides as well as on the amount and quality of the erodible material in the catchment area. Of the few studies that have linked lake sediment deposits to the intensity of historic earthquake events, seismic event-linked deposits in lake sediment cores from the European Alps ranged between 1 to >20 cm thickness [47,51], in a Chilean lake between 0.5 to 7 cm [48] and in alpine lakes of New Zealand between 0.2 to 20 cm [50]. This demonstrates that each lake needs to be analyzed individually to allow any deduction of the impact of the seismic event on lake sediments.

The 2016 Kaikoura earthquake had a magnitude of 7.8, which was higher or equivalent to the seismic event related changes in the sediments described for the European Alps, Chile and earlier seismic events in New Zealand. Thus, it is likely that the disturbed upper 3–4 cm in Lake Rotorua resulted from a mixing effect as well an increased deposition of erosion material [51]. The finding that other cyanobacterial taxa were detected in the upper 3 cm and intact biofilms were found at the sediment surface (Figure S3), may appear contradictory to the disturbed sediment hypothesis. However, *Cyanobium* and *Snowella* are common freshwater picocyanobacteria [53]. They are known to be very adaptive to changing environmental conditions [54,55] and may represent the pioneering species occupying new niches after major environmental changes induced by increased sediment fluxes. *Aphanizomenon* and *Dolichospermum* both produce akinetes, thick walled cells that remain dormant until environmental conditions are favorable. It is possible that these were more resilient to sediment mixing than *Microcystis* vegetative cells or survived in shoreline terrestrial habitats, allowing these genera to rapidly recolonize following the sediment disturbance.

4. Conclusions

In this study, we demonstrate that bacterial communities of Lake Rotorua surface sediments significantly differed between lake sites (mid, littoral and embayment). However, the majority of ASV were shared between the mid and littoral sites, indicating that in small to medium sized lakes, the sampling approach used by most paleolimnological studies will also capture dominant microbial communities. The absence of *Microcystis* 16S rRNA and *mcyE* gene copies and MCs in surface sediments, in addition to the disturbed ^{210}Pb profile, suggests that allochthonous processes may have impacted the surface sediment, changing the cyanobacterial communities. The 2016 Kaikoura earthquake may have played a role in this, but further research is required to confirm this hypothesis. The absence of *Microcystis* 16S rRNA and *mcyE* gene copies in the sediment layers prior to about 1950's indicates that toxin producing cyanobacterial blooms in Lake Rotorua are a relatively recent occurrence.

5. Materials and Methods

5.1. Study Site

Lake Rotorua (42°24′05″ S, 173°34′57″ E) is a small (0.55 km^2), shallow (max. depth 3 m), hypertrophic lake in the northeast of the South Island of New Zealand [34] (Figure 8). The average temperature in the nearby town of Kaikoura is 12.7 °C and precipitation averages 881 mm per year. In the early 1900s the native podocarp-mixed hardwood forest that dominated the catchment was removed. The catchment now has a mixed land use of low intensity grazing, native scrub and bushland, and exotic weeds and trees. Inflow comes largely from surface run-off during rainfall events. There is a single small outflow at the southern end of the lake, and the lake has an estimated residence of 0.86 y^{-1} [56]. During a two year monitoring study, Lake Rotorua was shown to experience annual cyanobacterial blooms between spring and autumn, with seasonal succession from nitrogen fixers (*Dolichospermum* and *Aphanizomenon*) to *Microcystis* later in summer [38]. During *Microcystis* blooms, peak concentrations of >2000 µg L^{-1} total MC and 4 × 106 cells mL^{-1} have been measured [38,57].

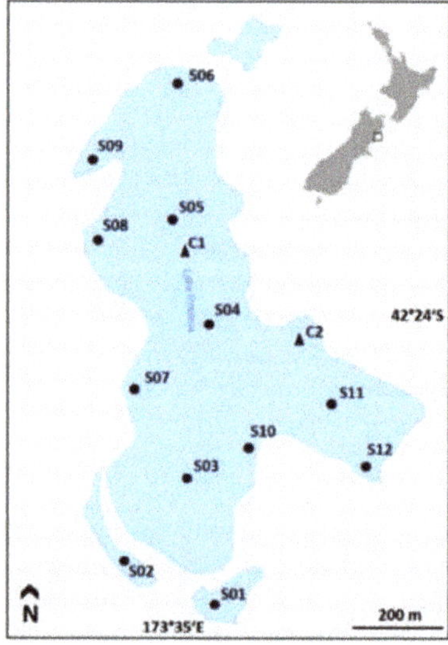

Figure 8. Map of Lake Rotorua (Kaikoura, New Zealand) and the 14 sediment sampling sites (12 surface sediments (S01–S12) and two sediment cores (C1 and C2)). Inset: map of New Zealand showing site of the lake.

5.2. Sediment Sampling

Surface sediment samples were collected at 12 sites in January 2018 using a sediment core sampler (UWITEC, Mondsee, Austria; Figure 8, Table 1). Surface samples were collected in embayments (S01 and S02), mid-lake (S03-05, S11) and littoral zones (S06-10, S12) from the top layer (0.5 cm) of the core. At each site, three separate surface samples were taken. At two sites, sediment cores were retrieved; (i) C1, collected mid-lake (40 cm length), and (ii) C2, collected in the littoral zone (32 cm length). Sediment cores were sectioned into 1 cm increments. Sediment subsamples for DNA extraction were stored at −20 °C until further processing. Subsamples for toxin extraction and age dating were sampled into 50 mL Falcon tubes and stored at −20 °C until analysis. At each sampling site the water temperature, pH, salinity and dissolved oxygen were measured using a YSI water quality sonde (YSI, Yellow Springs, OH, USA).

Table 1. Overview of sediment samples taken in this study.

	Surface Sediments												Sediment Cores	
Sample name	S01	S02	S03	S04	S05	S06	S07	S08	S09	S10	S11	S12	C1	C2
Sediment depth (cm)	0–0.5												0-40	0-32
Replicates	3												1	
Increments	1												40	32
Site location	embayment			mid-lake			littoral				mid-lake	littoral	mid-lake	littoral

5.3. DNA Extraction and Inhibition Test

Subsamples of sediment (0.2–0.4 g) were weighed into the first tube of a PowerSoil DNA Isolation Kit (Qiagen, Germantown, MD, USA) and DNA extracted according to the protocol supplied by the manufacturer. A random selection of DNA samples ($n = 10$) were screened in duplicates for inhibition using an internal control quantitative PCR (qPCR) assay [58].

5.4. Quantitative PCR for Microcystis and mcyE Enumeration in Surface Sediments

The number of toxin producing *Microcystis* cells present was determined by amplifying the *mcyE* gene (a single copy gene) and quantifying the number of amplicons, using a standard curve generated from MC-producing *Microcystis* sp. strain CAWBG617 [38]. The qPCR was performed in triplicate for each sample in a 12.5 µL of reaction mix containing 6.25 µL KAPA Probe Fast qPCR Kit Master Mix (2×), 1 µL of primers targeting a region within the *mcyE* open reading frame (0.4 µM, McyE-F2 and MicMcyE-R8) [59], 0.2 µL of *mcyE* probe [60] and 1 µL of template DNA per sample. The standard curve was constructed using a purified (AxyPrep PCR Clean-up Kit, Axygen Biosciences, Union City, CA, USA) PCR product generated with the primers described above. The number of copies in the PCR product used for the standard curves was determined using: (A × 6.022 × 1023)/(B × 1×109 × 650), with A being the concentration of the PCR product, 6.022 × 1023 (Avogadro's number), B being the length of the PCR product, 1 × 109 used to convert to ng, and 650 the average molecular weight per base pair (bp). Each point of the standard curve was analyzed in triplicate for each qPCR run conducted. The standard curve generated was linear ($R^2 > 0.99$) and PCR efficiency was >0.9. The results of copies per µL DNA were converted to copies per gram of sediment using: (D × 80)/E, with D being the number of copies per µL, × 80 since 1 µL DNA of the elution volume (80 µL) was used, and E being the weight of sediment (g) used for DNA extraction.

To detect and quantify *Microcystis* sp., a region of the 16S rRNA gene was amplified using *Microcystis* sp. specific primers MICR184F [61] and MICR431R [62]. qPCR and standard curve construction was performed as described for *mcyE* enumeration.

5.5. Droplet Digital PCR for mcyE Enumeration in Sediment Cores

Quantification of the *mcyE* copy numbers in sediment cores was carried out with droplet digital PCR (ddPCR) using a BioRad QX200 system. Each ddPCR reaction included 2 µL of 900 nM of each primer and 0.55 µL of 250 nM of primer probe (primers and probe as described for qPCR), 11 µL of BioRad ddPCR Supermix for probes, 1.5 µL DNA, and 4.95 µL of sterile water. The reaction mixture was divided into nanodroplets by joining 20 µL of the reaction mixture with 70 µL of BioRad droplet oil. After processing, the nanodroplet volume resulted in a total of 40 µL, which was transferred to a PCR plate for amplification at 96 °C for 10 min, followed by 40 cycles at 94 °C for 30 s, 60 °C for 60 s and a final extension step of 96 °C for 10 min. The ddPCR was run on one plate, with four negative controls (containing all reagents plus DNA/RNA-free water) and four positive controls (genomic DNA extracted from a sample known to contain the *mcyE* gene) included. The results were converted to copies per gram of sediment using: (D × 66)/E, with D being the number of copies per µL, × 66 since 1.5 µL DNA of the elution volume (100 µL) was used, and E being the weight of sediment (g) used for DNA extraction.

5.6. High Throughput Sequencing

5.6.1. 16S rRNA PCR

Surface and core sediments were analysed via Illumina™ sequencing. A region of the 16S rRNA gene approx. 400 bp long was amplified using bacteria-specific primers 341F: 5′-CCT ACG GGN GGC WGC AG-3′ and 805R: 5′-GAC TAC HVG GGT ATC TAA TCC-3′, modified to include Illumina™ adapters [63,64]. PCR reactions were performed in 50 μL volumes containing 25 μL of AmpliTaq Gold® 360 master mix (Life Technologies, Pleasanton, CA, USA), 12.5 μL GC inhibitor (Life Technologies, Pleasanton, CA, USA), 2 μL of each primer (10 μM, Integrated DNA Technologies, Coralville, IA, USA), 2 μL template DNA (between 60 and 220 ng) and 6.5 μL Milli-Q water. PCR cycling conditions were: 95 °C for 10 min, followed by 27 cycles of 95 °C for 30 s, 50 °C for 30 s, 72 °C for 45 s, and a final extension of 72 °C for 7 min. PCR products were visualized with 1.5% agarose gel electrophoresis with Red Safe DNA Loading Dye (iNtRON Biotechnology Inc, Kyungki-Do, Korea) and UV illumination to ensure amplification of a single 400 bp product. PCR products were purified (Agencourt® AMPure® XP Kit; Beckman Coulter, CA, USA), quantified (Qubit® 20 Fluorometer, Invitrogen), diluted to 10 ng μL^{-1} and submitted to New Zealand Genomics Limited (Auckland, New Zealand) for library preparation. Sequencing adapters and sample-specific indices were added to each amplicon via a second round of PCR using the NexteraTM Index kit (IlluminaTM, Melbourne, Australia). Amplicons were pooled into a single library and paired-end sequences (2 × 250) were generated on a MiSeq instrument using the TruSeqTM SBS kit (IlluminaTM, Melbourne, Australia). Sequence data were automatically demultiplexed using MiSeq Reporter (v2), and forward and reverse reads were assigned to samples.

5.6.2. Quality Control and Taxonomic Assignment of Illumina Sequences

Analysis of sequencing data was performed using the QIIME 2 2019.1 [65]. Raw sequence reads were demultiplexed and quality filtered using q2-demux plugin. Denoising of sequences was followed by chimera identification and removal using DADA2 [66]. To account for differential sequencing depth among samples, the number of reads per sample was rarefied to 28,700 (surface sediment samples) and 26,177 reads (sediment core samples). Five samples were excluded from further analysis because of low read number; two surface sediments (S01 and S02), two subsamples from the mid-lake sediment core (9–10 cm and 17–18 cm) and one subsample from the littoral core (29–30 cm). The implemented q2-feature-classifier-plugin (a scikit-learn naive Bayes machine-learning classifier) [67] was used for classification and pre-trained on the SILVA bacteria 16S rRNA gene reference database (version 132) [68], trimmed with the primer sequences prior to taxonomic assignment. Quality filtered and taxa assigned ASVs were further pre-processed and filtered using R (version 3.6.1, R Foundation for Statistical Computing, Vienna, Austria) [69] and the phyloseq package [70]. Chloroplast, rare ASVs accounting for less than 0.1% of the total bacterial community, and ASVs which could not be assigned further than kingdom level were excluded from further analysis.

5.6.3. Graphics and Data Analysis of High Throughput-Sequences

To visualize bacterial community dissimilarities between sampling sites and depth increments, non-metric-multidimensional-scaling (NMDS) was performed, using the R vegan package [71]. Filtered and pre-processed ASVs were previously fourth root transformed and a Bray–Curtis dissimilarity matrix was calculated. To test for significance of observed dissimilarities in bacterial community between sites, a PERMANOVA was performed using PRIMER (version 7, PRIMER-e, Auckland, New Zealand) with the PERMANOVA + add on [72]. Venn diagrams were designed using the Venny tool version 2.1 [73]. GraphPad prism (version 5, San Diego, CA, USA) for Windows was used for all other figure preparations.

5.7. Toxin Analysis

5.7.1. Toxin Extraction from Surface Sediments

Surface sediment samples (6–10 g wet weight) were freeze-dried and extracted with three consecutive extractions using 100% MeOH (5 mL each). During each extraction step, samples were vortexed for 1 min, sonicated for 15 min and centrifuged (3200× g, 10 min, 10 °C). The extracts were pooled and dried at 40 °C under a stream of nitrogen gas and dissolved in 1 mL 100% MeOH. Extracts were stored at −20 °C until measurement.

5.7.2. Toxin Analysis of Surface Sediments

Toxin extracts were analyzed by ultra-performance liquid chromatography-tandem mass spectrometry (UPLC-MS/MS) as described previously [74]. To assess the extraction method and check for matrix effects, the recovery rate of MC was determined by fortification with a semi-purified extract of *Microcystis* CAWBG11 which produces a range of MC congeners [75]. The MC-enriched extract (dissolved in 80% MeOH) was added to sediment samples and incubated at 4 °C for 30 min. Samples were extracted as described previously and analyzed via UPLC-MS/MS.

5.8. Sediment Chronology of Mid-Lake Sediment Deep Core

For age-dating of the mid-lake sediment core, subsamples representing six increments (4–5, 6–7, 10–11, 14–15, 24–25 and 32–33 cm) were analyzed using gamma spectrometry of the long-lived radionuclides ^{210}Pb and ^{137}Cs at the Institute of Environmental Science and Research (Christchurch, New Zealand). Activities were quantified using a gamma counter with a high-purity germanium well detector. Activities are reported in Bq per kg. Uncertainties are based on the combined standard uncertainty multiplied by a coverage factor (k) = 2, (providing a level of confidence of 95%) as described before [76,77]. Core chronologies from ^{210}Pb and ^{137}Cs measurements were calculated using the constant rate of supply (CRS) model in R by means of the mudata2 package [78]. Assuming that the supply of unsupported ^{210}Pb to the sediment is the same for each time interval, the CRS model is calculated as follows:

$$tz = \lambda^{-1} - 1 \ln(A_0 \, Az^{-1}) \tag{1}$$

where (t) is the age of any interval (z) [79].

Supplementary Materials: The following are available online at http://www.mdpi.com/2072-6651/12/9/580/s1. Table S1: abiotic parameters of surface sediment sampling sites, Table S2: recovery rates of microcystin (MC) congeners, Figure S1: Venn diagram illustrating the unique and shared amplicon sequence variant (ASV) between the two sediment cores from the mid-lake (C1) and the littoral (C2), Figure S2: bacterial community of the sediment core from the mid-lake (C2). Relative abundance of; (a) *Microcystis* and other cyanobacteria in the total bacterial community, and (b) Cyanobacteria genera in the total cyanobacteria community based on 16S rRNA sequences. Missing bars indicate that cyanobacterial taxa accounted for <0.1% of the total bacterial community in these samples, Figure S3: biofilm growing on the surface sediment of Lake Rotorua.

Author Contributions: Conceptualization, S.A.W., B.W., D.R.D.; methodology, S.A.W., J.P., B.W.; software, O.L., B.W.; validation, S.A.W., O.L., J.P., K.S., B.W., D.R.D.; formal analysis, B.W., S.A.W.; investigation, B.W., R.W.-W.; resources, S.A.W., J.P.; data curation, B.W., R.W.-W., K.S., J.P.; writing—original draft preparation, B.W.; writing—review and editing, all authors; visualization, B.W.; supervision, S.A.W., K.S., B.W., D.R.D.; project administration, B.W., S.A.W., D.R.D.; funding acquisition, D.R.D., S.A.W., J.P. All authors have read and agreed to the published version of the manuscript.

Funding: The work was financially supported by the Ministry of Science, Research and the Arts of the Federal State Baden-Württemberg, Germany (Water Research Network project 'Challenges of Reservoir Management—Meeting Environmental and Social Requirements'; 814/15), the Ministry of Business, Innovation and Employment, New Zealand (Programe: 'Our Lake's Health; Past, Present and Future'; C05X1707) and the Marsden Fund of the Royal Society of New Zealand (Blooming Buddies; CAW1601). The International Max Planck Research School for Organismal Biology provided a travel grant.

Acknowledgments: The authors thank Sean Waters (Cawthron Institute) for his support during field sampling.

Conflicts of Interest: The authors declare no conflict of interest.

References

1. Carpenter, S.R.; Stanley, E.H.; Vander Zanden, M.J. State of the World's Freshwater Ecosystems: Physical, Chemical, and Biological Changes. *Annu. Rev. Environ. Resour.* **2011**, *36*, 75–99. [CrossRef]
2. Vitousek, P.; Mooney, H.; Lubchenco, J.; Melillo, J. Human Domination of Earth's Ecosystems. *Science* **1997**, *277*, 494. [CrossRef]
3. Huisman, J.; Codd, G.A.; Paerl, H.W.; Ibelings, B.W.; Verspagen, J.M.H.; Visser, P.M. Cyanobacterial blooms. *Nat. Rev. Microbiol.* **2018**, *16*, 471. [CrossRef] [PubMed]
4. Taranu, Z.E.; Gregory-Eaves, I.; Leavitt, P.R.; Bunting, L.; Buchaca, T.; Catalan, J.; Domaizon, I.; Guilizzoni, P.; Lami, A.; McGowan, S.; et al. Acceleration of cyanobacterial dominance in north temperate-subarctic lakes during the Anthropocene. *Ecol. Lett.* **2015**, *18*, 375–384. [CrossRef]
5. Sivonen, K. Cyanobacterial Toxins. In *Encyclopedia of Microbiology*; Schaechter, M., Ed.; Elsevier: Oxford, UK, 2009; pp. 290–307.
6. Legrand, B.; Lamarque, A.; Sabart, M.; Latour, D. Benthic archives reveal recurrence and dominance of toxigenic cyanobacteria in a eutrophic lake over the last 220 years. *Toxins* **2017**, *9*, 271. [CrossRef]
7. Monchamp, M.-E.; Walser, J.-C.; Pomati, F.; Spaak, P. Sedimentary DNA reveals cyanobacterial community diversity over 200 years in two perialpine lakes. *Appl. Environ. Microbiol.* **2016**, *82*, 6472–6482. [CrossRef]
8. Pilon, S.; Zastepa, A.; Taranu, Z.E.; Gregory-Eaves, I.; Racine, M.; Blais, J.M.; Poulain, A.J.; Pick, F.R. Contrasting histories of microcystin-producing cyanobacteria in two temperate lakes as inferred from quantitative sediment DNA analyses. *Lake Reserv. Manag.* **2019**, *35*, 102–117. [CrossRef]
9. Zastepa, A.; Taranu, Z.E.; Kimpe, L.E.; Blais, J.M.; Gregory-Eaves, I.; Zurawell, R.W.; Pick, F.R. Reconstructing a long-term record of microcystins from the analysis of lake sediments. *Sci. Total Environ.* **2017**, *579*, 893–901. [CrossRef]
10. Zastepa, A.; Pick, F.; Blais, J.; Saleem, A. Analysis of intracellular and extracellular microcystin variants in sediments and pore waters by accelerated solvent extraction and high performance liquid chromatography-tandem mass spectrometry. *Anal. Chim. Acta* **2015**, *872*, 26–34. [CrossRef]
11. Dietrich, D.R.; Fischer, A.; Michel, C.; Hoeger, S.J. Toxin mixture in cyanobacterial blooms—A critical comparison of reality with current procedures employed in human health risk assessment. Advances in Experimental Medicine and Biology. In *Cyanobacterial Harmful Algal Blooms: State of the Science and Research Needs*; Hudnell, H.K., Ed.; Springer: New York, NY, USA, 2008; pp. 885–912.
12. *Handbook of Cyanobacterial Monitoring and Cyanotoxin Analysis*; Meriluoto, J.; Spoof, L.; Codd, G.A. (Eds.) John Wiley & Sons: Hoboken, NJ, USA, 2017.
13. Tillett, D.; Dittmann, E.; Erhard, M.; von Döhren, H.; Börner, T.; Neilan, B.A. Structural organization of microcystin biosynthesis in *Microcystis aeruginosa* PCC7806: An integrated peptide–polyketide synthetase system. *Chem. Biol.* **2000**, *7*, 753–764. [CrossRef]
14. Chorus, I. *Current Approaches to Cyanotoxin Risk Assessment, Risk Management and Regulations in Different Countries*; Federal Environment Agency: Dessau-Roßlau, Germany, 2012; p. 63.
15. Medlin, L.K.; Orozco, J. Molecular techniques for the detection of organisms in aquatic environments, with emphasis on harmful algal bloom species. *Sensors* **2017**, *17*, 1184. [CrossRef] [PubMed]
16. Paerl, H.W. Mitigating harmful cyanobacterial blooms in a human- and climatically-impacted world. *Life* **2014**, *4*, 988–1012. [CrossRef]
17. Ibelings, B.W.; Bormans, M.; Fastner, J.; Visser, P.M. CYANOCOST special issue on cyanobacterial blooms: Synopsis—a critical review of the management options for their prevention, control and mitigation. *Aquat. Ecol.* **2016**, *50*, 595–605. [CrossRef]
18. Rastogi, R.P.; Madamwar, D.; Incharoensakdi, A. Bloom dynamics of cyanobacteria and their toxins: Environmental health impacts and mitigation strategies. *Front. Microbiol.* **2015**, *6*, 1254. [CrossRef] [PubMed]
19. Giguet-Covex, C.; Ficetola, G.F.; Walsh, K.; Poulenard, J.; Bajard, M.; Fouinat, L.; Sabatier, P.; Gielly, L.; Messager, E.; Develle, A.L.; et al. New insights on lake sediment DNA from the catchment: Importance of taphonomic and analytical issues on the record quality. *Sci. Rep.* **2019**, *9*, 14676. [CrossRef]
20. Wörmer, L.; Cirés, S.; Quesada, A. Importance of natural sedimentation in the fate of microcystins. *Chemosphere* **2011**, *82*, 1141–1146. [CrossRef]

21. Borges, H.; Wood, S.A.; Puddick, J.; Blaney, E.; Hawes, I.; Dietrich, D.R.; Hamilton, D.P. Intracellular, environmental and biotic interactions influence recruitment of benthic *Microcystis* (Cyanophyceae) in a shallow eutrophic lake. *J. Plankton Res.* **2016**, *38*, 1289–1301. [CrossRef]

22. Preston, T.; Stewart, W.D.P.; Reynolds, C.S. Bloom-forming cyanobacterium *Microcystis aeruginosa* overwinters on sediment surface. *Nature* **1980**, *288*, 365–367. [CrossRef]

23. Capo, E.; Debroas, D.; Arnaud, F.; Perga, M.-E.; Chardon, C.; Domaizon, I. Tracking a century of changes in microbial eukaryotic diversity in lakes driven by nutrient enrichment and climate warming. *Environ. Microbiol.* **2017**, *19*, 2873–2892. [CrossRef]

24. Rühland, K.; Priesnitz, A.; Smol, J.P. Paleolimnological evidence from diatoms for recent environmental changes in 50 lakes across Canadian arctic treeline. *Arct. Antarct. Alp. Res.* **2003**, *35*, 110–123. [CrossRef]

25. Hunter, P.D.; Tyler, A.N.; Willby, N.J.; Gilvear, D.J. The spatial dynamics of vertical migration by *Microcystis aeruginosa* in a eutrophic shallow lake: A case study using high spatial resolution time-series airborne remote sensing. *Limnol. Oceanogr.* **2008**, *53*, 2391–2406. [CrossRef]

26. Vander Woude, A.; Ruberg, S.; Johengen, T.; Miller, R.; Stuart, D. Spatial and temporal scales of variability of cyanobacteria harmful algal blooms from NOAA GLERL airborne hyperspectral imagery. *J. Gt. Lakes Res.* **2019**, *45*, 536–546. [CrossRef]

27. Wang, H.; Zhang, Z.; Liang, D.; du, H.; Pang, Y.; Hu, K.; Wang, J. Separation of wind's influence on harmful cyanobacterial blooms. *Water Res.* **2016**, *98*, 280–292. [CrossRef] [PubMed]

28. Holst, T.; Jørgensen, N.O.G.; Jørgensen, C.; Johansen, A. Degradation of microcystin in sediments at oxic and anoxic, denitrifying conditions. *Water Res.* **2003**, *37*, 4748–4760. [CrossRef]

29. Christoffersen, K.; Lyck, S.; Winding, A. Microbial activity and bacterial community structure during degradation of microcystins. *Aquat. Microb. Ecol. Aquat. Microb. Ecol.* **2002**, *27*, 125–136. [CrossRef]

30. Monchamp, M.-E.; Spaak, P.; Pomati, F. High dispersal levels and lake warming are emergent drivers of cyanobacterial community assembly in peri-Alpine lakes. *Sci. Rep.* **2019**, *9*, 7266. [CrossRef]

31. der Gucht, K.V.; Cottenie, K.; Muylaert, K.; Vloemans, N.; Cousin, S.; Declerck, S.; Jeppesen, E.; Conde-Porcuna, J.-M.; Schwenk, K.; Zwart, G.; et al. The power of species sorting: Local factors drive bacterial community composition over a wide range of spatial scales. *Proc. Natl. Acad. Sci. USA* **2007**, *104*, 20404–20409. [CrossRef]

32. Lindström, E.S.; Langenheder, S. Local and regional factors influencing bacterial community assembly. *Environ. Microbiol. Rep.* **2012**, *4*, 1–9. [CrossRef]

33. de Vargas, C.; Audic, S.; Henry, N.; Decelle, J.; Mahé, F.; Logares, R.; Lara, E.; Berney, C.; Bescot, N.L.; Probert, I.; et al. Eukaryotic plankton diversity in the sunlit ocean. *Science* **2015**, *348*, 1261605. [CrossRef]

34. Flint, E. Phytoplankton in some New Zealand lakes. In *New Zealand Lakes*; Jolly, V.H., Brown, J.M.A., Eds.; Auckland University Press: Auckland, New Zealand, 1975; pp. 63–192.

35. Monchamp, M.-E.; Spaak, P.; Domaizon, I.; Dubois, N.; Bouffard, D.; Pomati, F. Homogenization of lake cyanobacterial communities over a century of climate change and eutrophication. *Nat. Ecol. Evol.* **2018**, *2*, 317–324. [CrossRef]

36. O'Donnell, J.L.; Kelly, R.P.; Shelton, A.O.; Samhouri, J.F.; Lowell, N.C.; Williams, G.D. Spatial distribution of environmental DNA in a nearshore marine habitat. *PeerJ* **2017**, *5*, e3044. [CrossRef]

37. Savichtcheva, O.; Debroas, D.; Kurmayer, R.; Villar, C.; Jenny, J.P.; Arnaud, F.; Perga, M.E.; Domaizon, I. Quantitative PCR Enumeration of total/toxic *Planktothrix rubescens* and total cyanobacteria in preserved DNA isolated from lake sediments. *Appl. Environ. Microbiol.* **2011**, *77*, 8744–8753. [CrossRef] [PubMed]

38. Wood, S.A.; Borges, H.; Puddick, J.; Biessy, L.; Atalah, J.; Hawes, I.; Dietrich, D.R.; Hamilton, D.P. Contrasting cyanobacterial communities and microcystin concentrations in summers with extreme weather events: Insights into potential effects of climate change. *Hydrobiologia* **2017**, *785*, 71–89. [CrossRef]

39. Fallon, R.D.; Brock, T.D. Overwintering of *Microcystis* in Lake Mendota. *Freshw. Biol.* **1981**, *11*, 217–226. [CrossRef]

40. Reynolds, C.S.; Jaworski, G.H.M.; Cmiech, H.A.; Leedale, G.F.; Lund, J.W.G. On the annual cycle of the blue-green alga *Microcystis aeruginosa*. Kütz. Emend. Elenkin. *Philos. Trans. R. Soc. Lond. B Biol. Sci.* **1981**, *293*, 419–477.

41. Brunberg, A.-K.; Blomqvist, P. Recruitment of *Microcystis* (cyanophyceae) from lake sediments: The importance of littoral inocula. *J. Phycol.* **2003**, *39*, 58–63. [CrossRef]

42. Kitchens, C.M.; Johengen, T.H.; Davis, T.W. Establishing spatial and temporal patterns in *Microcystis* sediment seed stock viability and their relationship to subsequent bloom development in Western Lake Erie. *PLoS ONE* **2018**, *13*, e0206821. [CrossRef] [PubMed]

43. Ministry for the Environment & Stats NZ. *New Zealand's Environmental Reporting Series: Environment Aotearoa 2019*; Ministry for the Environment & Stats NZ: Wellington, New Zealand, 2019.

44. Dellow, S.; Massey, C.I.; McColl, S.T.; Townsend, D.B.; Villeneuve, M. *Landslides Caused by the 14 November 2016 Kaikoura Earthquake, South Island, New Zealand. Proceedings 20th NZGS Geotechnical Symposium*; Alexander, G.J., Chin, C.Y., Eds.; New Zealand Geotechnical Society: Napier, New Zealand, 2017.

45. Massey, C.; Townsend, D.; Rathje, E.; Allstadt, K.E.; Lukovic, B.; Kaneko, Y.; Bradley, B.; Wartman, J.; Jibson, R.W.; Petley, D.N.; et al. Landslides triggered by the 14 November 2016 Mw 7.8 Kaikōura Earthquake, New Zealand. *Bull. Seismol. Soc. Am.* **2018**, *108*, 1630–1648. [CrossRef]

46. Schillereff, D.N.; Chiverrell, R.C.; Macdonald, N.; Hooke, J.M. Flood stratigraphies in lake sediments: A review. *Earth-Sci. Rev.* **2014**, *135*, 17–37. [CrossRef]

47. Wilhelm, B.; Nomade, J.; Crouzet, C.; Litty, C.; Sabatier, P.; Belle, S.; Rolland, Y.; Revel, M.; Courboulex, F.; Arnaud, F.; et al. Quantified sensitivity of small lake sediments to record historic earthquakes: Implications for paleoseismology. *J. Geophys. Res. Earth Surf.* **2016**, *121*, 2–16. [CrossRef]

48. Moernaut, J.; Daele, M.V.; Heirman, K.; Fontijn, K.; Strasser, M.; Pino, M.; Urrutia, R.; Batist, M.D. Lacustrine turbidites as a tool for quantitative earthquake reconstruction: New evidence for a variable rupture mode in south central Chile. *J. Geophys. Res. Solid Earth* **2014**, *119*, 1607–1633. [CrossRef]

49. Avşar, U.; Jónsson, S.; Avşar, Ö.; Schmidt, S. Earthquake-induced soft-sediment deformations and seismically amplified erosion rates recorded in varved sediments of Köyceğiz Lake (SW Turkey). *J. Geophys. Res. Solid Earth* **2016**, *121*, 4767–4779. [CrossRef]

50. Howarth, J.D.; Fitzsimons, S.J.; Norris, R.J.; Jacobsen, G.E. Lake sediments record high intensity shaking that provides insight into the location and rupture length of large earthquakes on the Alpine Fault, New Zealand. *Earth Planet. Sci. Lett.* **2014**, *403*, 340–351. [CrossRef]

51. Arnaud, F.; Lignier, V.; Revel, M.; Desmet, M.; Beck, C.; Pourchet, M.; Charlet, F.; Trentesaux, A.; Tribovillard, N. Flood and earthquake disturbance of ^{210}Pb geochronology (Lake Anterne, NW Alps). *Terra Nova* **2002**, *14*, 225–232. [CrossRef]

52. Arnaud, F.; Magand, O.; Chapron, E.; Bertrand, S.; Boës, X.; Charlet, F.; Mélières, M.-A. Radionuclide dating (^{210}Pb, ^{137}Cs, ^{241}Am) of recent lake sediments in a highly active geodynamic setting (Lakes Puyehue and Icalma—Chilean Lake District). *Sci. Total Environ.* **2006**, *366*, 837–850. [CrossRef]

53. Callieri, C.; Coci, M.; Corno, G.; Macek, M.; Modenutti, B.; Balseiro, E.; Bertoni, R. Phylogenetic diversity of nonmarine picocyanobacteria. *FEMS Microbiol. Ecol.* **2013**, *85*, 293–301. [CrossRef]

54. Callieri, C. Single cells and microcolonies of freshwater picocyanobacteria: A common ecology. *J. Limnol.* **2010**, *69*, 257–277. [CrossRef]

55. Callieri, C.; Cronberg, G.; Stockner, J.G. Freshwater picocyanobacteria: Single cells, microcolonies and colonial forms. In *Ecology of Cyanobacteria II: Their Diversity in Space and Time*; Whitton, B.A., Ed.; Springer: Dordrecht, The Netherlands, 2012; pp. 229–269.

56. Kelly, D.; Shearer, K.; Schallenberg, M. *Nutrient Loading to Shallow Coastal Lakes in Southland for Sustaining Ecological Integrity Values*; Cawthron Institute: Nelson, New Zealand, 2013.

57. Wood, S.A.; Rueckert, A.; Hamilton, D.P.; Cary, S.C.; Dietrich, D.R. Switching toxin production on and off: Intermittent microcystin synthesis in a *Microcystis* bloom. *Environ. Microbiol. Rep.* **2011**, *3*, 118–124. [CrossRef]

58. Haugland, R.A.; Siefring, S.C.; Wymer, L.J.; Brenner, K.P.; Dufour, A.P. Comparison of *Enterococcus* measurements in freshwater at two recreational beaches by quantitative polymerase chain reaction and membrane filter culture analysis. *Water Res.* **2005**, *39*, 559–568. [CrossRef]

59. Vaitomaa, J.; Rantala, A.; Halinen, K.; Rouhiainen, L.; Tallberg, P.; Mokelke, L.; Sivonen, K. Quantitative real-time PCR for determination of microcystin synthetase E copy numbers for *Microcystis* and *Anabaena* in lakes. *Appl. Environ. Microbiol.* **2003**, *69*, 7289–7297. [CrossRef]

60. Rueckert, A.; Cary, S.C. Use of an armored RNA standard to measure microcystin synthetase E gene expression in toxic *Microcystis* sp. by reverse-transcription QPCR. *Limnol. Oceanogr. Methods* **2009**, *7*, 509–520. [CrossRef]

61. Neilan, B.A.; Jacobs, D.; Therese, D.D.; Blackall, L.L.; Hawkins, P.R.; Cox, P.T.; Goodman, A.E. rRNA sequences and evolutionary relationships among toxic and nontoxic cyanobacteria of the genus *Microcystis*. *Int. J. Syst. Evol. Microbiol.* **1997**, *47*, 693–697. [CrossRef] [PubMed]

62. Rinta-Kanto, J.M.; Ouellette, A.J.A.; Boyer, G.L.; Twiss, M.R.; Bridgeman, T.B.; Wilhelm, S.W. Quantification of toxic *Microcystis* spp. during the 2003 and 2004 blooms in western Lake Erie using quantitative real-time PCR. *Environ. Sci. Technol.* **2005**, *39*, 4198–4205. [CrossRef]

63. Herlemann, D.P.; Labrenz, M.; Jürgens, K.; Bertilsson, S.; Waniek, J.J.; Andersson, A.F. Transitions in bacterial communities along the 2000 km salinity gradient of the Baltic Sea. *ISME J.* **2011**, *5*, 1571–1579. [CrossRef] [PubMed]

64. Klindworth, A.; Pruesse, E.; Schweer, T.; Peplies, J.; Quast, C.; Horn, M.; Glöckner, F.O. Evaluation of general 16S ribosomal RNA gene PCR primers for classical and next-generation sequencing-based diversity studies. *Nucleic Acids Res.* **2013**, *41*, e1. [CrossRef] [PubMed]

65. Bolyen, E.; Rideout, J.R.; Dillon, M.R.; Bokulich, N.A.; Abnet, C.; Al-Ghalith, G.A.; Alexander, H.; Alm, E.J.; Arumugam, M.; Asnicar, F.; et al. *QIIME 2: Reproducible, Interactive, Scalable, and Extensible Microbiome Data Science*; PeerJ Inc.: San Diego, CA, USA, 2018.

66. Callahan, B.J.; McMurdie, P.J.; Rosen, M.J.; Han, A.W.; Johnson, A.J.A.; Holmes, S.P. DADA2: High-resolution sample inference from Illumina amplicon data. *Nat. Methods* **2016**, *13*, 581–583. [CrossRef] [PubMed]

67. Pedregosa, F.; Varoquaux, G.; Gramfort, A.; Michel, V.; Thirion, B.; Grisel, O.; Blondel, M.; Prettenhofer, P.; Weiss, R.; Dubourg, V.; et al. Scikit-learn: Machine Learning in Python. *J. Mach. Learn. Res.* **2011**, *12*, 2825–2830.

68. Quast, C.; Pruesse, E.; Yilmaz, P.; Gerken, J.; Schweer, T.; Yarza, P.; Peplies, J.; Glöckner, F.O. The SILVA ribosomal RNA gene database project: Improved data processing and web-based tools. *Nucleic Acids Res.* **2013**, *41*, D590–D596. [CrossRef]

69. The R Core Team. *R: A Language and Environment for Statistical Computing*; The R Core Team: Vienna, Austria, 2019.

70. McMurdie, P.J.; Holmes, S. Phyloseq: An R Package for reproducible interactive analysis and graphics of microbiome census data. *PLoS ONE* **2013**, *8*, e61217. [CrossRef]

71. Oksanen, J.; Blanchet, F.G.; Friendly, M.; Kindt, R.; Legendre, P.; McGlinn, D.; Minchin, P.R.; O'Hara, R.B.; Simpson, G.L.; Solymos, P.; et al. *Vegan: Community Ecology Package*. 2019. Available online: https://CRAN.R-project.org/package=vegan (accessed on 8 September 2020).

72. Anderson, M.J. A new method for non-parametric multivariate analysis of variance. *Austral Ecol.* **2001**, *26*, 32–46.

73. Oliveros, J.C. *Venny. An Interactive Tool for Comparing Lists with Venn's Diagrams*. 2015. Available online: https://bioinfogp.cnb.csic.es/tools/venny_old/venny.php (accessed on 8 September 2020).

74. Puddick, J.; Wood, S.A.; Hawes, I.; Hamilton, D.P. Fine-scale cryogenic sampling of planktonic microbial communities: Application to toxic cyanobacterial blooms. *Limnol. Oceanogr. Methods* **2016**, *14*, 600–609. [CrossRef]

75. Puddick, J.; Prinsep, M.R.; Wood, S.A.; Kaufononga, S.A.F.; Cary, S.C.; Hamilton, D.P. High levels of structural diversity observed in microcystins from *Microcystis* CAWBG11 and characterization of six new microcystin congeners. *Mar. Drugs* **2014**, *12*, 5372–5395. [CrossRef] [PubMed]

76. FAO/IAEA. *Use of ^{137}Cs for Soil Erosion Assessment*; Fulajtar, E., Mabit, L., Renschler, C.S., Lee, Z., Yi, A., Eds.; Food and Agriculture Organization of the United Nations: Rome, Italy, 2017; p. 64.

77. Sikorski, J. A new method for constructing Pb-210 chronology of young peat profiles sampled with low frequency. *Geochronometria* **2019**, *46*, 1–14. [CrossRef]

78. Dunnington, D.W.; Spooner, I.S. Using a linked table-based structure to encode self-describing multiparameter spatiotemporal data. *Facets* **2018**, *3*, 326–337. [CrossRef]

79. Appleby, P.G.; Oldfield, F. The calculation of lead-210 dates assuming a constant rate of supply of unsupported ^{210}Pb to the sediment. *Catena* **1978**, *5*, 1–8. [CrossRef]

 *toxins*

Article

Insight into Unprecedented Diversity of Cyanopeptides in Eutrophic Ponds Using an MS/MS Networking Approach

Andreja Kust [1,2], Klára Řeháková [2,3], Jaroslav Vrba [2,4], Vincent Maicher [5], Jan Mareš [1,2,4], Pavel Hrouzek [1,4], Maria-Cecilia Chiriac [2], Zdeňka Benedová [6], Blanka Tesařová [6,7] and Kumar Saurav [1,*]

[1] Laboratory of Algal Biotechnology-Centre Algatech, Institute of Microbiology of the Czech Academy of Sciences, 37901 Třeboň, Czech Republic; kust@alga.cz (A.K.); mares@alga.cz (J.M.); hrouzek@alga.cz (P.H.)

[2] Biology Centre of the Czech Academy of Sciences, Institute of Hydrobiology, 37005 České Budějovice, Czech Republic; klara.rehakova@hbu.cas.cz (K.Ř.); jaroslav.vrba@prf.jcu.cz (J.V.); cecilia.chiriac@icbcluj.ro (M.-C.C.)

[3] Institute of Botany of the Czech Academy of Sciences, 37901 Třeboň, Czech Republic

[4] Faculty of Science, University of South Bohemia, 37005 České Budějovice, Czech Republic

[5] Nicholas School of the Environment, Duke University, Durham, NC 27710, USA; vincent.maicher@duke.edu

[6] ENKI, o.p.s. Třeboň, Dukelská 145, 37901 Třeboň, Czech Republic; benedova@enki.cz (Z.B.); blanka.tesarova@prirodou.cz (B.T.)

[7] Faculty of Agriculture, University of South Bohemia, Applied Ecology Laboratory, 37005 České Budějovice, Czech Republic

* Correspondence: sauravverma17@gmail.com; Tel.: +420-3-8434-0469

Received: 7 August 2020; Accepted: 28 August 2020; Published: 31 August 2020

Abstract: Man-made shallow fishponds in the Czech Republic have been facing high eutrophication since the 1950s. Anthropogenic eutrophication and feeding of fish have strongly affected the physicochemical properties of water and its aquatic community composition, leading to harmful algal bloom formation. In our current study, we characterized the phytoplankton community across three eutrophic ponds to assess the phytoplankton dynamics during the vegetation season. We microscopically identified and quantified 29 cyanobacterial taxa comprising non-toxigenic and toxigenic species. Further, a detailed cyanopeptides (CNPs) profiling was performed using molecular networking analysis of liquid chromatography-tandem mass spectrometry (LC-MS/MS) data coupled with a dereplication strategy. This MS networking approach, coupled with dereplication, on the online global natural product social networking (GNPS) web platform led us to putatively identify forty CNPs: fourteen anabaenopeptins, ten microcystins, five cyanopeptolins, six microginins, two cyanobactins, a dipeptide radiosumin, a cyclooctapeptide planktocyclin, and epidolastatin 12. We applied the binary logistic regression to estimate the CNPs producers by correlating the GNPS data with the species abundance. The usage of the GNPS web platform proved a valuable approach for the rapid and simultaneous detection of a large number of peptides and rapid risk assessments for harmful blooms.

Keywords: cyanobacteria; cyanopeptides; harmful bloom; liquid chromatography-tandem mass spectrometry; global natural product social networking (GNPS); dereplication strategy

Key Contribution: The combination of non-targeted HRMS/MS and GNPS has been proved to be a valuable approach for simultaneous, rapid, and early detection of bioactive and potentially harmful peptides, such as microcystins, anabaenopeptins, microginins, and cyanopeptolins.

---

1. Introduction

Cyanobacteria are important primary producers in the food chain with high nutritional value [1] and tend to proliferate, forming dense blooms, scums, and floating mats under favorable environmental conditions [2,3]. Eutrophication and climatic changes have led to increases in the geographical extent, population densities, and duration of cyanobacterial blooms in fresh, brackish, and marine waters [4]. These blooms can be hazardous to humans, animals, and plants due to the production of cyanotoxins apart from disrupting the ecosystem functions, such as nutrient cycles, light availability, dissolved oxygen levels/content, and consequent community reorganization and reduced biodiversity [5]. The most frequently reported cyanotoxins can be classified as cyclic oligopeptides (i.e., microcystins (MCs) and nodularins (NODs)) or alkaloids (i.e., anatoxins and cylindrospermopsin) based on their chemical structures, and as hepatotoxins, neurotoxins, and dermatotoxins based on their mechanism of toxic action in vertebrates [6,7]. The most extensively studied cyanotoxins are cyclic heptapeptides, MCs, produced most often by *Microcystis, Planktothrix*, and *Dolichospermum* (former *Anabaena*) [4]. NODs, cyclic pentapeptides, are structurally related to MCs and are produced mainly by *Nodularia spumigena*. Both MCs and NODs are hepatotoxins, inhibiting serine/threonine protein phosphatases [8]. To date, about forty cyanobacterial genera have been described as potential cyanotoxins producers [9,10], of which the most common bloom-forming genera include *Microcystis, Aphanizomenon, Cylindrospermopsis, Dolichospermum, Nodularia, Planktothrix, Oscillatoria*, and *Trichodesmium* [4,9]. Moreover, more than six hundred peptides or peptidic metabolites (hereafter "cyanopeptides (CNPs)") have been isolated from cyanobacteria [11], many of which are unknown with regard to their toxic potential and are not regularly monitored during the cyanobacterial bloom events. The co-occurrence of CNPs has been reported during the cyanobacterial proliferation events, and the necessity for extending their regular monitoring has been addressed [12–15]. CNPs, such as aeruginosins, microginins, cyanopeptolins, anabaenopeptilides, microviridins, anabaenopeptins, and nostophycins, with numerous structural variants are regularly found in cyanobacterial blooms [13,16–18]. Recent findings have suggested that metabolomic profiles consisting of different CNPs affect the cohabiting invertebrates and fish populations differently [19–21], underlining the need for the expansion of the number of regularly monitored and studied CNPs.

Early methods for the detection of toxins are based mostly on animal assays using intraperitoneal or intravenous injections on mice [22–24]. However, recent advancements in the field of fast and accurate methods, such as high-performance liquid chromatography connected to tandem mass spectrometry with high-resolution mass spectrometry (HPLC-HRMS/MS) and introduction of the global natural product social (GNPS) molecular networking platform, have gained considerable attention towards its application in the field of identification of novel compounds [25]. Further introduction of an in-silico annotation tool (such as Dereplicator+) at the GNPS online workflow has revolutionized the detection of known/unknown natural products by comparing experimental MS/MS spectra against chemical structure databases. These tools enable the analyses and curation of hundreds to thousands of obtained MS/MS data from analytes within the extract, which is almost impossible to analyze manually [26]. Recent application of these tools in the field of annotating metabolites from cyanobacterial bloom has led to the discovery of various novel compounds as well as unknown analogs [27–30].

Hence, the current study was focused on three eutrophic shallow ponds in the South Bohemia region of the Czech Republic and the determination of their phytoplankton composition and metabolomic profiles during the vegetation season. The metabolic composition was determined by leveraging the GNPS online workflow in silico tools and molecular networking to obtain a complete CNP profile of cyanobacterial proliferation of the studied ponds.

2. Results and Discussion

The studied ponds have been in use for fish production since the 16th century. During the 20th century, natural eutrophication and the intensification of fish production increased and led to heavy eutrophication of these water bodies at present [31], resulting in intensified cyanobacterial proliferation during the summer months. We sampled three ponds, KL (Klec), DH (Dehtář), and KV (Kvítkovický), located in South Bohemia, Czech Republic, once per month during the whole vegetation season (six months in total), to investigate their phytoplankton, CNPs composition, and dynamics. The chemical background data of the studied ponds indicated high concentrations of total nitrogen (TN), total phosphorus (TP), and chlorophyll-a (chl-a), illustrating the hypertrophic status of ponds (Table 1) [32]. All three ponds included in the current study showed a high content of chl-a, the primary and dominant photosynthetic pigment used as a proxy for phytoplankton biomass [33], with lowest concentration in KL-Apr (61.0 µg/L) and highest in KL-Jul (376.1 µg/L). Overall, the increase of water temperature and of total nitrogen resulted in higher cyanobacterial proliferation, while chl-a concentrations were correlated with the increase of cyanobacteria and/or diatoms biomass. To study seasonal dynamics of phytoplankton with emphasis on cyanobacterial species composition, we quantified phytoplankton at the species level (wherever clear taxonomical identification was possible) and statistically correlated cyanobacterial taxa with detected CNPs.

Table 1. Physicochemical characteristics of water of investigated lakes during sampling season. Sampling dates, water temperature, pH, conductivity, transparency, dissolved organic carbon (DOC), total nitrogen (TN), total phosphorus (TP), dissolved organic phosphorus (DRP), and chlorophyll-a (Chl-a) during each sampling. KL stands for Klec, DH for Dehtář, and KV for Kvítkovický.

Locality	Sampling Date	Temperature	pH	Conductivity	Secchi Depth	DOC	TN #200	TP #200	DRP	Chl-a
		(°C)		(µS/cm)	(cm)	(mg/L)	(mg/L)	(mg/L)	(mg/L)	(µg/L)
KL	24 April 2018	19.3		214	50	12.5	1.84	0.18	0.001	61.0
KL	15 May 2018	19.3	6.9	229	90	16.1	1.57	0.18	0.001	67.3
KL	19 June 2018	21.0	8.5	216	40	14.4	2.34	0.13	0.011	106.1
KL	17 July 2018	21.7	9.4	204	20	16.7	4.37	0.31	0.011	376.1
KL	14 August 2018	23.7	8.9	201	25	19.7	6.61	0.37	0.015	270.2
KL	11 September 2018	18.7	9.4	214	20	20.7	7.77	0.39	0.021	351.7
DH	26 April 2018	18.0		330	50	17.8	1.79	0.22	0.001	65.0
DH	17 May 2018	18.9	8.6	338	40	19.6	2.03	0.24	0.031	68.0
DH	21 June 2018	22.7	8.9	337	65	18.9	1.86	0.20	0.009	71.2
DH	19 July 2018	22.1	8.7	338	40	20.0	2.50	0.30	0.026	104.0
DH	16 August 2018	23.5	9.2	329	35	22.3	3.44	0.29	0.013	161.5
DH	13 September 2018	21.0	9.6	323	40	22.0	4.35	0.15	0.014	114.4
KV	26 April 2018	18.8		313	30	15.6	2.22	0.36	0.094	128.0
KV	17 May 2018	17.3	7.7	372	30	19.4	2.63	0.61	0.422	94.3
KV	21 June 2018	21.8	8.3	349	35	19.3	2.03	0.32	0.060	89.0
KV	19 July 2018	20.9	9.0	334	20	17.7	3.45	0.41	0.022	327.7
KV	16 August 2018	22.1	8.6	347	15	22.7	4.20	0.26	0.041	254.7
KV	13 September 2018	19.7	8.6	341	25	22.0	4.87	0.20	0.014	152.4

2.1. Phytoplankton Composition and Seasonal Dynamics

Phytoplankton of the three studied sites was assigned to classes Chlorophyceae, Cyanophyceae, Cryptophyceae, Bacillariophyceae, Euglenophyceae, Dinophyceae, and Zygnematophyceae (Figure S1A–C). During April and May, phytoplankton of all the three studied sites (KL, DH, KV) was dominated by *Chlorophyceae*, while cyanobacterial biomass did not exceed 3 mg/L of total phytoplankton biomass (Table S1, Figure S1A–C). Total cyanobacterial biomass in KL-Apr was 2.2 mg/L, from which 90.7% was composed of toxigenic taxa *Cuspidothrix issatschenkoi, Microcystis aeruginosa, Dolichospermum circinale* and *viguieri, Aphanizomenon flos-aquae*, and *Planktothrix agardhii* (Figure 1A and Figure S1D). On the other hand, cyanobacterial taxa in DH-Apr and KV-Apr were composed mainly of picocyanobacteria (84.2% and 88.3%, respectively). During May, low cyanobacterial biomass with the dominance of picocyanobacteria was observed in all studied ponds, with the exception of 1.9 mg/L (68.1% of cyanobacterial biomass) of toxigenic *Microcystis aeruginosa* in DH-May. Even though the cyanobacterial biomass was lower during April and May, it still formed an important part of the total phytoplankton biomass in some of the samples, i.e., 17.1%, 6.0%, and 5.9% in KL-Apr, KV-May, and DH-Apr, respectively. The dominance of taxa, which have been reported as CNPs producers, was observed in KL-Jun: *Woronichinia naegeliana* 2.6 mg/L (34.7%) and *Microcystis aeruginosa* 2 mg/L (27.5%); in DH-Jun: *Aphanizomenon flos-aquae* 3.3 mg/L (52.8%) and *Dolichospermum circinale* 1.2 mg/L (19.9%), while KV-Jun was dominated by planktic picocyanobacteria 1.5 mg/L (73.7%) and, in general, had the lowest cyanobacterial biomass (~2 mg/L) in comparison with the other two ponds (Figure S1D,E).

In July, we detected onset of summer phytoplankton peak (except for DH-Jul), with a record of total phytoplankton biomass of 104.4 mg/L in KL-Jul. KL-Jul exhibited the highest diversity of cyanobacterial taxa without clear dominance of a single cyanobacterial taxon, while the cyanobacteria from DH-Jul (12.6 mg/L) were co-dominated by 5.4 mg/L of *Aphanizomenon floss-aquae*, 2.3 mg/L *Dolichospermum circinale*, and 1.4 mg/L *Planktothrix agardhii*. Unlike the other two ponds, KV-Jul was dominated by Bacillariophyceae with only 3.6 mg/L (9.2%) of total phytoplankton biomass belonging to cyanobacteria, out of which the most abundant were the harmful taxa *Dolichospermum circinale* (1.2 mg/L) and *Aphanizomenon floss-aquae* (1.2 mg/L) (Figure 1A). All three studied sites exhibited the highest cyanobacterial biomass (total phytoplankton was composed of more than 50% cyanobacteria) during August, with the dominance of a single or two toxigenic taxa. While toxigenic cyanobacteria formed a major part of KL-Sep phytoplankton, Bacillariophyceae took over cyanobacteria in DH-Sep and KV-Sept, however, with a still high abundance of toxigenic cyanobacteria (i.e., *Aphanizomenon flos-aquae, Planktothrix agardhii, Microcystis aeruginosa*, and *Dolichospermum circinale*). DH and KV showed similar phytoplankton dynamics to the previous studies with an early spring maximum, followed by phytoplankton depression, with a final summer peak, while KL had its phytoplankton depression in spring months with a summer maximum (Figure S1). Observed phytoplankton development corresponds to previously reported plankton dynamics in shallow eutrophic ponds [34,35].

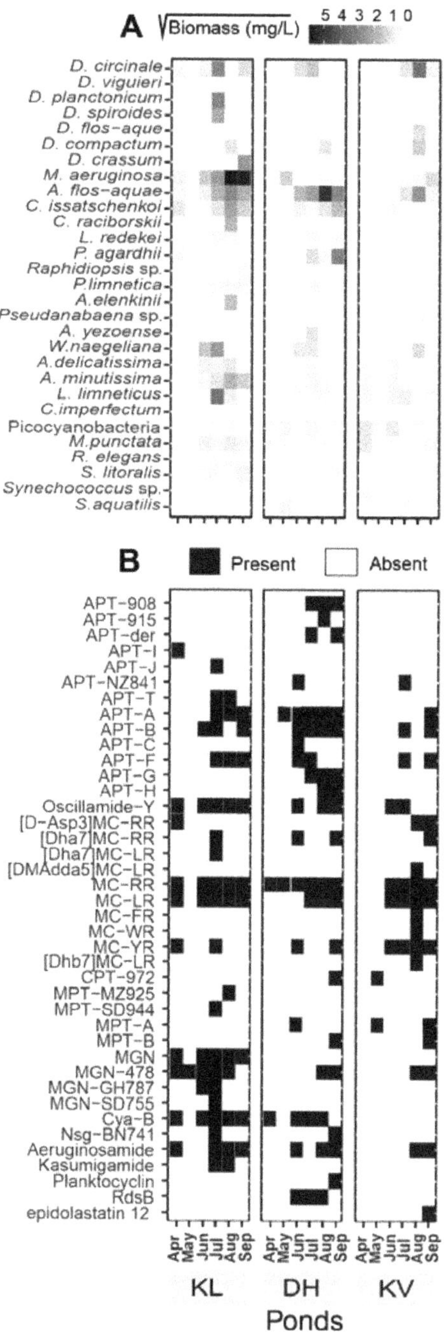

Figure 1. Heat maps showing (**A**) the square root of the biomass in mg/L of different cyanobacterial species in all ponds during all sampled months, and (**B**) the presence/absence of the different cyanopeptides (CNPs) detected in all ponds during all sampled months. The full names of cyanobacterial species can be found in Figure S1.

2.2. CNPs Diversity: Molecular Networking

Diverse cyanobacterial communities among studied ponds were reflected in the production of a wide array of CNPs. Abundance in CNPs diversity in a given ecosystem could affect any co-existing organisms, especially due to their inhibitory and toxic activities [11,18,36,37]. It has been hypothesized that physiological and ecological relevance of chemotype variability in a single strain, and even higher diversity in natural cyanobacterial bloom population, is advantageous for cyanobacterial dominance against other photoautotrophs and protection against grazing zooplankton [38–40]. Applying the online workflow of GNPS for high throughput screening, we detected forty CNPs (Figure 1B and Table S2). A molecular network of 87 clusters was generated using high-resolution mass spectrometry (HRMS) spectra data on the global natural product social molecular networking (GNPS) online workflow (Figure 2). GNPS algorithm automatically aligned and compared each spectrum against the spectra available in the database and then further grouped them by assigning cosine score (0 to 1).

Further, the obtained molecular network was annotated using an in-silico tool, Dereplicator+. With this tool, it was possible to search all the spectra in the GNPS launched in the molecular network and identify an order of magnitude more natural products than previous dereplication efforts [41]. Eighteen spectrum files were dereplicated, generating 26,220 spectrum scans. A total of 568 peptide-spectrum matches (PSMs) were identified with 321 PSMs, exhibiting a significant score of $\geq$11.0. The dereplication algorithm enabled us to facilitate natural product discovery by high-throughput peptide natural product identification among large-scale mass spectrometry-based screening platforms [42]. Different analytes were grouped in the same molecular clusters based on the similarity of their fragmentation patterns, with each cluster being potentially specific to the structure of the chemical families. These unidentified ions that belong to annotated clusters are then considered as potential new analogs of their respective molecular family [27]. Additionally, the availability of HRMS/MS spectral data of known cyanotoxins in the GNPS database facilitates the detection process [43–45]. This led to the detection of 14 anabaenopeptins (APTs), ten MCs, five cyanopeptolins (CPTs), six microginins (MGNs), two cyanobactins, a dipeptide radiosumin (RdsB), a cyclooctapeptide planktocyclin, and epidolastatin 12 from methanolic extracts of the biomass (Figure 1B, Table S2). Recently, new MC variants have been discovered using an MS-based molecular networking approach from the freshwater cyanobacterial harmful bloom at Green Lake, Seattle [28]. Similarly, numerous reports have been published, where molecular networking is employed to track changes in secondary metabolic profiles, including MCs and other peptides [46,47]. Fragmentation spectra of unknown variants discovered for APTs and MCs using molecular networking have been further manually curated for the identification of diagnostic ion peak [45,48–52].

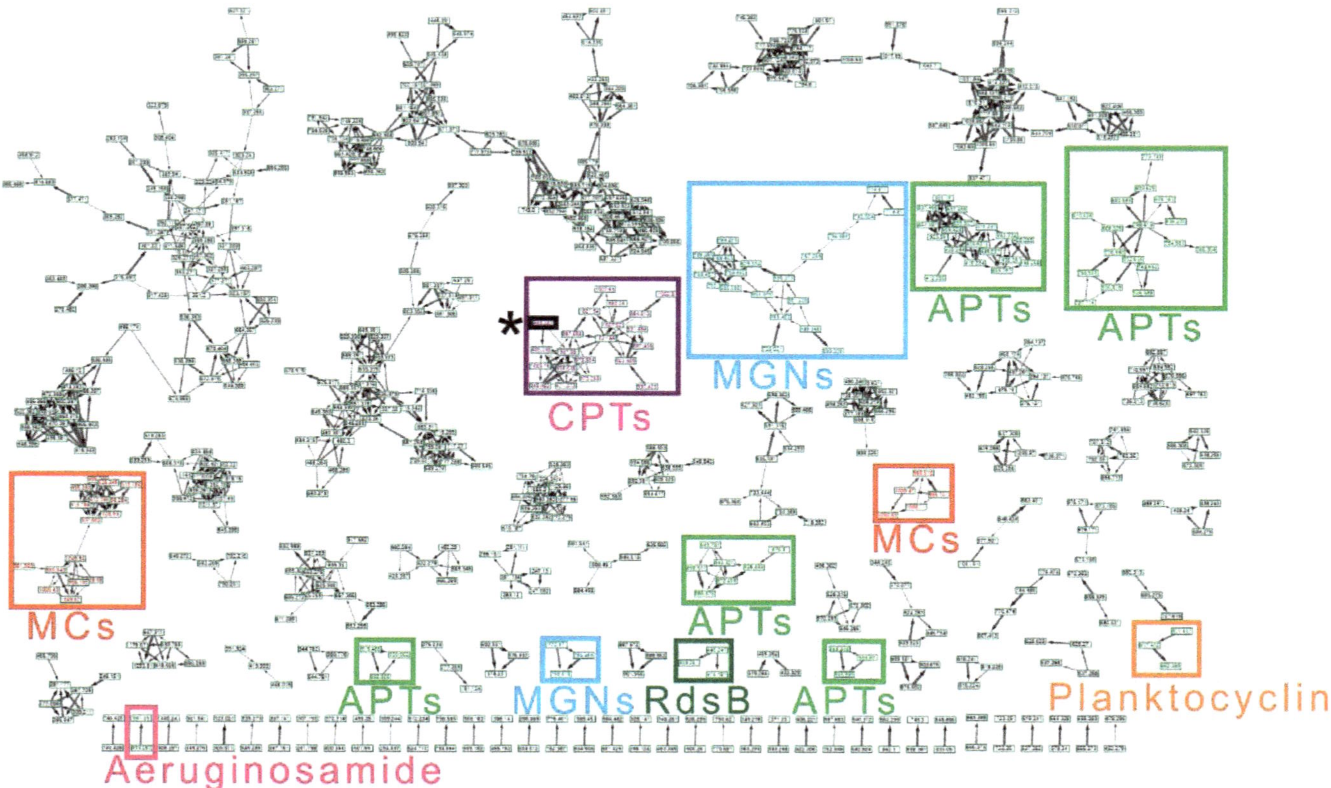

Figure 2. The molecular network generated from HRMS/MS spectra from all the samples of three ponds using global natural product social molecular networking (GNPS) tool. Analytes were compared with the components from the fragmentation pattern library available from the GNPS server. Only clusters of at least two nodes are represented. APTs: anabaenopeptins, MCs: microcystins, CPTs: cyanopeptolins, MGNs: microginins, RdsB: radiosumin_B, *: epidolastatin 12.

2.2.1. Anabaenopeptins (APTs)

Anabaenopeptins are a highly diverse family of cyclic hexapeptides, first described from *Anabaena flos-aquae* NRC 525-17 [53]. They exhibit diverse biological activities; however, studies are mostly focused on their serine protease and chymotrypsin inhibiting activity [54]. They are *N*-methylated and contain a conserved ureido linkage, connecting the side-chain amino acid residue to the D-lys [11]. In the current study, we detected the presence of 14 APTs variants; APT-908, APT-915, APT-I, APT-J, APT-NZ841, APT-T, APT-A, APT-B, APT-C, APT-F, APT-G, APT-H, Oscillamide-Y, and one defined as APT-derivative using a dereplication strategy (Figure 1B, Table S2). Oscillamide-Y (a serine protease inhibitor, isolated from *Planktothrix agardhii* NIES-610; [54]) was detected in ten samples, including KL-Apr with low cyanobacterial abundance (Figure 1). Apt-A and -B were detected nine times, while Apt-F seven times. Apt-A and -B inhibit carboxypeptidase A and protein phosphatase 1 with varying potency, but no inhibition against chymotrypsin, trypsin, and thrombin has been reported [55]. Detected APTs formed five clusters (Figure 2) corresponding to ions, presenting a match with the mass of previously described APTs, suggesting the presence of 34 structural variants, corresponding to potentially new analogs (Figures 2 and 3). Some of the compounds formed single nodes and were removed from the networking. However, we reported here their putative presence based on dereplication. Further, mass spectra of these putative new variants were evaluated for the presence of characteristic ion peak at m/z 84.081 (an immonium fragment ion of lysine) together with other fragment ions of amino acids (Figures S2 and S3) [45,49,50]. The recent increased reports of APTs co-occurrence with MCs raise the attention to this class of CNPs since their impact on the cohabiting aquatic organisms remains unclear, and their ecological role is uncertain [46].

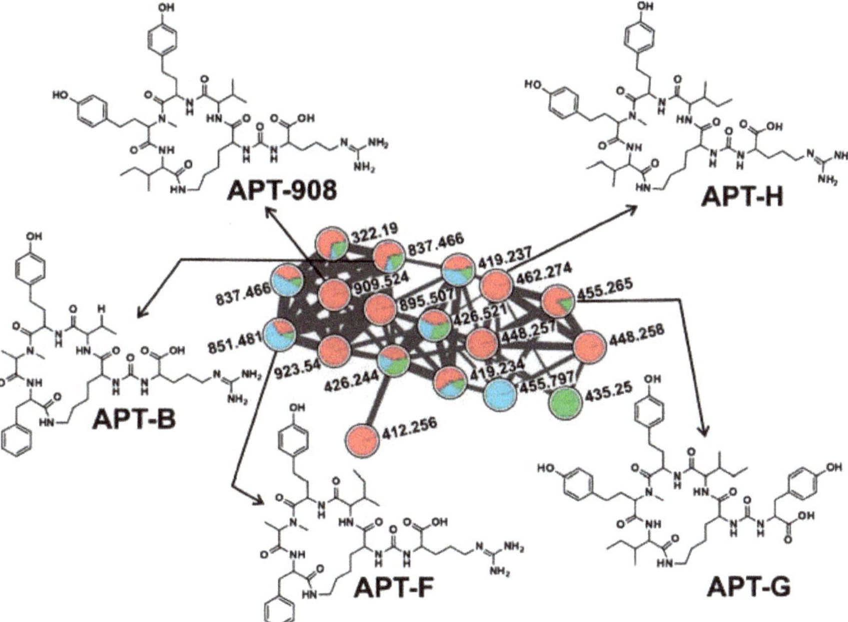

Figure 3. Anabaenopeptin (APT) cluster, formed by the GNPS analysis based on the MS/MS fragmentation spectra obtained from all three sampling sites (Red: DH, blue: KL, green: KV). Here depicted are APTs congener chemical structures detected in this respective cluster, with fragmentation patterns available in the library of the GNPS server. Note that (M + H)$^+$ and (M + 2H)$^{2+}$ ions are in the molecular cluster.

2.2.2. Microcystins (MCs)

Microcystins are cyclic heptapeptides produced, among others, by *Microcystis, Anabaena/Dolichospermum, Nodularia,* and *Oscillatoria*. While the liver is the primary target of MCs, they are also a skin, eye, and throat irritants and immunomodulating agents [56,57]. Microcystin-LR was the first identified cyanotoxin and is the most studied. The WHO has established a provisional guideline value of 1 ug/L for microcystin-LR in drinking water [58].

We were able to detect putatively ten microcystin congeners—MC-LR, MC-RR, MC-FR, MC-WR, MC-YR, [D-Asp3]MC-RR, [Dha7]MC-RR, [Dha7]MC-LR, [DMAdda5]MC-LR, and [Dhb7]MC-LR (Figure 1B)—using dereplication and molecular networking, forming two clusters (Figure 2). Potentially eleven new analogs were also observed in these clusters, showing distinct but similar fragmentation patterns to those of other known variants spectra present in the GNPS library (Figures 2 and 4). Further, the manual curation of the spectra was performed to identify the most common fragments characterizing MC together with the characteristic ion peak originating from Adda moiety at *m/z* 135.0804 Da, *m/z* 121.0653 for DMAdda, and m/z 163.0759 for ADMAdda (Figure S4 and Figure S5) [48]. MC-RR and MC-LR were detected in all studied ponds, MC-RR was detected in 15 samples, while MC-LR in 12. However, the presence of other congeners was more scattered, with the highest diversity in KV-Aug (eight MC variants).

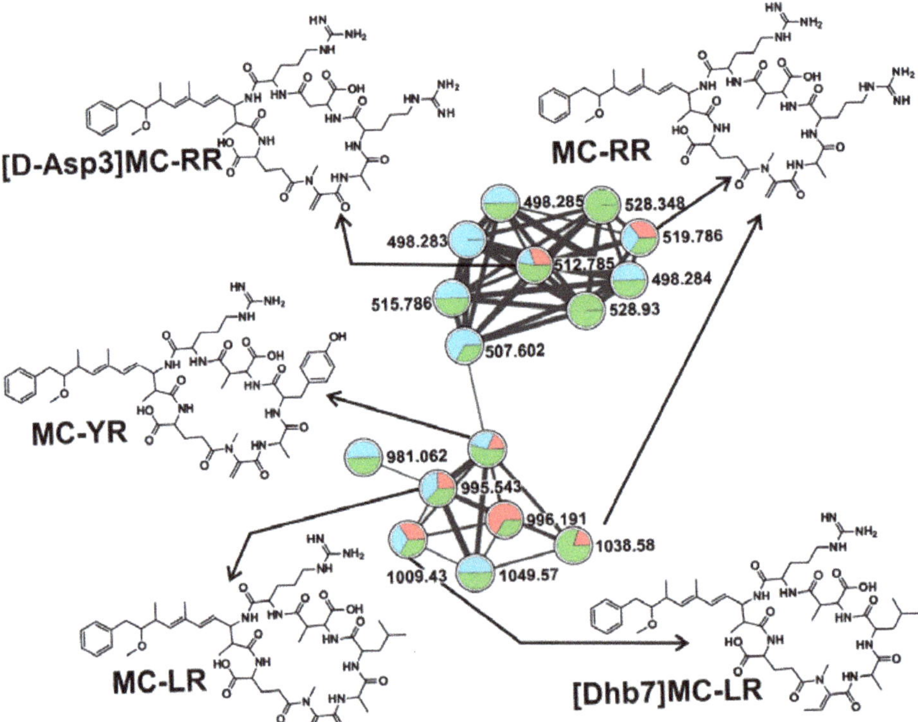

Figure 4. A microcystin (MC) cluster, formed by the GNPS analysis based on the MS/MS fragmentation spectra obtained from all three sampling sites (Red: DH, blue: KL, green: KV). MCs congener chemical structures detected in this respective cluster is depicted here, whose fragmentation patterns are available in the library of the GNPS server. Note that (M + H)$^+$ and (M + 2H)$^{2+}$ ions are in the molecular cluster.

2.2.3. Cyanopeptolins (CPTs)

CPTs are a diverse class of cyclic depsipeptides previously isolated from *Microcystis* sp. PCC 7806 [59], composed of a six amino acid residue ring structure, a conserved 3-amino-6-hydroxy-2-piperidone (AHP) residue, and a side chain of variable length [59,60]. We detected five CPTs variants (micropeptin (MPT) -MZ925, -SD944, -A, -B, and CPT 972 (Figure 1B) exclusively from summer samples), with a high abundance of cyanobacteria in the phytoplankton community (total nine hits). Two micropeptins and an epidolastatin12 formed a cluster together with compounds that very likely corresponded to potentially eighteen new analogs (Figures 2 and 5). The general activities reported from CPTs are protease inhibitory, fungicidal, cytotoxic, and antitumor activities [61], and the recent elucidation of the molecular basis of AHP-cyclodepsipeptides has opened new possibilities for customizing them as serine protease-specific inhibitors [60].

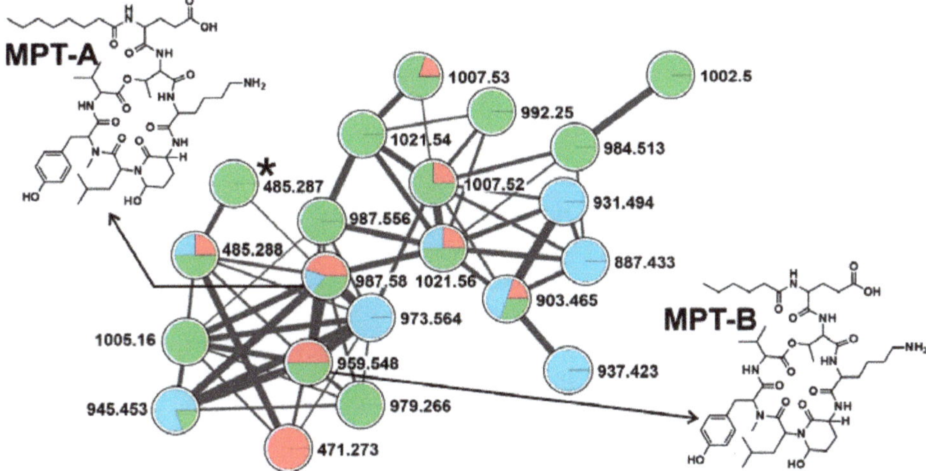

Figure 5. Cyanopeptolins (CPTs) cluster formed by the GNPS analysis based on the MS/MS fragmentation spectra obtained from all three sampling sites (Red: DH, blue: KL, green: KV). CPTs congener chemical structures (MPT: micropeptin) detected in this respective cluster is depicted here together with epidolastatin 12 (*), whose fragmentation patterns are available in the library of the GNPS server. Note that $(M + H)^+$ and $(M + 2H)^{2+}$ ions are in the molecular cluster.

Another peptide, epidolastatin 12, detected in the current study, formed a cluster together with CPTs (Figure 5). Dolastatins were originally reported from the mollusk *Dolabella auricularia* [62]; however, their structural variants were found to be produced by axenic cyanobacteria, implying the possibility that even the first reported dolastatin is produced by cyanobacteria [63,64]. It has been reported as an epimer of dolastatin 12 isolated from marine *Lyngbya majuscula/Schizothrix calcicola* cyanobacterial assemblages [63]. The detection of dolastatins epimer in a single sample (KV-Sep), is to our knowledge, the first report of the dolastatin variant detected from a freshwater source.

2.2.4. Microginins (MGNs)

Microginins are linear pentapeptides originally isolated from *Microcystic aeruginosa* NIES-100 as an angiotensin-converting enzyme inhibitor [65]. We detected six MGNs variants (MGN, MGN-478, -GH787, -SD755, cyanostatin B (Cya-B), nostoginin BN741 (NSG-BN741)) throughout the sampling season (Figure 1B). In KL-Jul, the most cyanobacterial diverse sampling point (16 taxa, Figure 1), we detected six MGNs variants matching with the library spectra match together with seventeen putative variants forming two clusters (Figures 2 and 6). One cluster comprising three variants is depicted in Figure 6. Similarly, to MCs and APTs, MGNs were detected, also in samples with low cyanobacterial biomass (KL-Apr, -May, DH-Apr). The most frequently detected variants were MGN-478 and Cya-B, both detected in nine samples. The biological activity among MGN variants also varies widely; for example, Cya-B is reported as an aminopeptidase M inhibitor [54], whereas MGN-478 has not exhibited any protease inhibitory activity [66].

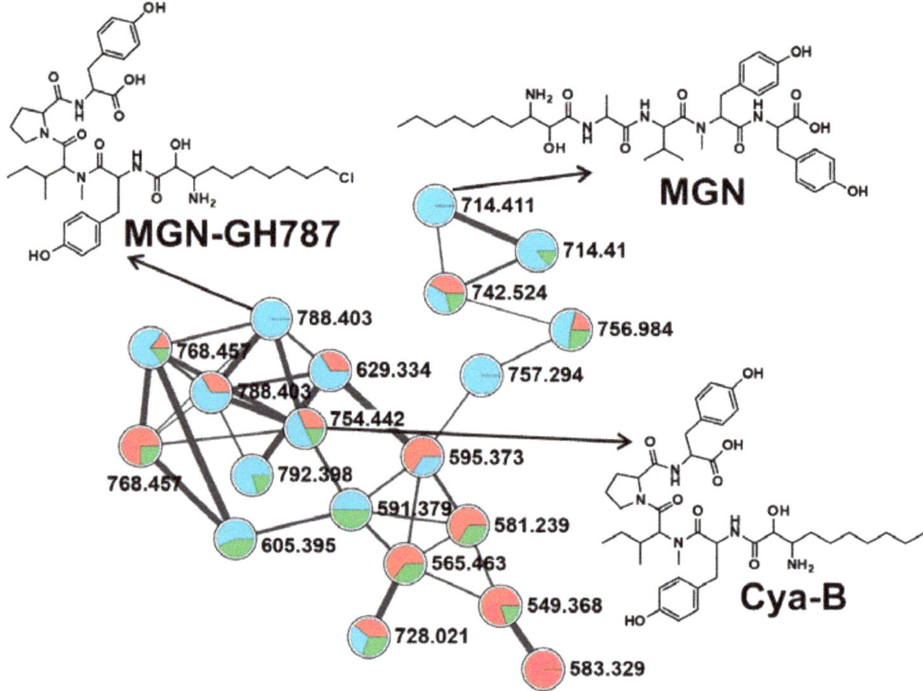

Figure 6. Microginin (MGN) compounds clustered together by the GNPS analysis based on the MS/MS fragmentation spectra obtained from all three sampling sites (Red: DH, blue: KL, green: KV). Selective known MGNs congener chemical structures detected in this respective cluster is depicted here, whose fragmentation patterns are available in the library of the GNPS server. Note that $(M + H)^+$ and $(M + 2H)^{2+}$ ions are in the molecular cluster.

2.3. CNPs Composition and Seasonal Dynamics

Cyanobacterial blooms are formed by diverse coexisting cyanobacterial species, resulting in the production of a wide array of CNPs, altering the natural habitat with their toxic activities [11,18,36,37,67]. Recent studies have reported the co-production of diverse CNPs with MCs [14], suggesting high chemotype variability in a single strain and even higher in natural cyanobacterial blooms [68,69].

The obtained array of CNPs (MCS, APTs, CPTs, MGNs) in our study, in general, corresponds to the expected chemical composition of cyanobacterial blooms dominated by commonly reported toxigenic planktic taxa, such as *Microcystis*, *Dolichospermum*, *Aphanizomenon*, and *Planktothrix* (Table S1) [14,55,70]. However, the amino-protease inhibitor Nsg-BN741 has been previously reported only from periphytic and terrestrial heterocytous cyanobacteria *Nostoc* [71]; thus, it was surprising to find it in the planktic environment (ponds KL and DH). Despite the absence of *Plectonema radiosum* and *Planktothrix rubescens* in the phytoplankton in the DH pond, we detected the presence of RdsB and planktocyclin [72,73].

In the ponds, we detected also two cyanobactins, unlike other CNPs reported here, produced ribosomally [74]. Cyanobactin kasumigamide, a ribosomal tetrapeptide isolated originally from *Microcystis aeruginosa* [66], occurred only in one pond (KL-Jul and -Aug) in sampling points, which showed the highest cyanobacterial diversity. On the other hand, second cyanobactin, aeruginosamide, reported originally from *Microcystis* [75], occurred in all three ponds in a variety of sampling points with different abundances of *Microcystis aeruginosa* co-occurring with diverse cyanotaxa (Figure 1).

As mentioned above, spring samples of all three ponds were dominated by Chlorophyceae with low cyanobacteria abundance. While no CNPs were detected in KV-Apr, we detected nine CNPs in KL-Apr, where the common toxigenic taxa were present, and two CNPs in DH-Apr without the presence of common CNPs producers. Although detection of the CNPs already in samples with low cyanobacterial biomass has been reported previously [76], they are often neglected for detailed monitoring. The detection of diverse CNPs in our samples further addressed a need for detailed monitoring of water bodies, even in samples with low cyanobacterial abundance.

In the following month, all three ponds were dominated mainly by picocyanobacteria, where we found one CNPs in KL-May and two in DH-May. The increase in cyanobacterial abundance late in the sampling season was reflected by an increase in CNPs. An increase in the diversity of common toxigenic cyanobacterial taxa in June resulted in a higher number of detected CNPs (Figure 1). Particularly, KL-Jun and DH-Jun exhibited higher cyanobacterial biomass and diversity of common toxigenic taxa compared with previous samples; thus, in these sampling points, we detected eight and 13 CNPs, respectively. KV-Jun was less abundant in cyanobacteria, resulting in the detection of only four CNPs.

The highest diversity of the CNPs was detected during the cyanobacterial proliferation in the months of July, August, and September (Figure 1). The most prolific sample in both cyanobacterial and CNP diversity was KL-Jul. This sampling point was characterized by the presence of 16 cyanotaxa and detection of 20 CNPs, referring that the co-occurrence of several toxigenic taxa would result in higher metabolic diversity [14,15].

As mentioned above, *Aphanizomenon flos-aquae* was the most abundant species in DH-Jul, where we detected nine CNPs, while KV-Jul was unlike the other ponds dominated by Bacillariophyceae with the lowest cyanobacterial biomass. Accordingly, we detected lower CNP diversity. Generally, the KV pond exhibited lower cyanobacterial and CNP diversity in comparison with the other two studied ponds.

All ponds showed increased cyanobacterial abundance and dominance of a single taxon in August. *Microcystis aeruginosa*, reported diverse CNPs producer [77,78], was the most abundant cyanobacterial taxon in KL-Aug (55%), where we detected 12 diverse CNPs. Thirteen CNPs were detected in DH-Aug dominated by *Aphanizomenon flos-aquae* (90%), while *Dolichospermum* dominated KV-Aug (50%), showing the presence of 12 CNPs, with the highest detected diversity of MCs congeners (eight) among all samples. Both the dominating taxa have been demonstrated as rich secondary metabolite producers [79,80]. KL-Sept had the highest abundance of *Microcystis aeruginosa* (65% of cyanobacterial biomass) with the presence of 11 more taxa, resulting in the detection of nine CNPs. Known for their high CNPs potential, *Aphanizomenon flos-aquae* and *Planktothrix agardhii* [80,81] were the most abundant species in DH-Sep, where 17 CNPs were found. KV-Sep showed a lower diversity of cyanobacteria when compared with the previous sampling point (KV-Aug) and the other two ponds. Dominated with *Microcystis*, KV-Sep exhibited the highest CNPs diversity (13) when compared with previous sampling points of the same pond.

Among all the samples, four samples (DH-May, KL-Aug, KL-Sep, KV-Sep) were dominated mainly by *Microcystis aeruginosa*, where we detected the presence of different variants of CNPs found in other samples not dominated by *Microcystis aeruginosa*; only MPT-MZ925 and epidolastatin 12 were detected exclusively in KL-Aug and KV-Sep, respectively.

Since high cyanobacterial and CNPs diversity co-occurred throughout the sampling campaign, a binary logistic regression was performed in order to correlate specific cyanobacterial taxa with individual CNP occurrence. A number of common toxigenic cyanobacterial taxa exhibited a strong correlation with some of the CNPs; on the other hand, we also observed previously unreported associations (Figure 7). The presence of several reported APTs was significantly correlated with a certain cyanobacterial taxon, previously reported as a producer [55]. APT-915 and APT-H were associated with *Aphanizomenon flos-aquae*, APT-der with *Planktothrix agardhii*, and APT-I with *Dolichospermum viguieri*. Three APTs showed correlation with two taxa, oscillamide Y with *Cuspidothrix issatschenkoi* and *Limnococcus limneticus*, while APT-908 and APT-G were correlated with *Planktothrix agardhii* and *Aphanizomenon flos-aquae*. Other APTs were correlated with three or more taxa, or not correlated to any.

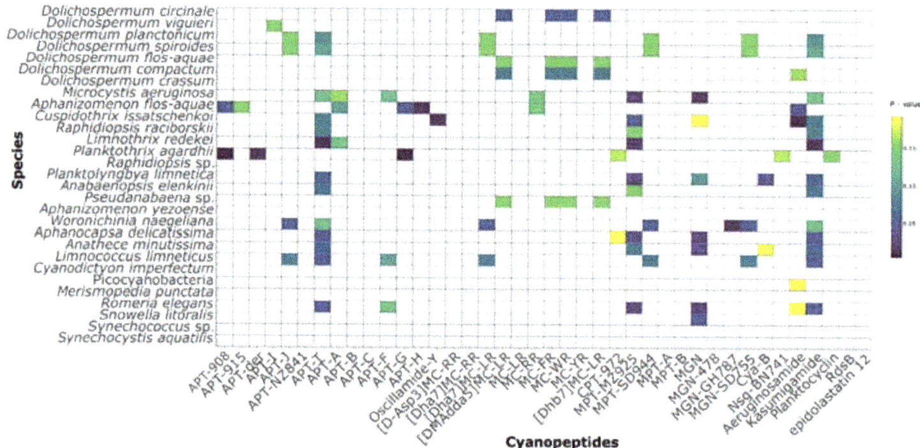

Figure 7. Binary logistic regression of cyanobacterial taxa with distinct CNP production. The full names of CNPs are in Table S2.

Six out of 10 MCs were correlated with two or more taxa; four MCs (MC-FR, -WR, [DMAdda5]MC-LR, and [Dhb7]MC-LR) detected only in one sample, KV-Aug, showed the same correlation pattern with four cyanobacteria (*Dolichospermum circinale*, *Dolichospermum flos-aquae*, *Dolichospermum compactum*, and *Pseudanabaena* sp.). Similarly, three CPTs showed a correlation with more than one taxon, including not previously reported producers (*Aphanocapsa delicatissima*, *Anathece minutissima*, *Romeria elegans*). MGN-GH787 was correlated with a single taxon, *Woronichinia naegeliana*, previously reported as MGNs producing species [81]. Nsg-BN741, previously reported from *Nostoc*, exhibited a correlation with *Planktothrix agardhii*, while MGN and MGN-SD755 showed a correlation with seven and four taxa, respectively. Furthermore, Cya-B showed association not only with *Planktolyngbya limnetica* but also with *Anathece minutissima* (Figure 7), often reported in cyanobacterial blooms, but never directly associated with CNPs production so far [68,82]. Aeruginosamide was correlated with five cyanobacteria, while planktocyclin, previously reported from *Planktothirx rubescens*, exhibited a correlation with *Planktothrix agardhii*. APT-T and kasumigamide were both detected only in KL-Jul and KL-August and were significantly correlated to the same thirteen cyanotaxa, including non-toxigenic taxa. While *Limnococcus limneticus* (formerly *Chroococcus limneticus*) is generally not considered as toxigenic, it was associated with MGN-SD755, APT-I, -T, -F, [Dha7]MC-LR, Mpt-SD944, MGN-SD755, and kasumigamide. Similarly, picocyanobacterial taxa exhibited a correlation with several CNPs. Picocyanobacteria have been, in general, considered as cyanobacteria with low secondary metabolite potential, although their correlation with MCs production has been repeatedly reported since the eighties [83–85].

Jakubowska [82] suggested more in-depth toxicological studies on picocyanobacteria, in general, since CNPs have been already detected in bloom samples with high picocyanobacterial abundance. Only several recent studies have investigated picocyanobacterial's capability for bioactive secondary metabolites production, such as hepatotoxins, β-N-methylamino-L-alanine (BMAA), lipopolysaccharides, and other bioactive metabolites, i.e. bacteriocins [86]. However, the direct proofs of picocyanobacteria toxicity are still scarce, and authors tend to consider them as non-toxic.

3. Conclusions

In the current study, we implied a non-targeted mass spectrometry approach to determine the metabolome profile of three ponds used for fish farming in the Czech Republic. We detected a range of harmful MCs variants and other potentially harmful CNPs during the entire sampling season and especially (but not exclusively) in samples where cyanobacterial proliferation occurred, which raises concerns on the high presence of harmful CNPs. Usage of the online workflow at GNPS enabled us to identify several classes of CNPs beyond MC. In addition, we were able to putatively determine the presence of several unknown variants of CNPs, which was further evaluated manually by targeting their respective diagnostic ion peak. There is no such study, which investigates/monitor a broad range of CNPs in fish farming ponds, which is regularly used in the fishmarket for human consumption [14]. The current study also aimed to develop this minimal sample treatment method and apply regularly on the freshwater sample monitoring process. Furthermore, detected CNPs (i.e., APT, CPT, MGN) were reported as a co-product along with the MCs; thus, the possible synergistic effect of several compounds produced was addressed. We also introduced a rapid and efficient monitoring approach, combining the GPNS approach and binary logistic regression, for the detection of a wide range of CNPs, even in the samples with low cyanobacterial biomass, which could help to understand the early development and dynamics of CNPs production in aquaculture ponds.

4. Material and Methods

4.1. Study Sites and Sampling

Three nutrient-rich shallow eutrophic ponds were sampled monthly to cover the growth season, from April until September 2018. The investigated ponds are used for fish production in the Czech Republic: KL (Klec 49.090N, 14.767E, max. depth 2 m, area 0.64 km^2), DH (Dehtář) 49.006 N, 14.294 E, max. depth 4 m, area 2.28 km^2), and KV (Kvítkovický 48.963N, 14.337E max. depth 3 m, area 0.24 km^2). During each sampling point, temperature, pH, conductivity, and transparency were measured (Table 1). Water samples for plankton and background physicochemical analysis were collected, as described previously [32]. Briefly, horizontally integrated mixed water samples from surface water were collected from seven different points with van Dorn sampler (length of 1 m, 6.4 L volume). Chlorophyll a (Chl a) was determined spectrophotometrically after the extraction of samples collected on GF/C filters (Merck KGaA, Darmstadt, Germany), as described elsewhere [87] (Table 1). A subsample (3–5 L) was taken for chemical analysis, and 100 mL was preserved with Lugol's solution for the analysis of phytoplankton. For the CNPs' analysis, surface water samples were repeatedly collected with the plankton net (20 μm mesh) until obtaining dense biomass, refrigerated on the way, transferred to the lab, and kept at −80 °C until the analysis.

4.2. Phytoplankton Analysis

Biomass of individual phytoplankton taxa was determined in Lugol preserved samples using Utermöhl's sedimentation method [88] and the inverted microscope (Olympus IMT2, Hamburg, Germany). The abundance of each taxon was multiplied by their respective biovolume calculated from mean cell dimensions using an approximation to geometrical solids [89]. For the taxonomic determination of cyanobacteria, the taxonomic keys by Komárek and Anagnostidis were used [90–92].

4.3. Crude Extracts Preparation and HPLC-MS/MS Analysis

Crude extracts were prepared following the pre-established protocol [93]. Briefly, freeze-dried biomass of collected pond samples (~20 mg) was ground (with the sea sand) and extracted three times with 75% MeOH in water, followed by bath sonication. Extracts were evaporated under vacuum using a rotary vacuum evaporator (Heidolph, Schwabach, Germany) and dissolved with DMSO to get a final concentration of 4 mg/mL prior to analysis. Thermo Scientific DionexUltiMate 3000 UHPLC (Thermo Fischer Scientific, Waltham, MA, USA) equipped with a diode array detector (DAD) and high-resolution mass spectrometry with electrospray ionization source (ESI-HRMS; Impact HD Mass Spectrometer, Bruker Billerica, MA, USA) was used for the analysis of the crude extracts. HPLC separation was performed on reversed-phase Kinetex Phenomenex C_{18} column (150 × 4.6 mm, 2.6 μm; Phenomenex, Aschaffenburg, Germany) with H_2O/acetonitrile containing 0.1% HCOOH as a mobile phase. The flow rate during the analysis was 0.6 mL/min. The gradient was as follows: H_2O/MeOH 85/15 (0 min), 85/15 (in 1 min), 0/100 (in 20 min), 0/100 (in 25 min), and 85/15 (in 30 min). The mass spectrometer settings were as follows: dry temperature 200 °C; drying gas flow 12 L/min; nebulizer 3 bar; capillary voltage 4500 V; endplate offset 500 V. The spectra were collected in the range 20–2000 *m/z*, with the spectra rate 4 Hz. A ramp was set with collision-induced dissociation from 20 to 60 eV on successive *m/z* 200–1200. Data were collected by an initial precursor ion survey scan, followed by product ion generation from precursor ions selected in small isolation windows (≈4 Da wide). Calibration was performed using LockMass 622 (abcr GmbH, Karlsruhe, Germany) as an internal calibration solution and CH_3COONa clusters at the beginning of each analysis.

4.4. Molecular Networking

The raw data files obtained from HPLC-HRMS/MS analysis were converted to mzXML format using MSConvert from the ProteoWizard suite [94]. A molecular network was created using the online workflow on the GNPS website [26]. The data were filtered by removing all MS/MS fragment ions within +/− 17 Da of the precursor m/z. MS/MS spectra were window filtered by choosing only the top 6 fragment ions in the +/− 50Da window throughout the spectrum. The precursor ion mass tolerance was set to 2 Da and an MS/MS fragment ion tolerance of 0.1 Da. A network was then created where edges were filtered to have a cosine score above 0.65 and more than 6 matched peaks. Further, edges between the two nodes were kept in the network if and only if each of the nodes appeared in each other's respective top 10 most similar nodes. Finally, the maximum size of a molecular family was set to 100, and the lowest-scoring edges were removed from molecular families until the molecular family size was below this threshold. The spectra in the network were then searched against GNPS spectral libraries. The library spectra were filtered in the same manner as the input data. All matches, kept between network spectra and library spectra, were required to have a score above 0.65 and at least 4 matched peaks. Further, the network was annotated using dereplicator+ to putatively identify the structural details of the compounds present. For annotation using dereplication+, precursor ion mass tolerance of 0.1 Da, fragment ion mass tolerance of 0.01 Da, max charge of 2, min score to consider a PSM as 8.25, and fragmentation mode applied as general_6_1_6.

4.5. Statistical Analysis

Statistical analyses were performed in R v. 3.6.1 [95]. The association of specific cyanobacterial species abundance (continuous variable) with distinct cyanotoxin presence/absence (nominal variable) was evaluated using a binary logistic regression in R ('*glm*' function from 'stats' package) [96]. An asymptotic chi-square statistic based on the deviance was used to assess the goodness-of-fit of each model. *p*-values were adjusted to reduce the number of false positives using the Benjamini–Hochberg procedure [97], with a false discovery rate (FDR) threshold of 0.2. Heatmaps generated using the 'heatmaply' function in R ('heatmaply' package). The R code developed for the entire analysis is available in Supplementary material as Code 1.

4.6. Data Deposition

The mass spectrometry data was deposited in MassIVE public repository (MSV000085840). The molecular networking job can be publicly accessed with the task ID: task=c2034223333641b3a06a72b40d27b2e4.

Supplementary Materials: The following are available online at http://www.mdpi.com/2072-6651/12/9/561/s1, Figure S1: Phytoplankton composition of studied ponds throughout the sampling season, Table S1: Biomass of cyanobacterial species and phytoplankton classes in all three studied ponds during the sampling season, Table S2: Detected cyanopeptides (CNPs) in all three studied ponds during the sampling season, Figure S2: HR-MS/MS product ion spectra APT-B in comparison with two unknown variants, highlighting the presence of diagnostic ion peak at m/z 84.0810 (Lys immonium ion) together with other fragment ions of amino acids, Figure S3: HR-MS/MS product ion spectra of protonated known/unknown APTs, forming five clusters (Figure 2), Figure S4: HR-MS/MS product ion spectra MC-RR in comparison with two unknown variants, highlighting the presence of diagnostic ion peak originating from Adda moiety at m/z 135.0804 Da, Figure S5: HR-MS/MS product ion spectra of protonated known/unknown MCs, forming two clusters (Figure 2), Code 1: The R code developed for the entire analysis.

Author Contributions: Conceptualization, K.S.; Data curation, J.V., V.M., J.M., P.H., M.-C.C. and B.T.; Formal analysis, K.Ř., J.V., J.M., P.H., C.C. and B.T.; Funding acquisition, J.V. and K.S.; Investigation, A.K.; Methodology, A.K. and Z.B.; Visualization, V.M.; Writing–original draft, A.K. and K.S.; Writing–review and editing, P.H. and K.S. All the authors have read and agreed to the published version of the manuscript. All authors have read and agreed to the published version of the manuscript.

Funding: This study was supported by the Ministry of Education, Youth and Sports of the Czech Republic MSCA IF II project (CZ.02.2.69/0.0/0.0/18_070/0010493), and Czech Science Foundation (GAČR)-project no. 19-17868Y and 17-09310S.

Conflicts of Interest: The authors declare no conflict of interest.

References

1. Gantar, M.; Svirčev, Z. Microalgae and cyanobacteria: Food for thought. *J. Phycol.* **2008**, *44*, 260–268. [CrossRef] [PubMed]
2. Schindler, D.W. Recent advances in the understanding and management of eutrophication. *Limnol. Oceanogr.* **2006**, *51*, 356–363. [CrossRef]
3. Weisse, T. Limnoecology: The ecology of lakes and streams. *J. Plankton Res.* **2008**, *30*, 489–490. [CrossRef]
4. Buratti, F.M.; Manganelli, M.; Vichi, S.; Stefanelli, M.; Scardala, S.; Testai, E.; Funari, E. Cyanotoxins: Producing organisms, occurrence, toxicity, mechanism of action and human health toxicological risk evaluation. *Arch. Toxicol.* **2017**, *91*, 1049–1130. [CrossRef] [PubMed]
5. Whitton, B.A. *Ecology of Cyanobacteria ii: Their Diversity in Time and Space*; Springer: Dordrecht, The Netherlands, 2012; p. 760.
6. Corbel, S.; Mougin, C.; Bouaïcha, N. Cyanobacterial toxins: Modes of actions, fate in aquatic and soil ecosystems, phytotoxicity and bioaccumulation in agricultural crops. *Chemosphere* **2014**, *96*, 1–15. [CrossRef]
7. Chorus, I.; Bartram, J. *Toxic Cyanobacteria in Water: A Guide to Their Public Health Consequences, Monitoring and Management*; Ingrid, C., Jamie, B., Eds.; World Health Organization: Geneva, Switzerland, 1999.
8. Runnegar, M.; Berndt, N.; Kong, S.M.; Lee, E.Y.; Zhang, L. In Vivo and In Vitro binding of microcystin to protein phosphatases 1 and 2A. *Biochem. Biophys. Res. Commun.* **1995**, *216*, 162–169. [CrossRef]
9. Bernard, C.; Ballot, A.; Thomazeau, S.; Maloufi, S.; Furey, A.; Mankiewicz-Boczek, J.; Pawlik-Skowrońska, B.; Capelli, C.; Salmaso, N. *Appendix 2: Cyanobacteria associated with the production of cyanotoxins. Handbook of Cyanobacterial Monitoring and Cyanotoxin Analysis*; John Wiley & Sons: Hoboken, NJ, USA, 2017; pp. 501–525.
10. Carmichael, W.W. Health effects of toxin-producing cyanobacteria: "The CyanoHABs". *Hum. Ecol. Risk Assess. HERA* **2001**, *7*, 1393–1407. [CrossRef]
11. Welker, M.; Von Döhren, H. Cyanobacterial peptides–Nature's own combinatorial biosynthesis. *FEMS Microbiol. Rev.* **2006**, *30*, 530–563. [CrossRef]
12. Beversdorf, L.J.; Weirich, C.A.; Bartlett, S.L.; Miller, T.R. Variable cyanobacterial toxin and metabolite profiles across six eutrophic lakes of differing physiochemical characteristics. *Toxins* **2017**, *9*, 62. [CrossRef]
13. Fastner, J.; Erhard, M. Determination of oligopeptide diversity within a natural population of *Microcystis* spp. (cyanobacteria) by typing single colonies by matrix-assisted laser desorption ionization–time of flight mass spectrometry. *Appl. Environ. Microbiol.* **2001**, *67*, 5069–5076. [CrossRef]

14. Janssen, E.M.L. Cyanobacterial peptides beyond microcystins–A review on co-occurrence, toxicity, and challenges for risk assessment. *Water Res.* **2019**, *151*, 488–499. [CrossRef] [PubMed]

15. Roy-Lachapelle, A.; Vo Duy, S.; Munoz, G.; Dinh, Q.T.; Bahl, E.; Simon, D.F.; Sauvé, S. Analysis of multiclass cyanotoxins (microcystins, anabaenopeptins, cylindrospermopsin and anatoxins) in lake waters using on-line SPE liquid chromatography high-resolution Orbitrap mass spectrometry. *Anal. Methods* **2019**, *11*, 5289–5300. [CrossRef]

16. Mazur-Marzec, H.; Bertos-Fortis, M.; Toruńska-Sitarz, A.; Fidor, A.; Legrand, C. Chemical and genetic diversity of *Nodularia spumigena* from the Baltic sea. *Mar. Drugs* **2016**, *14*, 209. [CrossRef] [PubMed]

17. Namikoshi, M.; Rinehart, K.L. Bioactive compounds produced by cyanobacteria. *J. Ind. Microbiol. Biotechnol.* **1996**, *17*, 373–384. [CrossRef]

18. Welker, M.; Maršálek, B.; Šejnohová, L.; von Döhren, H. Detection and identification of oligopeptides in *Microcystis* (cyanobacteria) colonies: Toward an understanding of metabolic diversity. *Peptides* **2006**, *27*, 2090–2103. [CrossRef] [PubMed]

19. Palíková, M.; Krejčí, R.; Hilscherová, K.; Babica, P.; Navrátil, S.; Kopp, R.; Bláha, L. Effect of different cyanobacterial biomasses and their fractions with variable microcystin content on embryonal development of carp (*Cyprinus carpio* L.). *Aquat. Toxicol.* **2007**, *81*, 312–318. [CrossRef]

20. Pawlik-Skowrońska, B.; Toporowska, M.; Mazur-Marzec, H. Effects of secondary metabolites produced by different cyanobacterial populations on the freshwater zooplankters *Brachionus calyciflorus* and *Daphnia pulex*. *Environ. Sci. Pollut. Res.* **2019**, *26*, 11793–11804. [CrossRef]

21. Wejnerowski, L.; Falfushynska, H.; Horyn, O.; Osypenko, I.; Kokocinski, M.; Meriluoto, J.; Jurczak, T.; Poniedzialek, B.; Pniewski, F.; Rzymski, P. In Vitro Toxicological screening of stable and senescing cultures of *Aphanizomenon, Planktothrix, and Raphidiopsis. Toxins (Basel)* **2020**, *12*, 400. [CrossRef]

22. Carmichael, W.W.; Biggs, D.F.; Gorham, P.R. Toxicology and pharmacological action of *Anabaena flos-aquae* toxin. *Science* **1975**, *187*, 542–544. [CrossRef]

23. Sivonen, K.; Kononen, K.; Esala, A.L.; Niemelä, S.I. Toxicity and isolation of the cyanobacterium *Nodularia spumigena* from the southern Baltic Sea in 1986. *Hydrobiologia* **1989**, *185*, 3–8. [CrossRef]

24. Devlin, J.P.; Edwards, O.E.; Gorham, P.R.; Hunter, N.R.; Pike, R.K.; Stavric, B. Anatoxin-a, a toxic alkaloid from *Anabaena flos-aquae* NRC-44h. *Can. J. Chem.* **1977**, *55*, 1367–1371. [CrossRef]

25. Nothias, L.F.; Nothias-Esposito, M.; Da Silva, R.; Wang, M.; Protsyuk, I.; Zhang, Z.; Sarvepalli, A.; Leyssen, P.; Touboul, D.; Costa, J.; et al. Bioactivity-based molecular networking for the discovery of drug leads in natural product bioassay-guided fractionation. *J. Nat. Prod.* **2018**, *81*, 758–767. [CrossRef] [PubMed]

26. Wang, M.; Carver, J.J.; Phelan, V.V.; Sanchez, L.M.; Garg, N.; Peng, Y.; Nguyen, D.D.; Watrous, J.; Kapono, C.A.; Luzzatto-Knaan, T.; et al. Sharing and community curation of mass spectrometry data with Global Natural Products Social Molecular Networking. *Nat. Biotechnol.* **2016**, *34*, 828–837. [CrossRef] [PubMed]

27. Ding, C.Y.G.; Pang, L.M.; Liang, Z.X.; Goh, K.K.K.; Glukhov, E.; Gerwick, W.H.; Tan, L.T. MS/MS-Based molecular networking approach for the detection of aplysiatoxin-Related compounds in environmental marine cyanobacteria. *Mar. Drugs* **2018**, *16*, 505. [CrossRef] [PubMed]

28. Teta, R.; Sala, G.D.; Glukhov, E.; Gerwick, L.; Gerwick, W.H.; Mangoni, A.; Costantino, V. A combined LC-MS/MS and molecular networking approach reveals new cyanotoxins from the 2014 cyanobacterial bloom in Green Lake, Seattle. *Environ. Sci. Technol.* **2017**, *176*, 139–148. [CrossRef]

29. Via, C.W.; Glukhov, E.; Costa, S.; Zimba, P.V.; Moeller, P.D.R.; Gerwick, W.H.; Bertin, M.J. The metabolome of a cyanobacterial bloom visualized by MS/MS-Based molecular networking reveals new neurotoxic smenamide analogs (C, D, and E). *Front. Chem.* **2018**, *6*, 316. [CrossRef]

30. Fox Ramos, A.E.; Evanno, L.; Poupon, E.; Champy, P.; Beniddir, M.A. Natural products targeting strategies involving molecular networking: Different manners, one goal. *Nat. Prod. Rep.* **2019**, *36*, 960–980. [CrossRef]

31. Pechar, L. Impacts of long-term changes in fishery management on the trophic level water quality in Czech fish ponds. *Fish. Manag. Ecol.* **2000**, *7*, 23–31. [CrossRef]

32. Šimek, K.; Grujčić, V.; Nedoma, J.; Jezberová, J.; Šorf, M.; Matoušů, A.; Pechar, L.; Posch, T.; Bruni, E.P.; Vrba, J. Microbial food webs in hypertrophic fishponds: Omnivorous ciliate taxa are major protistan bacterivores. *Limnol. Oceanogr.* **2019**, *64*, 2295–2309. [CrossRef]

33. Graham, J.L.; Loftin, K.A. Book review: Handbook of cyanobacterial monitoring and cyanotoxin analysis. *Limnol. Oceanogr. Bull.* **2018**, *27*, 61–62. [CrossRef]

34. Fott, J.; Kořínek, V.; Pražáková, M.; Vondruš, B.; Forejt, K. Seasonal development of phytoplankton in fish ponds. *Int. Rev. Gesamten Hydrobiol. Hydrogr.* **1974**, *59*, 629–641. [CrossRef]

35. Sommer, U.; Gliwicz, Z.M.; Lampert, W.; Duncan, A. The PEG-model of seasonal succession of planktonic events in fresh waters. *Arch. Hydrobiol.* **1986**, *106*, 433–471.

36. Fastner, J.; Erhard, M. Determination of oligopeptide diversity within a natural population of. *Society* **2001**, *67*, 5069–5076.

37. Welker, M.; Brunke, M.; Preussel, K.; Lippert, I.; von Döhren, H. Diversity and distribution of *Microcystis* (cyanobacteria) oligopeptide chemotypes from natural communities studies by single-colony mass spectrometry. *Microbiology* **2004**, *150*, 1785–1796. [CrossRef] [PubMed]

38. Holland, A.; Kinnear, S. Interpreting the possible ecological role(s) of cyanotoxins: Compounds for competitive advantage and/or physiological aide? *Mar. Drugs* **2013**, *11*, 2239–2258. [CrossRef] [PubMed]

39. Smith, G.D.; Thanh Doan, N. Cyanobacterial metabolites with bioactivity against photosynthesis in cyanobacteria, algae and higher plants. *J. Appl. Phycol.* **1999**, *11*, 337–344. [CrossRef]

40. Von Elert, E.; Agrawal, M.K.; Gebauer, C.; Jaensch, H.; Bauer, U.; Zitt, A. Protease activity in gut of *Daphnia magna*: Evidence for trypsin and chymotrypsin enzymes. *Comp. Biochem. Physiol. B Biochem. Mol. Biol.* **2004**, *137*, 287–296. [CrossRef]

41. Gerwick, W.H. The face of a molecule. *J. Nat. Prod.* **2017**, *80*, 2583–2588. [CrossRef]

42. Mohimani, H.; Gurevich, A.; Mikheenko, A.; Garg, N.; Nothias, L.F.; Ninomiya, A.; Takada, K.; Dorrestein, P.C.; Pevzner, P.A. Dereplication of peptidic natural products through database search of mass spectra. *Nat. Chem. Biol.* **2017**, *13*, 30–37. [CrossRef]

43. Bertin, M.J.; Schwartz, S.L.; Lee, J.; Korobeynikov, A.; Dorrestein, P.C.; Gerwick, L.; Gerwick, W.H. Spongosine production by a *Vibrio harveyi* strain associated with the sponge *Tectitethya crypta*. *J. Nat. Prod.* **2015**, *78*, 493–499. [CrossRef]

44. Ishaque, N.M.; Burgsdorf, I.; Limlingan Malit, J.J.; Saha, S.; Teta, R.; Ewe, D.; Kannabiran, K.; Hrouzek, P.; Steindler, L.; Costantino, V.; et al. Isolation, genomic and metabolomic characterization of *Streptomyces tendae* VITAKN with quorum sensing inhibitory activity from southern india. *Microorganisms* **2020**, *8*, 121. [CrossRef] [PubMed]

45. Saha, S.; Esposito, G.; Urajová, P.; Mareš, J.; Ewe, D.; Caso, A.; Macho, M.; Delawská, K.; Kust, A.; Hrouzek, P.; et al. Discovery of unusual cyanobacterial tryptophan-containing anabaenopeptins by MS/MS-based molecular networking. *Molecules* **2020**, *25*, 3786. [CrossRef] [PubMed]

46. Kim Tiam, S.; Gugger, M.; Demay, J.; Le Manach, S.; Duval, C.; Bernard, C.; Marie, B. Insights into the diversity of secondary metabolites of *Planktothrix* using a biphasic approach combining global genomics and metabolomics. *Toxins* **2019**, *11*, 498. [CrossRef] [PubMed]

47. Briand, E.; Bormans, M.; Gugger, M.; Dorrestein, P.C.; Gerwick, W.H. Changes in secondary metabolic profiles of *Microcystis aeruginosa* strains in response to intraspecific interactions. *Environ. Microbiol.* **2016**, *18*, 384–400. [CrossRef]

48. Bouaïcha, N.; Miles, C.O.; Beach, D.G.; Labidi, Z.; Djabri, A.; Benayache, N.Y.; Nguyen-Quang, T. Structural diversity, characterization and toxicology of microcystins. *Toxins* **2019**, *11*, 714. [CrossRef]

49. Racine, M.; Saleem, A.; Pick, F.R. Metabolome variation between strains of *Microcystis aeruginosa* by untargeted mass spectrometry. *Toxins* **2019**, *11*, 723. [CrossRef]

50. Roy-Lachapelle, A.; Solliec, M.; Sauvé, S.; Gagnon, C. A Data-independent methodology for the structural characterization of microcystins and anabaenopeptins leading to the identification of four new congeners. *Toxins* **2019**, *11*, 619. [CrossRef]

51. Benke, P.I.; Vinay Kumar, M.C.; Pan, D.; Swarup, S. A mass spectrometry-based unique fragment approach for the identification of microcystins. *Analyst* **2015**, *140*, 1198–1206. [CrossRef]

52. Sanz, M.; Andreote, A.P.; Fiore, M.F.; Dorr, F.A.; Pinto, E. Structural characterization of new peptide variants produced by cyanobacteria from the brazilian atlantic coastal forest using liquid chromatography coupled to quadrupole time-of-flight tandem mass spectrometry. *Mar. Drugs* **2015**, *13*, 3892–3919. [CrossRef]

53. Harada, K.I.; Fujii, K.; Shimada, T.; Suzuki, M.; Sano, H.; Adachi, K.; Carmichael, W.W. Two cyclic peptides, anabaenopeptins, a third group of bioactive compounds from the cyanobacterium *Anabaena flos-aquae* NRC 525-17. *Tetrahedron Lett.* **1995**, *36*, 1511–1514. [CrossRef]

54. Sano, T.; Kaya, K. Oscillamide Y, a chymotrypsin inhibitor from toxic *Oscillatoria agardhii*. *Tetrahedron Lett.* **1995**, *36*, 5933–5936. [CrossRef]

55. Spoof, L.; Błaszczyk, A.; Meriluoto, J.; Cegłowska, M.; Mazur-Marzec, H. Structures and activity of new anabaenopeptins produced by Baltic Sea cyanobacteria. *Mar. Drugs* **2016**, *14*, 8. [CrossRef] [PubMed]

56. Adamovsky, O.; Moosova, Z.; Pekarova, M.; Basu, A.; Babica, P.; Svihalkova Sindlerova, L.; Kubala, L.; Blaha, L. Immunomodulatory potency of microcystin, an important water-polluting cyanobacterial toxin. *Environ. Sci. Technol.* **2015**, *49*, 12457–12464. [CrossRef] [PubMed]

57. Moosova, Z.; Hrouzek, P.; Kapuscik, A.; Blaha, L.; Adamovsky, O. Immunomodulatory effects of selected cyanobacterial peptides in vitro. *Toxicon* **2018**, *149*, 20–25. [CrossRef] [PubMed]

58. World Health Organization. International Programme on Chemical, S. Health criteria and other supporting information. In *Guidelines for Drinking-Water Quality*, 2nd ed.; World Health Organization: Geneva, Switzerland, 1996; Volume 2.

59. Martin, C.; Oberer, L.; Ino, T.; Konig, W.A.; Busch, M.; Weckesser, J. Cyanopeptolins, new depsipeptides from the cyanobacterium *Microcystis* sp. PCC 7806. *J. Antibiot. Tokyo* **1993**, *46*, 1550–1556. [CrossRef] [PubMed]

60. Köcher, S.; Resch, S.; Kessenbrock, T.; Schrapp, L.; Ehrmann, M.; Kaiser, M. From dolastatin 13 to cyanopeptolins, micropeptins, and lyngbyastatins: The chemical biology of Ahp-cyclodepsipeptides. *Nat. Prod. Rep.* **2020**, *37*, 163–174. [CrossRef]

61. Weckesser, J.; Martin, C.; Jakobi, C. Cyanopeptolins, depsipeptides from cyanobacteria. *Syst. Appl. Microbiol.* **1996**, *19*, 133–138. [CrossRef]

62. Pettit, G.R.; Kamano, Y.; Herald, C.L.; Tuinman, A.A.; Boettner, F.E.; Kizu, H.; Schmidt, J.M.; Baczynskyi, L.; Tomer, K.B.; Bontems, R.J. The isolation and structure of a remarkable marine animal antineoplastic constituent: Dolastatin 10. *J. Am. Chem. Soc.* **1987**, *109*, 6883–6885. [CrossRef]

63. Harrigan, G.G.; Yoshida, W.Y.; Moore, R.E.; Nagle, D.G.; Park, P.U.; Biggs, J.; Paul, V.J.; Mooberry, S.L.; Corbett, T.H.; Valeriote, F.A. Isolation, structure determination, and biological activity of dolastatin 12 and lyngbyastatin 1 from *Lyngbya majuscula/Schizothrix calcicola* cyanobacterial assemblages. *J. Nat. Prod.* **1998**, *61*, 1221–1225. [CrossRef]

64. Ishida, K.; Okita, Y.; Matsuda, H.; Okino, T.; Murakami, M. Aeruginosins, protease inhibitors from the cyanobacterium *Microcystis aeruginosa*. *Tetrahedron* **1999**, *55*, 10971–10988. [CrossRef]

65. Okino, T.M.H.; Murakami, M.; Yamaguchi, K. Microginin, an angiotensin-converting enzyme inhibitor from the blue-green alga *Microcystis aeruginosa*. *Tetrahedron Lett.* **1993**, *34*, 501–504. [CrossRef]

66. Ishida, K.; Murakami, M. Kasumigamide, an antialgal peptide from the cyanobacterium *Microcystis aeruginosa*. *J. Org. Chem.* **2000**, *65*, 5898–5900. [CrossRef]

67. Welker, M.; Christiansen, G.; von Dohren, H. Diversity of coexisting *Planktothrix* (cyanobacteria) chemotypes deduced by mass spectral analysis of microystins and other oligopeptides. *Arch. Microbiol.* **2004**, *182*, 288–298. [CrossRef] [PubMed]

68. Christophoridis, C.; Zervou, S.K.; Manolidi, K.; Katsiapi, M.; Moustaka-Gouni, M.; Kaloudis, T.; Triantis, T.M.; Hiskia, A. Occurrence and diversity of cyanotoxins in Greek lakes. *Sci. Rep.* **2018**, *8*, 1–22. [CrossRef] [PubMed]

69. Gkelis, S.; Lanaras, T.; Sivonen, K.; Taglialatela-Scafati, O. Cyanobacterial toxic and bioactive peptides in freshwater bodies of Greece: Concentrations, occurrence patterns, and implications for human health. *Mar. Drugs* **2015**, *13*, 6319–6335. [CrossRef] [PubMed]

70. Preece, E.P.; Hardy, F.J.; Moore, B.C.; Bryan, M. A review of microcystin detections in Estuarine and Marine waters: Environmental implications and human health risk. *Harmful Algae* **2017**, *61*, 31–45. [CrossRef]

71. Ploutno, A.; Carmeli, S. Modified peptides from a water bloom of the cyanobacterium *Nostoc* sp. *Tetrahedron* **2002**, *58*, 9949–9957. [CrossRef]

72. Matsuda, H.; Okino, T.; Murakami, M.; Yamaguchi, K. Radiosumin, a trypsin inhibitor from the blue-green alga *Plectonema radiosum*. *J. Org. Chem.* **1996**, *61*, 8648–8650. [CrossRef]

73. Baumann, H.I.; Keller, S.; Wolter, F.E.; Nicholson, G.J.; Jung, G.; Sussmuth, R.D.; Juttner, F. Planktocyclin, a cyclooctapeptide protease inhibitor produced by the freshwater cyanobacterium *Planktothrix rubescens*. *J. Nat. Prod.* **2007**, *70*, 1611–1615. [CrossRef]

74. Martins, J.; Vasconcelos, V. Cyanobactins from cyanobacteria: Current genetic and chemical state of knowledge. *Mar. Drugs* **2015**, *13*, 6910–6946. [CrossRef]

75. Lawton, L.A.; Morris, L.A.; Jaspars, M. A bioactive modified peptide, aeruginosamide, isolated from the cyanobacterium *Microcystis aeruginosa*. *J. Org. Chem.* **1999**, *64*, 5329–5332. [CrossRef]

76. Oh, H.M.; Lee, S.J.; Kim, J.H.; Kim, H.S.; Yoon, B.D. Seasonal variation and indirect monitoring of microcystin concentrations in Daechung reservoir, Korea. *Appl. Environ. Microbiol.* **2001**, *67*, 1484–1489. [CrossRef] [PubMed]

77. Martins, J.; Saker, M.L.; Moreira, C.; Welker, M.; Fastner, J.; Vasconcelos, V.M. Peptide diversity in strains of the cyanobacterium *Microcystis aeruginosa* isolated from Portuguese water supplies. *Appl. Microbiol. Biotechnol.* **2009**, *82*, 951–961. [CrossRef] [PubMed]

78. Le Manach, S.; Duval, C.; Marie, A.; Djediat, C.; Catherine, A.; Edery, M.; Bernard, C.; Marie, B. Global metabolomic characterizations of *Microcystis* spp. Highlights clonal diversity in natural bloom-forming populations and expands metabolite structural diversity. *Front. Microbiol.* **2019**, *10*, 791. [CrossRef]

79. Osterholm, J.; Popin, R.V.; Fewer, D.P.; Sivonen, K. Phylogenomic analysis of secondary metabolism in the toxic cyanobacterial genera *Anabaena, Dolichospermum* and *Aphanizomenon*. *Toxins* **2020**, *12*, 248. [CrossRef] [PubMed]

80. Natumi, R.; Janssen, E.M. Cyanopeptide co-production dynamics beyond mirocystins and effects of growth stages and nutrient availability. *Environ. Sci. Technol.* **2020**, *54*, 6063–6072. [CrossRef]

81. Bober, B.; Lechowski, Z.; Bialczyk, J. Determination of some cyanopeptides synthesized by *Woronichinia naegeliana* (Chroococcales, Cyanophyceae). *Phycol. Res.* **2011**, *59*, 286–294. [CrossRef]

82. Jakubowska, N.; Szeląg-Wasielewska, E. Toxic picoplanktonic cyanobacteria–Review. *Mar. Drugs* **2015**, *13*, 1497–1518. [CrossRef]

83. Lincoln, E.P.; Carmichael, W.W. Preliminary tests of toxicity of *Synechocystis* sp. grown on wastewater medium. In *The Water Environment: Algal Toxins and Health*; Carmichael, W.W., Ed.; Springer: Boston, MA, USA, 1981; pp. 223–230.

84. Ludvek, B.; Blahoslav, M. Microcystin production and toxicity of picocyanobacteria as a risk factor for drinking water treatment plants. *Algol. Stud.* **1999**, *92*, 95–108.

85. Mitsui, A.; Rosner, D.; Goodman, A.; Reyes-Vasquez, G.; Kusumi, T.; Kodama, T.; Nomoto, K. Hemolytic Toxins in a Marine Cyanobacterium Synechococcus sp. In Proceedings of the International Red Tide Symposium, Takamatsu, Japan, November 1987.

86. Jasser, I.; Callieri, C. Picocyanobacteria: The smallest cell-size cyanobacteria. In *Handbook of Cyanobacterial Monitoring and Cyanotoxin Analysis*; Meriluoto, J., Spoof, L., Codd, G.A., Eds.; John Wiley & Sons, Ltd.: Hoboken, NJ, USA, 2017; pp. 19–27.

87. Pechar, L. Photosynthesis of natural populations of *Aphanizomenon flos-aquae*: Some ecological implications. *Int. Rev. Gesamten Hydrobiol. Hydrogr.* **1987**, *72*, 599–606. [CrossRef]

88. Utermöhl, H. Methods of collecting plankton for various purposes are discussed. *SIL Commun. 1953–1996* **1958**, *9*, 1–38. [CrossRef]

89. Hillebrand, H.; Dürselen, C.-D.; Kirschtel, D.; Pollingher, U.; Zohary, T. Biovolume calculation for pelagic and benthic microalgae. *J. Phycol.* **1999**, *35*, 403–424. [CrossRef]

90. Komárek, J. *Cyanoprokaryota, Teil 1/Part 1: Chroococcales*, 1st ed.; Spektrum Akademischer Verlag GmbH: Heidelberg/Berlin, Germany, 1999; p. 548.

91. Komárek, J.; Anagnostidis, K. Cyanoprokaryota 2. Teil/2nd Part: Oscillatoriales. In *Süsswasserflora von Mitteleuropa*; Elsevier: Heidelberg, Germany; Spektrum: Champaign, IL, USA, 2005; p. 759.

92. Komárek, J. Cyanoprokaryota,3. Teil/ 3rd Part: Heterocytous genera. In *Süsswasserflora von Mitteleuropa*; Elsevier: Heidelberg, Germany; Spektrum: Champaign, IL, USA, 2005; p. 1130.

93. Saurav, K.; Macho, M.; Kust, A.; Delawska, K.; Hajek, J.; Hrouzek, P. Antimicrobial activity and bioactive profiling of heterocytous cyanobacterial strains using MS/MS-based molecular networking. *Folia Microbiol. Praha* **2019**, *64*, 645–654. [CrossRef] [PubMed]

94. Chambers, M.C.; Maclean, B.; Burke, R.; Amodei, D.; Ruderman, D.L.; Neumann, S.; Gatto, L.; Fischer, B.; Pratt, B.; Egertson, J.; et al. A cross-platform toolkit for mass spectrometry and proteomics. *Nat. Biotechnol.* **2012**, *30*, 918–920. [CrossRef]

95. Team, RDC. A language and environment for statistical computing. R Foundation for Statistical Computing. 2006. Available online: http://www.R-project.org/ (accessed on 30 August 2020).

96. Tolles, J.; Meurer, W.J. Logistic Regression: Relating Patient Characteristics to Outcomes. *JAMA* **2016**, *316*, 533–534. [CrossRef]

97. Benjamini, Y.; Hochberg, Y. Controlling the false discovery rate: A practical and powerful approach to multiple testing. *J. R. Stat. Soc. Ser. B Methodol.* **1995**, *57*, 289–300. [CrossRef]

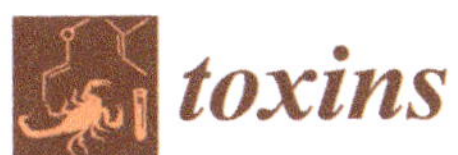 

Article

Cylindrospermopsin- and Deoxycylindrospermopsin-Producing *Raphidiopsis raciborskii* and Microcystin-Producing *Microcystis* spp. in Meiktila Lake, Myanmar

Andreas Ballot [1,*], **Thida Swe** [1,2,3], **Marit Mjelde** [1], **Leonardo Cerasino** [4], **Vladyslava Hostyeva** [1] **and Christopher O. Miles** [5]

1   Norwegian Institute for Water Research, Gaustadalléen 21, N-0349 Oslo, Norway; thida.swe@niva.no (T.S.); marit.mjelde@niva.no (M.M.); vladyslava.hostyeva@niva.no (V.H.)
2   Forest Research Institute, 15013 Yezin, Myanmar
3   Department of Natural Sciences and Environmental Health, University of South- Eastern Norway, Gullbringvegen 36, N-3800 Bø, Norway
4   Department of Sustainable Agro-ecosystem and Bioresources, Research and Innovation Centre, Fondazione Edmund Mach, Via E. Mach 1, 38010 San Michele all'Adige, Italy; leonardo.cerasino@fmach.it
5   National Research Council, 1411 Oxford Street, Halifax, NS B3H 3Z1, Canada; christopher.miles@nrc-cnrc.gc.ca
*   Correspondence: andreas.ballot@niva.no

Received: 12 March 2020; Accepted: 3 April 2020; Published: 7 April 2020

Abstract: Meiktila Lake is a shallow reservoir located close to Meiktila city in central Myanmar. Its water is used for irrigation, domestic purposes and drinking water. No detailed study of the presence of cyanobacteria and their potential toxin production has been conducted so far. To ascertain the cyanobacterial composition and presence of cyanobacterial toxins in Meiktila Lake, water samples were collected in March and November 2017 and investigated for physico-chemical and biological parameters. Phytoplankton composition and biomass determination revealed that most of the samples were dominated by the cyanobacterium *Raphidiopsis raciborskii*. In a polyphasic approach, seven isolated cyanobacterial strains were classified morphologically and phylogenetically as *R. raciborskii*, and *Microcystis* spp. and tested for microcystins (MCs), cylindrospermopsins (CYNs), saxitoxins and anatoxins by enzyme-linked immunosorbent assay (ELISA) and liquid chromatography–mass spectrometry (LC–MS). ELISA and LC–MS analyses confirmed CYNs in three of the five *Raphidiopsis* strains between 1.8 and 9.8 µg mg^{-1} fresh weight. Both *Microcystis* strains produced MCs, one strain 52 congeners and the other strain 20 congeners, including 22 previously unreported variants. Due to the presence of CYN- and MC-producing cyanobacteria, harmful effects on humans, domestic and wild animals cannot be excluded in Meiktila Lake.

Keywords: Meiktila Lake; *Raphidiopsis*; *Microcystis*; cylindrospermopsin; deoxycylindrospermopsin; microcystin

Key Contribution: This study confirmed the production of CYN and deoxyCYN by *Raphidiopsis raciborskii* strains and numerous MCs by *Microcystis* strains isolated from Meiktila Lake in Myanmar. The MCs included many novel congeners demonstrated by LC–MS and chemical derivatization methods. Among these were the rarely reported L-Glu and L-dihydrotyrosine-containing congeners. This is the first finding of toxin-producing cyanobacteria in a Myanmar waterbody.

---

1. Introduction

Many lakes and reservoirs worldwide are affected by periodic cyanobacterial dominance or even cyanobacterial blooms. Such mass developments of cyanobacteria are typical for eutrophic conditions and are often induced by nutrient enrichment caused by increased agricultural, urban and industrial activities and are also expected to increase due to regional and global climate change [1]. Various cyanobacterial species forming such blooms are potential producers of hepatotoxic or neurotoxic compounds and their presence is often associated with animal poisonings and a threat to human health [2].

Myanmar is characterized by the presence of several natural lakes and numerous man-made reservoirs. Meiktila Lake is one of the numerous reservoirs in Myanmar and was built in ancient times, dating from an unknown period [3] but most likely in the reign of King Narapathisithu (1173–1210) [4]. Today the lake is divided by a dam into a northern and a southern part (Figure 1) [5]. Meiktila Lake is exposed to sedimentation due to deforestation in the catchment and especially the northern part has been partially filled with sediment over a period of more than 100 years [6]. The priority use of water from Meiktila Lake is drinking water, water for domestic purposes and for irrigation, although the lake is also polluted with domestic waste water, street runoff and solid waste [6,7].

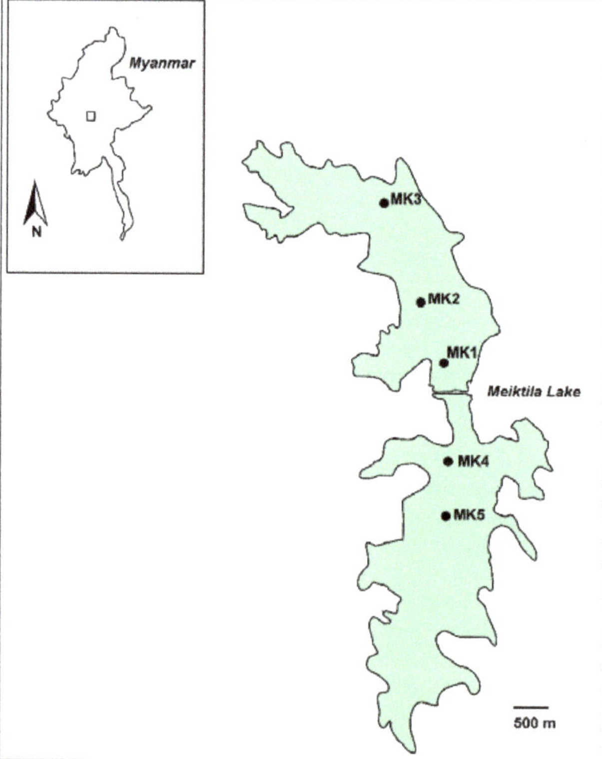

Figure 1. Map of Meiktila Lake. The map shows the locations of water sampling (Stations MK1-MK5). The location of Meiktila Lake in Myanmar is shown in the inset.

Only limited information is available about the limnological characteristics of Meiktila Lake and other freshwater habitats in Myanmar. In 1995, a study of the algal flora of Meiktila Lake was reported [4]. A recent study described the investigation of physical parameters, macrophyte and phytoplankton composition in the period 2011–2014 in Meiktila Lake [5]. Twenty taxa of aquatic macrophytes including helophytes have been documented in Meiktila Lake [5]. Several heterocytous cyanobacterial taxa, e.g., *Anabaena*, *Anabaenopsis* and *Calothrix* and a few nonheterocytous cyanobacterial taxa e.g., *Aphanocapsa*, *Chroococcus*, *Microcystis*, *Arthrospira* and *Oscillatoria* have been reported but not further investigated [4,5]. Neither study mentioned the presence of the cyanobacterium *Raphidiopsis raciborskii* (formerly *Cylindrospermopsis raciborskii*) (Woloszynska) Aguilera, Berrendero Gómez, Kastovsky, Echenique & Salerno nor the presence of the microcystin (MC) and cylindropsermopsin (CYN) groups of cyanobacterial toxins, which are documented from many lakes in Asia [8–10].

We suspect that the number of cyanobacterial species documented to date in Meiktila Lake was underestimated and that various toxin-producing cyanobacteria were present in the cyanobacterial community. There is clearly a lack of information about cyanobacteria and the production of cyanobacterial toxins in Meiktila Lake and the recently described potentially toxic cyanobacterium *Microcystis* is most likely not the only potential toxin-producing cyanobacterium. The ongoing pollution of the lake suggests the potential for occurrence of more frequent severe cyanobacterial blooms, which would have a negative impact on the use of the lake for drinking water and domestic purposes by the residents. This study aimed therefore to investigate the presence of cyanobacteria and their potential toxins in Meiktila Lake, applying modern analytical methods in a polyphasic approach to elucidate in detail the cyanobacterial composition, phylogeny and toxin production and toxin profiles.

2. Results

2.1. Physico-Chemical Parameters

At both sampling dates in March and November 2017, all sampling points in Meiktila Lake were characterized by water temperatures of 26.2–28.0 °C, pH of 8.5–9.3 and conductivities of 580–729 μS cm^{-1}. Secchi depth was between 0.8 m (at sampling point MK1) and 1.8 m (at MK3). Total phosphorus and total nitrogen concentrations were 12–23 and 360–570 μg L^{-1}, respectively.

2.2. Phytoplankton Community

At all sampling stations, and at both sampling dates, cyanobacteria were the dominant group in the phytoplankton in both parts of Meiktila lake together with diatoms (Bacillariophyceae) Cryptophyceae, Chlorophyceae and Euglenophyceae (Table 1). The most dominant cyanobacterium was *R. raciborskii*, which comprised biomasses between 0.2 and 1.9 mg L^{-1} fresh weight (FW), or 27%–91% of the cyanobacterial biomass at the sampling points MK1, MK2, MK4 and MK5. Other cyanobacteria present in the samples belonged to the genera *Aphanocapsa*, *Aphanothece*, *Chroococcus*, *Merismopedia*, *Limnothrix*, *Microcystis*, *Planktolyngbya*, *Planktothrix*, *Sphaerospermopsis* and *Synechococcus*. They together comprised biomasses between 0.06 and 1.2 mg L^{-1} cyanobacterial wet weight. At MK3, the biomass of *R. raciborskii* at both sampling dates (0.02–0.08 mg L^{-1}) was lower than at the other sampling points MK1, MK2, MK4 and MK5 (0.20–1.94 mg L^{-1}) (data not shown). The *Microcystis* biomass was lower than the *Raphidiopsis* biomass at all sampling points and sampling dates and ranged from 0.003 to 0.16 mg L^{-1} (data not shown).

Table 1. Biomass (mg L^{-1} FW) of phytoplankton groups at sampling points MK1–MK5 in Meiktila Lake in March and November of 2017.

Sampling Point	MK1	MK 1	MK 2	MK 2	MK 3	MK 3	MK 4	MK 4	MK 5	MK 5
Sampling Date / **Phytoplankton Group**	Mar	Nov	Mar	Nov	Mar	Nov	Mar	Nov	Mar	Nov
Bacillariophyceae	0.525	0.149	1.316	0.101	0.035	0.301	0.083	0.064	0.088	0.059
Chlorophyceae	0.143	0.055	0.182	0.146	0.102	0.045	0.101	0.102	0.107	0.165
Chrysophyceae	0.011	0	0.002	0	0.045	0.021	0.006	0.007	0	0.004
Conjugatophyceae	0.099	0.006	0.141	0	0.013	0	0.008	0.017	0.045	0
Cryptophyceae	0.023	0.132	0.007	0.170	0.155	0.299	0.127	0.275	0.106	0.138
Cyanobacteria	1.063	1.965	1.640	1.050	0.606	0.078	1.064	1.990	0.751	2.397
Dinophyceae	0.186	0.005	0.222	0	0.285	0	0.084	0.006	0.078	0
Euglenophyceae	0.255	0.029	0.159	0.005	0.014	0.050	0.009	0.002	0.053	0.021
Eustigmatophyceae	0	0	0	0	0	0	0	0.002	0	0
Klebsormidiophyceae	0	0	0	0	0	0.023	0	0	0.002	0
Prymnesiophyceae	0	0.002	0.000	0.002	0.003	0.004	0.002	0.003	0.008	0.007
Trebouxiophyceae	0	0	0.007	0	0	0.005	0.005	0	0	0.001
Xanthophyceae	0	0	0.008	0	0.005	0	0.003	0.001	0	0
Total	2.306	2.344	3.684	1.474	1.264	0.827	1.493	2.467	1.238	2.790

2.3. Morphological and Phylogenetic Characterization

Seven potentially toxin-producing cyanobacterial strains were isolated from Meiktila Lake (Table 2). Based on morphological features, e.g., presence and form of colonies or filaments, vegetative cells and heterocytes, five of the isolated cyanobacterial strains were identified as *R. raciborskii* and two strains as *M. aeruginosa* and *M. novacekii*, respectively (Figure 2). However, Harke et al. [11] suggested all *Microcystis* warrant placement into the same species complex. Therefore, we use "*Microcystis*" instead of species names in the following parts of the manuscript.

Table 2. Strains isolated from Meiktila Lake, strain codes and European Nucleotide Archive (ENA) accession numbers.

Species	Strain	Accession nr. 16S rRNA Gene
Raphidiopsis		
R. raciborskii	AB2017/05	LR590626
R. raciborskii	AB2017/09	LR590627
R. raciborskii	AB2017/12	LR590628
R. raciborskii	AB2017/13	LR590629
R. raciborskii	AB2017/16	LR746263
Microcystis		
Microcystis	AB2017/14	LR590630
Microcystis	AB2017/15	LR590631

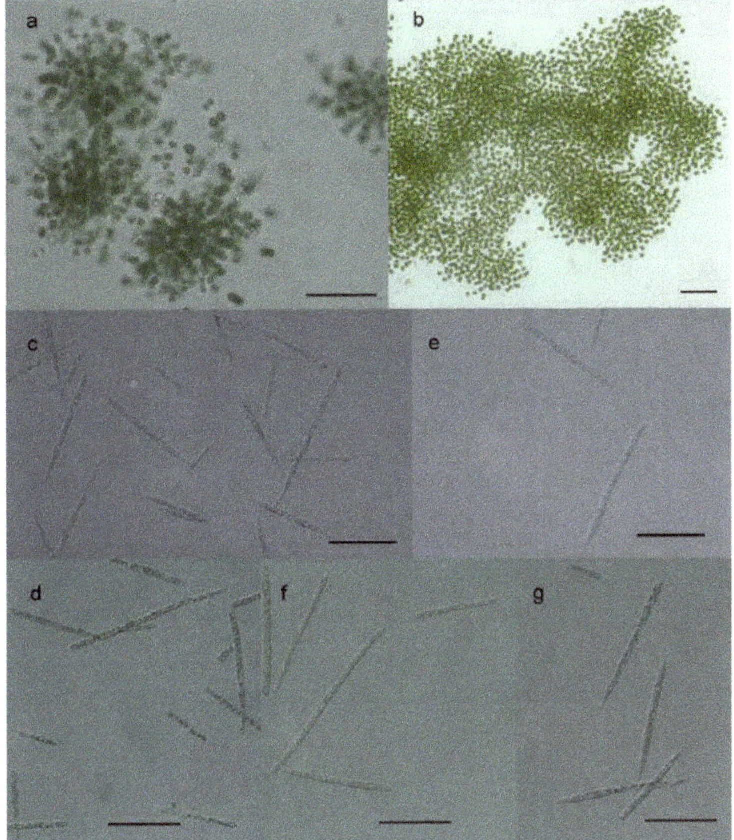

Figure 2. Micrographs of cyanobacteria investigated in this study. (**a**) *Microcystis novacekii* (AB2017/14); (**b**) *Microcystis aeruginosa* (AB2017/15); (**c**) *Raphidiopsis raciborskii* (AB2017/05); (**d**) *Raphidiopsis raciborskii* (AB2017/09); (**e**) *Raphidiopsis raciborskii* (AB2017/12); (**f**) *Raphidiopsis raciborskii* (AB2017/13); (**g**) *Raphidiopsis raciborskii* (AB2017/16). Scale bars indicate 50 μm.

The *Raphidiopsis* strains were mostly characterized by straight tapered filaments. The filament length and width varied between 8.8–90 × 1.9–5.8 μm. Heterocytes were observed in some filaments of all isolated strains. Akinetes were not observed in any of the investigated strains. As in the cultured strains, only a few of the filaments possessed heterocytes in the environmental samples. The two *Microcystis* strains were characterized by cell diameters ranging from 4.2 to 6.6 μm (strain AB2017/15) and from 3.7 to 5.8 μm (strain AB2017/14) (data not shown).

The morphological determination of the isolated strains was supported by phylogenetic analyses (Figure 3; Figure 4). Phylogenetic relationships of the investigated strains are presented in the maximum-likelihood (ML) tree of the 16S rRNA gene of *Cylindrospermopsis/Raphidiopsis* (Figure 3) and a separate ML tree of the *Microcystis* 16S rRNA gene (Figure 4). In the ML tree in Figure 3, the *Raphidiopsis* strains from Meiktila Lake grouped together with 16S rRNA gene sequences derived from *Cylindrospermopsis* and *Raphidiopsis* strains from Asia, Europe, Africa, Australia and North America (cluster I). The CYN-producing and nonCYN-producing *Raphidiopsis* strains from Meiktila Lake could not be distinguished phylogenetically using 16S rRNA gene and had similar 16 rRNA gene sequences (Figure 3). In cluster II, strains from North and South America, (USA, Mexico, Brazil), North

Africa (Tunisia), Southwest Europe (Spain) and New Zealand, grouped together. Both *Microcystis* strains from Meiktila Lake possessed similar 16S rRNA gene sequences and clustered together with 16S rRNA gene sequences of *Microcystis* from Europe, Asia, Africa and South America (Figure 4).

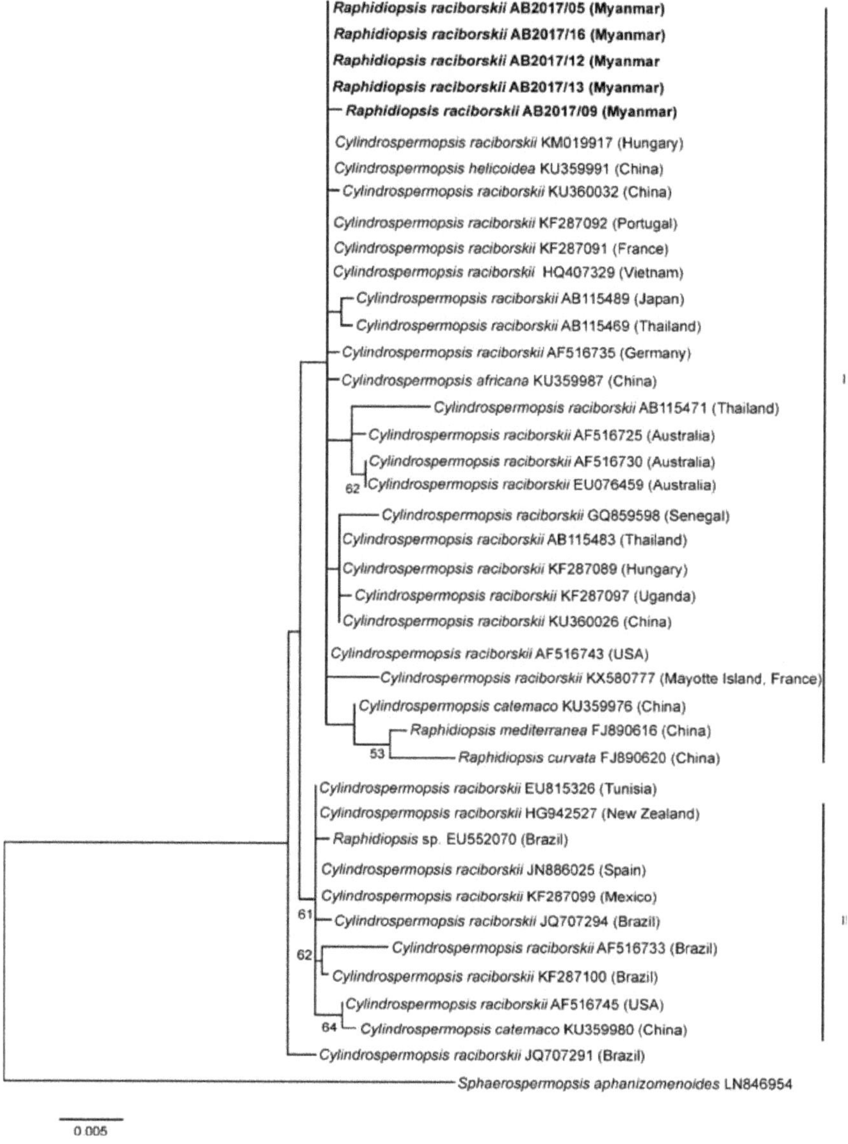

Figure 3. ML tree based on partial 16S rRNA gene sequences of 40 *Raphidiopsis/Cylindrospermospis* strains. Outgroup = *Sphaerospermopsis aphanizomenoides* (LN846954). Cluster I includes *Cylindrospermopsis* and *Raphidiopsis* strains from Asia, Europe, Africa, Australia and North America, cluster II includes strains from North and South America (USA, Mexico, Brazil), North Africa (Tunisia), Southwest Europe (Spain) and New Zealand. Strains from this study are marked in bold. Bootstrap values above 50 are included. The scale bar indicates 0.5% sequence divergence.

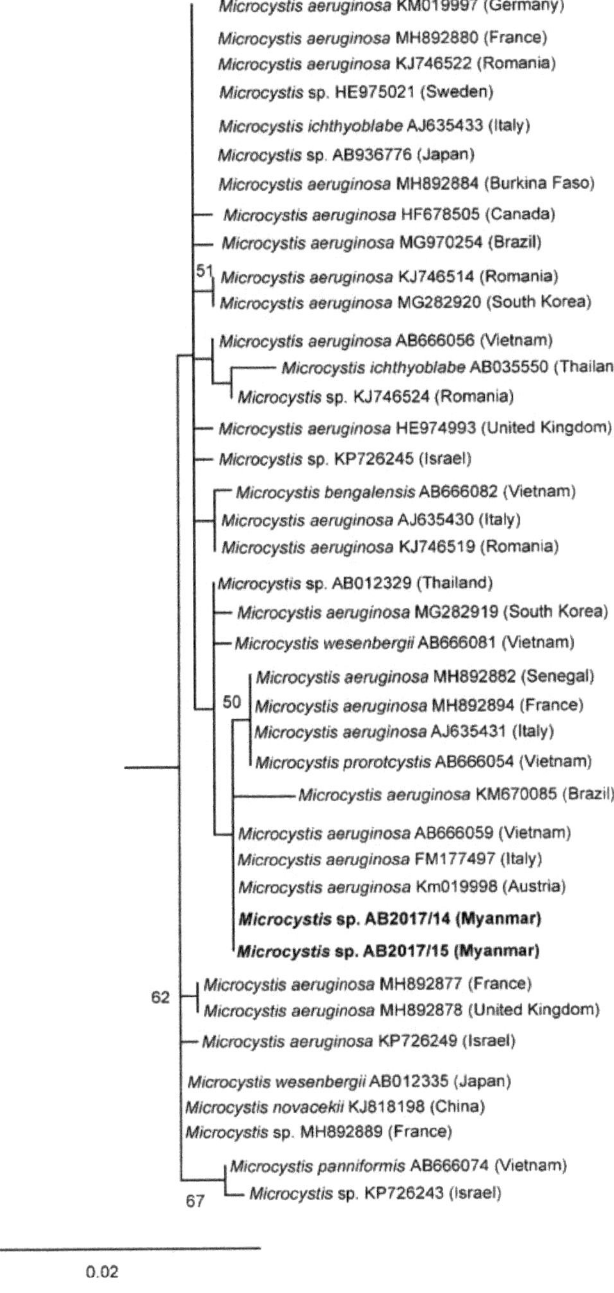

Figure 4. ML tree based on partial 16S rRNA gene sequences of 40 *Microcystis* strains. Outgroup = *Chroococcus subviolaceus* (MF072353). Strains from this study are marked in bold. Bootstrap values above 50 are included. The scale bar indicates 2% sequence divergence.

Table 4. Identities of microcystins detected by LC-HRMS/MS analysis in *Microcystis* strains AB2017/14 and /15 isolated from Meiktila Lake, their retention times (t_R), concentrations, relative abundances (%) and observed m/z values in negative ionisation mode[a].

	m/z	Compound Name	Confidence	t_R (min)	Concentration[b]			
					AB2017/14		AB2017/15	
					µg g^{-1}	%	µg g^{-1}	%
1	1022.5443	[D-Asp^3]MC-RR[c]	Confirmed	4.18	2.4	0.21	1002.4	7.21
2	1022.5454	[Dha^7]MC-RR[c]	Probable	4.53	0.1	0.01	31.8	0.23
3	1036.5597	MC-RR[c]	Confirmed	4.55	19.4	1.74	1458.0	10.49
4	995.4858	[D-Asp^3]MC-ER[c]	Probable	6.14	ND	-	119.6	0.86
5	1114.5657	MC-LR–Cys[d]	Probable	6.16	0.1	0.01	ND	-
6	1045.5378	MC-(H2)YR[c]	Probable	6.18	32.4	2.90	ND	-
7	1031.5224	[D-Asp^3]MC-(H2)YR[c]	Probable	6.20	32.9	2.95	ND	-
8	1033.5381	[D-Asp^3]MC-(H4)YR[c]	Probable	6.27	0.6	0.05	148.1	1.07
9	1009.5015	MC-ER[c]	Probable	6.55	0.6	0.05	37.8	0.27
10	1029.5072	[D-Asp^3]MC-YR[c]	Probable	6.59	ND	-	144.1	1.04
11	1047.5540	MC-(H4)YR[c]	Probable	6.64	6.4	0.57	310.8	2.24
12	1011.5530	[Mser^7]MC-LR	Probable	6.80	0.4	0.04	13.3	0.10
13	979.5273	[D-Asp^3]MC-LR[c]	Confirmed	6.89	40.8	3.65	1870.3	13.45
14	1043.5224	MC-YR[c]	Confirmed	6.99	9.1	0.81	317.4	2.28
15	1011.4999	MC-MR[c,d,e]	Probable	6.99	3.6	0.32	ND	-
16	1073.5329	MC-Y(OMe)R[c]	Probable	7.07	7.2	0.64	ND	-
17	993.5435	MC-LR[c]	Confirmed	7.12	183.0	16.38	3332.5	23.97
18	979.5289	[Dha^7]MC-LR[c]	Confirmed	7.13	0.8	0.07	37.4	0.27
19	1080.5170	Unidentified MC[c]	Unidentified	7.15	59.0	5.28	ND	-
20	1070.5332	MC-KynR[c]	Probable	7.35	4.5	0.40	ND	-
21	1007.5582	MC-HilR[c]	Confirmed	7.38	41.8	3.74	17.9	0.13
22	1080.5169	Unidentified[c]	Unidentified	7.38	59.6	5.34	ND	-
23	1082.5311	MC-OiaR[c]	Tentative	7.40	102.4	9.17	ND	-
24	995.4857	[D-Asp^3]MC-RE[c]	Probable	7.43	ND	-	110.1	0.79
25	1027.5269	MC-FR[c]	Confirmed	7.48	255.2	22.85	ND	-
26	1066.5380	MC-WR[c]	Confirmed	7.64	0.7	0.06	ND	-
27	1009.5015	MC-RE[c]	Probable	7.79	3.8	0.34	1281.0	9.21
28	951.4955	MC-RA[c]	Probable	8.24	13	1.16	ND	-
29	965.5116	MC-RAbu[c]	Probable	8.83	7.1	0.64	ND	-
30	946.4577	MC-(H2)YG[c]	Probable	9.06	5.4	0.48	ND	-
31	1011.4966	MC-RM[c,d,e]	Probable	9.23	0.2	0.02	ND	-
32	960.4738	MC-(H2)YA[c]	Probable	9.34	10.1	0.90	ND	-
33	953.4783	Unidentified MC[c]	Unidentified	9.50	12	1.07	ND	-
34	967.4942	Unidentified MC[c]	Unidentified	9.78	2.3	0.21	ND	-
35	984.4735	Unidentified MC[c]	Unidentified	9.96	3.7	0.33	ND	-
36	968.4272	[D-Asp^3]MC-EE[c]	Probable	11.43	ND	-	403.3	2.90
37	982.4432	MC-EE[c]	Probable	12.58	0.9	0.08	1533.5	11.03
38	952.4690	[D-Asp^3]MC-LE[c]	Probable	12.82	1.2	0.11	358.8	2.58
39	894.4631	[D-Asp^3]MC-LA[c]	Probable	13.32	6.5	0.58	ND	-
40	966.4849	MC-LE[c]	Probable	13.69	12.9	1.15	1376.1	9.90
41	908.4788	MC-LA[c]	Confirmed	14.87	44.8	4.01	ND	-
42	922.4942	MC-HilA[c]	Probable	15.29	5.9	0.53	ND	-
43	981.4737	MC-WA[c]	Probable	15.88	4.4	0.39	ND	-
44	942.4629	MC-FA[c]	Probable	15.95	42.3	3.79	ND	-
45	922.4943	MC-LAbu[c]	Probable	16.09	16.6	1.49	ND	-
46	936.5110	MC-HilAbu[c]	Probable	16.55	2.2	0.20	ND	-
47	995.4898	MC-WAbu[c]	Probable	17.20	13.4	1.20	ND	-
48	956.4784	MC-FAbu[c]	Probable	17.39	26.4	2.36	ND	-
49	936.5106	MC-LV[c]	Probable	17.62	3.6	0.32	ND	-
50	950.5254	MC-LL[c]	Probable	17.93	3.9	0.35	ND	-

Table 4. *Cont.*

	m/z	Compound Name	Confidence	t_R (min)	Concentration[b]			
					AB2017/14		AB2017/15	
					µg g^{-1}	%	µg g^{-1}	%
51	970.4944	MC-FV[c]	Probable	18.40	0.7	0.06	ND	-
52	950.5262	*iso*-MC-LL[c]	Tentative	18.49	1.2	0.11	ND	-
53	1009.5041	MC-WV[c]	Probable	18.54	2.9	0.26	ND	-
54	1023.5210	MC-WL[c]	Probable	18.74	2.8	0.25	ND	-
55	970.4942	*iso*-MC-FV[c]	Tentative	18.79	2.1	0.19	ND	-
56	1023.5216	*iso*-MC-WL[c]	Tentative	19.16	1.3	0.12	ND	-

[a] A comprehensive version of this table, including positive and negative ionisation MS data, reactivity towards thiols and mild oxidising agents, number of rings plus double-bond equivalents (RDBE) and presence of characteristic ions observed in positive and negative ionisation MS/MS spectra, is in the Supporting Information (Table S1) together with LC–HRMS/MS spectra (Figures S1–S59). [b] Concentration expressed per weight of biomass (FW) and as a percentage of total microcystins detected in each culture); ND = not detected; [c] Reacted with mercaptoethanol; [d] Oxidised by Oxone/DMSO; [e] HRMS/MS spectrum of oxidation product obtained.

In culture AB2017/14, 52 microcystin congeners were detected by LC–HRMS, with a total concentration of 1100 µg g^{-1} FW. Twenty-one of these were unidentified or previously unreported variants. In culture AB2017/15, 20 microcystin variants (of which six were previously unreported) were detected, with a total concentration of 14000 µg g^{-1} FW. The microcystin variants and the concentrations found in each strain are shown in Table 4 and Table S1.

3. Discussion

This study clearly demonstrates for the first time the presence of CYN- and deoxyCYN-producing *R. raciborskii* and MC-producing *Microcystis* in the phytoplankton community of Meiktila Lake in Myanmar. The relatively high biomass of *R. raciborskii*, up to 1.9 mg L^{-1} in the phytoplankton community of Meiktila Lake, is expected to cause elevated concentrations of CYNs in the lake water. The results suggest a higher risk for humans and animals to be affected by CYNs than by MCs, although this could be affected by variations in the biomass of CYN-producing *R. raciborskii* versus MC-producing *Microcystis*. Variations in cyanobacterial bloom composition and toxin production are influenced by abiotic factors such as nutrients, temperature and light and by biotic factors such as grazing, parasitism and predation [16,17]. The distribution of CYN/deoxyCYN-producing and nonproducing *Raphidiopsis* strains in Meiktila Lake is likely to vary over time, and dominance by a *Raphidiopsis* strain such as AB2017/13 would lead to CYN and deoxyCYN concentrations up to 20 µg L^{-1} for the highest *Raphidiopsis* biomasses measured in this study. It is therefore expected that CYN/ deoxyCYN concentrations in the lake water will at times exceed the guideline value for CYN in drinking water of 1 µg L^{-1} [18]. The tolerable daily intake value of 0.03 µg kg^{-1} for a person of 70 kg body weight would be exceeded after the intake of slightly more than 100 mL of lake water if CYN and deoxyCYN are similarly toxic. However, the toxicity of deoxyCYN to humans is not yet clear. According to Norris et al. [19], deoxyCYN does not contribute significantly to the toxicity of *R. raciborskii*. In contrast, cell viability assays showed that deoxyCYN was only slightly less toxic than CYN and most likely operates by similar toxicological mechanisms [20]. The potential risk of deoxyCYN for humans needs therefore to be clarified [20]. The use of Meiktila Lake water for drinking water, irrigation, domestic purposes or animal consumption is complicated by the fact that an unknown proportion of CYNs can be extracellular and is therefore not eliminated by filtration. Lake water contaminated by CYN and other toxins like MC-LR can lead to morphological and physiological changes and potential loss of productivity by agricultural plants, and bioaccumulation of cyanotoxins in the tissues of edible terrestrial plants in a concentration-dependent manner has been reported [21].

Griffith and Saker [22] have shown that in stationary phase of cultures, more than 50% of CYN can be extracellular. In environmental samples, the same authors found that extracellular CYN could exceed 90%. Boiling in water does not significantly degrade CYN within 15 min [23]. The removal of extracellular CYN/deoxyCYN therefore needs other methods, like the use of activated carbon, membrane filtration or chemical inactivation (Ultraviolet (UV), or oxidants) [24]. The presence of *Raphidiopsis* strains in Meiktila Lake that do not produce CYN makes it likely that CYN concentrations in the lake will vary considerably depending on the ratio of the two chemotypes in the phytoplankton community. As Meiktila Lake water is used for domestic purposes (drinking water, irrigation, washing of clothes, and personal hygiene), regular monitoring of cyanobacterial biomass and CYNs is recommended.

R. raciborskii has not been described from Myanmar water bodies and was not observed in a phytoplankton community study conducted in Meiktila Lake from 2011 to 2012 [5]. *Raphidiopsis* (and *Cylindrospermopsis*) spp., however, have been described from various other Southeast Asian freshwater habitats [25–27] and other water habitats worldwide [8]. *R. raciborskii* is only known to produce CYNs in Australia and the Asian countries China, Japan, Vietnam and Thailand and to produce STXs in Brazil [26,28–32]. The prime radiation centre of *R. raciborskii* is thought to be in Africa, with a second radiation centre in Australia [8]. Our 16S rRNA gene analysis confirms the close relationship of the *Raphidiopsis* strains from Meiktila Lake to other *Raphidiopsis* strains from Asia, Europe and Australia. Our 16S rRNA gene tree also clearly supports the suggested movement of *Raphidiopsis* from the American continents to Southwest Europe and North Africa and probably further to Greece and China, as has been described [33,34]. The close relationship of *Raphidiopsis* strains from Australia, Asia and Europe does not, however, explain why CYN- and deoxyCYN-producing strains have only been found in Asia and Australia but not in Europe, Africa or the Americas. Parts of, or the whole, CYN gene cluster could have been lost during the spread from Asia westwards, or only nonCYN-producing strains may have spread to Europe. Our finding of nontoxic *Raphidiopsis* strains in Meiktila Lake supports the latter hypothesis.

Both *Microcystis* strains AB 2017/14 and AB2017/15 isolated from Meiktila Lake are confirmed microcystin producers and are closely related to *Microcystis* strains from Africa, Europe and Asia based on 16S rRNA gene phylogeny. Both strains had identical 16S rRNA gene sequences but were clearly distinguished chemically by their MC congener profiles. Fifty-six microcystin variants were found in *Microcystis* strains AB2017/14 and AB2017/15 isolated from Meiktila Lake (Figure 6).

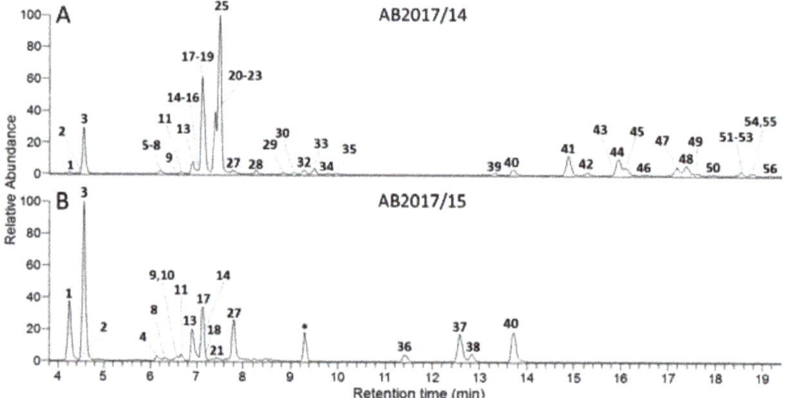

Figure 6. LC–HRMS full scan extracted ion chromatograms (3.8–19.4 min, positive ionization mode) of extracts from (**A**) *Microcystis* culture AB2017/14 and (**B**) *Microcystis* culture AB2017/15. Chromatograms were produced by extracting at *m/z* (± 5 ppm) for all MCs listed in Table 4 (see Table S1 for positive ionisation *m/z* values). Note that some of the smaller peaks are not labelled on the chromatograms, and the peak marked with an asterisk is not from a MC.

In order to reliably estimate the quantities of the MCs in the extracts by LC-HRMS, it was necessary to characterize, and if possible, identify all of them. The reason for this is that the response in LCMS can be expected to vary from congener-to-congener, primarily due to variations in the number of easily ionisable amino acid residues, especially Arg, present in the MC's structure. Only the identification of previously unreported MCs (see Bouaïcha et al. [35]) in the cultures is discussed further (i.e., **7, 9, 19, 22, 24, 27, 30–35, 37, 38, 40, 42, 46, 51–53, 55** and **56**) but spectra of all compounds for which adequate MS/MS spectra were obtained are available in the Supporting Information.

Three of the compounds were sulfide-containing variants (**5, 15,** and **31**) which reacted when the extract was oxidised with Oxone/DMSO (Table 4). The first two have been reported in cultures and blooms [15,36], and their characteristics were fully consistent with those reported for **5** and **15** here, and in the case of **15** its oxidation product (MC-M(O)R) showed characteristic product ions including neutral loss of CH_4OS and displayed an MS/MS spectrum (Figures S8, S11 and S14) identical to that reported previously for MC-M(O)R [15]. The third of sulfide-containing MC was identified as MC-RM (**31**) based on its physical and chemical properties (Table 4), which were essentially identical to those of **15** except for its longer t_R and that its MS/MS spectrum closely paralleled that of MC-RA (**28**) and displayed product ions characteristic of an MC with one Arg at position-2 rather than at the more common position-4 (Figures S8, S14 and S15). For example, fragments at m/z 440.2263 ($C_{18}H_{30}O_6N_7^+$, $\Delta m = 2.2$ ppm, from $Mdha^7$–D-Ala^1–Arg^2–D-$Masp^3$) and 731.3716 ($C_{34}H_{51}O_{10}N_8^+$, $\Delta m = -0.9$ ppm, from $Adda^5$–D-Glu^6–$Mdha^7$–D-Ala^1–Arg^2–D-$Masp^3$), together with the complete absence of a product ion at m/z 599.3552 (from Arg^4–$Adda^5$–D-Glu^6), confirmed Arg at position-2 and Met at position-4 (see Okello et al. [37] for assigned product ions from MC-YR) of **31**.

Eight of the compounds (**4, 9, 24, 27, 36–38** and **40**) showed characteristics of MCs containing one or more Glu residues at position-2 or -4. Two of these (**9** and **27**) had formulae consistent with MC-RE or MC-ER (Table 4). Compound **9** gave product ions (Figures S2 and S5–S7) typical of a MC with Arg at position-4, including m/z 599.3536 ($C_{31}H_{47}O_6N_6^+$, $\Delta m = -2.6$ ppm, from Arg^4–$Adda^5$–D-Glu^6), 284.1238 ($C_{12}H_{18}O_5N_3^+$, $\Delta m = -1.1$ ppm, from $Mdha^7$–D-Ala^1–Glu^2), and 286.1497 ($C_{11}H_{20}O_4N_5^+$, $\Delta m = -4.3$ ppm, from D-$Masp^3$–Arg^4), indicating **9** to be MC-ER. The MS/MS spectrum of **27** (Figures S2 and S15) included product ions at m/z 440.2248 ($C_{18}H_{30}O_6N_7^+$, $\Delta m = -0.9$ ppm, from $Mdha^7$–D-Ala^1–Arg^2–D-$Masp^3$), and 731.3678 ($C_{34}H_{51}O_{10}N_8^+$, $\Delta m = -6.1$ ppm, from $Adda^5$–D-Glu^6–$Mdha^7$–D-Ala^1–Arg^2–D-$Masp^3$) which, together with the complete absence of a product ion at m/z 599.3552, confirmed Arg at position-2 and Glu at position-4 (see Okello et al. [37]) for assigned product ions from MC-RY), showing **27** to be MC-RE. The characteristics of **37** (Table 4) were consistent with MC-EE. In addition, **37** gave product ions (Figures S27–S33) including m/z 276.1189 ($C_{10}H_{18}O_6N_3^+$, $\Delta m = -0.4$ ppm, from D-$Masp^3$–Glu^4), 405.1605 ($C_{15}H_{25}O_9N_4^+$, $\Delta m = -2.9$. ppm, from Glu^2–D-$Masp^3$–Glu^4) and 575.2703 ($C_{28}H_{39}O_9N_4^+$, $\Delta m = -1.5$ ppm, from $Adda^5$–D-Glu^6–$Mdha^7$–D-Ala^1–Glu^2) that confirmed **37** as MC-EE. Compound **40** was identified as MC-LE based on the characteristics presented in Table 4, as well as product ions (Figures S27–S33) observed in its MS/MS spectra, including m/z 460.2397 ($C_{19}H_{34}N_5O_8^+$, $\Delta m = -1.2$, from D-Ala^1–Leu^2-D-$Masp^3$–Glu^4) and 397.2073 ($C_{18}H_{29}N_4O_6^+$, $\Delta m = -2.2$, from $Mdha^7$–DAla^1–Leu^2-D-$Masp^3$), confirmed its identity as MC-LE (**40**). Earlier-eluting desmethylated D-variants of **9, 27, 37** and **40** were similarly identified as the corresponding D-Asp^3-congeners [D-Asp^3]MC-ER (**4**), [D-Asp^3]MC-RE (**24**), [D-Asp^3]MC-EE (**36**) and [D-Asp^3]MC-LE (**38**) based on analysis of their MS/MS spectra (Figures S21–S27) and characteristics presented in Table 4. Furthermore, the **4** and **36** in this sample coeluted with, and gave identical product ion spectra to, [D-Asp^3]MC-ER (**4**) and [D-Asp^3]MC-EE (**36**) identified [12] in an extract of a culture of *Planktothrix prolifica* NIVA-CYA544.

Twenty-one conventional late-eluting nonArg-containing MCs (**30, 32** and **36–56**) were detected. Of these, the identities of four that contained Glu^2 or Glu^4 (**36–38** and **40**) were discussed above. The remaining previously unreported nonArg MCs were **30, 32, 42, 46, 51–53, 55** and **56**. Compound **42** had the same characteristics as MC-LAbu (**45**) (Table 4), however, its MS/MS spectrum (Figures S34–S36) was consistent with MC-HilA. In particular, product ions at m/z 573.3270 ($C_{30}H_{45}N_4O_7^+$,

$\Delta m = -2.3$ ppm, from Adda^5–D-Glu^6–Mdha^7–D-Ala^1–Hil^2 minus $C_9H_{10}O$ (cf m/z 559.3126 for **41** and **45**)) and 411.2231 ($C_{19}H_{31}N_4O_6^+$, $\Delta m = -1.6$ ppm, from Mdha^7–D-Ala^1–Hil^2-D-Masp^3 (cf m/z 397.2082 for **41** and **45**)) as well as a range of other ions indicated the identity as MC-HilA (**42**), although the actual connectivity of the carbons in the amino acid side-chain at position-2 cannot be determined by mass spectrometry. A related compound (**46**) had characteristics (Table 4) and gave product ions (Figures S40–S42) that were consistent with MC-HilAbu. Product ions included m/z 573.3260 ($C_{30}H_{45}N_4O_7^+$, $\Delta m = -3.9$, from Adda^5–D-Glu^6–Mdha^7–D-Ala^1–Hil^2 minus $C_9H_{10}O$), 232.1291 ($C_9H_{18}N_3O_4^+$, $\Delta m = -0.4$, from D-Masp^3–Abu^4) and 430.2651 ($C_{19}H_{36}N_5O_6^+$, $\Delta m = -2.1$, from D-Ala^1–Hil^2-D-Masp^3–Abu^4 (cf m/z 402.2347 for **41** and 416.2504 **42**). This data unambiguously shows the presence of an extra CH_2 group in both amino acid-2 and -4 in **46**, relative to MC-LA (**41**) and is consistent with MC-HilAbu (**46**). Compound **51** displayed characteristics consistent with MC-FV (Table 4), as well as product ions (Figures S49–S51) at m/z 246.1456 ($C_{10}H_{20}N_3O_4^+$, $\Delta m = -0.4$, from D-Masp^3–Val^4), 593.2960 ($C_{32}H_{41}N_4O_7^+$, $\Delta m = -1.6$, from Adda^5–D-Glu^6–Mdha^7–D-Ala^1–Phe^2 minus $C_9H_{10}O$ (cf. 559.3126 for **41**)) and 464.2494 ($C_{22}H_{34}N_5O_6^+$, $\Delta m = -3.1$, from D-Ala^1–Phe^2 –D-Masp^3–Val^4 (cf. 402.2347 for **41**)). This establishes an extra C_5H_2 and 4 RDBE in amino acid-2 and C_2H_4 in amino acid-4, relative to MC-LA (**41**), consistent with MC-FV (**51**). Compound **53** had characteristics consistent with MC-WV (Table 4). This was supported by its MS/MS spectra (Figures S49–S51), which included product ions at m/z 246.1456 ($C_{10}H_{20}N_3O_4^+$, $\Delta m = 2.9$, from D-Masp^3–Val^4), 632.3063 ($C_{34}H_{42}N_5O_7^+$, $\Delta m = -2.5$, from Adda^5–D-Glu^6–Mdha^7–D-Ala^1–Trp^2 minus $C_9H_{10}O$ (cf. 559.3126 for **41**)) and 503.2605 ($C_{24}H_{35}N_6O_6^+$, $\Delta m = -2.6$, from D-Ala^1–Trp^2–D-Masp^3–Val^4 (cf. 402.2347 for **41**)). This indicates the presence of an extra C_5HN and 6 RDBE in amino acid-2 and C_2H_4 in amino acid-4, relative to MC-LA (**41**), consistent with MC-WV (**53**). Later-eluting isomers of **50**, **51** and **54** were also present (i.e., **52**, **55** and **56**), with identical characteristics (Table 5) and product ion spectra (Figures S52–S56). These compounds all contain branching amino acids at the variable position-2 (nominally Leu for **52** and **56**) or -4 (Val for **55**), and most likely the isomers present result from changes to this branching (e.g., Ile or 2-aminohexanoic acid at position-2, and 2-aminopentanoic acid or isovaline at position-4).

Table 5. Sampling points and sampling depth in Meiktila Lake for chemical and biological measurements.

Sampling Point	Water Depth (m)	Depth of Integrated Sample (m)	Geographical Position
MK1	3.3	0–1	N 20° 52′ 59.196, E 95° 51′ 12.204
MK2	2	0–1	N 20° 53′ 21.48, E 95° 51′ 2.124
MK3	2.5	0–1	N 20° 54′ 16.38, E 95° 50′ 40.092
MK4	4.4	0–2	N 20° 52′ 21.468, E 95° 51′ 12.528
MK5	7.1	0–3	N 20° 51′ 58.752, E 95° 51′ 18.936

In addition, **30** and **32** differed from each other by CH_2 and had characteristics consistent with MC-(H2)YA (**32**) and MC-(H2)YG (**30**), respectively (Table 4). These fragmented somewhat differently from typical Arg-free MCs such as MC-LA (**41**) (Figure S17). Both compounds showed weak product ions at m/z 155.0815 and 580.3017, indicating that **30** and **32** both contained Adda^5–D-Glu^6–Mdha^7–D-Ala^1. However, both **30** and **32** also gave product ions at m/z 320.1605, 611.3075 and 745.3807 (cf. 268.1650, 559.3117 and 693.3854 in MC-LA (**41**)), indicating the presence of an extra C_3O and 3 RDBE at amino acid-2 relative to **41**, consistent with the presence of the unusual amino acid L-dihydrotyrosine ((H2)Y) at position-2. Compounds **30** and **32** gave product ions as m/z 449.2033 and 449.2015 ($C_{21}H_{29}N_4O_7^+$, Δ 0.5 and -3.6 ppm, respectively, from Mdha^7–D-Ala^1–(H2)Tyr^2–D-Masp^3; cf. m/z 397.2082 for **41**, from Mdha^7–D-Ala^1–Leu^2–D-Masp^3). Thus, the difference in mass (14.0157, i.e., CH_2) between **30** and **32** lies not in residue-3 (D-Masp^3 vs D-Asp^3) as might be expected but in residue-4. Thus, **32** is identified as MC-(H2)YA, and **30** as MC-(H2)YG, which appears to be the first MC so far reported [35] with Gly at position-4.

Compounds **6** and **7** coeluted but differed by a mass corresponding to CH_2, with **6** having the same accurate mass, MS/MS spectrum (Figure S3) and retention time as MC-(H2)YR (**6**) identified in a sample from a recent study [13]. The MS of **7** was consistent with [D-Asp^3]MC-(H2)YR, and the MS/MS spectra of **6** and **7** were very similar. The MS/MS spectrum of **7** included product ions (Figure S4) at m/z 155.0813 ($\Delta m = -1.6$ ppm, from Mdha^7–D-Ala^1) and 599.3525 ($C_{31}H_{47}O_6N_6^+$, $\Delta m = -4.5$ ppm, from Arg^4–Adda^5–D-Glu^6), 120.0806 ($C_8H_{10}N^+$, $\Delta m = -1.6$ ppm, from (H2)Tyr), and 320.1611 ($C_{16}H_{22}O_4N_3^+$, $\Delta m = 2.0$ ppm, from Mdha^7–D-Ala^1–(H2)Tyr^2), 272.1353 ($C_{10}H_{18}O_4N_5^+$, $\Delta m = 2.0$ ppm, from D-Asp^3–Arg^4) and 714.3802 ($C_{35}H_{52}O_9N_7^+$, $\Delta m = -2.7$ ppm, from D-Asp^3–Arg^4–Adda^5–D-Glu^6), showing that **7** is [D-Asp^3]MC-(H2)YR.

Five MCs were present (**19**, **22** and **33–35**) whose structures could not be identified from their characteristics (Table 4 and Table S1) or product ion spectra (Figures S12 and S18–S20).Compounds **33** and **34** gave apparent m/z values that did not correspond to known or plausible MC variants and differed from each other by a mass corresponding to CH_2. They did not contain Arg but contained one extra nitrogen atom, 5 extra RDBE and, more surprisingly, one less oxygen atom than MC-LA (**41**). Compound **33** gave product ions at m/z 375.1911 ($C_{20}H_{27}O_5N_2^+$, $\Delta m = -0.9$ ppm, from Adda^5–D-Glu^6–Mdha^7 minus $C_9H_{10}O$), 446.2284 ($C_{23}H_{32}N_3O_6^+$, $\Delta m = -0.4$ ppm, from Adda^5–D-Glu^6–Mdha^7–D-Ala^1 minus $C_9H_{10}O$) and 580.2986 ($C_{32}H_{42}N_3O_7^+$, $\Delta m = -5.4$ ppm, from Adda^5–D-Glu^6–Mdha^7–D-Ala^1), however, fragments containing amino acids 2–4 were either shifted or absent. Although the data appear to be consistent with analogues containing Orn and Phe at positions 2 and 4 with an amide linkage to the neighbouring D-Asp^3/Masp^3 residue, further data are required for even a tentative structural assignment. The apparent pseudomolecular ion isotope envelopes of **33** and **34** (Figure S60) displayed unusual patterns that suggest that these compounds may be reacting during ionisation, possibly including dehydration, further complicating mass spectral analysis.

As with **30** and **32**, compound **35** showed product ions at m/z 213.0866, 320.1583, 375.1909, 446.2279, 449.2013, 509.2680 and 611.3040, consistent with the presence of Adda^5–D-Glu^6–Mdha^7–D-Ala^1–(H2)Tyr^2–D-Masp^3, which would require an MC with an amino acid side-chain at position-4 possessing an unprecedented C_3H_3 and 2 RDBE. This could possibly be due to the presence of a larger fragile amino acid at this position that undergoes ready elimination during MS, and the full structure of **35** remains undetermined.

Compounds **19** and **22** had identical product ion spectra including ions at m/z 135.0803, 213.0866, 269.1235, 375.1907, 446.2281, 599.3533, 640.2807 and 710.3886, identical to those from MC-LR, indicating the presence of D-Masp^3–Arg^4–Adda^5–D-Glu^6–Mdha^7–D-Ala^1 and that these compounds differed from MC-LR (**17**) and MC-WR (**26**) only in the amino acid at position-2. Product ions at 469.1815 and 486.2111 (both from X^2–D-Masp^3–Arg^4) and 173.0709 and 144.0455 (from X^2 and 173.0709–CH_3N) were consistent with this and indicated the presence of a side chain at position-2 containing C_9H_6NO and 7 RDBE (cf. C_9H_8N and 6 RDBE for the side chain of Trp in **26**), suggesting the presence of an unidentified oxidised variant of Trp present in the two isomeric MCs, **19** and **22**.

Between them, the two cultures contained 34 known and 22 previously unreported MCs of which 14 were assigned probable structures based on their chemical and mass spectrometric properties. In all cases, the number of Arg residues present in the MCs were reliably determined from their elemental compositions and charge-state, even for congeners for which definitive structures could not be established. Quantitation of the individual congeners was then performed from negative (Table 4) and positive mode full scan chromatograms, which gave essentially identical results, relative to 3-point calibration curves of the appropriate MC reference materials (RMs) containing no, one or two Arg residues ([D-Leu^1]MC-LY [38], MC-LR (**17**) and MC-RR (**3**), respectively). AB2017/14 contained 52 MCs, of which only 31 had been previously reported, and the newly reported MCs constituted nearly 20% of the total identified MC content. AB2017/15 contained fewer MCs (20) of which only six had not been reported previously, but these constituted nearly 34% of the total MC content in this culture. Notable amongst the newly reported MCs are those containing Glu at positions 2 and 4 (**4**, **9**, **24**, **27**, **36–38**, and **40**), which constituted 1.7% of the MCs in AB2017/14 and 37.5% of the MCs in AB2017/15. MCs of this type have only been reported previously from two sources [12,39]. AB2017/14 also contained a number of congeners that appeared to be derived from MC-WR (**19, 20, 22, 23** and **26**) that together constituted 20.2% of its MC content, and an unusually high number (25) of nonArg-containing MCs (**30, 32–35** and **37–56**) together constituting 20.5% of the total MC content. These results underscore both the diversity of MCs that may be present in a single sample and the potential difficulty of reliably quantitating the total MCs using traditional targeted LC–MS/MS methods. These factors may contribute to the reported apparent overestimation of MC levels when using less-targeted methods such as ELISA and PP2A inhibition, relative to highly congener-targeted LC–MS/MS approaches [40].

Although both strains produce a variety of MC variants, the risk of harmful effects caused by microcystins is likely to be low due to the low *Microcystis* biomasses observed in Meiktila Lake during this study. An increase in *Microcystis* biomass cannot, however, be excluded due to the pollution from various sources, e.g., waste water and street runoff [6]. This will most likely lead to an increase in microcystin concentrations in the lake, with potential harmful effects on humans, domestic and wild animals using the untreated lake water. The toxicity of microcystins found in the strains isolated from Meiktila Lake varies from highly toxic variants like MC-LR or MC-LA to less toxic variants such as MC-RR [41]. The toxicity of most of the MC variants found in this study has not yet been described, which makes a full risk assessment difficult. The production of a high number of MC variants (up to 47) has also been shown for two other *Microcystis* strains isolated from South African Haartbeestpoort Dam and Japanese Lake Kasumigaura [42,43].

Shallow lakes like Meiktila Lake are often characterized by competition between macrophytes and phytoplankton. High nutrient loading and phytoplankton growth lead to turbid water conditions and prevent the growth of macrophytes [44]. In Meiktila Lake the present turbid conditions are explained by deforestation and erosion in the catchment area, pollution with domestic waste water, street runoff and solid waste [6]. The relatively strong growth of *R. raciborskii* most likely is an additional reason for the observed turbidity at sampling points MK1, MK2, MK4 and MK5 in Meiktila Lake. The *Potamogeton* belt which separates MK2 and MK3 seems to act as a kind of filter or barrier because the Secchi depth at MK3 (1.8 m) was considerably higher than at MK1 (0.8 m). According to Van Donk and Van de Bund [45], macrophytes significantly modify the composition of the phytoplankton community and lead to a decrease in its abundance and biomass. It has been shown that certain macrophyte species exhibit allelopathic activity against certain phytoplankton species [46]. The obvious decrease of *Raphidiopsis* biomass from MK2 to MK3 may be therefore attributable to allelopathic effects of the macrophyte species in the belt. Allelopathy has been suggested by several authors to be responsible for observed phytoplankton patterns in whole-lake studies of vegetated, shallow lakes, but evidence for or against allelopathy has not been provided [47–49].

4. Conclusions

In conclusion, this is the first report of *R. raciborskii* and *Microcystis* in Meiktila Lake, Myanmar. Three of five *Raphidiopsis* isolates produced CYN and deoxyCYN, like other *Raphidiopsis/Cylindrospermopsis* isolates described from other Asian countries and Australia. Both *Microcystis* strains isolated from Meiktila Lake produced at least 56 MC variants (52 for AB2017/14 and 20 for AB2017/15), including 22 previously undescribed congeners. Harmful effects on humans and animals using Meiktila Lake as a water source cannot therefore be excluded.

5. Materials and Methods

5.1. Study Area, Measurements and Sampling

Meiktila Lake is a shallow reservoir, located close to Meiktila city in central Myanmar in the Mandalay region at an altitude of 230 m (Figure 1). During the rainy season, from April/May to October/November, it receives water from Mondaing Dam, located ca. 15 km west of Meiktila at an altitude of 245 m. Depending on the season, Meiktila Lake covers an area of around 54 km^2 with a maximum water depth of 10 m. It is divided by a dam into a northern and a southern lake [5]. Five sampling points were selected, three in the northern part of Meiktila Lake and two in the southern part (Figure 1, Table 5).

Sampling points 2 and 3 in the Northern Part of Meiktila Lake were separated by a broad *Potamogeton* belt. At all five sampling points in March and November 2017, in situ measurements of water temperature, pH, conductivity and dissolved oxygen were conducted, and integrated water samples were taken (1 m steps up to max. 3 m water depth) for the analysis of chemical parameters (ammonium, nitrate, total nitrogen, soluble reactive phosphorous, total phosphorous, Ca, phytoplankton composition and biomass and for the isolation of cyanobacterial strains. For quantitative phytoplankton analysis, a 50 mL subsample was removed from a sample taken from integrated samples and preserved with Lugol's solution and a concentrated net sample (mesh size 20 μm) was taken and preserved by addition of formaldehyde (4% final concentration) for qualitative analysis. A 50 mL water sample for isolation of cyanobacteria was taken at each sampling point and kept in a cool shady place and gently shaken twice per day before further treatment in Norway.

5.2. Phytoplankton Analysis

The Lugol-fixed phytoplankton samples were counted in sedimentation chambers (Hydro-Bios Apparatebau GmbH Kiel, Germany) using an inverted microscope (Leica DMi8; Ortomedic, Oslo, Norway) according to Utermöhl [50]. Phytoplankton biomass was calculated by geometrical approximations using the computerized counting programme Opticount (SequentiX - Digital DNA Processing, Klein Raden, Germany). The specific density of phytoplankton cells was calculated as $1\ \mathrm{g\ cm^{-3}}$.

5.3. Isolation of Strains and Morphological Characterization

Using a microcapillary, single colonies of *Microcystis* and filaments of *Raphidiopsis* were isolated. They were washed five times and placed in wells on microtiter plates containing 300 μL Z8 medium [51]. After successful growth, the samples were placed in 50 mL Erlenmeyer flasks containing 20 mL Z8 medium and maintained at 22 °C. Strains were classified based on morphological traits [52,53]. Morphological examination was conducted using a Leica DM2500 light microscope, Leica DFC450 camera and Leica Application Suite software (LAS) (Leica, Oslo, Norway). The morphological identification was based on the following criteria: (i) size of vegetative cells and heterocytes and (ii) nature and shape of filaments or colonies. Length and width of 50–250 vegetative cells or filaments and of 20–50 heterocytes were measured. Akinetes were not detected in the samples. All strains used in this study (Table 2) are maintained at the Norwegian Institute for Water Research, Oslo, Norway.

5.4. Genomic DNA Extraction, PCR Amplification and Sequencing

Genomic DNA was extracted according to Ballot et al. [54]. All PCRs were performed on a Bio-Rad CFX96 Real-Time PCR Detection System (Bio-Rad Laboratories, Oslo, Norway) using the iProof High-Fidelity PCR Kit (Bio-Rad Laboratories, Oslo, Norway). The 16S rRNA gene of the isolated strains from Meiktila Lake was amplified using the primers as described by Ballot et al. [54]. PCR products were visualized by 1% agarose gel electrophoresis with GelRed staining (GelRed Nucleic Acid Gel Stain, Biotium, Fremont, CA, USA) and UV illumination.

Amplified 16S rRNA gene products were purified through Qiaquick PCR purification columns (Qiagen, Hilden, Germany). Sequencing of the purified 16S rRNA gene products was performed using the same primers as for PCR and intermediate primers as described in Ballot et al. [54]. For each PCR product, both strands were sequenced on an ABI 3730 Avant genetic analyser using the BigDye terminator V.3.1 cycle sequencing kit (Applied Biosystems, Thermo Fisher Scientific Oslo, Norway) according to the manufacturer's instructions.

5.5. Phylogenetic Analysis

Sequences of the 16S rRNA gene of the cyanobacterial strains were analysed using the Seqassem software package (version 07/2008) and the Align MS Windows-based manual sequence alignment editor (version 03/2007) (SequentiX - Digital DNA Processing, Klein Raden, Germany). Segments with highly variable and ambiguous regions and gaps making proper alignment impossible were excluded from the analyses.

A 16S rRNA gene set containing 1135 positions was used in the phylogenetic tree for *Cylindrospermopsis/Raphidiopsis*. *Sphaerospermopsis aphanizomenoides* (LN846954) was employed as the outgroup, five *Raphidiopsis* strains from Meiktila Lake and 35 additional *Cylindrospermopsis/Raphidiopsis* sequences derived from GenBank were included in the analyses. A set containing 1426 positions was used for the *Microcystis* 16S rRNA gene analysis. *Chroococcus subviolaceus* (MF072353) was employed as the outgroup, two strains from Meiktila Lake and 38 additional *Microcystis* sequences derived from GenBank were included in the analysis. Phylogenetic trees for 16S rRNA genes were constructed using the ML algorithm in Mega v. 7 [55]. In the ML analyses, evolutionary substitution models were evaluated using Mega v. 7 [55]. The HKY+G+I evolutionary model was found to be the best-fitting evolutionary model for the Nostocales 16S rRNA gene tree and T92+G+I for the *Microcystis* 16S rRNA gene tree. ML analyses of both trees were performed with 1000 bootstrap replicates using Mega v. 7 [55]. The sequence data were submitted to the European Nucleotide Archive (ENA) under the accession numbers listed in Table 2.

5.6. Toxin Analysis

5.6.1. Material

LC–MS/MS utilised standards of ATX-a (Tocris Bioscience, Bristol, UK), homoATX-a (Novakits, Nantes, France) and CYN (Vinci Biochem, Vinci, Italy) and certified reference materials (CRMs) of STX, dcSTX, NeoSTX, GTX1, GTX4, GTX5 and C1 and C2 toxins (National Research Council of Canada, Halifax, NS, Canada (NRC)). LC–HRMS utilised CRMs of MC-RR (**3**), MC-LR (**17**) and [Dha^7]MC-LR (**18**) (NRC) and an RM of [D-Leu^1]MC-LY [38]. Additional RMs of [D-Asp^3]MC-RR (**1**), D-Asp^3]MC-LR (**13**), MC-YR (**14**), MC-HilR (**21**) containing traces of MC-FR (**25**), MC-WR (**26**) and MC-LA (**41**) were prepared at NRC from commercial samples (Enzo Life Sciences, Farmingdale, NY, USA), and extracts containing an array of other identified MCs were available from previous work [12,13,15]. Standards for the Adda-ELISA and for the CYN, ATX and STX ELISAs were as provided with the kits (Abraxis LLC, Warminister, PA, USA).

5.6.2. ELISA for MCs, CYNs, ATXs and STXs

Fresh culture material of two *Microcystis* and five *Raphidiopsis* strains was frozen and thawed three times. The *R. raciborskii* strains were tested for CYNs using the Abraxis Cylindrospermopsin ELISA kit (Abraxis LLC, Warminister, PA, USA) following the manufacturer's instructions. The test is a direct competitive ELISA that detects cylindrospermopsin but also recognizes deoxycylindrospermopsin and 7-*epi*-cylindrospermopsin. The ELISA results do not distinguish between dissolved and cell-bound toxins. Both *Microcystis* strains were tested for microcystins using the Abraxis Microcystins/Nodularins (ADDA) ELISA kits (Abraxis LLC, Warminister, PA, USA). The test is an indirect competitive ELISA designed to detect Adda, (3-amino-9-methoxy-2,6,8-trimethyl-10-phenyldeca-4,6-dienoic acid), based on specific recognition of the Adda moiety [56]. ADDA is a nonprotein amino acid and is the most common side chain at position-5 in microcystins (Figure 5).

All strains were also tested for saxitoxins and anatoxin-a using the Abraxis Saxitoxins (PSP) and Abraxis Anatoxin (VFDF) ELISA kits (Abraxis LLC, Warminister, PA, USA). The saxitoxin ELISA is a direct competitive ELISA that detects saxitoxin based on specific antibody recognition but also recognizes other saxitoxins (e.g., dcSTX, GTXs, lyngbyatoxin, NeoSTX) to varying degrees according to the manufacturer's instructions. The test for anatoxin-a is a direct competitive ELISA that detects anatoxin-a based on specific antibody recognition but also recognizes homoanatoxin according to the manufacturer's instructions The colour reaction of all ELISA tests was evaluated at 450 nm on a Perkin Elmer1420 Multilabel counter Victor3 (Perkin Elmer, Waltham, MA, USA), and concentrations were evaluated by manual analysis of the absorbance data as recommended by the vendor.

5.6.3. Microcystin Analysis by LC–HRMS

Fresh culture material of both *Microcystis* strains was prepared for LC–HRMS by freeze-thawing (3 times), diluting with an equal volume of MeOH and filtering (0.22 μm) [57]). LC–HRMS/MS analysis was performed on a Q Exactive-HF Orbitrap mass spectrometer equipped with a HESI-II heated electrospray ionization interface (ThermoFisher Scientific, Waltham, MA, USA) using an Agilent 1200 LC system including a binary pump, autosampler and column oven (Agilent, Santa Clara, CA, USA). Analyses were performed with SymmetryShield 3.5 μm C18 column (100×2.1 mm; Waters, Milford, MA, USA) held at 40 °C with mobile phases A and B of H_2O and CH_3CN, respectively, each of which contained formic acid (0.1% v/v). Gradient elution (0.3 mL min^{-1}) was from 20% to 90% B over 18 min, then to 100% B over 0.1 min and a hold at 100% B (2.9 min), then returned to 20% B over 0.1 min with a hold at 20% B (3.9 min) to equilibrate the column (total run time 25 min). Injection volume was typically 1–5 μL.

The MS was operated in positive ion mode and calibrated from *m/z* 74 to 1622. The spray voltage was 3.7 kV, the capillary temperature was 350 °C and the sheath and auxiliary gas flow rates were 25 and 8 units, respectively, with MS data acquired from 2 to 20 min. Mass spectral data were collected using a combined full scan (FS) and data independent acquisition (DIA) method. FS data was collected from *m/z* 500 to 1400 using the 60,000-resolution setting, an AGC target of 1×10^6 and a max IT of 100 ms. DIA data was collected using the 15,000 resolution setting, an AGC target of 2×10^5, maxIT set to "auto" and a stepped collision energy of 30, 60 and 80 eV. Precursor isolation windows were 62 *m/z* wide with centering at *m/z* 530, 590, 650, 710, 770, 830, 890, 950, 1010, 1070, 1130, 1190, 1250, 1310 and 1370. Mass spectral data were also collected using a combined full scan (FS) and top-10 data-dependent acquisition (DDA) method. Data was acquired as for DIA but with an exclusion list generated from a blank injection and an inclusion list (both at ± 5 ppm) from a publicly available database of MC *m/z* values [58], except that maxIT was set to 100 ms and dynamic exclusion 5.0 s and "if idle pick others" were selected. Putative MCs detected using the above FS/DIA method were further probed in a targeted manner using the parallel reaction monitoring scan (PRM) mode with a 0.7 *m/z* precursor isolation window, typically using the 30,000-resolution setting, an AGC target of 5×10^5 and a max IT of 400 ms. Typical collision energies were: stepped CE at 30 and 35 eV for MCs with no Arg; stepped CE at 60, 65 and 70 eV for MCs with one Arg; and CE at 65 eV for [M+H]$^+$ and stepped CE at 20, 25 and 30 eV for [M+2H]$^{2+}$ of MCs with two Arg groups. Full scan chromatograms were obtained in MS-SIM mode as for DIA but with resolution 120,000 and max IT 300 ms.

In negative mode, the mass spectrometer was calibrated from *m/z* 69 to 1780 and the spray voltage was −3.7 kV, while the capillary temperature, sheath and auxiliary gas flow rates were the same as for positive mode. Mass spectrometry data were collected in FS/DIA scan mode as above using a scan range of *m/z* 750–1400, a resolution setting of 60,000, AGC target of 1×10^6 and a max IT of 100 ms. For DIA, MS/MS data was collected from *m/z* 93 to 1400 using a resolution setting of 15,000, AGC target of 2×10^5, max IT set to "auto" and stepped collision energy 65 and 100 eV. Isolation windows were 45 *m/z* wide and centered at *m/z* 772, 815, 858, 902, 945, 988, 1032, 1075, 1118, 1162, 1205, 1248, 1294, 1335 and 1378. Mass spectral data were also collected using a combined full scan (FS) and top-10 data dependent acquisition (DDA) method. Data were acquired as for DIA but with an exclusion list generated from a blank injection and an inclusion list from a publicly available list (both at ± 5 ppm) of MC *m/z* values [58], except that maxIT was set to 100 ms and dynamic exclusion 5.0 s and "if idle pick others" were selected. Full scan chromatograms were obtained over a scan range *m/z* 750–1400 at a resolution setting of 120,000 using an AGC target of 1×10^6 and a max IT of 300 ms.

Thiol derivatizations were performed by addition of $(NH_4)_2CO_3$ (0.1 M, 200 µL) to the filtered extract (200 µL), with 200 µL transferred to two LC-MS vials. To one vial was added 1 µL of a 1:1 mixture of mercaptoethanol and d_4-mercaptoethanol (Sigma–Aldrich, St. Louis, MO, USA), while 1 µL of water was added to the other vial as a control. Oxidations were performed by addition of DMSO (5 µL) and Oxone (10 mg/mL in water; 25 µL) to 50 µL of extract [15]. Samples and reactions were placed in the sample tray (held at 15 °C) for analysis, and the reactions were monitored periodically until completion and then analysed.

5.6.4. CYN, deoxyCYN, ATXs and STXs Analysis by LC-MS/MS

Extraction was performed on freeze-dried cultures (40 mL), according to the protocol in [59]. In brief, dry material was treated with 6 mL of 50% methanol and sonicated (Omniruptor4000 probe sonicator, Omni-Inc., Kennesaw, MA, USA) for 10 min in pulsed mode (50%) using 160 W power. An aliquot of the solution was then filtered on Phenex RC syringe filters (0.2 µm; Phenomenex, Castel Maggiore, Italy) and analyzed by LC–MS/MS.

LC-MS/MS analysis were performed using a Waters Acquity UPLC system (Waters, Milford, MA, USA) coupled to a SCIEX 4000 QTRAP mass spectrometer (AB Sciex Pte. Ltd., Singapore). Chromatographic separation of analytes was performed using a HILIC column (Ascentis Express OH5, 2.7 µm, 50 × 2.1 mm; Merck Life Science S.r.l., Milan, Italy), while MS detection was performed using positive electrospray ionization using scheduled Multiple Reaction Monitoring. Details of the experimental set up are as described by Cerasino et al. [60], and the method was suitable for the detection and quantification of the following toxins: ATX-a, homoATX-a, CYN, STX, dcSTX, NeoSTX, GTX1, GTX4, GTX5 and C1 and C2 [60]. Quantification limits were 0.2–200 µg L^{-1}. Other toxic alkaloids not available as pure standards were also screened but only for tentative analysis (hydroxy- and epoxy- and homo-ATXs, deoxyCYN, dcNeoSTX, GTX2/3, dcGTX2/3 and C3 and C4 toxins) using equivalent detection settings to their most similar analogs.

Supplementary Materials: The following are available online at http://www.mdpi.com/2072-6651/12/4/232/s1, Figures S1–S59: LC-HRMS/MS spectra of **1**, **3**, **4**, **6**, **7**, **9**, **11**–**15**, **17**, **19**–**25**, and **27**–**56**, Figure S60: LC-HRMS spectra of unidentified microcystins **33**–**35**, Table S1: Identities of microcystins detected by LC-HRMS/MS analysis in *M. aeruginosa* strains AB2017/14 and /15, their retention times (t_R), concentrations, elemental compositions, observed m/z values in positive and negative ionisation modes, whether they matched retention times in reference samples and whether characteristic microcystin product ions were observed.

Author Contributions: Conceptualization, A.B.; methodology, A.B., C.O.M., L.C., V.H.; validation, A.B., C.O.M., L.C.; investigation, A.B., C.O.M., L.C., M.M., T.S.; data curation, A.B., C.O.M., L.C.; writing—original draft preparation, A.B., T.S., C.O.M.; writing—review and editing, A.B., C.O.M., M.M., L.C., T.S., V.H.; supervision, A.B.; project administration A.B. All authors have read and agreed to the published version of the manuscript.

Funding: This research was supported by the Project "Integrated Water Resources Management (IWRM) – Institutional Building and Training in Myanmar (funded by the Norwegian Ministry of Foreign Affairs and the Royal Norwegian Embassy in Myanmar).

Acknowledgments: We thank K. Thomas for preparation of RMs of [D-Asp^3]MC-RR (**1**), D-Asp^3]MC-LR (**13**), MC-YR (**14**), MC-HilR (**21**) containing traces of MC-FR (**25**), MC-WR (**26**) and MC-LA (**41**) at National Research Council, Halifax, Canada in partnership with the Norwegian Veterinary Institute (Oslo, Norway).

Conflicts of Interest: The authors declare no conflict of interest

References

1. Paerl, H.W.; Huisman, J. Climate change: a catalyst for global expansion of harmful cyanobacterial blooms. *Environ. Microbiol. Rep.* **2009**, *1*, 27–37. [CrossRef] [PubMed]

2. Sivonen, K. Cyanobacterial toxins. In *Encyclopedia of Microbiology*; Schaechter, M., Ed.; Elsevier: Oxford, UK, 2009; pp. 290–307.

3. Harvey, G.E. *History of Burma from the Earliest Times to 10 March 1824: The Beginning of the English Conquest, 1*; Longmans, Green and Company: London, UK, 1925.

4. Nyunt, K.K. Algal flora of Meiktila Lake and its surrounding area. Master's Thesis, Mandalay University, Mandalay, Myanmar, 1995.

5. Hlaing, N.K. Limnological study of Meiktila Lake, Mandalay Division. Ph.D. Thesis, University of Mandalay, Mandalay, Myanmar, 2014.

6. Klink, T.; de Jongh, I.; Leushuis, M.; Bijlmakers, L.; Speets, R.; Brinkman, L. *Scoping Study Meiktila Lake Area IWRM*; Royal HaskoningDHV, Arcadis, Rebel Group, Dutch Water Authorities: Den Haag, The Netherlands, 2015; pp. 1–50.

7. Aung, T.T.; Zin, T. *Zooplankton community in Meiktila Lake, Meiktila, central Myanmar*; Universities Research Journal; The Government of The Republic of the Union of Myanmar Ministry of Education: Yangon, Myanmar, 2015; Volume 7, pp. 1–11.

8. Antunes, J.T.; Leao, P.N.; Vasconcelos, V.M. *Cylindrospermopsis raciborskii*: Review of the distribution, phylogeography, and ecophysiology of a global invasive species. *Front. Microbiol.* **2015**, *6*, 13. [CrossRef] [PubMed]

9. Burford, M.A.; Beardall, J.; Willis, A.; Orr, P.T.; Magalhaes, V.F.; Rangel, L.M.; Azevedo, S.; Neilan, B.A. Understanding the winning strategies used by the bloom-forming cyanobacterium *Cylindrospermopsis raciborskii*. *Harmful Algae* **2016**, *54*, 44–53. [CrossRef] [PubMed]

10. Mowe, M.A.D.; Mitrovic, S.M.; Lim, R.P.; Furey, A.; Yeo, D.C.J. Tropical cyanobacterial blooms: A review of prevalence, problem taxa, toxins and influencing environmental factors. *J. Limnol.* **2015**, *74*, 205–224. [CrossRef]

11. Harke, M.J.; Steffen, M.M.; Gobler, C.J.; Otten, T.G.; Wilhelm, S.W.; Wood, S.A.; Paerl, H.W. A review of the global ecology, genomics, and biogeography of the toxic cyanobacterium, *Microcystis* spp. *Harmful Algae* **2016**, *54*, 4–20. [CrossRef]

12. Mallia, V.; Uhlig, S.; Rafuse, C.; Meija, J.; Miles, C.O. Novel microcystins from *Planktothrix prolifica* NIVA-CYA 544 identified by LC-MS/MS, functional group derivatization and ^{15}N-labeling. *Mar. Drugs* **2019**, *17*, 643. [CrossRef]

13. Yilmaz, M.; Foss, A.J.; Miles, C.O.; Özen, M.; Demir, N.; Balci, M.; Beach, D.G. Comprehensive multi-technique approach reveals the high diversity of microcystins in field collections and an associated isolate of *Microcystis aeruginosa* from a Turkish lake. *Toxicon* **2019**, *167*, 87–100. [CrossRef]

14. Miles, C.O.; Sandvik, M.; Haande, S.; Nonga, H.; Ballot, A. LC-MS analysis with thiol derivatization to differentiate [Dhb^7]- from [Mdha^7]-microcystins: analysis of cyanobacterial blooms, *Planktothrix* cultures and European crayfish from Lake Steinsfjorden, Norway. *Environ. Sci. Technol.* **2013**, *47*, 4080–4087. [CrossRef]

15. Miles, C.O.; Melanson, J.E.; Ballot, A. Sulfide oxidations for LC-MS analysis of methionine-containing microcystins in *Dolichospermum flos-aquae* NIVA-CYA 656. *Environ. Sci. Technol.* **2014**, *48*, 13307–13315. [CrossRef]

16. Lürling, M.; Van Oosterhout, F.; Faassen, E. Eutrophication and warming boost cyanobacterial biomass and microcystins. *Toxins* **2017**, *9*, 64. [CrossRef]

17. Guedes, I.A.; Rachid, C.T.C.C.; Rangel, L.M.; Silva, L.H.S.; Bisch, P.M.; Azevedo, S.M.F.O.; Pacheco, A.B.F. Close link between harmful cyanobacterial dominance and associated bacterioplankton in a tropical eutrophic reservoir. *Front. Microbiol.* **2018**, *9*, 13. [CrossRef] [PubMed]

18. Humpage, A.R.; Falconer, I.R. Oral toxicity of the cyanobacterial toxin cylindrospermopsin in male Swiss albino mice: determination of no observed adverse effect level for deriving a drinking water guideline value. *Environ. Toxicol.* **2003**, *18*, 94–103. [CrossRef] [PubMed]

19. Norris, R.L.; Eaglesham, G.K.; Pierens, G.; Shaw, G.R.; Smith, M.J.; Chiswell, R.K.; Seawright, A.A.; Moore, M.R. Deoxycylindrospermopsin, an analog of cylindrospermopsin from *Cylindrospermopsis raciborskii*. *Environ. Toxicol.* **1999**, *14*, 163–165. [CrossRef]

20. Neumann, C.; Bain, P.; Shaw, G. Studies of the comparative in vitro toxicology of the cyanobacterial metabolite deoxycylindrospermopsin. *J. Toxicol. Environ. Health, Part A* **2007**, *70*, 1679–1686. [CrossRef]

21. Machado, J.; Campos, A.; Vasconcelos, V.; Freitas, M. Effects of microcystin-LR and cylindrospermopsin on plant–soil systems: a review of their relevance for agricultural plant quality and public health. *Environ. Res.* **2017**, *153*, 191–204. [CrossRef]

22. Griffiths, D.J.; Saker, M.L. The Palm Island mystery disease 20 years on: A review of research on the cyanotoxin cylindrospermopsin. *Environ. Toxicol.* **2003**, *18*, 78–93. [CrossRef]

23. Chiswell, R.K.; Shaw, G.R.; Eaglesham, G.; Smith, M.J.; Norris, R.L.; Seawright, A.A.; Moore, M.R. Stability of cylindrospermopsin, the toxin from the cyanobacterium, *Cylindrospermopsis raciborskii*: Effect of pH, temperature, and sunlight on decomposition. *Environ. Toxicol.* **1999**, *14*, 155–161. [CrossRef]

24. Environmental Protection Agency. *Cyanobacteria and Cyanotoxins: Information for Drinking Water Systems*; United States Environmental Protection Agency: Washington, DC, USA, 2014; pp. 1–11.

25. Li, R.H.; Carmichael, W.W.; Brittain, S.; Eaglesham, G.K.; Shaw, G.R.; Liu, Y.D.; Watanabe, M.M. First report of the cyanotoxins cylindrospermopsin and deoxycylindrospermopsin from *Raphidiopsis curvata* (cyanobacteria). *J. Phycol.* **2001**, *37*, 1121–1126. [CrossRef]

26. Nguyen, T.T.L.; Hoang, T.H.; Nguyen, T.K.; Duong, T.T. The occurrence of toxic cyanobacterium *Cylindrospermopsis raciborskii* and its toxin cylindrospermopsin in the Huong River, Thua Thien Hue province, Vietnam. *Environ. Monit. Assess.* **2017**, *189*, 11. [CrossRef]

27. Dao, T.S.; Cronberg, G.; Nimptsch, J.; Do-Hong, L.-C.; Wiegand, C. Toxic cyanobacteria from Tri An Reservoir, Vietnam. *Nowa Hedwigia* **2010**, *90*, 433–448. [CrossRef]

28. Hawkins, P.R.; Chandrasena, N.R.; Jones, G.J.; Humpage, A.R.; Falconer, I.R. Isolation and toxicity of *Cylindrospermopsis raciborskii* from an ornamental lake. *Toxicon* **1997**, *35*, 341–346. [CrossRef]

29. Saker, M.L.; Neilan, B.A. Varied diazotrophies, morphologies, and toxicities of genetically similar isolates of *Cylindrospermopsis raciborskii* (Nostocales, Cyanophyceae) from northern Australia. *Appl. Environ. Microbiol.* **2001**, *67*, 1839–1845. [CrossRef] [PubMed]

30. Li, R.; Carmichael, W.W.; Brittain, S.; Eaglesham, G.K.; Shaw, G.R.; Mahakhant, A.; Noparatnaraporn, N.; Yongmanitchai, W.; Kaya, K.; Watanabe, M.M. Isolation and identification of the cyanotoxin cylindrospermopsin and deoxy-cylindrospermopsin from a Thailand strain of *Cylindrospermopsis raciborskii* (Cyanobacteria). *Toxicon* **2001**, *39*, 973–980. [CrossRef]

31. Chonudomkul, D.; Yongmanitchai, W.; Theeragool, G.; Kawachi, M.; Kasai, F.; Kaya, K.; Watanabe, M.M. Morphology, genetic diversity, temperature tolerance and toxicity of *Cylindrospermopsis raciborskii* (Nostocales, Cyanobacteria) strains from Thailand and Japan. *FEMS Microbiol. Ecol.* **2004**, *48*, 345–355. [CrossRef] [PubMed]

32. Zarenezhad, S.; Sano, T.; Watanabe, M.M.; Kawachi, M. Evidence of the existence of a toxic form of *Cylindrospermopsis raciborskii* (Nostocales, Cyanobacteria) in Japan. *Phycol. Res.* **2012**, *60*, 98–104. [CrossRef]

33. Cirés, S.; Wörmer, L.; Ballot, A.; Agha, R.; Wiedner, C.; Velazquez, D.; Casero, M.C.; Quesada, A. Phylogeography of cylindrospermopsin and paralytic shellfish toxin-producing Nostocales cyanobacteria from Mediterranean Europe (Spain). *Appl. Environ. Microbiol.* **2014**, *80*, 1359–1370. [CrossRef]

34. Panou, M.; Zervou, S.K.; Kaloudis, T.; Hiskia, A.; Gkelis, S. A Greek *Cylindrospermopsis raciborskii* strain: missing link in tropic invader's phylogeography tale. *Harmful Algae* **2018**, *80*, 96–106. [CrossRef]

35. Bouaïcha, N.; Miles, C.O.; Beach, D.G.; Labidi, Z.; Djabri, A.; Benayache, N.Y.; Nguyen-Quang, T. Structural diversity, characterization and toxicology of microcystins. *Toxins* **2019**, *11*, 714. [CrossRef]

36. Foss, A.J.; Miles, C.O.; Samdal, I.A.; Lovberg, K.E.; Wilkins, A.L.; Rise, F.; Jaabaek, J.A.H.; McGowan, P.C.; Aubel, M.T. Analysis of free and metabolized microcystins in samples following a bird mortality event. *Harmful Algae* **2018**, *80*, 117–129. [CrossRef]

37. Okello, W.; Portmann, C.; Erhard, M.; Gademann, K.; Kurmayer, R. Occurrence of microcystin-producing cyanobacteria in Ugandan freshwater habitats. *Environ. Toxicol.* **2010**, *25*, 367–380. [CrossRef]

38. LeBlanc, P.; Merkley, N.; Thomas, K.; Lewis, N.I.; Békri, K.; Renaud, S.L.; Pick, F.R.; McCarron, P.; Miles, C.O.; Quilliam, M.A. Isolation and characterization of [D-Leu^1]microcystin-LY from *Microcystis aeruginosa* CPCC-464. *Toxins* **2020**, *12*, 77. [CrossRef] [PubMed]

39. Namikoshi, M.; Yuan, M.; Sivonen, K.; Carmichael, W.W.; Rinehart, K.L.; Rouhiainen, L.; Sun, F.; Brittain, S.; Otsuki, A. Seven new microcystins possessing two L-glutamic acid units, isolated from *Anabaena* sp. strain 186. *Chem. Res. Toxicol.* **1998**, *11*, 143–149. [CrossRef] [PubMed]

40. Foss, A.J.; Aubel, M.T. Using the MMPB technique to confirm microcystin concentrations in water measured by ELISA and HPLC (UV, MS, MS/MS). *Toxicon* **2015**, *104*, 91–101. [CrossRef] [PubMed]

41. Rinehart, K.L.; Namikoshi, M.; Choi, B.W. Structure and biosynthesis of toxins from blue–green algae (cyanobacteria). *J. Appl. Phycol.* **1994**, *6*, 159–176. [CrossRef]

42. Ballot, A.; Sandvik, M.; Rundberget, T.; Botha, C.J.; Miles, C.O. Diversity of cyanobacteria and cyanotoxins in Hartbeespoort Dam, South Africa. *Mar. Freshwat. Res.* **2014**, *65*, 175–189. [CrossRef]

43. Fewer, D.P.; Rouhiainen, L.; Jokela, J.; Wahlsten, M.; Laakso, K.; Wang, H.; Sivonen, K. Recurrent adenylation domain replacement in the microcystin synthetase gene cluster. *BMC Evol. Biol.* **2007**, *7*, 183. [CrossRef]

44. Scheffer, M.; Carpenter, S.; Foley, J.A.; Folke, C.; Walker, B. Catastrophic shifts in ecosystems. *Nature* **2001**, *413*, 591–596. [CrossRef]

45. van Donk, E.; van de Bund, W.J. Impact of submerged macrophytes including charophytes on phyto- and zooplankton communities: allelopathy versus other mechanisms. *Aquat. Bot.* **2002**, *72*, 261–274. [CrossRef]

46. Gao, Y.N.; Dong, J.; Fu, Q.Q.; Wang, Y.P.; Chen, C.; Li, J.H.; Li, R.; Zhou, C.J. Allelopathic effects of submerged macrophytes on phytoplankton. *Allelopathy J.* **2017**, *40*, 1–22.

47. Mjelde, M.; Faafeng, B.A. *Ceratophyllum demersum* hampers phytoplankton development in some small Norwegian lakes over a wide range of phosphorus concentrations and geographical latitude. *Freshwat. Biol.* **1997**, *37*, 355–365. [CrossRef]

48. Blindow, I.; Hargeby, A.; Andersson, G. Seasonal changes of mechanisms maintaining clear water in a shallow lake with abundant *Chara* vegetation. *Aquat. Bot.* **2002**, *72*, 315–334. [CrossRef]

49. Lombardo, P. Applicability of littoral food-web biomanipulation for lake management purposes: snails, macrophytes, and water transparency in northeast Ohio shallow lakes. *Lake Reserv. Manage.* **2005**, *21*, 186–202. [CrossRef]

50. Utermöhl, H. Zur Vervollkommnung der quantitativen Phytoplankton-Methodik. *Mitt. Int. Ver. Theor. Angew. Limnol.* **1958**, *9*, 1–38. [CrossRef]

51. Kotai, J. *Instructions for Preparation of Modified Nutrient Solution Z8 for Algae.—Publication B-11/69*; Norwegian Institute for Water Research: Oslo, Norway, 1972.

52. Komárek, J.; Anagnostidis, K. Cyanoprokaryota 1. Chroococcales.—19/1. In *Süsswasserflora von Mitteleuropa*; Ettl, H., Gärtner, G., Heynig, H., Mollenhauer, D., Eds.; Gustav Fischer: Jena-Stuttgart-Lübeck-Ulm, Germany, 1999; p. 548.

53. Komárek, J. Cyanoprokaryota. 3. Heterocytous genera. In *Süsswasserflora von Mitteleuropa/Freshwater flora of Central Europe*; Büdel, B., Krienitz, L., Gärtner, G., Schagerl, M., Eds.; Springer Spektrum: Berlin-Heidelberg, Germany, 2013; p. 1130.

54. Ballot, A.; Cerasino, L.; Hostyeva, V.; Cirés, S. Variability in the sxt gene clusters of PSP toxin producing *Aphanizomenon gracile* strains from Norway, Spain, Germany and North America. *PLoS One* **2016**, *11*, e0167552. [CrossRef] [PubMed]

55. Kumar, S.; Stecher, G.; Tamura, K. MEGA7: molecular evolutionary genetics analysis version 7.0 for bigger datasets. *Mol. Biol. Evol.* **2016**, *33*, 1870–1874. [CrossRef] [PubMed]

56. Fischer, W.J.; Garthwaite, I.; Miles, C.O.; Ross, K.M.; Aggen, J.B.; Chamberlin, A.R.; Towers, N.R.; Dietrich, D.R. Congener-independent immunoassay for microcystins and nodularins. *Environ. Sci. Technol.* **2001**, *35*, 4849–4856. [CrossRef] [PubMed]

57. Miles, C.O.; Sandvik, M.; Nonga, H.E.; Rundberget, T.; Wilkins, A.L.; Rise, F.; Ballot, A. Thiol derivatization for LC-MS identification of microcystins in complex matrices. *Environ. Sci. Technol.* **2012**, *46*, 8937–8944. [CrossRef]

58. Miles, C.O.; Stirling, D. Toxin mass list, version 15b. Available online: https://www.researchgate.net/publication/324039408_Toxinmasslist_COM_v15b (accessed on 11 October 2018).

59. Cerasino, L.; Meriluoto, J.; Bláha, L.; Carmeli, S.; Kaloudis, T.; Mazur-Marzec, A. SOP 6. Extraction of cyanotoxins from cyanobacterial biomass. In *Handbook of Cyanobacterial Monitoring and Cyanotoxin Analysis*; Meriluoto, J., Spoof, L., Codd, G.A., Eds.; Wiley: Chichester, UK, 2017; pp. 350–353.

60. Cerasino, L.; Capelli, C.; Salmaso, N. A comparative study of the metabolic profiles of common nuisance cyanobacteria in southern perialpine lakes. *Adv. Oceanogr. Limnol.* **2017**, *8*, 22–32. [CrossRef]

Review

Determination of Cyanotoxins and Prymnesins in Water, Fish Tissue, and Other Matrices: A Review

Devi Sundaravadivelu [1], Toby T. Sanan [2,*], Raghuraman Venkatapathy [1], Heath Mash [2], Dan Tettenhorst [2], Lesley DAnglada [3], Sharon Frey [3], Avery O. Tatters [4] and James Lazorchak [5,*]

[1]   Pegasus Inc., U.S. EPA, Cincinnati, OH 45268, USA; sundaravadivelu.devi@epa.gov (D.S.); venkatapathy.raghuraman@epa.gov (R.V.)
[2]   Office of Research and Development, Center for Environmental Solutions and Emergency Response, U.S. EPA, Cincinnati, OH 45268, USA; mash.heath@epa.gov (H.M.); tettenhorst.dan@epa.gov (D.T.)
[3]   Office of Water, Science and Technology, U.S. EPA, Washington, DC 20004, USA; danglada.lesley@epa.gov (L.D.); frey.sharon@epa.gov (S.F.)
[4]   Center for Environmental Measurement and Modeling, U.S. EPA, Gulf Breeze, FL 32561, USA; tatters.avery@epa.gov
[5]   Center for Environmental Measurement and Modeling, U.S. EPA, Cincinnati, OH 45268, USA
*   Correspondence: sanan.toby@epa.gov (T.T.S.); lazorchak.jim@epa.gov (J.L.); Tel.: +1-513-569-7076 (J.L.)

Abstract: Harmful algal blooms (HABs) and their toxins are a significant and continuing threat to aquatic life in freshwater, estuarine, and coastal water ecosystems. Scientific understanding of the impacts of HABs on aquatic ecosystems has been hampered, in part, by limitations in the methodologies to measure cyanotoxins in complex matrices. This literature review discusses the methodologies currently used to measure the most commonly found freshwater cyanotoxins and prymnesins in various matrices and to assess their advantages and limitations. Identifying and quantifying cyanotoxins in surface waters, fish tissue, organs, and other matrices are crucial for risk assessment and for ensuring quality of food and water for consumption and recreational uses. This paper also summarizes currently available tissue extraction, preparation, and detection methods mentioned in previous studies that have quantified toxins in complex matrices. The structural diversity and complexity of many cyanobacterial and algal metabolites further impede accurate quantitation and structural confirmation for various cyanotoxins. Liquid chromatography–triple quadrupole mass spectrometer (LC–MS/MS) to enhance the sensitivity and selectivity of toxin analysis has become an essential tool for cyanotoxin detection and can potentially be used for the concurrent analysis of multiple toxins.

Keywords: harmful algal blooms; cyanotoxins; cyanobacteria; fish tissue; shellfish; detection methods

Key Contribution: This review article examines the available methodologies for the detection and quantification of cyanotoxins in complex matrices, such as fish tissue. An understanding of toxin concentrations using a standardized and reliable methodology is crucial to accurately assess the threat of these toxins to ecosystems and human health.

Citation: Sundaravadivelu, D.; Sanan, T.T.; Venkatapathy, R.; Mash, H.; Tettenhorst, D.; DAnglada, L.; Frey, S.; Tatters, A.O.; Lazorchak, J. Determination of Cyanotoxins and Prymnesins in Water, Fish Tissue, and Other Matrices: A Review. *Toxins* **2022**, *14*, 213. https://doi.org/10.3390/toxins14030213

Received: 25 January 2022
Accepted: 13 March 2022
Published: 16 March 2022

Publisher's Note: MDPI stays neutral with regard to jurisdictional claims in published maps and institutional affiliations.

1. Introduction

Harmful algal blooms (HABs) have been observed in freshwater, estuarine, and marine waters in the U.S. and around the globe. Cyanobacteria frequently contribute to HABs in freshwater systems and are able to produce highly potent toxins, known as cyanotoxins. A large number of cyanotoxins have been reported from different species of cyanobacteria, including microcystins (*Microcystis, Anabaena, Hapalosiphon, Dolichospermum, Gloeotrichia, Nostoc, Oscillatoria, Phormidium/Microcoleus,* and *Synechocystis*), nodularins (*Nodularia, Nostoc,* and *Iningainema*), anatoxins (*Anabaena, Aphanizomenon, Cylindrospermum, Dolichospermum Planktothrix, Oscillatoria, Geitlerinema, Phormidium/Microcoleus,* and *Tychonema*), cylindrospermopsins (*Oscillatoria* and *Raphidiopsis*), lyngbyatoxin-a (*Lyngbya*), saxitoxins (*Anabaena*, *Aphanizomenon*, *Cylindrospermum*, *Lyngbya*, *Phormidium*, and *Raphidiopsis*), and aplysiatoxins (*Lyngbya*, *Schizothrix*, and *Oscillatoria*) [1–11].

Microcystins (MCs) are the most frequently studied and the most widespread cyanotoxins [12], with approximately 50% of publications on cyanotoxins focusing on MCs, 25% on saxitoxins, and 25% on other toxins (such as nodularin, anatoxin-a, β-N-methylamino-L-alanine, cylindrospermopsins, and prymnesins). Of these publications, a majority focus on detection in water matrices. There is limited information on the presence of MCs and other cyanotoxins in matrices such as aquatic food webs, phytoplankton, zooplankton, periphyton, macroinvertebrates, forage fish, bottom feeders, and top carnivore fish. Since research on cyanotoxins has mostly focused on MCs, especially in water matrices, the development and standardization of better monitoring methods for other cyanotoxins and complex matrices, such as those mentioned above, are of utmost importance.

Various techniques have been developed to analyze cyanotoxins in water. These methods include enzyme-linked immunosorbent assay (ELISA), protein phosphatase inhibition assay (PPIA), oxidation of MCs to produce erythro-2-methyl-3-methoxy-4-phenylbutyric acid (MMPB) [13], and chromatography coupled with various detection methods [14]. A survey of chromatographic methods for the detection of cyanotoxins in water could be divided into the following main categories: liquid chromatography with ultraviolet detection (LC–UV), LC with fluorescence detection (LC–FLD), LC–photo-diode array (LC–PDA) or similar, LC with tandem mass spectrometry (LC–MS/MS), gas chromatography with mass spectrometry (GC–MS), ion mobility spectrometry (IMS), and immunological assays. LC–MS, or single quadrupole mass spectrometry, has largely been replaced by LC–MS/MS. Among these methods, a few LC–MS/MS and ELISA-based standard methods developed by the U.S. Environmental Protection Agency (EPA) and the International Organization for Standardization (ISO) are available, are considered standard, are widely used for the measurement of cyanotoxins in water, and can potentially be adapted for use with other matrices, such as fish tissue [15–18]. However, there is no standardized analytical method to detect these cyanotoxins in other matrices [19]. The surveyed studies herein show that ELISA and LC–MS/MS analysis have most frequently been used for detection/quantitation. While specific extraction procedures used vary considerably, overall method performance metrics, such as matrix spike recoveries and replicate precision, were sometimes poor or unreported. This raises questions about the quality of the data in the case of fish tissue and organs, shellfish and similar matrices exposed to HABs.

Most methods developed to date depend on extracting toxins in some form of an aqueous matrix, typically water and one or a combination of organic solvents. For example, ELISA methods should be carried out directly in water with only limited amounts of organic solvent, while LC–MS/MS methods typically require samples to be prepared with significant amounts of organic solvent in water [20]. In addition to the presence of organic solvents, some methods are sensitive to low pH, lipid content, or other matrix effects due to the properties of the target toxins or the nature of the matrix [21]. Hence, analytical methods for cyanotoxins in complex matrices, such as tissue, organs, and plants, frequently require efficient extraction, clean-up, and transfer of the toxin into a suitable solvent for analysis. Use of inefficient extraction methods may lead to poor matrix spike recoveries and poor precision among replicate samples.

The prevalence of HAB events globally and the potential for exposure and bioaccumulation of toxins necessitate monitoring and detecting cyanotoxins in tissues associated with HAB-related mortality events. There is a need for the development of reliable extraction and analytical methods capable of analyzing edible fish/shellfish, plant, and animal tissues for multiple toxins and for such methods to be broadly applicable to all inland waters and coastal food webs. The overarching goal of this study is to summarize available methodologies for the measurement of toxins, identify gaps in the detection of toxins in complex matrices, and highlight the complexity of quantifying groups of compounds with diverse chemistries.

2. Cyanotoxins and Prymnesins

Cyanotoxins can be categorized based on (1) their mechanism of action on terrestrial vertebrates, especially mammals, e.g., hepatotoxins, neurotoxins, and dermatotoxins, and (2) their chemical structure, e.g., cyclic peptides, alkaloids, and lipopolysaccharides [22–24]. Table 1 lists the names, chemical groups, and mammalian targets of some common cyanotoxins. Physicochemical properties of cyanotoxins not only determine which extraction procedures or detection techniques are suitable but also impact the bioaccumulative and toxicological properties of the toxins on several taxa of aquatic invertebrates and vertebrates [25].

Table 1. Characteristics of the most commonly studied cyanotoxins and prymnesins.

Common Name	Chemical Group	Effect/Target in Mammals
Microcystins	Heptapeptide	Cytotoxicity; genotoxic effects in liver
Nodularins	Pentapeptide	Cytotoxicity; liver
Anatoxin-a	Bicyclic amine alkaloid	Neurotoxicity
Saxitoxins	Tricyclic perhydropurine alkaloids	Neurotoxicity
β-N-methylamino-L-alanine	Amino acid	Neurotoxicity
Cylindrospermopsins	Polycyclic uracil with guanidine and sulfate group	Multitarget alkaloids
Prymnesins	Polyether polycyclic core with several conjugate double and triple bonds	Hemolytic activity and ichthyotoxicity

MCs, the most studied class of cyanotoxin, are produced several cyanobacterial genera including *Microcystis*, *Anabaena*, *Hapalosiphon*, *Nostoc*, *Dolichospermum*, *Gleootrichia*, and *Oscillatoria* [12]. These compounds are short cyclic heptapeptides synthesized by non-ribosomal pathways [26]. MCs are characterized by significant structural variability in amino acid composition, including residue substitutions, methylations, and demethylations. In total, more than 248 congeners/variants have been reported in the environment. Among them, MC-LR and MC-LA are among the most toxic [27,28]. Only a limited subset of MCs has commercially available analytical standards, which limits the scope of quantitative methods, such as LC–MS/MS, which rely on these for accurate quantification. MCs are cytotoxic and inhibit protein phosphatases, which leads to several subsequent harmful effects at concentrations over 10 µg/L [29].

Nodularins (NODs) are a class of cyclic pentapeptides structurally similar to MCs. Currently, 10 variants of this water-soluble and stable toxin have been identified [30,31], the most common being the variant with arginine as the variable amino acid (NOD-R). Similar to MCs, NODs are hepatotoxins acting through the inhibition of protein phosphatases and are potential tumor promoters. While NODs have historically been considered to be less common than MCs, their occurrence has been observed in bloom events corresponding to the species *Nodularia spumigena*, and animal deaths have been reported [31].

Anatoxin-a (ANA-a) and its structural variants have been associated with the genus *Anabaena* (now *Anabaena/Dolichospermum*) and also to *Aphanizomenon*; *Planktothrix*; *Cylindrospermum*; *Microcystis*; and the benthic *Oscillatoria*, *Phormidium*, *Tychonema*, and *Geitlerinema* [32]. ANA-a is an alkaloid known to exhibit acetylcholinesterase inhibition activity, which is its primary mode of toxicity. Of the structural variants of ANA-a, homoanatoxin-a, resulting from the methylation of the carbon at the extremity of the ketone function, is one of the more commonly observed [33]. Several other derivatives of ANA-a have been identified, including 2,3-epoxy-anatoxin-a; 4-hydroxy- and 4-oxo-derivatives; dihydroanatoxin-a; and dihydrohomoanatoxin-a [34,35]. Toxicity indications for these other variants are mixed, with some reports indicating lower toxicity while others suggesting higher oral toxicity from the dihydroanatoxin variant specifically, suggesting that monitoring of these variants is of interest [36].

Saxitoxins (STXs) are highly polar, nonvolatile, tricyclic perhydropurine alkaloids. While the overall molecular structure is largely conserved within the group, substitutions and variations amount toat least 57 analogs in the environment. They are also referred to as paralytic shellfish toxins (PSTs). PSTs can be non-sulfated, singly sulfated, or doubly sulfated [33,37]. These water-soluble toxins can persist for over 90 days in freshwater [38], but they are altered by high temperatures and may be degraded into more toxic variants [32]. STX is one of the most potent natural neurotoxins, and a dose of approximately 1 mg of the toxin from a single serving of contaminated shellfish is fatal to humans [39].

The cyanotoxin β-N-methylamino-L-alanine (BMAA) is a non-protein amino acid reportedly produced by the majority of cyanobacterial isolates [40]. Some literature reports have suggested that cellular exposure to BMAA may lead to neurological damage in the brain and the central nervous system of humans and animals, potentially contributing to one of several neurodegenerative diseases [41]. However, this remains uncertain as while other studies have illustrated a wide spectrum of effects to exposure, the neurodegenerative disease relationship was not universally observed [42].

Cylindrospermopsins (CYNs) are another class of alkaloid cyanotoxins produced by a variety of freshwater cyanobacteria. Production of CYNs have been reported from Nostocalean species mostly, as well as recently from one Oscillatoriale [43]. CYN and its variants are highly polar, polycyclic uracil derivatives containing guanidino and sulfate groups, and their detection has been reported worldwide. It is a potent hepatotoxin and is likely to be taken up by a variety of aquatic organisms, suggesting potential bioaccumulation risks [22,44].

The invasive algae *Prymnesium parvum* (golden algae) is a species of haptophyte (Prymnesiophyta). The species is of concern because of its ability to produce a suite of complex polyether toxins, known as prymnesins. Prymnesins may refer to Prymnesin-1, Prymnesin-2, or Prymnesin-B1 (to name a few of the presently identified species). These variants differ primarily in the length of the carbon backbone and are recognized hemolytic and ichthyotoxic agents that have been associated with massive fish kills in at least 14 countries [45]. Because of the uncertainty as to the identity of potentially toxic species, and the unavailability of analytical standards, the detection methods used for other cyanotoxins are often not suitable for the detection of prymnesins, which often require more complex analyses, such as high-resolution mass spectrometry (HRMS), quadrupole-time of flight mass spectrometry (qTOF), and nuclear magnetic resonance (NMR) [46,47].

3. Current Detection Methods for Cyanobacterial Toxins

A wide range of detection methods is available for the analysis of cyanotoxins. The most commonly used methods for different toxins are listed in Table 2. These methods are primarily targeted for water matrices; however, some can potentially be adapted for use with other matrices, such as fish tissue, shellfish, or organs.

Table 2. Analytical methods available for the detection of cyanotoxins and prymnesins (primarily used for water but adapted for use with fish tissue and other matrices).

Common Name	Commonly Used Analytical Detection Techniques
Microcystins	Immunoassay, LC–PDA, LC–MS *, GC–MS, PPIA
Nodularins	Immunoassay, LC–PDA, LC–MS *, PPIA
Anatoxin-a	Immunoassay, LC–UV, LC–MS *, IMS
Saxitoxins	Immunoassay, LC–FLD, LC–MS *
BMAA	Immunoassay, LC–FLD, LC–MS *
Cylindrospermopsins	Immunoassay, LC–UV, LC–MS *
Prymnesins	MS–MS, HRMS, qTOF, NMR

* Includes LC–MS, LC–MS/MS, and high-resolution mass spectrometric technologies.

The quantification of cyanotoxins in water matrices is at this point a well-studied problem, with multiple EPA standard methods developed for drinking water and an extensive literature presence. However, studies on the extraction and quantification of cyanotoxins in more complex matrices, such as shellfish, fish tissue, and organs, are more limited in quantity and scope. This review will present a summary of studies evaluating cyanotoxins in various matrices, and their extraction, preconcentration, and clean-up stages will be discussed, along with the specific quantitation techniques employed.

4. Sample Preparation and Analytical Methods for Cyanotoxins and Prymnesins Detection

Various methods are available for the quantification of cyanotoxins, as described above, and are primarily split between immunoassays and chromatographic separation (i.e., GC or LC), followed by quantification using mass spectrometry, UV, and photodiode array detectors. The matrix suitability varies by procedure, with a variety of interferences inherent to specific analysis; for example, mass spectrometric methodologies are susceptible to matrix interferences related to the ionization technique employed, with ion suppression/enhancement commonly observed in more challenging sample matrices. For each method, then, the samples often require specific preparation before analysis. This is especially common for tissue and other more complex matrices than water.

4.1. Sample Preparation

Sample preparation procedures differ according to the toxin analyzed and the nature of the tissue, organ, or biomass analyzed. Typically, this includes extraction, clean-up, and any post-extraction modifications (e.g., derivatization and solvent exchange) prior to analysis. Table 3 lists a variety of sample extraction and preparation procedures from a survey of the literature. The typical workflow being followed is summarized in Figure 1.

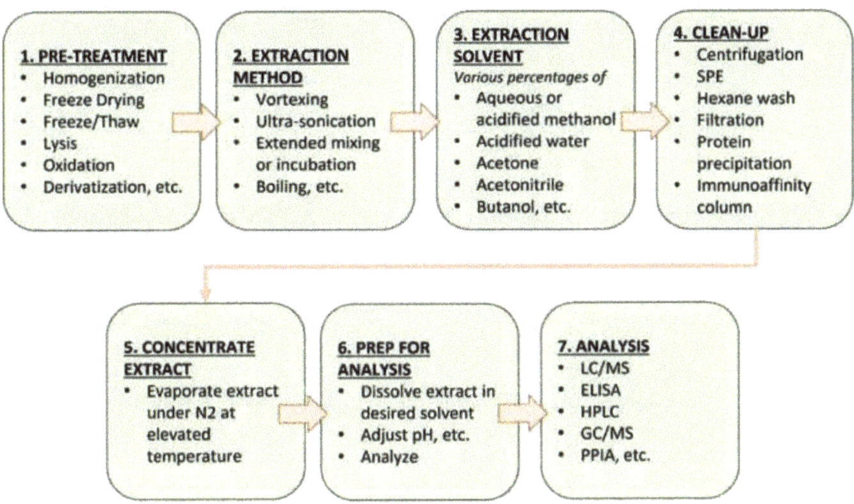

Figure 1. Common sample extraction and preparation procedures for measuring cyanotoxins in tissue.

Procedures for extracting toxins from complex matrices vary widely (Table 3). In most cases, heterogeneous samples, including tissues, organs, and food supplements, are mechanically homogenized or lyophilized to break down any solid structures and produce a suspension or emulsion amenable to extraction by solvents containing various percentages of aqueous and/or organic solvents (acidified methanol, acidified water, acetone, acetonitrile, butanol, etc.). In samples containing intact cyanobacterial cells, lysis to release intracellular toxins is often facilitated by ultrasonication, extended mixing, or incubation at a specific temperature and application of heat [48]. Following lysis/homogenization and extraction, a variety of clean-up techniques are available (Figure 1), including centrifugation, solid-phase extraction (SPE), protein precipitation, hexane washes, filtration, and immunoaffinity columns [49]. The clean-up steps attempt to remove matrix components, after which the toxin is eluted/extracted with a solvent and reduced in volume by evaporation prior to analysis. The efficiency of extraction techniques and the clean-up steps in the recovery of cyanotoxins from complex matrices vary widely based on the toxin physicochemical properties, matrix effects, holding times, collection methods, and other site-specific variables. In addition, each of these extraction techniques and clean-up steps may result in some loss of the target toxins during sample processing.

Extraction of cyanotoxins from complex matrices requires use of appropriate solvent mixtures to remove them, ideally without also extracting interferences to the analytical techniques being employed. The references cited in Table 3 are, broadly, either attempting to extract only individual types of toxins or attempting to quantify multiple classes with a single extraction workflow. The former can potentially allow for optimal conditions to be obtained, while in the case of more general extractions, compromises may result from attempting to measure chemicals less compatible with the extraction conditions. The primary differentiating step is the extraction procedure, solvents, and conditions used, which are elaborated upon below.

4.1.1. MCs and NODs

There are numerous literature examples of procedures for the extraction of MCs from a variety of matrices, such as tissue, listed in Table 3, including some that have systematically evaluated various extraction conditions. Because of their similar structure and chemical properties to MCs, there is little to differentiate extraction conditions for NODs, and as seen in Table 3, the conditions are comparable. The most common methods for the extraction of MCs have used methanol:water mixtures of varying ratios, in some cases with added acid, typically acetic acid. Because MCs are a diverse class of species with varied amino acid substitutions, one complicating factor is the variability in hydrophobicity/hydrophilicity of MCs that could be present in a sample. This has been found to be an issue in extractions with pure methanol, for which lower recovery of more hydrophilic congeners, such as MC-RR, has been reported. Instead, most procedures have used 75% to 90% methanol [50], the remainder being water, which has been seen to be generally appropriate for a broad suite from hydrophobic to hydrophilic congeners (See Table 3). An additional complication, however, has been observed in extractions using methanol when used in conjunction with some ELISA based toxin measurements. In at least some studies, interferences resulting in false positive detections were observed and confirmed by comparison with LC–MS/MS, where no MCs were observed [49]. While it is possible that the targeted methods were missing unknown MCs, this was further evaluated by comparison with a direct monoclonal ELISA with known higher specificity to MCs, and the absence of MCs via that method was confirmed.

One alternative to attempting to extract individual MC congeners has been developed using Lemieux oxidation, which has been demonstrated to react with a variety of MCs and convert them to a common product from all MCs with "adda", producing MMPB, which can be extracted instead. Quantification of MMPB can then serve as a proxy measurement for the total of MC congeners present in the sample. This does have limitations in that MMPB oxidation is known to generate positive results when parent MCs may be partially metabolized or otherwise transformed, as in drinking water treatment where oxidized MCs give a false positive signal, but recent publications indicate correlations seen between solution phase MC detections and detections in fish tissue [51].

4.1.2. ANA-a

Examples in Table 3 show three different procedures for the extraction of ANA-a from a solution. The first uses an acidic ethanol:water mixture (80:20) that is largely comparable to that for the majority of MC detecting methods above [52]. An alternative approach used immunoaffinity beads to pull the toxin from waters and slurries, but the practicality of this for heterogeneous systems was less clear [53]. Finally, for water samples, SPE is a viable avenue for extracting ANA-a, but this is again of limited utility for more heterogeneous matrices, where cartridge clogging will be more of a challenge [54]. More recent examples for ANA-a extraction in mixed methods are discussed below.

4.1.3. STXs

STXs are universally prepared using acidic or buffered extractions owing to their chemical instability under basic conditions. As these are a hydrophilic class of toxins, the extraction solvents used have generally been acidic, typically aqueous acetic [55–57] or hydrochloric acid [58], with buffered water being used in one instance [59]. Elevated temperatures are frequently used to extract STXs, including boiling of the solvent/sample mixture to facilitate extraction efficacy. Interestingly, in one study [59], acidic water and acidic methanol:water mixtures were evaluated and found to produce significantly lower recoveries of STX specifically, while the use of neutral pH phosphate buffer increased recoveries to >50%, which they speculated was due to reduced solubility in low pH.

4.1.4. CYNs

Cylindrospermopsins are a significantly more hydrophilic toxin than MCs, without as much structural diversity, and as a result extraction procedures have generally been more straightforward. In Table 3, three procedures are listed for the analysis of CYNs, and these all use mechanical homogenization or lyophilization, followed by extraction with methanol or water.

4.1.5. BMAA

Extraction of BMAA is typically performed under acidic conditions following lyophilization [60–62]. Because it is an amino acid, it is highly water soluble and does not require the use of organic solvents for extraction from tissue. However, removal of proteins and/or lipids is part of the process in some cases on extraction with chloroform [61].

4.1.6. Prymnesins

Evaluation of extraction techniques for prymnesins and other toxins associated with *P. parvum* is difficult owing to the many questions related to the toxic species, their structure, and the lack of available analytical standards. The protocols listed in Table 3 illustrate these challenges, with [46,47,63] repeatedly re-extracting the water/algal cell lysates with solvents of varying polarity in an attempt to isolate and characterize potential toxic species. In [46], cold acetone was used as a pre-extraction solvent to remove chlorophyll, followed by methanol extraction, in which prymnesins were observed by LC–HRMS. In [47], *P. parvum* lysates were extracted with dichloromethane, ethyl acetate, methanol, and water and while prymnesins were not observed under LC–HRMS analysis, cytotoxic activity was seen in the ethyl acetate and methanol extracts and a variety of fatty acid amides and one hydroxamic acid were observed. Finally, in [63], cold acetone was again used as a chlorophyll-removal step before sequential methanol and *n*-propanol extraction of the cellular material, which was then followed by solvent exchange to water. The aqueous material was then extracted with ethyl acetate four times to remove fatty material, before a final SPE step for clean-up prior to analysis. In this study, two prymnesins and a variety of related ions and fragments were characterized by LC–HRMS.

4.1.7. Extraction of Multiple Cyanotoxins

Workflows to extract multiple toxin classes typically attempt either to group compatible species together or to accept compromises in recovery arising from the impossibility of trying to recover diverse species simultaneously. As described above, procedures for the extraction of MCs and NODs have typically settled on 75–90% methanolic water as an extraction solvent. In studies such as [20,64–66], extraction of both MCs and NODs is described and the solvent mixtures used were uniformly in that range, showing the ease of extracting similar classes of toxins in tandem. However, studies attempting to extract those toxins as well as the more hydrophilic species, such as STXs or CYNs, as in [67–70], typically reduced the percentage of organic solvent to ~50% in order to improve solubility of the more hydrophilic constituents of the methods. An alternative approach was followed in [71], where two separate extraction procedures were devised to recover collectively STX, ANA-a, and CYN with 25% acetonitrile in water, while MCs and NOD were extracted with 75% acetonitrile in water from a split sample. The latter approach attempts to reduce the impact of chemical incompatibility, but at a cost of doubling extraction and analysis requirements. Once the samples are prepared based on the extraction and clean-up techniques described above, the many analytical methods discussed in Section 4.2 may be used for detection and quantification.

4.2. Analytical Methods

4.2.1. Immunological Assays

Cyanotoxins can be detected through recognition and binding to specific antibodies, either monoclonal or polyclonal. For example, ELISA kits are commercially available for the detection of MCs in water based upon either the common "adda" moiety present or specific recognition of a single MC congener [49,72–74]. Depending upon the antibody and the procedure employed, these kits can achieve a detection limit (DLs) as low as 4 ng/L, with an upper quantitation limit (due to saturation) of 5 µg/L for MC-LR [73]. While ELISA is frequently employed for the detection of MCs, particularly in drinking water, ELISA kits have also been made commercially available for the detection of ANA-a, CYN, and STX [75,76]. The most significant advantage of ELISA methods is that they do not require expensive and high-upkeep analytical instrumentation to be maintained, as they typically rely only on colorimetric assays for quantification.

However, detection methods based on ELISA have some limitations of varying severity by target compound. The measurement of a variety of MCs by ELISA is possible because the antibody assay is broadly cross-reactive over different MC congeners; however, this cross-reactivity is not uniform, and in some cases, MCs may be measured with greater or lesser responses relative to MC-LR [20]. Beyond that, only a single signal for a general MC concentration will be obtained, even for a sample that might contain a variety of MC congeners of varying potential toxicity levels. As a result, ELISA-based measurements should be considered to be semi-quantitative in that the measured concentrations are influenced by a number of variables that may or may not be known at the time of analysis. In addition to this limitation, cross-reactivity of the assay with other compounds in the sample and matrix may lead to over- or underestimation of the concentration of toxins, for both MCs and other cyanotoxins. This was observed in one paper, mentioned in Table 3, where ELISA kits targeting the "adda" moiety common to most MCs were found to be generating false positive results when compared with LC–MS/MS and an alternative, monoclonal antibody specific to MC-LR [77]. Conversely, it has also been demonstrated that ELISA-based screening may fail to measure protein-bound or glutathione-conjugated species in tissue matrices, perhaps due to the molecules being inaccessible to the antibody. When ELISA extracts were compared with a chemical oxidation/derivatization technique for measuring total MCs in tissue, a significant enhancement in the measured concentrations was observed in the latter [78]. These types of issues are not specific only to ELISA, but rather a general complication for any measurement technique. The sequestration/transformation of toxins through metabolic/removal processes may make them unavailable for many types of assays.

ELISA-based techniques can also be sensitive to the presence of solvents such as methanol and acetonitrile that are often used for tissue extraction. Commercially available ELISA kits typically recommend <5–10% solvent, but this value may be as low as <2.5% for ANA-a and as high as <20% for CYN and STX, based on standard protocols included in the various ELISA kits; therefore, in many studies relying on ELISA to quantify toxins from tissue extracts (see Table 3), the organic extracts are evaporated to dryness under nitrogen at 30–60 °C and reconstituted in DI water or an appropriate diluent prior to analysis [20]. Low pH can also affect ELISA performance, which could be an issue for STX extracted under acidic conditions. In many of the reviewed studies, pH was adjusted with 0.1 N NaOH to improve compatibility with the ELISA assay.

4.2.2. Mass Spectrometry

Following sample extraction and clean-up, separation of compounds typically employs either liquid or gas chromatographic (LC or GC) techniques. Typically, LC methods use a reversed phase C18 or a hydrophilic interaction liquid chromatography (HILIC) column and either methanol:water or water:acetonitrile gradients for separation, as these allow for flexibility, speed, and adaptability to a wide range of detectors relying on UV absorbance, fluorescence, or mass spectrometry. GC can be used as a separation method for cyanotoxins [79]. However, many cyanotoxins, including MCs, are larger molecules and are either not volatile or do not ionize well without chemical derivatization techniques. In addition, GC-based methods might require a solvent-exchange step to a nonpolar solvent such as hexane or ethyl acetate prior to analysis, which also typically involves blowing samples down to dryness prior to reconstitution with the nonpolar solvent. As such, GC-based separation requires more complex and time-consuming sample preparation, while also presenting more avenues for toxin loss from samples. While there are examples in the literature of analytical methods for cyanotoxins using GC–MS methods, these are a clear minority [78]. The single example in Table 3 using GC–MS involved MMPB oxidation of MCs, and in this instance, using GC–MS as the detector required an added derivatization step in the workflow.

Mass spectrometric methods for quantifying cyanotoxins rely on comparison with specific analytical standards for target compounds and can provide greater specificity than is possible with ELISA. In the LC–MS/MS techniques typically employed for toxin measurement, both a starting mass and a characteristic fragment ion specific to a given chemical are used to ensure specificity of assignment, along in some cases with a separate confirming ion fragment and a unique ratio of confirming to main ion. This is advantageous in that it means that false positive identifications are less common (although not impossible; ANA-a is isobaric with the common amino acid phenylalanine and both parent and product masses are identical; confusion of these two species can occur when MS-based detection is used [53]).

An additional useful feature of LC–MS/MS methods is that surrogate and internal standards, usually consisting of isotopically labeled analogs of target compounds, can be added to samples prior to extraction and analysis. This enables the losses encountered during extraction, clean-up, and analysis steps to be accounted for through comparison with expected recoveries of surrogates, while isotopic dilution techniques can allow for compensation for some matrix interferences during all of these processes, improving overall quantitation accuracy [77].

Quantitative mass spectrometric methods have been developed for the majority of the cyanobacterial toxins, including MCs, ANA-a, CYNs, STXs, and BMAA, either as a native compound or after chemical modification [80]. This allows samples to be separated using conventional C18 phases, although with ANA-a(S), HILIC methods work well for the assessment of this very hydrophilic toxin [81]. Due to the specificity, sensitivity, and rapidity of the analysis, mass spectrometric methods are now the physicochemical method of choice for the quantitative analysis of most cyanobacterial toxins in complex matrices for labs with access to appropriate instrumentation.

Cyanotoxins can also be detected by MS without preliminary chromatographic separation, particularly with time of flight (TOF) mass spectrometers in which many compounds can be identified and quantified concurrently. For example, matrix-assisted laser desorption/ionization coupled with TOF (MALDI-TOF) can be used to perform toxin analysis in even individual cell colonies [79]. In a typical workflow, target compounds enclosed in the dried and solid sample are ionized by a laser beam and accurately identified through the high mass resolution provided by the TOF instrument. However, TOF mass spectrometers usually tend to be less sensitive than other mass spectrometers of the same generation, and these methods are less commonly used for routine sample screening and quantitation than LC–MS/MS methods. However, TOF- and HRMS-based instruments can be used to qualitatively identify toxins that lack analytical standards, unknown toxins, and/or their metabolites or quantify using standards of other structurally similar toxins [82–84].

LC–MS/MS is the most commonly used chromatographic method for cyanotoxin detection in general, but it is limited by the need for analytical standard material to be available. For example, LC–MS/MS methods for the measurement of prymnesins and other toxins produced by *P. parvum* have not been commonly reported in the literature; these represent an emerging class of contaminants for which there is presently little information. The key issue preventing their measurement is the absence of a pure analytical reference standard for these toxins, which could be used for calibration. The ability to incorporate an internal standard would be ideal for highest precision for an analytical method, ideally a stable, isotopically labeled form of the analyte, but along with the native (unlabeled) prymnesins, these are not available. To date, only a few studies have reported isolating potential toxins, including prymnesin molecules, from *P. parvum*, [85–89], and replication of this work is yet to be reported in the peer-reviewed literature. Analytical standards for the prymnesins, either normal or isotopically labeled, along with any other structurally similar molecule are not currently available. A few studies have used UTEX strain 2797 as the source material for the preparative isolation of toxin material for analytical method development [46,63,89–92], but these preparations are not quantitatively exact enough to allow for use as true analytical standards. As a result, there are currently no published validated methods for the quantitative analysis of prymnesins [93].

4.2.3. LC–UV and LC–FLD

Monitoring UV absorbance was historically one of the first techniques to detect cyanotoxins after LC separation. MCs and CYNs have specific UV spectra with maximum absorbances at 240 and 262 nm, respectively [94,95]. Analytical workflows using LC–UV and/or LC–PDA allow for measuring the concentration of MCs by using these characteristic absorbances. In conjunction with good chromatographic separation, samples can not only have toxin concentrations measured but also give some information on the congeners of the MCs present. The key limitation of UV-absorbance-based techniques, however, is that the absorbance is not limited to only MCs and that background interferences may also impact the baseline signal. This can result in both lower DLs and potentially false positive signals, depending on the matrices being studied. Literature results for MC monitoring using LC–UV and LC–PDA have allowed for measurement of up to seven MC variants through comparison with analytical standards [96,97].

Detection by fluorescence is also commonly used after LC separation. However, cyanotoxins do not naturally fluoresce and, therefore, require the addition of a derivatization process during the sample preparation. For example, in the high-performance liquid chromatography with fluorescence detection (HPLC–FLD) system used by [98], post-column derivatization was performed using a solution containing 10 mmol/L of periodic acid and 550 mmol/L of ammonia in water (flow rate 0.3 mL/min). Nitric acid (0.75 mol/L in water; flow rate 0.4 mL/min) was used for reducing the pH value to 2–3. Fluorescence detection was applied for the determination of STX oxidation products at an excitation and emission wavelength of 330/395 nm.

While most studies that have used LC–UV and LC–FLD techniques have done so in water matrices, several prior studies have used fluorescence and/or UV-based methods to measure cyanotoxins (and in particular MCs) in fish, sediments, and plants [96,99]. The results of these studies show sensitivities comparable to those of LC–MS/MS methods, but with reduced specificity, due to which they can be more susceptible to matrix interferences, resulting in higher practical DLs. While LC–UV, LC–FLD, and LC–PDA methods can be simple and cost effective, misidentification can occur due to the non-specific nature of these methods. LC–MS/MS methods are becoming more prevalent in environmental laboratories now and are the preferred method for toxin identification as they can precisely and accurately identify toxins based on a specific mass/charge precursor ion that is unique to each toxin. In addition, despite typically offering only a unit mass resolution, LC–MS/MS methods can improve confidence in the identification of target compounds through screening for confirmation ions and the ratio of target and confirmation ions being produced.

4.2.4. Biochemical Assays

MCs and NODs are potent inhibitors of protein phosphatases (PPs), and in addition to antibody screening, these toxins can also be detected and measured using a protein phosphatase inhibition assay (PPIA) [74,100–103]. This assay measures the rate of formation of *p*-nitrophenol (pNP; yellow color) through hydrolysis of *p*-nitrophenol phosphate by PPs over time and measures all PP inhibitors present in a sample. The colorimetric PPIA has been optimized for MC and NOD detection in cyanobacteria extracts using 96-well plates and has shown acceptable correlation with HPLC data [104]. However, PPIA cannot distinguish co-occurring variants of MCs and cannot distinguish MCs from NODs. Therefore, results are often expressed as equivalent dosages relative to MC-LR, which is used as a reference standard. In addition, when analyzing bloom-containing water, interferences with unknown compounds can occur, leading to overestimation or underestimation of toxin concentration. Few studies have documented the use of PPIA for toxin measurement in tissue because of the complexity of tissue extracts and potential for interferences [64,105]. Note that because PPIA detects only inhibitors of that enzyme, within the realm of cyanotoxins it is known to quantify only MCs and NODs and alternative analysis should be undertaken to detect other cyanotoxins if required.

5. Discussion

Methods for the extraction and measurement of toxins from complex matrices are influenced by the chemistry of both the toxins and the composition of various matrices. MCs are a particularly diverse class of cyanotoxins, with more than 248 structurally diverse congeners identified in the environment, with chemical properties (including hydrophobicity) varying accordingly. Analysis of MCs in tissue and organs primarily relies on the extraction of homogenized tissue with various fractions of methanol:water, typically 75–80%, or conversion of MCs to MMPB to enable measurements of a single species rather than an ensemble [78,106–108]. In both pathways, the extracts are often cleaned via centrifugation [106], hexane wash [108], or SPE [107]. The methanol:water extracts are compatible with LC–MS/MS or other chromatographic methods directly or with ELISA analysis after dilution or solvent evaporation and reconstitution (to reduce the organics in the final sample) to avoid matrix effects [109]. While only a limited number of individual congeners can be quantified using LC–MS/MS methods due to the need for matching analytical standards (in limited availability) and difficulty with isomer separation, both ELISA and the MMPB method can be used to measure total MCs, with the latter potentially also including bound, conjugated, or partially degraded MCs. DLs for ELISA and LC–MS/MS methods for tissue matrices in the literature are generally as low as a few nanograms/gram. DLs for PPIA [106] in water matrices were higher, on the order of 1 µg/L in water. LC–PDA methods also generally have higher detections for MCs in water, with 2.9 µg/L as the lowest level detected, in [97]. Total MCs quantified using the MMPB method, as a proxy for total MCs, have been shown to have DLs of 2.18 ng/g wet weight [51]. One concern for anti-adda-based ELISA methods in the measurement of MCs is the potential for false positives, discussed above; confirmation of ELISA results with LC–MS/MS or direct monoclonal antibodies has been found to be effective at confirming or refuting such results [49].

Due to their structural similarity to MCs, NODs are also frequently extracted using methanol or methanol:water mixtures. If only pure methanol is used, as in [110], if employed for a simultaneous analysis for MCs, it would be expected to be susceptible to the same under-extraction of more hydrophilic MCs, as seen in [72]. More broadly, there is cross-reactivity in ELISA analysis between NODs and MCs, which can lead to difficulties differentiating the specific toxins at a site. LC–MS/MS methods should be used to confirm the specific toxins present if this information is needed. In the reports discussed herein, detection of NODs in tissue varied significantly from organ to organ, from 21 µg/kg to 1.4 mg/kg, with ELISA being the predominant detection method used for tissue samples.

In this review, methods for the analysis of solely ANA-a or CYN were limited and some are more than a decade old. These studies employed extraction of ANA-a using solvents (ethanol:acetic acid, 80:20), immunoaffinity beads, or SPE, followed by analysis using ELISA, IMS, or HPLC–UV [52–54]. While DLs for these methods were comparable to extraction/analysis procedures reviewed for other toxins (25 ng/L in water and 4 µg/g in tissue), they are rarely used. Similarly, the extraction of only CYN was achieved using methanol or water, followed by analysis using ELISA, LC–MS/MS, or LC–UV methods, achieving detection varying from 2.7 ng/g for the highly sensitive MS/MS methods to 2.5 µg/g using LC–UV [111–113]. LC–MS/MS is the most common methodology currently employed to detect ANA-a and CYN concurrently in the studies reviewed. Since most recent studies optimize the extraction of multiple toxins in one workflow, the multi-toxin methods reviewed herein provide the path forward and best practices for ANA-a and CYN detection in complex matrices. For example, the use of 75:25 water:acetonitrile for the extraction and use of chemically similar isotopically labeled surrogates, such as ANA-a-$^{13}C_4$ or CYN-$^{15}N_5$, which have similar extraction recovery, column retention, and ionization efficiency, can help with improved compensation for recovery bias and matrix effects [71,114]. These highly sensitive methods provide method detection limits (MDLs) as low as 0.14 ng/g for ANA-a and 0.12 ng/g for CYN [71].

The methods for the analysis of only BMAA in tissue were also limited in scope. Because it is a small amino acid and highly hydrophilic, studies that concurrently analyze BMAA with these other toxins are limited because of the need for either chemical derivatization or specialized HILIC LC–MS/MS methods, as seen in [60–62]. In general, the extraction of BMAA from tissue has been accomplished with an acidic solvent, but the need for hydrolysis, derivatization, and clean-up steps to reduce matrix interferences have made the process lengthy and potentially incompatible with other toxins. Following extraction, analysis was performed using HPLC–FLD or LC–MS/MS, with measurements ranging from micrograms/gram to milligrams/gram. The data from both analysis methods are generally comparable, indicating that the use of HPLC–FLD could be a beneficial first step or screening method. Studies support the use of LC–MS/MS as a necessary confirmation tool [115,116].

Most methods for STX detection in complex matrices usually involve extraction under acidic conditions (with hydrochloric acid or acetic acid, for instance) and often at an elevated temperature [55–58]. Detection was accomplished with ELISA, HPLC–FLD, LC–MS/MS, or LC–qTOF MS, with DLs mostly in the micrograms/kilogram range. The ELISA method is sensitive and allows for rapid screening of a large number of samples, and LC–MS/MS serves as a useful confirmation tool. STX is stable at high temperatures, which is an important consideration for food safety; it also plays a role in many extraction procedures that require the sample to be boiled [55–59,117]. STX is known to persist for over 90 days and is considered extremely stable even at a high temperature and low pH [38]. Although it can be stored in acid without loss for many decades, some studies suggest that it may not be stable at pH > 8, even at ambient temperature [118]. This suggests that STX may not be stable in water and bivalves unless it is stabilized with acid. The stability also depends on the chemical structure of specific compounds; GTX1, GTX4, and NEO variants are less stable at acidic pH than GTX2, GTX3, and STX [119]. Overall, the stability of various STX variants is an unsolved analytical problem that is in need of continued study.

Significant questions and uncertainties remain with regard to the toxins associated with *P. parvum* at this time. Sequential extraction of cultures with various solvents in [47] identified material with cytotoxic properties, including fatty acids and fatty acid amides, but no prymnesins. However, in [63], two toxins classified as prymnesins were observed by LC–MS/MS, following multiple preparative and clean-up steps, including chlorophyll removal and post-extraction solid-phase extraction. At this point, it remains unclear which toxins are the specific causative agents for *P. parvum* toxicity, a problem exacerbated by a lack of commercially available standards from which to perform toxicity assessments.

One of the emerging needs in analytical laboratories is for methods suitable for quantifying multiple cyanotoxins in a single workflow. As cyanobacterial blooms become more common and severe, the need for rapid analysis is only going to increase and methods suitable for screening and quantifying multiple classes of toxin at once will greatly improve sample throughput. In Table 3 are listed 20 studies relying on multi-toxin methods.

One approach for extracting multiple toxins is to attempt a compromise extraction mixture that is suitable for both hydrophilic and more hydrophobic toxins, as described in [68–70]. In these studies, the percentage of organic solvent in the extraction mixture was reduced to ~50%, rather than the 75–90% range typical for MCs, to improve the recovery of the more hydrophilic classes of contaminants, including CYNs, ANA-a, and STXs. In [68], an acidic methanol:water (50:50) mixture was used to extract benthic algae and analysis was performed using LC–MS/MS and LC–qTOF. In this study, two STX variants were observed in field samples at 209–270 ng/g, but no detections were seen for MCs, NODs, CYN, or ANA-a. The matrix fortification of dried tissue prior to extraction found that all five classes of contaminant were recovered with 80–105% efficiency, with the caveat that the only MC congeners used were MC-LR and MC-RR, two of the most hydrophilic MCs. In [69], in contrast to a single ~50% organic extraction, sequential extractions were performed in fish and shellfish, first with 100% methanol, followed by an acidic water:acetonitrile (55:45) mixture, with the two pooled before further processing. This study was designed to be as broadly suitable as possible for screening purposes, with two chromatographic methods used, HILIC for hydrophilic toxins and reverse phase for the more lipophilic toxins. As a part of method validation, an extensive matrix spike evaluation was performed for numerous MCs, STX, and other toxins outside the scope of this review, such as domoic acid, and okadaic acid. The authors found that recoveries ranged significantly, with many within an 80–120% window but others, particularly ANA-a, showing around 200% recovery of the spiked amount. According to them, this was mostly likely the result of matrix interferences in the analysis procedure, which is not surprising given the exhaustive extraction procedure and tissue matrix, but without isotopically labeled materials, confirmation of this hypothesis is impossible. In addition, the DLs in this study were somewhat higher than typical, with DLs of 150 ng/g for MCs and 600 ng/g for the more hydrophilic toxins.

In another multi-toxin study, [71], a method to measure MCs, NOD, ANA-a, CYN, and STX in tissue using LC–MS/MS was reported. Because of differences in chemical properties, the study relied on two separate extraction procedures, one for ANA-a, CYN, and STX and the other for NOD and MCs (Table 3). Once prepared, the samples were subject to one of two LC–MS/MS methods to separate and quantify the compounds, with C18-based chromatography for NOD/MCs and HILIC for ANA/CYN/STX. As described above, the technique of isotope dilution was used in this procedure to track recovery and quantification of ANA-a, CYN, and a subset of MCs through extraction and analysis. Matrix effects for MCs were seen in fish tissue, with magnitudes ranging from −26% to −58%, which were corrected through the internal standards to −16% to 10%. Comparable matrix suppressions of −44% to −50%, corrected to 1.4 to 3.4%, were observed for ANA-a and CYN, respectively. Similar magnitudes were seen in water extraction. Extensive method validation was performed using matrix spikes to determine DLs, from which MDLs were determined to be 0.12 to 0.70 µg/kg in tissue and 4 to 80 ng/L in water. However, validation of this procedure in fish exposed to exogenous MCs in [71] was not successful in producing positive results, which the authors attributed to the short duration of exposure, but MCs were detected in environmental samples. Overall, this study illustrates the utility of the use of isotopically labeled materials in these multi-toxin methods to improve the tracking of interferences and extraction efficiencies.

Method development and optimization of techniques to extract and analyze toxins from different matrices rely on spiking the matrix with the target analytes, processing these samples through the optimized workflow, and reporting recoveries of the spiked compounds and precision as a metric of method performance. While this process may work in the case of simple matrices, such as water or algal cells, it may only partially mimic toxin-containing environmental samples from more complex matrices, such as zooplankton or tissue, as noted in [71]. For example, in spike recovery studies involving tissue, either the sample is spiked with the target toxins using a syringe needle inserted into the tissue prior to homogenization or the homogenate is spiked with the target toxins, followed by mixing prior to the processing of these samples through the optimized workflow. However, these spiked toxins may not entirely mimic the way that natural toxins are bound in tissue samples in the environment. This may lead to biases in estimated recoveries, especially if the spiked toxins are more readily amenable to the extraction process than the natural toxins, leading to incorrect quantifications of these toxins in complex matrices. While spike recovery studies are a good first step for developing and optimizing methods for toxin detection, method validation with different types of samples collected from HAB sites will provide an insight into the reliability of the methods.

Table 3. Selected list of publications summarizing methods for toxin detection in various matrices.

Reference	Toxins Measured	Matrix	Extraction/Sample Preparation Procedure	Analysis Method	Toxin Detection
			Microcystins		
[106]	MCs	Dissolved fractions of the water table, silversides, and common carp	For tissue extraction, 75% methanol and then 75% methanol with 0.05% acetic acid was used. Samples were centrifuged, and the supernatant was blown down to dryness and resuspended in a suitable solvent for ELISA (phosphate-buffered saline), LC (methanol), and PPIA (water). The ELISA did not detect MCs within the limits of the assay, but the PPase showed that bioactive variants are present.	ELISA, LC–MS/MS, and PPIA	0.02 to 0.36 µg/L in water/sestonic, 0.16 to 0.87 µg/g in fish by ELISA, >1 µg/L PPIA
[107]	MCs	Muscle, liver, fish tissue, and lake water samples	Tissue samples were homogenized, mixed with 10 mL methanol:acidified water (90:10, v/v), and sonicated. The extracts were then centrifuged, diluted with water (not to exceed 5% methanol), and acidified. For samples with lipid >1%, an extra SPE step was included.	LC–MS/MS	349–450 ng/g in tissue
[96]	MCs	Common carp and silver carp	Tissue samples were homogenized with methanol, sonicated, and centrifuged. The supernatant was analyzed, blown down to dryness, and resuspended in a suitable solvent for analysis.	LC–PDA and ELISA	PDA: 13.8–539 µg/L in water and dry biomass. ELISA: 1.4–29 ng/g in tissue
[120]	MCs	Crab tissue	Samples were analyzed following the protocol included in the kit. This comprised tissue homogenization, extraction with methanol, sonication, and centrifugation. The supernatant was blown down prior to analysis (<5% methanol).	ELISA	Up to 1.42 µg/L in water; 65–820 µg/kg in tissue, including liver and viscera
[108]	MCs	Fish tissue	Tissue was homogenized with 3 mL of methanol, sonicated in an ultrasonic bath, and centrifuged. Supernatants were pooled and extracted with 1 mL of hexane to remove lipids. The extract was evaporated at 50 °C and reconstituted in methanol for analysis.	LC–MS/MS	<DL (1.2 ng/g) to 50.3 ng/g in tissue
[121]	MCs	Fish and crustacean tissue	Fish and crustaceans were treated with 100% methanol and then with hexane. The obtained methanolic fraction was concentrated/cleaned using SPE. The eluent was dried, redissolved in methanol, filtered (nylon filter), and analyzed by ELISA.	ELISA	0.25 to 103.3 µg/kg in tissue
[122]	MCs	Fish tissue (common carp)	Tissues were homogenized, extracted in 100% methanol, stirred overnight at room temperature, and then centrifuged. The supernatants were collected and concentrated under a N_2 stream to 350 µL to remove the organic solvent. A 100 µL aliquot of the concentrated sample extract was diluted with 900 µL of distilled water, filtered (pore size of 0.45 µm and diameter of 4 mm), and analyzed.	ELISA	114 to 732 µg/kg in muscle, kidney, and liver

Table 3. *Cont.*

Reference	Toxins Measured	Matrix	Extraction/Sample Preparation Procedure	Analysis Method	Toxin Detection
[109]	MCs	Fish tissue	Tissues were homogenized and extracted with 75% methanol. Extracts were centrifuged, the supernatant was removed, and the solids were resuspended in 75% methanol for two more extractions. The supernatant from all extractions was pooled and diluted to one-quarter strength with deionized water. The resulting solution was concentrated/cleaned with SPE (C18 column) and eluted with 5 mL of 100% methanol. The sample was then diluted to <5% methanol and analyzed.	ELISA	<7.5 to 203 ng/g in tissue
[78]	MCs	Fish tissue	Lemieux oxidation reactions were performed to convert MCs to MMPB. After termination of reactions, the samples were centrifuged at 3000 rpm for 5 min to remove tissue. Aliquots of oxidation products in the supernatant were dried and dissolved in a 5% HCl methanol solution, followed by heating and neutralization with silver carbonate. Total MC content was measured by headspace by polydimethylsiloxane/divinylbenzene (PDMS-DVB) solid-phase microextraction (SPME) GC/MS/MS analysis.	SPME–GC–MS/MS	0.018 to 0.87 µg/g by ELISA; 0.84 to 4.7 µg/g by MMPB oxidation
[51]	MCs	Rainbow trout tissue, liver, kidney	MCs were oxidized to MMPB. Total MCs were quantified using isotope dilution with d3-MMPB by LC–qTOF MS.	LC–qTOF–MS	MDL 2.18 ng/g (wet wt); tissue < MDL to 8.3 ± 6.9 ng/g; liver < MDL to 173.1 ± 97.8; kidney < MDL to 209.9 ± 42.3 ng/g
[49]	MCs	Water, fish, and mussels	A gram of water or fish was added to a centrifuge tube, spiked with MC-LR (standard addition), mixed with solvent (methanol, aqueous methanol, or aqueous acetonitrile), vortexed, centrifuged, cleaned as per [20,107,123,124], evaporated to dryness, re-constituted, and analyzed.	ELISA	0.1 µg/L (MDL)–0.2 µg/L in fish
[97]	MCs	Fish tissue	Tissue was extracted with 80% (v/v) aqueous methanol and centrifuged and the supernatant was filtered through a 0.2 µm filter. The supernatants were separated and evaporated to dryness at 40 °C. The samples were reconstituted in 300 µL of ultra-pure water for ELISA and the same amount of 100% methanol for HPLC.	ELISA and LC–PDA	0.043–1.72 mg/kg in tissue, 7.0–17.6 mg/kg in sediment, 2.9–13.5 µg/L in water

Table 3. *Cont.*

Reference	Toxins Measured	Matrix	Extraction/Sample Preparation Procedure	Analysis Method	Toxin Detection
			Nodularins		
[110]	NOD	Flounder, mussel, and clam tissue	Tissue was homogenized, frozen at −30 °C, and freeze-dried. Dry samples were extracted with 100% methanol and centrifuged. The supernatant was collected, concentrated at 50 °C, centrifuged, diluted 10x with Milli-Q water, filtered, and analyzed.	LC–MS/MS, MALDI–TOF–MS, LC–UV–MS/MS, and ELISA	Up to 1.490 mg/kg MCs and/or NOD in tissue by ELISA; NOD confirmed but not quantified by LC–MS
[105]	NOD	Sediments, mussels, and fish	A freeze-dried sample was ground with a mortar and pestle, extracted with 75% methanol, sonicated, and centrifuged. The supernatant was evaporated to dryness and dissolved in Milli-Q water. The sample was then vortexed, sonicated, centrifuged, and cleaned up with SPE. The cartridge was eluted with 100% methanol, dried, and re-suspended in Milli-Q water prior to analysis.	ELISA and LC–MS/MS	2.3–75 µg/kg in sediment; up to 139 µg/kg in mussels. 489 µg/kg in liver, 21 µg/kg in guts, and 21 µg/kg in flounder
[125]	NOD	Flounder and cod	Samples were extracted in water:methanol:n-butanol 75:20:5, *v:v:v*, in an ultrasonic bath for 8 h at 50–60 °C. Then, the samples were centrifuged, supernatant was evaporated to 1.2 mL, and (LC-PDA only) cleaned/concentrated with SPE and eluted with methanol. The eluent was evaporated at 50 °C to dryness, dissolved in 150 µL of methanol, filtered, and diluted with water for analysis.	LC–PDA, ELISA, and PPIA	30–70 µg/kg in liver by ELISA and PPIA, <DL of LC–PDA
			Anatoxin-a		
[52]	ANA-a	Phytoplankton, stomach contents of birds, and blooms	Samples were extracted with ethanol:acetic acid (20:80) and centrifuged. The supernatant was used for assay.	ELISA	ACE inhibition equivalent to 4 µg/g ANA-a in extracts; 0.1 to 0.9 µg/g MCs by ELISA
[53]	ANA-a	Water samples/ slurry	Immunoaffinity beads were employed for the extraction of ANA-a from water. Sample pH was adjusted to 10, then a magnetic immunosorbent was added to the sample and mixed for 10 min. The magnetic particles were separated rapidly from the solution by an external magnet, and the water sample was gently removed. Then, ANA-a was completely eluted with 2-propanol. The solution was separated from the magnetic particles by an external magnet and directly analyzed by IMS.	IMS	0.02 to 5 µg/L linear range by IMS
[54]	ANA-a	Water	SPE cartridges were conditioned with 2-propanol, followed by HPLC-grade water. The samples were applied to the cartridges dried under vacuum and the analyte eluted with methanol containing 0.1% v/v trifluoroacetic acid. The extracts were blown down to dryness at 40 °C, re-dissolved in 5 % v/v aqueous acetonitrile containing 0.1% v/v trifluoroacetic acid, and analyzed.	LC–UV	DL of 25 ng/L

Table 3. *Cont.*

Reference	Toxins Measured	Matrix	Extraction/Sample Preparation Procedure	Analysis Method	Toxin Detection
			Saxitoxin		
[55]	STX	Shellfish tissue	Samples were extracted using 0.1 M HCl with ultrasonication, cleaned/concentrated with SPE (C18 cartridge), and analyzed.	LC–qTOF MS	0.1–1.6 µg/kg recovery from spiked tissues
[58]	STX	Sheep intestine and blood	Samples were sonicated with 0.1 M acetic acid and incubated for 2 h at 4 °C. Clean-up was performed with a C18 cartridge; 1 mL of 0.05 M acetic acid was used to elute the toxin fraction.	LC with spectrofluorometric detector	STX detected in intestine but not in blood; exact concentration not reported
[56]	STX	Seabird tissues, forage fish, and invertebrates	Seabird tissues and whole forage fish and invertebrates were extracted for STX analysis using the procedure of [117]. Tissue was homogenized, extracted in 3 mL of 1 % acetic acid, vortexed, boiled, allowed to cool to room temperature, vortexed again, and centrifuged. The remaining supernatant was poured into a vessel; the procedure was repeated. The combined supernatant was filtered and diluted in Milli-Q water for analysis.	ELISA and LC–FLD	0.14–1.08 µg/kg in liver by ELISA; no detection using LC–FLD with DL of 1 µg/kg
[59]	STX	Bivalves	A tissue homogenate (1.0 g) was mixed with 5.0 mL of phosphate buffer solution in a 50 mL plastic centrifugal tube and then placed in a boiling water bath for 5 min, cooled, extracted in ultrasonic water bath at room temperature, and centrifuged. The supernatant was collected, and the residue extracted once more. The supernatant was combined and filtered by microfiber filters. The filtrate was cleaned using immunoaffinity column (IAC). The eluent from the IAC was blown dry with N_2 at 55 °C, redissolved with 1 mL of water, and filtered by a 0.22 µm membrane before determination by LC–MS/MS.	LC–MS/MS	DL 0.1 µg/kg
[57]	STX	Abalone	About 2 g of abalone tissue (epipodium, viscera, or foot muscle) was mixed with 18 mL of 1% acetic acid (v/v). The mixture was vortexed, boiled, cooled, vortexed again, and centrifuged. This was followed by the addition of 5 µL of ammonium hydroxide before SPE clean-up. STX was eluted using 2 mL of acetonitrile:water:acetic acid (20:80:1, v/v/v) and diluted with acetonitrile before analysis.	LC–MS/MS	High detection in muscle/epipodium (up to 1.085 mg/kg) exposed to STX producing cultures

Table 3. *Cont.*

Reference	Toxins Measured	Matrix	Extraction/Sample Preparation Procedure	Analysis Method	Toxin Detection
			BMAA		
[60]	BMAA	Cyanobacterial samples	The lyophilized sample was hydrolyzed using 6 N HCl liquid hydrolysis for 20 h at 105 °C in the absence of oxygen. After hydrolysis, samples for derivatized analysis were dissolved in 500 μL of hot 20 mM HCl and subsequently diluted 10 times in HCl to obtain a protein concentration below 0.1 g/L. Hydrolyzed samples for underivatized LC–MS/MS analysis were dissolved in 1 mL of 65% acetonitrile, 35% Millipore water, and 0.1% formic acid (*v:v:v*).	LC–FLD and LC–MS/MS	ND by LC–MS/MS, false positives by HPLC–FLD
[61]	BMAA	Water samples and tissue samples (crustacean, mollusk, and fish)	Centrifuged, homogenized tissue was suspended in trichloroacetic acid and washed with chloroform for the removal of residual lipids. Samples (5 mL) and standards were derivatized with 6-aminoquinolyl-N-hydrosuccinimidyl carbamate (AQC), and BMAA was separated from the protein amino acids by reverse-phase elution (Waters Nova-Pak C18 column). Identification of a BMAA peak detected by reverse-phase HPLC was verified by LC–MS/MS using product ion mode in a triple quadrupole system.	LC–FLD	<DL to 7 mg/g by FLD; confirmed by LC–MS/MS
[62]	BMAA	Freshwater surface samples, mollusks, crustaceans, and fishes	The lyophilized sample was extracted with 2 mL of 0.1 M trichloroacetic acid by sonication in an ice bath. The extract was centrifuged, and the supernatant was N_2 dried for the collection of free BMAA. The precipitated protein pellets were subsequently hydrolyzed in 6 M HCl and filtered. The hydrolysate was then N_2 dried for protein-associated BMAA collection. The free and protein-associated BMAA fractions were reconstituted in 20 mM HCl. Samples were derivatized by adding 60 μL of borate buffer and 20 μL of AQC. The mixture was incubated in a water bath at 55 °C for derivatization and was prepared for LC analysis.	LC–MS/MS	0.45–6.05 μg/g dry weight
			Cylindrospermopsin		
[111]	CYN	Fish tissue and liver	Tissue and liver were homogenized in 10 mL of 100% methanol, sonicated, and centrifuged. The supernatant was decanted and filtered. The extraction was repeated on the pellet, and the two extracts were collected together and then dried by rotavapor at 40 °C; the residue was re-suspended in 2 mL of distilled water and analyzed.	LC–MS/MS and ELISA	2.6 to 126 μg/L in water; up to 2.7 ng/g in fish tissue
[112]	CYN	Crayfish tissue	Freeze-dried samples of cyanobacteria and tissue were taken up in distilled water with sonication, filtered, and diluted to a concentration within the linear range of the method. Water samples were filtered and diluted when necessary.	LC–MS/MS	589 μg/L in water; up to 4.3 and 0.9 μg/g in liver and muscle tissue, respectively

Table 3. *Cont.*

Reference	Toxins Measured	Matrix	Extraction/Sample Preparation Procedure	Analysis Method	Toxin Detection
[113]	CYN	Mussel	Lyophilized tissues and samples for the analysis of intra- and extracellular CYN were extracted in 100% methanol with ultrasonication on ice. Tissue and cell debris was removed by centrifugation and the supernatant dried at 50 °C under N_2 and re-suspended in Milli-Q water. The samples were centrifuged again to remove insoluble materials and analyzed.	LC–UV	Up to 2.52 µg/g in tissue
			Prymnesin(s)		
[126]	*P. parvum* strains	Water and algal cells	Samples were placed in 15 °C and incubated at an irradiance of 5–7 mmol photons $m^2 s^1$ for 2 h. After 2 h, the in vivo fluorescence of the samples was measured on a Turner Design Trilogy1 Laboratory Fluorometer.	Relative fluorescence	Toxin extracts highly unstable when extracellular; storage at −80 °C with no headspace indicated
[47]	*P. parvum* strains	Water and algal cells	Water and cultured and field-collected algal cell mass was lyophilized. An elutropic extraction scheme using solvents with increasing polarity (dichloromethane, ethyl acetate, methanol, and water) was used to fractionate toxic compounds in samples by polarity. Individual compounds were obtained via semi-preparative HPLC–MS purification. Isolated compounds were then structurally characterized by MS/MS and NMR.	LCMS/MS, LC–HRMS, and NMR	Structural identification of potentially toxic compounds in extracts
[92]	*P. parvum* strains	Water and algal cells	Liquid–liquid partitioning of the whole cultures (medium plus cells) using ethyl acetate was performed. The ethyl acetate layers from partitioning against 50 L of *P. parvum* cultures were combined, and the organic extract was subjected to gradient MPLC.	GC–MS and NMR	Additional structural characterization of potential toxins
[127]	*P. parvum* strains	Water and algal cells	Samples were preserved with acid Lugol's solution, and cells were counted using a particle counter.	Cell density	Characterization of parameters influencing toxicity of *P. parvum* cells
[46]	*P. parvum* strains	Water and algal cells	The biomass pellets were thawed and extracted twice with cold acetone for removing, among other, chlorophylls. After vortexing and centrifugation, the supernatants were collected (acetone). After chlorophyll extraction, the biomass was extracted twice with methanol and sonicated. Both extracts (acetone and methanol) were concentrated to dryness under N_2 at 35 °C, reconstituted in 1 mL methanol, and analyzed.	LC–DAD–HRMS	Prymnesins characterized and identified

Table 3. *Cont.*

Reference	Toxins Measured	Matrix	Extraction/Sample Preparation Procedure	Analysis Method	Toxin Detection
[63]	*P. parvum* strains	Water	The samples were extracted with cold acetone, methanol, and isopropanol. This was followed by pooling of samples, SPE, and evaporating the eluent to dryness. The dried methanol:isopropanol fraction was resuspended in water. An equal volume of ethyl acetate was added to the sample and placed on the vortex mixer, followed by centrifugation. The aqueous portion was recovered and defatted with ethyl acetate three more times. After the last phase of partitioning, the aqueous layer was transferred back to the methanol:isopropanol vial and evaporated to dryness. This was followed by SPE and analysis.	Thin-layer chromatography (TLC) and LC–HRMS	Prymnesins characterized and identified
			Concurrent Analysis of Multiple Cyanotoxins		
[67]	MC NOD ANA CYN STX	Fish from aquaculture	MCs, NOD, ANA, and CYN: Toxins were extracted twice with a water:methanol mixture (50:50, v/v), followed by 10 min sonication in an ultrasonic bath, and then treated for 2 min in an ultrasonic homogenizer. The extracts were centrifuged (14,000 rpm), and the supernatant was and analyzed via LC–ESI–MS. STX: By adding 1 mL of acetic acid (0.03 N), 50 mg of lyophilized samples was extracted, sonicated in an ice bath, and centrifuged. The supernatant was then filtered, and the extract was analyzed via LC–FLD.	LC–MS/MS and LC–FLD	No detection of MCs, NODs, ANA, or CYN by LC–MS/MS; up to 350 ng/g STXs by LC–FLD
[68]	STX ANA NOD MCs CYN	Benthic *Lyngbya wollei* algae samples	Dry algae were mixed with 1 mL of methanol and water (1:1) with 0.1 M acetic acid. Samples was vortexed, sonicated, and centrifuged. The supernatant was collected and filtered through a 0.45 μm PTFE filter. This procedure was repeated three times in total. All aliquots were combined, evaporated to dryness under N_2, and resuspended in 0.5 mL of acetonitrile:water (9:1) with 5 mM ammonium acetate and 3.6 mM formic acid (pH 3.5) for analysis.	HILIC and RPLC–MS/MS, LC–QqQMS, and LC–QqTOFMS ǂ	209–279 μg/g of two STX analogs (LWTX-1 and LWTX-6) in algae; no other cyanotoxins detected
[128]	MC ANA	Lake water samples and freeze-dried bloom material	Lyophilized cells (about 100 mg) were extracted three times with 10 mL of 0.05 M acetic acid for 30 min while stirring. The extract was centrifuged, and the supernatant was adjusted to pH 10 with 7% ammonium hydroxide. This pH 10 extract was directly applied to 0.2 g of a reversed-phase ODS disposable extraction column.	LC–PDA	20–1500 μg/g of various MCs; up to 1444 μg/g ANA
[64]	MCs NOD	Cyanobacterial bloom material obtained from freshwater lakes	Samples were lyophilized, and extracts were prepared using 70% (v/v) methanol and centrifuged. The resulting supernatants were analyzed.	PPIA and LC–PDA	DLs of as low as to 1 μg/L in drinking water

Table 3. *Cont.*

Reference	Toxins Measured	Matrix	Extraction/Sample Preparation Procedure	Analysis Method	Toxin Detection
[65]	MCs NOD	Otter tissue, digesta, and water	Tissue samples were first homogenized, mixed with methanol:water (90:10), sonicated, and analyzed. Sample preparation and analysis followed protocols from previously published studies [107].	LC–MS/MS	1.36–348 µg/kg in otter liver; up to 1324 µg/kg in clams, mussels, and oysters
[66]	MCs NOD	Water, algal cells, algal supplement tablets, and mussels	Water samples were analyzed directly by LC without any extraction steps. Algal samples were centrifuged to isolate cells. The cells, tablets, and mussel tissue were extracted using a variety of solvents in different proportions (aqueous methanol, isopropyl alcohol, and 1% acetic acid). It was found that 80% aqueous methanol enabled the optimum extraction of toxins. Samples were extracted by vortex mixing.	LC–MS/MS	Limit of detection ranging from 0.01 and 0.19 ng/mL for water, 0.4 and 3.6 pg/mL for algal cells, 0.12 to 1.18 µg/kg for algal supplement tablet powders, and 0.01 and 0.21 µg/kg for mussels
[129]	MCs NOD	Bottlenose dolphin liver	Samples were oxidized to convert MCs/NODs to MMPB. Samples were cleaned with SPE and 12 cc Novum simplified liquid extraction (SLE) tubes and analyzed with LC. Individual variants were extracted with 75% methanol in 0.1 M acetic acid, followed by a butanol rinse. Supernatants were blown to dryness using N$_2$ at 60 °C, reconstituted in deionized water, clarified using SPE, eluted with acetonitrile, blown to dryness (60 °C, N$_2$), reconstituted (1 mL of 5% methanol), filtered (0.2 µm polyvinylidene fluoride), and analyzed using LC. The final extract was also diluted 10-fold for ELISA analysis.	LC–MS/MS and ELISA	MDL 1.3 ng/g for the MMPB method and 1.6–11.5 ng/g for the variants
[69]	MCs STXs CYNs Others	Fish, shellfish tissue, and food supplements	A gram of tissue homogenate was extracted with 4 mL of methanol with a vortex mixer and centrifuged and the supernatant decanted. To the pellet, 5 mL of water/acetonitrile/ammonium formate/formic acid (55:45 v/v, 2 mM, 0.5 mM) was added and extracted with a pulse mixer and centrifuged. The supernatant was combined with the previously obtained methanol extract. The tube was filled with 10 mL of acetonitrile. The aliquot of the extract was filtered with a 0.2 µm filter and used for analysis with LC–HRMS. The supplements followed a similar procedure with an additional clean-up step using a Strata-X polymeric reversed-phase cartridge.	LC–HRMS	DLs of 150 ng/g for MCs and 600 ng/g for the more hydrophilic toxins; 80–200% recoveries

Table 3. *Cont.*

Reference	Toxins Measured	Matrix	Extraction/Sample Preparation Procedure	Analysis Method	Toxin Detection
[130]	MCs NOD ANA CYN	Water, fish tissue, and liver	Water: Of the sample, 150 mL was filtered using a glass fiber filter, adjusted to pH 11, and cleaned with SPE. Eluents were evaporated to dryness under N_2, reconstituted with 150 µL of 5% (v/v) methanol, and then analyzed with LC. PIPPA and ELISA analysis was performed using commercial kits as per manufacturer-provided guidelines. Tissue/liver: With 5 mL of 80% methanol containing 0.5% formic acid, 0.2 g of lyophilized powdered flesh or 0.25 g of liver was extracted by stirring for 15 min, followed by ultrasonication for 30 min. The mixture was centrifuged, and the supernatant was washed three times with 1 mL of hexane. The extract was cleaned by SPE. The eluents were dried in a water bath at 40 °C under N_2, reconstituted with 200 µL of 5% methanol, and analyzed with LC.	LC–MS/MS, PPIA, and ELISA	25.8–429.3 µg/L MCs in water; no detection in tissue
[114]	MCs ANA-a CYN	Fish tissue	Tissue (500 mg wet weight) was amended with 150 µL of a mixture of isotope-labeled internal standards. After a 1 h equilibration time, 4 mL of methanol was added, vortexed, ultrasonicated, and centrifuged. The supernatant was removed, and the tissue was re-extracted twice as previously described. The combined supernatants were concentrated to 4 mL (N_2, 40 °C). The samples were then frozen and centrifuged (defatting step), and the supernatants were evaporated to dryness (N_2, 40 °C), reconstituted in 2 mL water, vortexed, ultrasonicated, filtered (0.2 µm), and analyzed.	Online SPE–LC–MS/MS	MDL of 0.1 to 10 µg/kg; 0.16–7.8 µg/kg of MCs and 46 µg/kg of CYN detected in field samples
[131]	MCs NOD CYN STX	Carp, otter, dalmatian pelican tissue, and liver and stomach contents	Samples were freeze-dried, ground using a pestle and mortar, and extracted three times at 60 °C in 0.5 mL of 75:25 methanol:water (*v:v*). Extracts were dried in SpeedVac and reconstituted in 600 µL of methanol. The reconstituted samples were transferred to 2 mL of Eppendorf vials with a cellulose acetate filter and centrifuged for 5 min. Filtrates were transferred to amber glass vials for MC analysis.	LC–MS/MS	0.8–1.9 µg/g of MCs in carp liver; 0.7 µg/g MCs in otter liver; 0.4–1.5 µg/g MCs in pelican liver, tissue, and stomach sample
[71]	MCs ANA STX NOD CYN	Fish tissue	Homogenized whole fish, 2 g, was lyophilized in a freeze dryer for 72 h. ANA, CYN, and SAX were extracted with 10 mL of 25:75 (*v:v*) acetonitrile:water added to each vial. MCs and NOD were extracted using 10 mL of 75:25 (*v:v*) acetonitrile:aqueous 0.1% formic acid added to each vial. Samples were sonicated and centrifuged. The supernatant was collected, syringe-filtered, blown down under N_2, and re-suspended in 20 mL of water. SPE was performed for clean-up (ANA, CLD, and SAX were extracted on a Supelclean ENVI-carb. MCs and NOD were extracted using an Oasis HLB). Analysis was performed using LC–MS/MS.	LC–MS/MS	Non-detection in fish exposure study method; MDLs from 80 to 960 ng/L in water and 0.12 to 0.70 µg/kg in tissue

Table 3. *Cont.*

Reference	Toxins Measured	Matrix	Extraction/Sample Preparation Procedure	Analysis Method	Toxin Detection
[132]	STX ANA	Phytoplankton samples	Freeze-dried material (10 mg) and 2 mL of 0.03 N acetic acid were mixed, frozen and thawed three times, sonicated, and centrifuged. The supernatant was filtered and stored at −20 °C until analysis.	LC–FLD	5.9–224.1 ng/g STX equivalents
[20]	MCs NOD	Fish	Homogenized fish tissue was weighed, extracted with a 3:1 methanol:water solution with 1% formic acid, vortexed, centrifuged, extracted with hexane clean-up to reduce lipid content, centrifuged, and analyzed.	LC–MS/MS and ELISA	10 ng/g DLs in tissue
[133]	STX NeoSTX GTX(1,2,3,4,5)	*Hoplias malabaricus* (wolf fish)	The samples were homogenized in HCl (0.1 N) and centrifuged at 10,000 ×g at 19 °C for 10 min. The supernatants were filtered with cellulose filters and analyzed.	LC–FLD	No detections of STXs in the tissue after exposure
[116]	BMAA DABA ANA-a	Water fish aquatic plants	Samples were mixed with 1 mL of 0.1 N TCA by vortexing for 1 min and washed with 100% purified water, 50% methanol in water, and 100% methanol. The mixture was then vortexed and centrifuged to separate solids from the aqueous extract. The extract containing unbound or "free" amino acids, including BMAA and DABA (2,4-diaminobutyric acid), were transferred to a microcentrifuge filter tube for removal of suspended proteins and centrifuged before analysis.	LC–FLD and LC–MS/MS	BMAA between 8 and 59 ng/g in tissue; no ANA-a detections reported; BMAA, DABA, and ANA-a detected in plants
[134]	BMAA DABA ANA-a	Lake water, fish, and aquatic plants	Freeze-dried samples were ground into a fine powder and extracted with 0.1 N TCA. The mixture was sonicated for 30 s, refrigerated for 16 h, and centrifuged and the supernatants retained. The process was repeated once more. The supernatants were combined, filtered, and analyzed with HPLC–FLD for preliminary analysis of all extracts; confirmation was performed using LC–MS/MS.	LC–FLD and LC–MS/MS	0.8–3.2 µg/L DL for LC–MS/MS; 5–7 µg/L for FLD
[135]	STX NOD MCs	Fish tissue	Freeze-dried muscle tissue was extracted with methanol, sonicated, and centrifuged and the supernatants retained. For lipid removal, hexane was added to the supernatants and then discarded after phase separation. Samples were evaporated and 10% methanol was added, followed by sonication and passage of the material through reversed-phase cartridges (OASIS HLB Cartridge 200 mg, Waters). Cartridges were eluted with 100% methanol, followed by evaporation and dissolving the residues in 75% aqueous methanol. After vortexing, the samples were filtered and centrifuged. The supernatants were then diluted 10-fold with 75% methanol for analysis.	LC–MS/MS	Detection of STXs, NODs, and MCs in water, while only MC-RR detection in tissue

Table 3. *Cont.*

Reference	Toxins Measured	Matrix	Extraction/Sample Preparation Procedure	Analysis Method	Toxin Detection
[77]	MCs ANA-a	Fish tissue	Two types of ELISA kits were used for samples: Envirologix™ anti MC-LR and Abraxis LLC anti-adda. The samples were extracted with methanol:water followed by SPE clean-up, similar to [107]. LC–MS/MS was used for confirmation.	ELISA and LC–MS/MS	2.2 to 132 µg/kg by anti-adda ELISA; 0.2–2.4 by anti-MC-LR ELISA; 2.5–14 µg/L MC-LA by LC–MS/MS; potential false positive detection by adda-ELISA
[70]	MCs ANA	Cyanobacterial biomass and fish tissue	Cyanobacterial biomass and fish tissues were prepared in acidified (0.002 M HCl) 50% methanol. Both biomass and tissues were homogenized, ultrasonicated (3 times), and treated with n-hexane to remove lipids (hexane layers were discarded). The obtained methanol extracts were analyzed.	LC–PDA	Up to 18.4 µg/g ANA and up to 4.4 µg/g MCs in liver tissue

ⱡ Hydrophilic interaction liquid chromatography (HILIC), reverse-phased liquid chromatography (RPLC) coupled to triple quadrupole mass spectrometry (LC–QqQMS), and quadrupole–time of flight mass spectrometry (LC–QqTOFMS).

6. Conclusions and Recommendations

This review documents the methodologies currently used to measure cyanotoxins and prymnesins in complex matrices and to assess the advantages and limitations of the various techniques summarized in this paper.

Due to the diversity in chemical structures and properties among cyanotoxins, it is a significant challenge to develop and validate procedures for consistent and high-yield extraction from various matrices, including tissues. In individual toxin methods, optimum conditions can typically be achieved through validation experiments, such as the consensus for extractions using 75–90% methanol:water for MCs/NODs consistently seen in the literature, although variations in matrix constituents such as lipid or protein content could potentially result in varied outcomes. The use of matrix spikes, where known amounts of toxins are fortified to sample matrices, can help further quantify the various matrix effects and interferences in extraction and analysis of cyanotoxins and add confidence to an analytical workflow. This is particularly important in cases where ambient detection of toxins was negative, as in studies such as [55], where only spiked samples showed detectable levels of toxins to allow for evaluation of recovery efficiencies.

As the breadth of cyanotoxins known to be present in the environment continues to increase, it is clear that methods for quantifying single classes of toxin may eventually overwhelm analytical facilities. A single method capable of extracting and detecting many or all cyanotoxins would be an ideal goal for the future. Several publications describe such approaches using mass spectrometric detectors for a subset of contaminants of interest, but as expected from a group of compounds with diverse chemistry, there are significant limitations in recoveries during sample processing, chromatographic performance, and sensitivity [104–113]. For example, in [69], MCs, NOD, ANA-a, CYN, and STXs were all analyzed using LC–MS/MS following extraction, with sequential extraction first with methanol, then a mixture of acetonitrile:water (45:55), which were pooled for analysis. Method recoveries in this study varied between 80 and 200% and the reported detection limits were high, most likely due to matrix interferences. In other cases, multiple complementary workflows for hydrophilic and hydrophobic toxins were developed, as in [71], and showed improved performance.

To control for variation in extraction efficiencies, one approach is to add isotopically labeled internal standards to the sample prior to extraction, allowing for quantification of recovery percentages through the extraction and analysis procedures. This can help compensate for recovery bias and matrix suppression/enhancement and is a recommended best practice but is limited in scope to those compounds for which isotopically labeled materials are readily available and when analytical methods can differentiate the native and labeled material (e.g., mass spectrometric methods). Recently, labeled materials have become available for a broader set of cyanotoxins, including MCs, STX, ANA-a, and CYN, which has made this approach feasible. Examples of this as applied to toxin measurement can be found in reference [71], where recoveries were assessed and corrected for both extraction efficiency and matrix effects using labeled toxin analogs.

Analysis with LC–MS/MS has become an essential tool for cyanotoxin detection and it can potentially be used for the concurrent analysis of multiple classes of toxins due to its rapid scan rate and ability to cycle polarity from positive to negative ion modes. LC–MS/MS is best employed for targeted screening for toxins, particularly where both native and isotopically labeled standards are available. In contrast, LC coupled with HRMS or TOF is better suited to detect unknown toxins or those for which standards are not available, but a more complete discussion is beyond the scope of this article [82].

An alternative to LC–MS/MS methods could be an ELISA microplate test strip that contains antibodies for multiple toxins. This would require design in a way that similar incubation periods for the binding step could be achieved, as the typical ELISA workflow has specific duration of each step of binding and rinsing of the plate. If this was achievable, it would allow simultaneous testing for multiple toxins without requiring expensive and bulky experimental apparatus, such as mass spectrometers. Because ELISA methods tend to be semi-quantitative for reasons described above, it is a best practice that detections be confirmed by LC–MS/MS or some other more specific technique where possible. One potential alternative in the future could be electrochemical biosensors that contain a biological recognition element that specifically reacts with the target of interest; these are in active development for use with water samples. The suitability of these biosensors to fish tissue and other matrices still needs to be determined.

Another complicating factor for assessment of toxins in tissue matrices is common to any analytical workflow, namely, the amenability of these bound toxins to be extracted into a solution for measurement. Literature results suggest a significant fraction of cyanotoxins in tissue samples could be bound to organs or otherwise unavailable through the extraction procedure [106]. Ref. [78] showed significant differences in MC measurements in tissue using ELISA and GC–MS/MS following Lemieux oxidation, which they attributed to the oxidation technique freeing bound analytes for measurement. Ref. [106] and references therein suggest that the total concentration could potentially be an order of magnitude higher. The level of underestimation and impact on risk assessments and health outcomes should be investigated in future studies.

An additional challenge is that many emerging cyanotoxins do not have commercially available standards, and, in other cases, toxin standards are prepared from minute amounts of natural sources or unidentified sources of unknown purity, making it difficult to accurately quantify toxin concentrations. In this review, prymnesins and other potential toxins produced by *P. parvum* are one such class of toxins, which at present are only available through laborious culturing, extraction, and isolation, as described above, and no commercial sources exist. Even in cases where materials can be procured, the use of certified reference standards (CRMs) with exact concentrations is recommended to improve confidence in the absolute concentrations; however, these are not available for many contaminants, and even for those with CRMs available, such as MCs, only a subset of variants may be covered. It is recommended that laboratories monitor their cyanotoxin standards over time for variations in purity/concentrations and where possible obtain standards from multiple vendors if certified materials are not available.

There is a great deal of variability in the analytical procedures presently being used to prepare, extract, and analyze for cyanotoxins associated with harmful algal blooms in diverse sample matrices. In this review, the most common procedures were highlighted, and best practices were identified. It is clear that there is a compelling need for more standardized, reliable, and affordable screening methods compatible with tissue and similar matrices exposed to cyanotoxins, particularly as new toxins are continually being identified in the environment. For LC–MS/MS methods, the best approach is to ensure the use of extensive quality control procedures, including evaluating matrix interferences, though matrix spikes where possible and using labeled surrogate and internal standards to monitor method performance across both the extraction and analysis phases (e.g., [71]). For the commonly used immunological methods, because they are incompatible with the use of labeled standards, researchers should instead ensure they perform similar studies of matrix performance to determine cross-reactivity parameters and potential interferences as part of their method validation procedures.

Author Contributions: Conceptualization, D.S., T.T.S., R.V., and J.L.; resources, J.L.; data curation, D.S. and J.L.; writing—original draft preparation, D.S., T.T.S., R.V., L.D., A.O.T., and J.L.; writing—review and editing, D.S., T.T.S., R.V., H.M., D.T., L.D., S.F., A.O.T., and J.L.; visualization, D.S.; supervision, R.V. and J.L.; project administration, J.L.; funding acquisition, J.L. All authors have read and agreed to the published version of the manuscript.

Funding: This research received no external funding.

Institutional Review Board Statement: Not applicable.

Informed Consent Statement: Not applicable.

Data Availability Statement: Not applicable.

Conflicts of Interest: The authors declare no conflict of interest. **Disclaimer:** The views expressed in this article are those of the author(s) and do not necessarily reflect the views or policies of the U.S. Environmental Protection Agency. This research was supported in part by U.S. EPA's Office of Research and Development, Cincinnati, OH, under Contract No. 68HERC20D0029. The use of trade names or commercial products is for identification only and does not imply endorsement by any agency of the U.S. government.

References

1. Sivonen, K. Cyanobacterial toxins and toxin production. *Phycologia* **1996**, *35*, 12–24. [CrossRef]
2. Zanchett, G.; Oliveira-Filho, E.C. Cyanobacteria and cyanotoxins: From impacts on aquatic ecosystems and human health to anticarcinogenic effects. *Toxins* **2013**, *5*, 1896–1917. [CrossRef] [PubMed]
3. Rastogi, R.P.; Madamwar, D.; Incharoensakdi, A. Bloom Dynamics of Cyanobacteria and Their Toxins: Environmental Health Impacts and Mitigation Strategies. *Front. Microbiol.* **2015**, *6*, 1254. [CrossRef] [PubMed]
4. Hawkins, P.R.; Runnegar, M.T.; Jackson, A.; Falconer, I. Severe hepatotoxicity caused by the tropical cyanobacterium (blue-green alga) Cylindrospermopsis raciborskii (Woloszynska) Seenaya and Subba Raju isolated from a domestic water supply reservoir. *Appl. Environ. Microbiol.* **1985**, *50*, 1292–1295. [CrossRef] [PubMed]
5. Gehringer, M.M.; Adler, L.; Roberts, A.A.; Moffitt, M.C.; Mihali, T.K.; Mills, T.J.; Fieker, C.; Neilan, B.A. Nodularin, a cyanobacterial toxin, is synthesized in planta by symbiotic *Nostoc* sp. *ISME J.* **2012**, *6*, 1834–1847. [CrossRef] [PubMed]
6. McGregor, G.B.; Sendall, B.C. *Iningainema pulvinus* gen nov., sp nov.(Cyanobacteria, Scytonemataceae) a new nodularin producer from Edgbaston Reserve, north-eastern Australia. *Harmful Algae* **2017**, *62*, 10–19. [CrossRef]
7. Gantar, M.; Sekar, R.; Richardson, L.L. Cyanotoxins from black band disease of corals and from other coral reef environments. *Microb. Ecol.* **2009**, *58*, 856–864. [CrossRef]
8. Tatters, A.O.; Howard, M.D.; Nagoda, C.; Fetscher, A.E.; Kudela, R.M.; Caron, D.A. Heterogeneity of toxin-producing cyanobacteria and cyanotoxins in coastal watersheds of southern California. *Estuaries Coasts* **2019**, *42*, 958–975. [CrossRef]
9. Shams, S.; Capelli, C.; Cerasino, L.; Ballot, A.; Dietrich, D.R.; Sivonen, K.; Salmaso, N. Anatoxin-a producing Tychonema (Cyanobacteria) in European waterbodies. *Water Res.* **2015**, *69*, 68–79. [CrossRef]
10. Bormans, M.; Lengronne, M.; Brient, L.; Duval, C. Cylindrospermopsin accumulation and release by the benthic cyanobacterium Oscillatoria sp. PCC 6506 under different light conditions and growth phases. *Bull. Environ. Contam. Toxicol.* **2014**, *92*, 243–247. [CrossRef]
11. Vico, P.; Bonilla, S.; Cremella, B.; Aubriot, L.; Iriarte, A.; Piccini, C. Biogeography of the cyanobacterium Raphidiopsis (Cylindrospermopsis) raciborskii: Integrating genomics, phylogenetic and toxicity data. *Mol. Phylogenetics Evol.* **2020**, *148*, 106824. [CrossRef] [PubMed]
12. Merel, S.; Walker, D.; Chicana, R.; Snyder, S.; Baurès, E.; Thomas, O. State of knowledge and concerns on cyanobacterial blooms and cyanotoxins. *Environ. Int.* **2013**, *59*, 303–327. [CrossRef] [PubMed]
13. Williams, D.E.; Craig, M.; McCready, T.L.; Dawe, S.C.; Kent, M.L.; Holmes, C.F.; Andersen, R.J. Evidence for a covalently bound form of microcystin-LR in salmon liver and Dungeness crab larvae. *Chem. Res. Toxicol.* **1997**, *10*, 463–469. [CrossRef] [PubMed]
14. Meriluoto, J. Chromatography of microcystins. *Anal. Chim. Acta* **1997**, *352*, 277–298. [CrossRef]
15. *ISO 22104:2021 (en)*; Water Quality—Determination of Microcystins—Method Using Liquid Chromatography and Tandem Mass Spectrometry (LC-MS/MS). International Organization for Standardization: Geneva, Switzerland, 2021.
16. EPA. *Method 544—Determination of Microcystins and Nodularin in Drinking Water by Solid Phase Extraction and Liquid Chromatography/Tandem Mass Spectrometry (LC/MS/MS)*; National Exposure Research Laboratory, Office of Research and Development: Cincinnati, OH, USA, 2015.
17. EPA. *Method 545—Determination of Cylindrospermopsin and Anatoxin-a in Drinking Water by Liquid Chromatography Electrospray Ionization Tandem Mass Spectrometry (LC/ESI-MS/MS)*; National Exposure Research Laboratory, Office of Research and Development: Cincinnati, OH, USA, 2015.
18. EPA. *Method 546—Determination of Total Microcystins and Nodularins in Drinking Water and Ambient Water by Adda Enzyme-Linked Immunosorbent Assay*; Standards and Risk Management Division—Technical Support Center, Office of Ground Water and Drinking Water: Cincinnati, OH, USA, 2016.
19. Yuan, M.; Carmichael, W.W.; Hilborn, E.D. Microcystin analysis in human sera and liver from human fatalities in Caruaru, Brazil 1996. *Toxicon* **2006**, *48*, 627–640. [CrossRef]
20. Geis-Asteggiante, L.; Lehotay, S.J.; Fortis, L.L.; Paoli, G.; Wijey, C.; Heinzen, H. Development and validation of a rapid method for microcystins in fish and comparing LC-MS/MS results with ELISA. *Anal. Bioanal. Chem.* **2011**, *401*, 2617–2630. [CrossRef]
21. Mol, H.G.; Plaza-Bolaños, P.; Zomer, P.; de Rijk, T.C.; Stolker, A.A.; Mulder, P.P. Toward a generic extraction method for simultaneous determination of pesticides, mycotoxins, plant toxins, and veterinary drugs in feed and food matrixes. *Anal. Chem.* **2008**, *80*, 9450–9459. [CrossRef]
22. Carmichael, W.W. Cyanobacteria secondary metabolites—The cyanotoxins. *J. Appl. Bacteriol.* **1992**, *72*, 445–459. [CrossRef]
23. Chorus, I.; Bartram, J. *Toxic Cyanobacteria in Water: A Guide to Their Public Health Consequences, Monitoring and Management*; E & FN Spon: London, UK, 1999.

24. Codd, G.; Bell, S.; Kaya, K.; Ward, C.; Beattie, K.; Metcalf, J. Cyanobacterial toxins, exposure routes and human health. *Eur. J. Phycol.* **1999**, *34*, 405–415. [CrossRef]

25. Ferrão-Filho Ada, S.; Kozlowsky-Suzuki, B. Cyanotoxins: Bioaccumulation and effects on aquatic animals. *Mar. Drugs* **2011**, *9*, 2729–2772. [CrossRef]

26. Kumar, J.; Singh, D.; Tyagi, M.B.; Kumar, A. Cyanobacteria: Applications in Biotechnology. In *Cyanobacteria*; Elsevier: Amsterdam, The Netherlands, 2019; pp. 327–346.

27. Chernoff, N.; Hill, D.; Lang, J.; Schmid, J.; Le, T.; Farthing, A.; Huang, H. The comparative toxicity of 10 microcystin congeners administered orally to mice: Clinical effects and organ toxicity. *Toxins* **2020**, *12*, 403. [CrossRef] [PubMed]

28. Catherine, A.; Bernard, C.; Spoof, L.; Bruno, M. Microcystins and nodularins. In *Handbook of Cyanobacterial Monitoring and Cyanotoxin Analysis*; Wiley: Hoboken, NJ, USA, 2017; Volume 1, pp. 107–126.

29. Krishnan, A.; Mou, X. A Brief Review of the Structure, Cytotoxicity, Synthesis, and Biodegradation of Microcystins. *Water* **2021**, *13*, 2147. [CrossRef]

30. Codd, G.A.; Lindsay, J.; Young, F.M.; Morrison, L.F.; Metcalf, J.S. Harmful cyanobacteria. In *Harmful Cyanobacteria*; Springer: Berlin/Heidelberg, Germany, 2005; pp. 1–23.

31. Chen, G.; Wang, L.; Wang, M.; Hu, T. Comprehensive insights into the occurrence and toxicological issues of nodularins. *Mar. Pollut. Bull.* **2021**, *162*, 111884. [CrossRef] [PubMed]

32. Sivonen, K.; Jones, G. Cyanobacterial toxins. In *Encyclopedia of Microbiology*, 3rd ed.; Elsevier: Amsterdam, The Netherlands, 2009; pp. 290–307. [CrossRef]

33. Van Apeldoorn, M.E.; van Egmond, H.P.; Speijers, G.J.; Bakker, G.J. Toxins of cyanobacteria. *Mol. Nutr. Food Res.* **2007**, *51*, 7–60. [CrossRef] [PubMed]

34. Namikoshi, M.; Murakami, T.; Watanabe, M.F.; Oda, T.; Yamada, J.; Tsujimura, S.; Nagai, H.; Oishi, S. Simultaneous production of homoanatoxin-a, anatoxin-a, and a new non-toxic 4-hydroxyhomoanatoxin-a by the cyanobacterium Raphidiopsis mediterranea Skuja. *Toxicon* **2003**, *42*, 533–538. [CrossRef]

35. Mann, S.; Cohen, M.; Chapuis-Hugon, F.; Pichon, V.; Mazmouz, R.; Mejean, A.; Ploux, O. Synthesis, configuration assignment, and simultaneous quantification by liquid chromatography coupled to tandem mass spectrometry, of dihydroanatoxin-a and dihydrohomoanatoxin-a together with the parent toxins, in axenic cyanobacterial strains and in environmental samples. *Toxicon* **2012**, *60*, 1404–1414.

36. Puddick, J.; van Ginkel, R.; Page, C.D.; Murray, J.S.; Greenhough, H.E.; Bowater, J.; Selwood, A.I.; Wood, S.A.; Prinsep, M.R.; Truman, P. Acute toxicity of dihydroanatoxin-a from Microcoleus autumnalis in comparison to anatoxin-a. *Chemosphere* **2021**, *263*, 127937. [CrossRef]

37. Nicholson, B.C.; Shaw, G.R.; Morrall, J.; Senogles, P.J.; Woods, T.A.; Papageorgiou, J.; Kapralos, C.; Wickramasinghe, W.; Davis, B.C.; Eaglesham, G.K.; et al. Chlorination for degrading saxitoxins (paralytic shellfish poisons) in water. *Environ. Technol.* **2003**, *24*, 1341–1348. [CrossRef]

38. Jones, G.J.; Negri, A.P. Persistence and degradation of cyanobacterial paralytic shellfish poisons (PSPs) in freshwaters. *Water Res.* **1997**, *31*, 525–533. [CrossRef]

39. Wiese, M.; D'agostino, P.M.; Mihali, T.K.; Moffitt, M.C.; Neilan, B.A. Neurotoxic alkaloids: Saxitoxin and its analogs. *Mar. Drugs* **2010**, *8*, 2185–2211. [CrossRef]

40. Esterhuizen, M.; Downing, T. β-N-methylamino-L-alanine (BMAA) in novel South African cyanobacterial isolates. *Ecotoxicol. Environ. Saf.* **2008**, *71*, 309–313. [CrossRef] [PubMed]

41. Delcourt, N.; Claudepierre, T.; Maignien, T.; Arnich, N.; Mattei, C. Cellular and Molecular Aspects of the β-N-Methylamino-l-alanine (BMAA) Mode of Action within the Neurodegenerative Pathway: Facts and Controversy. *Toxins* **2018**, *10*, 6. [CrossRef] [PubMed]

42. Chernoff, N.; Hill, D.; Diggs, D.; Faison, B.; Francis, B.; Lang, J.; Larue, M.; Le, T.-T.; Loftin, K.A.; Lugo, J. A critical review of the postulated role of the non-essential amino acid, β-N-methylamino-L-alanine, in neurodegenerative disease in humans. *J. Toxicol. Environ. Health Part B* **2017**, *20*, 183–229. [CrossRef] [PubMed]

43. Seifert, M.; McGregor, G.; Eaglesham, G.; Wickramasinghe, W.; Shaw, G. First evidence for the production of cylindrospermopsin and deoxy-cylindrospermopsin by the freshwater benthic cyanobacterium, Lyngbya wollei (Farlow ex Gomont) Speziale and Dyck. *Harmful Algae* **2007**, *6*, 73–80. [CrossRef]

44. Ohtani, I.; Moore, R.E.; Runnegar, M.T. Cylindrospermopsin: A potent hepatotoxin from the blue-green alga Cylindrospermopsis raciborskii. *J. Am. Chem. Soc.* **1992**, *114*, 7941–7942. [CrossRef]

45. Barkoh, A.; Fries, L.T. Aspects of The Origins, Ecology, and Control of Golden Alga *Prymnesium parvum*: Introduction to the Featured Collection. *JAWRA J. Am. Water Resour. Assoc.* **2010**, *46*, 1–5. [CrossRef]

46. Binzer, S.B.; Svenssen, D.K.; Daugbjerg, N.; Alves-de-Souza, C.; Pinto, E.; Hansen, P.J.; Larsen, T.O.; Varga, E. A-, B- and C-type prymnesins are clade specific compounds and chemotaxonomic markers in *Prymnesium parvum. Harmful Algae* **2019**, *81*, 10–17. [CrossRef]

47. Bertin, M.J.; Zimba, P.V.; Beauchesne, K.R.; Huncik, K.M.; Moeller, P.D. Identification of toxic fatty acid amides isolated from the harmful alga *Prymnesium parvum* carter. *Harmful Algae* **2012**, *20*, 111–116. [CrossRef]

48. Smith, J.L.; Boyer, G.L. Standardization of microcystin extraction from fish tissues: A novel internal standard as a surrogate for polar and non-polar variants. *Toxicon* **2009**, *53*, 238–245. [CrossRef]

49. Preece, E.P.; Moore, B.C.; Swanson, M.E.; Hardy, F.J. Identifying best methods for routine ELISA detection of microcystin in seafood. *Environ. Monit. Assess.* **2015**, *187*, 12. [CrossRef]

50. Fastner, J.; Flieger, I.; Neumann, U. Optimised extraction of microcystins from field samples—A comparison of different solvents and procedures. *Water Res.* **1998**, *32*, 3177–3181. [CrossRef]

51. Shahmohamadloo, R.S.; Ortiz Almirall, X.; Simmons, D.B.; Lumsden, J.S.; Bhavsar, S.P.; Watson-Leung, T.; Eyken, A.V.; Hankins, G.; Hubbs, K.; Konopelko, P. Cyanotoxins within and Outside of Microcystis aeruginosa Cause Adverse Effects in Rainbow Trout (*Oncorhynchus mykiss*). *Environ. Sci. Technol.* **2021**, *55*, 10422–10431. [CrossRef] [PubMed]

52. Henriksen, P.; Carmichael, W.W.; An, J.; Moestrup, Ø. Detection of an anatoxin-a (s)-like anticholinesterase in natural blooms and cultures of cyanobacteria/blue-green algae from Danish lakes and in the stomach contents of poisoned birds. *Toxicon* **1997**, *35*, 901–913. [CrossRef]

53. Le, T.; Esteve-Turrillas, F.A.; Armenta, S.; de la Guardia, M.; Quiñones-Reyes, G.; Abad-Fuentes, A.; Abad-Somovilla, A. Dispersive magnetic immunoaffinity extraction. Anatoxin-a determination. *J. Chromatogr. A* **2017**, *1529*, 57–62. [CrossRef] [PubMed]

54. Powell, M. Analysis of anatoxin-a in aqueous samples. *Chromatographia* **1997**, *45*, 25–28. [CrossRef]

55. Fang, X.; Fan, X.; Tang, Y.; Chen, J.; Lu, J. Liquid chromatography/quadrupole time-of-flight mass spectrometry for determination of saxitoxin and decarbamoylsaxitoxin in shellfish. *J. Chromatogr. A* **2004**, *1036*, 233–237. [CrossRef] [PubMed]

56. Van Hemert, C.; Schoen, S.K.; Litaker, R.W.; Smith, M.M.; Arimitsu, M.L.; Piatt, J.F.; Holland, W.C.; Hardison, D.R.; Pearce, J.M. Algal toxins in Alaskan seabirds: Evaluating the role of saxitoxin and domoic acid in a large-scale die-off of Common Murres. *Harmful Algae* **2020**, *92*, 101730. [CrossRef] [PubMed]

57. Seger, A.; Hallegraeff, G.; Stone, D.A.; Bansemer, M.S.; Harwood, D.T.; Turnbull, A. Uptake of paralytic shellfish toxins by blacklip abalone (Haliotis rubra rubra leach) from direct exposure to Alexandrium catenella microalgal cells and toxic aquaculture feed. *Harmful Algae* **2020**, *99*, 101925. [CrossRef]

58. Negri, A.P.; Jones, G.J.; Hindmarsh, M. Sheep mortality associated with paralytic shellfish poisons from the cyanobacterium *Anabaena circinalis*. *Toxicon* **1995**, *33*, 1321–1329. [CrossRef]

59. Yue, Y.; Zhu, B.; Lun, L.; Xu, N. Quantifications of saxitoxin concentrations in bivalves by high performance liquid chromatography-tandem mass spectrometry with the purification of immunoaffinity column. *J. Chromatogr. B* **2020**, *1147*, 122133. [CrossRef]

60. Faassen, E.J.; Gillissen, F.; Lürling, M. A comparative study on three analytical methods for the determination of the neurotoxin BMAA in cyanobacteria. *PLoS ONE* **2012**, *7*, e36667. [CrossRef] [PubMed]

61. Brand, L.E.; Pablo, J.; Compton, A.; Hammerschlag, N.; Mash, D.C. Cyanobacterial blooms and the occurrence of the neurotoxin, beta-N-methylamino-l-alanine (BMAA), in South Florida aquatic food webs. *Harmful Algae* **2010**, *9*, 620–635. [CrossRef] [PubMed]

62. Jiao, Y.; Chen, Q.; Chen, X.; Wang, X.; Liao, X.; Jiang, L.; Wu, J.; Yang, L. Occurrence and transfer of a cyanobacterial neurotoxin β-methylamino-l-alanine within the aquatic food webs of Gonghu Bay (Lake Taihu, China) to evaluate the potential human health risk. *Sci. Total Environ.* **2014**, *468*, 457–463. [CrossRef] [PubMed]

63. Manning, S.R.; La Claire II, J.W. Isolation of polyketides from *Prymnesium parvum* (Haptophyta) and their detection by liquid chromatography/mass spectrometry metabolic fingerprint analysis. *Anal. Biochem.* **2013**, *442*, 189–195. [CrossRef] [PubMed]

64. Metcalf, J.S.; Bell, S.G.; Codd, G.A. Colorimetric immuno-protein phosphatase inhibition assay for specific detection of microcystins and nodularins of cyanobacteria. *Appl. Environ. Microbiol.* **2001**, *67*, 904–909. [CrossRef]

65. Miller, M.A.; Kudela, R.M.; Mekebri, A.; Crane, D.; Oates, S.C.; Tinker, M.T.; Staedler, M.; Miller, W.A.; Toy-Choutka, S.; Dominik, C. Evidence for a novel marine harmful algal bloom: Cyanotoxin (microcystin) transfer from land to sea otters. *PLoS ONE* **2010**, *5*, e12576. [CrossRef]

66. Turner, A.D.; Waack, J.; Lewis, A.; Edwards, C.; Lawton, L. Development and single-laboratory validation of a UHPLC-MS/MS method for quantitation of microcystins and nodularin in natural water, cyanobacteria, shellfish and algal supplement tablet powders. *J. Chromatogr. B* **2018**, *1074*, 111–123. [CrossRef] [PubMed]

67. Galvao, J.A.; Oetterer, M.; do Carmo Bittencourt-Oliveira, M.; Gouvêa-Barros, S.; Hiller, S.; Erler, K.; Luckas, B.; Pinto, E.; Kujbida, P. Saxitoxins accumulation by freshwater tilapia (*Oreochromis niloticus*) for human consumption. *Toxicon* **2009**, *54*, 891–894. [CrossRef]

68. Lajeunesse, A.; Segura, P.A.; Gélinas, M.; Hudon, C.; Thomas, K.; Quilliam, M.A.; Gagnon, C. Detection and confirmation of saxitoxin analogues in freshwater benthic Lyngbya wollei algae collected in the St. Lawrence River (Canada) by liquid chromatography–tandem mass spectrometry. *J. Chromatogr. A* **2012**, *1219*, 93–103. [CrossRef]

69. Klijnstra, M.D.; Faassen, E.J.; Gerssen, A. A generic LC-HRMS screening method for marine and freshwater phycotoxins in fish, shellfish, water, and supplements. *Toxins* **2021**, *13*, 823. [CrossRef]

70. Pawlik-Skowrońska, B.; Kalinowska, R.; Skowroński, T. Cyanotoxin diversity and food web bioaccumulation in a reservoir with decreasing phosphorus concentrations and perennial cyanobacterial blooms. *Harmful Algae* **2013**, *28*, 118–125. [CrossRef]

71. Haddad, S.P.; Bobbitt, J.M.; Taylor, R.B.; Lovin, L.M.; Conkle, J.L.; Chambliss, C.K.; Brooks, B.W. Determination of microcystins, nodularin, anatoxin-a, cylindrospermopsin, and saxitoxin in water and fish tissue using isotope dilution liquid chromatography tandem mass spectrometry. *J. Chromatogr. A* **2019**, *1599*, 66–74. [CrossRef] [PubMed]

72. Carmichael, W.W.; An, J. Using an enzyme linked immunosorbent assay (ELISA) and a protein phosphatase inhibition assay (PPIA) for the detection of microcystins and nodularins. *Nat. Toxins* **1999**, *7*, 377–385. [CrossRef]

73. Lindner, P.; Molz, R.; Yacoub-George, E.; Dürkop, A.; Wolf, H. Development of a highly sensitive inhibition immunoassay for microcystin-LR. *Anal. Chim. Acta* **2004**, *521*, 37–44. [CrossRef]
74. Rapala, J.; Erkomaa, K.; Kukkonen, J.; Sivonen, K.; Lahti, K. Detection of microcystins with protein phosphatase inhibition assay, high-performance liquid chromatography–UV detection and enzyme-linked immunosorbent assay: Comparison of methods. *Anal. Chim. Acta* **2002**, *466*, 213–231. [CrossRef]
75. Bláhová, L.; Oravec, M.; Marsálek, B.; Sejnohová, L.; Simek, Z.; Bláha, L. The first occurrence of the cyanobacterial alkaloid toxin cylindrospermopsin in the Czech Republic as determined by immunochemical and LC/MS methods. *Toxicon* **2009**, *53*, 519–524. [CrossRef]
76. Campbell, K.; Huet, A.C.; Charlier, C.; Higgins, C.; Delahaut, P.; Elliott, C.T. Comparison of ELISA and SPR biosensor technology for the detection of paralytic shellfish poisoning toxins. *J. Chromatogr. B* **2009**, *877*, 4079–4089. [CrossRef]
77. Hardy, F.J.; Johnson, A.; Hamel, K.; Preece, E. Cyanotoxin bioaccumulation in freshwater fish, Washington State, USA. *Environ. Monit. Assess.* **2015**, *187*, 667. [CrossRef]
78. Suchy, P.; Berry, J. Detection of total microcystin in fish tissues based on lemieux oxidation and recovery of 2-methyl-3-methoxy-4-phenylbutanoic acid (MMPB) by solid-phase microextraction gas chromatography-mass spectrometry (SPME-GC/MS). *Int. J. Environ. Anal. Chem.* **2012**, *92*, 1443–1456. [CrossRef]
79. Kaushik, R.; Balasubramanian, R. Methods and approaches used for detection of cyanotoxins in environmental samples: A review. *Crit. Rev. Environ. Sci. Technol.* **2013**, *43*, 1349–1383. [CrossRef]
80. Spoof, L.; Neffling, M.R.; Meriluoto, J. Separation of microcystins and nodularins by ultra performance liquid chromatography. *J. Chromatogr. B* **2009**, *877*, 3822–3830. [CrossRef] [PubMed]
81. Dörr, F.A.; Rodríguez, V.; Molica, R.; Henriksen, P.; Krock, B.; Pinto, E. Methods for detection of anatoxin-a(s) by liquid chromatography coupled to electrospray ionization-tandem mass spectrometry. *Toxicon* **2010**, *55*, 92–99. [CrossRef] [PubMed]
82. Bogialli, S.; Bortolini, C.; Di Gangi, I.M.; Di Gregorio, F.N.; Lucentini, L.; Favaro, G.; Pastore, P. Liquid chromatography-high resolution mass spectrometric methods for the surveillance monitoring of cyanotoxins in freshwaters. *Talanta* **2017**, *170*, 322–330. [CrossRef] [PubMed]
83. Flores, C.; Caixach, J. High Levels of Anabaenopeptins Detected in a Cyanobacteria Bloom from NE Spanish Sau-Susqueda-El Pasteral Reservoirs System by LC–HRMS. *Toxins* **2020**, *12*, 541. [CrossRef]
84. Filatova, D.; Jones, M.R.; Haley, J.A.; Núñez, O.; Farré, M.; Janssen, E.M.-L. Cyanobacteria and their secondary metabolites in three freshwater reservoirs in the United Kingdom. *Environ. Sci. Eur.* **2021**, *33*, 29. [CrossRef]
85. Igarashi, T.; Satake, M.; Yasumoto, T. Prymnesin-2: A potent ichthyotoxic and hemolytic glycoside isolated from the red tide alga *Prymnesium parvum*. *J. Am. Chem. Soc.* **1996**, *118*, 479–480. [CrossRef]
86. Igarashi, T.; Aritake, S.; Yasumoto, T. Biological activities of prymnesin-2 isolated from a red tide alga *Prymnesium parvum*. *Nat. Toxins* **1998**, *6*, 35–41. [CrossRef]
87. Igarashi, T.; Satake, M.; Yasumoto, T. Structures and partial stereochemical assignments for prymnesin-1 and prymnesin-2: Potent hemolytic and ichthyotoxic glycosides isolated from the red tide alga *Prymnesium parvum*. *J. Am. Chem. Soc.* **1999**, *121*, 8499–8511. [CrossRef]
88. Morohashi, A.; Satake, M.; Oshima, Y.; Igarashi, T.; Yasumoto, T. Absolute configuration at C14 and C85 in prymnesin-2, a potent hemolytic and ichthyotoxic glycoside isolated from the red tide alga *Prymnesium parvum*. *Chirality* **2001**, *13*, 601–605. [CrossRef]
89. Rasmussen, S.A.; Meier, S.; Andersen, N.G.; Blossom, H.E.; Duus, J.; Nielsen, K.F.; Hansen, P.J.; Larsen, T.O. Chemodiversity of Ladder-Frame Prymnesin Polyethers in *Prymnesium parvum*. *J. Nat. Prod.* **2016**, *79*, 2250–2256. [CrossRef]
90. Blossom, H.E.; Rasmussen, S.A.; Andersen, N.G.; Larsen, T.O.; Nielsen, K.F.; Hansen, P.J. *Prymnesium parvum* revisited: Relationship between allelopathy, ichthyotoxicity, and chemical profiles in 5 strains. *Aquat. Toxicol.* **2014**, *157*, 159–166. [CrossRef] [PubMed]
91. Bertin, M.J.; Zimba, P.V.; Beauchesne, K.R.; Huncik, K.M.; Moeller, P.D. The contribution of fatty acid amides to *Prymnesium parvum* Carter toxicity. *Harmful Algae* **2012**, *20*, 117–125. [CrossRef]
92. Henrikson, J.C.; Gharfeh, M.S.; Easton, A.C.; Easton, J.D.; Glenn, K.L.; Shadfan, M.; Mooberry, S.L.; Hambright, K.D.; Cichewicz, R.H. Reassessing the ichthyotoxin profile of cultured *Prymnesium parvum* (golden algae) and comparing it to samples collected from recent freshwater bloom and fish kill events in North America. *Toxicon* **2010**, *55*, 1396–1404. [CrossRef] [PubMed]
93. Schug, K.A.; Skingel, T.R.; Spencer, S.E.; Serrano, C.A.; Le, C.Q.; Schug, C.A.; Valenti, T.W., Jr.; Brooks, B.W.; Mydlarz, L.D.; Grover, J.P. Hemolysis, Fish Mortality, and LC-ESI-MS of Cultured Crude and Fractionated Golden Alga (*Prymnesium parvum*) 1. *JAWRA J. Am. Water Resour. Assoc.* **2010**, *46*, 33–44. [CrossRef]
94. Merel, S.; LeBot, B.; Clément, M.; Seux, R.; Thomas, O. Ms identification of microcystin-LR chlorination by-products. *Chemosphere* **2009**, *74*, 832–839. [CrossRef]
95. Merel, S.; Clément, M.; Mourot, A.; Fessard, V.; Thomas, O. Characterization of cylindrospermopsin chlorination. *Sci. Total Environ.* **2010**, *408*, 3433–3442. [CrossRef]
96. Adamovský, O.; Kopp, R.; Hilscherová, K.; Babica, P.; Palíková, M.; Pašková, V.; Navrátil, S.; Maršálek, B.; Bláha, L. Microcystin kinetics (bioaccumulation and elimination) and biochemical responses in common carp (*Cyprinus carpio*) and silver carp (Hypophthalmichthys molitrix) exposed to toxic cyanobacterial blooms. *Environ. Toxicol. Chem. Int. J.* **2007**, *26*, 2687–2693. [CrossRef]

97. Gurbuz, F.; Uzunmehmetoğlu, O.Y.; Diler, Ö.; Metcalf, J.S.; Codd, G.A. Occurrence of microcystins in water, bloom, sediment and fish from a public water supply. *Sci. Total Environ.* **2016**, *562*, 860–868. [CrossRef]

98. Diener, M.; Erler, K.; Hiller, S.; Christian, B.; Luckas, B. Determination of paralytic shellfish poisoning (PSP) toxins in dietary supplements by application of a new HPLC/FD method. *Eur. Food Res. Technol.* **2006**, *224*, 147–151. [CrossRef]

99. Lebo, J.A.; Smith, L.M. Determination of fluorene in fish, sediment, and plants. *J. Assoc. Off. Anal. Chem.* **1986**, *69*, 944–951. [CrossRef]

100. Almeida, V.P.S.; Cogo, K.; Tsai, S.M.; Moon, D.H. Colorimetric test for the monitoring of microcystins in cyanobacterial culture and environmental samples from southeast—Brazil. *Braz. J. Microbiol.* **2006**, *37*, 192–198. [CrossRef]

101. Bouaícha, N.; Maatouk, I.; Vincent, G.; Levi, Y. A colorimetric and fluorometric microplate assay for the detection of microcystin-LR in drinking water without preconcentration. *Food Chem. Toxicol.* **2002**, *40*, 1677–1683. [CrossRef]

102. Heresztyn, T.; Nicholson, B.C. Determination of cyanobacterial hepatotoxins directly in water using a protein phosphatase inhibition assay. *Water Res.* **2001**, *35*, 3049–3056. [CrossRef]

103. Ortea, P.M.; Allis, O.; Healy, B.M.; Lehane, M.; Shuilleabháin, A.N.; Furey, A.; James, K.J. Determination of toxic cyclic heptapeptides by liquid chromatography with detection using ultra-violet, protein phosphatase assay and tandem mass spectrometry. *Chemosphere* **2004**, *55*, 1395–1402. [CrossRef] [PubMed]

104. Ward, C.J.; Beattie, K.A.; Lee, E.Y.; Codd, G.A. Colorimetric protein phosphatase inhibition assay of laboratory strains and natural blooms of cyanobacteria: Comparisons with high-performance liquid chromatographic analysis for microcystins. *FEMS Microbiol. Lett.* **1997**, *153*, 465–473. [CrossRef] [PubMed]

105. Mazur-Marzec, H.; Tymińska, A.; Szafranek, J.; Pliński, M. Accumulation of nodularin in sediments, mussels, and fish from the Gulf of Gdańsk, southern Baltic Sea. *Environ. Toxicol. Int. J.* **2007**, *22*, 101–111. [CrossRef]

106. Berry, J.P.; Lee, E.; Walton, K.; Wilson, A.E.; Bernal-Brooks, F. Bioaccumulation of microcystins by fish associated with a persistent cyanobacterial bloom in Lago de Patzcuaro (Michoacan, Mexico). *Environ. Toxicol. Chem.* **2011**, *30*, 1621–1628. [CrossRef]

107. Mekebri, A.; Blondina, G.; Crane, D. Method validation of microcystins in water and tissue by enhanced liquid chromatography tandem mass spectrometry. *J. Chromatogr. A* **2009**, *1216*, 3147–3155. [CrossRef]

108. Kopp, R.; Palíková, M.; Adamovský, O.; Ziková, A.; Navrátil, S.; Kohoutek, J.; Mareš, J.; Bláha, L. Concentrations of microcystins in tissues of several fish species from freshwater reservoirs and ponds. *Environ. Monit. Assess.* **2013**, *185*, 9717–9727. [CrossRef]

109. Wituszynski, D.M. Variation of Microcystin concentrations in fish related to algae blooms in Lake Erie, and public health impacts. Ph.D. Thesis, The Ohio State University, Columbus OH, USA, 2014.

110. Sipiä, V.; Kankaanpää, H.; Pflugmacher, S.; Flinkman, J.; Furey, A.; James, K. Bioaccumulation and detoxication of nodularin in tissues of flounder (*Platichthys flesus*), mussels (*Mytilus edulis, Dreissena polymorpha*), and clams (*Macoma balthica*) from the northern Baltic Sea. *Ecotoxicol. Environ. Saf.* **2002**, *53*, 305–311. [CrossRef]

111. Messineo, V.; Melchiorre, S.; Di Corcia, A.; Gallo, P.; Bruno, M. Seasonal succession of Cylindrospermopsis raciborskii and Aphanizomenon ovalisporum blooms with cylindrospermopsin occurrence in the volcanic Lake Albano, Central Italy. *Environ. Toxicol. Int. J.* **2010**, *25*, 18–27.

112. Saker, M.L.; Eaglesham, G.K. The accumulation of cylindrospermopsin from the cyanobacterium Cylindrospermopsis raciborskii in tissues of the Redclaw crayfish Cherax quadricarinatus. *Toxicon* **1999**, *37*, 1065–1077. [CrossRef]

113. Saker, M.L.; Metcalf, J.S.; Codd, G.A.; Vasconcelos, V.M. Accumulation and depuration of the cyanobacterial toxin cylindrospermopsin in the freshwater mussel Anodonta cygnea. *Toxicon* **2004**, *43*, 185–194. [CrossRef] [PubMed]

114. Skafi, M.; Duy, S.V.; Munoz, G.; Dinh, Q.T.; Simon, D.F.; Juneau, P.; Sauvé, S. Occurrence of microcystins, anabaenopeptins and other cyanotoxins in fish from a freshwater wildlife reserve impacted by harmful cyanobacterial blooms. *Toxicon* **2021**, *194*, 44–52. [CrossRef]

115. Mondo, K.; Hammerschlag, N.; Basile, M.; Pablo, J.; Banack, S.A.; Mash, D.C. Cyanobacterial neurotoxin β-N-methylamino-L-alanine (BMAA) in shark fins. *Mar. Drugs* **2012**, *10*, 509–520. [CrossRef]

116. Al-Sammak, M.A.; Hoagland, K.D.; Cassada, D.; Snow, D.D. Co-occurrence of the cyanotoxins BMAA, DABA and anatoxin-a in Nebraska reservoirs, fish, and aquatic plants. *Toxins* **2014**, *6*, 488–508. [CrossRef]

117. Lawrence, J.F.; Niedzwiadek, B.; Menard, C. Quantitative determination of paralytic shellfish poisoning toxins in shellfish using prechromatographic oxidation and liquid chromatography with fluorescence detection: Collaborative study. *J. AOAC Int.* **2005**, *88*, 1714–1732. [CrossRef]

118. Kodama, M. Ecobiology, classification, and origin. In *Seafood and Freshwater Toxins: Pharmacology, Physiology, and Detection*; Marcel Dekker, Inc.: New York, NY, USA, 2000; pp. 125–150.

119. Rey, V.; Alfonso, A.; Botana, L.M.; Botana, A.M. Influence of different shellfish matrices on the separation of PSP toxins using a postcolumn oxidation liquid chromatography method. *Toxins* **2015**, *7*, 1324–1340. [CrossRef]

120. Garcia, A.C.; Bargu, S.; Dash, P.; Rabalais, N.N.; Sutor, M.; Morrison, W.; Walker, N.D. Evaluating the potential risk of microcystins to blue crab (*Callinectes sapidus*) fisheries and human health in a eutrophic estuary. *Harmful Algae* **2010**, *9*, 134–143. [CrossRef]

121. Magalhaes, V.d.; Marinho, M.M.; Domingos, P.; Oliveira, A.; Costa, S.M.; Azevedo, L.O.d.; Azevedo, S.M. Microcystins (cyanobacteria hepatotoxins) bioaccumulation in fish and crustaceans from Sepetiba Bay (Brasil, RJ). *Toxicon* **2003**, *42*, 289–295. [CrossRef]

122. Mitsoura, A.; Kagalou, I.; Papaioannou, N.; Berillis, P.; Mente, E.; Papadimitriou, T. The presence of microcystins in fish *Cyprinus carpio* tissues: A histopathological study. *Int. Aquat. Res.* **2013**, *5*, 8. [CrossRef]

123. De Magalhães, V.F.; Soares, R.M.; Azevedo, S.M. Microcystin contamination in fish from the Jacarepaguá Lagoon (Rio de Janeiro, Brazil): Ecological implication and human health risk. *Toxicon* **2001**, *39*, 1077–1085. [CrossRef]

124. Wilson, A.E.; Gossiaux, D.C.; Höök, T.O.; Berry, J.P.; Landrum, P.F.; Dyble, J.; Guildford, S.J. Evaluation of the human health threat associated with the hepatotoxin microcystin in the muscle and liver tissues of yellow perch (*Perca flavescens*). *Can. J. Fish. Aquat. Sci.* **2008**, *65*, 1487–1497. [CrossRef]

125. Sipiä, V.; Kankaanpää, H.; Lahti, K.; Carmichael, W.W.; Meriluoto, J. Detection of nodularin in flounders and cod from the Baltic Sea. *Environ. Toxicol. Int. J.* **2001**, *16*, 121–126. [CrossRef]

126. Blossom, H.E.; Andersen, N.G.; Rasmussen, S.A.; Hansen, P.J. Stability of the intra-and extracellular toxins of *Prymnesium parvum* using a microalgal bioassay. *Harmful Algae* **2014**, *32*, 11–21. [CrossRef]

127. Johansson, N.; Granéli, E. Influence of different nutrient conditions on cell density, chemical composition and toxicity of *Prymnesium parvum* (Haptophyta) in semi-continuous cultures. *J. Exp. Mar. Biol. Ecol.* **1999**, *239*, 243–258. [CrossRef]

128. Park, H.D.; Kim, B.; Kim, E.; Okino, T. Hepatotoxic microcystins and neurotoxic anatoxin-a in cyanobacterial blooms from Korean lakes. *Environ. Toxicol. Water Qual. Int. J.* **1998**, *13*, 225–234. [CrossRef]

129. Brown, A.; Foss, A.; Miller, M.A.; Gibson, Q. Detection of cyanotoxins (microcystins/nodularins) in livers from estuarine and coastal bottlenose dolphins (*Tursiops truncatus*) from Northeast Florida. *Harmful Algae* **2018**, *76*, 22–34. [CrossRef]

130. Hammoud, N.A.; Zervou, S.-K.; Kaloudis, T.; Christophoridis, C.; Paraskevopoulou, A.; Triantis, T.M.; Slim, K.; Szpunar, J.; Fadel, A.; Lobinski, R. Investigation of the Occurrence of Cyanotoxins in Lake Karaoun (Lebanon) by Mass Spectrometry, Bioassays and Molecular Methods. *Toxins* **2021**, *13*, 716. [CrossRef]

131. Maliaka, V.; Lürling, M.; Fritz, C.; Verstijnen, Y.J.; Faassen, E.J.; Van Oosterhout, F.; Smolders, A.J. Interannual and spatial variability of cyanotoxins in the Prespa lake area, Greece. *Water* **2021**, *13*, 357. [CrossRef]

132. Kaas, H.; Henriksen, P. Saxitoxins (PSP toxins) in Danish lakes. *Water Res.* **2000**, *34*, 2089–2097. [CrossRef]

133. Da Silva, C.A.; Oba, E.T.; Ramsdorf, W.A.; Magalhães, V.F.; Cestari, M.M.; Ribeiro, C.A.O.; de Assis, H.C.S. First report about saxitoxins in freshwater fish Hoplias malabaricus through trophic exposure. *Toxicon* **2011**, *57*, 141–147. [CrossRef] [PubMed]

134. Al-Sammak, M.A.; Hoagland, K.D.; Snow, D.D.; Cassada, D. Methods for simultaneous detection of the cyanotoxins BMAA, DABA, and anatoxin-a in environmental samples. *Toxicon* **2013**, *76*, 316–325. [CrossRef] [PubMed]

135. Drobac, D.; Tokodi, N.; Lujić, J.; Marinović, Z.; Subakov-Simić, G.; Dulić, T.; Važić, T.; Nybom, S.; Meriluoto, J.; Codd, G.A. Cyanobacteria and cyanotoxins in fishponds and their effects on fish tissue. *Harmful Algae* **2016**, *55*, 66–76. [CrossRef] [PubMed]

Review

Anabaenopeptins: What We Know So Far

Patrick Romano Monteiro [1,2,*], **Samuel Cavalcante do Amaral** [1], **Andrei Santos Siqueira** [2], **Luciana Pereira Xavier** [1] **and Agenor Valadares Santos** [1,*]

[1] Laboratory of Biotechnology of Enzymes and Biotransformation, Biological Sciences Institute, Federal University of Pará, Belém 66075-110, Brazil; samuel.amaral@icb.ufpa.br (S.C.d.A.); lpxavier@ufpa.br (L.P.X.)

[2] Laboratory of Biomolecular Technology, Biological Sciences Institute, Federal University of Pará, Belém 66075-110, Brazil; andrei.siqueira@icb.ufpa.br

* Correspondence: patrick.monteiro@icb.ufpa.br (P.R.M.); avsantos@ufpa.br (A.V.S.); Tel.: +55-91-98086-9433 (P.R.M.); +55-91-99177-3164 (A.V.S.)

Abstract: Cyanobacteria are microorganisms with photosynthetic mechanisms capable of colonizing several distinct environments worldwide. They can produce a vast spectrum of bioactive compounds with different properties, resulting in an improved adaptive capacity. Their richness in secondary metabolites is related to their unique and diverse metabolic apparatus, such as Non-Ribosomal Peptide Synthetases (NRPSs). One important class of peptides produced by the non-ribosomal pathway is anabaenopeptins. These cyclic hexapeptides demonstrated inhibitory activity towards phosphatases and proteases, which could be related to their toxicity and adaptiveness against zooplankters and crustaceans. Thus, this review aims to identify key features related to anabaenopeptins, including the diversity of their structure, occurrence, the biosynthetic steps for their production, ecological roles, and biotechnological applications.

Keywords: cyanobacteria; peptide; NRPS; anabaenopeptin

Key Contribution: The present work approach various features of Anabaenopeptins, including structural, biosynthetic and regulatory aspects as well as their ecological and biotechnological importance.

Citation: Monteiro, P.R.; do Amaral, S.C.; Siqueira, A.S.; Xavier, L.P.; Santos, A.V. Anabaenopeptins: What We Know So Far. *Toxins* **2021**, *13*, 522. https://doi.org/10.3390/toxins13080522

Received: 29 March 2021
Accepted: 25 May 2021
Published: 27 July 2021

Publisher's Note: MDPI stays neutral with regard to jurisdictional claims in published maps and institutional affiliations.

1. Introduction

Cyanobacteria are photosynthetic microorganisms widely distributed in the world. They can inhabit several types of ecosystems, including aquatic and terrestrial. These microorganisms produce a great variety of bioactive compounds, which have been investigated mainly due to their biotechnological potential and environmental relevance [1–3]. Cyanotoxins are among the most studied compounds originated from cyanobacteria since they are capable of negatively affecting human and animal health [4,5]. These metabolites can vary drastically concerning their action mechanism and chemical structure, which include peptides, alkaloids, and lipopolysaccharides [6–8]. The majority of publications related to peptides from cyanobacteria have mainly focused on the class of microcystins with over 300 characterized variants [9,10]. However, cyanobacteria usually do not exclusively produce a single class of compounds, given that specific strains are co-producing different groups of secondary metabolites [11].

Other peptides beyond microcystins have been poorly explored, lacking information mainly in the environmental sciences [11]. These several metabolites are known for their potent inhibitory properties against several enzymes in nanomolar concentrations, resulting in toxic effects [12,13]. Moreover, similar to microcystins, they have been regularly detected in diverse environments [14]. In certain regions, their occurrence is more pronounceable than microcystins themselves [15]. However, information about the concentrations which are encountered is rarely reported [11]. Cyanobacteria have developed different peptides as a protection mechanism against parasites [16]. Concerning their origin, some peptides as microviridins and cyanobactins are produced via ribosomal whereas others as microginins and aeruginosins are synthesized by non-ribosomal pathways [13,17,18].

Among the most recurrent peptides encountered in the environment are anabaenopeptins (APs), a family of cyclic peptides containing six amino acid residues [19]. They have been found in an enormous variety of cyanobacteria isolated from both the aquatic and terrestrial environments, including *Anabaena, Nostoc, Microcystis, Planktothrix, Lyngbya,* and *Brasilonema* [12,20–24]. In their general structure is a well-conserved Lysine (Lys) residue in D-configuration, which is responsible for the ring formation and five additional variable amino acids, either proteinogenic or non-proteinogenic, resulting in 124 described AP variants from cyanobacteria (Supplementary Table S1). [19]. Besides their structural variety, molecules belonged to this group exhibit an impressive functional diversity, which includes inhibitory activity for proteases, phosphatases, and carboxypeptidases [22,25,26].

The enormous structural diversification of anabaenopeptins can be attributed to the low substrate specificity of some enzymes involved in their synthesis as well as the presence of alternative starter modules [16]. Their production is strongly influenced by environmental factors [27]. Besides that, because of their diversified bioactive properties, they exhibit an elevated biotechnological potential. This review aims at presenting the main researches on anabaenopeptins, emphasizing their general characteristics, biosynthesis as well as ecological and biotechnological relevance.

2. Structures of Anabaenopeptins

Being non-ribosomally synthesized, anabaenopeptin structures comprise a ring of five amino acids connected through an ureido linkage to an exocyclic amino acid. Thus, its general structure is represented by X^1-CO-[Lys^2-X^3-X^4-MeX^5-X^6], where the bracket represents the cyclic region of this peptide and X are variable amino acids according to their positions represented by the superscript numbers (Figure 1). Its ring is formed by cyclization of the C-terminal carboxyl of the amino acid at position 6 to the ε-NH$_3$ of the well-conserved D-lysine at position 2. Furthermore, the α-amino group of Lys is connected to the exocyclic amino acid X^1 via an ureido bridge. Due to its non-ribosomal nature, proteinogenic and non-proteinogenic amino acids are usually detected in this hexapeptide [19].

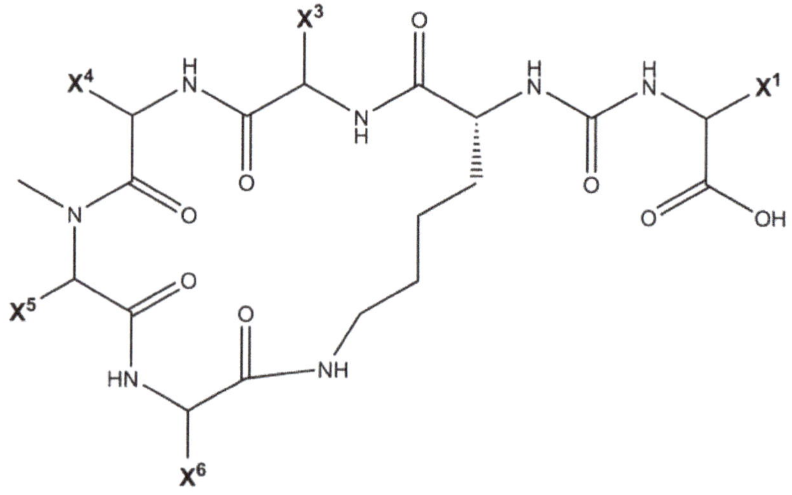

Figure 1. The general structure of the class of Anabaenopeptins. X corresponds to different amino acids in their respective positions represented by the superscript numbers.

This family of peptides is predominantly found in cyanobacteria, but they were also detected in some sponges [28–33]. However, Anabaenopeptins from cyanobacterial origins demonstrate a well-conserved D-Lys, while the other amino acids are in L configuration and vary in residues and modifications (e.g., acetylation, methylation) [19,34]. In comparison, Anabaenopeptins derived from sponges can have both D- and L-configuration of Lys residues at the second position. Besides these differences, some features in common are also encountered, such as the frequently N-methylated amino acids at position 5 and homo-amino acids at position 4. However, exceptions are also found, for example, Paltolides A–C, a subgroup of anabaenopeptin-like peptides that have in common a tryptophan residue at the C-terminus linked to ε-amine of the N-terminal Lys, and Leucine (Leu) in L-configuration at positions 4 and 5, and an L-Alanine (Ala) residue at position 3. Furthermore, Paltolide A is the first example of this class of peptides where the amino acid at position 5 lacks an N-methyl group [28–33].

The first Anabaenopeptins detected were Anabaenopeptins A and B (Figure 2; Supplementary Table S1) by Harada and co-workers in 1995 [20]. These peptides were isolated from *Anabaena flos-aquae* NRC 525-17, where they were co-produced with Microcystins (MCs) and the neurotoxic alkaloid anatoxin-A. Both peptides share the same cyclic sequence and structure, differing only at the exocyclic position: (Arginine/Tyrosine)-Lys-Valine-Homotyrosine-NMethylAla-Phenylalanine. Due to their origin, these peptides were then named after their producer. Following the first detection of this new peptide, Fujii and co-workers [35] identified Anabaenopeptins A-D from different *Anabaena* and *Oscillatoria* strains, as well as from *Nodularia spumigena*. APs C and D differ solely by the exocyclic amino acid, harboring a Lys and a Phenylalanine (Phe), respectively, and sharing the same pentapeptide ring with APs A and B. Furthermore, in the same year, Sano and Kaya [36] identified a peptide named Oscillamide Y, which was obtained from *Oscillatoria agardhii* NIES-610. Following the same nomenclature, Sano and colleagues [25] further characterized both Oscillamide B and C (known as Anabaenopeptin F) from *Planktothrix agardhii* CCAP 1459/11A and *P. rubescens* CCAP 1459/14. Besides their different nomenclature, Oscillamides peptides also possess the same common features of anabaenopeptin-peptides.

The cyanobacteria *Oscillatoria agardhii* NIES-204 had been assessed regarding AP production by two different research studies. During the first approach, only Anabaenopeptin B had been detected [37]. Later, Shin and co-workers [38] were able to characterize two new structures from the same organism: Anabaenopeptins E and F, which differ at those residues in positions 3 and 4. Later, two new AP structures were identified in *O. agardhii* NIES-595, then named Anabaenopeptins G and H, diverging by Tyrosine (Tyr) and Arginine (Arg) in position 1, respectively: (Tyr/Arg)-Lys-Isoleucine-Homotyrosine-NMethylhomotyrosine-Isoleucine [26].

The first unicellular cyanobacterium strain to be identified as an Anabaenopeptin producer was *Microcystis aeruginosa* Kutz. This freshwater strain was able to biosynthesize the anabaenopeptin-type Ferintoic Acids A and B [39]. Another *M. aeruginosa* strain and an environmental sample containing mostly *Microcystis* cells demonstrated to contain the Non-Ribosomal Peptide Synthetase (NRPS) apparatus for Anabaenopeptins B and F production. Also, the same work concluded that the filamentous cyanobacteria *Planktothrix agardhii* HUB011 produced the Anabaenopeptin G [40].

Kodani and co-workers [41] evaluated the presence of anabaenopeptins in an environmental sample from Lake Teganuma. Besides the identification of microginins and micropeptins, a newly found AP was characterized: Anabaenopeptin T. However, this nomenclature did not follow any specific order, as Anabaenopeptin I and J had been only used for the new peptides obtained from *Aphanizomenon flos-aquae* NIES-81 and identified by Muramaki and co-workers [42], one year later from this previous work.

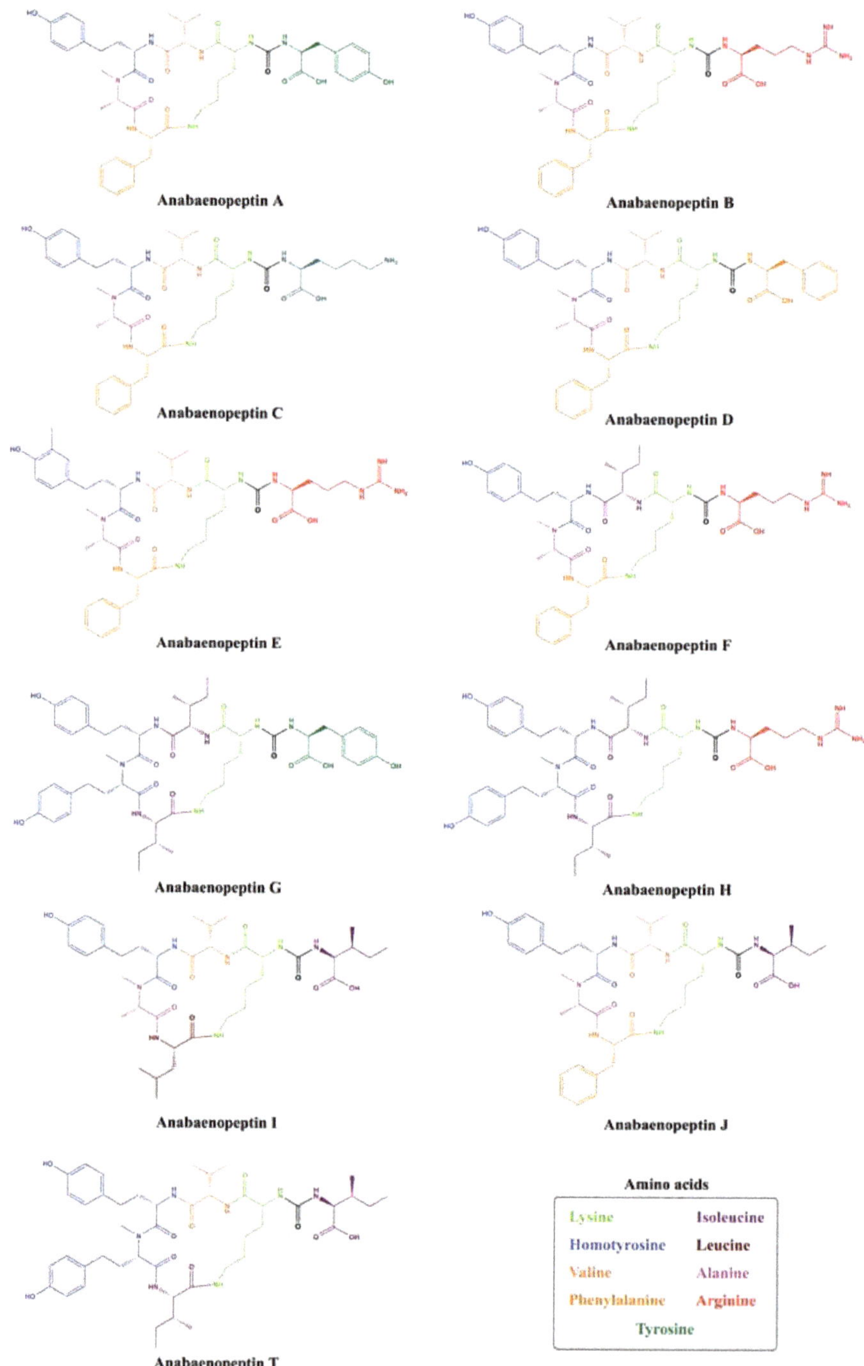

Figure 2. Structures of anabaenopeptins A–J [20,26,35,38,42] and T [41].

Some of those conserved features from APs can also be visualized in other cyanopeptides. Veraguamides A-G are cyclic hexadepsipeptides, and they do not possess any exocyclic residue. Lyngbyastatin peptides demonstrate elastase, trypsin, and chymotrypsin inhibitory properties. Their structures consist of a 6-member ring coupled to a chain of 2 exocyclic residues and can bear modified and unusual residues. Also possessing a 6-member ring structure and 2 exocyclic amino acids, Tiglicamides A-C were obtained from *Lyngbya confervoides* [43].

In addition, there are several classes of toxic peptides frequently detected in cyanobacteria, each one presents the main structure codified by a set of NRPS genes. Besides Anabaenopeptins, Microcystins, Cyanopeptolins, Aerucyclamides, Aeruginosis and Microginins are examples of well-characterized cyanopeptides. Microcystins also share some resemblances to APs as the former possesses a ring structure, but it is comprised of 7 residues and no exocyclic amino acid. Also produced by the NRPSs apparatus, MCs bear D-amino acids, and the unusual residue (2S,3S,8S,9S)-3-amino-9-methoxy-2,6,8-trimethyl-10-phenyldeca-4,6-dienoic acid, also known as Adda, which is present in all MCs. One example of MC variant is Microcystin-LR, which is composed by (1)-Ala, (2)-Leu, (3)-N-methyl-Asp, (4)-Arg, (5)-Adda, (6)-Glu, and (7)-N-methyl-dehydro-Ala, where LR refers to 2 and 4 variable positions among MCs. It was observed in different cyanopeptides inside this class that there are different variants according to these positions, contributing to their structural diversity [10,11].

Cyanopeptolins are depsipeptides containing a 6-amino acid ring bearing a side chain with 1–2 residues and modified residues, such as 3-amino-6-hydroxy-2-piperidone. Cyanopeptolin A is one example of this class of cyanopeptides and is composed by (1)-fatty acid, (2)-Arg, (3)-Ahp, (4)-Leu, (5)-methyl-Phe, (6)-Val, and (7)-Thr, in this case, the β-lacton ring is formed between Arg and Thr residues and positions 2, 4, 5 and 6 are variable. Using Anabaenopeptin A as reference (Figure 2), its structure is (1)-Tyr, (2)-D-Lys, (3)-Valine, (4)-Homotyrosine, (5)N-methyl-Alanine, (6)-Phenylalanine [44]. Positions 1, 3, 4, 5, and 6 are variable concerning APs (Figure 1) and the ureido bond is formed between 1 and 2 residues. Aerucyclamides are entirely cyclic peptides, Aerucyclamide A is composed by (1)-dehydro-Thr, (2)-Gly, (3)-thiozole, (4)-Ile, (5)-dhCys, and (6)-Ile, in this case, variations were reported in positions 2, 3, 4 and 6. Different from the cyanopeptides listed until now, Aeruginosins and Microginins are linear peptides. Aeruginosin KB 676 is formed by (1)-Hpla, (2)-Ile, (3)-Choi and (4)-Arg, only position 2 presents variation with amino acid substitution, and radical changes occur in positions 3 and 4. Finally, Microginin 713 is formed by (1)-Ahda, (2)-Ala, (3)-Val, (4)-N-methyl-Tyr, and (5)-Tyr, in this case, positions 2, 3, 4, and 5 had substitutions reported [11].

Structurally, despite the major amino acid variability, Microcystins, Cyanopeptolins, and Anabaenopeptins are most similar. Microcystins and Cyanopeptolins are heptapeptides and Anabaenopeptins are hexapeptides and comparting these structures, it is possible to distinguish a ring core and a linear region. Although Microcystins are technically cyclic peptides, the Adda moiety projected outside the ring may act like the fatty acid in Cyanopeptolin A or Tyrosine in Anabaenopeptin A. The Adda moiety is crucial for MCs inhibition towards phosphatases, as its long linear chain can penetrate the enzyme active site together with other side chains, having a similar role as the exocyclic residue of APs (Tyrosine from Anabaenopeptin A), as it will be further discussed in Section 7 [11,45]. This exocyclic or even protuberant residue was not observed in Aerucyclamides that only present a cyclic structure or Aeruginosins and Microginins, which are linear structures [11]. Therefore, cyclic peptides bearing exocyclic residues and unusual and D-configuration amino acids are also found in cyanobacteria, however, the ureido linkage in cyanopeptides is, so far, an exclusive characteristic of Anabaenopeptins.

All these AP variants named here so far are structurally related, differing only by amino acid substitutions responsible for their diversity (Supplementary Table S1). However, these peptides do not possess a fully systematic nomenclature, which can make it difficult to identify them as a member of a certain group of oligopeptides with similar structure. This fact is not specific to Anabaenopeptins, but cyanopeptides in general, as their denominations are frequently referring to the taxon or geographic locality from which the oligopeptide had been isolated, and also information regarding molecular weight, specific residues, or even the strain number can be used as a suffix, and some example can be seen applied to APs [11]. One example of a variant with a distinct name is the Schizopeptin 791 (Figure 3), which was named after the terrestrial cyanobacteria *Schizothrix* sp. IL-208-2-2 (Schizo-), its peptide nature (-peptin) and its molecular weight of 791 Da (791) [46]. Lyngbyaureidamides A and B are Anabaenopeptins named after their isolation from the filamentous freshwater cyanobacterium *Lyngbya* sp. SAG 36.91. These anabaenopeptin-like peptides also have an uncommon feature due to the presence of a D-Phenylalanine in the exocyclic position, being the only APs bearing an amino acid in D-configuration in this position [47]. Obtained from the marine *Lyngbya confervoides*, Pompanopeptin B is an anabaenopeptin-type peptide bearing in the fifth position the *N*-methyl-2-amino-6-(4′-hydroxyphenyl)hexanoic acid (N-Me-Ahpha), a methylated form of a residue found in Largamide C [23]. Nodulapeptins are also anabaenopeptin-like peptides and they were first identified by Fujii and co-workers [48] in the toxic *Nodularia spumigena* AV1. Among the different nomenclature of this class of cyclic hexapeptide, Nodulapeptin is one of the most used and it is often associated with the presence of Methionine (Met) or Serine (Ser) residues in position 6 of anabaenopeptin-like structures [49]. Isolated from the cyanobacteria *Tychonema* sp., Brunsvicamides A-C share a high resemblance to anabaenopeptin-like peptides obtained from sponges, thus indicating their possible cyanobacterial origin. These peptides obtained from a *Tychonema* sp. strain did not possess any homoamino acid and have a L-Lys besides D-Lys, in addition, Brunsvicamide C has an *N*-methyl-N′-formyl-D-kynurenine unit in position 5 [50].

Besides these distinct nomenclatures and structures for Anabaenopeptins obtained from cyanobacteria, this class of peptides can also be found in sponges, which were the initial organisms to be identified the first anabaenopeptin-related compound, not in a cyanobacterium [31,32]. Konbamide and Keramide A (Table 1 and Figure 4) were isolated from the marine sponge *Theonella* sp., which showed distinct features from cyanobacterial anabaenopeptins having a cyclic hexapeptide structure and the presence of an ureido bond. Both variants have L-Lys residue and also they contain a modified Tryptophan (Trp) residue at position 6. Konbamide had 2-bromo-5-hydroxytryptophan (2′Br-Trp) in position 6; in comparison, Keramide A possessed a 6-chloro-5-hydroxy-*N*-methyltryptophan (5′OH-6′ClTrp) in position 5 [31,32]. Keramide L was detected in *Theonella* sp. SS-342 together with Keramide K (a thiazole-containing cyclic peptide not belonging to anabaenopeptin-class). Keramide L shared similar features to Konbamide and Keramide A, having a modified Trp residue in position 5: a 6-chloro-*N*-methyltryptophan (NMe-6′ClTrp) residue [30]. Besides, the marine sponge *Theonella swinhoei* demonstrated to produce a similar class of peptides, named Paltolides A-C, and 3 additional peptides which were previously detected in the Australian sponge *Melophlus* sp. [51]. Unlike Konbamide and Keramide A, Paltolides A-C and the 3 unnamed peptides have a D-Lys residue in position 2, but 4 of these structures had modified Trp residues, and 3 of these modifications were also related to the presence of a halogen element (Bromo or Chlorine) similar to Konbamide and Keramides A and L [28]. Other examples from sponges are Mozamides A and B produced by *Theonella* sp. from Mozambique. Similar to other anabaenopeptin-like peptides from sponges, both Mozamides have the same modified Trp (*N*-methyl-L-5′-hydroxytryptophan) residue in position 5 and an L-Lysine at position 2. However, both have an amino acid in D-configuration in position 3: D-valine in Mozamide A; and D-Isoleucine in Mozamide B [29].

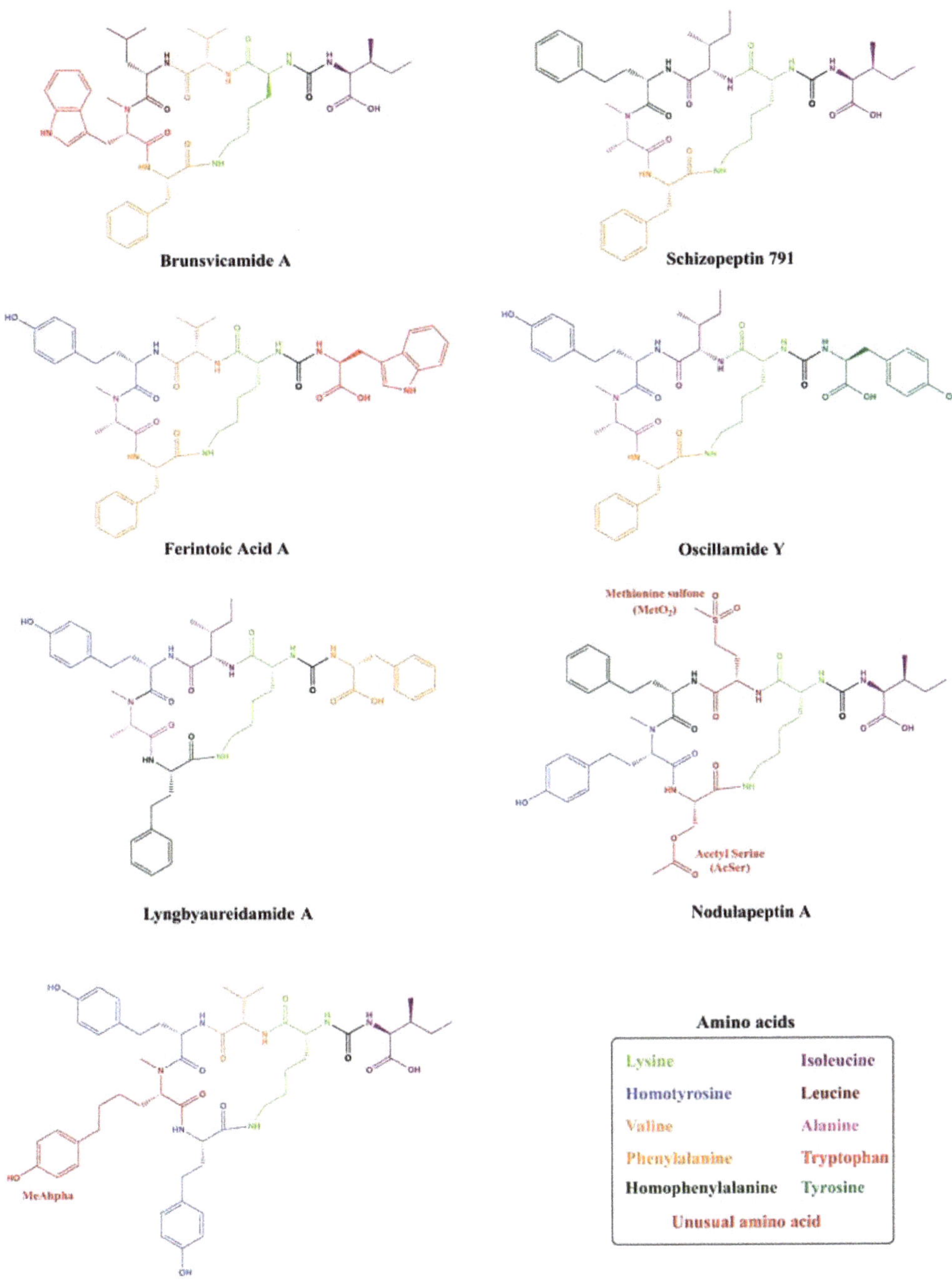

Figure 3. Example of different nomenclatures to anabaenopeptin-like structures [20,23,39,46–48,50].

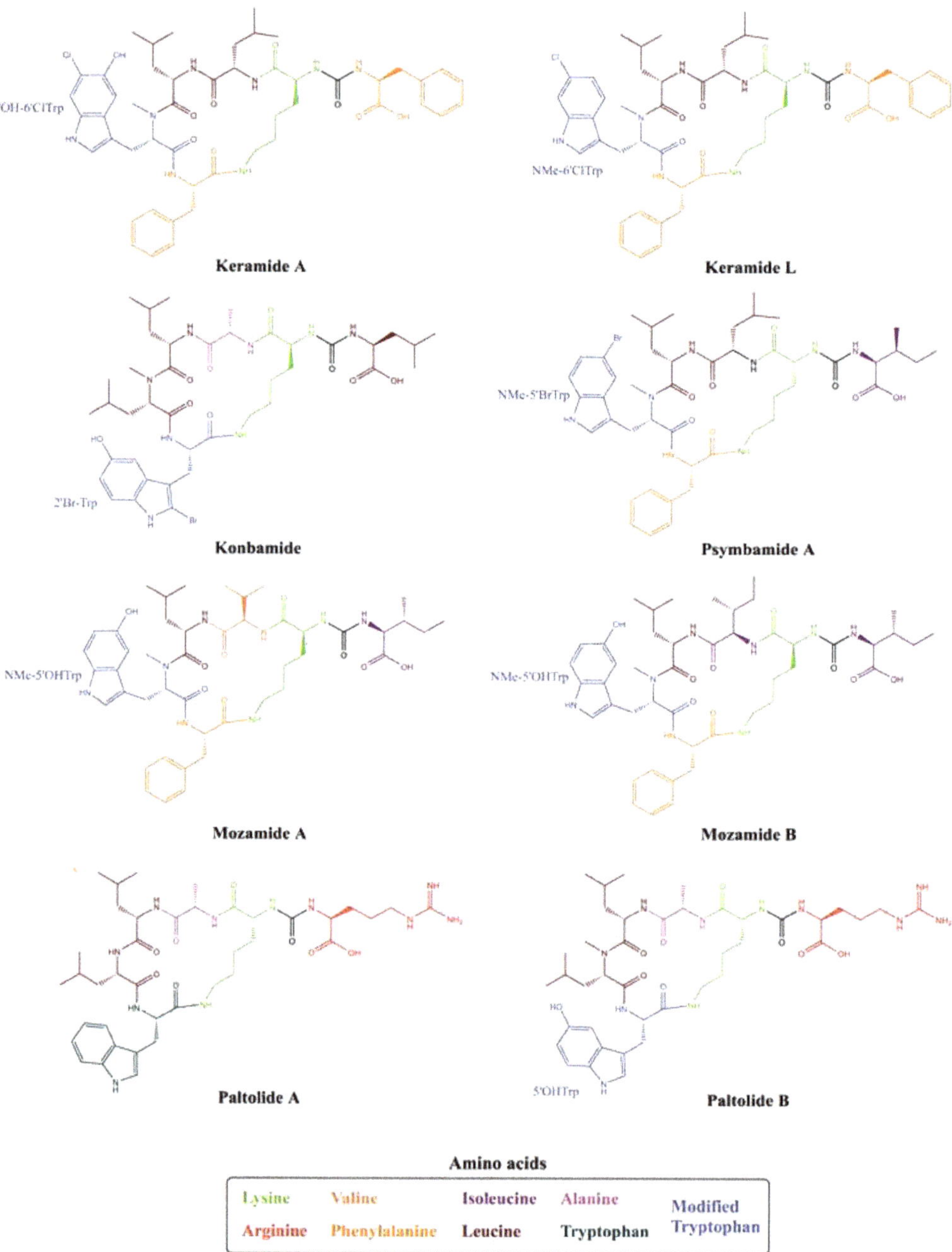

Figure 4. Structures of anabaenopeptin-like peptides obtained from sponges [28–33].

Table 1. Amino acid composition of anabaenopeptin-like peptides obtained from sponges. Amino acids are considered in L-configuration unless otherwise defined. Ala: Alanine; Arg: Arginine; Ile: Isoleucine; Leu: Leucine; Lys: Lysine; MeLeu: N-methyl-Leucine; Phe: Phenylalanine; Trp: Tryptophan; Allo-Ile: Allo-Isoleucine; 2′BrTrp: 2-bromo-5-hydroxytryptophan; NMe-6′ClTrp: 6-chloro-N-methyltryptophan; 5′OHTrp: 5-hydroxytryptophan; 6′BrTrp: 6′-bromotryptophan; 5′OH-6′Cl Trp: 6′-chloro-5-hydroxytryptophan; 6′ClTrp: 6′-chloro-tryptophan; NMe-5OHTrp: N-methyl-5′-hydroxytryptophan; NMe-5′BrTrp: 5′-Bromo-N-methyltryptophan.

Nomenclature		Position						Reference
		1	**2**	**3**	**4**	**5**	**6**	
Konbamide	-	Leu	L-Lys	Ala	Leu	MeLeu	2′BrTrp	[31]
Keramide	A	Phe	L-Lys	Leu	Leu	5′OH-6′ClTrp	Phe	[32]
	L	Phe	L-Lys	Leu	Leu	NMe-6′ClTrp	Phe	[30]
Paltolide	A	Arg	D-Lys	Ala	Leu	Leu	Trp	[28]
	B	Arg	D-Lys	Ala	Leu	MeLeu	5′OHTrp	[28]
	C	Arg	D-Lys	Ala	Leu	MeLeu	6′BrTrp	[28]
Unnamed	1	Arg	D-Lys	Ala	Leu	MeLeu	5′OH-6′ClTrp	[28,51]
	2	Arg	D-Lys	Ala	Leu	MeLeu	6′ClTrp	[28,51]
	3	Arg	D-Lys	Ala	Leu	MeLeu	Trp	[28,51]
Mozamide	A	Allo-Ile	L-Lys	D-Val	Leu	NMe-5OHTrp	Phe	[29]
	B	Allo-Ile	L-Lys	D-Ile	Leu	NMe-5OHTrp	Phe	[29]
Psymbamide	A	Ile	D-Lys	Leu	Leu	NMe-5′BrTrp	Phe	[33]

To this date, 124 Anabaenopeptins variants have been identified (Supplementary Table S1 and Table 1), making it difficult to compile a systematic nomenclature and to evaluate every individual feature. To overcome this problem, several works have named new Anabaenopeptins according to their mass or even mass and specific information regarding the strain or location in which it was isolated [34]. However, some Anabaenopeptins have unusual features, typically not found in most of these peptides belonging to this class. The previously cited Ferintoic acid B has an allo-isoleucine (Allo-Ile) residue in position 6, similar to both Mozamides, which have the same residue in position 1 [29,39]. None of the Brunsvicamides have a homoamino acid and they are also the only examples of this class of peptide with L-Lys in position 2 from cyanobacteria [50]. A group of Anabaenopeptins named SA demonstrated to possess some uncommon features. Anabaenopeptin SA1, SA4, SA5, and SA7 (Figure 5) have a 5-phenylnorvaline (PNV) in position 4, which has not been previously found in any other peptide of this class. Also, Anabaenopeptin SA8 was the only one to have a 6-phenylnorleucine (PNL) residue in position 4. Additionally, APs SA2 and SA13 have a Ser residue in position 5, which has not been previously detected in the same position, but in position 6, as in Nodulapeptins [12]. Anabaenopeptins 877B, 905, 862, and 896 are some of the few examples of N-ethylated peptides [24]. The only example of homoarginine in Anabaenopeptins is from AP KT864, which has this residue in position 1 [52]. A residue of glutamate at the exocyclic position has been only described in one Anabaenopeptin: the variant MM823 [22]. Besides its common presence in position 3, Valine (Val) has been only detected in position 4 in Nodulapeptin 855C [34]. Thus, demonstrating that Anabaenopeptin peptides have huge structural diversity.

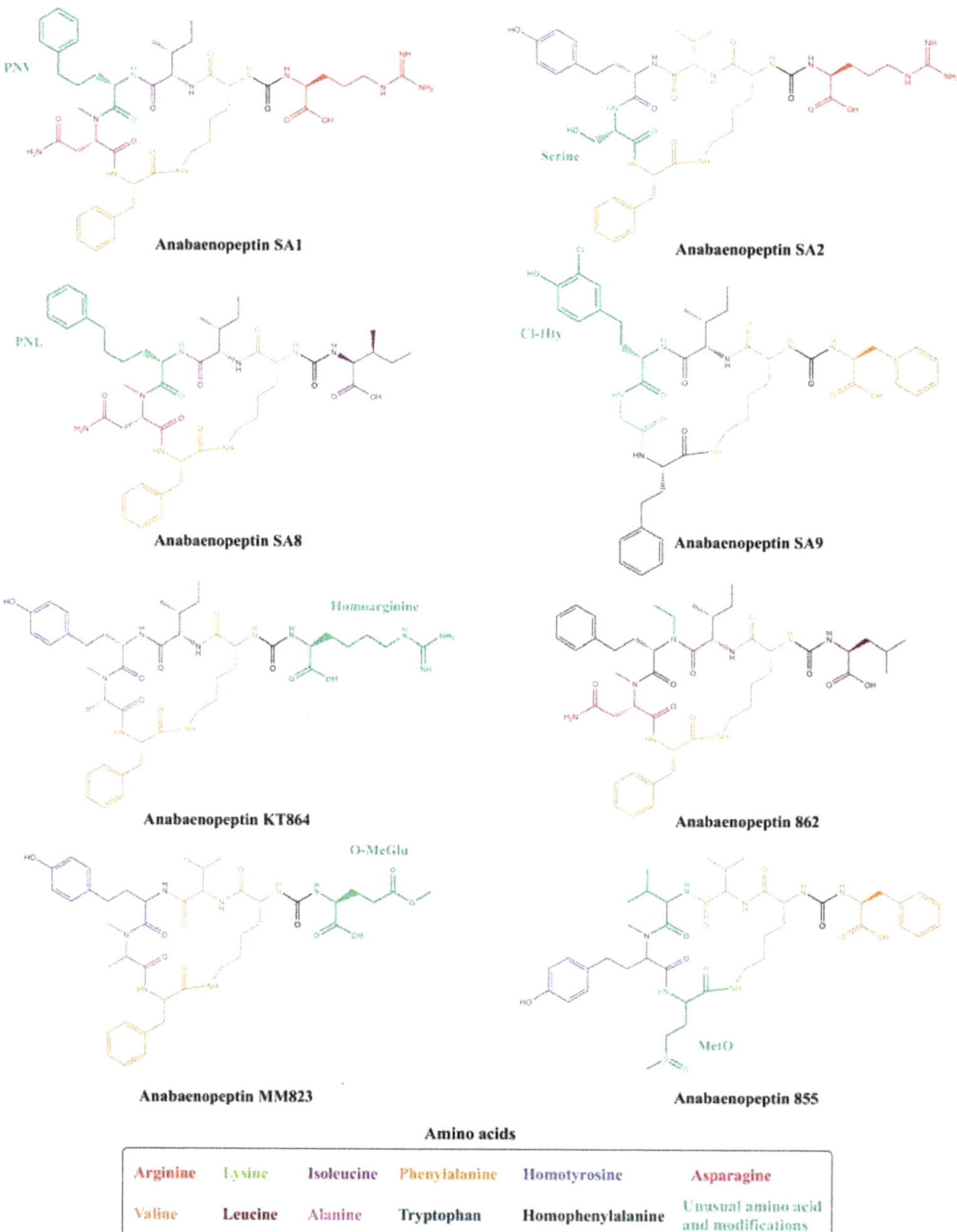

Figure 5. Examples of untypical features of anabaenopeptins from cyanobacteria [12,22,24,34,52,53].

Unusual Anabaenopeptins lacking residues in their structures are also visualized. Anabaenopeptin 679 is the only example of an anabaenopeptin-like peptide where the exocyclic residue is absent (Figure 6). This anabaenopeptin possesses solely the ring structure, which shares the same amino acid sequence as anabaenopeptins A, B, C, D, and

J [53]. Additionally, Namalides are anabaenopeptins with an atypical structure lacking two amino acids from the macrocycle. They are cyclic tetrapeptides firstly identified in the marine sponge *Siliquariaspongia mirabalis* [54] and then detected in cyanobacteria, such as *Sphaerospermopsis torques-reginae* ITEP-024 [55] and *Nostoc* sp. CENA543 [56], producing namalides B and C, and namalides B, D, E, and F, respectively.

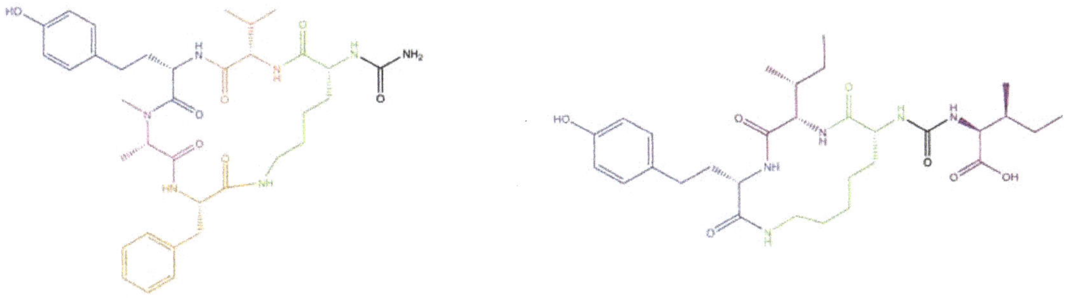

Anabaenopeptin 679

Namalide B

Amino acids

Lysine Isoleucine Valine Alanine
Homotyrosine Phenylalanine

Figure 6. Example of Anabaenopeptins with unusual structures lacking one amino acid (Anabaenopeptin 679) and two amino acids (Namalide B) residues [53,55,56].

3. Occurrence of Anabaenopeptins and factors involved in their expression

Besides its great structural diversity, it appears that those peptides are usually detected in some specific genera of cyanobacteria. As can be seen in Table 2, the majority of cyanobacteria able to biosynthesize anabaenopeptins belong to genera such as *Anabaena, Microcystis, Nodularia, Oscillatoria,* and *Planktothrix*. Except for *Microcystis*, those genera are filamentous cyanobacteria belonging to the order Nostocales and Oscillatoriales. Regarding the unicellular genus, as will be discussed later (Section 4), the Anabaenopeptin NRPS cluster seems to be horizontally transferred to *Microcystis* [57]. Also, Anabaenopeptins have been detected in genera *Aphanizomenon, Brasilonema,* and *Desmonostoc* belonging to Nostocales order. Similar to *Oscillatoria* and *Planktothrix*, the genus *Lyngbya* belonging to Oscillatoriales demonstrated to produce anabaenopeptins. In addition, two strains of unicellular genera of cyanobacteria that belonged to Synechococcales, *Schizothrix* and *Woronichinia*, proved to have the ability to produce Anabaenopeptins. Strains belonging to filamentous cyanobacteria tend to present a higher quantity of gene clusters than the unicellular strains [58]. The heterocyst presence in some members of the order of Nostocales can also confer some advantages in the Anabaenopeptin production since this differentiated cell provides the propitious microenvironment for the nitrogen fixation, which is an element required in large amount for the production of cyanopeptides [59].

Table 2. Occurrence of anabaenopeptins in different cyanobacteria genera and species.

Strains	Anabaenopeptin	Reference
Freshwater		
Anabaena flos-aquae 202 A 1	Anabaenopeptins B and D	[44]
Anabaena flos-aquae CYA 83/1	Anabaenopeptins B and D	[44]
Anabaena lemmermannii 202 A2/41	Anabaenopeptins B and C	[44]
Aphanizomenon flos-aquae NIES-81	Anabaenopeptins I and J	[42]
Lyngbya sp. (SAG 36.91)	Lyngbyaureamide A and B	[47]
Microcystis aeruginosa HUB 063	Anabaenopeptins B and F	[40]
Microcystis aeruginosa Kutz	Ferintoic acids A and B	[39]
Microcystis aeruginosa PCC7806	Anabaenopeptins A, B, E/F and Oscillamide Y	[60]
Microcystis aeruginosa TAU IL-342	Anabaenopeptin HU892	[61]
Microcystis sp. (MB-K)	Anabaenopeptin KT864	[52]
Microcystis sp. TAU IL-306	Anabaenopeptin F and Oscillamide Y	[61]
Microcystis sp. TAU IL-362	Anabaenopeptins MM823, MM850, MM913 and B	[61]
Microcystis spp.	Anabaenopeptin KB905, KB899, G, H, 908A, 915, HU892, MM913	[62]
Nodularia spumigena Node 2	Nodulapeptins B, C, 855B, 871, 879, 897 and 915A	[14,49]
Nodularia spumigena Nodg 3	Nodulapeptins B, C, 855B, 871, 879, 897 and 915A	[14]
Nodularia spumigena Nodh 2	Nodulapeptins B, C, 855B, 871, 879, 897 and 915A	[14]
Nodularia spumigena NSBL-05	Anabaenopeptin 807	[14]
Nodularia spumigena NSBL-06	Anabaenopeptin 807	[14]
Nodularia spumigena NSBR-01	Anabaenopeptin 807	[14]
Nodularia spumigena NSGL-01	Anabaenopeptin 807	[14]
Nodularia spumigena NSKR-07	Anabaenopeptin 807	[14]
Nodularia spumigena NSLA-01	Anabaenopeptin 807	[14]
Nodularia spumigena NSOR-02	Anabaenopeptin 807	[14]
Nodularia spumigena NSPH-02	Anabaenopeptin 807	[14]
Oscillatoria agardhii CYA 128	Anabaenopeptins A and C	[44]
Oscillatoria agardhii NIES-204	Anabaenopeptins B, E and F	[38]
Oscillatoria agardhii NIES-595	Anabaenopeptin G and H	[26]
Planktothrix agardhii CCAP 1459/11A	Anabaenopeptin F and Oscillamide B	[25]
Planktothrix agardhii CYA126/8	Anabaenopeptin 908A and 915	[63]
Planktothrix agardhii HUB 011	Anabaenopeptin G	[40]
Planktothrix agardhii NIVA CYA 15	Anabaenopeptins A and B	[64]
Planktothrix agardhii NIVA CYA 34	Anabaenopeptins A, B, F and Oscillamide Y	[64]
Planktothrix mougeotii NIVA CYA 405	Anabaenopeptins A, B, F and Oscillamide Y	[64]
Planktothrix mougeotii NIVA CYA 56/3	Anabaenopeptins C, 822 *, B, and F	[64]
Planktothrix prolifica NIVA CYA 406	Anabaenopeptins A, B, F and Oscillamide Y	[64]
Planktothrix prolifica NIVA CYA 540	Anabaenopeptins A, B, F and Oscillamide Y	[64]
Planktothrix prolifica NIVA CYA 98	Anabaenopeptins A, B, F and Oscillamide Y	[18,64]
Planktothrix rubescens	Anabaenopeptins A, B, F and Oscillamide Y	[65]
Planktothrix rubescens	Anabaenopeptins A, B, C, F and Oscillamide Y	[66]
Planktothrix rubescens	Anabaenopeptins B and F	[67]
Planktothrix rubescens	Anabaenopeptin A, B, and F	[68]
Planktothrix rubescens BGSD-500	Anabaenopeptins B and F	[69]
Planktothrix rubescens NIES-610	Anabaenopeptin F	[25]
Planktothrix rubescens NIVA CYA 407	Anabaenopeptins C, 822 *, B, and F	[64]
Woronichinia naegeliana	Anabaenopeptin 899	[70]
Marine		
Anabaena sp. TAU NZ-3-1	Anabaenopeptins NZ841, NZ825 and NZ857	[71]
Coelosphaeriaceae cyanobacterium 06S067	Anabaenopeptins A, B, F, 802 *, 827 *, 809 * and Oscillamide Y	[72]
Nodularia spumigena AV1	Nodulapeptins A, B, C, 871, 821, 839, 849, 855A, 863, 865, 867, 879, 881A, 881B, 883A, 897, 899A, 915A, 931	[14,48,49]
Nodularia spumigena B15a	Anabaenopeptins 841 and D	[14]
Nodularia spumigena BY1	Anabaenopeptin B and Nodulapeptins B, C, 821, 839, 855A, 855B, 871, 879, 881A, 881B, 883A, 897, 899A, 915A, 931	[14,48,49]

Table 2. *Cont.*

Strains	Anabaenopeptin	Reference
Nodularia spumigena CCNP 1401	Anabaenopeptins 841A and D	[14,49]
Nodularia spumigena CCNP 1423	Nodulapeptins 883B, 899B, 901, 915B, 917, 933	[14,49]
Nodularia spumigena CCNP 1424	Nodulapeptins 883B, 899B, 901, 915B, 917, 933	[14,49]
Nodularia spumigena CCNP 1425	Nodulapeptins 883B, 899B, 901, 915B, 917, 933	[14,49]
Nodularia spumigena CCNP 1402	and Nodulapeptins A, B, C, 821, 839, 855A, 855B, 871, 879, 881A, 881B, 883A, 897, 899A, 915A, 931	[14,49]
Nodularia spumigena CCNP 1403	Anabaenopeptins 841A and D	[14,49]
Nodularia spumigena CCNP 1426	Anabaenopeptins D and 841A	[49]
Nodularia spumigena CCNP 1427	Nodulapeptins B, C, 821, 855A, 855B, 871, 879, 881A, 881B, 883A, 897, 899A, 915A and 931	[49]
Nodularia spumigena CCNP 1428	Nodulapeptins 883B, 899B, 901, 915B, 917 and 933	[49]
Nodularia spumigena CCNP 1430	Anabaenopeptins D and 841A	[49]
Nodularia spumigena CCNP 1431	Nodulapeptins 883B, 885, 899B, 901, 915B, 917 and 933	[49]
Nodularia spumigena CCNP 1436	Nodulapeptins B, C, 839, 855A, 855B, 871, 879, 881A, 881B, 883A, 897, 899A, 915A, 921 and 931	[49]
Nodularia spumigena CCNP 1440	Nodulapeptins 883B, 885, 899B, 901, 915B, 917 and 933	[49]
Nodularia spumigena CCY 9414	Nodulapeptins A, B, C, 839, 855A, 855B, 871, 879, 881A, 881B, 883A, 897, 899A, 915A, 931	[14,49,73]
Nodularia spumigena KAC 11	Anabaenopeptins J and 807	[49]
Nodularia spumigena KAC 13	Anabaenopeptins D and 841A	[49]
Nodularia spumigena KAC 64	Nodulapeptins 883B, 885, 899B, 901, 915B, 917 and 933	[49]
Nodularia spumigena KAC 66	Nodulapeptins 883B, 885, 857, 899B, 901, 915B, 917 and 933	[14,49]
Nodularia spumigena KAC 68	Nodulapeptins 883B, 885, 857, 899B, 901, 917 and 933	[49]
Nodularia spumigena KAC 7	Nodulapeptins B, C, 921, 839, 855A, 855B, 871, 879, 881A, 881B, 883A, 897, 899A, 915A and 931	[49]
Nodularia spumigena KAC 70	Nodulapeptins 807, 823, 851, 865, 867 and 883C	[49]
Nodularia spumigena KAC 71	Nodulapeptins A, B, C, 921, 823, 839, 855A, 855B, 871, 879, 881A, 881B, 883A, 897, 899A, 915A and 931	[49]
Nodularia spumigena KAC 87	Nodulapeptins 807, 823, 849, 851, 865, 867 and 883C	[49]
Nodularia spumigena UHCC0039	Nodulapeptins A, B, C, 839, 849, 855A, 863, 865, 867, 871, 879, 881A, 881B, 897, 899A, 915A and 933	[73]
Terrestrial		
Anabaena circinalis 90	Anabaenopeptins A, B, and C	[44]
Anabaena flos-aquae NRC 525-17	Anabaenopeptins A and B	[20]
Brasilonema sp. 360	Anabaenopeptin 802A	[24]
Brasilonema sp. 382	Anabaenopeptin 802A	[24]
Brasilonema sp. CT11	Anabaenopeptins 788, 802A, 802B and 816	[74]
Desmonostoc sp. 386	Anabaenopeptins 848, 849, 862, 863, 877A, 877B, 891 and 905	[24]
Nostoc sp. 352	Anabaenopeptins 841B, 855, 857 and 871	[24]
Nostoc sp. 358	Anabaenopeptins 882 and 896	[24]
Nostoc sp. ASN_M	Anabaenopeptins 808 *, 828, 842 *, 844 * and 858 *,	[75]
Nostoc sp. ATCC 53789	Anabaenopeptin SA9, SA10, SA11 and SA12	[12]
Nostoc sp. KVJ2	Anabaenopeptins KVJ827, KVJ841, and KVJ811	[21]
Schizothrix sp. IL-208-2-2	Schizopeptin 791	[46]

* Anabaenopeptin variants with non-elucidated sequence.

A total of 45, 29, and 12 cyanobacteria strains from freshwater, marine and terrestrial environment have been analyzed for AP production, respectively. As seen in Table 2 and according to the literature [34,40,41,53,76–85], marine strains produced a total of 50 different variants of APs, in comparison to 43 and 34 variants from freshwater, and terrestrial strains, respectively (Figure 7). Thus, marine cyanobacteria demonstrate to produce a higher number of distinct APs variants in comparison to the remaining strains from different sources. However, APs from freshwater environments have the greatest diversity of amino acids in the majority of positions (Figure 8). Thus, these features could be associated to different obstacles faced in their respective environments as well as the fact that both belong to aquatic environments [86], however, this hypothesis requires further studies.

Some of those APs are shared among different strains isolated from distinct environments: 2 anabaenopeptins (A and B) variants were detected in all ecosystems; in comparison, strains from both aquatic habitats had 13 APs variants in common (D, F, J, 807, NZ841, Oscillamide Y, and Nodulapeptins B, C, 855B, 871, 879, 897 and 915A); in contrast, only anabaenopeptin C were produced by both terrestrial and freshwater, and none Anabaenopeptin variant was shared by both terrestrial and marine strains.

According to Table 2 and Figure 7, there are AP variants shared among cyanobacteria strains from different environments according to the previous discussion. Anabaenopeptins A and B are the only variants detected in all habitats analyzed, and the only difference between those variants reside at the exocyclic residue. AP B is still the most recurrent among these oligopeptides in cyanobacteria (Table 2), corroborating with the previously raised hypothesis that this variant was the first cyanotoxin of this class to be emerged. [57]. Furthermore, the number of common anabaenopeptins variants increases when a comparison is made among strains only from aquatic habitats (freshwater and marine): Anabaenopeptins D, F, J, 807, NZ841, Oscillamide Y, and Nodulapeptins B, C, 855B, 871, 879, 897 and 915A. Besides their production by both freshwater and marine cyanobacteria, these prevalent oligopeptides seem to be more recurrent in marine environments, given that a higher number of cyanobacteria strains from this habitat are able to produce these APs comparing to freshwater, except for Oscillamide Y, which is more recurring in the latter. Among those variants, Nodulapeptin B is the most frequent in marine microorganisms. Besides, the only difference between the AP C (produced by freshwater and terrestrial strains) and both A and B variants is the exocyclic amino acid, and the former was not detected in marine cyanobacteria.

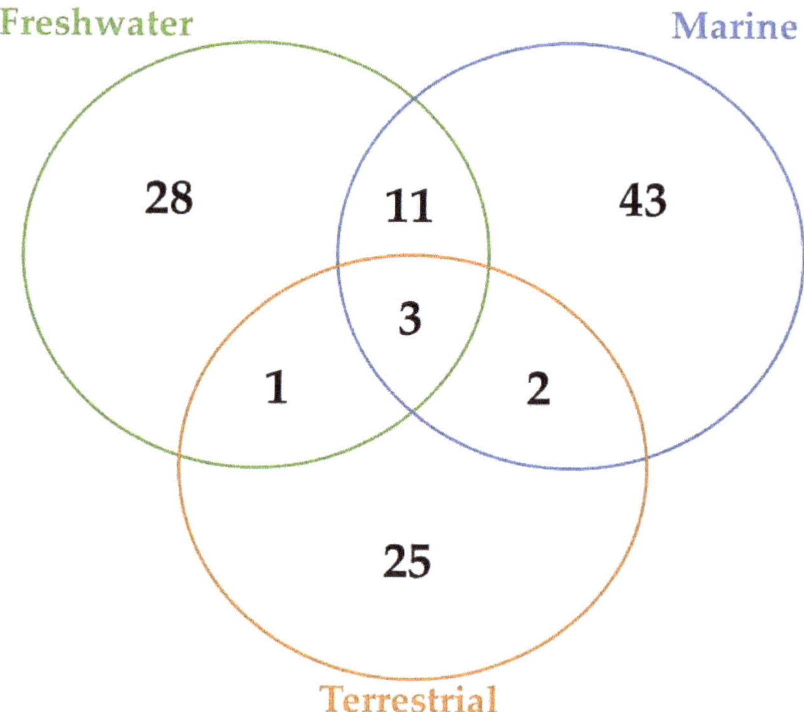

Figure 7. The number of Anabaenopeptins variants detected and shared among strains of cyanobacteria from different environments, including environmental samples.

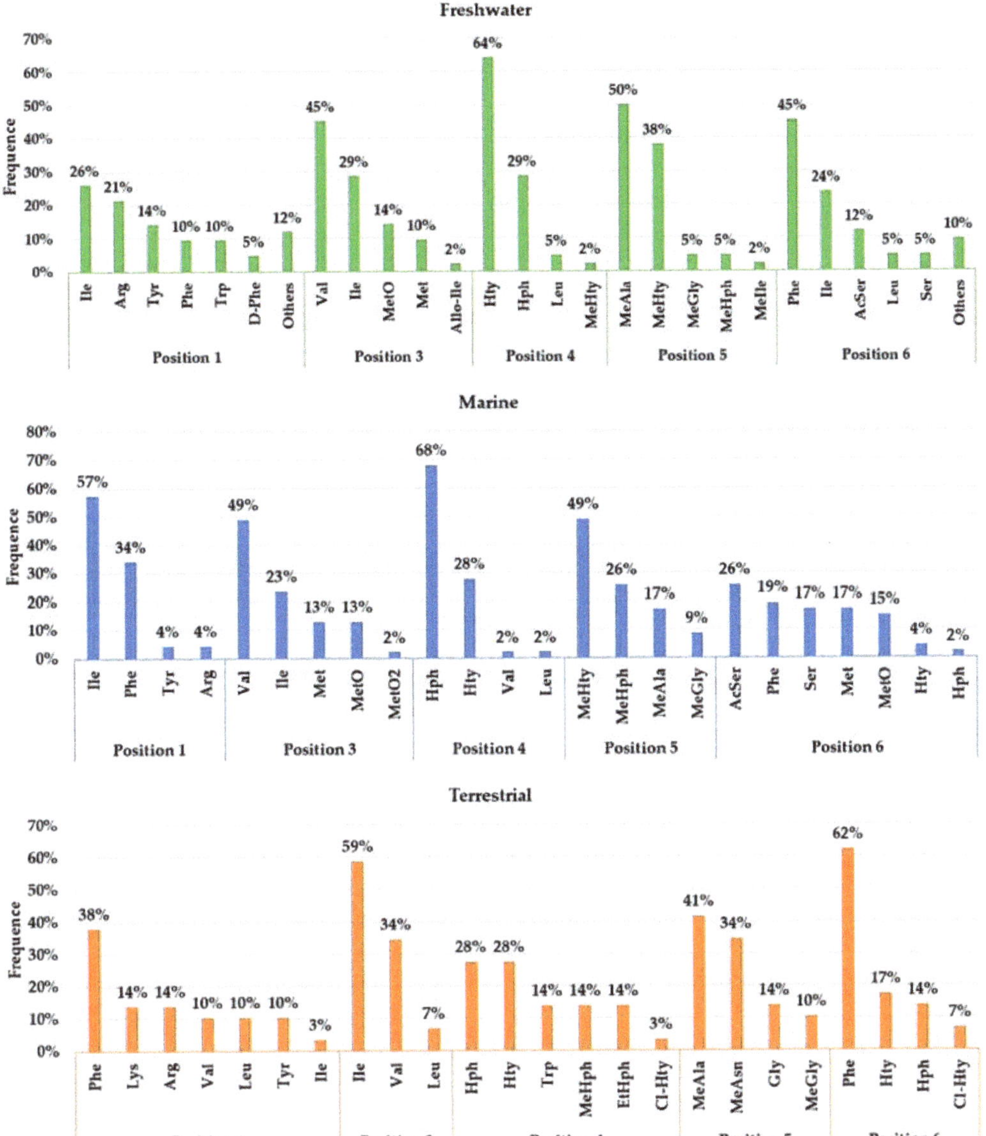

Figure 8. Relative frequency (%) of amino acids in positions 1 and 3–6 of variants of anabaenopeptins characterized according to their environment (freshwater, marine and terrestrial). The total number of variants with elucidated sequences were 42, 47 and 29 for freshwater, marine, and terrestrial environments, respectively. Position 2 was omitted as the D-Lys residue being conserved among AP variants.

As seen in Figure 7, the environment can exert a crucial role in the biosynthesis of different APs, justifying their distribution in certain locations. The presence and frequency of certain amino acids in Anabaenopeptin structures can vary according to their respective source environment. Anabaenopeptins from both aquatic environments demonstrate to have Isoleucine as the most recurrent amino acid in position 1, while this same amino acid was detected in only one AP variant in terrestrial strains (Figure 8). Phenylalanine was highly detected in position 1 of Anabaenopeptins isolated from terrestrial strains. Then, freshwater cyanobacteria may be promising biotechnological targets due to its highest diversity of amino acids in position 1, as the exocyclic residue is crucial for its inhibitory activity [12,34,45]. Regarding the variable position 3, Anabaenopeptins from freshwater and marine environments displayed a similar pattern of amino acid frequencies, Valine (Val) being the most frequent, followed by Ile and L-Methionine sulfone (MetO2). In contrast, terrestrial strains produce several AP variants with Ile in position 3, followed by Val and Leu, the latter being absent in this position on APs detected in aquatic environments. Homotyrosine (Hty) and Homophenylalanine (Hph) are the most found residues in position 4 among APs from all habitats analyzed, however, among terrestrial and marine strains Hph is more predominantly, while Hty is commonly observed in APs from freshwater strains. Except for Glycine (Gly) in some Anabaenopeptins from terrestrial strains, all the other residues in position 5 are N-methylated. APs from non-aquatic cyanobacteria do not harbor homoamino acids in the fifth position and, in addition, Asparagine is only detected in some of those variants in the respective position. Besides their detection in position 5, homoamino acids seem to be more persistent in position 4 from those APs analyzed. Position 6 has the highest richness of amino acids among AP variants obtained from marine environments, having incorporated 7 different residues, while this position in variants from freshwater habitats have assimilated 9 different amino acids, being the second most diverse site. Such heterogeneity in the last position in APs from aquatic strains is not clear, as the first amino acid residue demonstrated to be important in Anabaenopeptin interaction towards its enzyme target [12,34,45]. This array of several amino acids detected in position 6 is not visualized in Anabaenopeptins from terrestrial strains, where Phe was the amino acid more detected, similar to those APs from freshwater microorganisms.

The identification of the external physicochemical parameters involved in the regulation of these molecules can assist in controlling and assessing their risks [87]. Furthermore, this type of information can enable a better comprehension of their functions in producing organisms [27,88–91].

Microcystins, nodularins, and saxitoxins are among the most studied toxins from cyanobacteria, in which their gene clusters can operate independently, being, therefore, able to react oppositely when exposed to the same conditions [92,93]. The relationship between the gene cluster of Anabaenopeptins with these clusters within an individual strain is not well explored, demanding more detailed studies by a holistic approach, since enabling the study of various peptides at the same time [89,90].

Anabaenopeptins content per cell is strongly affected by environmental factors (Figure 9). Tonk and co-workers [27] investigated the effect of light intensity, temperature, and phosphorous concentration on the growth of the cyanobacterium *Anabaena* sp. 90 as well as its production of Anabaenopeptins A and C among some MC variants and with the anabaenopeptilide 90 B. This later belongs to an underexplored group of depsipeptides, which similar to the APs has the structure of a ring with a side chain, but without the ureido linkage. In the phosphorus-limited condition, all peptides were detected in a higher amount. These data match the result of Teikari and colleagues [94], who studied the same *Anabaena* strain and encountered higher quantities of transcripts belonged to anabaenopeptins, anabaenopeptilide, and microcystins gene clusters under low-phosphate conditions. Phosphate limitation also increases the content of protease inhibitors of the cyanobacterium *M. aeruginosa* NIVA CYA 43 [95].

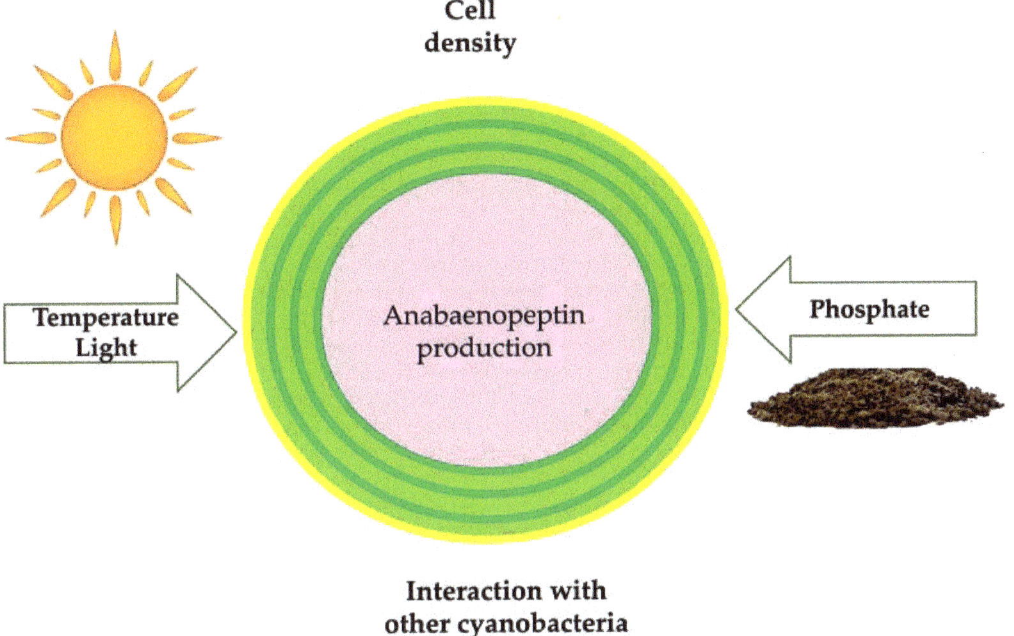

Figure 9. Major factors involved in anabaenopeptin regulation in cyanobacteria.

In contrast, anabaenopeptilide 90B responded in a different way of Anabaenopeptins A and C to light exposure. The former increases considerably with light intensity while the others had their production reduced. These two peptide groups exhibit a compensatory dynamic, where the reduction of one is accompanied by the increase of the other. This strategy employed by cyanobacteria ensures the constitutive production of peptides with similar functions in an unstable environment, increasing its survival change. For example, anabaenopeptilides are also serine protease inhibitors, having near functions to APs in the cell. In respect to the temperature, its increase favored anabaenopeptins production but resulted in a slower growth rate [27]. Such observations demonstrate that APs production is a constitutive process and is not always positively related to the growth of *Anabaena* sp. 90, contrasting, therefore, with the argument that the best growth conditions are more favorable to toxins production.

A compensatory mechanism has been described by other authors in cyanobacteria [88–90,96]. A comparison of the oligopeptide profile of *M. aeruginosa* PCC 7806 and its microcystin-deficient mutant revealed that the loss of this toxin has as consequence the increased production of cyanopeptolins and aeruginosins [90]. Pereira and co-workers [96] observed similar behavior for different variants of MCs in *Radiocystis fernandoii* 28. Microcystin-RR exhibited an opposite response to those observed for microcystins YR, FR, and WR under distinct light conditions in this strain. This type of modification can strongly affect its final toxicity since MC variants display different toxicity degrees. Another possible alteration consists of a change of antifeedant potential. APs significantly vary as substrate specificity, exhibiting different bioactivity. In a previous study, the inactivation of the genes involved in the synthesis of anabaenopeptilides in *Anabaena* sp. 90 resulted in a considerable increase in the production of anabaenopeptins [88]. One plausible explanation for this phenomenon is that the knockout of genes involved in the biosynthesis of some oligopeptides can lead to a credit of energy, which is allocated to the production of the remaining peptides [90].

The impact of the cell density on anabaenopeptin production and other oligopeptides have been investigated since it has been related to the increase in the production of some antibiotics. Moreover, this phenomenon can provide valuable data about probable alterations during bloom processes and mat formation [97,98]. High-density cultivation of photosynthetic microorganisms can be a challenge as the light availability decreases with cell density. Guljamow and co-workers [21] utilized a two-tier vessel developed by Bähr and co-workers [99] to cultivate *N. punctiforme* PCC 73102 and *Nostoc* sp. strain KVJ2 at high cell density and observed a higher diversity of secondary metabolites. Anabaenopeptin was absent in *Nostoc* sp. strain KVJ2 biomass obtained under conventional cultivation. High-density cultivation of this strain revealed the presence of three novel anabaenopeptins (KVJ827, KVJ841, and KVJ811). The increase of the content of these APs in the strain KVJ2 is attributed to a higher number of transcriptions among the cells. In the conventional cultivation, the distribution of the *aptA* transcripts (an NRPS gene related to AP production) was restricted only to a cell at (pre-)akinete state while in high-cell density culture, this transcript was widely distributed among the vegetative cells [21].

The interaction between different chemotypes of cyanobacteria in a water body can provoke significant alterations in their secondary metabolites profile. Consequently, differences are observed between laboratory culture and natural environments. In co-culture with *M. aeruginosa* PCC 7806, the non-microcystin-producing strain *M. aeruginosa* PCC 9432 enhanced its bioactive peptide content, including Ferintoic acids A and B [90]. These findings suggest the release of diffusible signals by cyanobacteria with the capacity of regulating the production of APs. The chemical nature of such metabolites was not determined in this study. However, certain oligopeptides can fill the signaling function since they are occasionally found in the extracellular compartment, acting as infochemicals. In addition to peptides, cyanobacterial exudate has also some nutrients, which affect the production of certain toxins and can be, consequently, responsible for the increase of Ferintoic acids A and B in *M. aeruginosa* PCC 9432 [100]. In a later study, the supplementation of the culture medium of a *P. agardhii* with two oligopeptides extracts from samples of *P. agardhii* as the predominant cyanobacterial species had different effects on the synthesis of the peptides of this strain. Both extracts showed a positive impact on biomass accumulation and chlorophyll-a production, being attributed to those nutrients and oligopeptides now present. The high nutritional content of the extracts is associated with the ability of cyanobacterial in fixing nitrogen and producing vitamins, phytohormones, and polysaccharides. Three out of four anabaenopeptins maintained constant (m/z 851, 844, and 837) while the variant with m/z 828 was substituted by other with m/z 923. One of the extracts increased the anabaenopeptin content of variants m/z 844, 851, and 837 while the other diminished the quantity of these last two [101]. The opposite responses to these extracts may be assigned to the content differences observed between them. The extract responsible for reducing the APs expression exhibits a superior concentration of nitrate and phosphate, which, as was previously mentioned has a negative effect on the production of anabaenopeptin [27].

In addition to interaction with other cyanobacteria, these microorganisms are capable to establish symbiotic associations with invertebrates, such as corals, mollusks, and sponges. Both organisms can be benefited during this consortium through secondary metabolite production, for example [102]. Sponges host an enormous quantity of microorganisms belonging to diverse phyla, where cyanobacteria are mainly represented by genera *Aphanocapsa*, *Synechocystis*, *Phormidium*, and *Oscillatoria* [103]. These photosynthetic microorganisms can occupy either extra- or intracellular spaces, aiding the host in the control of the redox potential, supplying pigments and energy through carbon fixation, and in the defense mechanism by the production of secondary metabolites. Published reports have demonstrated that as a consequence of these processes, cyanobacteria have their metabolic profile altered, resulting in the production of distinct variants of natural products. The compound 2-(2′,4′-dibromophenyl)-4,6-dibromophenol is solely biosynthesized by a cyanobacterium belonging to genus *Oscillatoria* in association with the sponge

Dysidea herbacea [104]. These factors corroborate with the hypothesis that anabaenopeptins primarily observed in sponges could be of cyanobacterial origin, as brominated APs variants were isolated only from sponges [28,31,33] and the *Oscillatoria* genus is known for APs production. For instance, the polyketide nosperin and some variants of oligopeptide nostopeptolide are encountered exclusively during symbiosis, which may be the same mechanism for anabaenopeptin variants production found in sponges.

4. Biosynthesis

The features of Anabaenopeptins are related to Non-Ribosomal Peptide Synthetases (NRPSs), which operate with a nucleic acid-free mechanism at the protein level and are structured as multifunctional proteins. NRPSs are organized as gene clusters in bacteria, usually possessing all the proteins required for proper biosynthesis of the secondary metabolites, from the generation of building blocks to product transport [105–107].

The variability of NRP structures, both cyclic and linear, reflects the concept of the complex modular system of NRPSs organized as an assembly line. Each module is responsible for the activation and coupling of an amino acid to the respective oligopeptide being synthesized. The principle known as the collinearity rule dictates that, for example, a hexapeptide requires six modules to be produced. Those modules are composed of enzymatic domains present in an NRPS, which are responsible for specific biosynthetic steps, as amino acid activation, bond formation, and oligopeptide liberation. Besides the initiation module, an elongation module from an NRPS requires, at least, an Adenylation-domain (A-domain) for amino acid recognition and activation; the Thiolation-domain (T-domain), required to carry the synthesized peptide; and a Condensation-domain (C-domain), responsible for the peptide bond formation. The last module of this assembly line requires the Thioesterase-domain (Te-domain) for the proper maturation of the peptide, also responsible for the cyclization step [18,105–108].

Similar to other peptides produced by NRPS, the biosynthesis of APs requires all the specific steps of the assembly line. Besides, due to some specific characteristics present in this cyclic hexapeptide and its variants, other proteins and domains can also be related to its synthesis, as the biosynthetic apparatus for homoamino acid production and domains for D-Lys formation (Epimerization-domain; E-domain) and N-methylation of specific residues (Methylation-domain; M-domain) [18,19,105,106,108,109].

Besides the fact that the anabaenopeptin structure's first detection in cyanobacteria occurred in 1995 [20], its gene cluster was only described ten years later in a *Planktothrix rubescens* strain [18]. The gene cluster detected in this cyanobacterium comprised of 5 genes (*ana*ABCDE): 4 NRPSs, and an ATP-Binding Cassette-transporter (ABC-transporter) protein. It was also visualized NRPSs possessing an epimerase domain (AnaA) and a methyltransferase domain (AnaC), which could be related to typical features encountered in APs, such as D-Lys and N-methylated amino acids, respectively. Only one cluster was detected in this organism, and it was attributed for the biosynthesis of all four peptides produced: Anabaenopeptin A, B, F, and Oscillamide Y, which differ by the combinatory of two residues in two distinct positions: (Tyr/Arg)-Lys-(Val/Ile)-Hty-MeAla-Phe. Thus, this phenomenon indicates that these NRPSs demonstrated a certain degree of promiscuity regarding their substrates and A-domains, as different amino acids can interact with the same catalytic site [18].

Rouhiainen and co-workers [110] detected gene clusters related to the production of APs in *Anabaena* sp. 90, *Nodularia spumigena* CCY9414, and *Nostoc punctiforme* PCC3102. In *Anabaena* sp. 90, five Open Reading Frame (ORF) were identified to be encoding NRPSs (*aptA1*, *aptA2*, *aptB*, *aptC*, and *aptD)* and two additional genes to be encoding proteins with similarity to HMGL-family (*aptE*) and ABC-transporter protein (*aptF*). When compared to the clusters identified in *N. spumigena* and *N. punctiforme*, 4 NRPS and two homolog proteins to AptE and -F were also detected, indicating that *Anabaena* sp. had an additional NRPS gene (*aptA1* and *aptA2*). Similar to AnaA from *Planktothrix rubescens* NIVA-CYA 98, AptA1 and AptA2 also have an epimerase domain indicating their role as initial enzymes, and AptC possessing the N-methyltransferase domain as AnaC [110]. The proteins AptA1/AptA2, AptB, AptC, and AptD are homologs to the NRPS proteins AnaA, AnaB, AnaC, and AnaD, sharing the same functions, respectively.

A genomic analysis of *Sphaerospermopsis torques-reginae* ITEP-024 accomplished by Lima and colleagues [107] demonstrated that the *apt* gene cluster is close to the spumigin cluster. Both AP and spumigin are peptides with protease inhibitory activity which usually possess Homophenylalanine and Homotyrosine residues, then indicating that both NRPS apparatus share a biosynthetic cluster related to the production of these nonproteinogenic residues. The *apt* gene cluster of *S. torques-reginae* strain has a similar organization to the anabaenopeptin clusters from *Anabaena, Nodularia, Nostoc,* and *Plaktothrix* [18,110]. Thus, its cluster also holds four genes encoding a six-module NRPS (*apt*ABCD), where the Te-domain is present at the last module, then being responsible for the final step of AP production, similarly to other NRPS products [107].

Entfellner and co-workers [57] suggested that the AP cluster could be transferred among cyanobacterial species due to horizontal gene transfer (HGT). This hypothesis is supported by the high similarity visualized between the *apnA-E* cluster from *Planktothrix* and *Microcystis* composed by *apnA, apnB, apnC, apnD* and *apnE*, which genes codified proteins homologs to AnaA/AptA, AnaB/AptB, AnaC/AptC, AnaD/AptD, and AnaE/AptF, respectively. Some strains belonging to the *Planktothrix* genus demonstrated to possess the same AP cluster, but not all of them, thus suggesting that the common ancestors of these organisms did not have the NRPS apparatus for AP biosynthesis, which could be visualized by a phylogenetic analysis using *apnA-E* clusters as biological markers. By phylogenetic analysis of different sequences of anabaenopeptin cluster, it could be inferred that an ancestral cluster was introduced into the chromosome of a *Planktothrix* strain and diversified into different variants, which could be grouped according to *apnA* sequences. Thus, the high frequency of AP producers belonging to *Microcystis* and *Planktothrix* in nature could be an indication of this mechanism of genetic transference by the AP cluster and its wide distribution among those genera, requiring further analysis of the same mechanism in other AP producers, such as *Anabaena, Aphanizomenon, Nodularia, Nostoc, Oscillatoria,* and *Lyngbya* (Table 2) [57].

It had been detected in *Nostoc* sp. CENA543 six variants of APs. Through genomic analysis, a gene cluster of 26 kb containing four NRPS and additional enzymes was visualized related to AP production. Following the same pattern, the NRPS proteins were AptABCD, and the additional enzymes were an ABC-transporter, 2-isopropylmalate synthase (HphA), and an ORF similar to Nuclear Transport Factor-2 (NFT2) proteins [56].

Thus, as discussed, several AP clusters have been identified (Figure 10) and their nomenclatures are not standardized, which usually are assigned according to the strains detected. For example: *ana* and *apn* for *Planktothrix* [18,57,111,112]; *apt* for *Anabaena, Microcystis, Nodularia, Nostoc* and *Sphaerospermosis* [56,107,110]; and even *kon* from *Candidatus Entotheonella* sp. TSY referencing the konbamide biosynthetic gene cluster [113]. Among these nomenclatures, *apt* is the most recurring, being applied to refer the AP gene cluster along this manuscript. However, all anabaenopeptin gene clusters from these different strains of cyanobacteria share common features. The first NRPS, AptA, is a bimodular initiation enzyme containing two A-domains, two T-domains, one C-domain, and one E-domain. The second NRPS enzyme, AptB, contains one elongation module (condensation, adenylation, and thiolation domains), followed by the third enzyme, AptC, which is an NRPS enzyme with two elongation modules, which commonly contains distinct domains related to peptide modification, such as N-methyltransferases. Finally, the termination module from AptD comprises an elongation module which also includes a Te-domain (Figures 10 and 11). Then, it totalizes 6 modules, following the collinearity rule and confirmed by bioinformatic analyses regarding the specificity of each module with its amino acid [18,56,57,107,110–112].

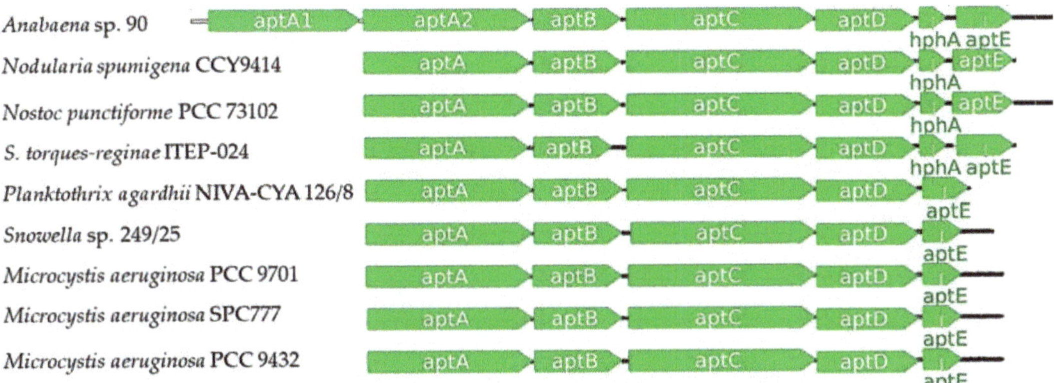

Figure 10. Anabaenopeptin cluster (*apt*) organization from different cyanobacteria strains. The genes *aptA1*, *aptA2*, *aptA*, *aptB*, *aptC*, *aptD* and *aptE* are Non-Ribosomal Peptide Synthetases (NRPSs) related to Anabaenopeptin production; *hphA* gene belongs to homoamino acid biosynthetic pathway and *hph*ABCD cluster. These clusters were obtained according to their accession codes (AC) from National Center for Biotechnology Information (NCBI): *Anabaena* sp. 90 (AC: GU174493), *Nodularia spumigena* CCY9414 (AC: CP007203), *Nostoc punctiforme* PCC 73102 (AC: CP001037), *Sphaerospermopsis torques-reginae* ITEP-024 (AC: KX788858), *Planktothrix agardhii* NIVA-CYA 126/8 (AC: EF672686), *Snowella* sp. 249/25 (AC: MF741700), *Microcystis aeruginosa* PCC 9701 (AC: HE974200), *Microcystis aeruginosa* SPC777 (AC: PRJNA205171), *Microcystis aeruginosa* PCC 9432 (AC: HE972547). This information is available on the public database NCBI (https://www.ncbi.nlm.nih.gov/; accessed on 16 March 2021).

The first adenylation domain from the NRPS apparatus belonging to the first module of AptA (Figures 10 and 11) had been analyzed by several works due to the inhibitory role of the first amino acid residue towards specific enzymes [57,111]. Evolutionary analysis coupled to molecular biology demonstrated that one of the first anabaenopeptin to be produced possessed Arg at position 1, such as anabaenopeptin B (Figure 2). This data corroborates with the inhibitory activity of carboxypeptidase B of AP variants bearing Arg at the exocyclic position, which is greater than Tyr, Phe, and Ile. In addition, analysis of *Planktothrix* producers strains demonstrated a high frequency of AP B producers (83 out of 89 strains), followed by AP A, AP F, and Oscillamide Y (55%, 45%, and 33% of the strains), corroborating with Table 2 [57]. Some wild-type adenylation domains from the first module of AptA demonstrated to be highly specific for arginine and tyrosine, and single point mutations within this domain can result in significant substrate promiscuity [57,110–112]. Due to its high frequency, inhibition towards carboxypeptidase B, and the possibility to be the first oligopeptide of its group to be originated, the biosynthesis of Anabaenopeptin B is outlined in Figure 11 and will be used as a standard for APs production.

Through a search of APs biosynthetic clusters in several cyanobacteria, Shishido and colleagues [56] detected that the majority of strains of cyanobacteria contained only one *apt*A gene. However, ten cyanobacteria and the tectomicrobia *Candidatus Entotheonella* sp. TSY1 possessed two alternative *apt*A genes. Thus, under other works [18,56,57,107,110,111,113], the biosynthesis initiation of APs has two different approaches. The first one is the NRPS with the presence of two starter modules with distinct substrate specificities that can produce different variants of APs. The second mechanism is due to the promiscuity of the first adenylation domain of AptA, producing different variants at position one [112]. Both mechanisms can increase the chemical diversity of Anabaenopeptins produced.

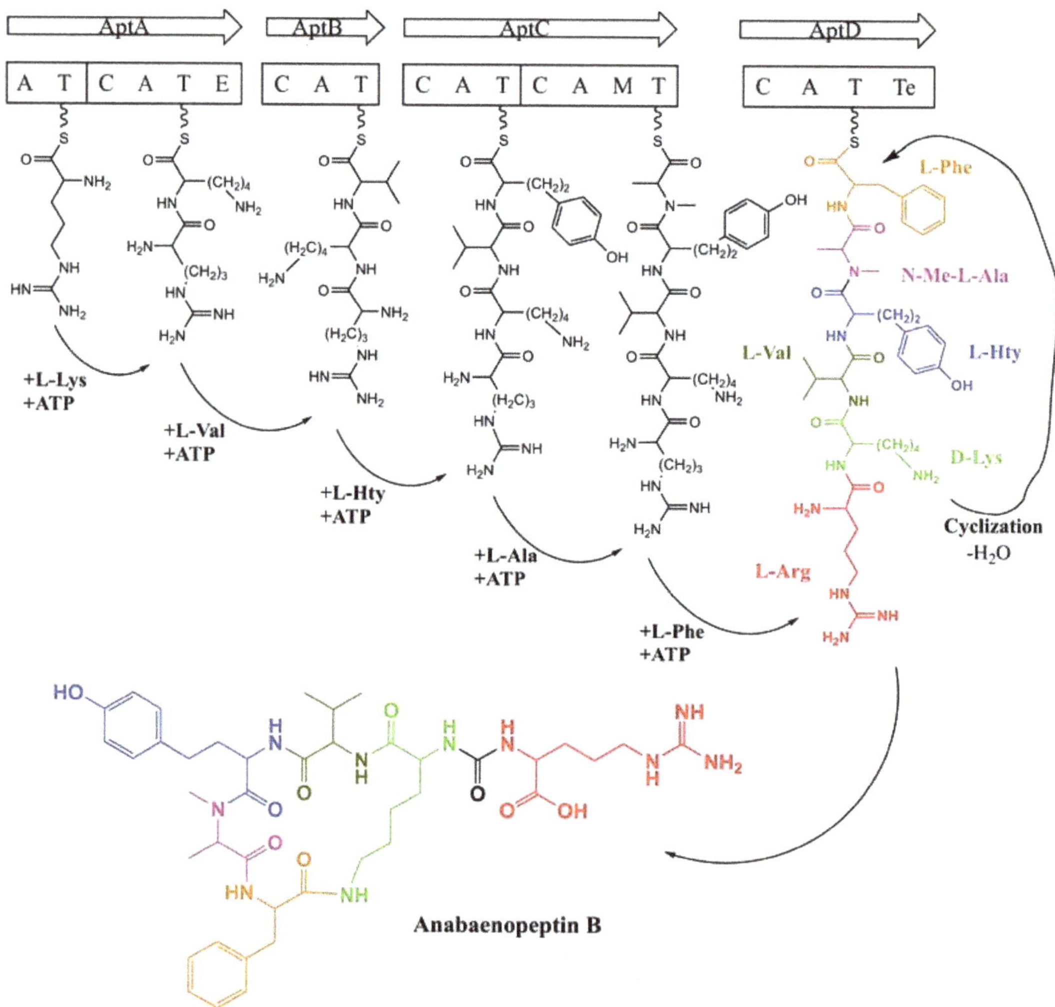

Figure 11. Scheme of biosynthesis of anabaenopeptin B in *Anabaena* sp. 90 by NRPS apparatus [107,110]. A: adenylation domain; T: thiolation domain; C: condensation domain; E: epimerization domain; M: N-methylation domain; Te: thioesterase domain.

As discussed previously, Rouhiainen and co-workers [110] identified an anabaenopeptin cluster from *Anabaena* sp. 90, possessing one additional NRPS enzyme with two modules (AptA1 and/or AptA2). This cyanobacterium was able to produce 3 different AP variants differing at position one. Through sequence comparison and substrate specificity analysis, it had been demonstrated that the first adenylation domain of AptA1 had an affinity to L-Lys and L-Arginine (Arg), while AptA2 demonstrated to interact with L-Tyr. Both adenylation domains from the second module of AptA1 and AptA2 incorporated D-Lys. Thus, demonstrating that *Anabaena* sp. 90 carried two distinct initiations NRPS producing different variants of anabaenopeptin, which a similar mechanism could also be visualized for puwainaphycins and minutissamides [110,114]. However, in Figure 11, only AptA1 is represented due to its specificity towards Arg.

Regarding the promiscuity of the adenylation domains aiming to understand the production of distinct AP variants, the adenylation domain of AptA from *Plaktothrix agardhii* PCC 7821 had been evaluated and concluded that it demonstrated to be bispecific for two different amino acids: Arg and Tyr. This feature corroborates with the variants produced by this strain of *P. agardhii*: Anabaenopeptins 908A and 915, which differs solely in the exocyclic residue (Arg or Tyr) [111,112]. A similar pattern had been visualized in *Planktothrix rubescens* NIVA-CYA 98, which possesses only one AP cluster, but it was able to biosynthesize different variants of anabaenopeptin differing at the exocyclic position (Tyr and Arg) and the third position (Val and Ile) [18].

One important feature encountered only in Anabaenopeptin among cyanobacterial peptides is the ureido linkage between the first and second residues [34,49]. However, this linkage can also be found in other natural products, including pacidomycins, mureidomycins, napsmycins, and syringolin A. This configuration is not common due to the mechanism present in NRPSs, which assembles amide bonds in an approach where the chain polarity remains unidirectional. The presence of ureido linkage alters this polarity due to the presence of N-to-N terminal condensation. Then, a specific enzyme and/or domain must be present in NRPSs involved at the ureido linkage formation, suggesting a possible role of the first elongation module in their formation [115].

When comparing the initial NRPSs genes encoding both modules of AptA from *Anabaena* sp. 90, *Nodularia spumigena* CCY9414, and *Nostoc punctiforme* PCC 73102, they all contained typical adenylation and condensation domains, also demonstrating highly conserved motifs. Besides their conservation, one hypothesis was that both modules of initiation and elongation of AptA would be related to ureido bond formation, similar to the SylC protein, from *Pseudomonas syringae*, which role is the catalysis of the ureido linkage between two Val residues from syringolin A [110,115]. Though the SylC protein possesses a domain with structural similarity to acetyltransferase between the A- and C-domains from the NRPSs, which is responsible for the ureido linkage formation and no homologous is present in the anabaenopeptins synthetases, suggesting a different mechanism for this step during AP biosynthesis [107,115].

Besides the initiation step and the formation of the ureido bond between the first residue and the conserved Lys, several steps of elongation of the peptide are required to produce a fully mature peptide. The signature sequences analyzed of the A-domains of these NRPS enzymes, such as AptB and AptC, are consistent with the respective amino acid residue of the final product and confirmed in vitro by biochemical methods. Also, usually, the fifth module bears an *N*-methyltransferase domain, as seen in AptC and their homologs, responsible for the *N*-methyl in Ala in position 5 of Anabaenopeptin B, as seen in Figure 11 [110].

Unlike the initiation enzyme related to residues at position 1 and 2, clusters related to AP production has not been shown to possess more than one NRPS for each residue. Thus, the variants produced by the cyanobacterial differing at positions 3–6 are biosynthesized due to the promiscuity of the adenylation domains of AptBCD. This phenomenon can be visualized by innumerous AP variants differing at those positions with only one correspondent gene cluster in the genome, for example, *Nostoc* sp. CENA 543 producing six variants [56].

Anabaenopeptins usually have homoamino acids at positions 4 and 5, which are added by AptC during elongation steps, as visualized in Figure 11 by the additional Hty added in position 4. The AptE, now known as HphA, was first suggested to be responsible for ureido linkage formation and is related to homoamino acid synthesis [110]. Succeeding previous works, it has been elucidated that AptE belongs to a biosynthetic cluster named *hphABCD*. Genes from *hph* cluster are frequently detected in the same genomic region as *apt* and *spu* clusters, which both products, Anabaenopeptins and Spumigins, are peptides displaying protease inhibitory activity and homoamino acids. A genomic analysis of *Sphaerospermopsis torques-reginae* ITEP-024 demonstrated that both Spumigin and Anabaenopeptin clusters were present in proximity in the genome. In between both clusters, the *hphABCD* biosynthetic cluster and additional genes were detected in this region, which a similar organization was also visualized in *Nodularia spumigena* CCY9414 [107]. The *hph* genes are responsible for the biosynthesis of Hph and Hty, nonproteinogenic amino acids commonly found in both anabaenopeptin and spumigin [116]. Thus, indicating that HphA is not responsible for ureido linkage formation but behind the supply of both Hph and Hty. In addition, the presence of the homophenylalanine and homotyrosine biosynthetic enzymes in this region could suggest that this cluster is supplying both homoamino acids for APs and Spumigins [107]. In accordance with Lima and co-workers [107], Shishido and colleagues [56] also visualized that from 56 genomes analyzed containing the *apt* cluster all demonstrated to possess the *hph* biosynthetic cluster, except for *Scytonema hofmanii* PCC 7110 and *Candidatus Entotheonella* sp. TSY. The genes encoding the proteins HphABCD were frequently found upstream or downstream of the AP cluster, supporting the hypothesis about their roles in providing homoamino acids to APs [107].

Thus, homoamino acids are produced by the HphABCD enzymes and then incorporated by the NRPS apparatus. In addition, these non-proteinogenic amino acids can also be further modified by the NRPS enzymes, considering that residues at position 5 are mostly methylated by the N-methylation domain in the second module of AptC. However, methylation of residues at position 4 was also visualized, such in Ferintoic acids A and B [39], Anabaenopeptin E [37], 863, 891, 848, and 882 [24].

The final step for Anabaenopeptin production is mediated by a Te-domain, which is commonly associated with the termination process of the biosynthesis of NRPS peptides. Thus, after the incorporation of the last residue, for example, L-Phenylalanine in AP B (Figure 11), these domains can be involved with the release of the peptide by hydrolysis, or even cyclization involving peptidic or ester bonds [19,106]. The last NRPS enzymes AptD and its homologs [18,111] bear the thioesterase domain, suggesting then their role as the termination step.

Besides those typical alterations to the amino acid residues discussed, several variants of APs have been found with different modifications, such as ethylated (Figure 2, Figure 3, and Figure 5), acetylated, and oxidized residues [22,24,34]. In addition to such modifications during the elongation steps by the NRPS, an analysis of cytochrome P450 monooxygenases from cyanobacteria revealed that some proteins of this class may be related to anabaenopeptin modifications. In *Synechococcus* sp. PCC 7502, it had been suggested that a P450 belonging to CYP110 is involved in the production of Anabaenopeptin NZ857. *Anabaena* sp. TAU NZ-3-1 was capable to coproduce this anabaenopeptin and APs NZ 825 and NZ841. Anabaenopeptin NZ857 differs from AP NZ825 and AP NZ841 by the number of oxidized residues at positions 4 and 6. Anabaenopeptin NZ857 has in both positions 4 and 6 the homotyrosine residue, while the other peptides have at least one homophenylalanine. Besides the possible relation of cytochrome P450 in anabaenopeptin production, its possible catalytic role has not been demonstrated [117].

Regarding the unusual anabaenopeptins lacking residues in their structure, the biosynthesis of Anabaenopeptin 679 (Figure 6) has not been described so far [53], requiring further analysis of its production. Due to Namalide similarity to APs, it has been suggested that the biosynthesis of this tetrapeptide is realized by the *apt* cluster, as during a genomic screening of both namalides-producing cyanobacteria no exclusive cluster related to the production of these peptides have been found. The prediction of amino acids incorporation of adenylation domains of AptABCD is in accordance with both AP and Namalides. Thus, the preliminary results obtained by Shishido and co-workers [56] strongly suggested that Namalides are biosynthesized by *apt* cluster through a module skipping event. During synthesis, the second domain of AptC and the C-domain of AptD (but not the thioesterase domain) are ignored resulting in the production of namalides, similar to the module-skipping process of Myxochromide from myxobacteria [56].

5. Ecology

Cyanopeptides confer a competitive advantage for their producing organisms due to their toxicity, which effect has been examined against parasites and grazers (Figure 12) [118,119]. Other strategies, such as colony formation and filaments aggregation with low nutrition content have also been documented as a defensive mechanism [120]. However, they cannot, on some occasions, be sufficient to explain the different susceptibility levels encountered among cyanobacterial populations [121].

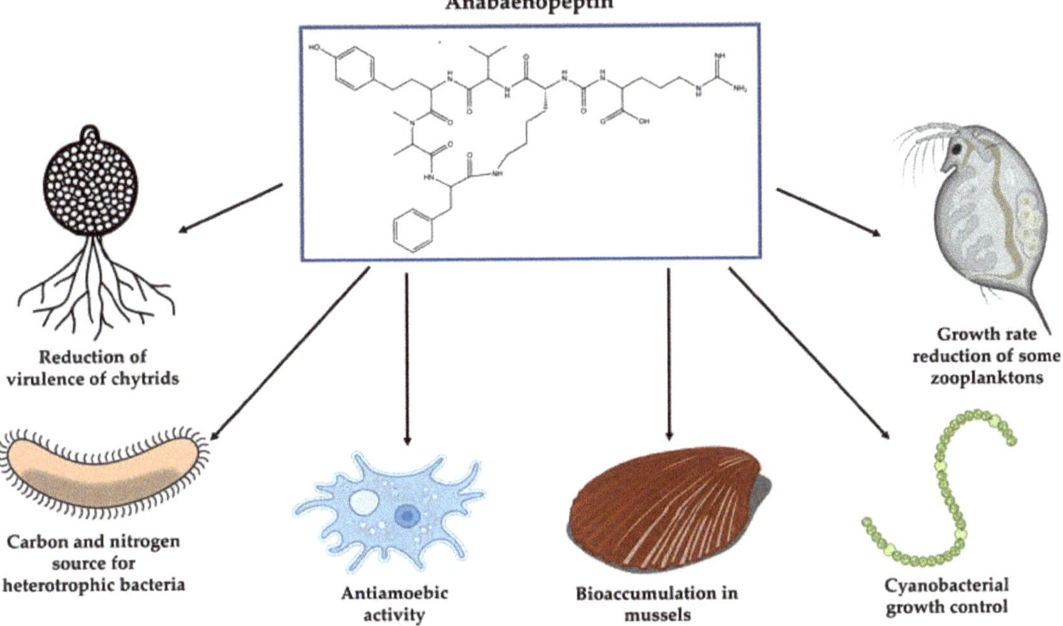

Figure 12. Ecological relevance of anabaenopeptins.

Anabaenopeptin presence in the cyanobacterial extract can confer a certain level of protection against some predators but is not a determining factor in the process as illustrated by the work developed by Urrutia-Cordero and coworkers [122]. These authors attested anti-amoeba activity against *Acanthamoeba castellanii* by *Microcystis* strains capable of producing either APs or MCs. Among the tested strains, the anabaenopeptin-producing was the one that caused the highest mortality rate. In contrast, the existence of the same APs in the extract of *A. lemmermannii* NIVA-CYA 426 did not result in any type of activity for the protozoan. Due to APs and MCs inhibitory activities against phosphatase, the loss of cytoskeleton integrity of *A. castellanii* was associated with the action of these cyanopeptides, which led to impairment of crucial functions associated with cytoplasmic projections, including motility and feeding.

Deleterious effects in organisms belonging to aquatic fauna were also linked to APs production and other cyanopeptides [123–125]. The negative impact of these metabolites can partially justify the substitution of large-bodied zooplankton by small-bodied species during the blooming process since they affect differently these living beings [126,127]. The absorption of such molecules can occur by ingestion of cyanobacteria or through uptake of water. Like the filtration system of large-bodied zooplankton has a greater tendency to absorb these microorganisms, they are more susceptible to the effect of toxins [128].

Some published reports have focused solely on the effect on a determined organism by an individual oligopeptide, especially MCs [92,129]. However, this type of investigation is not sufficient to verify the real impact of cyanobacterial bloom in the environment. Studies that employ only one type of solvent to obtain the cyanobacterial extract can also provide limited information since these microorganisms harbor an enormous variety of metabolites with distinct polarity, which are not, therefore, totally isolated and investigated during this type of analyses. Even though APs concentration in the aquatic environment can exceed 1 g.L^{-1}, they pose unknown consequences for human health [15]. Furthermore, their full effects on other animals are largely unknown. In Zebrafish (*Danio rerio*), an animal model very close to the human being, APs B and F as well as Oscillamide Y do not have any significant effect on the mortality of their embryos [130]. Otherwise, another study demonstrated that the APs A, B, and F exhibited the greatest toxicity as compared to other cyanopeptides, such as microcystin-RR, microginin 690, and cyanopeptolins CYP-1007, CYP-1020, and CYP-1041 to the nematode *Caenorhabditis elegans*. The exposure to these understudied toxins was responsible for diminishing the reproduction potential of this worm, affecting the brood size, the hatching time of eggs and vulvar integrity. Moreover, lifespan was also reduced by nearly 5 days [131].

Concerning APs action in other animals, Pawlik-Skowrońska and colleagues [127] demonstrated that extracts containing Anabaenopeptins originated from bloom samples, where the predominant species were *P. agardhii* or *Microcystis* sp., caused different responses in the behavior of planktonic species *Daphnia pulex* and *Brachionus calyciflorus*. The unicellular cyanobacterium extract did not cause acute toxicity to any of the investigated zooplanktons. Otherwise, the *Planktothrix* extract strongly reduced the survivorship rate of *D. pulex*. This difference was attributed to the oligopeptide profile in the extract, which considerably varies as to their quantity and structure. Anabaenopeptin and Aeruginoside had a superior contribution in *P. agardhii* extract as compared to *Microcystis* one, suggesting that those oligopeptides act in synergism. A similar analysis harboring a larger number of organisms, verified the toxicity of *Nodularia spumigena* extracts against the crustaceans *Thamnocephalus platyurus* and *Artemia franciscana* and also the bioaccumulation of their oligopeptides in these invertebrates and some blue mussels [118]. Nine APs and nodularin were encountered in the mussels collected from a bloom formed by this cyanobacterium. In *T. platyurus* and *A. franciscana*, the exposure to *Nodularia spumigena* extract results in the accumulation of various Anabaenopeptins, one aeruginosin, and one spumigin. The cyclic structure of APs confers them chemical stability preventing their degradation by the mussels tested, as linear peptides were not detected. Moreover, it also led to an increased mortality rate for both organisms. Among the fractions obtained from *N. spumigena* biomass extract, that with APs and a demethylated form of nodularin exerted the highest acute toxicity effect.

Anabaenopeptins also participate in the defensive mechanism of *Planktothrix* allowing, therefore, its dominance towards pathogens in the same environment. A comparison between a wild-type strain of *P. agardhii* NIVA CYA 126/8 and their mutants with dysfunctions in the production of APs, microviridins or MCs, indicated that these oligopeptides reduce the virulence of the fungi belonged to the division of Chytridiomycota, known as chytrids. This contrast may partially explain the dominance of a cyanobacterial chemotype in a determined environment, serving as a great defensive strategy to retard the parasite adaptation and to increase its diversity [16].

Anabaenopeptins and microviridins are most likely involved with the inhibition of the protease released by rhizoids whereas MCs are most probably related to the inactivation of their phosphatases [16]. Both groups of enzymes occupy a significant position in cell metabolism, participating in regulatory processes and signaling [132]. Strategies utilized by chytrids during infection are still an open question, but it is known that they can infect akinetes, vegetative cells, and/or heterocyst. Resource available inside the host can be one of the reasons for variations encountered among the infectivity methods. Akinetes offer higher energy and organic material content than vegetative cells. Oligopeptides distribution among the vegetative cells and those that are differentiated, such as akinetes and heterocysts should also be investigated since these secondary metabolites can sometimes be restricted to a cellular subgroup [133].

Cyanopeptides could also act as carbon and nitrogen source for various heterotrophic bacteria. These microorganisms are capable of degrading an enormous quantity of molecules with variable structures [134,135]. Besides the oligopeptides, the phycosphere is rich in carbohydrates, proteins, and lipids that originated from the exudate of cyanobacteria or its cell lysis. The mineralization of these organic compounds leads to CO_2 production, which may contribute to the growth of cyanobacteria [136]. Briand and colleagues [137] observed that the supplementation of an axenic culture of *M. aeruginosa* PCC7806 with bacteria associated with the mucilage of *M. aeruginosa* colonies collected during a bloom eliminated all oligopeptide encountered in extracellular fraction, including MC, cyanopeptolin, and cyanobactin. In a previous investigation, Kato and co-workers [138] identified in the cell extract of the bacterium *Sphingomonas* sp. B-9 hydrolytic activity for AP A, microcyclamide, nostophycin, aeruginopeptin 95-A, and microviridin I. Anabaenopeptin A degradation was gradual and subproducts were not observed.

APs and planktopeptin BL1125, both isolated from the bloom-forming *Planktothrix rubescens*, were associated with the collapse of cyanobacterial populations during the bloom termination. These oligopeptides act as triggers, inducing cellular lysis by virus-like particles, most likely cyanophages. Such propriety can explain in part the dominance and the high invasive potential of this species [139]. Sedmak and colleagues [67] attested it when verified that the cell growth of a non-xenic culture of *M. aeruginosa* MA2-NIB was inhibited when treated with these oligopeptides whereas the axenic were not affected. Such activity was not attributed to the known property of these peptides since the serine protease inhibitor 4-(2-aminoethyl) benzenesulfonyl fluoride hydrochloride failed in inducing any type of effect in the growth of these strains. Enrichment of the medium with the bacteria isolated from the non-axenic culture also did not produce any type of alteration. In contrast, the addition of the planktopeptin in an axenic culture of *Microcystis* previously supplemented with particulate materials obtained from cell lysate of the non-axenic culture provokes cell disintegration.

According to the hypothesis raised by these authors [67], the release of the cyclic peptides, mediated by cell lysis, signalizes the presence of a determined host and consequently activates the lytic cycle. A small concentration of these oligopeptides in the environment causes limited lysis, confined solely to a specific region. Cyanobacterial blooms offer the ideal condition for collapse since their cells are exposed to an elevated quantity of infection agents and cyclic peptides [99]. Cell-lysis provoked by cyanophages can promote the release of oligopeptide to the extracellular matrix, feeding positively the cycle [140]. In a subsequent study, algaecide property was reported for Anabaenopeptin KVJ811, which was capable of jeopardizing the growth of the strain *Nostoc* sp. KVJ11, pointing out the importance of these oligopeptides in populational control [21].

Given the above importance of these agents in the environment, the ecological functions of APs in the aquatic environment are much more extensive than we have already known. They are remarkably diverse in structural and functional terms. Novel Anabaenopeptins have been constantly isolated and identified. From an ecological and evolutionary perspective, these cyclopeptides allow communication with different organisms and are decisive elements in natural selection.

6. Applications of Anabaenopeptins

Cyanopeptides such as APs have a well-demonstrated capacity of protease inhibition [141]. Protein Phosphatase 1 (PP1), Protein Phosphatase 2A, Carboxypeptidase-A (CPA), Human Serine Protease, Leucine Aminopeptidase, Trypsin, and Thrombin have already been tested against several cyanobacterial extracts and confirmed the catalysis blockage [11].

Among the enzymes listed above, APs were more effective inhibitors against CPA, PP1, and elastase. In a cyanobacteria bloom, it was isolated eight different Anabaenopeptins which showed activity towards CPA and protein phosphatase 1 [34,142]. PP1 inhibition may be influencing the HIV-1 transcription, cancer, or cardiac hypertrophy, for example [143,144]. Some APs half-maximal inhibitory concentration (IC_{50}) values are presented in Table 3.

Table 3. Detected IC_{50} values of some Anabaenopeptins in nM. (TAFIa: Thrombin Activatable Fibrinolysis Inhibitor; PP1: Protein Phosphatase 1).

Anabaenopeptin	TAFIa (nM)	Carboxipeptidase A (nM)	PP1 (nM)	References
Anabaenopeptin 679	-	6740	-	[53]
Anabaenopeptin 908A	1.8	>11,000	-	[12,53]
Anabaenopeptin 915	530	130	-	[12,53]
Anabaenopeptin A	440	-	104,300	[12,34]
Anabaenopeptin B	1.5	>60,000; 3900	119,500	[12,145,146]
Anabaenopeptin C	1.9	>100,000	-	[12,145]
Anabaenopeptin E	-	>60,000	-	[53]
Anabaenopeptin F	1.5	>60,000; 1100	44770	[12,25,53]
Anabaenopeptin G	-	2–8	-	[53]
Anabaenopeptin H	-	3700–10,200	-	[53]
Anabaenopeptin I	-	7	-	[53]
Anabaenopeptin J	-	10	-	[53]
Anabaenopeptin SA1	2.2	-	-	[12]
Anabaenopeptin SA10	7000	-	-	[12]
Anabaenopeptin SA11	15000	-	-	[12]
Anabaenopeptin SA12	4300	-	-	[12]
Anabaenopeptin SA13	2500	-	-	[12]
Anabaenopeptin SA2	16	-	-	[12]
Anabaenopeptin SA3	2.1	-	-	[12]
Anabaenopeptin SA4	3.4	-	-	[12]
Anabaenopeptin SA5	790	-	-	[12]
Anabaenopeptin SA6	51000	-	-	[12]
Anabaenopeptin SA7	13000	-	-	[12]
Anabaenopeptin SA8	4800	-	-	[12]
Anabaenopeptin SA9	31000	-	-	[12]
Anabaenopeptin T	-	3–2300	-	[53]
Oscillamide Y	400	-	72.26	[12,34]

Serine/threonine protein phosphatases inhibition was also reported [22,25]. Nevertheless, several other cyanopeptides presented more effective IC_{50} levels against elastase, such as some variants of lyngbyastatins, symplostatins, microvirins, and others. Concerning PP1, MCs remain the best inhibitor among all cyanopeptides [11]. IC_{50} reported values to MCs and nodularins are from 1.1 to 1.9 nM as PP1 inhibitors [147]. In this case, APs remain promising candidates in Carboxypeptidase inhibition.

Cyanopeptides blooms events may present the production of different classes of cyanopeptides like MCs, APs, and cyanopeptolins. A few studies quantified cyanopeptides beyond Microcystins, even so, in 10 eutrophic lakes in the United States and Europe the cyanopeptides concentration including these 3 types of cyanopeptides were from <4 µg/L to >40 µg/L [11]. In wet weight, 2.1 mg of AP and 7.4 mg of Microcystin-LR were obtained from 1.7 kg of biomass in a water bloom of lake Teganuma (Japan) [41].

In a study conducted by Spoof and coworkers [34], the production range of the APs measured in extracts from cyanobacteria sampled by plankton net was from 1.7 to 181.9 µg/mL in 22 isolated Anabaenopeptins. Bioactivity assays identified IC_{50} values from 16 to 435 ng/mL (Nodulapeptin 933 and Anabaenopeptin 813, respectively) against PP1 and from below 3 to 45 µg/mL against CPA (Anabaenopeptins A, D and Nodulapeptin 883C and 917: <3 µg/mL; Nodulapeptin 867: 45 µg/mL). The inhibition of elastase, trypsin, or thrombin does not occur independently of the exocyclic residue (Phe, Ile, and Tyr). The residues adjacent to the ureido bond have a major influence on CPA inhibition. Therefore, APs with Ile and Tyr in the exocyclic position presented the best IC_{50} values against this enzyme. Thus, hydrophobic aromatic or linear sidechain next to the ureido moiety presents more favorable interactions with CPA while positive amino acids such as Arg are unfavorable. It explains why Anabaenopeptin B presents IC_{50}: >20 µg/mL and Anabaenopeptin 679 (different only in this position) had an improved inhibitory activity IC_{50}: 4.6 µg/mL [53].

Anabaenopeptins B and F presented activity against human leukocyte elastase (HLE) and porcine pancreatic elastase (PE). Ki values of HLE inhibition were in the 0.1–1 µM range in a linear competitive model [148]. In another study, APs A and B were capable of relaxing rat aortic preparations in a concentration-dependent form using 10–400 µg/mL [20].

Some studies have been explored APs bioactive properties in a pharmaceutical/ biotechnological way. Despite APs ability to inhibit diverse proteases, other cyanopeptides present the best IC_{50} values than them in most cases. However, one application shows more promising results using APs: the inhibition of the Thrombin Activatable Fibrinolysis Inhibitor (TAFIa), which is a proteolytic enzyme that cleaves Arg and Lys residues on fibrin and may be a novel antithrombotic mechanism [149]. Anabaenopeptins B, C, and F, isolated from *Planktothrix rubescens*, presented high promising results inhibiting TAFIa selectively over other coagulation enzymes as Carboxypeptidases A, B and N, FXa, FVIIa, FIIa, and FXIa [12,145]. In this sense, Anabaenopeptin B showed the best values of IC_{50} (1.5 nM, in different studies, similar to PP1 inhibition by microcystins) on a screening performed with 20 APs isolated from *Nostoc* and *Planktothrix* strains. It was elucidated that Lys and Arg residues in the R1 position (considering Anabaenopeptin B as reference: Arg-Lys-Val-Hty-MeAla-Phe) are associated with high activity (IC_{50} values of 2.1 and 1.5 nM, respectively) since those structures presenting Tyr residue in this position showed a significant decline of activity by two orders of magnitude (IC_{50} of 400 nM). In the R3 position, it was observed that the substitution of Ile by Val does not affect activity. R5 position also presents a loss of potency when the residue Ala is replaced by Ser. It is also observed a high tolerance towards substitutions in the pentacyclic region [12]. The Val to Ile difference among APs B and F does not implicate in gain or loss of activity against TAFIa. The mechanism involved in TAFIa inhibition depends on the linear part of APs mimics the carboxy−terminus of fibrin which is able of penetrating the active site pocket. Hence, the circular fraction of APs blocks the channel's entrance, preventing the interaction with other molecules [12]. To compare the interactions of the Anabaenopeptin B-TAFIa complex with Microcystin LR-PP1 complex, TAFIa structure was obtained from Protein Data Bank (PDB), it was resolved by X-ray diffraction presenting 2,5 Å resolution (PDB code: 3LMS) (Figure 13). Also, 1.84 Å resolution PP1 in complex with Microcystin-LR was used to represent the binding mechanism (PDB code: 6OBQ). Microcystin MeAsp residue blocks the access to the PP1 active site, the long hydrophobic tail composed of Adda residues plays an important role in this inhibition due to its interaction with the hydrophobic groove region, adjacent to the catalytic site [45]. Yet, different from the linear part of Anabaenopeptin B which accesses a protein channel, Adda residue in Microcystin-LR makes contact with superficial residues of PP1 (Figure 13).

Besides some cyanopeptides presented anticancer activity, APs have been presented poor results in cytotoxic tests [150]. Anabaenopeptin B had been tested about its anticancer potential and did not demonstrate cytotoxic effects against N2a, MCF−7, and GH4 cells even at the 500 µg/mL concentration [151]. Despite anticancer activity was detected in *Aliinostoc* sp. CENA543 extract containing AP, it was not possible to attribute this effect exclusively to this class of oligopeptides because there were other cyanopeptides in the extract, and the exact AP was not identified [152]. No cytotoxic activity was presented by Nodulapeptins 883C, 869, 867, 865, and Anabaenopeptin 813 as well [34].

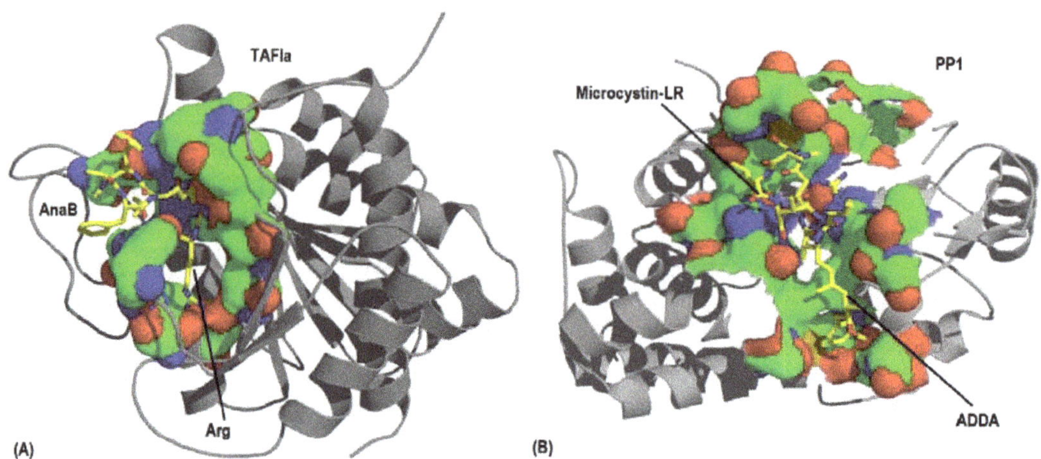

Figure 13. Interaction between (**A**) Anabaenopeptin B (AnaB) with Thrombin Activatable Fibrinolysis Inhibitor (TAFIa) and (**B**) the complex between Microcystin-LR and Protein Phosphatase 1 (PP1).

7. Final Considerations

Anabaenopeptins are structurally diverse molecules widely distributed in distinct ecosystems. Some structural features of these oligopeptides are shared with other cyanotoxins, such as the presence of modified residues, exocyclic amino acids, circular structure, and amino acids in D-configuration. However, among the cyanopeptides, the ureido linkage is exclusively found in APs. Despite their elevated occurrence and structural diversity, the majority of this group of peptides has been isolated from filamentous cyanobacteria, being commonly associated with specific genera. In freshwater environments, APs B and F are the most recurrent whereas the marine strains normally display a higher number of exclusive APs. Terrestrial cyanobacteria possess in common few APs with those from the aquatic environment.

The production of these toxins is influenced by environmental factors, which include nutrient concentration, temperature, light intensity, and association with other organisms. NRPS apparatus mediates the APs biosynthesis, following the collinearity rule of the modules. The low specificity of adenylation domains toward their substrate and the presence of additional modules are responsible for the production of various variants by a single strain. Homoamino acids present in the APs structure are supplied by HphABCD biosynthetic pathway. Some modifications are catalyzed by specific domains encountered in the NRPS modules. However, the mechanism of ureido bond formation is still unknown.

APs have been increasingly detected in reservoirs, lakes, and oceans in very elevated concentrations. Their toxicity for human beings has not yet been determined. However, assays employing animal models as well as other organisms have demonstrated their deleterious actions. In face of this fact, there is a dire need to further investigate the real impact of these oligopeptides on human health. The inhibitory activity of these molecules against proteases, phosphatases, and carboxylases makes them very promising for biotechnological use, but their mechanisms of action need to be investigated in detail to be properly applied.

However, Anabaenopeptins still require further studies to comprehend their behaviors in nature. Among AP producers, it must be evaluated the evolutionary relationship between terrestrial and aquatic strains as they do not share a high number of AP variants, similar to freshwater and marine. Besides, specific residues are more predominantly found in some environments, requiring additional analysis to comprehend the relationship between their frequency and habitats. Moreover, just a few variants have been analyzed regarding their inhibitory properties, then demanding more tests to discern the role of specific amino acids during the interaction with their targets. Another issue that must be investigated deeply is the APs amount produced by cyanobacteria. Their low yield can be a limited factor for industrial purposes. Such bottlenecks can be minimized through the use of heterologous expression, which has been well established for other cyanopeptides.

Supplementary Materials: The following are available online at https://www.mdpi.com/article/10.3390/toxins13080522/s1, Table S1: Sequence composition of Anabaenopeptins.

Author Contributions: Conceptualization, P.R.M., S.C.d.A., L.P.X., and A.V.S. Investigation, P.R.M. and S.C.d.A. Writing—original draft preparation, P.R.M., S.C.d.A. and A.S.S. Writing—review and editing, P.R.M., S.C.d.A., A.V.S. and L.P.X. Supervision, A.V.S. and L.P.X. All authors have read and agreed to the published version of the manuscript.

Funding: This study was financed in part by Coordenação de Aperfeiçoamento de Pessoal de Nível Superior—Brasil (CAPES)—Finance Code 001 and Fundação Amazônia Paraense de Amparo a Estudos e Pesquisas (FAPESPA)—03/2019.

Institutional Review Board Statement: Not applied.

Informed Consent Statement: Not applied.

Data Availability Statement: Not applied.

Acknowledgments: The authors would like to thank Pró–Reitoria de Pesquisa e Pós–Graduação da Universidade Federal do Pará (PROPESP/UFPA).

Conflicts of Interest: The authors declare no conflict of interest.

References

1. do Amaral, S.C.; Santos, A.V.; da Cruz Schneider, M.P.; da Silva, J.K.R.; Xavier, L.P. Determination of Volatile Organic Compounds and Antibacterial Activity of the Amazonian Cyanobacterium Synechococcus sp. Strain GFB01. *Molecules* **2020**, *25*, 4744. [CrossRef]
2. Gradíssimo, D.G.; Xavier, L.P.; Santos, A.V. Cyanobacterial Polyhydroxyalkanoates: A Sustainable Alternative in Circular Economy. *Molecules* **2020**, *25*, 4331. [CrossRef]
3. de Oliveira, D.T.; da Costa, A.A.F.; Costa, F.F.; da Rocha Filho, G.N.; do Nascimento, L.A.S. Advances in the Biotechnological Potential of Brazilian Marine Microalgae and Cyanobacteria. *Molecules* **2020**, *25*, 2908. [CrossRef] [PubMed]
4. Huisman, J.; Codd, G.A.; Paerl, H.W.; Ibelings, B.W.; Verspagen, J.M.H.; Visser, P.M. Cyanobacterial blooms. *Nat. Rev. Microbiol.* **2018**, *16*, 471–483. [CrossRef] [PubMed]
5. Bittner, M.; Štern, A.; Smutná, M.; Hilscherová, K.; Žegura, B. Cytotoxic and Genotoxic Effects of Cyanobacterial and Algal Extracts—Microcystin and Retinoic Acid Content. *Toxins* **2021**, *13*, 107. [CrossRef]
6. Chen, G.; Wang, L.; Li, W.; Zhang, Q.; Hu, T. Nodularin induced oxidative stress contributes to developmental toxicity in zebrafish embryos. *Ecotoxicol. Environ. Saf.* **2020**, *194*, 110444. [CrossRef]
7. Zhu, H.; Sonoyama, T.; Yamada, M.; Gao, W.; Tatsuno, R.; Takatani, T.; Arakawa, O. Co-Occurrence of Tetrodotoxin and Saxitoxins and Their Intra-Body Distribution in the Pufferfish Canthigaster valentini. *Toxins* **2020**, *12*, 436. [CrossRef]
8. Moosová, Z.; Šindlerová, L.; Ambrůzová, B.; Ambrožová, G.; Vašíček, O.; Velki, M.; Babica, P.; Kubala, L. Lipopolysaccharides from Microcystis Cyanobacteria-Dominated Water Bloom and from Laboratory Cultures Trigger Human Immune Innate Response. *Toxins* **2019**, *11*, 218. [CrossRef] [PubMed]
9. Massey, I.Y.; Yang, F. A Mini Review on Microcystins and Bacterial Degradation. *Toxins* **2020**, *12*, 268. [CrossRef]
10. Jones, M.R.; Pinto, E.; Torres, M.A.; Dörr, F.; Mazur-Marzec, H.; Szubert, K.; Tartaglione, L.; Dell'Aversano, C.; Miles, C.O.; Beach, D.G.; et al. CyanoMetDB, a comprehensive public database of secondary metabolites from cyanobacteria. *Water Res.* **2021**, *196*, 117017. [CrossRef]
11. Janssen, E.M.L. Cyanobacterial peptides beyond microcystins—A review on co-occurrence, toxicity, and challenges for risk assessment. *Water Res.* **2019**, *151*, 488–499. [CrossRef]
12. Schreuder, H.; Liesum, A.; Lönze, P.; Stump, H.; Hoffmann, H.; Schiell, M.; Kurz, M.; Toti, L.; Bauer, A.; Kallus, C.; et al. Isolation, Co-Crystallization and Structure-Based Characterization of Anabaenopeptins as Highly Potent Inhibitors of Activated Thrombin Activatable Fibrinolysis Inhibitor (TAFIa). *Sci. Rep.* **2016**, *6*, 32958. [CrossRef] [PubMed]
13. do Amaral, S.C.; Monteiro, P.R.; da Neto, J.S.P.; Serra, G.M.; Gonçalves, E.C.; Xavier, L.P.; Santos, A.V. Current Knowledge on Microviridin from Cyanobacteria. *Mar. Drugs* **2021**, *19*, 17. [CrossRef]
14. Mazur-Marzec, H.; Kaczkowska, M.; Blaszczyk, A.; Akcaalan, R.; Spoof, L.; Meriluoto, J. Diversity of Peptides Produced by Nodularia spumigena from Various Geographical Regions. *Mar. Drugs* **2013**, *11*, 1–19. [CrossRef] [PubMed]

15. Gkelis, S.; Lanaras, T.; Sivonen, K. Cyanobacterial Toxic and Bioactive Peptides in Freshwater Bodies of Greece: Concentrations, Occurrence Patterns, and Implications for Human Health. *Mar. Drugs* **2015**, *13*, 6319–6335. [CrossRef] [PubMed]

16. Rohrlack, T.; Christiansen, G.; Kurmayer, R. Putative Antiparasite Defensive System Involving Ribosomal and Nonribosomal Oligopeptides in Cyanobacteria of the Genus Planktothrix. *Appl. Environ. Microbiol.* **2013**, *79*, 2642–2647. [CrossRef] [PubMed]

17. Martins, J.; Vasconcelos, V. Cyanobactins from Cyanobacteria: Current Genetic and Chemical State of Knowledge. *Mar. Drugs* **2015**, *13*, 6910–6946. [CrossRef]

18. Rounge, T.B.; Rohrlack, T.; Nederbragt, A.J.; Kristensen, T.; Jakobsen, K.S. A genome-wide analysis of nonribosomal peptide synthetase gene clusters and their peptides in a Planktothrix rubescens strain. *BMC Genom.* **2009**, *10*, 396. [CrossRef]

19. Welker, M.; Von Döhren, H. Cyanobacterial peptides—Nature's own combinatorial biosynthesis. *FEMS Microbiol. Rev.* **2006**, *30*, 530–563. [CrossRef]

20. Harada, K.I.; Fujii, K.; Shimada, T.; Suzuki, M.; Sano, H.; Adachi, K.; Carmichael, W.W. Two cyclic peptides, anabaenopeptins, a third group of bioactive compounds from the cyanobacterium Anabaena flos-aquae NRC 525-17. *Tetrahedron Lett.* **1995**, *36*, 1511–1514. [CrossRef]

21. Guljamow, A.; Kreische, M.; Ishida, K.; Liaimer, A.; Altermark, B.; Bähr, L.; Hertweck, C.; Ehwald, R.; Dittmann, E. High-Density Cultivation of Terrestrial Nostoc Strains Leads to Reprogramming of Secondary Metabolome. *Appl. Environ. Microbiol.* **2017**, *83*, 1–15. [CrossRef] [PubMed]

22. Zafrir-Ilan, E.; Carmeli, S. Eight novel serine proteases inhibitors from a water bloom of the cyanobacterium *Microcystis* sp. *Tetrahedron* **2010**, *66*, 9194–9202. [CrossRef]

23. Matthew, S.; Ross, C.; Paul, V.J.; Luesch, H. Pompanopeptins A and B, new cyclic peptides from the marine cyanobacterium Lyngbya confervoides. *Tetrahedron* **2008**, *64*, 4081–4089. [CrossRef]

24. Sanz, M.; Andreote, A.P.D.; Fiore, M.F.; Dörr, F.A.; Pinto, E. Structural characterization of new peptide variants produced by cyanobacteria from the Brazilian Atlantic Coastal Forest using liquid chromatography coupled to quadrupole time-of-flight tandem mass spectrometry. *Mar. Drugs* **2015**, *13*, 3892–3919. [CrossRef]

25. Sano, T.; Usui, T.; Ueda, K.; Osada, H.; Kaya, K. Isolation of new protein phosphatase inhibitors from two cyanobacteria species, *Planktothrix* spp. *J. Nat. Prod.* **2001**, *64*, 1052–1055. [CrossRef]

26. Itou, Y.; Suzuki, S.; Ishida, K.; Murakami, M. Anabaenopeptins G and H, potent carboxypeptidase A inhibitors from the cyanobacterium *Oscillatoria agardhii* (NIES-595). *Bioorganic Med. Chem. Lett.* **1999**, *9*, 1243–1246. [CrossRef]

27. Tonk, L.; Welker, M.; Huisman, J.; Visser, P.M. Production of cyanopeptolins, anabaenopeptins, and microcystins by the harmful cyanobacteria Anabaena 90 and Microcystis PCC 7806. *Harmful Algae* **2009**, *8*, 219–224. [CrossRef]

28. Plaza, A.; Keffer, J.L.; Lloyd, J.R.; Colin, P.L.; Bewley, C.A. Paltolides A−C, Anabaenopeptin-Type Peptides from the Palau Sponge *Theonella swinhoei*. *J. Nat. Prod.* **2010**, *73*, 485–488. [CrossRef] [PubMed]

29. Schmidt, E.W.; Harper, M.K.; Faulkner, D.J. Mozamides A and B, cyclic peptides from a theonellid sponge from Mozambique. *J. Nat. Prod.* **1997**, *60*, 779–782. [CrossRef]

30. Uemoto, H.; Yahiro, Y.; Shigemori, H.; Tsuda, M.; Takao, T.; Shimonishi, Y.; Kobayashi, J. Keramamides K and L, new cyclic peptides containing unusual tryptophan residue from *Theonella sponge*. *Tetrahedron* **1998**, *54*, 6719–6724. [CrossRef]

31. Kobayashi, J.; Sato, M.; Murayama, T.; Ishibashi, M.; Wälchi, M.R.; Kanai, M.; Shoji, J.; Ohizumi, Y. Konbamide, a novel peptide with calmodulin antagonistic activity from the Okinawan marine sponge Theonella sp. *J. Chem. Soc. Chem. Commun.* **1991**, 1050–1052. [CrossRef]

32. Kobayashi, J.; Sato, M.; Ishibashi, M.; Shigemori, H.; Nakamura, T.; Ohizumi, Y. Keramamide A, a novel peptide from the Okinawan marine sponge *Theonella* sp. *J. Chem. Soc. Perkin Trans.* **1991**, 2609. [CrossRef]

33. Robinson, S.J.; Tenney, K.; Yee, D.F.; Martinez, L.; Media, J.E.; Valeriote, F.A.; van Soest, R.W.M.; Crews, P. Probing the Bioactive Constituents from Chemotypes of the Sponge *Psammocinia* aff. bulbosa. *J. Nat. Prod.* **2007**, *70*, 1002–1009. [CrossRef] [PubMed]

34. Spoof, L.; Błaszczyk, A.; Meriluoto, J.; Cegłowska, M.; Mazur-Marzec, H. Structures and Activity of New Anabaenopeptins Produced by Baltic Sea Cyanobacteria. *Mar. Drugs* **2015**, *14*, 8. [CrossRef] [PubMed]

35. Greunke, C.; Duell, E.R.; D'Agostino, P.M.; Glöckle, A.; Lamm, K.; Gulder, T.A.M. Direct Pathway Cloning (DiPaC) to unlock natural product biosynthetic potential. *Metab. Eng.* **2018**, *47*, 334–345. [CrossRef]

36. Sano, T.; Kaya, K. Oscillamide Y, a chymotrypsin inhibitor from toxic *Oscillatoria agardhii*. *Tetrahedron Lett.* **1995**, *36*, 5933–5936. [CrossRef]

37. Murakami, M.; Shin, H.J.; Matsuda, H.; Ishida, K.; Yamaguchi, K. A cyclic peptide, anabaenopeptin B, from the cyanobacterium *Oscillatoria agardhii*. *Phytochemistry* **1997**, *44*, 449–452. [CrossRef]

38. Shin, H.J.; Matsuda, H.; Murakami, M.; Yamaguchi, K. Anabaenopeptins E and F, two new cyclic peptides from the cyanobacterium *Oscillatoria agardhii* (NIES-204). *J. Nat. Prod.* **1997**, *60*, 139–141. [CrossRef]

39. Williams, D.E.; Craig, M.; Holmes, C.F.B.; Andersen, R.J. Ferintoic acids A and B, new cyclic hexapeptides from the freshwater cyanobacterium *Microcystis aeruginosa*. *J. Nat. Prod.* **1996**, *59*, 570–575. [CrossRef]

40. Erhard, M.; Von Döhren, H.; Jungblut, P.R. Rapid identification of the new anabaenopeptin G from *Planktothrix agardhii* HUB 011 using matrix-assisted laser desorption/ionization time-of-flight mass spectrometry. *Rapid Commun. Mass Spectrom.* **1999**, *13*, 337–343. [CrossRef]

41. Kodani, S.; Suzuki, S.; Ishida, K.; Murakami, M. Five new cyanobacterial peptides from water bloom materials of lake Teganuma (Japan). *FEMS Microbiol. Lett.* **1999**, *178*, 343–348. [CrossRef]

42. Murakami, M.; Suzuki, S.; Itou, Y.; Kodani, S.; Ishida, K. New Anabaenopeptins, Potent Carboxypeptidase-A Inhibitors from the Cyanobacterium Aphanizomenon flos-aquae. *J. Nat. Prod.* **2000**, *63*, 1280–1282. [CrossRef] [PubMed]

43. Mi, Y.; Zhang, J.; He, S.; Yan, X. New Peptides Isolated from Marine Cyanobacteria, an Overview over the Past Decade. *Mar. Drugs* **2017**, *15*, 132. [CrossRef]

44. Fujii, K.; Harada, K.; Suzuki, M.; Kondo, F.; Ikai, Y.; Oka, H.; Sivonen, K. Novel cyclic peptides together with microcystins produced by toxic cyanobacteria, *Anabaena* sp. *Symp. Chem. Nat. Prod. Symp. Pap.* **1995**, *37*, 445–450. [CrossRef]

45. Campos, A.; Vasconcelos, V. Molecular mechanisms of microcystin toxicity in animal cells. *Int. J. Mol. Sci.* **2010**, *11*, 268–287. [CrossRef] [PubMed]

46. Reshef, V.; Carmeli, S. Schizopeptin 791, a new anabeanopeptin-like cyclic peptide from the cyanobacterium *Schizothrix* sp. *J. Nat. Prod.* **2002**, *65*, 1187–1189. [CrossRef]

47. Zi, J.; Lantvit, D.D.; Swanson, S.M.; Orjala, J. Lyngbyaureidamides A and B, two anabaenopeptins from the cultured freshwater cyanobacterium *Lyngbya* sp. (SAG 36.91). *Phytochemistry* **2012**, *74*, 173–177. [CrossRef] [PubMed]

48. Fujii, K.; Sivonen, K.; Adachi, K.; Noguchi, K.; Sano, H.; Hirayama, K.; Suzuki, M.; Harada, K. Comparative study of toxic and non-toxic cyanobacterial products: Novel peptides from toxic *Nodularia spumigena* AV1. *Tetrahedron Lett.* **1997**, *38*, 5525–5528. [CrossRef]

49. Mazur-Marzec, H.; Bertos-Fortis, M.; Toruńska-Sitarz, A.; Fidor, A.; Legrand, C. Chemical and genetic diversity of nodularia spumigena from the baltic sea. *Mar. Drugs* **2016**, *14*, 209. [CrossRef]

50. Müller, D.; Krick, A.; Kehraus, S.; Mehner, C.; Hart, M.; Küpper, F.C.; Saxena, K.; Prinz, H.; Schwalbe, H.; Janning, P.; et al. Brunsvicamides A-C: Sponge-related cyanobacterial peptides with Mycobacterium tuberculosis protein tyrosine phosphatase inhibitory activity. *J. Med. Chem.* **2006**, *49*, 4871–4878. [CrossRef]

51. Bjoerquist, P.; Buchanan, M.; Campitelli, M.; Carroll, A.; Hyde, E.; Neve, J.; Polla, M.; Quinn, R. Use of cyclic anabaenopeptin-type peptides for the treatment of a condition wherein inhibition of carboxypeptidase U is beneficial, novel anabaenopeptin derivatives and intermediates thereof. U.S. Patent WO2005039617, 6 May 2005.

52. Beresovsky, D.; Hadas, O.; Livne, A.; Sukenik, A.; Kaplan, A.; Carmeli, S. Toxins and Biologically Active Secondary Metabolites of Microcystis sp. isolated from Lake Kinneret. *Isr. J. Chem.* **2006**, *46*, 79–87. [CrossRef]

53. Harms, H.; Kurita, K.L.; Pan, L.; Wahome, P.G.; He, H.; Kinghorn, A.D.; Carter, G.T.; Linington, R.G. Discovery of anabaenopeptin 679 from freshwater algal bloom material: Insights into the structure–activity relationship of anabaenopeptin protease inhibitors. *Bioorganic Med. Chem. Lett.* **2016**, *26*, 4960–4965. [CrossRef] [PubMed]

54. Cheruku, P.; Plaza, A.; Lauro, G.; Keffer, J.; Lloyd, J.R.; Bifulco, G.; Bewley, C.A. Discovery and Synthesis of Namalide Reveals a New Anabaenopeptin Scaffold and Peptidase Inhibitor. *J. Med. Chem.* **2012**, *55*, 735–742. [CrossRef]

55. Sanz, M.; Salinas, R.K.; Pinto, E. Namalides B and C and Spumigins K-N from the Cultured Freshwater Cyanobacterium Sphaerospermopsis torques-reginae. *J. Nat. Prod.* **2017**, *80*, 2492–2501. [CrossRef]

56. Shishido, T.K.; Jokela, J.; Fewer, D.P.; Wahlsten, M.; Fiore, M.F.; Sivonen, K. Simultaneous Production of Anabaenopeptins and Namalides by the Cyanobacterium *Nostoc* sp. CENA543. *ACS Chem. Biol.* **2017**, *12*, 2746–2755. [CrossRef] [PubMed]

57. Entfellner, E.; Frei, M.; Christiansen, G.; Deng, L.; Blom, J.; Kurmayer, R. Evolution of Anabaenopeptin Peptide Structural Variability in the Cyanobacterium Planktothrix. *Front. Microbiol.* **2017**, *8*. [CrossRef]

58. Wang, H.; Fewer, D.P.; Sivonen, K. Genome Mining Demonstrates the Widespread Occurrence of Gene Clusters Encoding Bacteriocins in Cyanobacteria. *PLoS ONE* **2011**, *6*, e22384. [CrossRef]

59. Hallenbeck, P.C. (Ed.) *Microbial Technologies in Advanced Biofuels Production*; Springer: Boston, MA, USA, 2012; Volume 1, ISBN 978-1-4614-1207-6.

60. Natumi, R.; Janssen, E.M.L. Cyanopeptide Co-Production Dynamics beyond Mirocystins and Effects of Growth Stages and Nutrient Availability. *Environ. Sci. Technol.* **2020**, *54*, 6063–6072. [CrossRef]

61. Gesner-Apter, S.; Carmeli, S. Protease Inhibitors from a Water Bloom of the Cyanobacterium Microcystis aeruginosa. *J. Nat. Prod.* **2009**, *72*, 1429–1436. [CrossRef]

62. Elkobi-Peer, S.; Carmeli, S. New Prenylated Aeruginosin, Microphycin, Anabaenopeptin and Micropeptin Analogues from a Microcystis Bloom Material Collected in Kibbutz Kfar Blum, Israel. *Mar. Drugs* **2015**, *13*, 2347–2375. [CrossRef]

63. Okumura, H.S.; Philmus, B.; Portmann, C.; Hemscheidt, T.K. Homotyrosine-Containing Cyanopeptolins 880 and 960 and Anabaenopeptins 908 and 915 from Planktothrix agardhii CYA 126/8. *J. Nat. Prod.* **2009**, *72*, 172–176. [CrossRef] [PubMed]

64. Tooming-Klunderud, A.; Sogge, H.; Rounge, T.B.; Nederbragt, A.J.; Lagesen, K.; Glöckner, G.; Hayes, P.K.; Rohrlack, T.; Jakobsen, K.S. From Green to Red: Horizontal Gene Transfer of the Phycoerythrin Gene Cluster between Planktothrix Strains. *Appl. Environ. Microbiol.* **2013**, *79*, 6803–6812. [CrossRef] [PubMed]

65. Welker, M.; Erhard, M. Consistency between chemotyping of single filaments ofPlanktothrix rubescens (cyanobacteria) by MALDI-TOF and the peptide patterns of strains determined by HPLC-MS. *J. Mass Spectrom.* **2007**, *42*, 1062–1068. [CrossRef]

66. Rohrlack, T.; Edvardsen, B.; Skulberg, R.; Halstvedt, C.B.; Utkilen, H.C.; Ptacnik, R.; Skulberg, O.M. Oligopeptide chemotypes of the toxic freshwater cyanobacterium Planktothrix can form sub-populations with dissimilar ecological traits. *Limnol. Oceanogr.* **2008**, *53*, 1279–1293. [CrossRef]

67. Sedmak, B.; Carmeli, S.; Eleršek, T. "Non-Toxic" Cyclic Peptides Induce Lysis of Cyanobacteria—An Effective Cell Population Density Control Mechanism in Cyanobacterial Blooms. *Microb. Ecol.* **2008**, *56*, 201–209. [CrossRef] [PubMed]

68. Grach-Pogrebinsky, O.; Sedmak, B.; Carmeli, S. Protease inhibitors from a Slovenian Lake Bled toxic waterbloom of the cyanobacterium *Planktothrix rubescens*. *Tetrahedron* **2003**, *59*, 8329–8336. [CrossRef]

69. Vasas, G.; Farkas, O.; Borics, G.; Felföldi, T.; Sramkó, G.; Batta, G.; Bácsi, I.; Gonda, S. Appearance of Planktothrix rubescens Bloom with [D-Asp3, Mdha7]MC–RR in Gravel Pit Pond of a Shallow Lake-Dominated Area. *Toxins* **2013**, *5*, 2434–2455. [CrossRef]

70. Bober, B.; Chrapusta-Srebrny, E.; Bialczyk, J. Novel cyanobacterial metabolites, cyanopeptolin 1081 and anabaenopeptin 899, isolated from an enrichment culture dominated by Woronichinia naegeliana (Unger) Elenkin. *Eur. J. Phycol.* **2020**, 1–11. [CrossRef]

71. Grach-Pogrebinsky, O.; Carmeli, S. Three novel anabaenopeptins from the cyanobacterium Anabaena sp. *Tetrahedron* **2008**, *64*, 10233–10238. [CrossRef]

72. Häggqvist, K.; Toruńska-Sitarz, A.; Błaszczyk, A.; Mazur-Marzec, H.; Meriluoto, J. Morphologic, Phylogenetic and Chemical Characterization of a Brackish Colonial Picocyanobacterium (Coelosphaeriaceae) with Bioactive Properties. *Toxins* **2016**, *8*, 108. [CrossRef]

73. Popin, R.V.; Delbaje, E.; de Abreu, V.A.C.; Rigonato, J.; Dörr, F.A.; Pinto, E.; Sivonen, K.; Fiore, M.F. Genomic and Metabolomic Analyses of Natural Products in Nodularia spumigena Isolated from a Shrimp Culture Pond. *Toxins* **2020**, *12*, 141. [CrossRef] [PubMed]

74. Saha, S.; Esposito, G.; Urajová, P.; Mareš, J.; Ewe, D.; Caso, A.; Macho, M.; Delawská, K.; Kust, A.; Hrouzek, P.; et al. Discovery of Unusual Cyanobacterial Tryptophan-Containing Anabaenopeptins by MS/MS-Based Molecular Networking. *Molecules* **2020**, *25*, 3786. [CrossRef]

75. Nowruzi, B.; Khavari-Nejad, R.-A.; Sivonen, K.; Kazemi, B.; Najafi, F.; Nejadsattari, T. Identification and toxigenic potential of a Nostoc sp. *Algae* **2012**, *27*, 303–313. [CrossRef]

76. Grabowska, M.; Kobos, J.; Toruńska-Sitarz, A.; Mazur-Marzec, H. Non-ribosomal peptides produced by Planktothrix agardhii from Siemianówka Dam Reservoir SDR (northeast Poland). *Arch. Microbiol.* **2014**, *196*, 697–707. [CrossRef] [PubMed]

77. de Carvalho, L.R.; Pipole, F.; Werner, V.R.; Laughinghous, H.D., IV; de Camargo, A.C.M.; Rangel, M.; Konno, K.; Sant' Anna, C.L. A toxic cyanobacterial bloom in an urban coastal lake, Rio Grande do Sul state, Southern Brazil. *Braz. J. Microbiol.* **2008**, *39*, 761–769. [CrossRef]

78. Teta, R.; Della Sala, G.; Glukhov, E.; Gerwick, L.; Gerwick, W.H.; Mangoni, A.; Costantino, V. Combined LC–MS/MS and Molecular Networking Approach Reveals New Cyanotoxins from the 2014 Cyanobacterial Bloom in Green Lake, Seattle. *Environ. Sci. Technol.* **2015**, *49*, 14301–14310. [CrossRef]

79. Beversdorf, L.; Weirich, C.; Bartlett, S.; Miller, T. Variable Cyanobacterial Toxin and Metabolite Profiles across Six Eutrophic Lakes of Differing Physiochemical Characteristics. *Toxins* **2017**, *9*, 62. [CrossRef]

80. Bartlett, S.L.; Brunner, S.L.; Klump, J.V.; Houghton, E.M.; Miller, T.R. Spatial analysis of toxic or otherwise bioactive cyanobacterial peptides in Green Bay, Lake Michigan. *J. Great Lakes Res.* **2018**, *44*, 924–933. [CrossRef]

81. Roy-Lachapelle, A.; Solliec, M.; Sauvé, S.; Gagnon, C. A Data-Independent Methodology for the Structural Characterization of Microcystins and Anabaenopeptins Leading to the Identification of Four New Congeners. *Toxins* **2019**, *11*, 619. [CrossRef]

82. Flores, C.; Caixach, J. High Levels of Anabaenopeptins Detected in a Cyanobacteria Bloom from N.E. Spanish Sau-Susqueda-El Pasteral Reservoirs System by LC–HRMS. *Toxins* **2020**, *12*, 541. [CrossRef] [PubMed]

83. Kust, A.; Řeháková, K.; Vrba, J.; Maicher, V.; Mareš, J.; Hrouzek, P.; Chiriac, M.-C.; Benedová, Z.; Tesařová, B.; Saurav, K. Insight into Unprecedented Diversity of Cyanopeptides in Eutrophic Ponds Using an MS/MS Networking Approach. *Toxins* **2020**, *12*, 561. [CrossRef]

84. Riba, M.; Kiss-Szikszai, A.; Gonda, S.; Boros, G.; Vitál, Z.; Borsodi, A.K.; Krett, G.; Borics, G.; Ujvárosi, A.Z.; Vasas, G. Microcystis Chemotype Diversity in the Alimentary Tract of Bigheaded Carp. *Toxins* **2019**, *11*, 288. [CrossRef] [PubMed]

85. Konkel, R.; Toruńska-Sitarz, A.; Cegłowska, M.; Eżerinskis, Ž.; Šapolaitė, J.; Mažeika, J.; Mazur-Marzec, H. Blooms of Toxic Cyanobacterium Nodularia spumigena in Norwegian Fjords During Holocene Warm Periods. *Toxins* **2020**, *12*, 257. [CrossRef] [PubMed]

86. Logares, R.; Bråte, J.; Bertilsson, S.; Clasen, J.L.; Shalchian-Tabrizi, K.; Rengefors, K. Infrequent marine-freshwater transitions in the microbial world. *Trends Microbiol.* **2009**, *17*, 414–422. [CrossRef]

87. Boopathi, T.; Ki, J.-S. Impact of Environmental Factors on the Regulation of Cyanotoxin Production. *Toxins* **2014**, *6*, 1951–1978. [CrossRef]

88. Repka, S.; Koivula, M.; Harjunpä, V.; Rouhiainen, L.; Sivonen, K. Effects of Phosphate and Light on Growth of and Bioactive Peptide Production by the Cyanobacterium Anabaena Strain 90 and Its Anabaenopeptilide Mutant. *Appl. Environ. Microbiol.* **2004**, *70*, 4551–4560. [CrossRef]

89. Tonk, L.; Visser, P.M.; Christiansen, G.; Dittmann, E.; Snelder, E.O.F.M.; Wiedner, C.; Mur, L.R. The Microcystin Composition of the Cyanobacterium Planktothrix agardhii Changes toward a More Toxic Variant with Increasing Light Intensity. *Appl. Environ. Microbiol.* **2005**, *71*, 5177–5181. [CrossRef]

90. Briand, E.; Bormans, M.; Gugger, M.; Dorrestein, P.C.; Gerwick, W.H. Changes in secondary metabolic profiles of Microcystis aeruginosa strains in response to intraspecific interactions. *Environ. Microbiol.* **2016**, *18*, 384–400. [CrossRef]

91. Hesse, K.; Dittmann, E.; Bornerm, T. Consequences of impaired microcystin production for light-dependent growth and pigmentation of Microcystis aeruginosa PCC 7806. *FEMS Microbiol. Ecol.* **2001**, *37*, 39–43. [CrossRef]

92. Ferrão-Filho, A.D.S.; Kozlowsky-Suzuki, B. Cyanotoxins: Bioaccumulation and Effects on Aquatic Animals. *Mar. Drugs* **2011**, *9*, 2729–2772. [CrossRef]

93. Pearson, L.A.; Crosbie, N.D.; Neilan, B.A. Distribution and conservation of known secondary metabolite biosynthesis gene clusters in the genomes of geographically diverse *Microcystis aeruginosa* strains. *Mar. Freshw. Res.* **2020**, *71*, 701–716. [CrossRef]

94. Teikari, J.; Österholm, J.; Kopf, M.; Battchikova, N.; Wahlsten, M.; Aro, E.-M.; Hess, W.R.; Sivonen, K. Transcriptomic and Proteomic Profiling of Anabaena sp. Strain 90 under Inorganic Phosphorus Stress. *Appl. Environ. Microbiol.* **2015**, *81*, 5212–5222. [CrossRef] [PubMed]

95. Burberg, C.; Petzoldt, T.; von Elert, E. Phosphate Limitation Increases Content of Protease Inhibitors in the Cyanobacterium *Microcystis aeruginosa*. *Toxins* **2020**, *12*, 33. [CrossRef] [PubMed]

96. Pereira, D.A.; Pimenta, A.M.C.; Giani, A. Profiles of toxic and non-toxic oligopeptides of *Radiocystis fernandoii* (Cyanobacteria) exposed to three different light intensities. *Microbiol. Res.* **2012**, *167*, 413–421. [CrossRef]

97. Seyedsayamdost, M.R.; Chandler, J.R.; Blodgett, J.A.V.; Lima, P.S.; Duerkop, B.A.; Oinuma, K.; Greenberg, E.P.; Clardy, J. Quorum-sensing-regulated bactobolin production by Burkholderia thailandensis E264. *Org. Lett.* **2010**, *12*, 716–719. [CrossRef]

98. Guo, X.; Liu, X.; Wu, L. The algicidal activity of *Aeromonas* sp. strain GLY-2107 against bloom-forming *Microcystis aeruginosa* is regulated by N-acyl homoserine lactone-mediated quorum sensing. *Environ. Microbiol.* **2016**, *18*, 3867–3883. [CrossRef]

99. Bähr, L.; Wüstenberg, A.; Ehwald, R. Two-tier vessel for photoautotrophic high-density cultures. *J. Appl. Phycol.* **2016**, *28*, 783–793. [CrossRef]

100. Wells, M.L.; Potin, P.; Craigie, J.S.; Raven, J.A.; Merchant, S.S.; Helliwell, K.E.; Smith, A.G.; Camire, M.E.; Brawley, S.H. Algae as nutritional and functional food sources: Revisiting our understanding. *J. Appl. Phycol.* **2017**, *29*, 949–982. [CrossRef]

101. Toporowska, M.; Mazur-Marzec, H.; Pawlik-Skowrońska, B. The Effects of Cyanobacterial Bloom Extracts on the Biomass, Chl-a, MC and Other Oligopeptides Contents in a Natural Planktothrix agardhii Population. *Int. J. Environ. Res. Public Health* **2020**, *17*, 2881. [CrossRef]

102. Mutalipassi, M.; Riccio, G.; Mazzella, V.; Galasso, C.; Somma, E.; Chiarore, A.; de Pascale, D.; Zupo, V. Symbioses of Cyanobacteria in Marine Environments: Ecological Insights and Biotechnological Perspectives. *Mar. Drugs* **2021**, *19*, 227. [CrossRef]

103. Carpenter, E.J.; Foster, R.A. Marine cyanobacteria symbioses. In *Cyanobacteria in Symbiosis*; Rai, A.N., Bergman, B., Rasmussen, U., Eds.; Kluwer Academic Publisher: Dordrecht, The Netherlands, 2002; pp. 10–17. ISBN 0306480050.

104. Shridhar, D.M.P.; Mahajan, G.; Kamat, V.; Naik, C.; Parab, R.; Thakur, N.; Mishra, P. Antibacterial Activity of 2-(2′,4′-Dibromophenoxy)-4,6-dibromophenol from *Dysidea granulosa*. *Mar. Drugs* **2009**, *7*, 464–471. [CrossRef]

105. Finking, R.; Marahiel, M.A. Biosynthesis of Nonribosomal Peptides. *Annu. Rev. Microbiol.* **2004**, *58*, 453–488. [CrossRef] [PubMed]

106. Süssmuth, R.D.; Mainz, A. Nonribosomal Peptide Synthesis-Principles and Prospects. *Angew. Chem. Int. Ed.* **2017**, *56*, 3770–3821. [CrossRef]

107. Lima, S.T.; Alvarenga, D.O.; Etchegaray, A.; Fewer, D.P.; Jokela, J.; Varani, A.M.; Sanz, M.; Dörr, F.A.; Pinto, E.; Sivonen, K.; et al. Genetic Organization of Anabaenopeptin and Spumigin Biosynthetic Gene Clusters in the Cyanobacterium *Sphaerospermopsis torques-reginae* ITEP-024. *ACS Chem. Biol.* **2017**, *12*, 769–778. [CrossRef] [PubMed]

108. Martínez-Núñez, M.A.; López, V.E.L. Nonribosomal peptides synthetases and their applications in industry. *Sustain. Chem. Process.* **2016**, *4*, 13. [CrossRef]

109. Siezen, R.J.; Khayatt, B.I. Natural products genomics. *Microb. Biotechnol.* **2008**, *1*, 275–282. [CrossRef]

110. Rouhiainen, L.; Jokela, J.; Fewer, D.P.; Urmann, M.; Sivonen, K. Two Alternative Starter Modules for the Non-Ribosomal Biosynthesis of Specific Anabaenopeptin Variants in *Anabaena* (Cyanobacteria). *Chem. Biol.* **2010**, *17*, 265–273. [CrossRef]

111. Christiansen, G.; Philmus, B.; Hemscheidt, T.; Kurmayer, R. Genetic Variation of Adenylation Domains of the Anabaenopeptin Synthesis Operon and Evolution of Substrate Promiscuity. *J. Bacteriol.* **2011**, *193*, 3822–3831. [CrossRef] [PubMed]

112. Kaljunen, H.; Schiefelbein, S.H.H.; Stummer, D.; Kozak, S.; Meijers, R.; Christiansen, G.; Rentmeister, A. Structural Elucidation of the Bispecificity of A Domains as a Basis for Activating Non-natural Amino Acids. *Angew. Chemie Int. Ed.* **2015**, *54*, 8833–8836. [CrossRef]

113. Wilson, M.C.; Mori, T.; Rückert, C.; Uria, A.R.; Helf, M.J.; Takada, K.; Gernert, C.; Steffens, U.A.E.; Heycke, N.; Schmitt, S.; et al. An environmental bacterial taxon with a large and distinct metabolic repertoire. *Nature* **2014**, *506*, 58–62. [CrossRef]

114. Mareš, J.; Hájek, J.; Urajová, P.; Kust, A.; Jokela, J.; Saurav, K.; Galica, T.; Čapková, K.; Mattila, A.; Haapaniemi, E.; et al. Alternative Biosynthetic Starter Units Enhance the Structural Diversity of Cyanobacterial Lipopeptides. *Appl. Environ. Microbiol.* **2018**, *85*, 1–17. [CrossRef]

115. Imker, H.J.; Walsh, C.T.; Wuest, W.M. SylC Catalyzes Ureido-Bond Formation During Biosynthesis of the Proteasome Inhibitor Syringolin A. *J. Am. Chem. Soc.* **2009**, *131*, 18263–18265. [CrossRef]

116. Koketsu, K.; Mitsuhashi, S.; Tabata, K. Identification of Homophenylalanine Biosynthetic Genes from the Cyanobacterium Nostoc punctiforme PCC73102 and Application to Its Microbial Production by *Escherichia coli*. *Appl. Environ. Microbiol.* **2013**, *79*, 2201–2208. [CrossRef]

117. Khumalo, M.J.; Nzuza, N.; Padayachee, T.; Chen, W.; Yu, J.-H.; Nelson, D.R.; Syed, K. Comprehensive Analyses of Cytochrome P450 Monooxygenases and Secondary Metabolite Biosynthetic Gene Clusters in *Cyanobacteria*. *Int. J. Mol. Sci.* **2020**, *21*, 656. [CrossRef]

118. Mazur-Marzec, H.; Sutryk, K.; Hebel, A.; Hohlfeld, N.; Pietrasik, A.; Błaszczyk, A. Nodularia spumigena Peptides—Accumulation and Effect on Aquatic Invertebrates. *Toxins* **2015**, *7*, 4404–4420. [CrossRef] [PubMed]

119. Agha, R.; Gross, A.; Rohrlack, T.; Wolinska, J. Adaptation of a chytrid parasite to its cyanobacterial host is hampered by host intraspecific diversity. *Front. Microbiol.* **2018**, *9*, 1–10. [CrossRef] [PubMed]

120. Ger, K.A.; Urrutia-Cordero, P.; Frost, P.C.; Hansson, L.A.; Sarnelle, O.; Wilson, A.E.; Lürling, M. The interaction between cyanobacteria and zooplankton in a more eutrophic world. *Harmful Algae* **2016**, *54*, 128–144. [CrossRef] [PubMed]
121. Kyle, M.; Haande, S.; Ostermaier, V.; Rohrlack, T. The Red Queen Race between Parasitic Chytrids and Their Host, Planktothrix: A Test Using a Time Series Reconstructed from Sediment DNA. *PLoS ONE* **2015**, *10*, e0118738. [CrossRef]
122. Urrutia-Cordero, P.; Agha, R.; Cirés, S.; Lezcano, M.Á.; Sánchez-Contreras, M.; Waara, K.-O.; Utkilen, H.; Quesada, A. Effects of harmful cyanobacteria on the freshwater pathogenic free-living amoeba *Acanthamoeba castellanii*. *Aquat. Toxicol.* **2013**, *130–131*, 9–17. [CrossRef]
123. Rohrlack, T.; Christoffersen, K.; Kaebernick, M.; Neilan, B.A. Cyanobacterial Protease Inhibitor Microviridin J Causes a Lethal Molting Disruption in *Daphnia pulicaria*. *Appl. Environ. Microbiol.* **2004**, *70*, 5047–5050. [CrossRef]
124. Schwarzenberger, A.; Sadler, T.; Von Elert, E. Effect of nutrient limitation of cyanobacteria on protease inhibitor production and fitness of *Daphnia magna*. *J. Exp. Biol.* **2013**, *216*, 3649–3655. [CrossRef]
125. Papadimitriou, T.; Kagalou, I.; Stalikas, C.; Pilidis, G.; Leonardos, I.D. Assessment of microcystin distribution and biomagnification in tissues of aquatic food web compartments from a shallow lake and evaluation of potential risks to public health. *Ecotoxicology* **2012**, *21*, 1155–1166. [CrossRef] [PubMed]
126. Leonard, J.A.; Paerl, H.W. Zooplankton community structure, micro-zooplankton grazing impact, and seston energy content in the St. Johns river system, Florida as influenced by the toxic cyanobacterium *Cylindrospermopsis raciborskii*. *Hydrobiologia* **2005**, *537*, 89–97. [CrossRef]
127. Pawlik-Skowrońska, B.; Toporowska, M.; Mazur-Marzec, H. Effects of secondary metabolites produced by different cyanobacterial populations on the freshwater zooplankters *Brachionus calyciflorus* and *Daphnia pulex*. *Environ. Sci. Pollut. Res.* **2019**, *26*, 11793–11804. [CrossRef] [PubMed]
128. Moustaka-Gouni, M.; Sommer, U. Effects of Harmful Blooms of Large-Sized and Colonial Cyanobacteria on Aquatic Food Webs. *Water* **2020**, *12*, 1587. [CrossRef]
129. Holland, A.; Kinnear, S. Interpreting the Possible Ecological Role(s) of Cyanotoxins: Compounds for Competitive Advantage and/or Physiological Aide? *Mar. Drugs* **2013**, *11*, 2239–2258. [CrossRef] [PubMed]
130. Roegner, A.; Truong, L.; Weirich, C.; Schirmer, M.P.; Brena, B.; Miller, T.R.; Tanguay, R. Combined Danio rerio embryo morbidity, mortality and photomotor response assay: A tool for developmental risk assessment from chronic cyanoHAB exposure. *Sci. Total Environ.* **2019**, *697*, 134210. [CrossRef]
131. Lenz, K.A.; Miller, T.R.; Ma, H. Anabaenopeptins and cyanopeptolins induce systemic toxicity effects in a model organism the nematode *Caenorhabditis elegans*. *Chemosphere* **2019**, *214*, 60–69. [CrossRef] [PubMed]
132. Muszewska, A.; Stepniewska-Dziubinska, M.M.; Steczkiewicz, K.; Pawlowska, J.; Dziedzic, A.; Ginalski, K. Fungal lifestyle reflected in serine protease repertoire. *Sci. Rep.* **2017**, *7*, 9147. [CrossRef]
133. Dehm, D.; Krumbholz, J.; Baunach, M.; Wiebach, V.; Hinrichs, K.; Guljamow, A.; Tabuchi, T.; Jenke-Kodama, H.; Süssmuth, R.D.; Dittmann, E. Unlocking the Spatial Control of Secondary Metabolism Uncovers Hidden Natural Product Diversity in Nostoc punctiforme. *ACS Chem. Biol.* **2019**, *14*, 1271–1279. [CrossRef]
134. Mazur-Marzec, H.; Toruńska, A.; Błońska, M.J.; Moskot, M.; Pliński, M.; Jakóbkiewicz-Banecka, J.; Węgrzyn, G. Biodegradation of nodularin and effects of the toxin on bacterial isolates from the Gulf of Gdańsk. *Water Res.* **2009**, *43*, 2801–2810. [CrossRef]
135. Kansole, M.; Lin, T.-F. Microcystin-LR Biodegradation by Bacillus sp.: Reaction Rates and Possible Genes Involved in the Degradation. *Water* **2016**, *8*, 508. [CrossRef]
136. Seymour, J.R.; Amin, S.A.; Raina, J.-B.; Stocker, R. Zooming in on the phycosphere: The ecological interface for phytoplankton–bacteria relationships. *Nat. Microbiol.* **2017**, *2*, 17065. [CrossRef]
137. Briand, E.; Humbert, J.; Tambosco, K.; Bormans, M.; Gerwick, W.H. Role of bacteria in the production and degradation of *Microcystis cyanopeptides*. *Microbiologyopen* **2016**, *5*, 469–478. [CrossRef] [PubMed]
138. Kato, H.; Imanishi, S.Y.; Tsuji, K.; Harada, K. Microbial degradation of cyanobacterial cyclic peptides. *Water Res.* **2007**, *41*, 1754–1762. [CrossRef]
139. Kurmayer, R.; Deng, L.; Entfellner, E. Role of toxic and bioactive secondary metabolites in colonization and bloom formation by filamentous cyanobacteria *Planktothrix*. *Harmful Algae* **2016**, *54*, 69–86. [CrossRef] [PubMed]
140. Šulčius, S.; Mazur-Marzec, H.; Vitonytė, I.; Kvederavičiūtė, K.; Kuznecova, J.; Šimoliūnas, E.; Holmfeldt, K. Insights into cyanophage-mediated dynamics of nodularin and other non-ribosomal peptides in *Nodularia spumigena*. *Harmful Algae* **2018**, *78*, 69–74. [CrossRef] [PubMed]
141. Mazur-Marzec, H.; Błaszczyk, A.; Felczykowska, A.; Hohlfeld, N.; Kobos, J.; Toruńska-Sitarz, A.; Devi, P.; Montalvão, S.; D'souza, L.; Tammela, P.; et al. Baltic cyanobacteria—A source of biologically active compounds. *Eur. J. Phycol.* **2015**, *50*, 343–360. [CrossRef]
142. Ahmad, S.; Saleem, M.; Riaz, N.; Lee, Y.S.; Diri, R.; Noor, A.; Almasri, D.; Bagalagel, A.; Elsebai, M.F. The Natural Polypeptides as Significant Elastase Inhibitors. *Front. Pharmacol.* **2020**, *11*, 1–19. [CrossRef]
143. Ammosova, T.; Jerebtsova, M.; Beullens, M.; Voloshin, Y.; Ray, P.E.; Kumar, A.; Bollen, M.; Nekhai, S. Nuclear Protein Phosphatase-1 Regulates HIV-1 Transcription. *J. Biol. Chem.* **2003**, *278*, 32189–32194. [CrossRef]
144. McConnell, J.L.; Wadzinski, B.E. Targeting Protein Serine/Threonine Phosphatases for Drug Development. *Mol. Pharmacol.* **2009**, *75*, 1249–1261. [CrossRef] [PubMed]

145. Halland, N.; Brönstrup, M.; Czech, J.; Czechtizky, W.; Evers, A.; Follmann, M.; Kohlmann, M.; Schiell, M.; Kurz, M.; Schreuder, H.A.; et al. Novel Small Molecule Inhibitors of Activated Thrombin Activatable Fibrinolysis Inhibitor (TAFIa) from Natural Product Anabaenopeptin. *J. Med. Chem.* **2015**, *58*, 4839–4844. [CrossRef]
146. Hameed, S. *Investigation of the Production and Isolation from Cyanobacteria*; Robert Gordon University: Aberdeen, UK, 2013.
147. Gulledge, B.; Aggen, J.; Huang, H.; Nairn, A.; Chamberlin, A. The Microcystins and Nodularins: Cyclic Polypeptide Inhibitors of PP1 and PP2A. *Curr. Med. Chem.* **2002**, *9*, 1991–2003. [CrossRef] [PubMed]
148. Bubik, A.; Sedmak, B.; Novinec, M.; Lenarčič, B.; Lah, T.T. Cytotoxic and peptidase inhibitory activities of selected non-hepatotoxic cyclic peptides from cyanobacteria. *Biol. Chem.* **2008**, *389*, 1339–1346. [CrossRef] [PubMed]
149. Vercauteren, E.; Gils, A.; Declerck, P. Thrombin Activatable Fibrinolysis Inhibitor: A Putative Target to Enhance Fibrinolysis. *Semin. Thromb. Hemost.* **2013**, *39*, 365–372. [CrossRef]
150. Shah, S.; Akhter, N.; Auckloo, B.; Khan, I.; Lu, Y.; Wang, K.; Wu, B.; Guo, Y.-W. Structural Diversity, Biological Properties and Applications of Natural Products from Cyanobacteria. A Review. *Mar. Drugs* **2017**, *15*, 354. [CrossRef]
151. Wahome, P.; Beauchesne, K.; Pedone, A.; Cavanagh, J.; Melander, C.; Zimba, P.; Moeller, P. Augmenting Anti-Cancer Natural Products with a Small Molecule Adjuvant. *Mar. Drugs* **2014**, *13*, 65–75. [CrossRef]
152. Shishido, T.K.; Popin, R.V.; Jokela, J.; Wahlsten, M.; Fiore, M.F.; Fewer, D.P.; Herfindal, L.; Sivonen, K. Dereplication of Natural Products with Antimicrobial and Anticancer Activity from Brazilian Cyanobacteria. *Toxins* **2019**, *12*, 12. [CrossRef]

Review

Co-Occurrence of Cyanobacteria and Cyanotoxins with Other Environmental Health Hazards: Impacts and Implications

James S. Metcalf [1],* and Geoffrey A. Codd [2,3]

1   Brain Chemistry Labs, Jackson, WY 83001, USA
2   School of Life Sciences, University of Dundee, Dundee DD1 5EH, UK; g.a.codd@stir.ac.uk
3   Biological and Environmental Sciences, University of Stirling, Stirling FK9 4LA, UK
*   Correspondence: james@ethnomedicine.org; Tel.: +1-307-734-1680

Received: 11 August 2020; Accepted: 15 September 2020; Published: 1 October 2020

Abstract: Toxin-producing cyanobacteria in aquatic, terrestrial, and aerial environments can occur alongside a wide range of additional health hazards including biological agents and synthetic materials. Cases of intoxications involving cyanobacteria and cyanotoxins, with exposure to additional hazards, are discussed. Examples of the co-occurrence of cyanobacteria in such combinations are reviewed, including cyanobacteria and cyanotoxins plus algal toxins, microbial pathogens and fecal indicator bacteria, metals, pesticides, and microplastics. Toxicity assessments of cyanobacteria, cyanotoxins, and these additional agents, where investigated in bioassays and in defined combinations, are discussed and further research needs are identified.

Keywords: cyanobacteria; co-occurrence; toxicity; plastics; metals; biocide

Key Contribution: This review article examines the potential for co-occurrence of cyanobacterial toxins with other toxic compounds, bacteria, and chemicals. An understanding of such risks is essential to accurately assess the threat of these combinations to human health.

---

1. Introduction

Research on the production, properties, monitoring, and analysis of toxigenic cyanobacteria, and of particular cyanotoxins, has increased greatly over the past 40 years [1–5]. Aspects of the environmental occurrence, biosynthesis, properties, and health significance of the most widely investigated individual classes of cyanotoxins, principally the microcystins, nodularins, saxitoxins, cylindrospermopsins, and anatoxins have been reviewed [6–12]. On the toxicity assessment of cyanobacterial cultures and environmental samples containing cyanobacteria, and the involvement of specific cyanotoxins, the research has passed through several discernible phases. The earliest known experimental investigations into the suspected toxicity of cyanobacteria, which followed animal poisonings, were performed by George Francis [13,14]. These involved the oral dosing of healthy animals, of the same species which had succumbed to intoxication, with the suspected toxic material, in this case *Nodularia* scum, and the consequent replication in the dosed animals of mortalities and gross signs of organ damage. Bioassays, a quantitative development of this logical approach, but initially without the chemical identification and quantification of the toxic substances in the test material, were then increasingly used to test for toxic principles for several decades [1–3]. These were gradually complemented, and then overtaken by the emerging methods for the identification and quantification of specific cyanotoxins in the test material and the quantitative characterization of toxicity using individual purified cyanotoxins originally obtained via bioassay-guided purification of cyanobacterial material. The increasing availability of physico-chemical methods for analysis and cyanotoxin purification, especially for microcystins, nodularin-R, saxitoxins and anatoxin-a, and later for cylindrospermopsins [1–5], appears to have led to an increased focus on the use of these methods, with a corresponding decline in the use of the whole animal bioassays. The move away from whole animal bioassays has been further influenced by increasing ethical and humanitarian concerns. Although non-mammalian in vivo and biochemical, enzyme-based, and cell-based in vitro bioassays are available [3,15] the continuing focus on the use of physico-chemical methods has also been influenced by the increasing need for validated, quantitative procedures to satisfy statutory regulations and guidelines [16]. In addition, genetic methods [17] may also contribute to a reduction in bioassay use through providing alternative testing to understand the potential for toxin production by cyanobacteria.

The further development and application of physico-chemical analytical methods for the detection, identification, and quantification of specific cyanotoxins continues to support and enable the application of policies for the risk management of water resources affected by cyanobacterial mass populations. These include specific methods for individual classes of cyanotoxins [5] and, increasingly, methods for the multiclass analysis of the toxins in single procedures [18–21]. However, whilst such methods alone can provide a partial indication of toxicity presented by axenic strains of cyanobacteria grown in the laboratory, they may not take into full account the toxicological significance of mass populations of cyanobacteria in open environments.

In marine and freshwater environments and in terrestrial and aerial habitats, whilst cyanobacteria can readily appear as "dominant", i.e., to account for the majority of the microbial biomass, they can be accompanied by a wide range of other microbes including microalgae, chemoheterotrophic bacteria, and protozoa [22,23]. Where such environments are subjected to intensive anthropogenic human pressures (e.g., wastewater and human sewage disposal and industrial discharges), the dominant cyanobacteria can co-occur with additional biological toxins (e.g., [19]), chemical pollutants e.g., metals [23], and with pathogenic microbes [24]. Examples of the co-occurrence of cyanotoxins, plus other chemical agents and microbial pathogens, and cases of co-exposure are reviewed (Figure 1). Research to further understand the health significance of cyanotoxins is discussed in a broader context of cyanotoxin co-occurrence and co-exposures with additional biological and anthropogenic toxic agents.

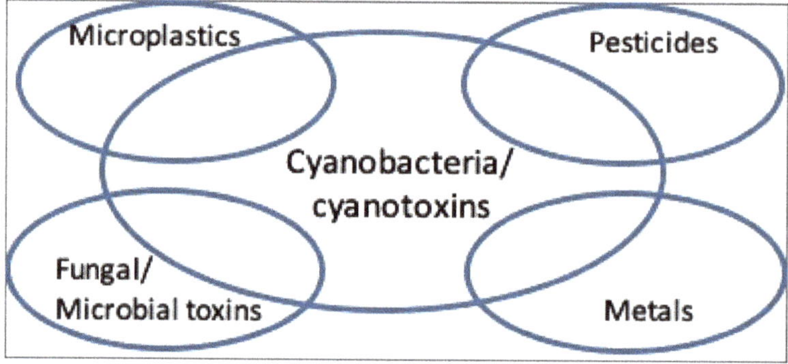

Figure 1. Examples of interactions of cyanobacteria with other potentially toxic components of water.

It is well established that when exposure to toxic compounds involves mixtures of two or more toxic substances, the effect of such mixtures on susceptible individuals or populations can be unpredictable, but different from the exposures to each of the toxins if applied separately. Examples of the three possible outcomes, namely additive, synergistic, and antagonistic effects [25] are increasingly emerging, where bioassays including toxigenic cyanobacteria, and specifically when purified cyanotoxins and non-cyanobacterial toxic agents, are performed (Table 1).

Table 1. Examples of toxicity assessment by bioassay of purified cyanotoxins in combinations and of cyanotoxins plus additional environmental toxins/pollutants.

Toxic Agents	Test Organism/Cell Line	Outcome [1]	Ref.
MC-LR plus Antx-a	Mouse, intranasal bioassay	SYN	[26]
MC-LR plus Antx-a	*Selenastrum capricornutum*	SYN	[27]
MC-LR plus Antx-a	*Cyprinus carpio*, carp cells	SYN (potential)	[28]
MC-LR plus Antx-a	*Vallisneria natans*, and microbial biofilm	ANTAG	[29]
MC-LR plus *Microcystis* LPS	*Artemia salina* and *Daphnia* sp.	ANTAG	[30]
MC-LR plus CYN	*Chlorella vulgaris*	SYN	[31]
MC-LR plus cyanobacterial λ-linolenic acid	*Daphnia magna*	ADDIT	[32]
MC-LR plus aflatoxin B1	Human hepatic cells	ANTAG	[33]
MC-LR plus aflatoxin B1, and plus fumonisin B1	HepG2, Caco2, MDBK cell lines	ADDIT, SYN, ANTAG	[34]
MC-LR and copper	*Danio rerio*	SYN	[35]
MC-LR plus copper	*Vallisneria natans*	SYN	[36]
MC-LR plus linear alkyl-benzene sulphonate	*Lactuca sativa*	SYN	[37]
MC-LR plus phenanthrene	*Lemna gibba*	ANTAG, SYN, ADDIT	[38]
CYN plus chloropyrifos	Human SH-SY5Y neuroblastoma cell line	ANTAG	[39]
CYN plus bisphenols	HepG2 cells	SYN, ADDIT	[40]

[1] SYN, synergistic; ADDIT, additive; ANTAG, antagonistic toxicological outcomes of applications of multiple toxic agents. MC-LR, microcystin-LR, Antx-a, anatoxin-a; CYN, cylindrospermopsin. Outcomes in several systems were influenced by relative concentrations, and in some cases by timing of applications of toxins/toxicants, see references.

Of the classes of toxins produced by cyanobacteria, with the exception of guanitoxin (formerly anatoxin-a(S) [41]), all encompass multiple chemical structures, ranging from a few (e.g., anatoxin-a and homoanatoxin-a) to hundreds, in the case of microcystins (>240, [42]). Similarly, analyses of blooms and cultures of cyanobacteria have shown that multiple variants can exist in extracts of individual strains, such as for *Microcystis* PCC7820 with at least 10 microcystin variants reported [43] and a Thai strain of *Cylindrospermopsis* that has been shown to produce two additional variants of cylindrospermopsin [44]. Therefore, when toxicity assessment is performed, then although classes of cyanotoxins may be identified, the differing amounts of variants, whether of e.g., microcystins or anatoxins, each with potentially differing toxicities when determined individually (e.g., [45]), may influence the risk assessment of such bloom material. On occasion, some strains have been reported to include multiple classes of toxins, such as microcystins and guanitoxin, as in *Anabaena* 525-17 [46]. In addition to these cyanotoxins, all cyanobacteria characteristically produce lipopolysaccharide endotoxins [47].

Fitzgeorge et al. [26] provided early evidence of the synergistic action of two purified cyanotoxins (anatoxin-a and microcystin-LR) when administered intranasally to mice. In this case, the administration of a sublethal dose of microcystin-LR 30 min before that of anatoxin-a lowered the LD_{50} of the latter four-fold. Perhaps the most high-profile human poisoning event occurred at Caruaru, Brazil [48,49]. Extensive investigations into the deaths of 50 people at a hemodialysis clinic identified microcystins as the most likely principal cause of the fatalities, although several symptoms were reported by those affected. Investigations continued into the intoxications and cylindrospermopsin was also implicated, suggesting the possibility of multiple cyanotoxin exposures [50].

Clearly, when bioassays are performed in vivo and in vitro with an individual purified cyanotoxin or defined combinations of purified cyanotoxins, then the responses can be unequivocally ascribed to the toxin(s) administered. However, since cyanobacterial cells, whether in axenic monocyanobacterial culture or in environmental samples, can produce a range of cyanotoxins and other bioactive secondary products, then bioassays with crude extracts from such sources unavoidably involve exposures to mixtures of toxic agents. These may be exclusively of cyanobacterial origin when derived from monocyanobacterial axenic cultures, or from cyanobacterial plus a wide range of toxins/toxicants from additional biological and anthropogenic sources in the case of environmental samples.

2. Environmental Intoxications Involving Toxigenic Cyanobacteria and Additional Agents

Some indications are available of contributory exposure to environmental health hazards in addition to toxigenic cyanobacteria or specific cyanotoxins in wildlife poisoning episodes. Thus, whilst a major role for microcystins and anatoxin-a in the mass mortalities of Lesser Flamingos (*Phoeniconaias minor*) at Kenyan lakes was identified, additional contributions due to heavy metals, pesticides, and mycobacterial infection were likely to have occurred [51,52]. A major role of microcystins was similarly deduced from cyanotoxin analyses and pathology in the deaths of Mute Swans (*Cygnus olor*) in the UK, although the additional contribution of lead was inferred from the abundance of fishermen's lead sinker-pellets in the birds' stomachs [53]. Microcystin-, saxitoxin-, and cylindrospermopsin-producing cyanobacterial blooms were indicated to have been early contributors to an extended major fish-kill in the Lower St. John's River, Florida. Later factors arising from cyanobacterial decomposition and lysis included lower dissolved oxygen concentrations, elevated ammonia concentrations, and hemolytic contribution from the dinoflagellate alga *Heterosigma akashiwo* [54].

Cases of human illness associated with exposure to toxigenic cyanobacteria also indicate the possible contribution of additional factors. In 1979, a severe hepato-enteritis outbreak occurred among consumers (mainly children) of drinking water from a Palm Island reservoir in Queensland, Australia, containing a bloom of *Cylindrospermopsis raciborskii* [55]. The subsequent isolation and characterization of cylindrospermopsin from *C. raciborskii* from the reservoir provided strong evidence for a contributing role of the cyanotoxin in the Palm Island illnesses. However, as pointed out by Hawkins et al. the evidence suggested that the cylindrospermopsin-producing bloom should be considered as only one possible cause [55]. It is unknown whether the copper concentration in the water, arising from the prior treatment to kill the *C. raciborskii* bloom, also contributed to the illness. An episode requiring hospitalization of army cadets undergoing swimming and canoeing exercises at Rudyard Lake, northern England, occurred after the cadets ingested microcystin-containing *Microcystis aeruginosa* scum [56]. Atypical pneumonia, liver damage, and blistering around the mouth were attributed to microcystin ingestion. However, whilst no evidence for enterovirus contamination of the water was apparent, *Escherichia coli* counts indicated that the water was unsuitable for bathing [56].

3. Additional Health Hazards: Their Co-Occurrence and Interactions with Cyanobacteria and Cyanotoxins

A wide range of microbial and chemical health hazards exists with which cyanobacteria can be associated. The associations range from the use of purified materials in experimental laboratory designs to investigate combined toxicological responses in vitro and in vivo, to the co-occurrence of multiple hazards in open environments.

3.1. Fungal and Algal Toxins

The health significance of human exposure to the fungal toxin aflatoxin B1 (via the diet), plus to microcystin via drinking water, was recognized almost 30 years ago in China [57–59]. A high incidence of primary liver cancer among local populations using surface drinking water containing microcystin-producing cyanobacteria was associated with additional chronic exposure to the hepatocellular carcinoma risk factor aflatoxin B1 via food consumption, including moldy maize. Further risk factors were presented by Hepatitis B virus and alcohol [60]. Whilst not characterized as a primary carcinogen itself, the tumor-promoting action of the microcystin(s) was expressed by the chronic exposure of the population to the cyanotoxin, and to the primary carcinogenicity of aflatoxin

B1 and/or the Hepatitis B virus. These milestone discoveries in the history of cyanotoxins research and risk management [6,8,9,57–59], not only illustrate the importance of exposure to cyanotoxins plus additional health hazards, but also provide an example of the hazards of such combined exposure via multiple exposure media: in this case, the media being water and food. Whether the timing of the exposure to microcystin-containing cyanobacterial blooms and to aflatoxin-B1, i.e., as co-exposure or sequential exposure, influenced the outcome among the human population, remains less certain. Trials using human hepatic cells in vitro and rats have indicated that whilst the tumor promoting actions of microcystin were confirmed, antagonistic action can occur when the cyanotoxin at low dose is co-applied together with the fungal toxin [33].

The co-occurrence of cyanotoxins with microalgal phycotoxins has received little attention. Blooms and shoreline scums of several species of toxigenic marine microalgae have long been recognized for their roles in the mass mortalities of fish, seabirds, and sea mammals and in human intoxications, from mild to fatal, via shellfish consumption [2]. For reviews concerning marine microalgal phycotoxins, including the saxitoxins, palytoxins, ciguatoxins, brevetoxins, domoic acid, okadaic acid, pinnatoxins, yessotoxins, and azaspiracids, see [61–63]. No evidence for the production of most classes of cyanotoxins by microalgae is available, namely of microcystins, nodularins, cylindrospermopsins, anatoxin-a, or guanitoxin. In addition, the lipopolysaccharide (LPS) endotoxins, being structural components of Gram-negative prokaryotes are characteristic of cyanobacteria [47] but are apparently lacking in eukaryotic microalgae. Thus, for these classes of cyanotoxins to co-occur with the potent phycotoxins of marine waters, it is necessary (i) for toxigenic cyanobacteria to also be present and growing in the environment, or (ii) for the cyanotoxins to be introduced into the microalgal environment. Indeed, such an introduction occurred in the Monterey Bay National Marine Sanctuary, California, where microcystin-producing *Microcystis* blooms entered the Bay waters from inland freshwaters via river inflows. Mass deaths of sea otters (*Enhydra lutris nereis*) occurred due to the consumption of marine shellfish which had accumulated the microcystins from the river inflows [64]. In this case, analyses for other candidate cyanotoxins (nodularin and anatoxin-a) and for marine phycotoxins (okadaic acid and yessotoxin) were negative. Another area of California, namely San Fransisco Bay, has shown the presence of multiple toxin classes, often detected in molluscs [65]. From an analysis of mussels, microcystins, domoic acid, diarrhetic shellfish toxins, and paralytic shellfish toxins were identified with all four toxin classes detected in 37% of mussels.

However, some classes of cyanotoxins are not exclusively the products of cyanobacteria [66]. Saxitoxins have long been known to be produced by marine dinoflagellates, including species of *Alexandrium*, *Gymnodinium*, and *Pyrodinium* [63] and by several strains of cyanobacterial species, including *Aphanizomenon* spp., *Dolichospermum circinale* (formerly *Anabaena circinalis*), *Cylindrospermopsis raciborskii*, *Raphidiopsis brookii*, *Lyngbya wollei* [67], and *Scytonema crispum* [68,69]. The neurotoxic diaminoacids, β-*N*-methylamino-L-alanine (BMAA), *N*-(2-aminoethyl)glycine (AEG), and 2,4-diaminobutyric acid (2,4-DAB), with LC-MS/MS used to confirm analytical specificity, also appear to have multiple origins, including diverse cyanobacteria, marine and freshwater diatoms, and a brackish coastal water dinoflagellate [70–77], and these toxins can also be present along with microcystins and brevetoxins [78]. Wider origins of these neurotoxins are also indicated by the presence of BMAA in the peptides found in chemoheterotrophic bacteria including environmentally widespread *Paenibacillus* spp. [79,80].

3.2. Microbial Pathogens

A close association of cyanobacteria, including toxigenic species, typically occurs with other microbes in aquatic and terrestrial environments [22,23]. The close proximity of the cyanobacteria and their non-phototrophic, mutualistic partners can enable the two-way exchange of metabolites and nutrients and provide a protective physical substrate for bacterial attachment and gene transfer. Cyanobacterial extracellular polysaccharides and glycoprotein sheath materials can provide a substrate for the bacterial attachment [23]. In addition to the association with a wide range of non-pathogenic bacteria (e.g., [81]), marine, estuarine, and freshwater cyanobacteria can form close associations with human pathogenic bacteria. When large blooms of cyanobacteria occur and people are exposed to such blooms, then health complaints and non-specific symptoms are often reported [82]. Such symptoms can include fever, pneumonia, and headaches as examples (reviewed in [82]). Therefore, exposure to cyanobacterial blooms may have the added issue of multiple insults affecting human health, especially in those individuals with underlying medical conditions.

In 1884, Robert Koch, the discoverer of the cholera bacillus, suggested from field studies that "aquatic flora" might serve as environmental reservoirs of cholera [83]. Indeed, searches for possible reservoirs of survival of toxigenic *Vibrio cholerae* 01 in a pond in Dhaka, Bangladesh, used for bathing, swimming, washing, and drinking, indicated the seasonal survival of the pathogen in the extracellular mucilage of the cyanobacterium *Anabaena variabilis* [84]. Whilst the specificity of this association is not extended to euglenoid microalgae, it does extend to the mucilaginous masses of other cyanobacteria, including *Microcystis* colonies [85] and microcystin-producing *Oscillatoria* filaments [86]. Without the application of more effective cyanobacterial risk management measures, the role of cyanobacteria in providing inter-epidemic reservoirs of *V. cholerae* is likely to increase as compounding anthropogenic pressures and climate change continue to result in increases in cyanobacterial population size, seasonal duration, and geographical spread [4]. Populations of other potentially pathogenic *Vibrio* spp. have increased from 2000 to 2018 in the Neuse River Estuary, North Carolina, USA but no correlations were apparent with changes in temperature, salinity, or dissolved oxygen concentration, factors which have influenced *Vibrio* abundance elsewhere [87]. The eutrophic estuary waters contain abundant blooms including microalgae and cyanobacteria [88] although whether the cyanobacteria are physically associated with the potentially pathogenic *Vibrio* spp. in the Neuse water is not apparent.

Although the waterborne human pathogenic protozoon *Cyclospora* was initially thought to be a cyanobacterium, this is not so, and the protzoon is able to cause diarrhea, sickness, and abdominal pain, similar to the protozoa *Cryptosporidium parvum* and *Giardia lamblia* [89]. An increased abundance of potential microcystin-producing cyanobacteria (*Aphanocapsa* and *Microcystis* spp.) was observed with *Cryptosporidium* and *Giardia* spp. in a reservoir supplying drinking water to the metropolitan Belo Horizonte area, southeastern Brazil [90]. It is notable that waterbody conditions which favor the persistence of *Cryptosporidium* and *Giardia* spp., including nutrient enrichment and high retention times, can also enhance cyanobacterial growth [23], increasing the potential for co-occurrence and co-exposure to pathogenic protozoa and cyanotoxins.

3.3. Metals

Relations between cyanobacteria and metals constitute a large field. Of toxicological relevance are (i) the chronic and acute effects of metals on the growth, metabolism, and survival of cyanobacteria and (ii) the ability of cyanobacteria to accumulate, detoxify, metabolize, and sequester metals [91–93]. Depending on growth conditions, the effects of dissolved metal ions (e.g., Ca, Cu, Pb, Cd) can include both the inhibition and stimulation of *Microcystis* blooms [94], and the presence of copper may affect the detoxication of e.g., nodularin by disrupting microbes that may degrade this cyanotoxin [95]. Copper is often used as a biocide to lyse cyanobacterial cells during blooms. Although successful, in the case of e.g., *Microcystis*, this has had the effect of releasing cyanotoxins from an intracellular pool to an extracellular pool with the result that liver damage may result after drinking water from which cyanobacterial cells and cell debris have been removed, but without the assured removal of extracellular microcystin [96]. Furthermore, if the copper used for lysis is not effectively removed during drinking water treatment, then this may subsequently also pose an additional toxicological burden on people and animals.

Research on metals and cyanotoxins has largely focused on the effects of iron on microcystin production. Pioneering studies on iron limitation in axenic *Microcystis aeruginosa* cultures revealed increases in microcystin-RR and -LR (MC-RR, MC-LR) production [97]. Early sampling and analytical procedures in the latter study are likely to have resulted in the combined analysis of the former intra- and extracellular pools. Further investigations into microcystin production by *M. aeruginosa*, *Microcystis novacekii*, and *Phormidium autumnale* cultures have included increases upon iron addition but also increases upon iron limitation during culture. Positive effects of copper, zinc, and manganese ions on microcystin production under conditions of metal enrichment of, and metal limitation of, cyanobacterial cultures have also been observed, possibly indicating physiological and biochemical roles of the metals in microcystin biosynthesis, in addition to a siderophore function for microcystins in metal acquisition [98]. No specific differences in anatoxin-a production by *P. autumnale* were found in response to growth under an environmentally encountered, low-to-high range of iron or copper concentrations [99].

Direct interaction between cyanotoxin molecules and metals has been investigated in vitro with two classes of cyanotoxins. The purified post-synaptic neuromuscular blocking neurotoxin, BMAA, is a potent chelator of divalent metal cations including copper and zinc [100]. Whether such chelation occurs in BMAA-producing cyanobacterial cells and, if so, if toxicity is thereby influenced, is not known. Divalent copper and zinc binding occur to at least three purified microcystins (MC-LR, MC-LW, MC-LF) with formation constants (K_i) indicating that all three cyanotoxins are medium-strength metal ligands. Single amino acid substitution in the heptapeptide microcystin ring (arginine versus tryptophan, or versus phenylalanine) did not influence the strength of the metal-microcystin association [101]. Whether metal-microcystin binding influences toxicity and whether this binding occurs in the cyanobacterial producer-cells and/or in the surrounding water after microcystin release are also unknown. However, this possibility is viewed alongside the increasing evidence for the combined toxicity of microcystins and metals: co-exposure bioassays involving MC-LR and copper, at environmentally encountered concentrations, have revealed synergistic toxicity against the early development of the zebrafish, *Danio rerio*. Uptake of the toxins by the fish involved microcystin and copper transporters [35]. Toxicity assessment of BMAA and methylmercury to primary neurocortical cells show a synergistic toxicity [102] and assessment of other cyanotoxins with such organic forms of metals is required.

3.4. Pesticides

One cyanotoxin, the phosphorylated cyclic *N*-hydroxyguanidine, guanitoxin (anatoxin-a(S)) exerts toxicity via the irreversible inhibition of acetylcholinesterases in common with organophosphorus pesticides [6,103]. No other cyanotoxins are known to have the same modes of action as those of synthetic pesticides. The co-occurrence of cyanobacteria, and of pesticides, including herbicides, fungicides, and insecticides, is a concern in water resources with a high human usage and dependency. For example, in paddy fields nitrogen-fixing cyanobacteria can serve as a valuable biofertilizer and contribute to rice production and an aim is to reduce the negative impacts of the pesticides on cyanobacterial growth [104]. Nevertheless, the inadvertent entry of pesticides into waterbodies from human activities, especially agriculture, appears to be a ubiquitous process and numerous investigations into the inhibitory impacts of pesticides on aquatic biota have occurred. A wide-ranging survey of investigations into the effects of pesticides in aquatic microbes included the effects of insecticides, herbicides, and fungicides on the viability of cyanobacteria, with overall dose-dependent growth inhibition occurring [105]. Most of the named cyanobacteria in this survey were members of cyanotoxin-producing taxa. However, investigations to date do not appear to have included relations between pesticides and the production and impacts of cyanotoxins. In addition to the possible contribution of pesticides (and metals) alongside cyanotoxins to the mass mortalities of Lesser Flamingos in East African lakes [51,52,106], both pesticides and cyanotoxins may contribute to the marked decline in the American Alligator (*Alligator mississippiensis*) population in eutrophic Lake Apopka, Florida, with high egg failure and anomalous endocrine function [107].

The adult invertebrate grazer *Daphnia pulicaria* was exposed to the purified pesticide carbaryl (1-naphthyl methylcarbamate) plus whole cells of microcystin-producing *M. aeruginosa* at a range of sublethal concentrations [108]. The actual dose of microcystin(s) assimilated per animal was estimated from the analysis of whole *D. pulicaria* by ELISA, although the immunoassay used does not distinguish between authentic microcystin(s) and a range of microcystin detoxification products [109]. However, at a range of sublethal carbaryl concentrations, adverse effects on egg numbers per female, delayed maturation, offspring mortality, and body malformations occurred with outcomes increased by the addition of *M. aeruginosa* cells. Additive and synergistic actions between carbaryl and the microcystin-containing cyanobacterial cells were indicated [108]. Hinojosa et al. [39] have recently provided a significant example of the needed, quantitative, baseline studies on the toxicology of pesticides in co-occurrence with cyanotoxins. In vitro bioassays using the human neuroblastoma cell line SH-SY5Y evaluated the effects of individual, versus combined exposure to purified cylindrospermopsin and chlorpyrifos [*O,O*-diethyl *O*-(3,5,6-trichloro-2-pyridinol) phosphorothionate]. Cytotoxicity and mechanistic endpoint comparisons after 24 and 48 h of exposure, at environmentally relevant concentrations, indicated antagonistic action between the pesticide and cyanotoxin [39].

The herbicide glyphosate has been used widely to control plants and it has been shown to occur in waterbodies, with toxicity demonstrated [110]. Its presence in waterways is well known and has been shown to have an adverse effect on the growth of cyanobacteria, including *Microcystis* [111]. Further effects of glyphosate on cyanobacteria include enhanced extracellular release of microcystins [112] with the potential for organisms to respond to combinations of microcystins and glyphosate, such as the mussel *Unio pictorium* [113].

3.5. Microplastic and Nanoplastic Particles and Contaminants

The global occurrence of microscopic plastic particles (microplastics) throughout the world's oceans has been recognized for several years [114] and their further occurrence in human feces, freshwater lakes and rivers, groundwater, and estuaries is becoming increasingly recognized [115,116]. Cyanobacteria in both marine and freshwaters are among the wide range of microbes which, with mineral particles, attach to microplastics, forming a biofilm. The adhesion of microbes, with subsequent development of an extracellular polysaccharide layer, may contribute to the sinking or buoyancy properties of these complexes, and to the further sorption of metals, particularly, iron and manganese [117]. Among toxicological hazards presented by the microplastics accumulating in water resources are the endocrine-disrupting bisphenol contaminants, including bisphenols A, S, F, and AF [118]. Cylindrospermopsin is a strong candidate for toxicity evaluation in combination with bisphenols since (i) its wide geographical occurrence in freshwaters is becoming more apparent; (ii) a high percentage of the total cylindrospermopsin pool is extracellular when the *C. raciborskii* producer cells are still intact, and (iii) the cyanotoxin exhibits a wide range of actions, including hepatotoxicity, inhibition of protein synthesis, genotoxicity, and potential carcinogenicity [6,9,40]. A complex of interactions between purified cylindrospermopsin and bisphenols was found in in vitro bioassays using HepG2 cells. Whilst bisphenols alone reduced cell viability or induced DNA double strand breaks, antagonistic activity of bisphenol against these actions by cylindrospermopsin was indicated. However, further possible additive or synergistic effects on HepG2 gene deregulation were also indicated by co-exposure to the bisphenols plus cylindrospermopsin [40]. No bioassays involving purified microcystins with micro- or nanoplastics are yet apparent, although toxicologically significant interactions have been reported between nanoplastics and *Microcystis aeruginosa* cells [118], and laboratory-ware plastics are also known to bind microcystins from solution [119,120] and plastic nanoparticles may also bind glyphosate [111]. Amino-modified polystyrene nanoplastics increase microcystin production by the cyanobacterial cells and also increase the extracellular release of the toxin(s) according to microcystin immunoassay. The prevalence and proximity of microplastics, nanoplastics, and cyanobacteria [115–117], and widespread ability of the latter to produce microcystins [5,8,9,11,12] require further research into the toxicological interactions of these ubiquitous synthetic and naturally occurring, biological health hazards.

4. Implications for Water Safety Guidelines, Legislation, and Water Treatment

The need to provide safe recreational and bathing waters and drinking water provides multiple challenges due to the potential for combinations of toxic compounds, with possible synergistic and additive effects, to occur at concentrations deemed unsafe for human health [121,122]. In order to prevent potential adverse toxicities arising from exposure to multiple toxins, to cyanotoxins plus microbial pathogens, and to cyanotoxins plus more recently recognized health hazards including microplastics, the ability of water safety guideline values (GV) and legislation to provide adequate safety margins in the event of potential multiple exposures need to be verified. In several cases, the same drinking water treatment technologies are used for the removal and/or destruction of cyanotoxins, and of other toxicants (Table 2).

Table 2. Examples of water treatment processes for the removal of potentially toxic compounds.

Toxicant	Treatment Process	References
Fungal/microbial toxins	SF, C	[123]
Pesticides	Oz, AC,	[124–126]
Microplastics	C, U	[127,128]
Metals	C, A	[129,130]
Microcystins	AC, Ox, Oz, Cl	[121,131]
Anatoxin-a	Oz, Ox, Cl	[132]
Saxitoxins	AC, Cl	[133]
Cylindrospermopsin	AC, Oz, Cl	[134,135]

AC, activated carbon; Ox, oxidation; Ozonation; Cl, chlorination; C, coagulation; A, adsorption; U, ultrafiltration; SF, sand filtration.

Consequently, in order to prevent or reduce toxicity during events when co-occurrence arises, then the efficacy of such multipurpose water treatment technologies needs to be verified, and increased where needed, with respect to each of the classes of cyanotoxins and of additional toxicants in the raw water. Furthermore, the development and implementation of advanced water treatment technologies should take into account the potential for real-world exposures to a wide range of toxicants and scenarios, from e.g., natural occurrence [121], to deliberate or man-made additions [136]. Such events can potentially occur alone or in combination and during times of high-water usage which may include cyanobacterial blooms in source waters, the potential for multiple exposures should be acknowledged and contingency plans should consider this issue. Furthermore, often contingencies may include the switching of source waters and if different, but potentially hazardous situations exist in these waters, then alternative or enhanced treatment technologies, or e.g., the temporary provision of bottled waters, may be required.

Understanding of the health risks presented by cyanotoxins has progressed considerably [6–9], but the risk management of microcystins, and of microcystin-producing cyanobacteria has been the focus of guideline derivation for health protection [137]. GVs in addition to those for microcystins, including for cylindrospermopsin [121,134] are also being derived. Although GVs for microplastics do not exist, the WHO has derived GVs for six monomers which can leach from the plastics, ranging from 0.3 to 300 µg/L [138]. Some metals and pesticides also have GVs for drinking water and are also generally in the low µg/L range [139]. Whether the GV values, with in-built safety margins, accommodate the additive and synergistic toxicities which can arise due to cyanotoxin co-occurrences and to the co-occurrence of cyanotoxins with other toxicants including microplastics derivatives (Section 3.5) merits investigation.

5. Concluding Remarks

Following wildlife-, domestic animal-, and human-intoxications due to exposure to cyanobacterial mass populations, the volume of research over recent years into the toxicology and toxinology of cyanobacteria and cyanotoxins has increased greatly and such growth continues e.g., [1,5,7,8].

In recognition of the co-occurrence of multiple variants within individual classes of cyanotoxins, of different cyanotoxin classes, and of cyanotoxins plus phycotoxins, it is encouraging that physico-chemical methods for the co-analysis of these combinations are being developed [5,19–21,140]. However, since cyanobacterial mass populations commonly develop in waterbodies which are under intensive anthropogenic use (e.g., for domestic, industrial, and agricultural wastewater discharge, abstraction for drinking water treatment, recreation, crop irrigation, and fisheries) it should be anticipated that toxigenic cyanobacteria can co-occur with a wide range of additional biological and chemical health hazards. Indeed, sufficient examples exist, reviewed here, of animal and human health incidents and intoxications associated with multiple health hazards including cyanobacteria and cyanotoxins, and other contributory biological toxins, microbial pathogens and anthropogenic chemical products. Whilst further research on the toxicity of the established and emerging cyanotoxins via bioassays is needed, including defined cyanotoxin combinations and environmental materials [15], more cyanotoxin bioassays including other environmental toxins and toxicants, as exemplified in Table 1, are required. Data on the toxicity of cyanotoxins in combination with other biotoxins and chemical toxic compounds, may then contribute to the assessment of whether guideline values and legislation for health protection can also accommodate multiple exposure to cyanotoxins plus the other biological and chemical agents. The route of administration of cyanotoxins may also affect the toxicological outcome or the mixtures of compounds that may be present, such as lead and particular matter ($PM_{2.5}$ and PM_{10}) that can occur as airborne components [141,142], potentially in addition to cyanotoxins [143].

The recognition of the multiple occurrence and combined toxicity of cyanotoxins, plus additional toxicants and pathogens, is a growing area of research and such co-occurrences may increase further with the growth of the human population, increasing demands upon water resources and climate change. However, such co-occurrence(s) may have already adversely affected human health in an earlier period. High concentrations of mercury, of phosphate consistent with eutrophic conditions, plus 16S rRNA amplicons indicating *Planktothrix* and *Microcystis* blooms (and thus potentially microcystins), have been found in dated sediment profiles from former reservoirs serving the ancient former Mayan city of Tikal in Guatemala. This potentially toxic combination may have contributed to the demise of the Mayan population and of the city in the ninth Century CE [144].

Funding: This research received no external funding.

Conflicts of Interest: The authors declare no conflict of interest.

References

1. Carmichael, W.W. (Ed.) *The Water Environment: Algal Toxins and Health*; Plenum Press: New York, NY, USA, 1981; p. 491.
2. Falconer, I.R. (Ed.) *Algal Toxins in Seafood and Drinking Water*; Academic Press: London, UK, 1993; p. 224.
3. Codd, G.A.; Jefferies, T.M.; Keevil, C.W.; Potter, E. (Eds.) *Detection Methods for Cyanobacterial Toxins*; Royal Society of Chemistry: Cambridge, UK, 1994; p. 191.
4. Huisman, J.; Matthijs, H.C.P.; Visser, P.M. (Eds.) *Harmful Cyanobacteria*; Springer: Dordrecht, The Netherlands, 2005; p. 241.
5. Meriluoto, J.; Spoof, L.; Codd, G.A. (Eds.) *Handbook of Cyanobacterial Monitoring and Cyanotoxin Analysis*; Wiley: Chichester, UK, 2017; p. 548.
6. Metcalf, J.S.; Codd, G.A. Cyanotoxins. In *Ecology of Cyanobacteria II: Their Diversity in Space and Time*; Whitton, B.A., Ed.; Springer: Dordrecht, The Netherlands, 2012; pp. 651–675.
7. Testai, E.; Scardala, S.; Vichi, S.; Buratti, F.M.; Funari, E. Risk to human health associated with the environmental occurrence of cyanobacterial neurotoxic alkaloids anatoxins and saxitoxins. *Crit. Rev. Toxicol.* **2016**, *46*, 385–418. [CrossRef] [PubMed]
8. Svirčev, Z.; Drobac, D.; Tokodi, N.; Mijović, B.; Codd, G.A.; Meriluoto, J. Toxicology of microcystins with reference to cases of human intoxications and epidemiological investigations of exposure to cyanobacteria and cyanotoxins. *Arch. Toxicol.* **2017**, *91*, 621–650. [CrossRef] [PubMed]

9. Buratti, F.M.; Manganelli, M.; Vichi, S.; Stefanelli, M.; Scardala, S.; Testai, E.; Funari, E. Cyanotoxins: Producing organisms, occurrence, toxicity, mechanism of action and human health toxicological risk evaluation. *Arch. Toxicol.* **2017**, *91*, 1049–1130. [PubMed]

10. Pichardo, S.; Caméan, A.M.; Jos, A. In vitro toxicological assessment of cylindrospermopsin: A review. *Toxins* **2017**, *16*, 402. [CrossRef]

11. Metcalf, J.S.; Souza, N.R. Cyanobacteria and their toxins. *Sep. Sci. Technol.* **2019**, *11*, 125–148.

12. Svirčev, Z.; Lalić, D.; Savić, G.B.; Tokodi, N.; Backovic, D.D.; Chen, L.; Meriluoto, J.; Codd, G.A. Global geographical and historical overview of cyanotoxins distribution and cyanobacterial poisonings. *Arch. Toxicol.* **2019**, *93*, 2429–2481. [CrossRef]

13. Francis, G. Poisonous Australian lake. *Nature* **1878**, *18*, 11–12. [CrossRef]

14. Codd, G.A.; Morton, H.; Baker, P.D. George Francis, a pioneer in the investigation of the quality of South Australia's drinking water sources (1878–1883). *Trans. R. Soc. S. Aust.* **2015**, *139*, 164–170. [CrossRef]

15. Blaha, L.; Camean, A.M.; Fessard, V.; Gutierrez-Praena, D.; Jos, A.; Marie, B.; Metcalf, J.S.; Pichardo, S.; Puerto, M.; Torokne, A.; et al. Bioassay use in the field of cyanobacteria. In *Handbook of Cyanobacterial Monitoring and Cyanotoxin Analysis*; Wiley: Chichester, UK, 2017; pp. 272–279.

16. Meriluoto, J.; Metcalf, J.S.; Codd, G.A. Selection of analytical methodology for cyanotoxin analysis. In *Handbook of Cyanobacterial Monitoring and Cyanotoxin Analysis*; Wiley: Chichester, UK, 2017; pp. 308–312.

17. Kurmayer, R.; Sivonen, K.; Wilmotte, A.; Salmaso, N. (Eds.) *Molecular Tools for the Detection and Quantitation of Toxigenic Cyanobacteria*; John Wiley and Sons: Chichester, UK, 2017; p. 402.

18. Turner, A.D.; Waack, J.; Lewis, A.; Edwards, C.; Lawton, L. Development and single-laboratory validation of a UHPLC-MS/MS method for quantitation of microcystins and nodularin in natural water, cyanobacteria, shellfish and algal supplement tablet powders. *J. Chromatogr. B* **2018**, *1074–1075*, 111–123. [CrossRef]

19. Gimenéz-Campillo, C.; Pastor-Belda, M.; Campillo, N.; Arroyo-Manzanares, N.; Hernández-Córdoba, M.; Viñas, P. Determination of cyanotoxins and phycotoxins in seawater and algae-based supplements using ionic liquids and liquid chromatography with time-of-flight mass spectrometry. *Toxins* **2019**, *11*, 610. [CrossRef]

20. Roy-Lachapelle, A.; Duy, S.V.; Munoz, G.; Dinh, Q.T.; Bahl, E.; Simon, D.F.; Sauvé, S. Analysis of multiclass cyanotoxins (microcystins, anabaenopeptins, cylindrospermopsin and anatoxins) in lake water using on-line SPE liquid chromatography high-resolution Orbitrap mass spectrometry. *Anal. Methods* **2019**, *11*, 5289. [CrossRef]

21. Di Pofi, G.; Favero, G.; di Gregorio, F.N.; Ferretti, E.; Viaggiu, E.; Lucentini, L. Multi-residue ultra performance liquid chromataography-high resolution mass spectrometry method for the analysis of 21 cyanotoxins in surface water for human consumption. *Talanta* **2020**, *211*, 120738. [CrossRef] [PubMed]

22. Canter-Lund, H.; Lund, J.W.G. *Freshwater Algae: Their Microscopic World Explored*; Biopress Limited: Bristol, UK, 1995; p. 360.

23. Whitton, B.A. (Ed.) *Ecology of Cyanobacteria II: Their Diversity in Space and Time*; Springer: Dordrecht, The Netherlands, 2012; p. 760.

24. Islam, M.S.; Miah, M.A.; Hasan, M.K.; Sack, R.B.; Albert, M.J. Detection of non-culturable *Vibrio cholera* 01 associated with a cyanobacterium from an aquatic environment in Bangladesh. *Trans. R. Soc. Trop. Med. Hyg.* **1994**, *88*, 298–299. [CrossRef]

25. Timbrell, J.A. *Introduction to Toxicology*; Taylor & Francis: Bristol, UK, 1995; p. 167.

26. Fitzgeorge, R.B.; Clark, S.A.; Keevil, C.W. Routes of intoxication. In *Detection Methods for Cyanobacterial Toxins*; Royal Society of Chemistry: Cambridge, UK, 1994; pp. 69–74.

27. Chia, M.A.; Kramer, B.J.; Jankowiak, J.G.; do Carmo Bittencourt-Oliveira, M.; Gobler, C.J. The individual and combined effects of the cyanotoxins, anatoxin-a and microcystin-LR, on the growth, toxin production and nitrogen fixation of prokaryotic and eukaryotic algae. *Toxins* **2019**, *11*, 43. [CrossRef] [PubMed]

28. Rymuszka, A.; Sieroslawska, A. Cytotoxic and immunotoxic effects of the mixture containing cyanotoxins on carp cells following in vitro exposure. *Cent. Eur. J. Immunol.* **2013**, *38*, 159–163. [CrossRef]

29. Li, Q.; Gu, P.; Zhang, C.; Luo, X.; Zhang, H.; Zhang, J.; Zheng, Z. Combined toxic effects of anatoxin-a and microcystin-LR on submerged macrophytes and biofilms. *J. Hazard. Mater.* **2020**, *389*, 122053. [CrossRef]

30. Lindsay, J.; Metcalf, J.S.; Codd, G.A. Protection against the toxicity of microcystin-LR and cylindrospermopsin in *Artemia salina*, and *Daphnia* spp. by pretreatment with cyanobacterial lipopolysaccharide (LPS). *Toxicon* **2006**, *48*, 995–1001. [CrossRef]

31. Pinheiro, C.; Azevedo, J.; Campos, A.; Vasconcelos, V.; Loureiro, S. The interactive effects of microcystin-LR and cylindrospermopsin on the growth rate of the freshwater alga *Chlorella Vulgaris*. *Ecotoxicology* **2016**, *25*, 745–758. [CrossRef]

32. Reinikainen, M.; Meriluoto, J.A.O.; Spoof, L.; Harada, K.-I. The toxicities of a polyunsaturated fatty acid and a microcystin to Daphnia Magna. *Environ. Toxicol.* **2001**, *16*, 444–448. [CrossRef]

33. Wang, L.; He, L.; Zeng, H.; Fu, W.; Wang, J.; Tan, Y.; Zheng, C.; Qiu, Z.; Luo, J.; Lv, C.; et al. Low dose microcystin-LR antagonises aflatoxin B1 induced hepatocarcinogenesis through decreased cytochrome P450 1A2 expression and aflatoxin B1-DNA adduct generation. *Chemosphere* **2020**, *248*, 126036. [CrossRef]

34. Meneely, J.P.; Hašlová, J.; Krska, R.; Elliott, C.T. Assessing the combined toxicity of the natural toxins, aflatoxin B_1, fumonisin B_1 and microcystin-LR by high content analysis. *Food Chem. Toxicol.* **2018**, *121*, 527–540. [CrossRef] [PubMed]

35. Wei, H.; Wang, S.; Xu, E.G.; Liu, J.; Li, X.; Wang, Z. Synergistic toxicity of microcystin-LR and Cu to zebrafish (*Danio rerio*). *Sci. Total Environ.* **2020**, *713*, 736393. [CrossRef] [PubMed]

36. Wang, Z.; Zhang, J.; Li, E.; Zhang, L.; Wang, X.; Song, L. Combined toxic effects and mechanisms of microcystin-LR and copper on *Vallisneria natans* (Lous.) Hara seedlings. *J. Hazard. Mater.* **2017**, *328*, 108–116. [CrossRef] [PubMed]

37. Wang, Z.; Xiao, B.; Song, L.; Wu, X.; Zhang, Y.; Wang, C. Effects of microcystin-LR, linear alkylbenzene sulfonate and their mixture on lettuce (*Lactuca sativa* L.) seeds and seedlings. *Ecotoxicology* **2011**, *20*, 803–814. [CrossRef] [PubMed]

38. Wan, X.; Steinman, A.D.; Shu, X.; Cao, Q.; Yao, L.; Xie, L. Combined toxic effects of microcystin-LR and phenanthrene on growth and antioxidant system of duckweed (*Lemna gibba* L.). *Ecotoxicol. Environ. Saf.* **2019**, *185*, 109668. [CrossRef] [PubMed]

39. Hinojosa, M.G.; Prieto, A.I.; Gutiérrez-Praena, D.; Moreno, F.J.; Caméan, A.M.; Jos, A. In vitro assessment of the combination of cylindrospermopsin and the organophosphate chlorpyrifos on the human neuroblastoma SH-SY5Y cell line. *Ecotoxicol. Environ. Saf.* **2020**, *191*, 110222. [CrossRef]

40. Hercog, C.; Štern, A.; Maisanaba, S.; Filipič, M.; Bojana, Ž. Plastics in cyanobacterial blooms—genotoxic effects of binary mixtures of cylindrospermopsin and bisphenols in HepG2 cells. *Toxins* **2020**, *12*, 219. [CrossRef]

41. Fiore, S.M.F.; de Lima, S.T.; Carmichael, W.W.; McKinnie, S.M.K.; Checkan, J.R.; Moore, B.S. Guanitoxin, renaming a cyanobacterial organophosphate toxin. *Harmful Algae* **2020**, *92*, 101937. [CrossRef]

42. Spoof, L.; Catherine, A. Appendix 3: Tables of microcystins and nodularins. In *Handbook of Cyanobacterial Monitoring and Cyanotoxin Analysis*; Wiley: Chichester, UK, 2017; pp. 526–537.

43. Robillot, C.; Vinh, J.E.; Puiseux-Dao, S.; Hennion, M.-C. Hepatotoxin production kinetics of the cyanobacterium *Microcystis aeruginosa* PCC 7820, as determined by HPLC-mass spectrometry and protein phosphatase bioassay. *Environ. Sci. Technol.* **2000**, *34*, 3372–3378. [CrossRef]

44. Wimmer, K.M.; Strangman, W.; Wright, J.L.C. 7-deoxy-desulfo-cylindrospermopsin and 7-dexoy-desulfo-12-acetylcylindrospermopsin: Two new cylindrospermopsin analogs isolated from a Thai strain of *Cylindrospermopsis raciborskii*. *Harmful Algae* **2014**, *37*, 203–206. [CrossRef]

45. Ward, C.J.; Codd, G.A. Comparative toxicity of four microcystins of different hydrophobicities to the protozoan, Tetrahymena Pyriformis. *J. Appl. Microbiol.* **1999**, *86*, 874–882. [CrossRef]

46. Harada, K.-I.; Ogawa, K.; Kimura, Y.; Murata, H.; Suzuki, M. Microcystins from *Anabaena flos-aquae* NRC 525-17. *Chem. Res. Toxicol.* **1991**, *4*, 535–540. [CrossRef]

47. Monteiro, S.; Santos, R.; Bláha, L.; Codd, G.A. Lipopolysaccharide endotoxins. In *Handbook of Cyanobacterial Monitoring and Cyanotoxin Analysis*; Wiley: Chichester, UK, 2017; pp. 165–172.

48. Pouria, S.; de Andrade, A.; Barbosa, J.; Cavalcanti, R.L.; Barreto, V.T.; Ward, C.J.; Preiser, W.; Poon, G.K.; Neild, G.H.; Codd, G.A. Fatal microcystin intoxication in haemodialysis unit in Caruaru, Brazil. *Lancet* **1998**, *352*, 21–26. [CrossRef]

49. Jochimsen, E.M.; Carmichael, W.W.; An, J.S.; Cardo, D.M.; Cookson, S.T.; Holmes, C.E.; Antunes, M.B.; de Melo Filho, D.A.; Lyra, T.M.; Barreto, V.S.; et al. Liver failure and death after exposure to microcystins at a hemodialysis center in Brazil. *N. Engl. J. Med.* **1998**, *338*, 873–878. [CrossRef] [PubMed]

50. Carmichael, W.W.; Azevedo, S.M.; An, J.S.; Molica, R.J.; Jochimsen, E.M.; Lau, S.; Rinehart, K.L.; Shaw, G.R.; Eaglesham, G.K. Human fatalities from cyanobacteria: Chemical and biological evidence for cyanotoxins. *Environ. Health Perspect.* **2001**, *109*, 663–668. [CrossRef] [PubMed]

51. Krienitz, L.; Ballot, A.; Kotut, K.; Wiegand, C.; Pütz, S.; Metcalf, J.S.; Codd, G.A.; Pflugmacher, S. Contribution of hot spring cyanobacteria to the mysterious deaths of Lesser Flamingos at Lake Bogoria, Kenya. *FEMS Microbiol. Ecol.* **2003**, *43*, 141–148. [CrossRef] [PubMed]

52. Krienitz, L.; Ballot, A.; Caspar, P.; Codd, G.A.; Kotut, K.; Metcalf, J.S.; Morrison, L.F.; Pflugmacher, S.; Wiegand, C. Contribution of toxic cyanobacteria to massive deaths of Lesser Flamingos at saline-alkaline lakes. *Proc. Int. Assoc. Theor. Appl. Limnol.* **2005**, *29*, 783–786. [CrossRef]

53. Pennycott, T.; Young, F.M.; Metcalf, J.S.; Codd, G.A. Necrotic enteritis in mute swans associated with cyanobacterial toxins. *Vet. Rec.* **2004**, *154*, 575–576.

54. Landsberg, J.H.; Hendrickson, J.; Tabuchi, M.; Kiryu, Y.; Williams, B.J.; Tomlinson, M.C. A large-scale sustained fish kill in the St John's River, Florida: A complex consequence of cyanobacterial blooms. *Harmful Algae* **2020**, *92*, 101771. [CrossRef]

55. Hawkins, P.R.; Runnegar, M.T.C.; Jackson, A.R.B.; Falconer, I.R. Severe hepatotoxicity caused by the tropical cyanobacterium (blue-green alga) *Cylindrospermopsis raciborskii* (Woloszynska) Seenaya and Subba Raju isolated from a domestic water supply. *Appl. Environ. Microbiol.* **1985**, *50*, 11292–11295. [CrossRef]

56. Turner, P.C.; Gammie, A.J.; Hollinrake, K.; Codd, G.A. Pneumonia associated with contact with cyanobacteria. *Br. Med. J.* **1990**, *300*, 1440–1441. [CrossRef] [PubMed]

57. Yu, S.-Z. Drinking water and primary liver cancer. In *Primary Liver Cancer*; Tang, Z.Y., Wu, M.C., Xia, S.S., Eds.; China Academic Publishers: New York, NY, USA, 1989; pp. 30–37.

58. Yu, S.-Z. Primary prevention of hepatocellular carcinoma. *J. Gastroenterol. Hepatol.* **1995**, *10*, 674–682. [CrossRef]

59. Ueno, Y.; Nagata, S.; Tsutsumi, T.; Hasegawa, A.; Watanabe, M.; Park, H.D.; Chen, G.-C.; Chen, G.; Yu, S.-Z. Detection of microcystins, a blue-green algal hepatotoxin, in drinking water sampled in Haimen and Fusui, endemic areas of primary liver cancer in China, by highly sensitive immunoassay. *Carcinogenesis* **1996**, *17*, 1317–1321. [CrossRef] [PubMed]

60. Zheng, C.; Zeng, H.; Lin, H.; Wang, J.; Feng, Z.; Chen, J.-A.; Luo, J.; Luo, Y.; Huang, Y.; Wang, L.; et al. Serum microcystin levels positively linked with risk of hepatocellular carcinoma: A case-control study in southwest China. *Hepatology* **2017**, *66*, 1519–1528. [CrossRef] [PubMed]

61. Rasmussen, S.A.; Andersen, A.J.C.; Andersen, N.G.; Nielsen, K.F.; Hansen, P.J.; Larsen, T.O. Chemical diversity, origin, and analysis of phycotoxins. *J. Nat. Prod.* **2016**, *79*, 662–673. [CrossRef]

62. Vilariño, N.; Louzao, M.C.; Abal, P.; Cagide, E.; Carrera, C.; Vieytes, M.R.; Botana, L.M. Human poisoning from marine toxins: Unknowns for optimal consumer protection. *Toxins* **2018**, *10*, 324. [CrossRef]

63. Akbar, M.A.; Yusof, N.Y.M.; Tahir, N.I.; Ahmad, A.; Usup, G.; Sahrani, F.K.; Bunawan, H. Biosynthesis of saxitoxin in marine dinoflagellates: An omics perspective. *Mar. Drugs* **2020**, *18*, 103. [CrossRef]

64. Miller, M.A.; Kudela, R.M.; Mekebri, A.; Crane, D.; Oates, S.C.; Tinker, M.T.; Staedler, M.; Miller, W.A.; Toy-Choutka, S.; Dominik, C.; et al. Evidence for a novel marine harmful algal bloom: Cyanotoxin (microcystin) transfer from land to sea otters. *PLoS ONE* **2010**, *5*, e12576. [CrossRef]

65. Peacock, M.B.; Gibble, C.M.; Senn, D.B.; Cloern, J.E.; Kudela, R.M. Blurred lines: Multiple freshwater and marine algal toxins at the land-sea interface of San Fransisco Bay, California. *Harmful Algae* **2018**, *73*, 138–147. [CrossRef]

66. Codd, G.A.; Nunn, P.B. Cyanotoxin production beyond the cyanobacteria. *Toxicon* **2019**, *168*, 93–94. [CrossRef]

67. Ballot, A.; Bernard, C.; Fastner, J. Saxitoxin and analogues. In *Handbook of Cyanobacterial Monitoring and Cyanotoxin Analysis*; Wiley: Chichester, UK, 2017; pp. 149–154.

68. Smith, F.M.J.; Wood, S.A.; Wilks, T.; Kelly, D.; Broady, P.A.; Williamson, W.; Gaw, S. Survery of *Scytonema* (Cyanobacteria) and associated saxitoxins in the littoral zone of recreational lakes in Canterbury, New Zealand. *Phycologia* **2012**, *51*, 542–551.

69. Cullen, A.; D'Agostino, P.M.; Mazmouz, R.; Pickford, R.; Wood, S.A.; Nielan, B.A. Insertions within the saxitoxin biosynthetic gene cluster result in differential toxin profiles. *ACS Chem. Biol.* **2018**, *13*, 3107–3114. [PubMed]

70. Cox, P.A.; Banack, S.A.; Murch, S.J.; Rasmussen, U.; Tien, G.; Bidigare, R.R.; Metcalf, J.S.; Morrison, L.F.; Codd, G.A.; Bergman, B. Diverse taxa of cyanobacteria produce β-*N*-methylamino-L-alanine, a neurotoxic amino acid. *Proc. Natl. Acad. Sci. USA* **2005**, *102*, 5074–5078. [PubMed]

71. Metcalf, J.S.; Banack, S.A.; Lindsay, J.; Morrison, L.F.; Cox, P.A.; Codd, G.A. Co-occurrence of β-*N*-methylamino-L-alanine a neurotoxic amino acid with other cyanobacterial toxins in British waterbodies, 1990–2004. *Environ. Microbiol.* **2008**, *10*, 702–708. [PubMed]

72. Faassen, E.J.; Gillissen, F.; Lürling, M. A comparative study of three analytical methods for the determination of the neurotoxin BMAA in cyanobacteria. *PLoS ONE* **2012**, *7*, e36667. [PubMed]

73. Jiang, L.; Eriksson, J.; Lage, S.; Jonasson, S.; Shams, S.; Mehine, M.; Ilag, L.L.; Rasmussen, U. Diatoms: A novel source for the neurotoxin BMAA in aquatic environments. *PLoS ONE* **2014**, *9*, e84578.

74. Violi, J.P.; Facey, J.A.; Mitrovic, S.M.; Colville, A.; Rodgers, K.J. Production of β-methylamino-L-alanine (BMAA) and its isomers in freshwater diatoms. *Toxins* **2019**, *11*, 1109–1138.

75. Jiang, L.; Ilag, L.L. Detection of endogenous BMAA in dinoflagellate (*Heterocapsa triquetra*) hints at evolutionary conservation and environmental concern. *Pubraw Sci.* **2014**, *2*, 1–8.

76. Metcalf, J.S.; Banack, S.A.; Wessel, R.A.; Lester, M.; Pim, J.G.; Cassani, J.R.; Cox, P.A. Toxin analysis of freshwater cyanobacterial and marine harmful algal blooms on the west coast of Florida and implications for estuarine environments. *Neurotox. Res.* **2020**. [CrossRef]

77. O'Neal, R.M.; Chen, C.-H.; Reynolds, C.S.; Meghal, S.K.; Koeppe, R.E. The 'neurotoxicity' of L-2,4-diaminobutyric acid. *Biochem. J.* **1968**, *106*, 699–706.

78. Schneider, T.; Simpson, C.; Desai, P.; Tucker, M.; Lobner, D. Neurotoxicity of isomers of the environmental toxin L-BMAA. *Toxicon* **2020**, *184*, 175–179. [PubMed]

79. Nunn, P.B.; Codd, G.A. Metabolic solutions to the biosynthesis of some diaminomonocarboxylic acids in nature: Formation in cyanobacteria of the neurotoxins 3-*N*-methyl-2,3-diaminopropanoic acid (BMAA) and 2,4-diaminobutanoic acid (2,4-DAB). *Phytochemistry* **2017**, *144*, 2530270. [CrossRef] [PubMed]

80. Nunn, P.B.; Codd, G.A. Environmental distribution of the neurotoxin L-BMAA in *Paenibacillus* species. *Toxicol. Res.* **2019**, *8*, 781–783. [CrossRef] [PubMed]

81. Tu, J.; Chen, L.; Gao, S.; Zhang, J.; Bi, C.; Lu, N.; Lu, Z. Obtaining genome sequences of mutualistic bacteria in single *Microcystis* colonies. *Int. J. Mol. Sci.* **2019**, *20*, 5047. [CrossRef] [PubMed]

82. Stewart, I.; Webb, P.M.; Schluter, P.J.; Shaw, G.R. Recreational and occupational field exposure to freshwater cyanobacteria—A review of anecdotal and case reports, epidemiological studies and the challenges for epidemiologic assessment. *Environ. Health* **2006**, *5*, 6. [CrossRef] [PubMed]

83. Koch, R. An address on cholera and its bacillus. *Br. Med. J.* **1884**, *2*, 453.

84. Islam, M.S.; Drasar, B.S.; Sack, R.B. Probable role of blue-green algae in maintaining endemicity and seasonality of cholera in Bangladesh: A hypothesis. *J. Diarrhoeal Dis. Res.* **1994**, *12*, 245–256. [PubMed]

85. Islam, M.S.; Zaman, M.H.; Islam, M.S.; Ahmed, M.; Clemens, J.D. Environmental reservoirs of *Vibrio cholerae*. *Vaccine* **2020**, *38*, A52–A62. [CrossRef]

86. Ahmed, M.S.; Raknussaman, M.; Akther, H.; Ahmed, S. The role of cyanobacteria blooms in cholera epidemic in Bangladesh. *J. Appl. Sci.* **2007**, *7*, 1785–1789.

87. Froehlich, B.; Gonzalez, R.; Blackwood, D.; Lauer, K.; Noble, R. Decadal monitoring reveals an increase in *Vibrio* spp. concentrations in the Neuse River Estuary, North Carolina, USA. *PLoS ONE* **2019**, *14*, e0215254. [CrossRef]

88. Paerl, H.W.; Valdes, L.M.; Peierls, B.J.; Adolf, J.E.; Harding, L.W. Anthropogenic and climate influences on the eutrophication of large estuarine ecosystems. *Limnol. Oceanogr.* **2006**, *51*, 448–462. [CrossRef]

89. Marshall, M.M.; Naumovitz, D.; Ortega, Y.; Sterling, C.R. Waterborne protozoan pathogens. *Clin. Microbiol. Rev.* **1997**, *10*, 67–85. [CrossRef]

90. Lopes, A.M.M.B.; Gomes, L.N.L.; de Cerqueira, M.F.; Filho, C.R.M.; von Sperling, E.; de Pádua, V.L. Dynamic of pathogenic protozoa and cyanobacteria in a reservoir used for water supply in southeastern Brazil. *Eng. Sanit. E Ambient.* **2017**, *22*, 25–43. [CrossRef]

91. Fiore, M.F.; Trevors, J.T. Cell composition and metal tolerance in cyanobacteria. *BioMetals* **2017**, *7*, 83–103. [CrossRef]

92. Baptista, M.S.; Vasconcelos, M.T. Cyanobacteria metal interactions: Requirements, Toxicity and ecological implications. *Crit. Rev. Microbiol.* **2006**, *32*, 127–137. [CrossRef] [PubMed]

93. Ramakrishnan, B.; Megharaj, M.; Venkateswarlu, K.; Naidu, R.; Sethunathan, N. The impacts of environmental pollutants on microalgae and cyanobacteria. *Crit. Rev. Environ. Sci. Technol.* **2010**, *40*, 699–821. [CrossRef]

94. Gu, P.; Qi, L.; Zhang, W.; Zheng, Z.; Luo, X. Effects of different metal ions (Ca, Cu, Pb, Cd) on formation of cyanobacterial blooms. *Ecotoxicol. Environ. Saf.* **2020**, *189*, 109976. [CrossRef] [PubMed]

95. Heresztyn, T.; Nicholson, B.C. Nodularin concentrations in Lake Alexandrina and Albert, South Australia, during a bloom of the cyanobacterium (blue-green alga) *Nodularia spumigena* and degradation of the toxin. *Environ. Toxicol. Water Qual.* **1997**, *12*, 273–282. [CrossRef]

96. Falconer, I.R.; Beresford, A.M.; Runnegar, M.T. Evidence of liver damage by toxin from a bloom of the blue-green alga *Microcystis aeruginosa*. *Med. J. Aust.* **1983**, *1*, 511–514. [CrossRef]

97. Utkilen, H.; Gjolme, N. Iron-stimulated toxin production in *Microcystis aeruginosa*. *Appl. Environ. Microbiol.* **1995**, *61*, 797–800. [CrossRef]

98. Facey, J.A.; Apte, S.C.; Mitrovic, S.M. A review of the effect of trace metals on freshwater cyanobacterial growth and toxin production. *Toxins* **2019**, *11*, 643. [CrossRef] [PubMed]

99. Harland, F.M.J.; Wood, S.A.; Moltchanova, E.; Williams, W.M.; Gaw, S. *Phormidium autumnale* growth and anatoxin-a production under iron and copper stress. *Toxins* **2013**, *5*, 2504–2521. [CrossRef] [PubMed]

100. Nunn, P.B.; O'Brien, P.; Pettit, L.D.; Pyburn, S.I. Complexes of zinc, copper and nickel with the non-protein amino acid L-α-amino-β-methylaminopropionic acid: A naturally occurring neurotoxin. *J. Inorg. Biochem.* **1989**, *37*, 175–183. [CrossRef]

101. Humble, A.V.; Gadd, G.M.; Codd, G.A. Binding of copper and zinc to three cyanobacterial microcystins quantified by differential pulse polarography. *Water Res.* **1997**, *31*, 1679–1686. [CrossRef]

102. Rush, T.; Liu, X.; Lobner, D. Synergistic toxicity of the environmental neurotoxins methylmercury and β-N-methylamino-ʟ-alanine. *Neuroreport* **2012**, *23*, 216–219. [CrossRef]

103. Metcalf, J.S.; Bruno, M. Anatoxin-a(S). In *Handbook of Cyanobacterial Monitoring and Cyanotoxin Analysis*; Wiley: Chichester, UK, 2017; pp. 155–159.

104. Singh, A.K.; Singh, P.P.; Tripathi, V.; Verma, H.; Singh, S.K.; Srivastarva, M.K.; Kumar, A. Distribution of cyanobacteria and their interactions with pesticides in paddy field: A comprehensive review. *J. Environ. Manag.* **2018**, *224*, 361–375. [CrossRef]

105. Staley, Z.R.; Harwood, V.J.; Rohr, J.R. A synthesis of the effects of pesticides on microbial persistence in aquatic ecosystems. *Crit. Rev. Toxicol.* **2015**, *45*, 813–836. [CrossRef]

106. Bettinetti, R.; Quadroni, S.; Crosa, G.; Harper, D.; Dickie, J.; Kyalo, M.; Mavuti, K.; Galassi, S. A preliminary evaluation of the DDT contamination of sediments in Lakes Natron and Bogoria (Eastern Rift Valley, Africa). *AMBIO* **2011**, *40*, 341–350. [CrossRef]

107. Woodward, A.R.; Percival, H.F.; Rauschenberger, R.H.; Gross, T.S.; Rice, K.G.; Conrow, R. Abnormal alligators and organochlorine pesticides in Lake Apopka, Florida. In *Wildlife Ecotoxicology Forensics Approached*; Elliott, J.E., Bishop, C.A., Morrissey, C.A., Eds.; Springer: New York, NY, USA, 2011; pp. 153–187.

108. Cerbin, S.; Kraak, M.H.S.; de Voogt, P.; Visser, P.M.; van Donk, E. Combined and single effects of pesticide carbaryl and toxic *Microcystis aeruginosa* on the life history of *Daphnia pulicaria*. *Hydrobiologia* **2010**, *643*, 129–138. [CrossRef]

109. Metcalf, J.S.; Beattie, K.A.; Ressler, J.; Gerbersdorf, S.; Pflugmacher, S.; Codd, G.A. Cross-reactivity and performance assessment of four microcystin immunoassays with detoxication products of the cyanotoxin, microcystin-LR. *J. Water Supply Res. Technol. AQUA* **2002**, *51*, 145–151. [CrossRef]

110. Gill, J.P.K.; Sethi, N.; Mohan, A.; Datta, S.; Girdhar, M. Glyphosate toxicity for animals. *Environ. Chem. Lett.* **2017**. [CrossRef]

111. Zhang, Q.; Qu, Q.; Lu, T.; Ke, M.; Zhu, Y.; Zhang, M.; Zhang, Z.; Du, B.; Pan, X.; Sun, L.; et al. The combined toxicity effect of nanoplastics and glyphosate on *Microcystis aeruginosa* growth. *Environ. Pollut.* **2018**, *243*, 1106–1112. [CrossRef] [PubMed]

112. Wu, L.; Qiu, Z.; Zhou, Y.; Du, Y.; Liu, C.; Ye, J.; Hu, X. Physiological effects of the herbicide glyphosate on the cyanobacterium *Microcystis aeruginosa*. *Aquat. Toxicol.* **2016**, *178*, 72–79. [CrossRef] [PubMed]

113. Malécot, M.; Guével, B.; Pineau, C.; Holbech, B.F.; Bormans, M.; Wiegand, C. Specific proteome response of *Unio pictorum* mussel to a mixture of glyphosate and microcystin-LR. *J. Proteome Res.* **2013**, *12*, 5281–5292.

114. Moore, C.J. Synthetic polymers in the marine environment: A rapidly increasing long-term threat. *Environ. Res.* **2008**, *108*, 131–139. [CrossRef]

115. Yokota, K.; Waterfield, H.; Hastings, C.; Davidson, E.; Kwietniewski, E.; Wells, B. Finding the missing piece of the aquatic pollution puzzle: Interaction between primary producers and microplastics. *Limnol. Oceanogr. Lett.* **2017**, *2*, 91–104. [CrossRef]

116. Yao, L.; Hui, L.; Yang, Z.; Chen, X.; Xiao, A. Freshwater microplastics pollution: Detecting and visualizing emerging trends based on Citespace II. *Chemosphere* **2020**, *245*, 125627. [CrossRef]

117. Leiser, R.; Wu, G.-M.; Neu, T.R.; Wendt-Potthoff, K. Biofouling, metal sorption and aggregation are related to sinking of microplastics in a stratified reservoir. *Water Res.* **2020**, *176*, 115748. [CrossRef]

118. Feng, L.J.; Sun, X.-D.; Zhu, F.-P.; Feng, Y.; Duan, J.-L.; Xiao, F.; Li, X.-Y.; Shi, Y.; Wang, Q.; Sun, J.-W.; et al. Nanoplastics promote microcystin synthesis and release from cyanobacterial *Microcystis aeruginosa*. *Environ. Sci. Technol.* **2020**, *54*, 3386–3394. [CrossRef]

119. Hyenstrand, P.; Metcalf, J.S.; Beattie, K.A.; Codd, G.A. Losses of the cyanobacterial toxin microcystin-LR from aqueous solution by adsorption during laboratory manipulations. *Toxicon* **2001**, *39*, 589–594. [CrossRef]

120. Hyenstrand, P.; Metcalf, J.S.; Beattie, K.A.; Codd, G.A. Effects of adsorption to plastics and solvent conditions in the analysis of the cyanobacterial toxin microcystin-LR by high performance liquid chromatography. *Water Res.* **2001**, *35*, 3508–3511. [CrossRef]

121. Hitzfeld, B.C.; Höger, S.; Dietrich, D.R. Cyanobacterial toxins: Removal during water treatment, and human risk assessment. *Environ. Health Perspect.* **2000**, *108*, 113–122. [PubMed]

122. Delgado, L.F.; Charles, P.; Glucina, K.; Morlay, C. The removal of endocrine disrupting compounds, pharmaceutically activated compounds and cyanobacterial toxins during drinking water preparation using activated carbon—A review. *Sci. Total Environ.* **2012**, *435–436*, 509–525. [CrossRef] [PubMed]

123. Hageskal, G.; Lima, N.; Skaar, I. The study of fungi in drinking water. *Mycol. Res.* **2009**, *113*, 165–172. [CrossRef] [PubMed]

124. Sanches, S.; Crespo, M.T.B.; Pereira, V.J. Drinking water treatment of priority pesticides using low pressure UV photolysis and advanced oxidation processes. *Water Res.* **2010**, *44*, 1809–1818. [CrossRef]

125. Griffini, O.; Bao, M.L.; Burrini, D.; Santianni, D.; Barbieri, C.; Pantani, F. Removal of pesticides during water treatment process at Florence water supply, Italy. *J. Water Supply Res. Technol. AQUA* **1999**, *48*, 177–185. [CrossRef]

126. Ignatowicz, K. Selection of sorbent for removing pesticides during water treatment. *J. Hazard. Mater.* **2009**, *169*, 953–957. [CrossRef]

127. Ma, B.; Xue, W.; Hu, C.; Liu, H.; Qu, J.; Li, L. Characteristics of microplastic removal via coagulation and ultrafiltration during drinking water treatment. *Chem. Eng. J.* **2019**, *359*, 159–167. [CrossRef]

128. Pivokonsky, M.; Cermakova, L.; Novotna, K.; Peer, P.; Cajthaml, T.; Janda, V. Occurrence of microplastics in raw and treated drinking water. *Sci. Total Environ.* **2018**, *643*, 1644–1651. [CrossRef]

129. Elliott, H.A.; Dempsey, B.A.; Maille, P.J. Content and fractionation of heavy metals in water treatment sludges. *J. Environ. Qual.* **1990**, *19*, 330–334. [CrossRef]

130. Chiang, Y.W.; Ghyselbrecht, K.; Santos, R.M.; Martens, J.A.; Swennen, R.; Cappuyns, V.; Meesschaert, B. Adsorption of multi-heavy metals onto water treatment residuals: Sorption capacities and applications. *Chem. Eng. J.* **2012**, *200–202*, 405–415. [CrossRef]

131. Lambert, T.W.; Holmes, C.F.B.; Hrudey, S.E. Adsorption of microcystin-LR by activated carbon and removal in full scale water treatment. *Water Res.* **1996**, *30*, 1411–1422. [CrossRef]

132. Vlad, S.; Anderson, W.B.; Peldszus, S.; Huck, P.M. Removal of the cyanotoxin anatoxin-a by drinking water treatment processes: A review. *J. Water Health* **2014**, *12*, 601–617. [CrossRef] [PubMed]

133. Ho, L.; Tanis-Plant, P.; Kayal, N.; Slyman, N.; Newcombe, G. Optimising water treatment practices for the removal of *Anabaena circinalis* and its associated metabolites, geosmin and saxitoxins. *J. Water Health* **2009**, *7*, 544–556. [CrossRef] [PubMed]

134. Senogles, P.; Shaw, G.; Smith, M.; Norris, R.; Chiswell, R.; Mueller, J.; Sadler, R.; Eaglesham, G. Degradation of the cyanobacterial toxin cylindrospermopsin, from *Cylindrospermopsis raciborskii*, by chlorination. *Toxicon* **2000**, *38*, 1203–1213. [CrossRef]

135. Falconer, I.R.; Humpage, A.R. Cyanobacterial (blue-green algal) toxins in water supplies: Cylindrospermopsins. *Environ. Toxicol.* **2006**, *21*, 299–304. [CrossRef]

136. Metcalf, J.S.; Codd, G.A. The status and potential of cyanobacteria and their toxins as agents of bioterrorism. In *Handbook on Cyanobacteria: Biochemistry, Biotechnology and Applications*; Gault, P.M., Marler, H.J., Eds.; Nova Science Publishers: New York, NY, USA, 2009; pp. 259–281.

137. Chorus, I.; Bartram, J. *Toxic Cyanobacteria in Water: A Guide to Their Public Health Consequences, Monitoring and Management*; E.&F.N. Spon: London, UK, 1999.

138. World Health Organization. *Microplastics in Drinking Water*; World Health Organization: Geneva, Switzerland, 2019.

139. World Health Organization. *Guidelines for Drinking Water Quality*, 4th ed.; World Health Organization: Geneva, Switzerland, 2017.

140. Merlo, F.; Maraschi, F.; Piparo, D.; Profumo, A.; Speltini, A. Simultaneous pre-concentration and HPLC-MS/MS quantification of phycotoxins and cyanotoxins in inland and coastal waters. *Int. J. Environ. Res. Public Health* **2020**, *17*, 4782. [CrossRef]

141. Manton, W.I. Total contribution of airborne lead to blood lead. *Br. J. Ind. Med.* **1985**, *42*, 168–172. [CrossRef]

142. Querol, X.; Alastuey, A.; Moreno, T.; Viana, M.M.; Castillo, S.; Pey, J.; Rodríguez, S.; Artiñano, B.; Salvador, P.; Sánchez, M.; et al. Spatial and temportal variation in airborne particulate matter (PM_{10} and $PM_{2.5}$) across Spain 1999–2005. *Atmos. Environ.* **2008**, *42*, 3964–3979. [CrossRef]

143. Metcalf, J.S.; Richer, R.; Cox, P.A.; Codd, G.A. Cyanotoxins in desert environments may present a risk to human health. *Sci. Total Environ.* **2012**, *421–422*, 118–123. [CrossRef]

144. Lentz, D.L.; Hamilton, T.L.; Dunning, N.P.; Scarborough, V.L.; Luxton, T.P.; Vonderheide, A.; Tepe, E.J.; Perfetta, C.J.; Brunemann, J.; Grazioso, L.; et al. Molecular genetic and geochemical assays reveal severe contamination of drinking water reservoirs at the ancient Maya city of Tikal. *Sci. Rep.* **2020**, *10*, 10316. [CrossRef]

MDPI

St. Alban-Anlage 66

4052 Basel

Switzerland

Tel. +41 61 683 77 34

Fax +41 61 302 89 18

www.mdpi.com

Toxins Editorial Office

E-mail: toxins@mdpi.com

www.mdpi.com/journal/toxins